DESIGN OF STEEL STRUCTURES

(VOLUME I)

[S.I. UNITS]

By

Dr. Ram Chandra

B.E., M.E. (Hons.), M.I.E., Ph.D. (Roorkee)
Professor and Head
Department of Structural Engineering
Faculty of Engineering
M.B.M. Engineering College
University of Jodhpur, Jodhpur

STANDARD BOOK HOUSE

unit of : **RAJSONS PUBLICATIONS PVT. LTD.**

1705-A, Nai Sarak, PB.No. 1074, Delhi-110006 Ph.: +91-(011)-23265506
Show Room: 4262/3, First Lane, G-Floor, Gali Punjabian, Ansari Road, Darya Ganj, New Delhi-110002 Ph.: +91-(011)43551085 Tel Fax : +91-(011)43551185, Fax: +91-(011)-23250212

E-mail: sbh10@hotmail.com www.standardbookhouse.com

Published by:
RAJINDER KUMAR JAIN
Standard Book House
Unit of: Rajsons Publications Pvt. Ltd.
1705-A, Nai Sarak, Delhi - 110006
Post Box: 1074
Ph.: +91-(011)-23265506 Fax: +91-(011)-23250212

Showroom:
4262/3, First Lane, G-Floor, Gali Punjabian
Ansari Road, Darya Ganj, New Delhi-110002
Ph.: +91-(011)-43751128, +91-(011)-43551185
E-mail: sbhl0@ hotmail.com
Web: www.standardbookhouse.in

First Published	:	1970
Second Edition	:	1975
Third Edition	:	1978
Fourth Edition	:	1981
Fifth Edition	:	1982
Sixth Edition	:	1984
Seventh Edition	:	1986
Eight Edition	:	1988
Ninth Edition	:	1989
Tenth Edition	:	1992
Eleventh Edition	:	1995
Twelveth Edition	:	1998
Thirteenth Edition	:	2001
Fourteenth Edition	:	2004
Fifteenth Edition	:	2007
Sixteenth Edition	:	2010
Seventeenth Edition	:	2013
Eighteenth Edition	:	2016
Nineteenth Edition	:	2022

₹ 495.00

ISBN: 978-81-89401-40-5

Typeset by:
C.S.M.S. Computers, Delhi.

Printed by:
R.K. Print Media Company, New Delhi

Foreword

Tables and clauses from the Indian Standard Specifications have been reproduced in the book with the kind permission of the Indian Standards Institution.

It is desirable that for complete detail, reference be made to the latest versions of the Standards Institution, Manak Bhavan, 9, Bahadur Shah Zafar Marg, New Delhi-1, or from its branch offices at Mumbai, Kolkata, Kanpur and Chennai.

SYSTEM INTERNATIONAL d' UNITES

(SI System of Units)

In order to avoid the conversion of results obtained by engineers working with the Foot Pound Second (FPS) System (gravitational) of units in terms of centimetre-gram second absolute system of units used by the scientists, a need of common system of units was realised. The General Conference on Weights and Measures held at Paris in 1960 finalised the System International *d'* Unites (SI). It is an absolute system of units. The mass is considered as fundamental unit and not the force. BIS has included a comment of transition in IS 3616–1966. 'Recommendation on the International System (SI) Units' that this system has begun to replace older system of units in several branches of science and technology. The SI is a universal system of units and it has been adopted in France as a legal system and it is likely to become common in many countries. SI units have the following six basic units.

Unit of length (metre, m)

The length equal to 1,650, 763.73 wavelengths, in vacuum, of the radiation corresponding to the transition between $2p^{19}$ and $5d^5$ levels of the krypton η atom of mass 86 is known as *one metre.*

Unit of mass (kilogram, kg)

The mass of platinum-indium cylinder deposited at the International Bureau of Weights and Measures and declared as the international prototype of the kilogram by the First General Conference of Weights and Measures is called as *one kilogram.*

Unit of time (second, s)

1131, 566, 925, 974.7 of the length of the tropical year for 1900, the year commencing at 1200 hours universal time on the first day of January, 1900 is termed as *one second.*

Unit of electric current (ampere, A)

The constant current which flows in two parallel straight conductors of infinite length of negligible circular cross-section and placed at a distance of one metre from each other in vacuum producing a force of 2×10^{-7} New tons per metre length between the conductors is defined as an *ampere.*

Unit of thermodynamic temperature (degree Kelvin, K)

The degree interval of the thermodynamic scale on which the temperature of triple point of water is 273.16 degrees, is known *as one degree Kelvin.*

Units of luminous intensity (candela, cd)

One sixtieth part of luminous intensity normally emitted by one hundred millimetre square of integral radiator (black body) at the temperature of solidification of platinum is called as *one candela.*

The SI units make the use of multiples and sub-multiples 1000 times or 1/1000 times the unit quantity and in powers of 10^3 (*kilo*) or 10^{-3} (*milli*) in respect of still larger and smaller quantities respectively. The lengths are measured usually in kilometre (1 km = 1000 m), metre and millimetre (1 mm = 10^{-3} m).The symbols of units are not to be suffixed with V for plural.

The force is a derived quantity and physical law connecting the quantity to the fundamental quantities or previously obtained derived quantities is force = mass × acceleration. It is defined as that force which produces unit acceleration i.e., 1 m per $\sec^2$ in a unit mass of 1 kg. Its unit is Newton (N). Though, the Newton is a small unit, a still larger unit kN may be used. The intensity of force (viz., stress) due to 1 Newton over a unit area of one metre square is known as one *pascal.* It is denoted by symbol, Pa. (1 Pa = 1 N/m^2 and 10^6. Pa = 1 N/mm^2, viz. 1 MPa = 1 N/mm^2).

SI system of units have many advantages. The units are very handy. The burden of non-decimal coefficients in foot-pound second system is avoided. It has relatively large main units in contrast to centimetre-gram-second system. At the same time, it is closely related to centimetre-gram-second system of units. In practice, it results in perfectly reasonable number when the value of g = 10 m/$\sec^2$ is used instead of 9.806 m/$\sec^2$.

(Professor V.S. Mokashi, Visvesvaraya Regional College of Engineering, Nagpur in his paper titled as International System (SI) Units and their Application to Engineering published in Journal of Institution of Engineers, India, Vol. 19, March 1970 has highlighted the advantages and discussed SI units. A reference has been made to this paper).

Structural Engineering is the science and art of planning, design, construction, operation, maintenance and rehabilitation of structures. The term "structures" includes bridges, buildings and all types of civil engineering structures (towers, shells, etc.) composed of any structural material.

(Reference: Brochure, International Association for Bridge and Structural Engineering, Final Invitation to the International Conference, **Structural Eurocodes, Daros, Switzerland** September 14–16, 1992).

Author

Preface to the Eleventh Edition

In the subsequent editions of this book, since first edition published in 1970 uptil now, the author enhanced the text by adding useful matter, fresh topics such as column formulae for axial stress in compression, design of built-up and perforated cover plate columns, modified and adjusted interaction formulae, equivalent axial load method of design of eccentrically loaded columns, approximate method of design of combined footing, graphical method of curtailment of flange plates, corrugated aluminium sheets used for roof covering and several examples. The author also added further text of design of high strength friction grip bolts.

The eleventh edition of the book itself is a fourth edition in S.I. system of units (viz., *system international d' unites*) and revised, rewritten and updated as per the latest code (viz., 'Code of Practice for General Construction in Steel. IS : 800–1984) incorporating the revision of permissible stresses, effective length of the columns with idealised support conditions and columns in framed structures and Merchant Rankine formula for the allowable stresses. The concept of shear lag, design of semi-rigid connections, their behaviour (linear and non-linear) and methods of analysis have also been included.

The abbreviated symbols for Rolled Steel Sections as recommended in IS: 808–1989 have been used throughout the text of the book. Various definitions relating to the new and rational concept of **Wind-Load as per IS: 875 (Part III)–1987** have been given in Chapter 2. Accordingly Chapter 9 (viz. Design of Roof Trusses) has been completely revised and determination of wind load has been thoroughly described and illustrated.

Author expresses his sincere thanks to his colleagues, members of staff in various engineering colleges and students for appreciating the efforts made by them. The author also expresses his personal thanks for the Publisher Shri Rajinder Kumar Jain and Shri Sandeep Jain for getting the book prepared by latest technique and bringing out the book in such a nice getup.

Author shall welcome the suggestions from the readers for the further improvement of the book in fothcoming editions.

August 2013 **Dr. Ram Chandra**
Jodhpur

Preface to the First Edition

In this book, the author with his long teaching experience in the subject has made an attempt to present the subject matter of design of steel structures in a way which lays emphasis on the fundamentals, keeping in view the difficulties experienced by the students. Every basic principle, method, equation or theory has been presented in simplified manner. Metric system of units has been used throughout the text. Indian Standards Specifications have been followed. The book is intended for the use of degree, diploma and A.M.I.E. students in various branches of engineering. The book deals with design of structural members and their connections.

Each topic introduced is thoroughly described. A number of design problems including problems for examinations of the University of Jodhpur, and A.M.I.E.. has been solved to illustrate the theory and practice. Slide-rule computation accuracy is adequate for the design and has been followed. The chapters have been so arranged that it facilitates self-understanding of the subject, during study. In spite of careful scrutiny of the manuscript, it is possible that some typographical and computational errors are still left. The author shall be highly obliged to any one who brings these errors to his notice.

The author is thankful to Shri J.N. Srivastava and other colleagues who have very generously helped with their suggestions. The author is also thankful to the University of Jodhpur, Jodhpur and the Institution of Engineers, India, for following the use of their examinations, problems.

Suggestions from the readers for the improvement of the book are welcome.

September 11,1970 **Ram Chandra**

Contents

PART 1 DESIGN OF STEEL STRUCTURES

CHAPTER 12 DESIGN OF ROUND TUBULAR STRUCTURES 765–784

PART 2 DESIGN OF TIMBER STRUCTURES

CHAPTER 13 DESIGN OF TIMBER STRUCTURES 787–871

PART 3 DESIGN OF MASONRY STRUCTURES

IS CODES

Useful IS codes used in the Text of the book

1.	226–1975	Structural Steel (Standard quality)
2.	227	Galvanised (Plain and Corrugated) sheets
3.	459	Unreinforced Corrugated Asbestos Cement Sheets
4.	723	Mild Steel Wire Nails
5.	800–1984	Code of Practice for General Construction in Steel
6.	806	Use of Steel Tubes in General Building Construction
7.	808	Rolled Steel Beams, Channels and Angle Sections
8.	812–1957	Glossary of Terms Relating to Welding and Cutting Metals
9.	813–1961	Scheme for Symbols of Welding
10.	816–1969	Use of Metal Arc Welding for General Construction in Mild Steel
11.	819–1957	Resistançe Spot Welding for Light Assemblies in Mild Steel
12.	875–1984	Loading Standard for Structural Safety of Buildings
13.	961–1975	Structural Steel (High Tensile)
14.	1148–1973	Rivets Bars for Structural Purposes
15.	1149–1982	High Tensile Rivet Bars for Structural Purposes
16.	1161	Steel Tubes for Structural Purposes
17.	1173	Rolled Steel Sections, Tee-Bars
18.	1252	Rolled Steel Sections, Bulb-Angles
19.	1261–1959	Seam Welding in Mild Steel
20.	1323–1962	Code of Practice for Oxy-Acetylene Welding for Structural Works in Mild Steel
21.	1730	Dimensions for Steel Plates, Sheets and Strips for Structural and General Engineering Purposes
22.	1731	Dimensions for Steel Flats for Structural General Engineering Purposes
23.	1732	Dimensions for Round and Square Steel Bars for Structural and General Engineering Purposes
24.	1911	Schedule of Unit Weight of Materials
25.	1977–1975	Structural Steel (Ordinary quality)
26.	2062–1984	Weldable Structural Steel
27.	2585	Black Square Bolts and Nuts and Black Square Screws
28.	3139	Dimensions for Screw Threads for Bolts and Nuts
29.	3757–1972	High Tensile Friction Grip Fasteners for Structural Engineering Purposes
30.	4000–1967	Assembly for Structural Joints using High Tensile Friction Grip Fasteners
31.	456–1978	Code of Practice for Plain and Reinforced Concrete (Third Revision)
32.	6623–1972	High Tensile Friction Grip Nuts
33.	6639–1972	Hexagon Bolts for Steel Structures
34.	6649–1972	High Tensile Friction Grip Washers

Tables 1.5 Contd.

(1)	(2)	(3)	(4)	(5)	(6)
24.	Roofing				
	Allahabad tiles (including battens)				
	Single	–	0.83	85	m^2
	Double	–	1.67	170	m^2
	Country tiles (with battens)				
	Single	–	0.69	70	m^2
	Double	–	1.18	120	m^2
	Mangalore tiles with battens	–	0.64	65	m^2
	(tiles with mortar)	–	1.08	110	m^2
	Slates (on battens)	–	0.34–0.49	35–50	m^2
	Glazed with aluminium alloy bars for span upto 3 m	64 mm	0.19	19.5	m^2
	Glazed with lead covered steel bars at 0.6 m centres	6.4 mm	0.25–0.28	26–29	m^2
25.	Wall				
	Engg. bricks (in masonry)	–	23.55	2400	m^3
	Stone (masonry)				
	Granite ashlar	–	25.90	2640	m^3
	nibble	–	23.55	2400	m^3
	Lime ashlar	–	25.70	2560	m^3
	Marble dressed	–	26.5	2700	m^3
	Sandstone	–	22	2240	m^3

The dead loads are also termed as permanent loads as regards to the duration of their action.

1.23 LIVE LOADS

Live loads are the loads which vary in magnitude and/or in positions. Live loads are also known as *imposed* or *transient loads.* Live loads include any external loads imposed upon the structure when it is serving its normal purpose. Live loads are assumed to be produced by the intended use of occupancy in buildings including distributed, concentrated, impact and vibration and snow loads. Live loads are expressed as *uniformly distributed static loads* (*U.D.L.*). Live loads include the weight of materials stored, furniture and movable equipments. For buildings in most cities, the loads imposed on floors, stairs and roofs are specified in codes. For design purpose, certain live loads on floors have been assumed by many countries and which are almost in agreement. Some efforts have been made at the international level do decide live loads on floors and these have been specified in the International Standards 2103 (Imposed floor loads in residential and public building and 2633 (Determination of imposed floor loads in production buildings and warehouses). These codes have been published in the International Organisation of Standardization.

Code IS : 875 (Part 2)–1987 defines the *principal occupancy* for which a building or part of a building is used or intended to be used. The buildings are classified according to occupancy as under as per IS : 875 (Part 2) 1987. *An occupancy shall be deemed to include subsidiary occupancies which are contingent upon.*

1. Assembly buildings. The assembly buildings including any building or part of a building where groups of people gather together for amusement, recreation, social, religious, patriotic, civil, travel and similar purposes (e.g., theatres, motion picture houses, assembly halls, city halls, marriage halls, townhalls, auditoria, exhibition halls, museums, skating rinks, gymnasiums, restaurants, places of worship, dance halls, club rooms, passenger stations and terminals of air, surface and other public transportation services, recreation piers and stadia, etc.).

2. Business buildings. The business buildings include any building or part of a building, which is used to conduct business (other than that covered in 6) for maintaining of accounts and records for similar purpose, offices, banks, professional establishments, court houses, and libraries shall be classified in this group so far as principal function of these is to deal with public business and the keeping of books and records.

The office buildings are primarily used for office purposes, (e.g., purpose of administration, clerical work, handling money, telephone and telegraph operating and operating computers, calculating machines). The clerical work includes writing, book-keeping, sorting papers, typing, filing, duplicating, punching cards or tapes, drawing of matter for publication and the editorial preparation of matter for publication).

3. Educational buildings. The educational buildings include any building used for school, college or day-care purposes involving assembly for instructions, education or recreation and which is not covered in 1).

4. Industrial buildings. The industrial buildings include any building or a part of a building or structure in which products or materials of various kinds and properties are fabricated, assembled or processed like assembly plants, power plants, refineries, gas plants, mills, diaries, factories, workshops, etc.

5. Institutional buildings. Institutional buildings include any building or a part thereof, which is used for purpose, such as medical or other treatment in case of persons suffering from physical and mental illness, disease or infirmity; care of infants, convalescents of aged persons and for penal or correctional detention in which the liberty of the inmates is restricted. Institutional buildings ordinarily provide sleeping accommodation for the occupants. These include hospitals, sanitoria, custodial institutions or panel institutions (e.g., jails, prisons and reformations).

6. Mercantile buildings. These buildings include building or a part of a building which is used as shops, stores, market for display and sale of merchandise either wholesale or retail. Office, storage and service facilities incidental to the sale of merchandise and located in the same building shall be included under this group.

7. Residential buildings. The residential buildings include any building in which sleeping accommodation is provided for normal residential purposes with or without cooking a dining or both facilities (except building covered in 5). These buildings include one or multi-family dwellings, apartment houses (flats), lodging or rooming houses, restaurants, hostels, dormitories and residential hotels.

The *dwellings* include any building or part occupies by members of single/multi-family units with independent cooking facilities. These also include apartment houses (flats).

8. Storage buildings. The storage buildings include any building or part of building used primarily for the storage or sheltering of goods, wares or merchandize, (e.g., warehouses, cold storages, freight depots, transity sheds, store houses, garages, hangers, truck terminals, grain elevators, barns and stables).

The *imposed loads* are the loads assumed to be produced by the intended use or occupancy of a building, including the weight of movable partitions, distributed loads, concentrated loads, loads due to impact and vibration, and dust load but excluding wind, seismic, snow and other loads due to temperature changes, creep, shrinkage, differential settlement, etc. The imposed loads are the largest loads those probably will be produced by the intended use or occupancy, but shall not be less than the equivalent minimum loads specified in Table 1.6 subjected to any reductions allowed by code IS : 875 (Part 2)–1987.

The floors are investigated for both the uniformly distributed loads (UDL) and the corresponding concentrated loads specified in Table 1.6 and designed for the most adverse effects but these shall not be considered to act simultaneously. The concentrated loads specified in Table 1.6 may be assumed to act over an area of 0.3 m × 0.3 m. However, the concentrated loads may not be considered where the floors are capable of effective lateral distribution of this load.

Table 1.6 *Imposed floor loads for different occupancies*

[*As per IS : 875 (Part 2)–1987*]

Sl. No.	*Occupancy Classification*	*Uniformly Distributed Load (UDL)*	*Concentrated Load*
(1)	(2)	(3)	(4)
		(kN/m²)	(kN)
(I)	**RESIDENTIAL BUILDINGS**		
	(a) Dwelling houses:		
	(1) All rooms and kitchens	2.0	1.8
	(2) Toilet and bath rooms	2.0	–
	(3) Corridors, passages, staircases including fire escapes and store rooms	3.0	4.5

Contd.

Table 1.6 Contd.

(1)	(2)	(3)	(4)
		(kN/m^2)	(kN)
	(4) Balconies	3.0	1.5 per metre run concentrated at the outer edge
	(b) Dwelling units planned and executed in accordance with IS : 8888–1989* only :		
	(1) Habitable rooms, kitchens, toilet and bathrooms	1.5	1.4
	(2) Corridors, passages and staircases including fire escapes	1.5	1.3
	(3) Balconies	3.0	1.5 per metre run concentrated at outer edges
	(c) Hotels, hostels, boarding houses, lodging houses, dormitories, residential clubs :		
	(1) Living rooms, bed rooms and dormitories	2.0	1.8
	(2) Kitchens and laundries	3.0	4.5
	(3) Billiards room and public lounges	3.0	2.7
	(4) Storerooms	5.0	4.5
	(5) Dining rooms, cafeterias and restaurants	4.0	2.7
	(6) Office rooms	2.5	2.7
	(7) Rooms for indoor games	3.0	1.8
	(8) Baths and toilets	2.0	–
	(9) Corridors, passages, staircases including fire escapes, lobbies – as per the floor serviced (excluding stores and the like) but not less than	3.0	4.5
	(10) Balconies	Same as rooms to which they give access but with a minimum of 4.0	1.5 per metre run concentrated at the outer edge

Contd.

Table 1.6 Contd.

(1)	(2)	(3)	(4)
		(kN/m²)	(kN)
	(d) Boiler rooms and plant room – to be calculated but not less than	5.0	6.7
	(e) Garages :		
	(1) Garage floors (including parking area and repair workshop) for passenger cars and vehicles not exceeding 25 kN gross weight, including access ways and ramps – to be calculated but not less than	2.5	9.0
	(2) Garage floors for vehicles not exceeding 40 kN gross weight (including access ways and ramps) – to be calculated but not less than	5.0	9.0
(II)	**EDUCATIONAL BUILDINGS**		
	(a) Class rooms and lecture rooms (not used for assembly purposes)	3.0	2.7
	(b) Dining rooms, cafterias and restaurants	3.0†	2.7
	(c) Offices, lounges and staff rooms	2.5	2.7
	(d) Dormitories	2.0	2.7
	(e) Projection rooms	5.0	–
	(f) Kitchens	3.0	4.5
	(g) Toilets and bathrooms	2.0	–
	(h) Store rooms	5.0	4.5
	(i) Libraries and archives :		
	(1) Stack room/stack area	6.0 kN/m² for a minimum height of 2.2 m + 2.0 kN/m² per metre height beyond 2.2 m	4.5
	(2) Reading rooms (without separate storage)	4.0	4.5
	(3) Reading rooms (with separate storage)	3.0	4.5
	(j) Boiler rooms and plant rooms–to be calculated but not less than	4.0	4.5
	(k) Corridors, passages, lobbies, staircases including fire escapes – as per the floor services (without accounting for storage and projection rooms) but not less than	4.0	4.5

Contd.

Table 1.6 Contd.

(1)	(2)	(3)	(4)
		(kN/m²)	(kN)
	(l) Balconies	Same as rooms to which they give access but with a minimum of 4.0	1.5 per metre run concentrated at the outer edge
(III)	**INSTITUTIONAL BUILDINGS**	2.0	1.8
	(a) Bed rooms, wards, dressing rooms, dormitories and lounges		
	(b) Kitchens, laundries and laboratories	30	4.5
	(c) Dining rooms, cafeterias and restaurants	3.0†	2.7
	(d) Toilets and bathrooms	2.0	–
	(e) X-ray rooms, operating rooms, general storage areas – to be calculated but not less than	30	4.5
	(f) Office rooms and OPD rooms	2.5	2.7
	(g) Corridors, passages, lobbies and staircases including fire escapes – as per the floor services but not less than	4.0	4.5
	(h) Boiler rooms and plant rooms–to be calculated but not less than	5.0	4.5
	(i) Balconies	Same as the rooms to which they give access but with a minimum of 4.0	1.5 per metre run concentrated at the outer edge
(IV)	**ASSEMBLY BUILDINGS**		
	(a) Assembly areas :		
	(1) with fixed seals ‡	4.0	–
	(2) without fixed seats	5.0	3.6
	(b) Restaurants (subject to assembly), museums and an galleries and gymnasia	4.0	4.5
	(c) Projection rooms	5.0	–
	(d) Stages	5.0	4.5
	(e) Office rooms, kitchens and laundries	3.0	4.5
	(f) Dressing rooms	2.0	1.8
	(g) Lounges and billiards rooms	2.0	2.7
	(h) Toilets and bathrooms	2.0	–

Contd.

Table 1.6 Contd.

(1)	(2)	(3)	(4)
		(kN/m²)	(kN)
	(i) Corridors, passages, staircases including fire escapes	4.0	4.5
	(j) Balconies	Same as rooms to which they give access but with a minimum of 4.0	1.5 per metre run concentrated at the outer edge
	(k) Boiler rooms and plant rooms including weight of machinery	7.5	4.5
	(l) Corridors, passages subject to loads greater than from crowds, such as wheeled vehicles, trolleys and the like. Corridors, staircases and passages in grandstands	5.0	4.5
(V)	**BUSINESS AND OFFICE BUILDINGS** (See also 3.1.2)		
	(a) Rooms for general use with separate storage	2.5	2.7
	(b) Rooms without separate storage	4.0	4.5
	(c) Banking halls	3.0	27
	(d) Business computing machine rooms (with fixed computers or similar equipment)	3.5	4.5
	(e) Records/files store rooms and storage space	5.0	4.5
	(f) Vaults and strong room – to be calculated but not less than	5.0	4.5
	(g) Cafeterias and dining rooms	3.0†	2.7
	(h) Kitchens	3.0	2.7
	(i) Corridors, passages, lobbies and staircases including fire escapes – as per the floor serviced (excluding stores) but not less than	4.0	4.5
	(j) Bath and toilet rooms	2.0	–

Contd.

Table 1.6 Contd.

(1)	(2)	(3)	(4)
		(kN/m^2)	(kN)
	(k) Balconies	Same as rooms to which they give access but with a minimum of 4.0	1.5 per metre run concentrated at the outer edge
	(l) Stationary stores	4.0 for each metre of storage height	9.0
	(m) Boiler rooms and plant rooms – to be calculated but not less than	5.0	6.7
	(n) Libraries	See Sl. No. (II)	
(VI)	**MERCANTILE BUILDINGS**		
	(a) Retail shops	4.0	3.6
	(b) Wholesale shops – to be calculated but not less than	6.0	4.5
	(c) Office rooms	2.5	2.7
	(d) Dining rooms, restaurants and cafeterias	3.0†	2.7
	(e) Toilets	2.0	–
	(f) Kitchens and laundries :	3.0	4.5
	(g) Boiler rooms and plant rooms – to be calculated but not less than	5.0	6.7
	(h) Corridors, passages, staircases including fire escapes and lobbies	4.0	4.5
	(i) Corridors, passages, staircases subject to loads greater than from crowds, such as wheeled vehicles, trolleys and the like	5.0	4.5
	(j) Balconies	Same as rooms to which they give access but with a minimum of 4.0	1.5 per metre concentrated at outer edge
(VII)	**INDUSTRIAL BUILDINGS**		
	(a) Work areas without machines/equipment	2.5	4.5
	(b) Work areas with machinery/equipment§		
	1. Light duty — To be calculated but not less than	5.0	4.5
	2. Medium duty — To be calculated but not less than	7.0	4.5
	3. Heavy duty — To be calculated but not less than	10.0	4.5

Contd.

Table 1.6 Contd.

(1)	(2)	(3)	(4)
		(kN/m²)	(kN)
	(c) Boiler rooms and plant rooms – to be calculated but not less than	5.0	6.7
	(d) Cafeterias and dining rooms	3.0†	2.7
	(e) Corridors, passages and staircases including fire escapes	4.0	4.5
	(f) Corridors, passages, staircases subject to machine loads, wheeled vehicles – to be calculated but not less than	5.0	4.5
	(g) Kitchens	3.0	4.5
	(h) Toilets and bathrooms	2.0	–
(VIII)	**STORAGE BUILDINGS ¶**		
	(a) Storage rooms (other than cold storage) warehouses – to be calculated based on the bulk density of materials stored but not less than	2.4 kN/m² per each metre of storage height with a minimum of 7.5 kN/m²	7.0
	(b) Cold storage – to be calculated but not less than	50 kN/m² per each metre of storage height with a minimum of 15 kN/m²	9.0
	(c) Corridors, passages and staircases including fire escapes – as per the floor serviced but not less than	4.0	4.5
	(d) Corridors, passages subject to loads greater than from crowds, such as wheeled vehicles, trolleys and the like	5.0	4.5
	(e) Boiler rooms and plant rooms	7.5	4.5

* Guide for requirements of low income housing.

† Where unrestricted assembly of persons is anticipated, the value of UDL should be increased to 40 kN/m².

‡ 'With fixed scats' implies that the removal of the sealing and the use of the space for other purposes is improbable. The maximum likely load in this case is, therefore, closely controlled.

§ The loading in industrial buildings (workshops and factories) varies considerably and so three loadings under the terms 'light', 'medium' and 'heavy' are introduced in order to allow for more economical designs but the terms have no special meaning in themselves other than the imposed load for which the relevant floor is designed. It is, however, important particularly in the case of heavy weight loads, to assess the actual loads to ensure that they are not in excess of 10 kN/m^2; in case where they are in excess, the design shall be based on the actual loadings.

¶ For various mechanical handling equipments which are used to transport goods, as in warehouses, workshops, store rooms, etc. the actual load coming from the use of such equipment shall be ascertained and design should cater to such loads.

Note 1. *Where no values are given (in Table 1.6) for concentrated load, it may be assumed that the tabulation distributed load is adequate for design purposes.*

Note 2. *The loads specified in Table 1.6 are equivalent uniformly distributed loads on the plan area and provide for normal effect of impact and acceleration. These loads do not take into consideration special concentrated loads and other loads.*

Note 3. *Where the use of an area or floor is not provided in the Table 1.6, the imposed load due to the use and occupancy of such an area shall be determined from the analysis of the loads resulting from :*

(i) Weight of the probable assembly of persons; (ii) Weight of the probable accumulation of equipment and furnishing; (iii) Weight of the probable storage materials; and (iv) Impact factor, if any.

While designing columns, abutments, piers, walls, their supports and foundations, the live loads on floors are reduced as in Table 1.7 as per IS : 800–1984.

Table 1.7 *Percent reduction of total live load*

Number of floors carried by member under consideration	*Percent reduction in total live load on all floors above the member under consideration*
1	0
2	10
3	20
4	30
5 or more	40

Reduction in live load shall not be made in the case of warehouses, garages and other buildings used for storage purposes and for factories and workshops designed for 5 kN/m^2. However, above reductions are made for buildings such as factories and workshops designed for a live load more than 5 kN/m^2 provided that the loading assumed for any column, etc., is not less it would have been if the floors had been designed for 5 kN/m^2 with no reduction.

As per IS : 875–1984, where a single span of beam or girder supports not less than 50 m^2 of floor at one general level, the live load may be reduced in the design of beam or girder by 5 percent, for each 50 m^2 supported, subjected to a maximum reduction of 25 percent. This reduction or that given in Table 1.7, whichever is greater, may be taken into account in the design of columns, supporting such a beam. These reductions are not applicable for the floors used for storage purposes, in the weight of any plant or machinery which is specifically allowed for.

1.24 WIND LOAD

The wind loads are the *transient loads.* Relative to the surface of earth, the natural air remains in motion, (i.e., a current of air flows near the surface of earth). The difference in solar and terrestrial radiations setting up irregularities in temperatures (which gives rise to convection either upwards or downwards. The wind usually blows horizontal to the ground at high wind speeds. The vertical components of atmospheric motion are relative small. The term wind denotes almost exclusive the horizontal wind. The vertical winds are always identified as such. The wind speeds are assessed with the aid of anemometers or anemographs which are installed at meteorological observatories at heights generally varying from 10 m to 30 m above ground. Very weak winds (i.e., winds at low speeds say in the range of 5 kmph to 10 kmph) are called breeze. The strong winds (i.e., winds at speed say, 50 kmph) are generally associated with cyclonic storms. Very strong winds (i.e., winds at speed say 80 kmph) are also associated with cyclonic storms, thunderstorms, dust storms or vigorous monsoons. The cyclonic storms weaken rapidly after crossing the coast and move as depressions in land. A severe storm may have wind speed of 60 kmph to 120 kmph. The winds of very high speeds and very short duration (hurricanes) are called **Kal Baisaki** or **Norwesters** occur fairly frequently during summer months over North East India.

In addition to wind at any time, there are effects of gusts (sudden blast of wind, which may last for few seconds) which should be considered. These gusts cause increase in air pressure but their effect on the stability of the building may not be so important. Often, the gusts influence only part of the building and the increased local pressures may be more than balanced by a momentary reduction in pressure elsewhere. Short period gusts may not cause any appreciable increase in stress in main components of buildings (because of inertia of the building), although the walls, roof sheetings and individual cladding units (glass panels) and their supporting members such as purlins, sheeting rails and glazing bars may be more seriously affected. Gusts may also be extremely important for design of structures with slenderness ratios.

The liability of a building or a structure to high wind pressures depends not only upon the geographical location and proximity of other obstructions to air flow but also upon the characteristics of the structure itself. In general, wind speed in the atmospheric boundary layer increases with height from zero at ground level to maximum at a height called the gradient height. The variation of wind with height depends primarily on the terrain conditions. However, the wind speed at any height never remains constant and it has been found convenient to resolve its instantaneous magnitude into an average or mean value and a fluctuating component around this average value. The average value depends on the average time employed in analysing the meteorological data, which varies from a few seconds to several minutes. The magnitude of fluctuating component of the wind speed is called gust, it depends upon averaging time. In general, smaller the averaging interval, greater is the magnitude of the gust speed.

The wind load depends upon terrain, height of the structure, and the shape and size of structure. For pitched roof, wind load has been considered and described further in detail in Chapter 9 (Design of Roof Trusses). However, it is essential to know the following terms to study the new concept of wind as described in IS: 875 (Part 3) – 1987.

1. Angle of attack. It is defined as the angle between the direction of wind and a reference axis of the structure

2. Breadth. It means horizontal dimension of the building measured normal to the direction of wind.

Note. *The breadth and depth are dimensions measured in relation to the direction of wind. Whereas length and width are dimensions related to the plan of the building.*

3. Depth. It means the horizontal dimension of the building measured in the direction of the wind.

4. Developed height. It is height of upward penetration of the velocity profile in a new terrain.

At larger fetch lengths, such penetration reaches the gradient height, above which the wind speed may be taken to be constant. At lesser fetch lengths, a velocity profile of a smaller height but similar to that of the fully developed profile of that terrain category has to be taken. The velocity at the top of this shorter profile equals that of the unpenetrated earlier velocity profile at that height.

5. Effective frontal area. The projected area of the structure normal to the direction of wind.

6. Element of surface area. The area of the surface over which the pressure coefficient is taken to be constant.

7. Force coefficient. It is a non-dimensional coefficient such that the total wind force on a body is the product of the force coefficient, the dynamic pressure of the incident design wind speed and the reference area over which the force is required.

In the direction of wind, this coefficient is called as *drag coefficient*. In the perpendicular direction of wind, it is called as *lift coefficient*.

8. Ground roughness. The nature of the earth's surface as affected by small scale obstructions such as trees and buildings (as distinct from topography) is called ground roughness.

9. Gust. A positive or negative departure of wind speed from its mean value, lasting not more than say, 2 minutes over a specified interval of time.

10. Peak gust. Peak gust or peak gust speed is the wind speed associated with the maximum amplitude.

11. Fetch lengths. It is the distance measured along the wind from a boundary at which a change in the type of terrain occurs.

When the changes in terrain types are encountered (such as, the boundary of a town or city, forest, etc.), the wind profile changes in character but such changes are gradual and start at ground level, spreading or penetrating upwards with increasing fetch length.

12. Gradient height. It is the height above the mean ground level at which the gradient wind blows as a result of balance among pressure gradient force, coriolis force and centrifugal force.

It is taken as height above the mean ground level, above which the variation of wind speed with height need not be considered.

13. Mean ground level. It is the level of average horizontal plane of the area enclosed by the boundaries of the structure.

14. Pressure coefficient. It is the ratio of the difference between the pressure acting at a point on a surface and the static pressure of the incident wind to the design wind pressure.

Where the static and design wind pressures are calculated at the height of the point considered after taking into consideration the geographical location, terrain conditions and shielding effect. Also

$$\text{Pressure coefficient} = \left[1 - \left(\frac{V_p}{V_z}\right)^2\right]$$

where, V_p = actual wind speed at any point on the structure at a height corresponding to that of V_z.

Note. *When the sign of pressure coefficient is positive, it shows that the pressure acts towards the surface and the negative sign shows that the pressure acts away from the surface.*

15. Return period. It is the number of years, the reciprocal exceeding a given wind speed in any one year.

16. Shielding effect. It refers to the condition where wind has to pass along some structure(s) or the structural element(s) located on the upstream wind side, before meeting the structure or structural element under consideration.

A factor called *shielding factor* is used to consider such effects in estimating the force on the shielded structures.

17. Suction. It means the pressure less than the atmospheric (static) pressure and it is taken to act away from the surface.

18. Solidity ratio. It is equal to the effective area (projected area of all the individual elements) of a frame normal to the direction of wind divided by the area enclosed by the boundary of the frame normal to the direction of wind.

Note. *It is to be calculated for the individual frame.*

19. Terrain category. It means the characteristics of the surface irregularities of an area which arise from natural or constructed features. The categories are numbered in the increasing order of roughness.

20. Velocity profile. The variation of horizontal component of the atmospheric wind speed at different heights above the mean ground level is termed as velocity ratio.

21. Topography. The nature of the earth's surface as influenced by the hill and valley configuration.

1.25 SNOW LOAD

The snow load depends upon latitude of place and atmospheric humidity. The snow load acts vertically and it is expressed in kilo-Newtons per square metre of plan area. The actual load due to snow depends upon the shape of the roof and its capacity to retain the snow. When actual data for snow load is not available, snow load may be assumed to be 25 N/m^2 per mm depth of snow.

It is usual practice to assume that snow load and maximum wind load will not be acting simultaneously on the structure.

1.26 SEISMIC LOAD (EARTHQUAKE FORCE)

It becomes essential to consider 'seismic load' in the design of structure, if the structure is situated in the seismic areas. The seismic areas are the regions which are geologically young and unstable parts and which have experienced earthquakes in the past and are likely to experience earthquakes in future. The Himalayan region, Indo Gangetic Plain, Western India, Cutch and Kathiawar are the places in our country which experience earthquakes frequently. Sometimes these earthquakes are violent also. Seismic load is caused by the shocks due to an earthquake. The earthquakes range from small tremors to severe shocks. The earthquake shocks cause movement of ground, as a result of which the structure vibrates. The vibrations caused because of earthquakes may be resolved in three perpendicular directions. The horizontal direction of vibration dominates over other directions. In some cases structures are designed for horizontal seismic forces only and in some cases horizontal seismic forces and vertical seismic forces both are taken into account. The seismic accelerations for the design may be arrived at from *seismic coefficient,* which is defined as the ratio of acceleration due to earthquakes and acceleration due to gravity. This country has been divided into seven zones for determining seismic coefficients. The seismic coefficients have also been recommended for different types of soils for the guidance of designers. IS : 1893–1962 Indian Standard Recommendations for Earthquake Resistant Design of Structure, may be referred to for actual design.

1.27 SOIL AND HYDROSTATIC PRESSURE

The pressure exerted by soil or water or both should be taken into consideration for the design of structures or parts of structures which are below ground level. The soil pressure and hydrostatic pressure may be calculated from established theories.

1.28 ERECTION EFFECTS

The erection effects include all effects to which a structure or a part of structure is subjected during transportation of structural members, and erection of structural member by equipments. Erection effects also take into account the placing or storage of construction materials. The proper provisions shall be made, e.g., temporary bracings, to take care of all stresses caused during erection. The stresses developed because of erection effects should not exceed allowable stresses.

1.29 DYNAMIC EFFECTS (IMPACT)

The moving loads on a structure cause vibrations and have also impact effect. The dynamic effects resulting from moving loads are accounted for, by impact factor. The live load is increased by adding to it the impact load. The impact load is determined by the product of impact factor and live load. The effect of impact has been described in Chapter 8 (Design of Plate Girder).

1.30 TEMPERATURE EFFECTS

The variations in temperature results in expansion and contraction of structural material. The range of variation in temperature varies from localities to localities, season to season and day to day. The temperature effects should be accounted for properly and adequately. The allowable stress should not be exceeded by stress developed because of design loads and temperature effects.

1.31 LOAD COMBINATIONS

All the parts of the steel structures shall be capable of sustaining the most adverse combination of the dead loads, prescribed live (imposed) loads, wind loads, earthquake loads where applicable and any other forces or loads to which the steel structure may reasonably be subjected without exceeding the stresses specified in this standard.

The load combinations for dsign purposes shall be the one that produces maximum forces and effects and consequently maximum stresses from the following combinations.

(i) Dead load + imposed (live) loads,

(ii) Dead load + imposed (live) loads + wind or earthquake loads, and

(iii) Dead load + wind load or earthquake loads.

1.32 STRESSES

When a structural member is loaded, deformation of the member takes place, and resistance is set up against deformation. This resistance to deformation is known as stress. The direct stress is defined as force per unit cross-sectional area. The nature of stress developed in the structural member depends upon nature of loading on the member. The following are the various types of stresses:

1. Axial stress (Direct stress) :

   (i) Tensile stress (ii) Compressive stress
2. Bearing stress
3. Bending stress
4. Shear stress.

The bending stress and shear stress have been discussed in Chapter 6, 'Design of Beams'.

A member may be subjected to combined direct and bending stresses. Such stress is known as combined stress.

Sign convention for stresses. The tensile stresses are taken as positive and compressive stresses as negative. This sign convention for stresses is convenient as a structural member elongates on application of tensile load and shortens on application of compressive load.

1.33 STRESS–STRAIN RELATIONSHIP FOR MILD STEEL

When a mild steel bar is subjected to a tensile load, it elongates. The *elongation* per unit length is known as *strain*. The stress is proportional to strain within limit of proportionality. The stress-strain relationship for mild steel can be studied by plotting stress-strain curve. The stress and load on some suitable scales may be plotted on y-axis, and strain on some suitable scale may be plotted on x-axis.

When the tensile load increases with increases in strain, stress–strain curve follows a straight line relationship upto *'Limit of proportionality'*. The limit of proportionality is defined as stress beyond which straight line relationship ceases between stress and strain. Beyond the limit of proportionality stress approaches the *elastic limit*. The elastic limit is defined as the maximum stress upto which a specimen regains its original length on the removal of the applied load. There is hardly any distinct difference in the positions of *limit of proportionality* and *elastic limit*. Practically, position of limit of proportionality coincides with the elastic limit. When the specimen is loaded beyond the elastic limit, the specimen does not resume its original length on the removal of applied load and a little strain is left in the specimen. This little strain is known as *residual strain* or *permanent set*. When the tensile load further increases the stress–reaches *'yield stress'* and material starts yielding. The stress–strain curve suddenly falls showing a decrease in stress. The distinct position from where sudden fall of curve occurs marks the *upper yield point* and the position upto which fall of curve occurs is *known* as *lower yield point*. The material of bar stretches suddenly at constant stress. The adjustment of stress takes place in the elements of material in between upper yield point and lower yield point. On further increase of load, stress increases with the increase of strain. However, strain increases more rapidly. Finally the load reaches the value of *'ultimate load'*. The ultimate load is defined as maximum load, which can be placed prior strain increases more rapidly. Finally the load reaches the value of *'ultimate toad'*. The ultimate load is defined as maximum load, which can be placed prior to the breaking of specimen. The stress corresponding to the ultimate load is known as *'ultimate stress'*. The stress–strain curve suddenly falls with rapid increase in strain and specimen breaks. The load corresponding to breaking position is known as "breaking load'. The cross-section of specimen decreases. If actual breaking stress is computed on the basis of decreased (reduced) cross-sectional area, the breaking stress will be found to be more than the ultimate stress.

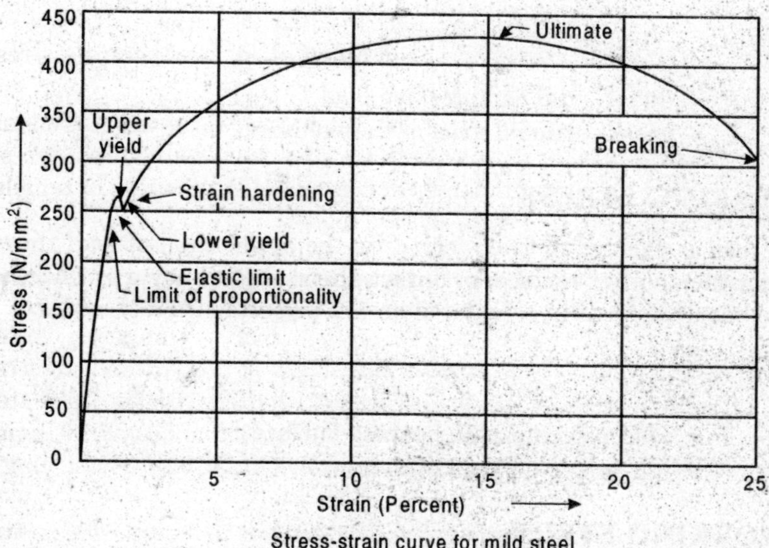

Fig. 1.8 *Stress–strain curve*

1.33.1 Significance of Upper Yield Point

The boundaries of grains of mild steel are composed of brittle material. This forms a rigid skeleton. The rigid skeleton prevents plastic deformation of the grains at low stress, and shows upper yield point in stress-strain curve. At upper yield point, this rigid skeleton breaks down. As a result of this, the stress in material drops down without elongation from upper yield point to lower yield point. This is followed by sudden stretching of the material at constant stress from lower yield point upto strain hardening.

1.34 TENSILE STRESS

When a structural member is subjected to direct axial tensile load, the stress is known as *tensile stress* (σ_{at}). The tensile stress is calculated on net cross-sectional area of the member

$$\sigma_{at} = \left(\frac{P_t}{A_n}\right)$$

where P_t is the direct axial tensile load and A_n is the net cross-sectional area of the member.

1.35 COMPRESSIVE STRESS

When a structural member is subjected to direct axial compressive load, the stress is known as *compressive stress,* (σ_{ac}). The compressive stress is calculated on gross cross-sectional area of the member

$$\sigma_{ac} = \left(\frac{P_c}{A_g}\right)$$

where, P_c = direct axial compressive load

A_g = gross-sectional area of the member.

1.36 BEARING STRESS

When a load is exerted or transferred by the application of load through one surface for the another surface in contact, the stress is known as *'bearing stress'* (σ_p). The bearing stress in calculated on net projected area of contact

$$\sigma_p = \left(\frac{P}{A}\right)$$

where, P = load placed on the bearing surface

A = net projected area of contact

1.37 WORKING STRESS

The working stress is also termed as *allowable stress* or *permissible stress*. The working stress is evaluated by dividing yield stress by factor of safety. For the purpose of computing safe load carrying capacity of a structural member, its strength is expressed in terms of working stress. The working stress is the stress which may be developed or set up in the member without causing structural damage to it. The actual stresses resulting in a structural member from design loads should not exceed working stresses. This ensures the safety of the structural member. The maximum working stresses are adopted from IS : 800–1984.

IS : 800–1984 specifies a number of grades of steel with different yield strengths, the design parameters and the geometrical properties. There are also developments in the design of steel structures. In view of these, the permissible stresses have been revised in general. These permissible stresses have been expressed to the extent possible in terms of the yield strength of the material. However, the specific values have also been given in IS : 800–1984 for steels commonly used.

1.38 INCREASE IN PERMISSIBLE STRESSES

A structure may be subjected to the different combinations of loads. The possible combinations of the loads with special reference to the roof trusses have been described in Chapter 9. These loads in such combinations do not act for long period. Most of the national codes allow some increase in permissible stresses. Increases in permissible stresses as per IS : 800–1984 is taken as follows :

(i) When the effect of wind or seismic load is taken into account, the permissible stresses in steel are increased by $33\frac{1}{3}$ per cent.

(ii) For rivets, bolts and tension rods, the permissible stresses are increased by 25 per cent, when the effect of wind or seismic load is taken into account.

The increased values of permissible stresses must not exceed yield stress of the material.

1.39 FACTOR OF SAFETY

The *factor of safety* is defined as the factor by which the yield stress of the material is divided to give the working stress (permissible stress) in the material. A greater value of factor of safety used to be adopted formerly and as such the working stresses computed were small. As a result of this a large cross-section of the member had to be adopted in design. From experience gained and experiments made in laboratories, it has been seen that such a large section of the structural member is not essential. Hence, factor of safety adopted now-a-days is comparatively small. This has resulted in appreciable saving in the material. The value of factor safety is decided keeping in view of the following considerations.

1. The average strength of materials is determined after making tests on number of specimens. The strengths of different specimens of given structural material are not identical. Some of the specimens may show relatively large deviations in strength from the average.
2. The values of design loads remains uncertain. The values of dead loads can of course be determined correctly. But live load, impact load, wind load, snow load etc. cannot be determined with certainty since these depend upon statistics available. The probable values of these loads are only determined.
3. The values of internal forces in many structures depend upon the methods of analysis. The degree of precision of different method varies. The methods involving detailed analysis are more precise. In case, analysis of the structure is done precisely, a small value of factor of safety may be adopted.
4. During fabrication, structural steel is subjected to different operations. The punching of a hole in a structural element distorts the surrounding material and causes high residual stresses. The warping and buckling of elements may take place during welding. The welding leaves high residual stresses. Structural elements are subjected to uncertain erection stress.
5. The variations in temperatures and settlement of supports are uncertain. Many times, a well designed structure is damaged because of these effects. The strength of materials decreases because of corrosion. The extent of corrosion is more, when a structure is located in industrial areas and exposed to chemical wastes.
6. The failure of some small structures or some elements of a structure is less serious and less disastrous than the failure of large structure or a main element of a structure.

The minimum value of factor of safety, m_{min} may be found by defining it in a rational method. The factor of safety may be defined as the ratio of computed strength, P of the structure or the structural member to the respective computed internal load, F. Therefore,

$$m = \left(\frac{P}{F}\right) \quad \text{...(1.1)}$$

The magnitude of lowest probable strength, P is influeneed by the uncertainties in the mechanism failure, properties of material and workmanship. The magnitude of highest probable internal load, F is influenced by the uncertainties in loading conditions and the structural behaviour. The probable values deviate from the computed value because of these uncertainties. The minimum probable value of strength is given by

$$P_{min} = (P - \Delta P) \quad \text{...(i)}$$

The maximum probable value of the internal load is given by

$$F_{max} = (F + \Delta F) \quad \text{...(ii)}$$

where, ΔP = Maximum probable deviation of actual value from the computed value of strength

ΔF = Maximum probable deviation of actual value from the computed value of internal load

When the minimum probable value of strength is atleast equal to or slightly more than the maximum probable value of the internal load, then the failure of structure is just prevented under most severe conditions. Therefore,

$$(P - \Delta P) \geq (F + \Delta F) \quad \text{...(iii)}$$

or

$$P\left(1 - \frac{\Delta P}{P}\right) \geq F\left(1 + \frac{\Delta F}{F}\right) \quad \text{...(iv)}$$

The minimum value of factor of safety is given by

$$m_{min} = \left(\frac{P}{F}\right) = \left(\frac{\left(1 + \frac{\Delta F}{F}\right)}{\left(1 - \frac{\Delta P}{P}\right)}\right) \quad \text{...(1.2)}$$

Assuming the values of maximum deviations ΔP and ΔF to 25 percent of the computed values of P anbd F respectively, the minimum value of factor of safety is given by

$$m_{min} = \left(\frac{P}{F}\right) = \left(\frac{1 + 0.25}{1 - 0.25}\right)$$

$$= 1.67 \quad \text{...(v)}$$

1.40 METHODS OF DESIGN

All parts of the steel framework of a structure shall be capable of sustaining the most adverse combination of the dead loads, the prescribed superimposed roof and floor loads, wind loads, seismic forces where applicable, supperimpose roof and floor loads, wind loads, seismic forces where applicable, and any other forces or loads to which the building may be reasonably subjected without exceeding the permissible stresses. The following methods may be employed for the design of the steel framework :

(i) Simple design
(ii) Semi-rigid design
(iii) Fully rigid designs, and
(iv) Plastic design.

1.40.1 Simple Design

This method is based on elastic theory and applies to structures in which the end connections between members are such that they will not develop restraint moments adversely affecting the members and the structures as a whole and in consequence the structure may, for the purpose of design, be assumed to be pin jointed. The method of simple design involves the following assumptions:

(a) The beams are simply supported.
(b) All connections of beams, girders, or trusses are virtually flexible and are proportioned for the reaction shears applied at the appropriate eccentricity.
(c) The members in compression are subjected to forces applied at the appropriate eccentricities.
(d) The members in tension are subjected to longitudinal forces applied over the net area of the section.
(e) The plane sections normal to the axis remain plane after bending. The stress strain relationship for steel is linear.

1.40.2 Semi-rigid Design

This method as compared with the simple design method permits a reduction in the maximum bending moment in beams suitably connected to their supports, so as to provide a degree of direction fixity. In the case of triangulated frames, it permits rotation account being taken of the rigidity of the connections and the moment of interaction of members. In cases where this method of design is employed, it is ensured that the assumed partial fixity is available and calculations based on general or particular experimental evidence shall be made to show that the stresses in any part of the structure are not in excess of those laid down in IS : 800–1984.

1.40.3 Filly-rigid Design

This method as compared to the methods of simple and semi-rigid design assumes that the end connections are fully rigid and are capable of transmitting moments and shears. It is also assumed that the angle between the members at the joint does not change, when it is subjected to loading. This method gives economy in the weight of steel used when applied in appropriate cases. The end connections of members of the frame shall have sufficient rigidity to hold virtually unchanged original angles between such members and the members they connect. The design should be based on accurate methods of elastic analysis and calculated stresses shall not exceed permissible stresses.

1.40.4 Plastic Design

The method of plastic analysis and design is recently developed. The method was thought of around 1935. However, even today all the problems related to this are not decided. In this method, the structural usefulness of the material is limited upto to ultimate load. This method has its main application in the analysis and design of statically indeterminate frame structures. This method provides striking economy as regards the weight of the steel. This method provides the margin of safety in terms of load factor which one is not less than provided in elastic design. A load factor of 1.85 is adopted for dead load plus live load and 1.40 is adopted for dead load, live load and wind or earthquake forces. The deflections under working loads should not exceed the limits prescribed in IS : 800–1984.

The method of plastic analysis and design has been described in Author's Vol. II (Design of Steel Structures).

1.41 DEFINITIONS

Following definitions concerning the mechanical properties of the steel are useful for the study of this subject.

1.41.1 Modulus of Elasticity

The modulus of elasticity is defined as the ratio of longitudinal stress to the longitudinal strain within the elastic region. It is denoted by letter *E*, the slope of elastic portion of the stress–strain diagram gives the value of modulus of elasticity, *E*.

1.41.2 Shear Modulus of Elasticity

The shear modulus of elasticity is defined as the ratio of shearing stress to the shearing strain within the elastic range. It is denoted by letter *G*. The shear modulus of elasticity is also known as modulus of rigidity. It is given by

$$G = \left(\frac{E}{2(1+\mu)}\right) \quad \text{...(1.3)}$$

where μ = Poisson' s ratio

1.41.3 Bulk Modulus of Elasticity

The bulk modulus of elasticity is defined as the ratio of hydrostatic stress to the volumetric strain within the elastic range. It is denoted by letter *K*. It is given by

$$K = \left(\frac{E}{3(1-2\mu)}\right) \quad \text{... (1.4)}$$

It is also given by

$$K = \left(\frac{2G(1+\mu)}{3(1-2\mu)}\right) \quad \text{... (1.5)}$$

1.41.4 Tangent Modulus of Elasticity

The tangent modulus of elasticity is defined as the slope of the tangent at a point on the stress–strain curve above the limit of proportionality. It is denoted by *Et*.

1.41.5 Poisson's Ratio

The Poisson's ratio is defined as the ratio of transverse strain to the longitudinal strain under an axial load. It is denoted by letter μ. The value of Poisson's ratio for steel within the elastic region ranges from 0.25 to 0.33.

1.41.6 Yield Point

The yield point on a stress–strain diagram is defined as the stress at which the strain increases to a large value without an increase of stress. For the mild steel, it is indicated by a flat portion of stress.strain diagram if the stress–strain diagram is plotted on the exaggerated scale. Some steels show upper yield point and lower yield point.

1.41.7 Yield Strength

Some steels do not show yield plateau, i.e., the flat portion on the stress–strain diagram at yield. The stress–strain diagram for such steels show continuous curve upto the tensile strength. Therefore, the yield strength of such steel is defined as a stress shown by a specific point on the curve obtained by drawing 0.2 per cent offset of the strain parallel to the initial elastic portion of the curve.

1.41.8 Tensile Strength

The tensile strength of a test piece is defined as maximum axial load divided by the original area of cross-section.

1.41.9 Fatigue Strength

The fatigue strength is defined as the stress at which steel fails under repeated applications of load.

1.41.10 Impact Strength

The impact strength of steel is the measure of ability of the steel to absorb energy at high rates of loading.

1.41.11 Strain Hardening Modulus

The strain hardening modulus is defined as the slope of the stress–strain curve in the strain–hardening range. It is denoted by E_{st}. It has the maximum value at the beginning of strain–hardening.

1.41.12 Ductility

The ductility is the unique property of steel. It is the property of steel by virtue of which, the deformation of structural member/structure occurs before failure. It is measured quantitatively by *percentage of elongation* of test specimen at failure in a uniaxial test. A gauge length, (L_{GL} = 5.65 $(A_0)^{1/2}$, where A_o is the original cross-sectional area of the bar) is marked in the middle of the bar. The bar is tested upto failure. Its elongated length, L_{EL} is measured. Then, the percentage elongation = $\left(\frac{L_{EL} - L_{GL}}{L_{GL}}\right) \times 100$. For a good ductile steel the percentage elongation should be 20 to 33 percent. The ductility is also measured by *percentage* of *reduction* in area. It is given by $\frac{(A_o - A_R)}{A_o} \times 100$ where A_o and A_R are the original and reduced cross-sectional areas of the bars. For good ductile steel bars, the percentage of reduction may be 65 percent.

1.42 STABILITY OF STRUCTURE

According to the *stability* requirement, the stability of a structure as a whole against overturning is ensured so that the *restoring moment* is greater than the maximum *over-turning moment.*

1.42.1 Overturning

The stability of a structure as a whole against overturning is ensured such that the restoring moment, M_R shall be not less than the sum of 12 times the maximum overturning moment due to the characteristic dead load, $M_{o \cdot CDL}$ and 1.4 times the maximum overturning moment due to characteristic imposed loads, $M_{o \cdot CIL}$. i.e.,

$$M_R \nless (1.2\, M_{o.CDL} + 1.4\, M_{o.CIL})$$

In cases, where dead load provides the restoring moment, only 0.9 times the characteristic dead load shall be considered. The restoring moment due to imposed loads shall be neglected.

The anchorages or counterweights against for overhanging members (during construction and service) should be such that the static equilibrium should be maintained, even when the overturning moment is doubled.

1.42.2 Sliding

The structure should have adequate factor of safety against *sliding* due to the most adverse combination of the applied loads.

The structure shall have a factor of safety against sliding not less than 1.4 under the most adverse combination of the applied characteristic forces. In case only dead loads are acting, only 0.9 times the characteristic dead load shall be taken into account.

To ensure stability at all times, account shall be taken of probable variations in dead load during construction, repair or other temporary measures. The wind and seismic loading shall be treated as imposed loading.

In designing the framework of a building, provisions shall be made by adequate moment connections or by a system of bracings to effectively transmit all the horizontal fores to the foundations.

1.43 STIFFNESS OF STRUCTURE

In order to meet the requirement of *stiffness* for the structure as a whole and for the structural members, the limitations for the deflections are specified. The deflection of the structure or part thereof should not adversely affect the appearance or efficiency of the structure or finishes or partitions.

Chapter 2

Design of Riveted, Bolted and Pin Connections

2.1 INTRODUCTION

The various types of connections used for connecting the structural members are given below :

1. Riveted connections
2. Bolted connections
3. Pin connections
4. Welded connections.

These connections are named after the type of fastening (viz., rivets, bolts and nuts, pins and welds) used for connecting the structural members. In the *riveted connections,* rivets are used. These rivets are the permanent fastenings. Once the rivets are driven, these rivets cannot be opened without considerable labour and without complete destruction.

The perfect theoretical analysis for stress distribution in riveted connections cannot be established. Hence a large factor of safety is employed in the design of riveted connections. The riveted connections should be as strong as the structural members. No part in the riveted connections should be so overstressed. These riveted connections should be so designed that there is neither any permanent distortion nor any wear. These should be elastic.

In general, the work of fabrication is completed in the workshops where the steel is fabricated.

2.2 RIVETS

A piece of round steel forged in place to connect two or more than two steel members together is known as a rivet.

The rivets for structural purposes are manufactured from mild steel and high tensile rivet bars. A rivet consists of a *head* and a *body* as shown in Fig. 2.1.

The body of rivet is termed as *shank*. The rivets are manufactured in different lengths to suit different purposes. The size of rivets is expressed by the diameter of the shank.

The most rivets used in structural steel work are heated uniformly throughout its length, without burning or excessive scaling, and shall be of sufficient length to provide a head of standard dimension when these rivets are driven, these rivets shall be completely fill the holes. These rivets are known as *hot driven rivets.*

For driving the rivets, they are heated till they become red hot and are then placed in the hole. Keeping the rivets pressed from one side, a number of blows are applied and a head at the other end is formed. The hot-driven rivets are divided in the following three types, according to the method of rivet-driving:

1. Power driven rivets
2. Hand driven rivets
3. Field rivets.

The rivets of the first category are more satisfactorily driven than the rivets of the other two categories and rivets of second category are more satisfactorily driven than rivets of the third category because of the difference in workmanship. Consequently strength of a rivet is different for distinctly seen in the working stresses for these rivets given in Table 2.1. The hot driven rivets of 16 mm, 18 mm, 20 mm and 22 mm diameter are used for the structural steel works.

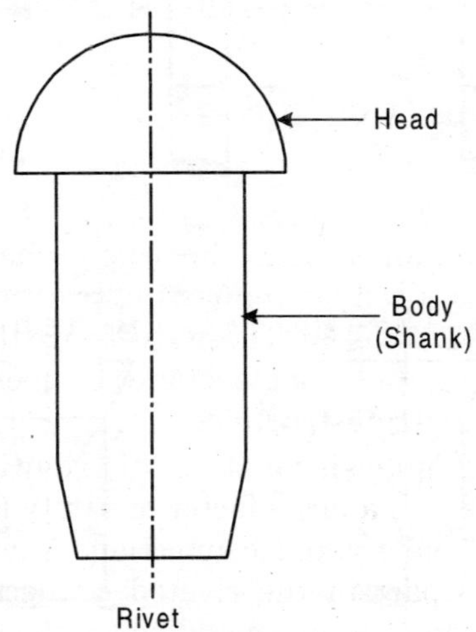

Fig. 2.1 *Rivet*

Some rivets are driven at atmospheric temperature. These rivets are known as *cold driven rivets.* The cold driven rivets need large pressure to form the head and complete the driving. The small size rivets ranging from 12 mm to 22 mm in diameter may be cold driven conveniently. The strength of rivet increases in the cold driving. The heating of rivet is not necessary. The use of cold driven rivets is limited because of equipment necessary and inconvenience caused in the field.

2.3 RIVET HEADS

The various types of rivet heads employed for different works are shown in Fig. 2.2.

The proportions of various shapes of rivet heads have been expressed in terms of diameter D of shank of rivet. The *snap head* is also termed as *round head* and *button head*. The snap heads are used for rivets connecting structural members. The countersunk heads are used to provide a flush surface.

2.4 RIVET HOLES

The rivet holes are made in the plates or structural members by one of the following two methods :

1. Punching
2. Drilling.

When the rivet holes are made by punching, the holes are not perfect, but taper. A punch damages the material around the hole. The operation known as reaming is done in the hole made by punching.

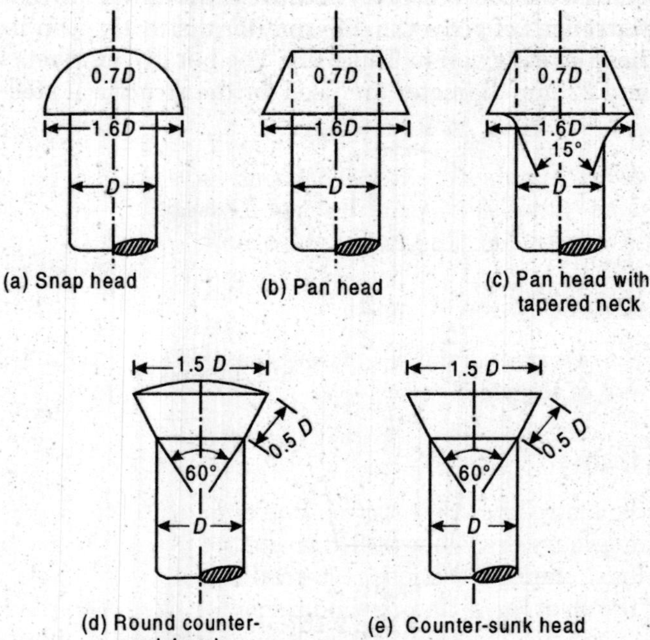

Fig. 2.2 *Rivet heads*

When the rivet holes are made by drilling, the holes are perfect and provide good alignment for driving the rivets.

The diameter of a rivet hole is made larger than the nominal diameter of the rivet by 15 mm of rivets less than or equal to 25 mm diameter and by 2 mm for diameters exceeding 25 mm.

2.5 DEFINITIONS OF TERMS USED IN RIVETING

2.5.1 Nominal Diameter of Rivet

The nominal diameter of rivet is the diameter of the cold rivet measured before driving.

2.5.2 Gross Diameter of Rivet

The gross diameter of rivet is the diameter of the rivet measured after driving and the diameter of the rivet hole is adopted as the gross diameter of a rivet.

2.5.3 Pitch of Rivet (p)

The pitch of rivets is the distance between two consecutive rivets measured parallel to the direction of the force in the structural member, lying on the same rivet line. It is the centre to centre distance between the individual fasteners. The pitch of the rivet is kept constant over as large a length as it is possible. The pitch of the rivet is not changed from rivet to rivet. It is better to keep a large number of pitches with an odd pitch cast to each end than to keep all the pitches the same with each pitch containing an odd figure. It facilitates the marking and checking the job. It saves time also.

2.5.4 Gauge Distance of Rivets (g)

The gauge distance is the transverse distance between two consecutive rivets of adjacent chains (parallel adjacent lines of fasteners) and is measured at right angles to the direction of the force in the structural member.

2.5.5 Gross Area of Rivet

The gross area of a rivet is the cross-sectional area of a rivet calculated from the gross diameter of the rivet.

2.5.6 Rivet Line

The *rivet line* is also known as *scrieve line* or *back line* or *gauge line.* The *rivet line* is the imaginary line along which rivets are placed. The rolled steel section (e.g., I-sections, channel sections, tee-sections) have been assigned *standard positions* of the rivet lines. These standard positions of rivet lines are conformed to whenever possible. The standard positions of the rivet lines depend upon the flange widths in the case if I-sections, channel sections and tee-sections. The standard positions of the rivet lines for the angle sections depend upon the length of legs. For the equal angle sections, the rivet lines are at equal distances from the heel. For the unequal angle sections, the rivet lines are also at unequal distances. The standard positions of rivet lines for the various sections may be noted from ISI Handbook No. 1 for the respective sections. The departure from standard position of the rivet lines may be done if necessary. The dimensions of rivet lines should be shown irrespective of whether the standard position have been followed or not.

2.5.7 Staggered Pitch

The *staggered pitch* is also known as *alternate pitch* or *reeled pitch*. The staggered pitch is defined as the distance measured along one rivet line from the centre of a rivet on it to the centre of the adjoining rivet on the adjacent parallel rivet line. One or both the legs of an angle section may have double rivet lines. The staggered pitch occurs between the double rivet lines.

2.6 WORKING STRESSES IN RIVETS

The working stresses in the rivets as per IS : 800–1984 are given in Table 2.1.

Table 2.1 *Working (maximum permissible) stresses in rivets*

Description	*Working stress*
In Tension, σ_{tf}	
Axial stress on gross area of rivets :	
(i) Power driven rivets	100 N/mm^2 (MPa)
(ii) Hand-driven rivets	80 N/mm^2 (MPa)
In Shear, τ_{vf}	
Shear stress on gross area of rivets :	
(i) Power-driven rivets	100 N/mm^2 (MPa)
(ii) Hand-driven rivets	80 N/mm^2 (MPa)
In Bearing, σ_{pf}	
Bearing stress on gross diameter of rivets :	
(i) Power-driven rivets	300 N/mm^2 (MPa)
(ii) Hand-driven rivets	250 N/mm^2 (MPa)

Note. *For high tensile steel rivets working stresses as given in IS : 800–1984 may be adopted. For the field rivets, the permissible stress are reduced by ten per cent.*

The calculated bearing stress of a rivet on the parts connected by it shall not exceed (i) value of the yield stress, f_y for hand driven rivets and (ii) value of 1.2 times yield stress of the connected parts for power driven rivets.

When the rivets are subjected to both shear and axial tension, these shall be so proportioned that the shear and axial stresses calculated shall not exceed the respective allowable stresses, τ_{sf} and σ_{tf} and the expression

$$\left(\frac{\tau_{vf\cdot\text{cal.}}}{\tau_{vf}} + \frac{\sigma_{tf\cdot\text{cal.}}}{\tau_{tf}}\right) \text{ shall not exceed 1.4.}$$

2.7 RIVETED JOINT

The riveted joints are of two types :

1. Lap joint 2. Butt joint.

2.7.1 Lap Joint

When one member is placed above the other and the two are connected by means of rivets the joint is known as *lap joint* as shown in Fig. 2.3 (a). In case, the lap joint, lines of forces are eccentric; this causes bending stress, and has the tendency to deform. The deformation of lap joint has been shown in Fig. 2.3 (b).

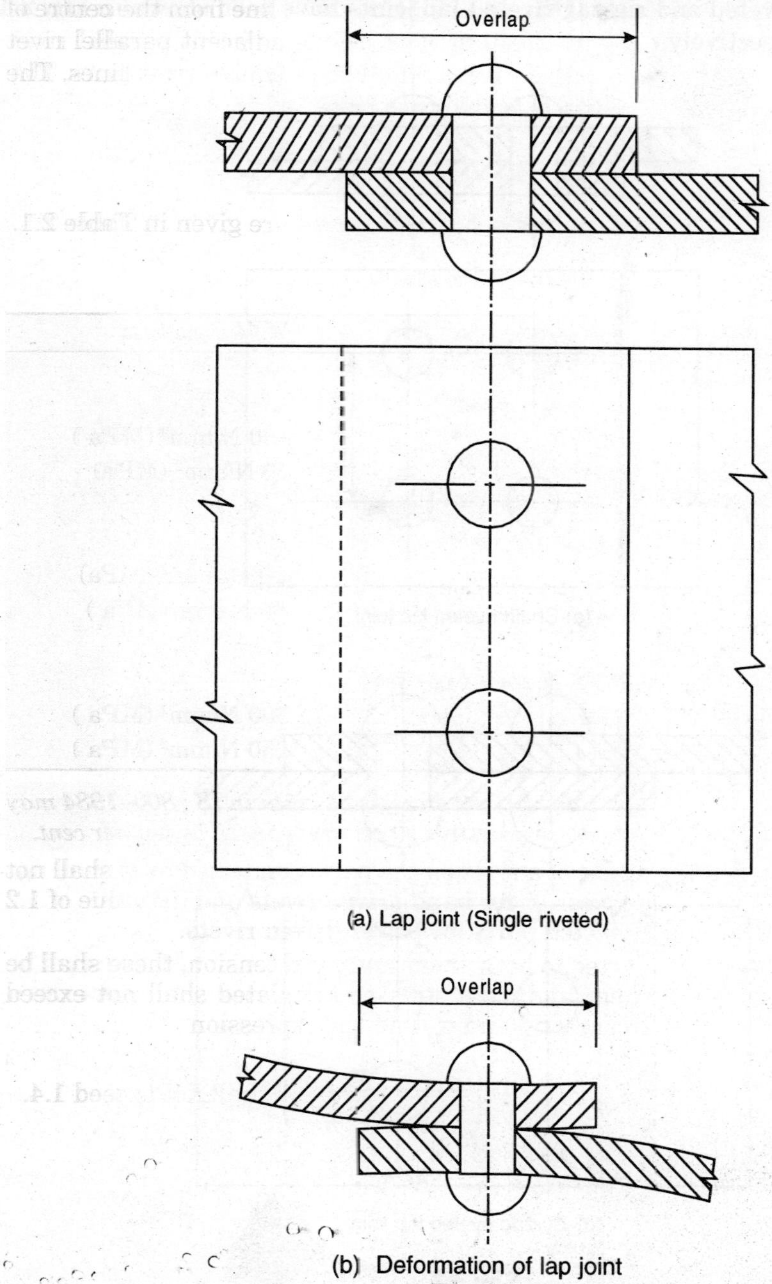

(a) Lap joint (Single riveted)

(b) Deformation of lap joint

Fig. 2.3

These joints are further classified according to the number of rivets used and the arrangement of rivets adopted. Following are the different types of lap joints :

1. Single riveted lap joint
2. Double riveted lap joint:
   (a) Chain riveted lap joint.
   (b) Zig-zag riveted lap joint.

The chain riveted and zig-zag riveted lap joints have been shown in Fig. 2.4 (a) and (b), respectively.

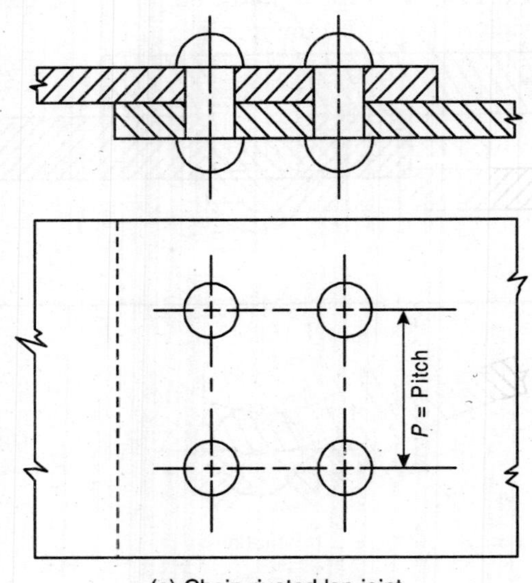

(a) Chain riveted lap joint

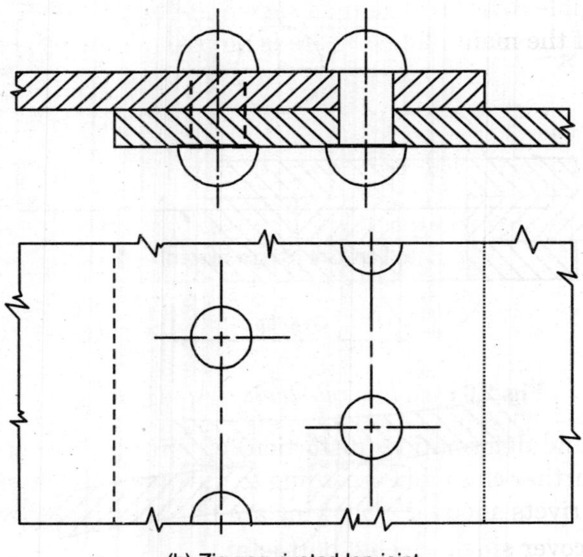

(b) Zig-zag riveted lap joint

Fig. 2.4

2.7.2 Butt Joint

When plates are placed end to end and flushed with each other and are joined by means of cover plates, the joint is known as *butt joint.* The butt joints are of two types :

(a) Single cover butt joint, (b) Double cover butt joint.

In single cover butt joint as shown in Fig. 2.5 (a), cover plate is provided on one side of main plates. In this type of joint, bending stress may develop which may cause deformation of joint as shown in Fig. 2.5 (b).

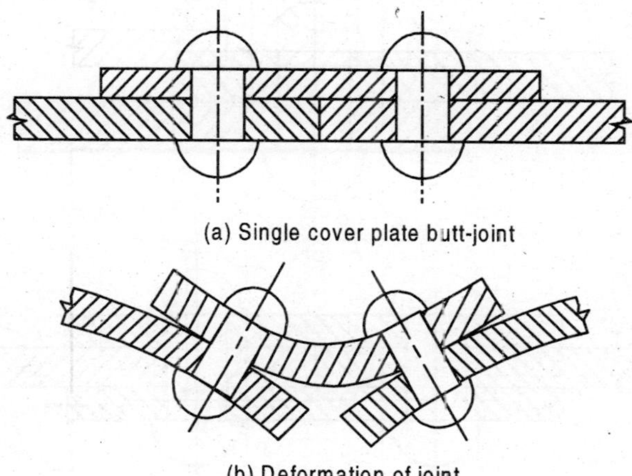

(a) Single cover plate butt-joint

(b) Deformation of joint

Fig. 2.5

In case of double cover butt joint as shown in Fig. 2.6, cover plates are used on either side of the main plates. There is no possibility of the development of

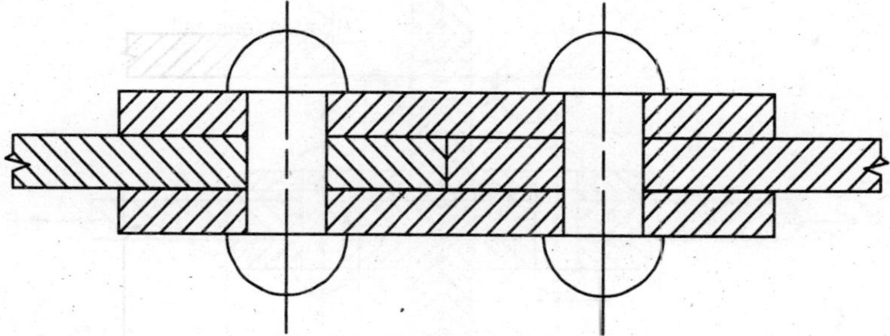

Fig. 2.6 *Double cover single riveted butt joint*

bending stress and deformation (distortion) of joint in this type of joint. Butt joints are also further classified according to the number of rivets used and the arrangement of rivets adopted. Following are the different types of butt joints :

1. Double cover single riveted butt joint
2. Double cover chain riveted butt joint
3. Double cover zig-zag riveted butt joint.

2.8 TRANSMISSION OF LOAD IN RIVETED JOINTS

There are two modes of transmission of load in riveted joints. When the load is transmitted by bearing between plates and shanks of rivets, the rivets are subjected to shear. When the shear of rivets is only across one cross-section of the rivet, it is known as single shear as shown in Fig. 2.7 (a). When the shear of rivet is across two cross-sections of the rivet, it is known as *double shear* as shown in Fig. 2.7 (b).

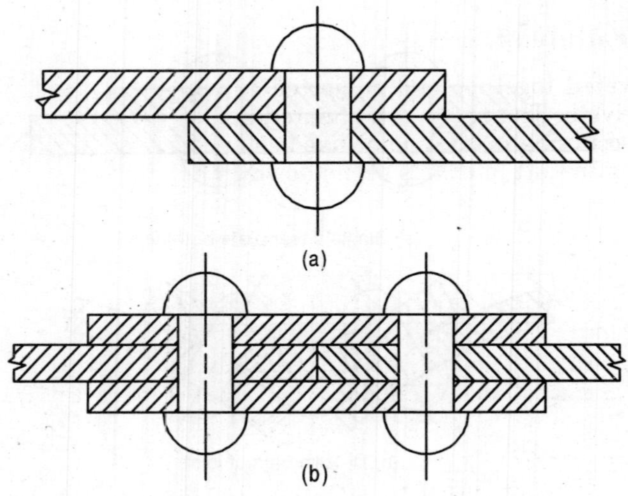

Fig. 2.7 *Rivets in shear*

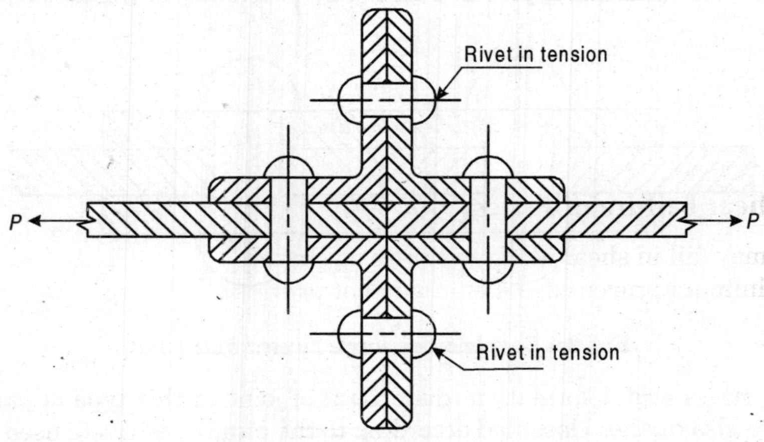

Fig. 2.8 *Rivet in tension*

When the load is transmitted by shearing between heads of rivets and plates, the rivets are subjected to tension as shown in Fig. 2.8.

2.9 FAILURE OF A RIVETED JOINT

The failure of a riveted joint may take place in any of the following ways :

1. Shear failure of rivets
2. Shear failure of plates
3. Tearing failure of plates
4. Bearing failure of plates
5. Splitting failure of plates at the edges
6. Bearing failure of rivets.

2.9.1 Shear Failure of Rivets

The plates riveted together and subjected to tensile loads may result in the shear of the rivets. The rivets are sheared across their cross-sectional areas. Single shear occurring in a lap joint has been shown in Fig. 2.9 (a) and double shear occurring in butt joint has been shown in Fig. 2.9 (b).

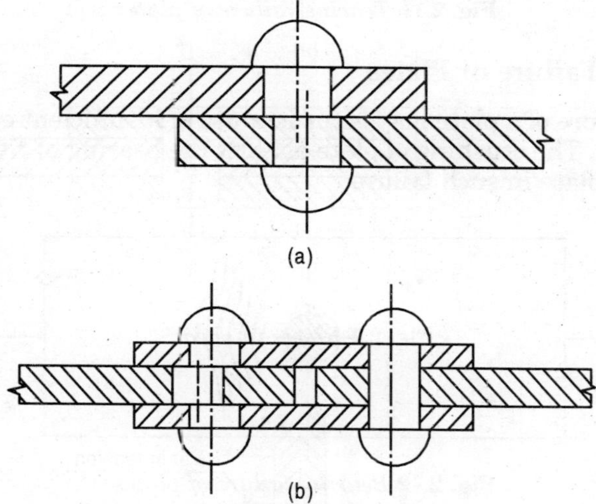

Fig. 2.9 *Shear failure of plates*

2.9.2 Shear Failure of Plates

A plate may fail in shear along two lines as shown in Fig. 2.10. This may occur when minimum proper edge distance is not provided.

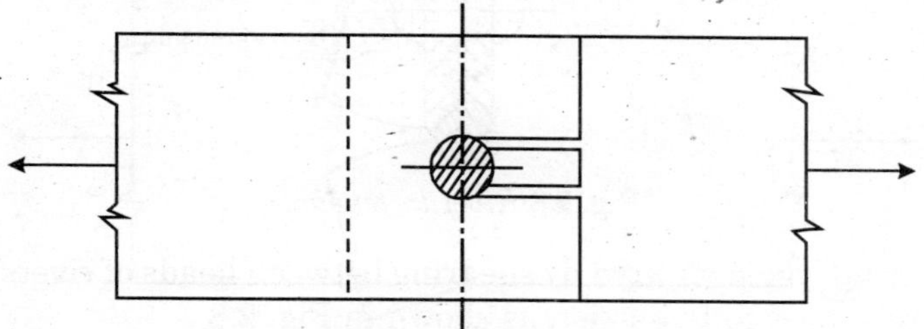

Fig. 2.10 *Shear failure of plates*

2.9.3 Tearing Failure of Plates

When plates riveted together are carrying tensile load, tearing failure of plate may occur, when strength of the plate is less than that of rivets. The tearing failure occurs at the net sectional area of plate as shown in Fig. 2.11.

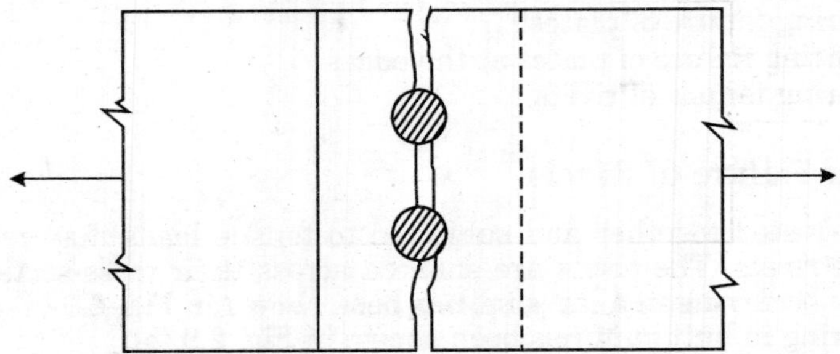

Fig. 2.11 *Tearing failure of plates*

2.9.4 Bearing Failure of Plates

The bearing failure of a plate may occur because of insufficient edge distance in the riveted joint. The crushing of plate against the bearing of rivet as shown in Fig. 2.12 takes place in such failure.

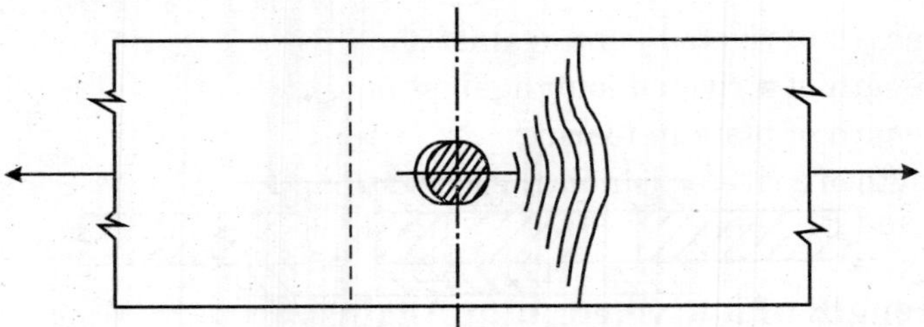

Fig. 2.12 *Bearing failure of plates*

2.9.5 Splitting Failure of Plates

The splitting failure of a plate may occur because of insufficient edge distance in the riveted joint. The splitting (cracking) of plate as shown in Fig. 2.13 takes place in such failure.

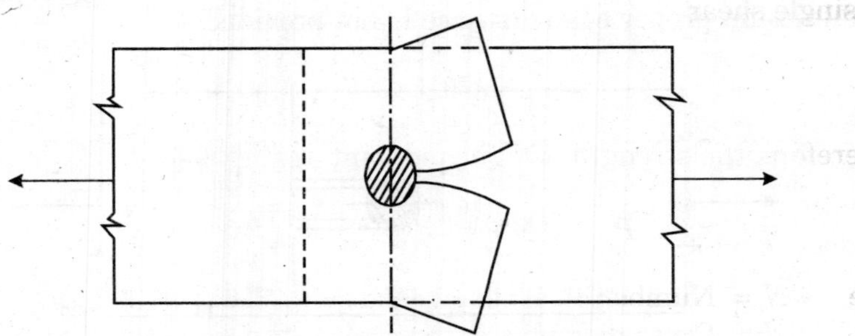

Fig. 2.13 *Splitting failure of plates*

2.9.6 Bearing Failure of Rivets

The bearing failure of a rivet occurs when the rivet is crushed by the plate as shown in Fig. 2.14.

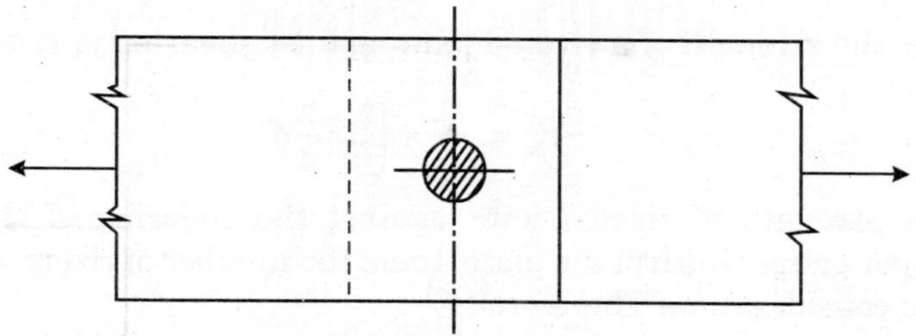

Fig. 2.14 *Bearing failure of rivets*

The bearing, shearing and splitting failure of plates may be avoided by providing adequate edge distance. To safeguard a riveted joint against other modes of failure, the joint should be designed properly.

2.10 STRENGTH OF RIVETED JOINT

The strength of a riveted joint is determined by computing the following strengths :

1. Strength of riveted joint against shearing of the rivets ... P_s
2. Strength of a riveted joint against bearing of the rivets ... P_b
3. Strength of plate in tearing ... P_t

The strength of a riveted joint is the least strength of the above three strengths viz., P_s, P_b, and P_t.

2.10.1 Strength of a Riveted Joint Against the Shearing of the Rivets

The strength of riveted joint against the shearing of rivets is equal to the product of strength of one rivet in shear and the number of rivets on each side of the joint. It is given by

$$P_s = \text{strength of a rivet in shearing} \times \text{Number of rivets on each side of joint} \quad \text{...(i)}$$

When the rivets are subjected to *single shear,* then, the strength of one rivet in single shear

$$= \left(\frac{\pi}{4} \times d^2 \times \tau_{vf}\right) \quad \text{...(ii)}$$

Therefore, the strength of a riveted joint against shearing of rivets

$$P_s = N \times \left(\frac{\pi}{4} d^2 \times \tau_{vf}\right) \quad \text{...(2.1)}$$

where N = Number of rivets on each side of the joint

d = Gross-diameter of the rivet

τ_{vf} = Maximum permissible shear stress in the rivet.

When the rivets are subjected to *double shear,* then, the strength of one rivet in double shear

$$= \left(2\times\frac{\pi}{4}\times d^2\times\tau_{vf}\right) \quad \text{...(iii)}$$

Therefore, the strength of a riveted joint against shearing of rivets

$$P_s = N\times\left(2\times\frac{\pi}{4}d^2\times\tau_{vf}\right) \quad \text{...(2.2)}$$

When the strength of riveted joint against the shearing of the rivets is determined per gauge width of the plate, then, the number of rivets, *n* per gauge is taken into consideration. Therefore

For single shear of rivets

$$P_{s_1} = n\times\left(\frac{\pi}{4}d^2\times\tau_{vf}\right) \quad \text{...(2.3)}$$

For double shear of rivets

$$P_{s_2} = n\times\left(2\times\frac{\pi}{4}d^2\times\tau_{vf}\right) \quad \text{...(2.4)}$$

2.10.2 Strength of Riveted Joint Against the Bearing of the Rivets

The strength of a riveted joint against the bearing of the rivets is equal to the product of strength of one rivet in bearing and the number of rivets on each side of the joint. It is given by

P_b = Strength of a rivet in bearing × Number of rivets on each side of the joint ...(iv)

In case of lap joint, the strength of one rivet in bearing

$$= (d\times t\times\sigma_{pb}) \quad \text{...(v)}$$

where t = Thickness of the thinnest plate

d = Gross diameter of the rivet

σ_{pb} = Maximum permissible stress in the bearing for the rivet.

In case of butt joint, the total thickness of both cover plates or thickness of main plate whichever is less is considered for determining the strength of a

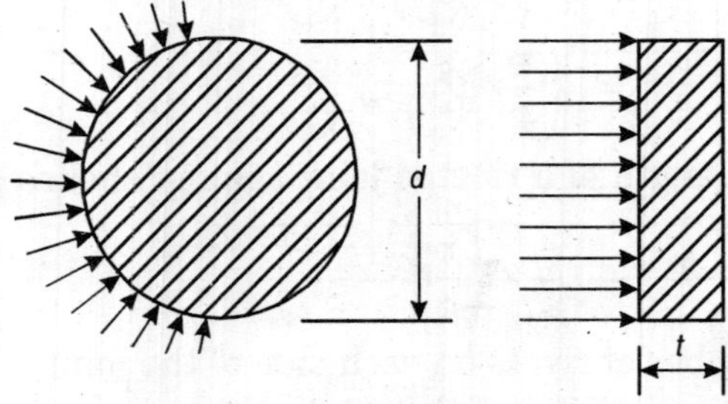

Fig. 2.15 *Bearing pressure on rivet*

rivet in the bearing. The bearing pressure acting on the rivet is radial as shown in Fig. 2.15. The projected area is used in determining the strength of rivet in bearing.

The strength of a riveted joint against the bearing of rivets

$$P_b = N \times (d \times t \times \sigma_{pb}) \quad ...(2.5)$$

When the strength of riveted joint against the bearing of rivets per gauge width of the plate is taken into consideration, then, the number of rivets n is also adopted per gauge. Therefore,

$$P_{b_1} = n \times (d \times t \times \sigma_{pb}) \quad ...(2.6)$$

2.10.3 Strength of Plate in Tearing

The strength of plate in tearing depends upon the resisting section of the plate. The strength of plate in tearing is given by

$$P_t = \text{Resisting section} \times \sigma_{tf} \quad ...(2.7)$$

When the strength of plate in tearing per gauge width of the plate is found, then, it is given by

$$P_{t_1} = (g - d) \times t \times \sigma_{tf} \quad ...(2.8)$$

where g = Width of plate equal to the gauge of the rivets.

The strength of a riveted joint is the least of P_s, P_b, and P_t. The strength of riveted joint per gauge width of plate is the least of P_{s_1}, P_{b_1} and P_{t_1}.

A riveted lap joint is shown in Fig. 2.16 (a). A riveted but joint is shown in Fig. 2.16 (b).

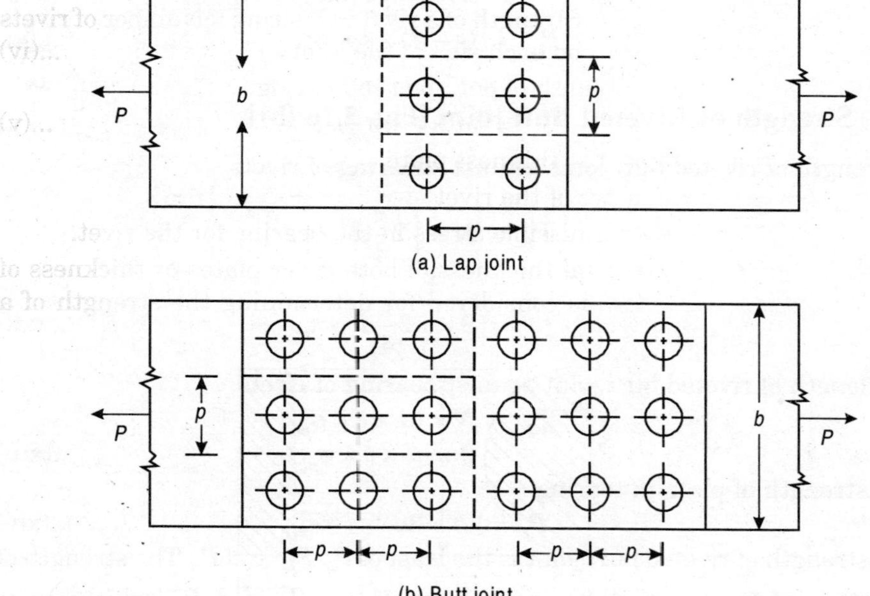

(a) Lap joint

(b) Butt joint

Fig. 2.16

2.10.4 Strength of Riveted Lap Joint [Fig. 2.16 (a)]

The strength of riveted lap joint against shearing of the rivets

$$P_s = N \times \left(\frac{\pi}{4} d^2 \times \tau_{vf}\right)$$

$$= 6 \times \left(\frac{\pi}{4} d^2 \times \tau_{vf}\right) \quad \text{...(vi)}$$

The strength of riveted lap joint against bearing of rivets

$$P_b = N \times (d \times t \times \sigma_{pb})$$
$$= 6 \times (d \times t \times \sigma_{pb}) \quad \text{...(vii)}$$

The strength of plate in tearing

$$P_t = (b - 3d) \times t \times \sigma_{tf} \quad \text{...(viii)}$$

The strength of riveted lap joint is the least *of* $P_s \cdot P_b$ and P_t. The strength of riveted lap joint per gauge width is the least *of* P_{s_1}, P_{b_1} and P_{t_1}, whichever is less, where

$$P_{s_1} = n \times \left(\frac{\pi}{4} d^2 \times \tau_{vf}\right)$$

$$= 2 \times \left(\frac{\pi}{4} d^2 \times \tau_{vf}\right) \quad \text{...(ix)}$$

$$P_{b_1} = n \times (d \times t \times \sigma_{pb})$$

$$= 2 \times (d \times t \times \sigma_{pb}) \quad \text{...(x)}$$

and

$$P_{t_1} = (g - d) \times t \times \sigma_{tf} \quad \text{...(xi)}$$

2.10.5 Strength of Riveted Butt Joint [Fig. 2.16 (b)]

The strength of riveted butt joint against shearing of rivets

$$P_3 = N \times \left(2\frac{\pi}{4} d^2 \times \tau_{vf}\right)$$

$$= 9 \times \left(2\frac{\pi}{4} d^2 \times \tau_{vf}\right) \quad \text{...(xii)}$$

The length of riveted butt joint against bearing of rivets

$$P_b = N \times (d \times t \times \sigma_{pb})$$
$$= 9 \times (d \times t \times \sigma_{pb}) \quad \text{...(xiii)}$$

The strength of plate in tearing

$$P_t = (b - 3d) \times t \times \sigma_{tf} \quad \text{...(xiv)}$$

The strength of riveted butt joint is the least of P_s, P_b, and P_t. The strength of rivet butt joint per gauge width is the least of P_{s_1}, P_{b_1} and P_{t_1}, whichever is less, where

$$P_{s_1} = n \times \left(\frac{\pi}{4}d^2 \times \tau_{vf}\right)$$

$$= 3 \times \left(\frac{\pi}{4}d^2 \times \tau_{vf}\right) \quad ...(xv)$$

$$P_{b_1} = n \times (d \times t \times \sigma_{pb})$$

$$= 3 \times (d \times t \times \sigma_{pb}) \quad ...(xvi)$$

and $$P_{t_1} = (g - d) \times t \times \sigma_{tf} \quad ...(xvii)$$

2.11 EFFICIENCY OR PERCENTAGE STRENGTH OF RIVETED JOINT

The efficiency of a joint is defined as the ratio of least strength of a riveted joint to the strength of solid plate. It is known as percentage strength of riveted joint as it is expressed in percentage.

Efficiency of riveted joint

$$n = \left(\frac{\text{Least strength of riveted joint}}{\text{Strength of solid plate}}\right) \times 100$$

$$= \left(\frac{\text{Least of } P_s, P_b \text{ or } P_t}{P}\right) \times 100 \quad ...(2.9)$$

where, P = Strength of solid plate.

2.12 RIVET VALUE

The strength of a rivet in shearing and in bearing is computed and the lesser is called the *Rivet Value (R)*.

2.13 ASSUMPTIONS FOR DESIGN OF RIVETED JOINT

The procedure for design of a riveted joint is simplified by making the following assumptions and by keeping in view the safety of the joint.

1. The load is assumed to be uniformly distributed among all the rivets.
2. The shear stress on a rivet is assumed to be uniformly distributed over its gross area.
3. The bearing stress is assumed to be uniform between the contact surfaces of plate and rivet.
4. The bending stress in a rivet is neglected.
5. The rivet hole is assumed to be completely filled by the rivet.
6. The stress in plate is assumed to be neglected.
7. The friction between plates is neglected.

2.14 ARRANGEMENT OF RIVETS

The rivets in a riveted joint are arranged into two forms :

1. Chain Riveting; 2. Diamond Riveting.

2.14.1 Chain Riveting

In chain riveting, the rivets are arranged as shown in Fig. 2.17.

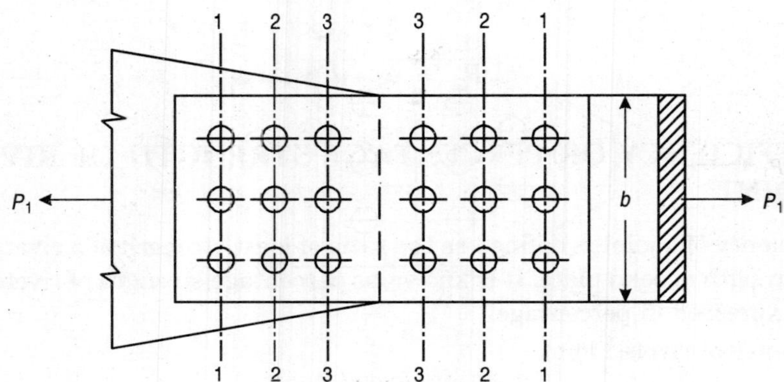

Fig. 2.17 *Chain riveting*

In Fig. 2.17, 1–1, 2–2 and 3–3 show sections on either side of the joint. Section 1–1 is the critical section as at the other section, strength of rivets prior to that section adds the strength of the joint at the section. The strength of plate in tearing can be computed at this section as under:

$$P_t = (b - 3d) \,.\, t \,.\, \sigma_t$$

where b = width of the plate

d = gross diameter of the rivet

t = thickness of the plate

When the safe load carried by the joint is known, width of plate can be found as follows :

$$b = \left(\frac{P_t}{\sigma_t \times t} + 3d \right)$$

2.14.2 Diamond Riveting

In diamond riveting, rivets are arranged as shown in Fig. 2.18.

All the rivets are arranged symmetrically about the centre line of the plate. Section 1–1 is the critical section. The strength of plate in tearing in diamond riveting at the section 1–1 can be computed as under :

$$P_t = (b - d) \,.\, t \,.\, \sigma_t$$

When the safe load carried by the joint is known, width of the plate can be found as follows :

$$b = \left(\frac{P_t}{\sigma_t \times t} + d\right)$$

where b = width of the plate
d = gross diameter of rivet
t = thickness of plate

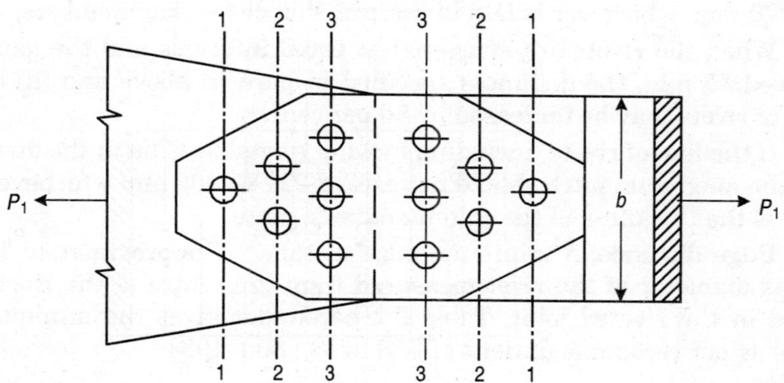

Fig. 2.18 *Diamond riveting*

The strength of the plate at all various sections can be found as follows :

At Section 2–2. All the rivets are stressed uniformly. Hence, strength of the plate at the section is

$$P_t = (b - 2d) \,.\, t \,.\, \sigma_t + \text{strength of one rivet}$$

At Section 3–3. $P_t = (b - 3d).\ t\ .\ \sigma_t$ + strength of three rivets.

The failure of plate in this form of riveted joint at any other section than section 1–1, can take place only when all rivets prior to that section have failed. Usually failure of plate takes place at the section 1–1.

The width of the plate required in case of chain riveting is more than that in diamond riveting by the twice the gross diameter of the rivet. In diamond riveting there is saving of material and efficiency is more ; hence it is preferred to chain riveting. The diamond riveting is used in bridge trusses generally.

2.15 SPECIFICATIONS FOR DESIGN OF RIVETED JOINT

1. **Members meeting at joint.** The centroidal axes of the members meeting at a joint should intersect at one point, and if there is any eccentricity, adequate resistance should be provided in the connection.

2. The centre of gravity of group of rivets should be on the line of action of load whenever practicable.

3. **Pitch.** *Minimum pitch.* The distance between centres of adjacent rivets should not be less than 2.5 times the gross diameter of the rivet.

Maximum pitch. (i) The maximum pitch should not exceed 12 t or 200 mm whichever is less in compression member, and 16 t or 200 mm whenever is less in case of tension member, when the line of rivet lies in the direction of stress. In the case of compression members in which the forces are transferred through the butting faces, this distance shall not exceed 45 times the diameter of the rivets for a distance from the abutting faces equal to 15 times the width of the member.

(ii) The distance between centres of any two consecutive rivets in a line adjacent and parallel to an edge of an outside plate shall not exceed (100 mm + 4 t) or 200 mm, whichever is less in compression or tension members.

(iii) When the rivets are staggered at equal intervals and the gauge does not exceed 75 mm, the distances specified in para (i) above and (ii) between centres of rivets may be increased by 50 per cent.

(iv) If the line of rivets (including tacking rivets) does lie in the direction of stress, the maximum pitch should not exceed 32 t or 300 mm whichever is less where t is the thickness of the thinner outside plate.

4. **Edge distance.** A minimum edge distance of approximately 15 times the gross diameter of the rivet measured from the centre of the rivet hole is provided in the riveted joint. Table 2.2 hereunder gives the minimum edge distance as per recommendations of BIS in IS : 800–1984.

Table 2.2 *Edge distance of holes*

Gross diameter of rivet	*Edge Distance of Hole*	
	Distance to sheared or hand flame cut edge	*Distance to rolled machine flame cut or planed edge*
mm	*mm*	*mm*
13.5 and below	19	17
15.5	25	22
17.5	29	25
19.5	32	29
21.5	32	29
23.5	38	32
25.5	44	38
29.0	51	44
32.0	57	51
35.0	57	51

5. **Rivets through packings.** The rivets carrying calculated shear stress through a packing greater than 6 mm thick shall be increased above number required by normal calculations by 2.5 per cent for each 2 mm thickness of packing. For double shear connections packed on both sides, the number of additional rivets required shall be determined from the thickness of the thicker packing. The additional rivets should preferably be placed in an extension of the packing. When the properly fitted packings are subjected to direct compression, then, the above mentioned specifications shall not apply.

6. **Long grip rivets.** When the grip of rivets carrying calculated loads exceeds 6 times the diameter of the holes, then, the rivets are subjected to bending in addition to shear and bearing. The number of rivets required by normal calculations shall be increased by not less than one percent for each additional 1.6 mm of grip, but the grip shall not exceed 8 times the diameter of the holes.

7. **Rivet line distance.** When two or more parts are connected together, a line of rivet shall be provided at a distance of not more than 37 mm + 4 t from the nearest edge where t is the thickness in mm of thinner outside plate. In case steel work is not exposed to weather, this may be increased to 12 t.

8. **Tacking rivets.** When the maximum distance between centres of two adjacent rivets connecting the members subjected to either compression or tension exceeds the maximum pitch, then, the tacking rivets not subjected to calculated stresses shall be used.

The tacking rivets shall have a pitch in line not exceeding 32 times the thickness of the outside plate or 300 mm whichever is less. Wherever the plates are exposed to the weather, the pitch in line shall not exceed 16 times the thickness of the outside plate or 200 mm, whichever is less. In both cases, the lines of rivets shall not be apart at a distance greater than these pitches.

For the design and construction composed of two flats, angles, channels or tees in contact back to back or separated back to back by a distance not exceeding the aggregate thickness of the connected parts, tacking rivets with solid distance pieces where the parts are separated, shall be provided at a pitch in line not exceeding 1000 mm.

2.16 DESIGN PROCEDURE FOR RIVETED JOINT

For the design of a lap joint or a butt joint the thickness of plates to be joined are known and the joint is designed for the full strength of the plate. For the design of a structural steel work, force (pull or push) to be transmitted by the joint is known and riveted joint can be designed. Following are the usual steps for the design of a riveted joint :

Step 1. The size of the rivet is determined for the *Unwin's formula* as

$$d = 6.04\,(t)^{1/2} \qquad \text{...(2.10)}$$

where t = thickness of plate in mm

d = nominal diameter of rivet.

The diameter of the rivet computed is rounded off to available size of rivets. The rivets are manufactured in nominal diameters of 12, 14, 16, 18, 20, 22, 24, 27, 30, 33, 36, 39, 42 and 48 mm.

In structural steel work, rivets of nominal diameter of 16, 18, 20 and 22 mm are used. The nomial diameter of rivets to be used in a joint is assumed.

Step 2. The strengths of rivets in shearing and bearing are computed. The working stresses in rivets and plates are adopted as per BIS. The rivet value R is found. For designing lap joint or butt joint tearing strength of plate is determined as under :

$$P_t = (g - d) \,.\, t \,.\, \sigma_{tf}$$

where g = gauge of rivets to be adopted
t = thickness of plate
σ_{tf} = working stress in direct tension for plate.

The tearing strength of plate should not exceed the rivet value R (P_s or P_b whichever is less) or

$$(g - d) \cdot t \cdot \sigma_{tf} \leq R$$

From this relation gauge of the rivets is determined.

In structural steel work, force to be transmitted by the riveted joint and the rivet value are known. Hence number of rivets required to be provided in the joint can be computed, as follows :

$$\text{No. of rivets required in the joint} = \left(\frac{\text{Force}}{\text{Rivet Value}}\right)$$

The number of rivets thus obtained is provided on one side of the joint and an equal number of rivets is provided on the other side of joint also.

For the design of joint in a tie member consisting of a flat, width/thickness of the flat is known. The section is assumed to be reduced by rivet holes, depending upon the arrangement of rivets to be provided. The strength of flat at the weakest section is equated to the pull transmitted by the joint. For example, assuming the section to be weakened by one rivet hole and also assuming that the thickness of the flat is known, we have

$$(b - d) \cdot t \cdot \sigma_{tf} = P$$

where b = width of flat
t = thickness of flat
P = pull to be transmitted by the joint
σ_{tf} = working stress in tension in plate.

From the above equation width of the flat can be determined.

Example 2.1. *A single riveted lap joint is used to connect plate 10 mm thick. If 20 mm diameter rivets are used at 55 mm gauge, determine the strength of joint and its efficiency.*

Working stress in shear in rivets = 80 N/mm² (MPa)

Working stress in bearing in rivets = 250 N/mm² (MPa)

Working stress in axial tension in plates = 0.6 f_y

$$f_y = 260 \text{ N/mm}^2.$$

Solution

Step 1 :

Assume that power driven field rivets are used

Nominal diameter of rivet = 20 mm
Gross diameter of rivet = 21.5 mm

Step 2 : Strength of rivet

$$\text{Strength of rivet in single shear} = \frac{\pi}{4} \times (21.5)^2 \times \frac{80}{1000} \text{ kN}$$

$$P_t = 29.029 \text{ kN}$$

Strength of rivet in bearing $= 21.5 \times 10 \times \dfrac{250}{1000}$ kN

$$P_b = 53.750 \text{ kN}$$

Strength of plate in tension per gauge length

$$P_t = (g - d) \,.\, t \,.\, \sigma_t$$

$$= (55 - 21.5) \times 10 \times 0.6 \times \frac{260}{1000} \text{ kN}$$

$$P_t = 52.260 \text{ kN}$$

Step 3 : Strength of joint

Strength of the joint is minimum of P_s, P_b, or P_t.

Strength of joint = **29.029 kN**

Step 4 : Efficiency of joint

Efficiency of joint

$$\eta = \left(\frac{\text{Strength of joint per pitch length}}{\text{Strength of solid plate}}\right) \times 100$$

$$\eta = \left(\frac{29.029 \times 10^3}{55 \times 10 \times 0.6 \times 260}\right) \times 100 = \mathbf{33.8\ \%}$$

Example 2.2. *A double riveted double cover butt joint is used to connect plates 12 mm thick. Using Unwin's formula, determine the diameter of rivet, rivet value, gauge and efficiency of joint. Adopt the following stresses :*

Working stress in shear in power driven rivets

$= 100$ *N/mm²* *(MPa)*

Working stress in bearing in power driven rivets

$= 300$ *N/mm²* *(MPa)*

For plates working stress in axis tension is 0.6 fv

$f_y = 260$ *N/mm²* *(MPa).*

Solution

Step 1 :

Nominal diameter of rivet from Unwin's formula

$$d = 6.04(t)^{1/2}$$

$$d = 6.04\,(t)^{1/2}$$

$$= 20.923 \text{ mm}$$

Adopt nominal diameter of rivet = 22 mm

Gross diameter of rivet = 23.5 mm

Step 2 : Rivet value. Strength of rivet in double shear

$$2 \times \frac{\pi}{4} \times \frac{(23.5)^2 \times 100}{1000} = 86.70 \text{ kN}$$

Strength of rivet in bearing

$$\left(\frac{23.5 \times 12 \times 300}{1000}\right) = 84.6 \text{ kN}$$

Rivet value R = **84.6 kN**.

Step 3: Gauge of rivets

Let g be the gauge of rivets. Strength of plate in tension per gauge length

$$P_t = (g - d) \times t \times p_t$$

$$= \left(\frac{(g - 23.5) \times 12 \times 0.6 \times 260}{1000}\right)$$

$$= 1.872\,(g - 23.5) \text{ kN}$$

In double riveted joint,

Strength of 2 rivets in shear $P_s = (2 \times 86.70) = 173.4$ kN

Strength of 2 rivets in bearing $P_b = (2 \times 84.6) = 169.2$ kN

The gauge of rivets can be computed by keeping $P_t = P_s$ or P_b whichever is less

$$1.872\,(g - 23.5) = 169.2$$

$$(g - 23.5) = \left(\frac{169.2}{1.872}\right) = 90.385$$

$$g = (90.35 + 23.5) = 113.85 \text{ mm}$$

Adopt gauge, $g = 100$ mm

Step 4 : Efficiency of joint :

Efficiency of joint $$\eta = \frac{(g - d)}{g} \times 100 = \frac{(100 - 23.5) \times 100}{100}$$

$$= 76.5 \text{ percent.}$$

Example 2.3. A *double cover butt joint is used to connect plates 16 mm thick. Design the riveted joint and determine its efficiency.*

Solution

Step 1 : Diameter of rivet :

Size of rivet, using Unwin's formula

$$d = 6.04\,(16)^{1/2} = 24.16 \text{ mm}$$

Unwin's formula gives higher value

Adopt nominal diameter of rivet = 22 mm

Gross diameter of rivet = 23.5 mm

Step 2 : Rivet value. In double cover butt joint, rivets are in double shear.

As per IS : 800–84

Shear stress for power driven rivet = 100 N/mm^2

Bearing stress for power driven rivet = 300 N/mm^2

Strength of plate in tension = 0.6 × 260 N/mm^2

Strength of rivet in double shear

$$P_s = \left(\frac{2\pi}{4}\times(23.5)^2\times\frac{100}{1000}\right) = 86.70 \text{ kN}$$

Strength of rivet in bearing.

$$P_b = \left(\frac{23.5\times16\times300}{1000}\right) = 112.8 \text{ kN}$$

Step 3 : Gauge distance of rivet

Let g be the gauge of rivets

Strength of plate per gauge length

$$P_t = (g-d)\,.\,t\,.\,\sigma_t$$

$$= \left((g-23.5)\times12\times0.6\times\frac{260}{1000}\right) \text{ kN}$$

$$= 2.496\,(g-23.5) \text{ kN}$$

Keep strength of plate $P_t \triangleq p_s$ or P_b whichever is less

$$2.496\times(g-23.5) = 86.70$$

or

$$g = \left(\frac{86.70}{2.496}\right) = 58.235 \text{ mm}$$

Adopt gauge, $g = 55$ mm

Adopt thickness of each cover plate

$$t = 10 \text{ mm} \left(t \triangleq \frac{5}{8}\times16\right) \text{ mm}$$

Step 4 : Efficiency of joint

$$\eta = \left(\frac{g-d}{g}\times100\right) = \left(\frac{55-23.5}{55}\right)\times100$$

$$= 57.273\ \%$$

Example 2.4. *Determine the strength of a double cover butt joint used to connect two flats 200 F 12. The thickness of each cover plate is 8 mm. Flats have been joined by 9 rivets in chain riveting at a gauge of 60 mm (Fig. 2.19). What is the efficiency of the joint ? Adopt working stresses in rivets and flats as per IS : 800–1984.*

Solution

Step 1 :

Size of flat used = 200 F 12

Width of flat = 200 mm

Thickness of flat = 12 mm

Use power driven rivets

Step 2: Strength of rivet

Strength of a rivet in double shear

$$2 \cdot \left(\frac{\pi}{4} \times \frac{(23.5)^2 \times 100}{1000} \right) = 86.70 \text{ kN}$$

Strength of a rivet in bearing

$$\left(\frac{12 \times 23.5 \times 300}{1000} \right) = 84.6 \text{ kN}$$

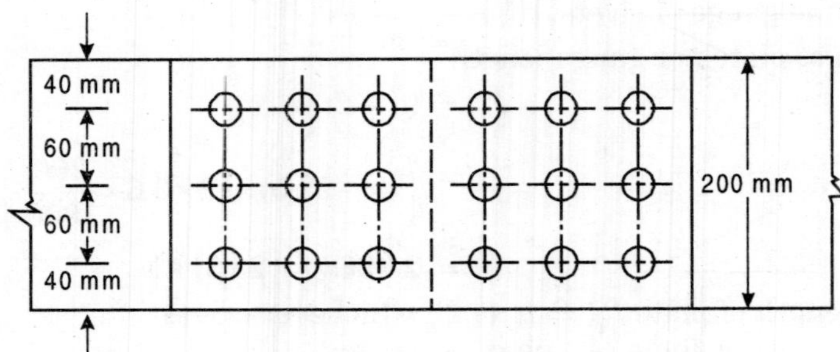

Fig. 2.19 *Double cover butt joint*

Step 3 : Strength of joint

Strength of joint in shear,

$$P_s = (9 \times 86.7) = 780.3 \text{ kN}$$

Strength of joint in bearing

$$\left(\frac{12 \times 23.5 \times 300}{1000} = 84.6 \text{ kN} \right)$$

$$p_b = (9 \times 84.6) = 761.4 \text{ kN}$$

Strength of plate in tearing

$$P_t = (b - 3d) \,.\, t \,.\, \sigma_t$$

$$P_t = \left(\frac{200 - 2 \times 23.5) \times 12 \times 6 \times 260}{1000} \right)$$

$$= 242.42 \text{ kN}$$

Strength of joint is P_s, P_b or P_t whichever is less

$$P_t = 242.42 \text{ kN}$$

Step 4 : Efficiency of joint

$$\eta = \left(\frac{P_s,\, P_b \text{ or } P_t \text{ whicnever is less}}{\text{Strength of solid plate}} \right) \times 100$$

$$\left(\frac{242.42 \times 100}{\frac{200 \times 12 \times 0.6 \times 260}{1000}} \right) = \mathbf{64.7489\ \%}$$

Example 2.5. *In a truss girder of a bridge, a diagonal consists of a 16 mm thick flat and carries a pull of 750 kN and is connected to a gusset plate by a double cover butt joint. The thickness of each cover plate is 8 mm. Determine the number of rivets necessary and the width of the flat required. What is the efficiency of the joint ? Sketch the joint.*

Take working stress in shear in rivet = 100 N/mm²

Working stresses in bearing in rivet = 300 N/mm²

Working stresses in tension in plate = 0.6 × 260 N/mm².

Solution

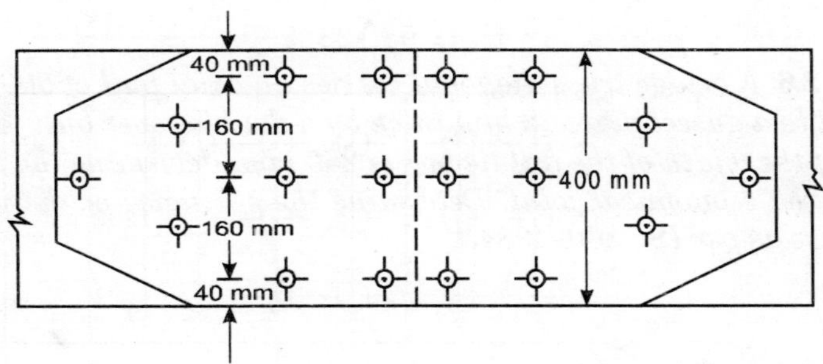

Fig. 2.20 *Double cover butt joint*

Step 1 : Rivet value

Use 22 mm diameter of rivets

Strength of power driven shop rivets in double shear

$$2\times\left(\frac{\pi}{4}\times(23.5)^2\times\frac{100}{1000}\right) = 86.70 \text{ kN}$$

Strength of power driven shop rivets in bearing

$$\left(23.5\times16\times\frac{300}{1000}\right) = 112.8 \text{ kN}$$

$$\text{Rivet value} = 86.7 \text{ kN}$$

Step 2 : Number of rivets

Number of rivets required to transmit pull of 750 kN

$$n = \left(\frac{750}{86.70}\right) = 8.7 \simeq 9 \text{ rivets}$$

Using diamond group of riveting, flat is weakened by one rivet hole.

Step 3 : Strength of plate

Strength of plate at Sec. 1–1 in tearing

$$P_t = (b - 23.5) \times \left(\frac{16 \times 0.6 \times 260}{1000}\right)$$

$$= 2.496\,(b - 23.5) \text{ kN}$$

$$P = 750 \text{ kN}$$

$$2.496\,(b - 23.5) = 750$$

$$\therefore \quad b = \left(\frac{750}{2.496} + 2.496 \times 23.5\right) = 359.136 \text{ mm}$$

Provide 400 mm width of diagonal member

Step 4 : Efficiency of joint

$$\eta = \left(\frac{(b-d)t \cdot p_t}{b \cdot t \cdot p_t}\right) = \left(\frac{400 - 23.5}{400}\right) \times 100$$

$$= 94.125\ \%$$

Example 2.6. *A bridge truss diagonal carries an axial pull of 500 kN. It is to be connected to a guseet plate 22 mm thick by a double cover butt joint with 22 mm rivets. If the width of the flat tie bar is 250 mm, determine the thickness of flat. Design the economical joint. Determine the efficiency of the joint. Adopt working stress as per IS : 800–1984.*

Solution

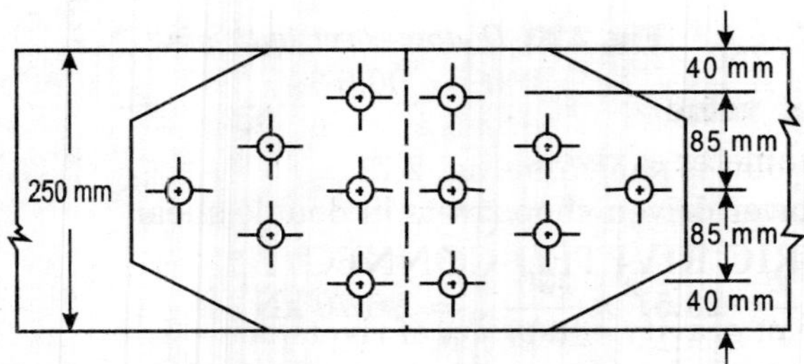

Fig. 2.21 *Diamond riveted double cover butt joint*

Step 1 : Diameter of rivet

Nominal diameter of rivet = 22 mm

Gross diameter of rivet = 23.5 mm

Step 2 : Rivet value

Adopt stresses from IS : 800–1984

Strength of power driven rivet in double shear

$$P_s = 2 \cdot \left(\frac{\pi}{4} \times (23.5)^2 \times \frac{100}{1000}\right) = 86.70 \text{ kN}$$

Strength of power driven rivet in bearing

$$P_b = \left(\frac{23.5 \times 22 \times 300}{1000}\right) = 155.1 \text{ kN}$$

$$\text{Rivet value, } R = 86.70 \text{ kN}$$

Step 3 : Number of rivets required

$$n = \left(\frac{500}{86.70}\right) = 5.75 \simeq 6$$

Provide 6 rivets in diamond group of riveting for efficient joint.

Step 4 : Thickness of flat

Let thickness of flat be t mm

Strength of plate at weakest section $(b - d) \times t \times \sigma_t = 500$

or $$\left(\frac{(250 - 23.5) \times t \times 6 \times 260}{1000}\right) = 500$$

$\therefore$ $$t = 14.151 \text{ mm}$$

Adopt 16 mm thickness of flat

Keep 40 mm edge distance from centre of rivet and 85 mm distance between centre to centre of rivet lines.

Step 5 : Efficiency of joint

$$\eta = \left(\frac{b-d}{b}\right) \times 100$$

$$\left(\frac{250 - 23.5}{250}\right) \times 100 = 90.6\ \%$$

The design of joint is shown in Fig. 2.21.

2.17 ECCENTRIC RIVETED CONNECTIONS

When the centre of gravity of a group of rivets does not lie on the line action of the load, the connections are known as *eccentric riveted connections.*

The eccentric riveted connections are of two types :

1. When the line of action of load the group of rivets, are in vertical plane and the centre of gravity of rivets is the centre of rotation, the eccentric riveted connections are of type one.
2. When the line of action of load does not lie in the plane of the group of rivets, and the line of rotation does not pass through the centre of gravity of the group of rivets, the eccentric riveted connections are of type two.

Type 1. Figure 2.22 shows an eccentric riveted connection. A vertical load P is acting at distance e from the centre of gravity of the group of rivets.

Consider a load P acting vertically downward and a load P acting vertically upward, both applied along and line passing dirough the C.G., of the group of rivets.

This is equivalent to

(i) an axial load (P) and

(ii) a couple ($P \times e$)

This couple causes *twisting moment.* This is also called as torsional moment :

Let n be the number of rivets in the bracket connection. The rivets are subjected to direct shear and a moment.

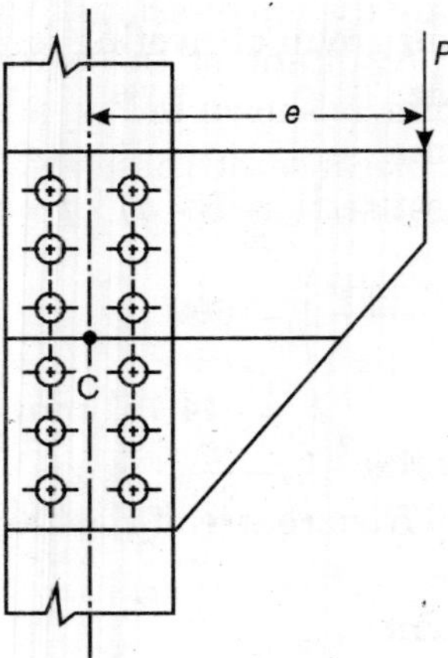

Fig. 2.22 *A bracket connection*

The direction of resisting shear force in a rivet is vertically upward. Shear force in any rivet is

$$F = \left(\frac{P}{n}\right)$$

The force resisting the twisting (torsional) moment in any rivet is proportional to the distance (r) of the centre of the rivet from the C.G. of the group of the rivets. The force acts in a direction perpendicular to the line joining the centre of the rivet to the C.G. of the group of the rivets and gives resisting moments about C.G. of group of rivets.

$$F_2 \propto r, \qquad \therefore F_2 = K \, . \, r \qquad \therefore K = \left(\frac{F_2}{r}\right)$$

Resisting moment because of force F_2 about C.G. of the group of rivets is

$$F_2 \, . \, r = K \, . \, r \, . \, r, \quad \text{or } F_2 \, . \, r = K \, . \, r^2$$

Total resisting moment is $\Sigma F_2 \, . \, r$ and is equal to the external moment $P \times e$.

$$\Sigma F_2 \, . \, r = \Sigma K \, . \, r^2, \quad \text{or } \Sigma F_2 \, . \, r = K \Sigma r^2$$

$$P \, . \, e = \frac{F_2}{r} \Sigma r^2, \quad \therefore F_2 = \left(\frac{P \cdot e \cdot r}{\Sigma r^2}\right)$$

The force resisting the twisting (torsional) moment is maximum in the rivet at the extreme distance (r_n) from the centre of gravity. The force F_2 in such a rivet is given by

$$F_2 = \left(\frac{P \cdot e \cdot r_n}{\Sigma r^2}\right)$$

The direct shear force F_1, and the force resisting the moment F_2 in a rivet are inclined at an angle θ. The resultant of these two forces can be found by determining their vectorial sum as shown in Fig. 2.23.

Resultant force in a rivet because of these two forces is obtained as the vectorial sum of two forces as under:

$$F = \left[F_1^2 + F_2^2 + 2F_1F_2 \cos\theta\right]^{1/2}$$

or $$F = \left[\left(\frac{P}{n}\right)^2 + \left(\frac{P \cdot e \cdot r_n}{\Sigma r^2}\right)^2 + 2\frac{P}{n}\frac{P \cdot e \cdot r_n}{\Sigma r^2} \cdot \cos\theta\right]^{1/2}$$

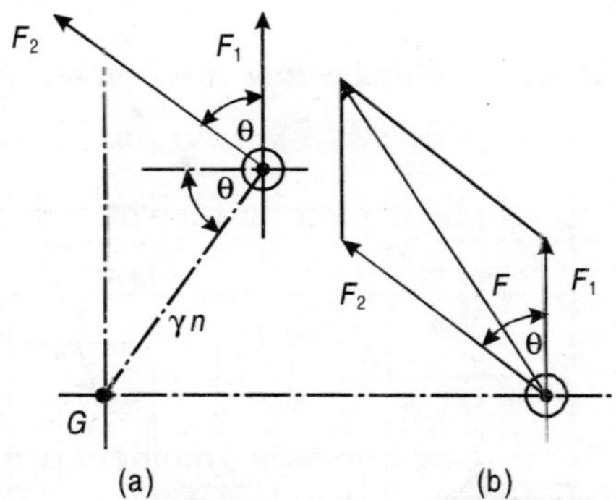

Fig. 2.23

The resultant force F in a rivet caused by an eccentric load should not exceed the rivet value R.

When the rivets are arranged in a narrow strip, the vertical distances may be taken as the distance to the centre of the rivets from the C.G. of the group of rivets.

Resultant force in a rivet under the above assumption is then

$$F = \left[\left(\frac{P}{n}\right)^2 + \left(\frac{P \cdot e \cdot y_n}{\Sigma y^2}\right)^2 + 2\frac{P}{n}\frac{P \cdot e \cdot y_n}{\Sigma y^2} \cdot \cos\theta\right]^{1/2} \quad ...(2.11)$$

where y_1, y_2 etc. are the vertical distances between the centres of rivets and C.G. of group of rivets. y_n is the distance to centre of extreme rivets from, the C.G. of group of rivets.

The direction of the applied load (P) may be at some inclination (θ) with the vertical as shown in Fig. 2.24.

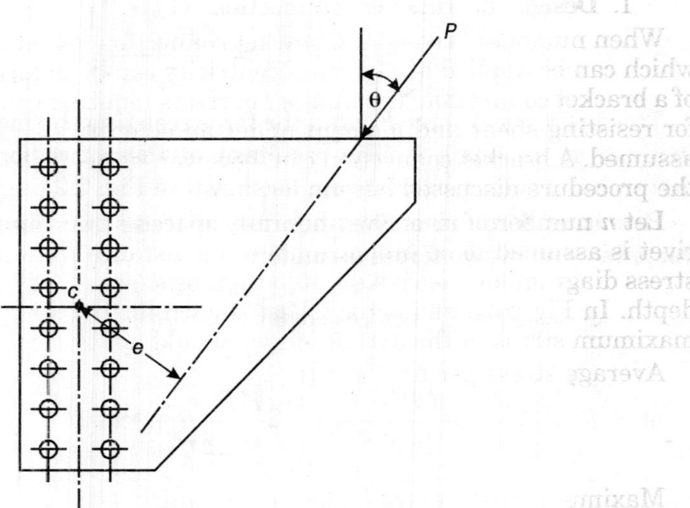

Fig. 2.24 *Inclined load acting in a bracket connection*

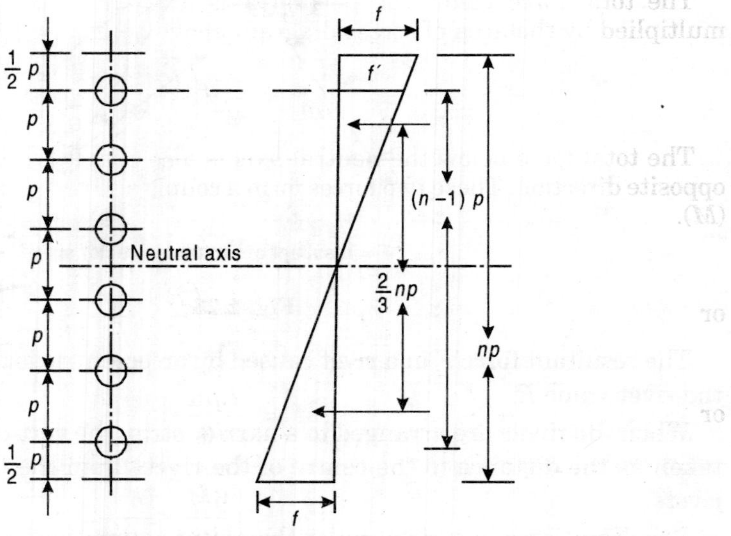

Fig. 2.25 *Stress diagram for rivets*

The eccentricity is computed. The direction of shear force in a rivet is parallel and opposite to the direction of the applied load. The direction of force in a rivet resisting the external moment (*P.e*) is in a direction perpendicular to the line joining the centre of rivet and centre of gravity *G* of the group of rivets. The maximum stress in the rivet most remotely placed from the C.G. of the group of rivets is found as already indicated. Its value should not exceed the rivet value *(R)*.

1. **Design of bracket connection (Type 1)**

When number of rivets in a bracket connection is known, the maximum load which can be applied at a given eccentricity can be determined. For the design of a bracket connection, the number of rivets required in the bracket connection for resisting shear and moment under an eccentrically applied load has to be assumed. A bracket connection can, however, be designed more conveniently by the procedure discussed hereunder:

Let n number of rivets be uniformly spaced at a distance 'p'. The stress in a rivet is assumed to be proportional to its distance from the neutral axis. Then stress diagram for the rivets would then be same as for rectangular beam of np depth. In Fig. 2.25 strips have been shown for the stresses in the rivets. The maximum stress in the extreme rivet should not exceed the rivet value R.

Average stress per unit depth

$$f' = \left(\frac{R}{p}\right)$$

Maximum stress on equivalent rectangular beam is

$$f = \frac{R}{P} \times \left(\frac{n}{n-1}\right)$$

The total force below the neutral axis is equal to the intensity of stress multiplied by the area of stress diagram above neutral axis.

$$F = \frac{1}{2} \times \left(\frac{R}{p}\right) \cdot \left(\frac{n}{(n-1)}\right) \cdot \left(\frac{np}{2}\right)$$

The total force below the neutral axis is also equal to this vatae and acts in opposite direction. These two forces form a couple and resist the external moment (M).

$$M = \text{Force} \times \text{Lever arm } n$$

or
$$M = \frac{1}{2} \times \left(\frac{R}{P} \times \frac{n}{(n-1)} \times \frac{np}{2}\right) \times \frac{2}{3} np$$

or
$$M = \frac{Rpn^2}{6} \times \left(\frac{n}{(n-1)}\right)$$

$$n = \left[\frac{6M}{Rp} \times \frac{(n-1)}{n}\right]^{1/2}$$

The factor $\left(\frac{n-1}{n}\right)$ may be assumed to be nearly 1. This assumption would be always on the safe side.

$$n = \left[\frac{6M}{Rp}\right]^{1/2} \qquad \text{...(2.12)}$$

The number of rivets computed from this expression is essential for resisting the external moment. For the purpose of resisting shear, a revision will seldom be found necessary. The size of rivets and pitch is may be assumed and a rivet value (R) can then be computed. The edge distance in the design of bracket connection is kept equal to one-half the pitch.

For two vertical lines of rivets, a rivet value of $2R$ is used and thus number of rivets is calculated for one row of rivets. These rivets are arranged for the bracket connection, and design is checked by general method.

2. Design of bracket connection (Type 2)

In this type of eccentric riveted connections, a rivet is subjected to tension, in addition to direct shear in the bracket connection as shown in Fig. 2.26.

A tee-section has been attached to a stanchion by riveting. The portion of tee-section above topmost rivet does not provide any resistance. The depth of the bracket is measured from bottom of the bracket upto the centre of the top rivet.

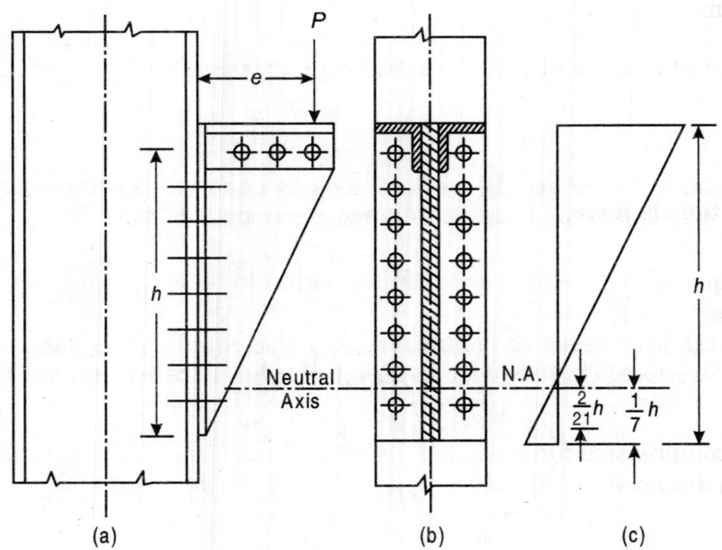

Fig. 2.26 *A bracket connection*

The position of the line of rotation depends upon the degree of rigidity of bracket connection. In practice, the line of rotation (neutral axis) is assumed to lie at a height of $\frac{1}{7}$th of the depth of the bracket measured from the bottom edge of the bracket. The rivets above the line of rotation are in tension in addition to being in direct shear, while below this line the tee-section is in compression against the column flange. The stress diagram has been shown in Fig. 2.26 (c).

The tensile force in a rivet F_t is proportional to its distance y from the line of rotation.

$$F_t \propto y, \qquad \therefore\ F_t = ky \qquad \therefore\ k = \left(\frac{F_t}{y}\right)$$

Moment of resistance due to this tensile force is

$$F_t . y = k . y . y, \;\therefore\; F_t . y = k . y^2$$

Total moment of resistance (M) provided by rivets in tension

$$M' = \Sigma k . y^2, \;\therefore\; M' \left(\frac{F_t}{y}\right) \Sigma y^2$$

or

$$F_t = \left(\frac{M'y}{\Sigma y^2}\right)$$

Let $y_1, y_2, ... y_n$ represent distances of the centres of the rivets from the line of rotation. The tensile forces in these rivets will be as given hereunder. It is obvious that tensile force in the topmost rivet which is at a distance y_n, is maximum.

$$F_{t_1} = \left(\frac{M'y_1}{\Sigma y^2}\right), F_{t_2} = \left(\frac{M'y_2}{\Sigma y^2}\right), F_{t_n} = \left(\frac{M'y_n}{\Sigma y^2}\right)$$

(max)

Total tensile force,

$$F_t = \left(\frac{M'\Sigma y}{\Sigma y^2}\right)$$

For equilibrium, total tensile force is equal to total compressive force. Total compressive force

$$C = \left(\frac{M'\Sigma y}{\Sigma y^2}\right)$$

This compressive force acts at the centroid of stress distribution diagram C viz., at a distance

$$\left(\frac{2}{3}\right) \times \left(\frac{h}{7}\right) = \left(\frac{2h}{21}\right)$$

Taking the moment about the line of rotation (neutral axis).

External moment = Moment resisted by rivets in tension
+ Moment resisted by area of tee-section in compression.

$$M = \left(M' + C \cdot \frac{2h}{21}\right), \;\therefore\; M = \left[M' + \frac{M'\Sigma y}{\Sigma y^2} \cdot \frac{2h}{21}\right]$$

or

$$M' = \left[\frac{M}{1 + \dfrac{2h}{21} \cdot \dfrac{\Sigma y}{\Sigma y^2}}\right] \qquad ...(2.13)$$

This equation gives the moment resisted by the rivets in tension from which maximum tensile force in the extreme rivet F_t can be calculated. Tensile stress:

$$\sigma_{tf.cal} = \left(\frac{F_t}{A}\right)$$

where A = gross cross-sectional area of rivet. Direct shear force in any rivet is

$$F = \left(\frac{P}{n}\right)$$

and shear stress $$f_s = \left(\frac{F}{A}\right)$$

The rivets subjected to shear and externally applied tensile force, should be so proportioned that the quantity

$$\left[\frac{\tau_{vf.cal}}{\tau_{vf}} + \frac{\tau_{tf.cal}}{\tau_{tf}}\right] \text{does not exceed 1.4}$$

where

$\tau_{vf.cal}$ = Actual shear stress in rivet
$\sigma_{tf.cal}$ = Actual tensile stress in rivet
τ_{vf} = Working shear stress in rivet.
τ_{tf} = Working tensile stress in rivet.

The stress in rivet is checked from the above equation.

Design procedure for bracket connection (Type 2)

In the design of bracket connection of this type, following are the usual steps for the design.

Step 1. The nominal diameter of rivet to be used in the bracket connection is assumed. Working stress in the rivet in tension is adopted as per IS : 800–1984 and the rivet value is calculated.

Step 2. The pitch of 2.5 times to 3 times the nominal diameter of rivet approximately is adopted for the rivets. The rivets are provided in two vertical rows. The number of rivets necessary for one row is computed from Eq. 2.12, by substituting rivet value of $2R$ in the expression, as the rivets are accommodated in two vertical rows. The rivet value 'R' is calculated in direct tension.

Step 3. The number of rivets provided in the design is checked by calculating actual shear and tensile stresses in the rivets and quantity

$$\left[\frac{\tau_{vf.cal}}{\tau_{vf}} + \frac{\sigma_{tf.cal}}{\sigma_{ft}}\right] \text{does not exceed 1.4.}$$

Example 2.7 *A riveted steel bracket connection has 22 mm diameter rivets 12 in number arranged as shown in Fig. 2.27. Determine the load P so that allowable stress in the extremely loaded rivet is just reached. The safe permissible stress in bearing in rivet is 300 N/mm^2 (MPa) and safe permissible stress in shearing in rivet is 100 N/mm^2 (MPa).*

Solution

Step 1. Diameter of rivets. Nominal diameter of rivet = 22 mm Gross diameter of rivet = 23.5 mm

Step 2. Rivet value (*R*). Strength of power driven shop rivet in single shear

$$\left(\frac{\pi}{4}\times\frac{(23.5)^2\times100}{1000}\right) = 43.35 \text{ kN}$$

Assume that strength of rivet in bearing is greater than strength of rivet in single shear

$$\text{Rivet value} = 43.35 \text{ kN}$$

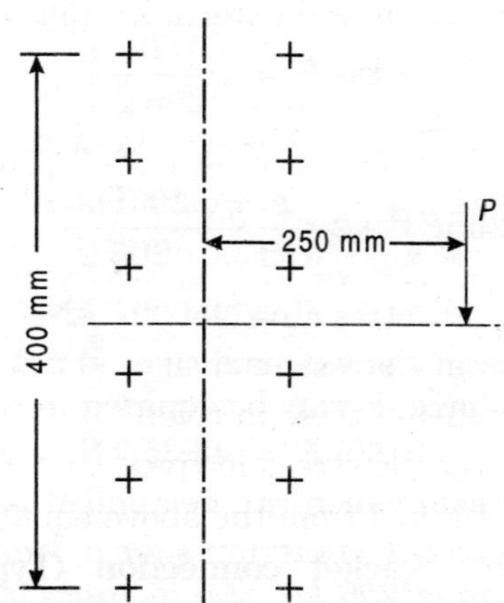

Fig. 2.27 *Rivets in a bracket connection*

Step 3. Direct shear in each rivet. Number of rivets in bracket connection is 12

$$F_1 = \left(\frac{P}{12}\right)\text{kN}$$

External moment acting on bracket connection

$$M = (P \times 250) \text{ kN-mm}$$

Force in extreme rivet due to twisting (torsional) moment

$$F_2 = \left(\frac{P\cdot 250\cdot r_n}{\Sigma r^2}\right)\text{kN}$$

Distance of centre of extreme rivet from C.G. of group of rivets

$$r_n = (50^2 + 200^2)^{1/2} = 206.155 \text{ mm}$$

Horizontal distance of each rivet from C.G. of group of rivtts

$$= 50 \text{ mm}$$

Vertical distances of rivets

Rivets in first row above C.G.= 40 mm

Rivets in second row C.G. = 120 mm

Rivets in topmost row C.G. = 200 mm

Other rivets are symmetrically placed

$\Sigma r^2 = [\,4\,(5^2 + 4^2) + 4\,(5^2 + 12^2) + 4\,(5^2 + 20^2]\times 100$ m .

$= 2540 \times 100$ mm^2

$$F_2 = \left(\frac{P \times 250 \times 206.2}{2540 \times 100}\right) = 0.205\, P \text{ kN}$$

Resultant of two forces

$$F = \left[F_1^2 + F_2^2 + 2F_1F_2\cos\theta\right]^{1/2}$$

From Fig. 2.27, $\cos\theta = \left(\dfrac{50}{206.2}\right)$

$$\therefore \quad F = \left[\left(\frac{P}{12}\right)^2 + (0.205P)^2 + 2\times\frac{P}{12}\times\frac{0.205P\times 50}{206.2}\right]^{1/2}$$

$$\therefore \quad F = 0.239P \text{ kN}$$

As the resultant force in the extreme rivet is not to exceed the rivet value. Therefore the resultant force, F may be equated to rivet, R

or $0.239\,P = 43.35$ kN, $P = 181.38$ kN

Hence the maximum load which can be applied is **181.38 kN.**

Example 2.8 *Design riveted connections for a bracket as shown in Fig. 2.28 carrying an eccentric load of 200 kN at a distance of 350 mm from the centre line. Adopt working stresses for rivets as per IS : 800–1984.*

Solution

Design :

Step 1. Rivet value. Use 22 mm normal diameter rivet. Gross diameter or rivet

$= 23.5$ mm

Adopt working stresses given for the power driven rivets from IS : 800–1984.

Strength of rivet in single shear

$$\left(\frac{\pi}{4}\times\frac{(23.5)^2\times 100}{1000}\right) = 43.35 \text{ kN}$$

Assume thickness of gusset plate as 14 mm

Thickness of flange of stanchion, HB 300, @ 0.588 kN/m is 10.6 mm.

Bearing strength of rivet

$$\left(\frac{23.5\times 10.6\times 300}{1000}\right) = 74.73 \text{ kN.}$$

Rivet value $R = 43.35$ kN

Step 2. Number of rivets. Provide rivet in two vertical rows and at 55 mm pitch then rivet value is $2R$

$$n = \left[\frac{6M}{P(2R)}\right]^{1/2} = \left[\frac{6\times200\times350}{55\times2\times43.35}\right]^{1/2} = 9.3$$

Provide 10 rivets in one vertical row.

Distance between centre to centre of rivets in vertical rows for HB 300, @ 0.588 kN/m from steel section tables in 140 mm.

Step 3. Check. C.G. of group of rivets is midway.

As rivets are in a narrow strip,consider the vertical distance as the distance of the centre of rivet from the C.G. of group of rivets.

$$\Sigma y^2 = [4(2.75^2 + 8.25^2 + 13.75^2 + 19.25 + 24.75^2] \times 10^2$$

or $$\Sigma y^2 = 4989.37\times10^2 \text{ mm}^2$$

Force in the extreme rivet, resisting the torsional moment

$$F_2 = \left(\frac{P\times e\cdot y_n}{\Sigma y^2}\right) = \left(\frac{200\times350\times24.75\times10}{4989.37\times100}\right)$$
$$= 34.72 \text{ kN.}$$

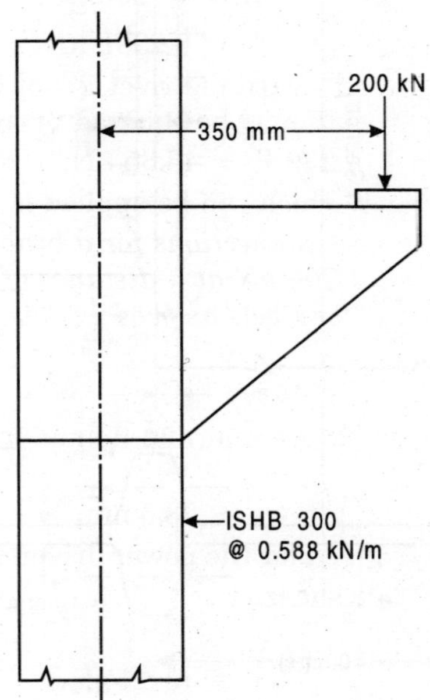

Fig. 2.28

Direct shear stress in the rivet

$$F_1 = \left(\frac{200}{20}\right) = 10 \text{ kN}$$

Resultant stress is obtained by vectorial sum of two stresses

$$F = \left[10^2 + 34.72^2 + 2 \times 10 \times 34.72 \times \frac{70}{206.2}\right]^{1/2}$$

$$= 39.258 \text{ kN} < 43.35 \text{ kN (Rivet value)}$$

Hence, the design is satisfactory.

Example 2.9 *The flange of a tee-section of 200 mm × 200 mm is riveted to the flange of a rolled steel column of I-section to form a bracket. It carries a vertical load of 280 kN at a distance of 200 mm from the face as shown in Fig. 2.29. Find a suitable depth for the bracket and design the riveted connections, connecting the tee-section with the flange of column.*

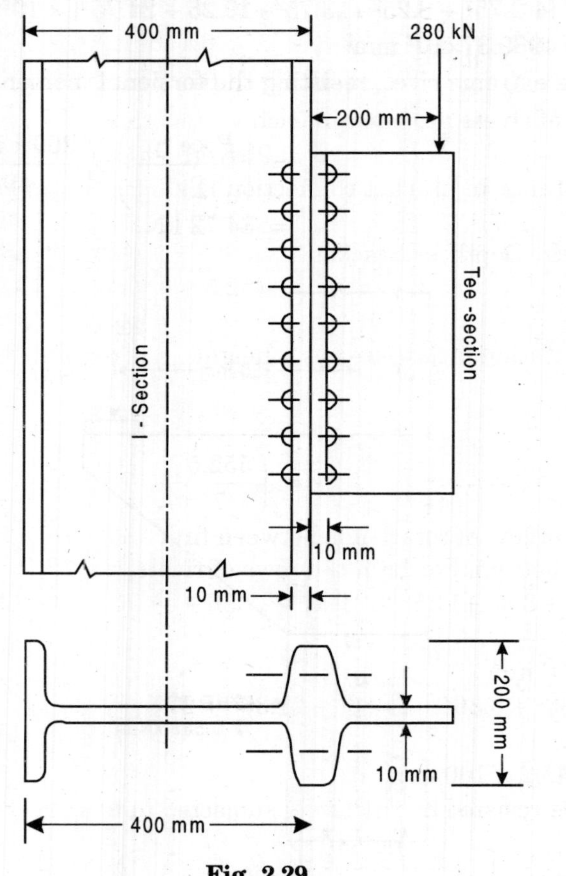

Fig. 2.29

Solution

Design :

Step 1. Rivet value. Use 22 mm nominal diameter power driven rivets. Gross diameter of rivet

$$= 23.5 \text{ mm} \quad \text{...(i)}$$

Rivets are subjected to tension and hence rivet value shall be the strength of rivet in tension. Gross area of rivet

$$\left(\frac{\pi}{4}\times 23.5^2\right) = 433.52 \text{ mm}^2 \qquad \text{...(ii)}$$

Allowable stress in axial tension for power driven rivets

$$= 100 \text{ N/mm}^2 \text{ (MPa)} \qquad \text{...(iii)}$$

Rivet value,

$$R = \left(\frac{433.52\times 100}{1000}\right) 43.35 \text{ kN} \qquad \text{...(iv)}$$

Step 2. Number of Rivets. The rivets in the bracket connection are used in the two vertical rows. Adopting a pitch of 65 mm the number of rivets in one vertical row is given by

$$n = \left[\frac{6M}{P\times 2R}\right]^{1/2} = \left[\frac{6\times 280\times 200}{65\times 2\times 43.35}\right]^{1/2} = 7.72 \qquad \text{...(v)}$$

The number of rivets provided in each vertical row

$$n = 9 \text{ rivets} \qquad \text{...(vi)}$$

The edge distance in bracket connection is kept equal to half the pitch, i.e., 32.5 mm.

Step 3. Check. Depth of bracket

$$h = (32.5 + 8 \times 65) = 552.5 \text{ mm} \qquad \text{...(vii)}$$

The line of rotation is assumed at a height $\frac{h}{7}$ above the bottom edge of the bracket

$$\frac{h}{7} = \left(\frac{552.5}{7}\right) = 78.9 \text{ mm} \qquad \text{...(viii)}$$

The position of line of rotation is between first and second horizontal rows of rivets from the bottom. For both the rows of rivets.

$$\Sigma y = [2\,(1.86 + 8.36 + 14.86 + 21.36 + 27.86 + 34.36 + 40.86 + 47.36)] \times 10 \text{ mm} \qquad \text{...(ix)}$$

or $\Sigma y = 3937.6$ mm

$$\Sigma y^2 = [2(1.86^2 + 8.36^2 + 14.86^2 + 21.36^2 + 27.86^2 + 34.36^2 + 40.86^2 + 47.36^2)] \times 10^2 \text{ mm}^2 \qquad \text{...(x)}$$

or $\Sigma y^2 = 13234.92 \times 100 \text{ mm}^2$

Moment to be resisted by the rivets subjected to tension

$$M' = \left[\frac{M}{1+\frac{2h}{21}\cdot\frac{\Sigma y}{\Sigma y^2}}\right] = \left(\frac{280\times 200}{1+\frac{2\times 552.5}{21}\times\frac{3937.6}{13234.92\times 100}}\right)$$

$$M' = 48500 \text{ mm-kN} \qquad \text{...(xi)}$$

Tensile stress in the topmost rivet

$$\sigma_{vf.cal} = \left(\frac{M'\times y_n}{A\times\Sigma y^2}\right) = \left(\frac{48500\times 473.6\times 1000}{433.52\times 13234.94\times 100}\right)$$

$$= 40.93 \text{ N/mm}^2 \quad \text{...(xii)}$$

Actual shear stress in the rivet

$$\tau_{vf.\text{cal}} = \left(\frac{280 \times 1000}{433.52 \times 2 \times 9}\right) = 35.88 \text{ N/mm}^2 \quad \text{...(xiii)}$$

The rivets are subjected to combined stress (i.e., tensile stress and shear stress). Substituting the values of these stresses

$$\left[\frac{\tau_{tf.\text{cal}}}{\sigma_{tf}} + \frac{\tau_{vf.\text{cal}}}{\tau_{vf}}\right] = \left[\frac{40.03}{100} + \frac{35.88}{100}\right] = 0.7591 < 1.4$$

Hence, the design is satisfactory.

Example 2.10 *A bracket connection is shown in Fig. 2.30. Check the design of all fasteners in this bracket.*

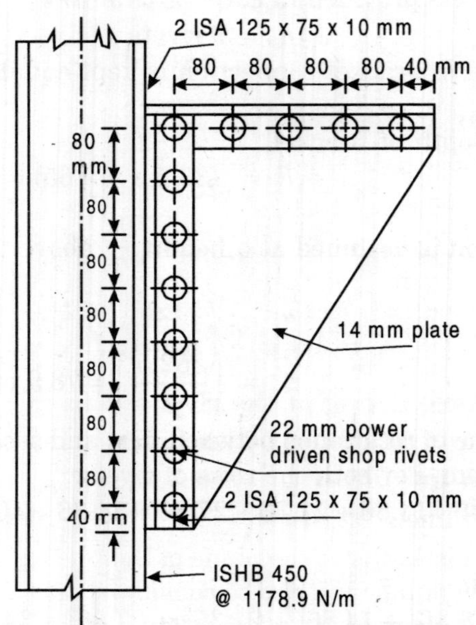

Fig. 2.30

Solution

Step 1. Rivet Value

Nominal diameter of rivets = 22 mm

Gross diameter of rivets = 23.5 mm

Strength of power-driven rivets in double shear

$$\left[2 \times \frac{\pi}{2} \times (23.5)^2 \times \frac{100}{1000}\right] = 86.70 \text{ kN}$$

Strength of power driven shop rivets in bearing

$$\left(23.5 \times 14 \times \frac{300}{1000}\right) = 98.70 \text{ kN}$$

Rivet value $R = 86.70$ kN

Step 2. Rivets in line *AA*

Consider rivets in line *AA*.

These rivets are subjected to direct shear and bending and can be considered as subjected to an axial load of $P = 125$ kN and a moment $M = 125 \times 80$ mm-kN as shown in Fig. 2.31.

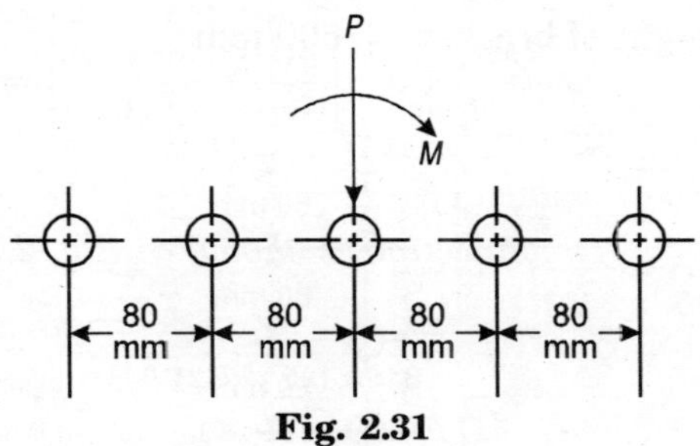

Fig. 2.31

1. **Direct shear in the rivet**

$$\frac{P}{5} = \left(\frac{125}{5}\right) = 25 \text{ kN}$$

2. **Maximum force in the extreme rivet due to torsional moment**

$$F_2 \left(\frac{P \cdot e \cdot r_n}{\Sigma r^2}\right) = \left(\frac{125 \times 80 \times 160}{2(8^2 + 16^2) \times 100}\right) = 25 \text{ kN}$$

3. **Resultant force acting in the rivet** (by vector sum of two forces)

$R = (25 + 25) = 50$ kN< 86.70 kN Rivet value. Hence satisfactory.

Step 3. Rivets in line *BB*

Consider rivets in line *BB*.

The eccentric load is acting in the plane of the rivets. The rivets are subjected to direct shear and bending, and can be considered as subjected to an axial load of $P_1 = 125$ kN and a moment $M_1 = 125 \times 240$ mm-kN as shown in Fig. 23.2.

1. **Direct shear force in the rivet**

$$F = \frac{125}{8} = 15.625 \text{ kN}$$

2. **Maximum force in the extreme rivet due to bending moment**

$$F = \left(\frac{P_1 \cdot e_1 \cdot r_n}{\Sigma r^2}\right)$$

$$F = \left(\frac{125 \times 240 \times 280}{2(4^2 + 12^2 + 20^2 + 28^2)} \times \frac{1}{100}\right)$$

$$= 31.25 \text{ kN}$$

3. **Resultant force in the rivet**

$$R = (15.625^2 + 31.25^2)^{1/2} = 34.94 \text{ kN}$$

Rivet value. Hence, satisfactory.

Consider rivets, connecting angle and the column section. These rivets are subjected to direct shear and tension. The rivets have been provided on both the sides.

Height of bracket = 600 mm

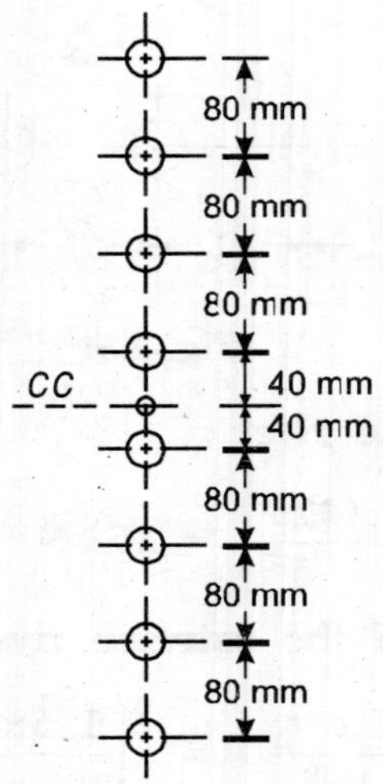

Fig. 2.32

Neutral axis of the rivet line is assumed at 1/7th the height of bracket above the bottom of the bracket.

Height of neutral axis

$$\frac{1}{7} \times 600 = 85.7 \text{ mm}$$

For both rows of rivets

$$\Sigma y = [2(3.43 + 11.43 + 19.43 + 27.43 + 35.43 + 43.43 + 51.43)] \times 10$$
$$= 2 \times 192.01 \times 10 \text{ mm}$$

$$\Sigma y^2 = [2(3.43^2 + 11.43^2 + 19.43^2 + 27.43^2 + 35.43^2 + 43.43^2 + 51.43^2)] \times 100$$
$$= 2 \times 7058.78 \times 100 \text{ mm}^2$$

Moment to be resisted by the rivets in tension from Eq. 2.13

$$M' = \frac{M}{\left(1 + \frac{2h}{21} \cdot \frac{\Sigma y}{\Sigma y^2}\right)}$$

The gauge distance for rivet for ISA 125 mm × 75 mm × 10 mm 125 mm length of leg from ISI Handbook No. 1 is 75 mm.

The eccentricity of the load measured from the face of column

$$e = (75 + 80 + 80 + 80) = 315 \text{ mm}$$

$$M' = \left(\frac{125 \times 315}{1 + \frac{2 \times 600}{21} \times \frac{2 \times 192.01 \times 10}{2 \times 7058.78 \times 100}} \right)$$

$$= 35650 \text{ mm-kN}$$

Tensile force resisted by top rivet

$$F_t = \left(\frac{M' \times y_n}{\Sigma y^2} \right) = \left(\frac{35650 \times 514.3}{2 \times 7058.78 \times 100} \right)$$

$$= 12.950 \text{ kN}$$

Actual tensile stress in the rivet

$$\sigma_{tf.cal} = \left(\frac{12.950 \times 1000}{\frac{\pi}{4} \times (23.5)^2} \right) = 29.872 \text{ N/mm}^2$$

Allowable axial tensile stress in the power driven rivets

$$\sigma_{tf} = 100 \text{ N/mm}^2 \text{ (MPa)}$$

Actual shear stress in the rivet

$$\tau_{vf.cal} = \left(\frac{125 \times 1000}{2 \times 8 \times \frac{\pi}{4} \times 23.5^2} \right) = 18.02 \text{ N/mm}^2$$

Allowable shear stress in the power driven rivets is

$$\tau_{vf} = 100 \text{ N/mm}^2$$

$$\therefore \quad \left(\frac{\sigma_{tf.cal}}{\sigma_{tf}} + \frac{\tau_{vf.cal}}{\tau_{vf}} \right) = \left(\frac{29.872}{100} \right) + \left(\frac{18.02}{100} \right) = 0.479 < 1.4$$

Hence, satisfactory.

Example 2.11 A *bracket connection is shown in Fig. 2.33. The thickness of plate attached to the flange of a column section MB 200, @ 25.4 kg/m is 10 mm. The plate carries a load of 20 kN inclined 60° with the horizontal at a distance 160 mm from the vertical through C.G. of group of the rivets. Design the bracket connection.*

Solution

Design :

Step 1. Distance of line of action of load from C.G. of rivets

The inclination of load P with the horizontal is 60°. Its inclination with the vertical is 30°. The prolonged direction of load P intersects the vertical through C.G. of group of the rivets at O. Consider the triangle ABC.

$$\angle ABC = 60°, \angle BAC = 90°, \angle ACB = 30°, AB = 160 \text{ mm}$$

$$\therefore \quad AC = AB \tan 60° = 160 \times \sqrt{3} = 276.8 \text{ mm} \quad \text{...(i)}$$

The eccentricity, e of the load P from the C.G. of group of rivets O, is OD. Consider the triangle OCD. The angle ODC is 90°

$$OC = AC - AO = 2768.40 = 236.8 \text{ mm} \quad \text{...(ii)}$$

$$OD = OC \sin 30° = 236.8 \times \tfrac{1}{2} = 118.4 \text{ mm} \quad \text{...(iii)}$$

Distance of line of action of load from C.G. of group of rivets

$$\therefore \quad e = 118.4 \text{ mm} \quad \text{...(iv)}$$

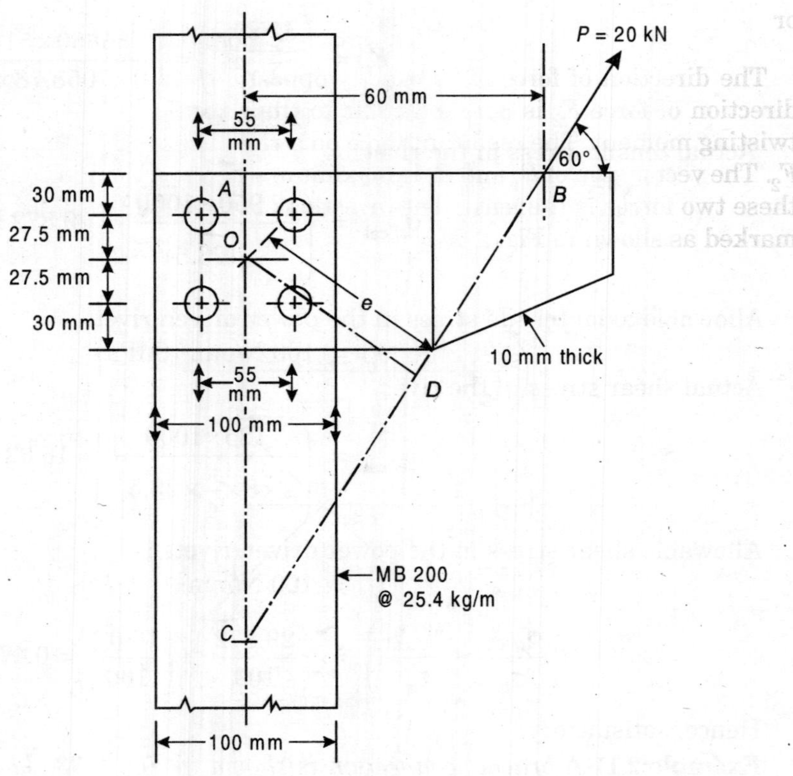

Fig. 2.33

The rivets are subjected to direct shear and twisting moment. The number of rivets in the bracket, N is 4.

Step 2. Force due to direct shear

Force in each rivet due to direct shear

$$\left(F_1 = \frac{P}{4} = \frac{20}{4}\right) = 5 \text{ kN} \quad \text{...(v)}$$

Step 3. Force due to twisting moment

All the corner rivets are at equal distance from the C.G. of group of rivets, O. The distance of corner rivet from O

$$r = 27.5\sqrt{2} = 27.5 \times 1.414 = 38.89 \text{ mm} \quad \text{...(vi)}$$

$$\Sigma\gamma^2 = (4 \times 38.89^2) = 6050 \text{ mm}^2 \quad \text{...(vii)}$$

Forces in the corner rivets due to twisting moment are equal. Force in each corner rivet due to twisting moment

$$F_2 = \left(\frac{M_r}{\Sigma r^2} = \frac{P \cdot e \cdot r}{\Sigma \gamma^2}\right)$$

or

$$F_2 = \left(\frac{20 \times 118.4 \times 38.89}{6050}\right) = 15.22 \text{ kN} \quad \text{...(viii)}$$

The direction of force F_1 is exactly opposite to the direction of load P. The direction of force F_2 is perpendicular to the radius vector and opposite to the twisting moment. The resultant force on a rivet is equal to vector sum of F_1 and F_2. The vector sum of F_1 and F_2 is maximum when the included angle θ, between these two forces is the least. The directions of force F_1 and F_2 on each rivet are marked as shown in Fig. 2.34.

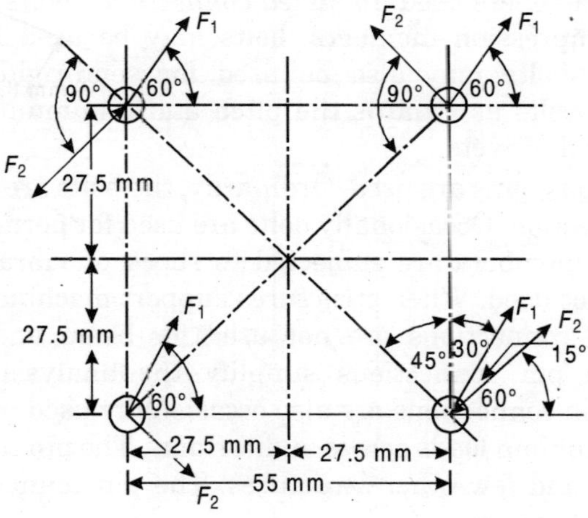

Fig. 2.34

Step 4. Resultant of forces

It is seen that the lower right rivet is subjected to maximum force. The angle θ_1 between F_1 and F_2 is least

$$\theta_1 = 15°, \cos\theta_1 = 0.9659$$

Resultant of the force F_1 and F_2 is given by

$$F = [5^2 + 15.22^2 + 2 \times 5 \times 15.22 \times 0.9659]^{1/2}$$
$$= 20.09 \text{ kN}$$

Strength of 16 mm nominal diameter rivet in single shear

$$\left(\frac{\pi}{4} \times \frac{(17.5)^2 \times 100}{1000}\right) = 24.05 \text{ kN}$$

Thickness of the plate is 10 mm. The thickness of flange of MB 200, @ 25.4 kg/m is 10.8 mm. Therefore, the strength of rivet in bearing

$$\left(17.5 \times 10 \times \frac{300}{1000}\right) = 52.5 \text{ kN}$$

Rivet value of 16 mm nominal diameter rivet

$$R = 24.05 \text{ kN} > 20.09 \text{ kN. Hence, unsafe.}$$

Provide 16 mm diameter power driven rivets for the back connection.

2.18 BOLT AND PIN CONNECTIONS

In Sec. 2.1, it was mentioned that various types of connections are used for connecting the structural members. The choice of type of fasteners depends on connection strength required, space limitations of the connections, available technicians to fabricate and erect the structure, service conditions and finally the total cost of installation. These fasteners serve essentially the same function in transferring loads from one component to another. The connections are named after the type of fasteners used. In *bolted connections,* bolts and nuts are used. In tension or compression members, bolts may be used like rivets for end connections. The bolts may also be used for semi-rigid connections. The requirements for bolts as regards the pitch and minimum edge distance are same as those for the rivets.

In *pin connections,* pins are used. Ordinarily, the bolts are used for temporary fastenings and erection. Occasionally bolts are used for permanent connections. When structural members are subjected to shock or vibrations, then, bolted connections are not used. When structures support machinery or rolling loads, then also, bolted connections are not used for beam and beam or column connections. The pin connections simplify the analysis of indeterminate structures. The pin connections are also occasionally used in buildings to carry extremely heavy column loads across auditoriums. The pin connections are used in hinged arches and few other structures. The pin connection acts as hinge where the moment is zero.

2.19 ADVANTAGES OF BOLTED CONNECTIONS

The advantages of bolted connections are as follows:

1. There is *silence in* preparing bolted connection.
   In riveting, hammering is done. The hammering causes noise in the riveting.
2. There is no *risk of fire* in bolted connection.
   The rivets are made red hot in riveting and there is risk of fire.
3. The bolted connections may be done quickly in comparison to the riveting.
4. Though the cost of bolts is more than the cost of rivets, the bolted connections are economical to use them than rivets because less persons are required for installation, and the work proceeds quickly.
5. The bolted connections facilitate the erection because of ease with which these connections can be done.

2.20 DISADVANTAGES OF BOLTED CONNECTIONS

The following are disadvantages of bolted connections:

1. If bolted connections become loose, their strength reduces considerably.
2. The unfinished bolts are not uniform in diameter and they have less strength.
3. The bolted connections have less strength when they are subjected to axial tension, because area at root of thread is less.
4. Generally, the diameter of hole is kept 16 mm more than the nominal diameter of black bolt. The bolt does not fill the hole and there remains a clearance in bolted connections.

2.21 BOLTS

A bolt is a metal pin with a head formed at one end and the shank threaded at the other in order to receive a nut.

The bolts for structural purposes are manufactured from mild steel and high strength steel. A bolt consists of head and body as shown in Fig. 2.35. The body of a bolt is termed as shank. The bolts are manufactured to different lengths to suit different purposes. The size of bolt is expressed by the diameter of shank. The shank is threaded at one end. A nut can be provided on the threaded end of the bolt. The threaded portion of each bolt shall project through the nut at least by one thread. In all cases, when the full bearing area of the bolt is to be developed, the bolt shall be provided with a washer of sufficient thickness under the nut to avoid any threaded portion of the bolt being within the thickness or the parts bolted together. The structural bolts are classified according to type of shank (unfinished or turned); material and strength (ordinary structural, or high strength steel); shape of head and nut (square or horizontal, regular or heavy duty); and pitch and fit of thread (standard, coarse or fine).

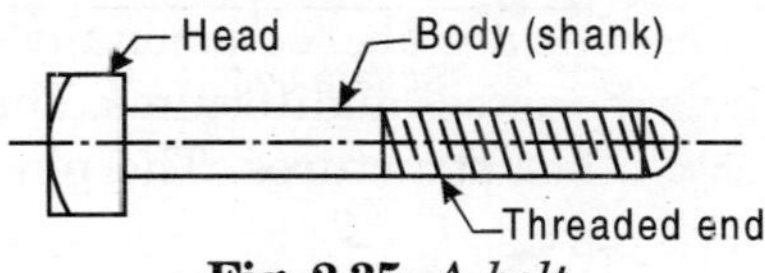

Fig. 2.35 *A bolt*

The following are the various types of bolts used for structural purposes.

2.21.1 Unfinished Bolts

The finished bolts are also termed as *ordinary bolts, common bolts, rough bolts* and *back bolts.* The unfinished bolts are manufactured from low carbon steel. The unfinished bolts have square heads. The heads of unfinished bolts are made by forging. The shanks of unfinished bolts are not uniform in diameter. The unfinished bolts are designed for lower permissible stresses in shear and bending than the rivets and turned bolts. The unfinished bolts are not used for structural connections of members, which are subjected to shock or vibrations. The unfinished bolts are used for ordinary field work and light load. The unfinished or black bolts are used with square nuts. These are also used with hexagonal nuts. These bolts, screws and nuts are designated as illustrated in Table 2.3 as per IS : 2585–1968.

Table 2.3 *Designation system for black square bolts, screws and nuts*

Description			*Designation*
Fasteners	*Thread size*	*Length mm*	
Square bolts with square nut	M 10	30	Square bolt M 10 × 30 N
Square bolt only	M 10	30	Square bolt M 10 × 30
Square screw	M 10	30	Square screw M 10 × 30
Square nut	M10	—	Square nut M 19
Square bolt with hexagonal nut	M 10	30	Square bolt M 10 × 30 NH

A large portion of site connections are made with *black bolts*. From the point of view of the erector, this makes for simplicity as the erection can be completed in almost one operation. The additional cost of special labour for riveting and the plant required is also provided.

The black bolts cannot, however, carry loads equal to those of a rivet with the same diameter. In such cases, the turned bolts are used.

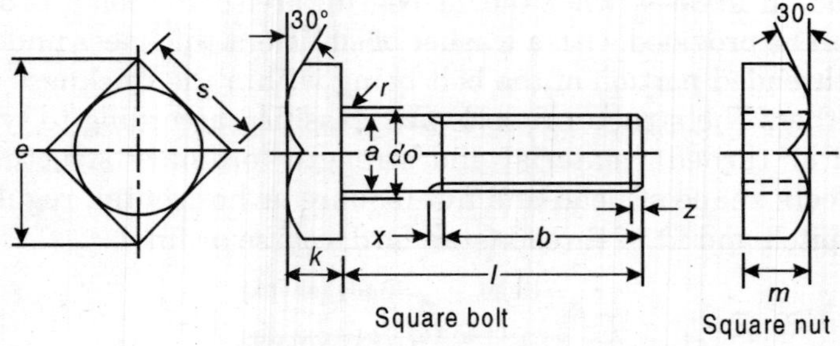

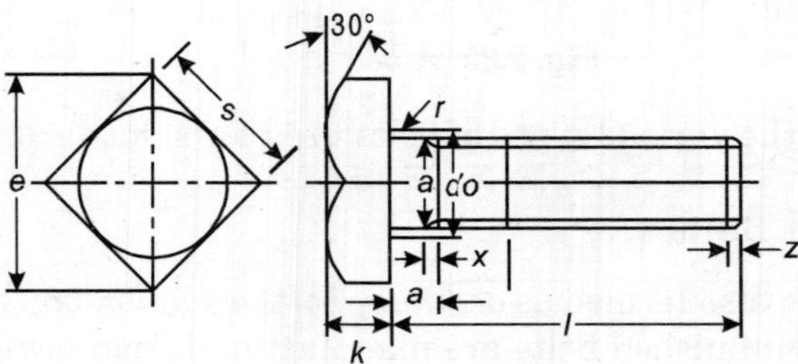

Fig. 2.36 *Square screw*

The various dimensions of bolts, screws and nuts are as shown in Fig. 2.36. The dimensions are as per IS : 2585–1968 and have been given in Tables 2.4 and 2.5. The diameters range from 6 to 39 mm.

Table 2.4 *Dimensions for black square bolts, screws and nuts (as per IS : 2585–1968)*

(Diamter range M-6 to M-18)

Size		*M6*	*M8*	*M10*	*M12*	*(M14)*	*M16*	*(M18)*	
		Nom.	6	8	10	13	14	16	18
d		Max.	6.48	8.90	10.90	12.10	15.10	17.10	19.10
		Min	5.70	7.64	9.64	11.57	13.57	15.57	17.57
s	<19-h14	Nom.	10	13	17	9	22	24	27
		Max.	10.00	13.00	17.00	19.00	22.00	24.00	27.00
	>19-h15	Min.	9.64	12.67	16.57	18.48	21.16	23.16	26.16
c		Min.	12.53	16.34	21.54	24.02	27.51	30.11	34.01
		Nom.	4	5.5	7	8	9	10	12
k	*js* 16	Max.	4.38	5.88	7.45	8.45	9.45	10.45	12.55
		Min.	3.62	5.12	6.55	7.55	8.55	9.55	11.45
da		Max.	7.2	10.2	12.2	15.2	17.2	19.2	21.2
r		Min.	0.25	0.4	0.4	0.6	0.6	0.6	0.6
a		Max.	4.0	4.5	5.0	6.0	7.5	7.5	9.0
	b_1	Min.	18	22	26	30	34	38	42
b	b_2	Min.	—	28	32	36	40	44	48
	b_3	Min.	—	—	—	40	53	57	61
		Nom.	5	5	8	10	11	13	15
m	*fs*16	Min.	5.38	6.95	7.55	10.45	11.55	13.55	15.55
		Max.	4.62	6.05	8.45	9.55	10.45	12.45	14.45

Notes. 1. Sizes shown in brackets are of second preference.

2. Use b_1 for l <130, b_2 130 < l < 200, and b_3, for l >200.

Table 2.5 *Dimensions for black square bolts, screws and nuts (as per is : 258 5–1968)*

(Diamter range 20 to M-39)

Size			*M 20*	*(M 22)*	*M 24*	*(M 27)*	*M 30*	*(M 33)*	*M 36*	*(M 39)*
d		Nom.	20	22	24	27	30	33	36	39
d		Max.	21.50	23.30	25.30	28.30	31.30	34.60	37.60	40.60
		Min.	19.481	21.48	23.48	26.48	29.48	32.38	35.38	38.38
s	< 19-*h*14	Nom.	30	32	36	41	46	50	55	60
		Max.	30.00	32.00	36.0	41.00	46.00	50.00	55.00	60.00
	> 19-*h*51	Min.	29.16	31.00	35.00	40.00	45.00	49.00	53.80	58.80
e		Min.	37.91	40.30	45.50	52.00	58.50	63.70	69.94	76.44
		Nom.	13	14	15	17	19	21	23	25
k	js 16	Max.	13.55	14.55	15.55	17.55	19.65	21.65	23.65	25.65
		Min.	12.45	13.45	14.45	16.45	18.35	20.35	22.35	24.35
da		Max.	24.4	26.4	28.4	32.4	35.4	38.4	42.4	45.4
r		Min.	0.8	0.8	0.8	1.0	1.0	1.0	1.0	1.0
a		Max.	9.0	9.0	1.1	—	—	—	—	—
b	b_1	Min.	46	50	54	60	66	72	78	84
	b_2	Min.	52	56	60	66	72	78	84	90
	b_3	Min.	65	69	73	79	85	91	97	100
		Nom.	16	18	19	22	24	26	29	31
m	*js*	Max.	16.55	18.55	19.55	22.65	24.65	26.65	29.65	31.80
		Min.	15.45	17.45	18.35	21.35	23.35	25.35	28.35	30.20

Notes. **1.** Sizes shown in brackets are of second preference.
2. Use b_1 for $l < 130$, b_2 for $130 < l < 200$.

2.21.2 Turned Bolts

Turned bolts are also termed as *turned and fitted bolts.* For these bolts, the holes are made slightly larger than the diameter of the bolts. These bolts are force fitted. Turned bolts are manufactured from hexagonal stock. The bars are turned to required diameters. The threads in turned bolts are usually cut with a die. If turned bolts are used in holes which are reamed, then turned bolts are equally satisfactory as rivets for many connections. The flat face of nut, inner face of head of bolt and both sides of washers are machined for better fitting.

The turned or black hexagonal bolts, nuts and lock nuts and screws, are designated as illustrated in Table 2.6 per IS : 1363–1967.

Table 2.6 *Designation system for bolts, screws, nuts and lock nuts*

Description			*Designation*
Fastener	*Size*	*Length*	
Belts with nuts and a lock nut	M 16	70	Hex. bolt M 16 × 70 NL
Bolts with a nut	M 16	70	Hex. bolt M 16 × 70 N
Bolt only	M 16	70	Hex. bolt M 16 × 70
Screw	M 16	70	Hex. screw M 16 × 70
Nuts	M 16	—	Hex. nut M 16
Lock nut	M 16	—	Hex. lock nut M 16
Double chamfered nut	M 16	—	Hex. nut DCM 16

The various dimensions of black hexagonal bolts, screws, nuts and lock nuts are as shown in Fig. 2.37. These dimensions are as per IS : 1363–1967 and have been given in Tables 2.7 and 2.8. The diameters range from 6 mm to 24 mm.

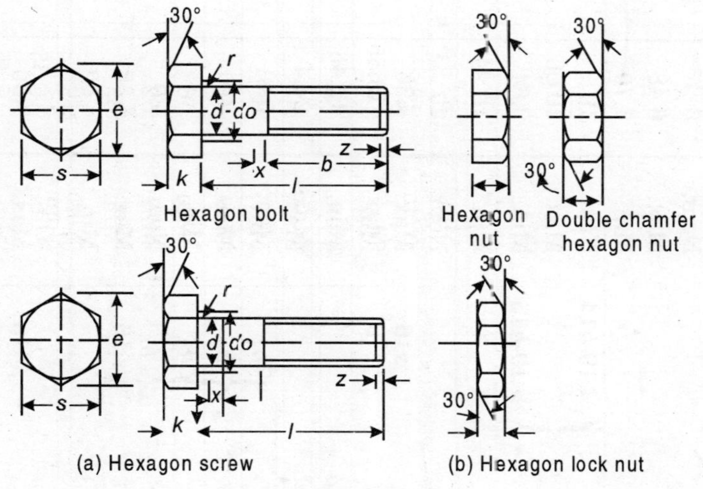

Fig. 2.37

Table 2.7 *Dimensions for black grade hexagonal bolts, screws, nuts and lock nuts (diamter range M-6 to m-18) as per is : 1363–1967*

Size			M6	M7	M 8	M10	M12	M14	M16	M18
d		Nom.	6	7	8	10	12	14	16	18
		Max.	6.48	7.58	8.90	10.90	13.10	15.10	17.10	19.10
		Min.	5.70	6.44	7.64	9.64	11.57	13.57	15.57	17.57
s	< 19-*h*14	Nom.	10	11	13	17	19	22	24	27
		Max.	10.00	11.00	13.00	17.00	19.00	22 .00	24.00	27.00
	> 19-*h*15	Min.	9.64	10.57	12.57	16.57	18.48	2.16	23.16	26.16
e		Min.	10.89	11.94	14.70	18.72	20.88	23.91	26.17	29.56
k	*js*16	Nom.	40	50	55	70	80	90	10	12
		Max.	4.38	5.38	5.88	7.45	8.45	945	10.45	12.55
		Min.	3.62	4 62	5.12	6.55	7.55	8.55	9.55	11.45
da		Max.	7.2	8.2	10.2	12.20	15.2	17.2	19.2	21.2
a		Max.	40	45	45	50	60	75	75	90
	b_1	Min	18	20	22	26	30	34	38	42
	b_2	Min.	—	—	28	32	36	40	44	48
	b_3	Min.	—	—	—	—	49	53	57	61
m	*js* 16	Nom.	5.0	5.5	6.5	8.0	10	11	13	15
		Max.	5.38	5.88	6.95	8.45	10.45	11.55	13.55	15.55
		Min.	4.62	5.12	6.05	7.55	9.55	10.45	12.45	14.45
f	*js* 16	Nom.	3	3	4	5	7	8	8	9
		Max.	3.30	3.30	4.38	5.38	7.45	8.55	8.45	9.45
		Min.	2.70	2.70	3.62	4.62	6.45	7.45	7.55	8.55
r		Min.	0.25	0.25	0.4	0.4	0.6	0.6	0.6	0.6

Notes. 1. Sizes shown in brackets are second preference.

2. Use b_1 for $l < 130$, b_2 for $130 < l < 200$, and b_3 for $l < 200$.

Table 2.8 *Dimensions for black grade hexagonal bolts, screws, nuts and lock nuts (Diamter range M-20 to M-30) as per IS : 1363–1967*

Size			*M20*	*M22*	*M 24*	*M27*	*M30*	*M33*	*M36*	*M39*
d		Nom.	20	22	24	27	30	33	36	39
		Max.	21.30	23.30	25.30	28.30	31.30	34.00	37.60	40.00
		Min.	19.48	21.48	23.48	26.48	29.48	32.38	35.38	38.38
s	< 19–*h*14	Nom.	30	32	36	41	46	50	55	60
		Max.	30.00	32.00	36.00	41.00	46.00	50.00	55.00	60.00
	> 19–*h*15	Min.	29.16	31.00	35.00	40.00	45.00	49.00	53.80	58.80
e		Min.	32.95	35.03	39.55	45.20	50.85	55.37	60.79	66.44
k	*js*16	Nom.	13	14	15	17	19	21	23	25
		Max.	13.55	14.55	15.5	17.55	19.65	21.65	23.65	25.65
		Min.	12.45	13.45	14.45	16.45	18.35	20.35	22.35	24.35
da		Max.	24.4	6.4	28.4	32.4	35.4	38.4	42.4	45.4
a		Max.	9.0	9.0	11	—	—	—	—	—
	b_1	Min.	46	50	54	60	66	72	78	84
	b_2	Min.	52	56	60	66	72	78	84	90
	b_3	Min.	65	69	73	79	85	91	97	103
m	*js* 16	Nom.	16	18	19	22	24	26	29	31
		Max.	16.55	18.55	19.65	22.65	24.65	26.65	29.65	31.80
		Min.	15.45	17.45	18.35	21.35	23.35	25.35	28.35	30.20
f	*js* 16	Nom.	9	10	12	12	12	14	14	15
		Max.	9.45	10.45	10.45	12.55	12.55	14.55	14.55	16.55
		Min.	8.55	9.55	11.45	11.45	13.45	13.45	13.45	15.45
r		Min.	0.8	0.8	1.0	1.0	1.0	1.0	1.0	1.0

Notes. 1. Sizes shown in brackets are of second preference.

2. Use b_1 for $l < 130$, b_2 for $130 < l < 200$.

Table 2.9 *Dimensions for high strength (tensile) friction grips bolts (as per IS : 3757–1966)*

d			*M12*	*M16*	*M20*	*M22*	*M 24*	*M27*	*M 30*	*M33*	*M 36*	*M39*
d_1		Min.	20	25	30	34	39	44	48	53	58	63
b		1	30	38	46	50	54	60	66	72	78	84
		2	—	44	52	56	60	66	72	78	84	90
c		Nom.	0.4	0.6	0.8	0.8	0.8	0.8	1.0	1.0	1.0	1.0
e		Max.	25.4	31.2	36.9	41.6	47.3	53.1	57.7	63.5	69.3	75.0
		Nom.	8	10	13	14	15	17	19	21	23	25
k	*j* 6	Max.	8.45	10.45	13.55	14.55	15.15	17.55	19.65	21.65	23.65	25.65
		Min.	7.55	9.55	12.45	13.45	14.45	16.45	18.35	20.35	22.35	24.35
		Max.	1.6	1.6	2	2	2	2.5	2.5	2.8	2.8	2.8
r		Min.	1.2	1.2	1.5	1.5	1.5	2.0	2.0	2.2	2.2	2.2
		Nom.	22	27	32	36	41	46	50	55	60	65
e	*h*15	Max.	22.00	27.00	32.00	36.00	41.00	46.00	50.00	55.00	60.00	65.00
		Min.	21.16	26.16	31.00	35.00	40.00	45.00	49.00	53.80	53.80	63.80

(1) For length upto 90 mm
(2) For length over 90 mm upto 200 mm.
Note. The dimensions d_1 shall not exceed the actual width across the flat.

2.21.3 High Strength Bolts

The *high strength bolts* are also called **high strength friction grip bolts.** Briefly, these are called HSFG bolts. These bolts are developed as a result of *recent developments* in the field of connections for the steel structures. These bolts are in use since 1938. Bureau of Indian Standards Institution has published a special publication SP : 6(4)–1969 (use of high strength friction bolts), which deals with the principles of design for such bolts. Use of high strength friction grip bolts has become very popular as a field fastener for structural connections. The term high strength friction grip bolts relates to bolts of high tensile steel. These bolts are used along with high tensile steel nuts and hardened steel washer. These bolts are tightened to predetermined shank tension which are equal to proof load. As a result of this, the structural elements connected together are clamped between the nuts and the head of bolt. The load is transferred from one structural element to other by friction between the parts as shown in Fig. 2.43 and not by shear in, and bearing on the bolts. These bolts do not allow slip to occur. When high strength friction grip bolts are used, all connected surface, including those adjacent to the bolts heads, nuts, or washers should be free of scale, burns, dirt and other foreign material that would prevent solid seating of parts. The contact surfaces of parts to be joined should be free of oil, paints, lacquer or galvanizing.

The high strength friction grip bolts have many advantages over rivets. Whereas the rivets are subjected to shear and bearing stresses, the bolts are subjected to uniform tensile stresses only. The high strength friction grip bolts have higher fatigue strength because there is no concentration of the stress in the hole. The bolts do not bear against the plates. Therefore, the uneven distribution of stress does not occur. These bolts are advantageously used in bridges and machine foundations subjected to vibrations. These bolts also simplify the problem of alterations and additions to structures as they can be assembled more easily than rivets. The various dimensions of high strength

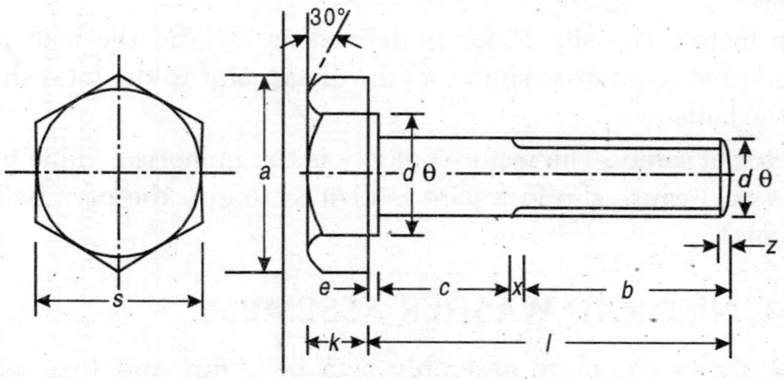

Fig. 2.38

friction grip bolts are as shown in Fig. 2.38. The dimensions are as per IS : 3757–1966 and have been given in Table 2.9. The lengths of these bolts are from 30 mm to 80 mm with variationn of 10 mm. These bolts are designated by expressing type of bolts, followed by the size, the length and symbol representing the mechanical properties and IS number e.g., friction grip bolts M 16 × 100 IS : 3757 : 10 K . The size of this bolt is M 16 (16 mm) and, the length of bolt is 100 mm. The bolt conforms to the mechanical properties of 10 K.

2.22 DEFINITIONS OF TERMS USED IN BOLTS AND BOLTING

1. • **Nominal diameters of bolt.** The nominal diameter of a bolt is diameter of unthreaded shank of bolt.

2. **Gross diameter of bolt.** The gross diameter of a bolt is the nominal diameter of the bolt.

3. **Gross area of bolt.** The gross area of a bolt is the nominal area of the bolt. It is the cross-sectional area of the bolt calculated from gross (or nominal) diameter of the bolt.

4. **Net area of the bolt.** The net area of a bolt is the area at the root of threaded part of cross-sectional area of the unthreaded part whichever is less. It is not cross-sectional area of the bolt calculated from diameter at the root of threads (or core diameter).

5. **Diameter of the bolt hole.** The diameter of a bolt hole is the nominal diameter of the bolt plus 1.5 mm unless specified otherwise.

The following definitions pertain to high strength friction grip bolts.

6. **Grip.** The grip of bolt is equal to total thickness of steel sections to be held together excluding washers.

7. **Ply.** The ply is equal to a single thickness of steel forming part of a structural joint.

8. **Length of bolt.** The length of bolt is equal to the distance from the underside of the bolt head to the extreme end of the shank, including any camber or radius.

9. **Effective interface.** The effective intersurface is a common contact surface between two load transmitting plies, excluding packing pieces, through which the bolt passes.

10. **Slip factor.** The slip factor is defined as ratio of the load per effective interface, required to produce slip in a pure shear joint to the total shank tension induced in the bolts.

11. **Factor of safety.** The factor of safety is the numerical value by which the load which would cause slip in a joint is divided to give the permissible working load on the joint.

2.23 BOLT, NUT AND WASHER ASSEMBLY

Figure 2.39 shows complete assembly of a bolt, nut and two washers. The complete assembly provides bolted connections.

The washers are used below the nut and inner face of a head of bolt. The washers distribute the load to be transmitted over a large area. The washers provide smooth seating for the nut to turn on. The washers prevent the nut from cutting into the plates.

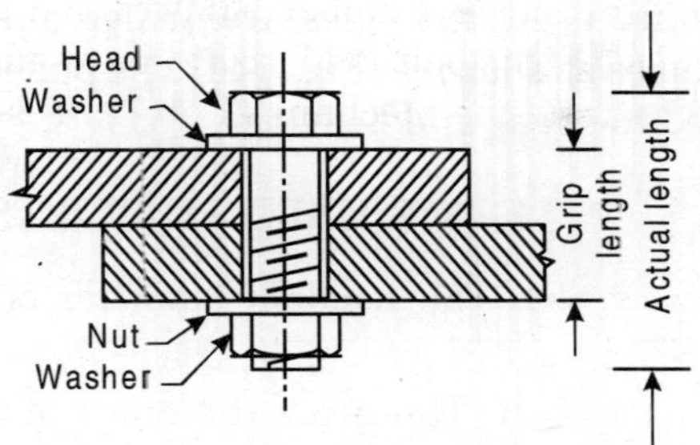

Fig. 2.39. *Bolt, nut and washers assembly*

Two plates are connected by bolts. The combined thickness of plates connected is called *grip length.* The actual length of bolt is kept greater than the grip length. The nut and washers are placed in length which is in excess over the grip length. A small threaded portion remains projecting outside the nut.

2.24 STRESSES IN BOLTS

Unfinished bolts are used in bolted connections in lap joint as shown in Fig. 2.40 (a) and in butt joint as shown in Fig. 2.40 (b). The plates are carrying an axial pull, *P*. The diameter of hole is 16 mm greater than the diameter of bolt. Firstly, there occurs a slip. Then plate elements come in contact with the bolts. The bolts are subjected to single shear in case of lap joint and double shear in case of butt joint. In addition to shear, bolts are also subjected to bearing.

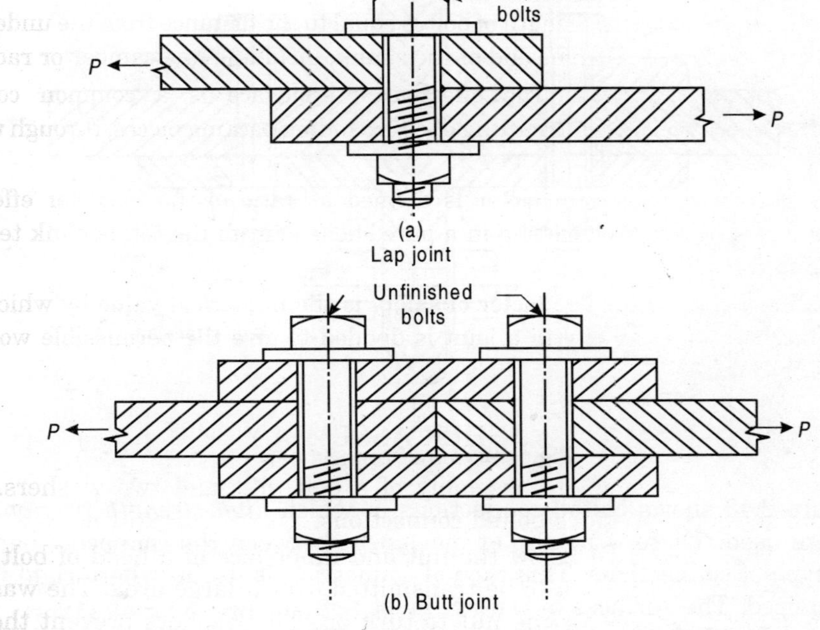

Fig. 2.40 *Bolted connections*

When turned bolts are used, the holes are drilled and reamed. The turned bolts fit tightly in the holes as shown in Fig. 2.41. The pull is transmitted directly. The bolts are subjected to shear and bearing.

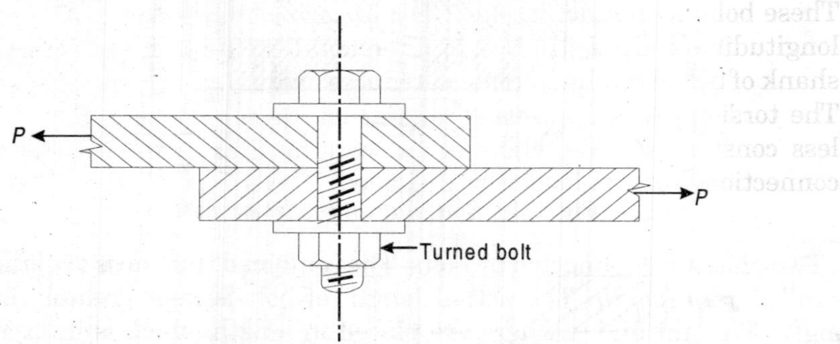

Fig. 2.41 *Bolted connection*

Figure 2.42 shows a bolted connection. Pull acts along the longitudinal axes of bolts *A–A*. The bolts are subjected to tension. The tension in bolt is resisted by the cross-section of bolt at the root of thread (i.e., the net area of bolt).

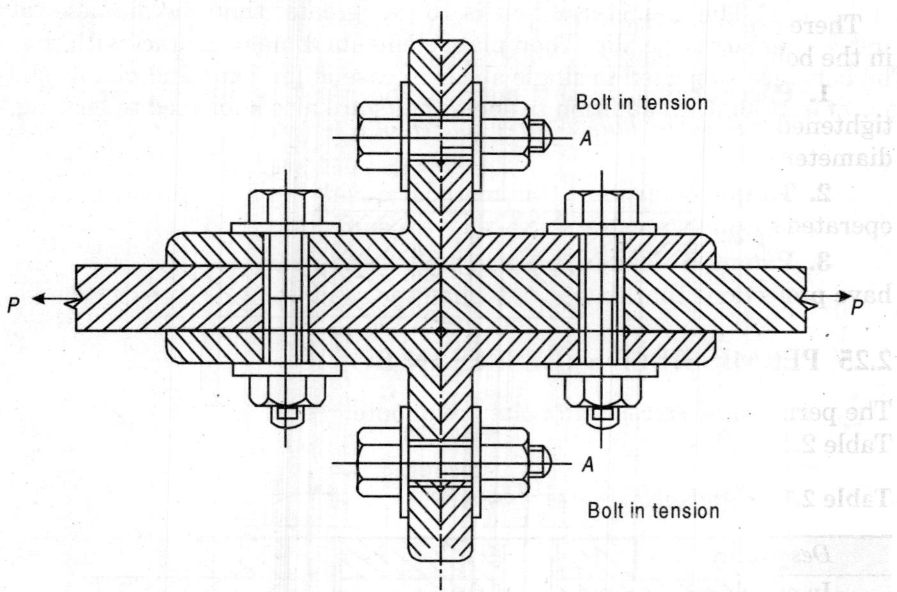

Fig. 2.42 *Bolted connections*

Figure 2.43 shows bolted connections in which *high strength friction grip bolts* are used. There is no slip or movement between the connected parts in these type of connections. This type of connection is useful where rigid joints are required. The surfaces in contact must remain free from oil, grease, scale and paint. H.S.F.G. bolts are tightened by applying a torque to the head of bolt. These are tightened to shank tension equal to the proof load *F*. In case, any part will remain loose, then slip will occur and the joint will act as an ordinary bolt joint. The plates are clamped together. As a result of this, friction μF, is developed between the plates as shown in Fig. 2.43. The load is transferred by friction. These bolts do not act in shear or in bearing. When these bolts are tightened, longitudinal tensile stresses, and torsional shear-stresses are developed in the shank of bolt. The shank tension reduces slightly after completion of tightening. The torsion drops appreciably. Later on, the shank tension remains more or less constant for the whole of its working life, irrespective of whether, the connections remain loaded or unloaded.

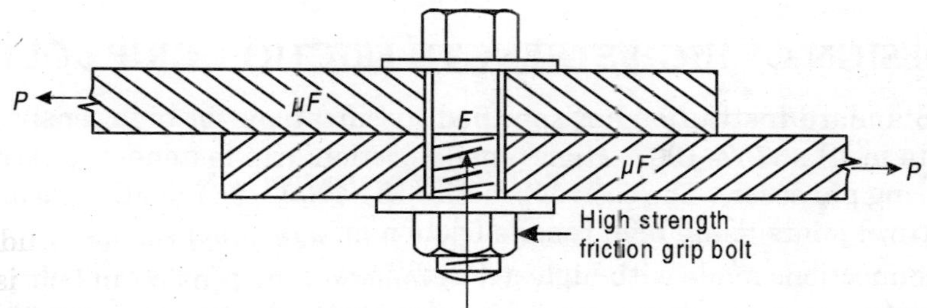

Fig. 2.43 *Bolted connection*

There are three methods to see that correct shank tension has been reached in the bolt. These methods are as follows :

1. Part turning. In this method, the nut is screwed up tight and then tightened further by *half* to *three-fourth* of a turn depending on the length and diameter of bolt.

2. Torque control. In this method, a calibrated power operated or hand operated torque wrench is used to give a suited torque to bolt.

3. Patented load-indicating bolts and washers. In this, the nuts have projections that flatten out when the required tension is reached.

2.25 PERMISSIBLE STRESSES IN BOLTS

The permissible stresses in bolts are adopted as per IS : 800–1984 as given in Table 2.10.

Table 2.10 *Permissible stresses in bolts*

Description		*Permissible stress N/mm² (MPa)*
In close tolerance and turned bolts		
(i) Axial tension	σ_{tf}	120
(ii) Shear	τ_{vf}	100
(iii) Bearing	σ_{pt}	300
In bolts in clearance holes		
(i) Axial tension	σ_{tf}	120
(ii) Shear	τ_{vf}	80
(iii) Bearing	σ_{pf}	250

Note. *The permissible stress in a bolt (other than a high strength friction grip bolt) of property class higher than 4.6 shall be those given in Table 2.10 multiplied by the ratio of its yield stress or 0.2 per cent proof stress or 0.7 times its tensile strength, whichever is less, to 235* N/mm^2 *(MPa).*

The calculated bearing stress of a bolt on the parts connected by it shall not exceed,

(i) the value of yield stress, f_y for bolts in clearance holes, and

(ii) the value of 1.2 f_y (f_y is the yield stress) for close tolerance and turned bolts.

2.26 DESIGN OF HIGH STRENGTH FRICTION GRIP BOLTS

Indian Standard Institution has specified specifications for high tensile friction grip bolts in IS : 3757–1967 (High tensile friction grip fasteners for structural engineering purposes). The code of practice IS : 4000–1967 deals with assembly of structural joints using high tensile friction grip fasteners.

The connections made with high strength friction grip bolts may be subjected to shear, tension in the direction of the axis of bolt, combined shear and tension or repeated variation of stresses. The principles for design of these connections are as following.

2.26.1 Shear Connections

When the bolt connections are subjected to shear in the plane of the friction faces, the force in the members, P as shown in Fig. 2.43 does not exceed the resulting frictional forces, that is, the maximum value of P may be expressed as follows :

$$P = \frac{1}{F \cdot S} \cdot (\mu \cdot F) \qquad \text{...(2.14)}$$

where, μ is the coefficient of friction between the interfaces and $F.S.$ is appropriate factor of safety. The coefficient of friction is also termed as slip factor. The slip factor is adopted as 0.45. For all loads except wind, the factor of safety is adopted as 1.4. When the effect of wind forces is considered, its value is adopted as 1.2. In case, the frictional forces are developed in N interfaces instead of one as shown in lap joint Fig. 2.43, Eq. 2.14 is modified as under

$$P = \frac{1}{F \cdot S} \cdot N \cdot \mu \cdot F \qquad \text{...(2.15)}$$

The common contact surfaces subjected to forces in the opposite directions give the number of effective interfaces. The value of proof load F depends on the diameter and the proof stress of the material.

The friction grip bolts are manufactured as conforming to mechanical properties of 10 K or 8 G. The proof loads F for the different diameters of bolts as per IS : 3757–1966 have been given in Table 2.11.

The values of proof loads for 10 *K* and 8 *G* bolts are based on 700 N/mm^2 and 600 N/mm^2, respectively.

Table 2.11 *Proof loads for bolts*

Diameter of bolts (mm)	*Proof load (kN)*	
	10 K bolts	*8 G bolts*
12	59.00	50.58
14	80.50	75.00
16	107.90	91.20
18	134.40	115.20
20	171.50	147.00
22	212.10	181.80
24	237.10	211.80
27	321.30	274.50
30	392.70	336.60
33	485.80	416.40
36	571.90	490.20
39	683.20	585.60

2.26.2 Tension Connections

In case, the bolt connections are subjected to external tension in the direction of the axis of bolts, then, the clamping force is likely to be decreased. Very large values of external tension may separate interfaces and the clamping force may be decreased to zero and cause the failure of the connections. Therefore, the maximum external tension of any bolt should not exceed 0.6 of the proof load of the bolt used.

2.26.3 Combined Shear and Tension

In beam to column connections and in some bracket connections, the connections are subjected to combined shear and tension. IS : 4000–1967 specifies that the bolts are so proportioned that the following expression is satisfied.

$$\left[\left(\frac{\text{Calculated shear}}{\text{Slip factor}}\right) + 1.2 \times (\text{calculated tension})\right] = \left(\frac{\text{Proof load}}{F \times S}\right)$$

The effective clamping action of a bolt will cease when the externally applied tension reaches 0.6 of its proof load.

2.26.4 Repeated Variation of Stress

The nut and washer of the bolt remain in position by spring action due to tension in the bolt. As such these nuts and washers do not become loose even if these are subjected to repeated variation of stress. Also there is very little vibrations in the tension of the bolt. The fatigue strength of friction grip connections has been found to be 25 per cent higher than the ordinary riveted joints. However, the external tension is limited to 0.5 of the proof load so that the separation of friction surfaces does not occur.

2.27 ADVANTAGES OF PIN CONNECTIONS

The following are the advantges of pin connections :

1. The pin connections simplify analysis of indeterminate structure. The moment at a pin-connection is zero.
2. The pin connections have freedom of rotation of connected members.
3. Only one pin is used in a pin-connection. Whereas in a rivet or bolted connection, more than one rivets or bolts are used.
4. The pin connections reduce secondary stress occurring at the joints.

2.28 DISADVANTAGES OF PIN CONNECTIONS

The following are the disadvantages of pin connections :

1. The pin cannot resist the longitudinal tension. Because, longitudinal tension produces friction which prevents free turning of pins. Free turning of pins is one of its principal function.
2. The pins and pin-holes require expensive machine work on them.
3. The pin connections have lack of rigidity. As a result of this, there is noise in bridges under traffic vibrations (specially in light structures).

2.29 PINS

The pins for structural connections are manufactured from mild steel. The pins can freely turn in a pin-connections. The sizes of pins range from 9 mm diameter, used for connecting strap, iron bars, to railway bridge pins 330 mm or more in diameter.

The following are various types of pins used for structural connections.

2.29.1 Forged Steel Pin

Figure 2.44 (a) shows a forged steel pin used for structural connections. When forged steel pins are used, these are secured by means of a cotter pin or a tape pin, or a thin nut screwed upto shoulder on the end of pin. Fig. 2.44 (b) and Fig. 2.44 (c) show washer and cotter pin, respectively.

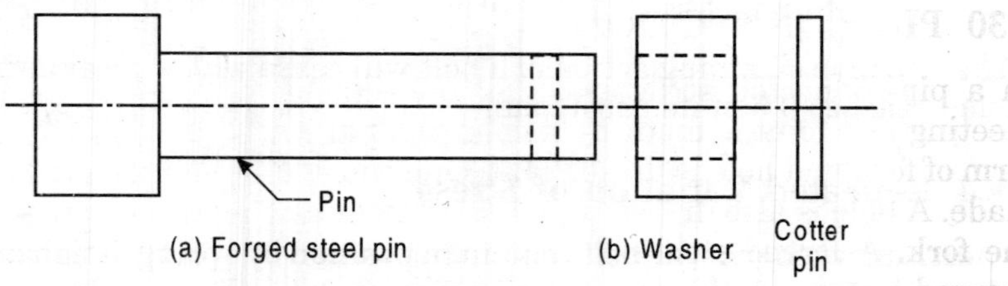

Fig. 2.44

2.29.2 Undrilled Pins

Figure 2.45 (a) shows undrilled pins. Undrilled pins are used in bridges. Pilot nut, driving nut and holding nut shown in Fig. 2.45 (b), (c) and (d), respectively, are used for securing the pin when it is placed in position.

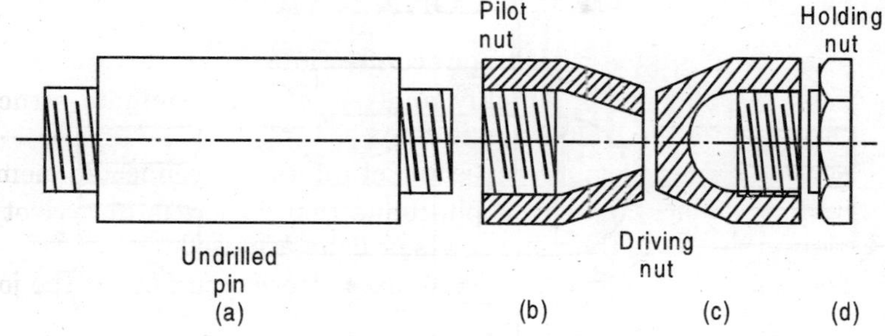

Fig. 2.45 *Undrilled pin*

2.29.3 Drilled Pins

When a diameter of pin is more than 200 mm, then drilled pin as shown in Fig. 2.46 (a) is used. A hole is drilled in this type of pin. Bolt and cap along with nut and washers as shown in Fig. 2.46 (b), (c) and (d) are used for securing the pin, when it is placed in position.

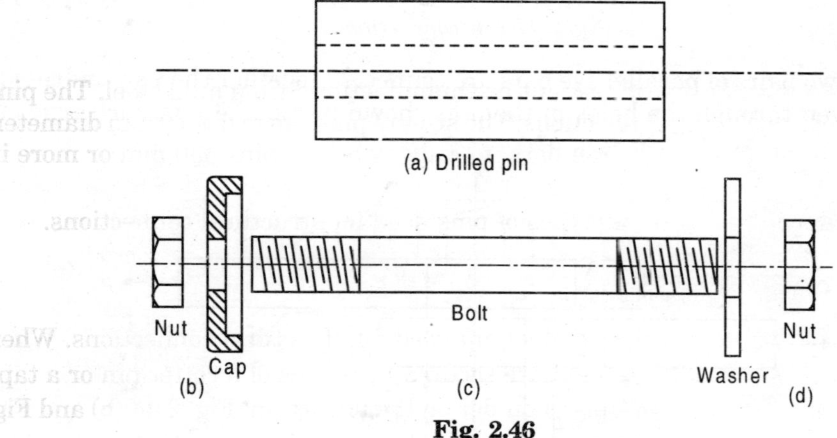

Fig. 2.46

2.30 PIN CONNECTION

In a pin-connected structure, the connection between the various members meeting at a joint is made by means of a pin. One end of a bar is forged in the form of fork and hole is drilled in it. One end of other bar is forged and an eye is made. A hole is also drilled in it. The eye-bar can be inserted within the jaws of the fork. A forged steel pin is driven through the holes in them. The pin is secured by means of a cotter pin or a screw. The ends of both parts are made octagonal for good grip. Such type of pin connection is shown in Fig. 2.47. This type of pin-connection is used in structures such as braced girders, links for suspension chains (chain-link cables of suspension bridges used for medium spans) etc.

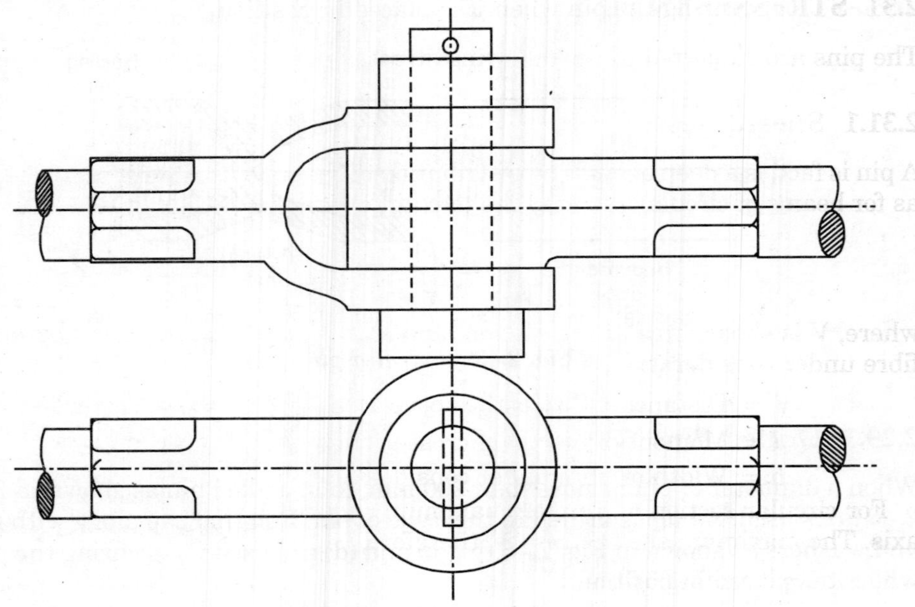

Fig. 2.47 *Pin connection*

When two pairs of parallel eye bars are connected together then an undrilled pin is driven through the holes in them as shown in Fig. 2.48. In such type of

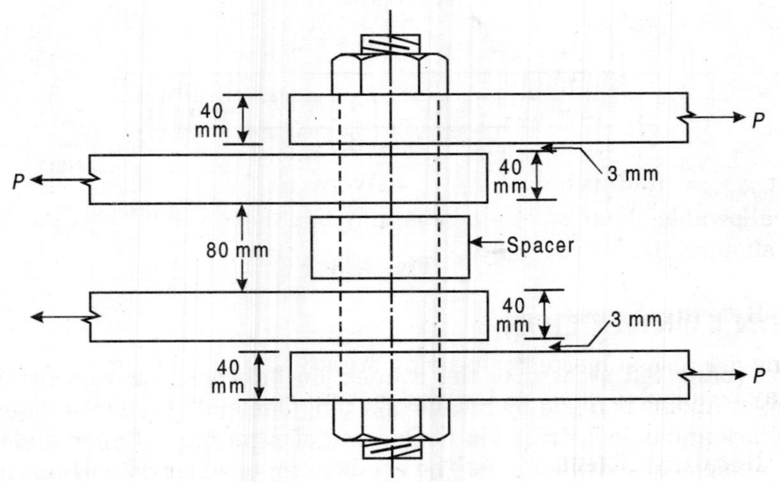

Fig. 2.48 *Pin connection*

connection, a spacer is inserted between inner pair of eye bars. Such type of connections are used in structures such as truss bridge girders. A clearance of 3 mm is kept between adjacent elements to allow free rotation at the joint.

2.31 STRESSES IN PINS

The pins are subjected to the following stresses in pin-connections.

2.31.1 Shear Stress

A pin is fact, is a deep beam The distribution of shear stress in a pin is the same as for beams of circular section. The maximum shear stress in a pin is

$$\tau_{vf\,(max)} = \left(\frac{V A \bar{y}}{I.b}\right)$$

where, V is shear force at the section, and A is area of cross-section above the fibre under consideration, and

$\bar{y}$ = Distance of C.G. of the above area from fibre under consideration

I = Moment of inertia of cross-section

b = Width of fibre under consideration.

For circular section of pin, the maximum shear stress occurs at the neutral axis. The maximum shear stress is, therefore,

$$\tau_{vf\,(max)} = \left[\frac{V\left(\frac{1}{2}\pi r^2\right)\left(\frac{4\,r}{3\,\pi}\right)}{\left(\frac{\pi}{4}\cdot r^4\right)(2\pi r)}\right]$$

$$= \frac{4}{3}\cdot\left(\frac{V}{\pi r^2}\right)$$

$$\tau_{vf\,(max)} = \left(\frac{4}{3}\right)\cdot \tau_{vf\,(ave)}$$

where $\tau_{vf\,(max)}$= Radius of pin, $\tau_{vf\,(ave)}$ = Average shear stress.

The allowable shear stress in pins may be adopted same as that for power driven shop rivets.

2.31.2 Bearing Stress

The pins are also subjected to bearing stress. The allowable bearing stress in pins may be adopted same as that for power driven rivets.

2.31.3 Bending Stress

Pins are subjected to bending stress. It may be necessary to proportions the pin for bending. The design of pin is likely to be governed by bending rather than by shear or by bearing. This is a significant difference between the design of a pin connection and the design of riveted joints.

2.32 PERMISSIBLE STRESSES IN PINS

Following values of permissible stress may be adopted for pins. The values of permissible stress in pins may be adopted as those for rivets.

Table 2.12 *Permissible stresses*

Description	*Permissible stress N/mm² (MPa)*
1. Shear	100
2. Bearing	300
3. Axial tension	100
4. Bending	
(a) Upto and including 20 mm diameter	$0.66 f_y$
(b) Over 20 mm diameter	$0.66 f_y$

Example 2.12 *The uplift on windward side of a steel chimney is 276 N/mm of periphery at the base. Design foundation bolts.*

Solution Design :

Step 1. Tensile strength of bolt

The bolts are subjected to axial tension

Tensile force at the base of steel chimney = 276 N/mm

Use 37 mm diameter bolts

Area at the root of threads = 840 mm²

Allowable stress in axial tension for bolts 37 mm diameter

= 120 N/mm²

$$\text{Strength of one bolt in tension} = \left(\frac{1.33 \times 120 \times 840}{1000}\right) = 134.064 \text{ kN}$$

Step 2 : Spacing of the bolts

$$= \left(\frac{134.064 \times 1000}{276}\right) = 485.74 \text{ mm}$$

Provide 37 mm diameter bolts at 480 mm spacing along the periphery at the base of chimney.

Example 2.13. *Two plates have been connected in bolted joint as shown in Fig. 2.49. The width of plate is 200 mm. Design bolted connection to transmit pull equal to the strength of plate. Thickness of plate is 10 mm.*

Solution Design :

Step 1. Bolt value

Let 20 mm diameter turned and fitted bolts are used. Two bolts are used in one row as shown in Fig. 2.49.

Strength of a bolt is double shear

$$2 \cdot \left(\frac{\pi}{4} \times \frac{(20)^2 \times 120}{1000}\right) = 75.36 \text{ kN}$$

When turned and fitted bolts are used, then, holes are drilled and reamed, the diameter of hole is kept equal to the diameter of bolt.

$$\text{Strength of a bolt in bearing} = \left(\frac{20 \times 10 \times 300}{1000}\right) = 65 \text{ kN}$$

$$\text{Bolt value} = 65 \text{ kN}$$

Step 2 : Strength of plate in axial tension

$$\left(\frac{0.6 \times 260 \,(200 - 2 \times 20) \times 10}{1000}\right) = 249.6 \text{ kN}$$

Number of bolts required to transmit this pull

$$\left(\frac{249.6}{65}\right) = 3.84$$

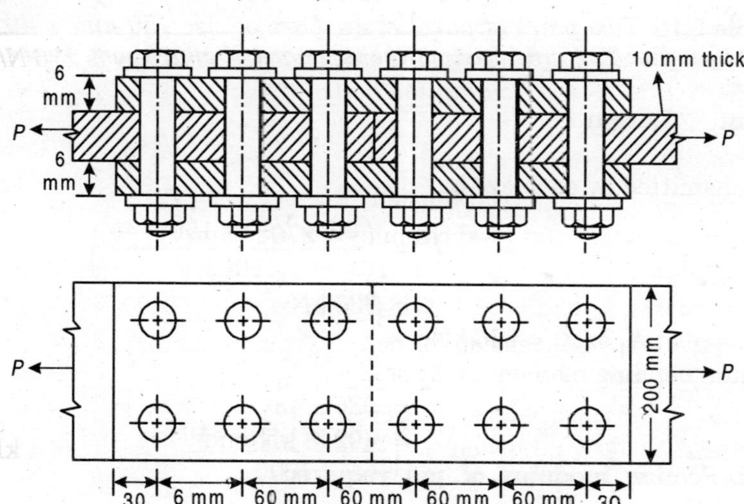

Fig. 2.49 *Bolted joint*

Provide 6 bolts 20 mm in diameter on each side as shown in Fig. 2.49.

Example 2.14. *Two plates of 16 mm thickness have been connected in a lap joint using high strength friction grip bolts. Design the joint so as to transmit a pull equal to full strength of the plate.*

Solution Design :

Step 1 : Strength of plate

In the lap joint connecting two plates, the number of interfaces N is equal to one only. Let p be the pitch of the bolts. The strength of plate per pitch.

$$F = p \times t \times \sigma_t$$

or

$$P = \left(\frac{p \times 16 \times 0.6 \times 260}{1000}\right) = 2.496\, p \text{ kN}$$

Step 2. Force in one bolt

Assuming that the plates are connected using bolts in two rows only. The force to be transmitted by one bolt

$$\frac{1}{2} \times 2.496\,p = 1.248\,p \text{ kN}$$

From Eq. 2.14

$$P = \frac{1}{F \cdot S}.(N \times \mu F) \qquad (\because \mu = 0.45)$$

$$\therefore \quad 1.248\,p = \frac{1}{1.4}(1 \times 0.45 \times F\,)$$

$$\therefore \quad F = \left(\frac{1.248 \times 1.4p}{0.45}\right) = 3.883\,p \text{ kN}$$

Assuming that the bolts are provided at 60 mm pitch. Then,

$$F = (3.883 \times 60) = 232.98 \text{ kN}$$

Example 2.15. *Two pairs of parallel eye bars of size 150 mm × 40 mm are connected by a pin. The inner pair of eye is spaced 80 mm apart. Design the pin connection.*

Solution **Design :**

Step 1

Pull transmitted by each eye bar

$$P = \left(\frac{0.6 \times 260 \times 150 \times 40}{1000}\right)$$

$$= 936 \text{ kN}$$

Shear force at X_1 or X_2 = 936 kN

Maximum bending moment at X_1 or X_2

$$= 936 \times 43$$

$$= 40248 \text{ kN-mm}$$

Step 2. Section modulus of pin required

$$Z = \left(\frac{40248 \times 1000}{0.66 \times 260}\right)$$

$$= 234.545 \times 1000 \text{ mm}^3$$

Step 3. Diameter of pin

Section modulus of a cylindrical pin

$$= \left(\frac{\pi}{32} \times d^3\right) \text{mm}^3$$

where, d = Diameter of pin in mm

$$\frac{\pi}{32} d^3 = 234.545 \times 1000$$

$$d = 133.70 \text{ mm}$$

Provide 140 mm diameter pin.

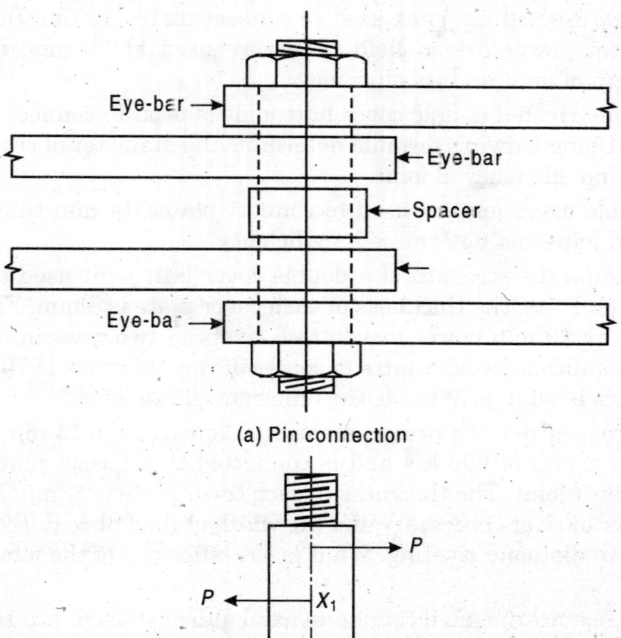

(a) Pin connection

(b)

Fig. 2.50

Step 4. Check :

1. Average shear stress in the pin

$$\tau_{va} = \left(\frac{936 \times 1000}{\frac{\pi}{4} \times (140)^2}\right) = 60.834 \text{ N/mm}^2$$

< 100 N/mm^2. Hence, satisfactory.

2. Bearing stress in pin

$$\sigma_p = \left(\frac{936 \times 1000}{140 \times 40}\right) = 167.143 \text{ N/mm}^2$$

< 300 N/mm^2. Hence satisfactory.

Provide 140 mm diameter undrilled pin for pin connection.

PROBLEMS

2.1. A single riveted lap joint used to connect plates 12 mm thick. If 22 mm diameter power driven field rivets are used at 70 mm, determine the strength of joint and its efficiency.

2.2. A double riveted double cover butt joint is used to connect plates 10 mm thick. Using Unwin's formula determine the diameter of rivet, rivet value, pitch and efficiency of joint.

2.3. A double cover joint is used to connect plates 14 mm thick. Design the riveted joint and determine its efficiency.

2.4. Determine the strength of a double cover butt joint used to connect two flats 150 F 10. The thickness of each cover plate is 8 mm. Flats have been joined by 22 mm power driven to 6 rivets in two rows in chain riveting. The distance between centre to centre of rows of rivets is 70 mm and edge distance is 40 mm. What is the efficiency of the joint ?

2.5. In a truss girder of a bridge, a diagonal consists of a 14 mm thick flat and carries a pull of 600 kN and is connected to a gusset plate by a double cover butt joint. The thickness of each cover plate is 8 mm. Determine the number of rivets necessary and the width of the flat required. Arrange the rivets in diamond riveting. What is the efficiency of the joint ? Sketch the joint.

2.6. A bridge truss diagonal carries an axial pull of 400 kN. It is to be connected to a gusset plate 22 mm thick by a double cover butt joint with 22 mm diameter power driven rivets. If the width of the flat tie bar is 200 mm, determine the thickness of flat. Design an economical joint. Determine the efficiency of the joint.

2.7. An engineer is investigating an exciting crane runway to see whether the crane load may be increased. The connection of the gantry girder

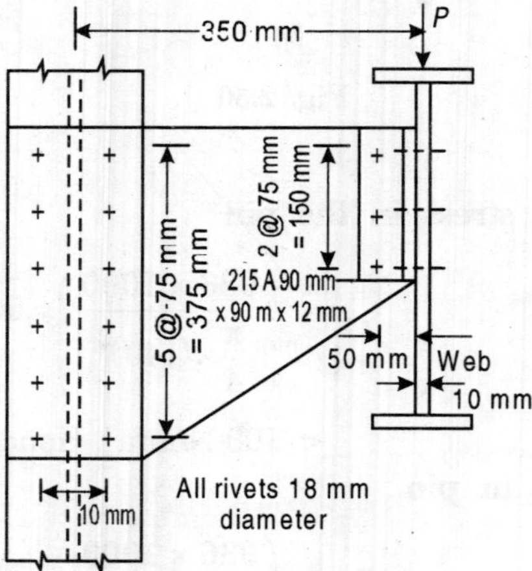

Fig. P 2.7

to the column is shown in Fig. P 2.8. Determine the maximum allowable load when the permissible shear stress in rivet is 90 N per mm square. Rivet in double shear may be allowed 2 × single shear value.

2.8. Design a bracket connection as shown in Fig. P 2.8 carrying an eccentric load of 150 kN at a distance of 300 mm from the centre line.

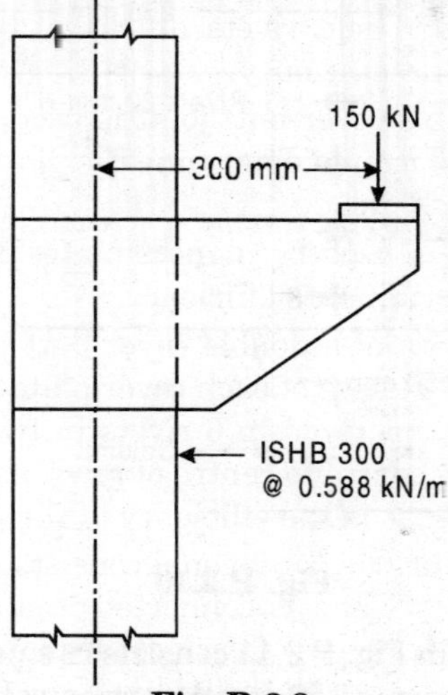

Fig. P 2.8

2.9. Design the riveted connections for the column bracket carrying an eccentric load of 100 kN at a distance 250 mm from the face of the column as shown in Fig. P 2.9.

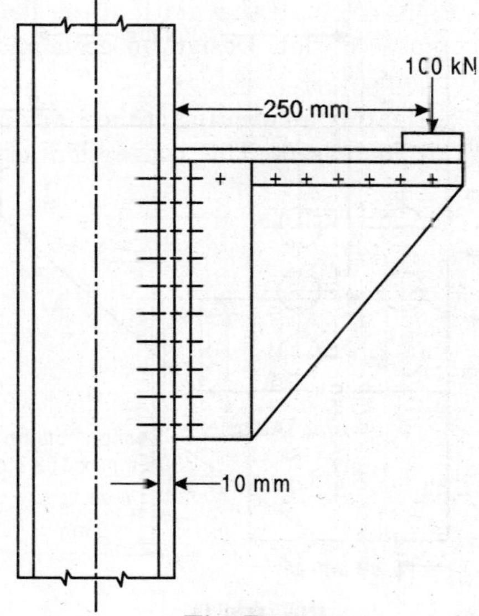

Fig. P 2.9

2.10. Determine the resultant stress on the rivet *B* of the eccentricted connections shown in Fig. P 2.9 and compare this stress with the allowable rivet value if 20 mm rivets are used. The bracket plate and column flange are both 10 mm thick and allowable stresses of 100 N/mm^2 in shear and 300 N/mm^2 in bearing may be assumed.

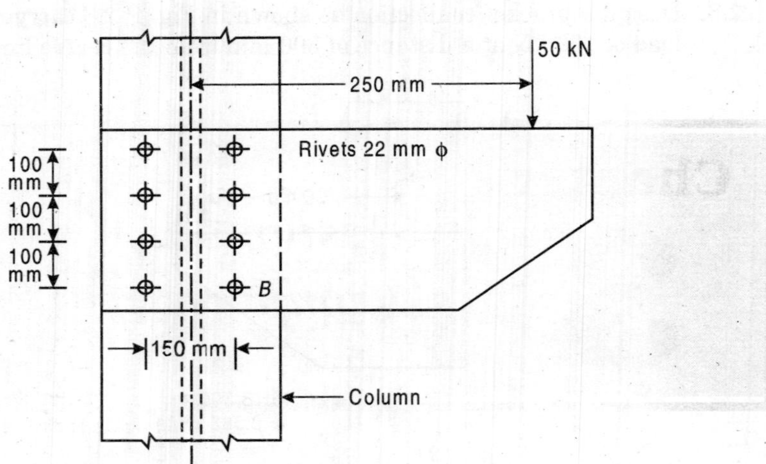

Fig. P 2.10

2.11. The bracket shown in Fig. P 2.11 consists of a pair of M.S. plates riveted to the flange of a 305 mm × 152 mm I–section column. If the resultant force on the critical rivet is limited to 45 kN, determine the load *P*, the bracket can support.

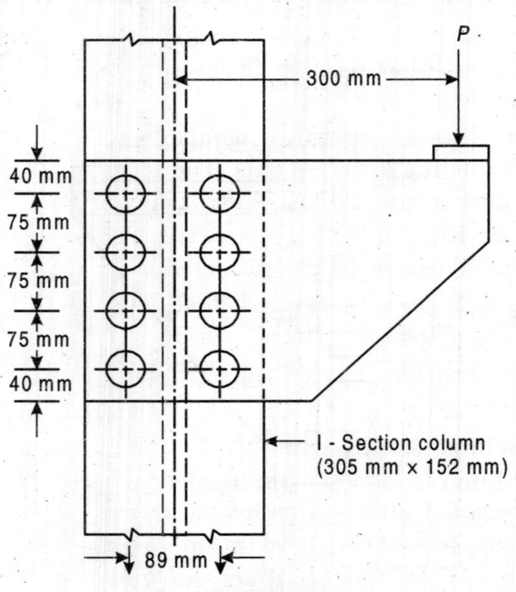

Fig. P 2.11

2.12. The uplift on a windward side of a steel chimney is 324 N/mm of periphery at the base. Design the foundation bolts.

2.13. Two pairs of parallel eye bars of size 240 mm × 60 mm are connected by a pin. The inner pair of eye bar is spaced 100 mm apart. Design pin-connections.

Chapter 3

Design of Columns and Compression Members

3.1 INTRODUCTION

A *column* is defined as a structural member subjected to compressive force in a direction parallel to its longitudinal axis. The terms *stanchions* and *posts* are also used for columns. The columns, stanchions and posts are general terms used in building construction. In truss bridge girders, end compression members are termed as *end posts*.

A *strut* is defined as a structural member subjected to compression in a direction parallel to its longitudinal axis. The term strut is commonly used for compression members in roof trusses. A strut may be used in a vertical position or in an inclined position in roof trusses. The compression members may be subjected to both axial compression and bending.

When compression members are overloaded then their failure may take place because of one of the following :

1. Direct compression
2. Excessive bending
3. Bending combined with twisting.

The failure of column depends upon its slenderness ratio. The load required to cause above mentioned failures decreases as the length of compression member increases, the cross-sectional area of the member being constant. Therefore, columns are commonly classified as *short* and *long columns*. This classification is arbitrary and there is no absolute way to determine the exact limits for each classification.

3.2 AXIALLY LOADED COLUMNS

In an axially loaded column, the load is applied at the centroid of the section and in a direction parallel to the longitudinal axis of the column. The terms *centrally loaded and concentrically loaded* are also used for *axially loaded* columns.

An axially loaded column as defined by the structural engineers transmits a compressive force without an explicit design requirement to carry lateral loads or end moments. There exist accidental end eccentricity, initial crookedness, initial curvature and the residual stresses within the tolerance limits in the columns. By having appropriate modifications in the analytical expressions and using suitable factor of safety, the design formula makes the account for these factors. Due to these factors, the strength of an actual column used to be less than that of a perfectly axially loaded column.

An ideal column is assumed initially to be perfectly straight. It is centrally loaded. Consider a case of a slender ideal column. The column is vertically fixed at the base and free at the upper end and subjected to an axial load P as shown in Fig. 3.1. The column is assumed to be perfectly elastic.

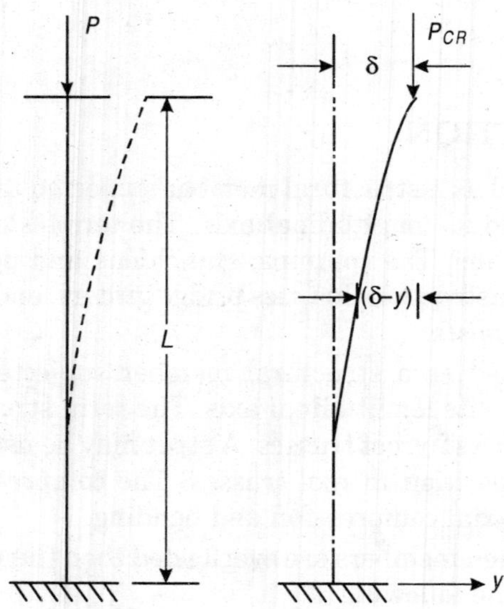

Fig. 3.1

When the value of load P is less than critical load, and stress is within the limit of proportionality, the column remains straight. The column is in *stable equilibrium* that is, if a small lateral load is applied at the free end, the column deflects. On withdrawal of the lateral load, the column resumes its vertical position and deflection vanishes. When the axial load P is gradually increased, a stage will be reached when the vertical position of the column is in the *unstable equilibrium* that is, if a small lateral load is applied, a deflection will be produced, which will not vanish on withdrawal of lateral load. The axial load which is sufficient to keep the column in such a slight deflected shape is called *critical load*. Critical load is also called as *buckling load* or *crippling load*. The *buckling load* is defined as the load at which a member or a structure as a whole collapses in service (or buckles in a load test). The *buckling* is defined as the *sudden bending,* warping, curling or crumpling of the elements or members under compressive stresses. *The direction of buckling* of a column depends upon flexural rigidity, *EI,* of the column. It buckles in a direction perpendicular to the axis, about which the moment of inertia of the section is minimum.

In about 1759 Prof. Leonhard Euler (a Swiss mathematician) derived the most popular column formula. The critical load for the column as shown in Fig. 3.1 was determined as under.

The differential equation of the deflected shape of the column is

$$\left(EI\frac{d^2y}{dx^2}\right) = +M \text{ (Hoging moment + ve)} \qquad \text{...(i)}$$

The bending moment at any point on the deflected shape

$$M = +P(\delta - y)$$

Therefore,

$$\left(EI\frac{d^2y}{dx^2}\right) = P(\delta - y) \qquad \text{...(ii)}$$

$$\frac{d^2y}{dx^2} - \frac{P}{EI}(\delta - y) = 0 \qquad \text{...(iii)}$$

Let

$$n = \left(\frac{P}{EI}\right)^{\frac{1}{2}}$$

Then the differential equation becomes

$$\frac{d^2y}{dx^2} + n^2y - n^2\delta = 0 \qquad \text{...(iv)}$$

The general solution of this equation is

$$y = A \,.\, \sin(nx) + B \,.\, \cos(nx) + \delta \qquad \text{... (v)}$$

At $\left(x = 0, y = 0, \frac{dy}{dx} = 0\right), B = -\delta, A = 0$

$\therefore \quad y = \delta\,(1 - \cos nx)$

At $\quad x = l, y = \delta$

This condition is satisfied when

$$\delta \cos(nl) = 0$$

From this, either $\quad \delta = 0$ or $\cos nl = 0$ ...(vi)

If δ is zero, buckling of column does not occur. If $\cos nl = 0$, then

$$nl = \frac{\pi}{2}(2\pi - 1) \text{ where } n = 1, 2, 3 \,.....$$

For $n = 1$, the values of nl is smallest. It is equal to $\frac{\pi}{2}$

$$\therefore \qquad P_{cr} = \left(\frac{\pi^2 EI}{4l^2}\right) \qquad ...(3.1)$$

The values of critical loads for other end conditions can be determined from this case. For a column, hinged at both ends,

$$P_{cr} = \left(\frac{\pi^2 EI}{l^2}\right) \qquad ...(3.2)$$

where P_{cr} = Critical load (or Buckling load).

Since each half of the column is in the same position as the whole of the column. This is called *fundamental case* of buckling of a bar.

3.3 EFFECTIVE LENGTH OF COMPRESSION MEMBER

The effective length of a compression member depends upon end restraint conditions. The end restraint conditions are of two types as given below :

1. Position restraint; 2. Direction restraint.

3.3.1 Position Restraint

In position restraint end of the column is not free to change its position but rotation about the end of the column can take place e.g., hinged end of column as shown in Fig. 3.2 (a).

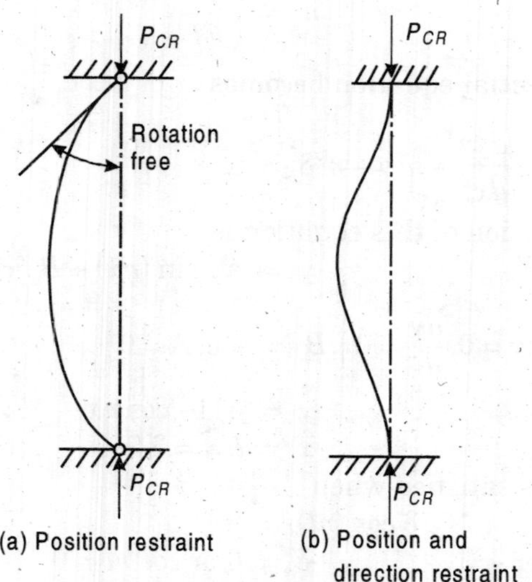

Fig. 3.2

3.3.2 Direction Restraint

In direction restraint, end of the column is free to change its position but rotation about the end of the column cannot take place.

When an end of a column is having restraint in position and direction both the end is not free to change its position, and the rotation about the end of the column also cannot take place as shown in Fig. 3.2 (b).

There are various possible combinations of restraints about either or both axes. The restraint conditions at the two ends of a column may be different or may be same. Following are the ideal cases of the end conditions :

1. Both ends of column hinged
2. Both ends of column fixed
3. One end of column fixed and the other end hinged
4. One end of column fixed and the other end free.

The deflected shapes of columns under critical loads have been shown in Figs. 3.3 (a), (b), (c) and (d) respectively. The actual lengths have been indicated by '*L*'. Cases one i.e., both ends of the column hinged has been considered as standard case. The effective length (l) of a column is expressed in terms of *equivalent length* of compression members, hinged at both the ends. It is the length of column between two adjacents points of zero moments, and is represented by '*l*'. It is also called as *unsupported length.*

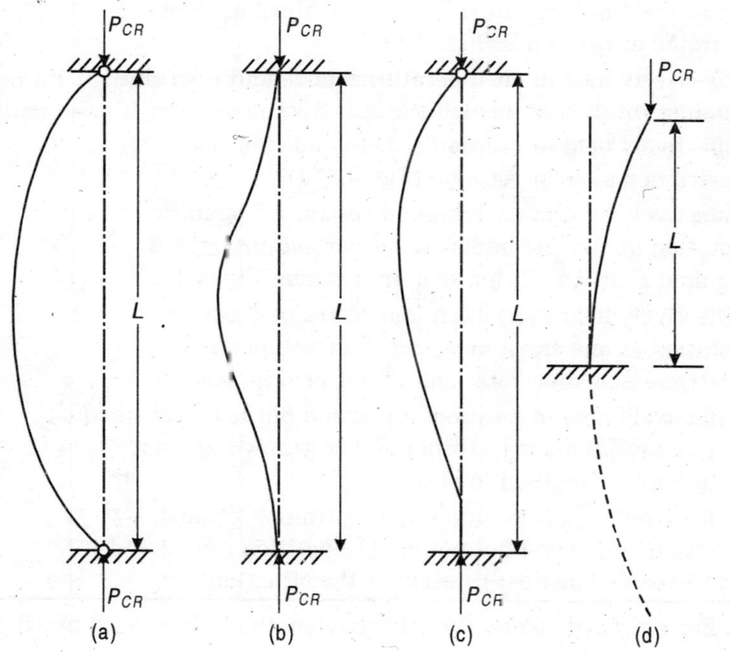

Fig. 3.3 *Ideal (standard) cases of end conditions*

The ideal end conditions cannot be achieved in actual practice. The effective length of a compression member is adopted as per Table 3.1 as recommended by BIS in IS : 800–1984 for different types of compression members. The effective length as given in Table 3.1 will be adequate in most of the cases. The effective length as given in this table may also be adopted where the columns directly form part of the frame structures.

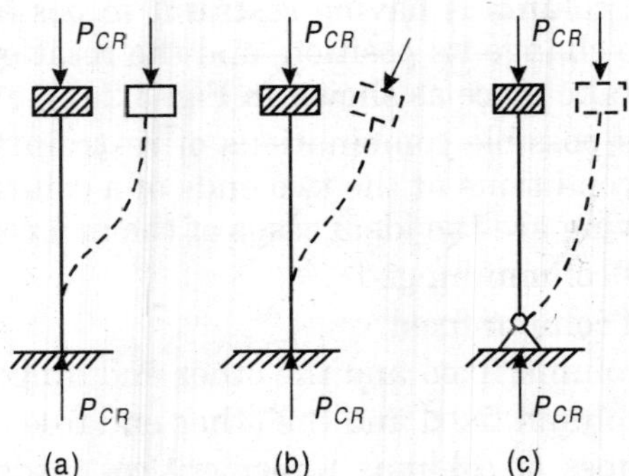

Fig. 3.4 *Ideal (Standard) cases of end conditions*

Table 3.1 *Effective length of compression members*

Sl. No.	*Type*	*Effective length of member (l)*
1.	Effectively held in position and restrained against rotation at both ends [Fig. 3.3 (b)]	0.65 *L*
2.	Effectively held in position at both ends and restrained against rotation at one ends [Fig. 3.3 (c)]	0.80 *L*
3.	Effectively held in position at both ends but not restrained against rotation [Fig. 3.3 (a)]	1.00 *L*
4.	Effectively held in position and restrained against rotation at one end and at the other end restrained against rotation but not held in position [Fig. 3.4 (a)]	1.00 *L*
5.	Effectively held in position and restrained against rotation at one end and at the other end partially restrained against rotation but not held in position	1.5 *L*
6.	Effectively held in position at one end but not restrained against rotation, at the other end restrained against rotation but not held in position	2.00 *L*
7.	Effectively held in position and restrained against rotation at one end but not held in position [Fig. 3.3 (d)] or restrained against rotation at the other end.	2.0 *L*

Note. For battened struts, the effective length *l* is increased by 10 per cent.

Where the exact frame analysis is not done, the effective length of columns in the frame structures my be found from the ratio of effective length to the unsupported length (*l*/*L*) from Fig. 3.5 when the relative displacements of the column is prevented (i.e., when there is no sway) and from Fig. 3.6 when the relative lateral displacement of the ends is not prevented (i.e., without restraint against sway viz., the sway occurs), when sway occurs, IS : 800–1984 recommends that the effective length ratio, (*l*/*L*) may not be taken to be less than 1.2.

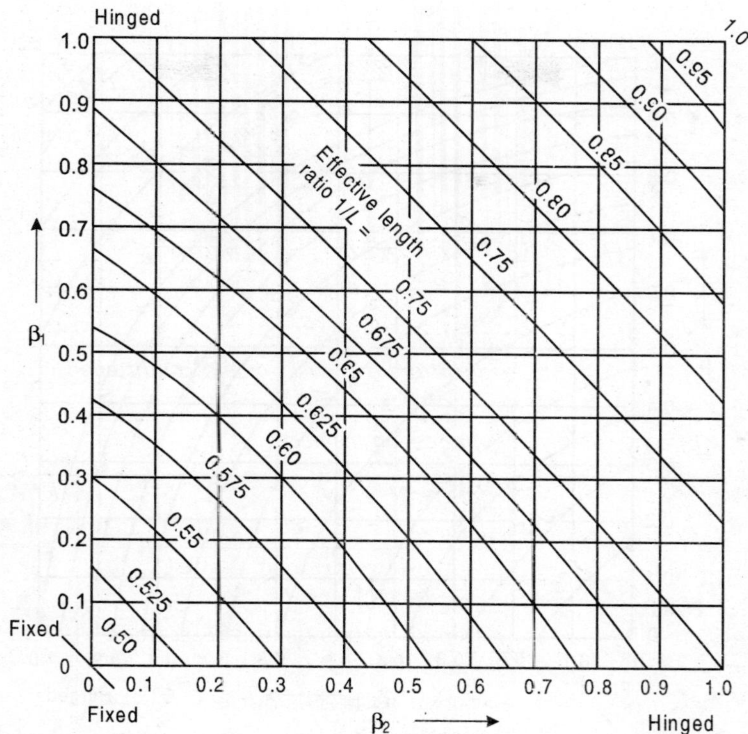

Fig. 3.5 *Effective length ratios for a column in a frame with no sway*

In Figs. 3.5 and 3.6.

$$\beta_1 = \left(\frac{\Sigma k_c}{\Sigma k_c + \Sigma k_b}\right) \quad \text{...(i)}$$

So also,

$$\beta_2 = \left(\frac{\Sigma k_c}{\Sigma k_c + \Sigma k_b}\right) \quad \text{...(ii)}$$

where, the summation is to be done for the members framing into a joint at *top* and *bottom* respectively, and

k_c = flexural stiffness of the column, and

k_b = flexural stiffness of the beam

Figures 3.6 and 3.7 are from the paper titled as Effective lengths of columns in multistorey buildings by Professor R.H. Wood, published in the Structural Engineer Vol. 52, No. 7 July 1974.

It is worthwhile to note that IS : 800–1984 'Code of Practice for General Construction in Steel' and IS : 456–1978 'Code of Practice for Plain and Reinforced Concrete' have recommended the same effective lengths for the columns with similar support conditions.

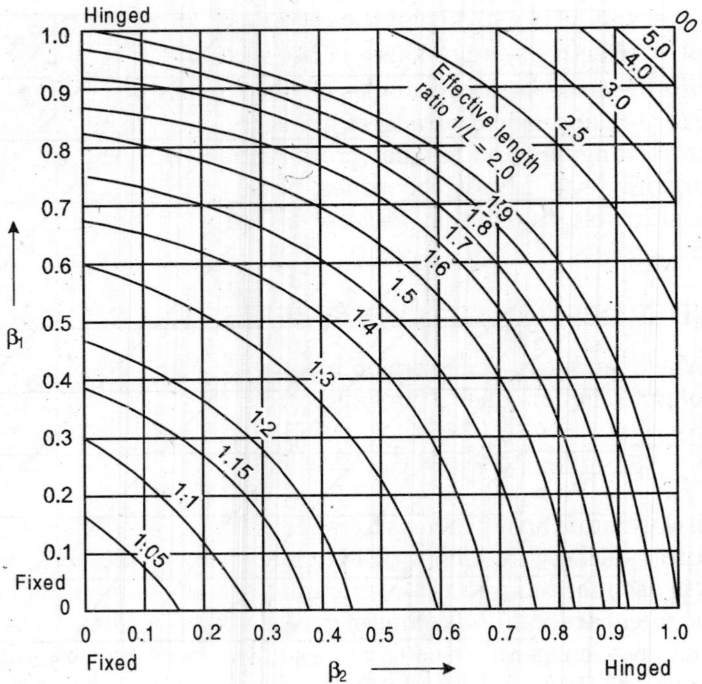

Fig. 3.6 *Effective length ratios for a column in a frame without restraint against sway*

3.4 EFFECTIVE SECTIONAL AREA

The gross cross-sectional area is the area as calculated from the specified size of the member or part thereof.

The effective sectional area of a compression member is the gross cross-sectional area of the member. The deduction is not made for members connected by rivets, bolts and pins. If the holes are not filled by the fastening material, then deduction is made for unfilled holes. The effective area is, however, modified when the ratio of the outstand to thickness exceeds the limits specified by BIS. This point has been discussed further in Sec. 3.13. The deduction is also made from the gross cross-sectional area for excessive effective plate width (if any) to determine the effective sectional area.

3.5 RADIUS OF GYRATION

The radius of gyration of a section is a geometrical property of the section and it is denoted by r

$$r = \left(\frac{I}{A}\right)^{1/2}$$

where I = moment of inertia of the section about the axis
r = radius of gyration of the section about the axis
A = effective sectional area of the section.

Along with the cross-sectional shapes of various columns used in practice as shown in Fig. 3.8, approximate radii of gyration for sections are given. In the cases of the rectangular and circular cross-sections, the values indicated are closely approximate to the correct values, but for the built-up sections, there may be considerable fluctuation because of the variation in relative cross-sectional dimensions.

3.6 SLENDERNESS RATIO OF COMPRESSION MEMBER

The slenderness ratio of a compression member is defined as ratio of effective length of compression member (l) to appropriate radius of gyration (r)

$$\text{Slenderness ratio, } \lambda = \frac{l}{r}.$$

The radii of gyration about various axes of rolled steel sections can be obtained from structural steel section tables (ISI Handbook No. 1). The minimum radius of gyration is used for computing the maximum slenderness ratio. For built-up compression members, value of radius of gyration is calculated. The slenderness ratio for a compression member should be as small as possible so that the material may be stressed to its greatest possible limit. The maximum slenderness ratio of compression members should not exceed the values given in Table 3.2. These limits have been laid down by BIS in IS : 800–1984.

The end restraints of columns arc often different in the two principal planes. The different moments of inertia of the column cross-section in these planes are sometimes desirable to achieve approximately equal slenderness ratios. The *intermediate supports* are provided to the columns for this purpose.

Table 3.2 *Maximum slenderness ratio of compression member* $\lambda = \frac{l}{r}$

Sl.No.	*Type of member*	$\lambda = \frac{l}{r}$
1.	A member carrying compressive loads resulting from dead loads and imposed loads	180
2.	A member subjected to compression forces resulting from wind/earthquake forces provided that the deformation of such members does not adversely affect the stress in any part of the structure	250
3.	A member normally acting as a tie in a roof truss or a bracing system but subjected to possible reversal of stress resulting from the action of wind or earthquake forces	350

The intermediate supports reduce the unsupported length of the columns. When the unsupported lengths of columns are reduced, then, the smaller sections may be used at a higher average stress.

Sometimes, the intermediate supports are furnished only in one direction. For example, a rolled steel I-section column is having its continuous length upto two storeys. At the level of one storey, intermediate support is provided by connecting beams with the web. The radius of gyration, r_{yy} of the section, about *yy*-axis (axis parallel to the web) is much smaller than the radius of gyration r_{xx} of the section, about *xx*-axis. By providing the intermediate support, the effective lengths of the column become different in two different directions. The effective length of columns, l_{yy} for bending about *yy*-axis is found by considering the length of column between one-storey only. The effective length of column l_{xx} for bending about *xx*-axis is found by considering the length of column between two storeys. It is seen that the effective length of column l_{yy} is much smaller than that of l_{xx}.

The values of slenderness ratio $\left(\frac{l_{yy}}{r_{yy}}\right)$ and $\left(\frac{l_{xx}}{r_{xx}}\right)$ about two directions may be made approximately equal. As such the use of sections with different values of radii of gyration in two directions may be made economical. When a column is subjected to different bending moments in two directions then, the greater value of r may be kept in the direction of greater moment. The intermediate supports in the weak direction make the use of I-section and channel section economical.

3.7 COLUMN FORMULAE FOR AXIAL STRESS IN COMPRESSION

The strength of a column depends upon large number of variables. The efforts are made to obtain a design formula by fitting a curve to experimentally found buckling loads for the intermediate range of the slenderness ratio. It is tried to draw a curve which may merge with the Euler hyperbola in the very slender column range on one side and with the material yield strength for the zero length on the other side.

A perfectly straight column of perfectly homogeneous material (i.e., an *ideal column*) is subjected to an axial load. The primary object is to find the average axial stress in compression, which corresponds to the allowable load. The average axial stress is uniform across the section. It is given by

$$\sigma_{ca} = \left(\frac{P_a}{A}\right) \qquad \text{...(i)}$$

where, P_a = allowable load
A = cross-sectional area of the column

The required cross-sectional area for a given design load may be found conveniently in case σ_c is known

$$A_{\text{reqd.}} = \left(\frac{P}{\sigma_c}\right) \qquad \text{...(ii)}$$

where P is the design load.

It is observed that the average axial stress in the column at the time of failure of column is less than the yield strength of the material. The difference in the stress depends on the type of failures of the column. The various types of failures of the column other than the direct compression or direct crushing are as follows :

1. Primary laterally buckling
   (a) Inelastic buckling
   (b) Elastic buckling
2. Gradual yielding due to excessive local stress
3. Local buckling, and
4. Torsional buckling.

The failure by *local buckling* and by *torsional buckling* are prevented by selecting the rolled steel sections of proper shape and proportions. Therefore, the failure of a column may occur by lateral buckling or by gradual yielding due to an excessive local stress. The average allowable axial stress for use in the design is influenced by amount of lateral buckling. It depends upon the *slenderness ratio* and the *compressive strength* of the material. Therefore, the columns may be classified in two categories viz., (i) those columns which fail in *elastic buckling*, and (ii) those columns which fail in *inelastic buckling*. The columns having very large slenderness ratios fail in elastic buckling. Lconhard Euler predicted that the lateral buckling of the column occurs when the value of load, P exceeds the critical or ultimate load, P_{cr} either in the elastic range or in the inelastic range, the critical stress in the column for *elastic buckling* is given by Euler's formula as follows :

$$\sigma_{cr} = \left(\frac{P_{cr}}{A}\right) = \left(\frac{\pi^2 E}{\left(\frac{KL}{r}\right)^2}\right) \quad \text{...(3.3)}$$

where, K = Factor, which depends upon the end conditions of the column
E = Modulus of elasticity
L = Actual length of the column
r = Radius of gyration of the section.

Equation 3.3 may be written as

$$\sigma_{cr} = \left(\frac{P_{cr}}{A}\right) = \left(\frac{\pi^2 E}{\left(\frac{l}{r}\right)^2}\right) \quad \text{...(3.4)}$$

where, l = K.L. = Effective length of the column.

1. Euler's formula is suitable for large slenderness ratio $\left(\text{i.e., } \frac{l}{r} > 200\right)$.

It was considered appropriate to apply a variable factor of safety to account for this stress level and additionally to consider the effect of eccentricity, residual stresses and the several other factors which complicate the column theory. The maximum average allowable stress in the axial compression may be found by dividing σ_{cr} by factor of safety (F.S.). Therefore,

$$f_a = \left(\frac{\sigma_{cr}}{F.S}\right) = \frac{1}{F.S.} \cdot \left(\frac{\pi^2 E}{\left(\frac{l}{r}\right)^2}\right) \quad ...(3.5)$$

The relationship between Euler's formula and A.I.S.C. (American Institute of Steel Construction) 1953 Specifications formula may be established.

The critical stress by Euler's formula may be expressed as

$$\sigma_{cr} = \left(\frac{\pi^2 E}{\left(\frac{l}{r}\right)^2}\right) \quad ...(i)$$

Differentiating (i) with respect to $\left(\frac{l}{r}\right)$

$$\left(\frac{d(\sigma_{cr})}{d\left(\frac{l}{r}\right)}\right) = \left(\frac{-2\pi^2 E}{\left(\frac{l}{r}\right)^3}\right) \quad ...(ii)$$

The maximum value of Euler's stress or any critical buckling stress is limited to yield stress f_y. The Euler's formula is not valid for small values of slenderness ratio. For such a region, the critical stress may be obtained by assuming the following expression (which one is also based on experimental column test data).

$$\sigma_{cr} = f_y - m.\left(\frac{l}{r}\right)^2 \quad ...(iii)$$

Differentiating (iii) with respect to $\left(\frac{l}{r}\right)$

$$\left(\frac{d(\sigma_{cr})}{d\left(\frac{l}{r}\right)}\right) = \left(-2m \cdot \left(\frac{l}{r}\right)\right) \quad ...(iv)$$

At $\frac{l}{r} = 0$, the slope may be assumed as zero. The two curves defined by the expressions (i) and (iii) will have a common tangent. At that point, the slopes given by (ii) and (iv) will be equal. Equating the two slopes at a common point.

$$\left(-2m\cdot\left(\frac{l}{r}\right)\right) = \left(\frac{-2\pi^2 E}{\left(\frac{l}{r}\right)^3}\right) \quad \text{...(v)}$$

The slenderness ratio at this common point be arbitrarily defined by a parameter

$$\left(\frac{l}{r}\right) = C_c \quad \text{...(vi)}$$

Substituting this in Eq. (v),

$$(-2m\,.\,(C_c) = \left(\frac{-2\pi^2 E}{(C_c)^3}\right)$$

$$\therefore \quad m = \left(\frac{\pi^2 E}{C_c^4}\right) \quad \text{...(vii)}$$

Thus, the expression (ii) may be written as

$$\sigma_{cr} = \left(f_y - \frac{\pi^2 E}{C_c^4}\left(\frac{l}{r}\right)^2\right) \quad \text{...(viii)}$$

By substituting, the values of σ_{cr} by Euler's formula (on left hand side)

$$\left(\frac{\pi^2 E}{\left(\frac{l}{r}\right)^2}\right) = \left[f_y - \frac{\pi^2 E}{(C_c^4)}\cdot\left(\frac{l}{r}\right)^2\right] \quad \text{...(ix)}$$

At $\left(\frac{l}{r}\right) = C_c$

$$\frac{\pi^2 E}{(C_c)^2} = f_y - \frac{\pi^2 E}{C_c^4}\cdot C_c^2 \quad \text{...(x)}$$

$$\therefore \quad C_c = \left(\frac{\pi^2\cdot E}{f_y}\right)^{1/2} \quad \text{(xi)}$$

Substituting the value of C_c in Eq. (x)

$$\sigma_{cr} = \left[f_y - \frac{\pi^2 E}{2\pi^2 E}\cdot f_y\right] = 0.5 f_y \quad \text{...(xii)}$$

Thus, the critical buckling stress becomes $0.5\,f_y$ at $\frac{l}{r} = C_c$.

In general, the expression (viii) may be written as

$$\sigma_{cr} = f_y \left[1.0 - \frac{0.5\left(\frac{l}{r}\right)^2}{C_c^2}\right] \quad ...(3.6)$$

The lowest of slenderness ratio upto which the elastic buckling occurs is denoted by C_c.

The minimum slenderness ratio $\left(\frac{l}{r_{min}} = C_c\right)$ distinguishes the elastic buckling from inelastic buckling. When the slenderness ratio is greater than C_c, then failure of column occurs by elastic buckling. A.I.S.C. Specifications recommended the following for the determination of allowable stress in axial compression for $\left(\frac{l}{r}\right) > C_c$.

$$\sigma_{cr} = \left[\frac{\pi^2 E}{\left(\frac{l}{r}\right)^2 \times (F.S.)}\right] = \left[\frac{9.87 \times 2.047 \times 10^6}{\left(\frac{l}{r}\right)^2 \times 1.92} \times \frac{10}{100}\right] \text{N/mm}^2$$

$$\sigma_{cr} = \left(\frac{10,52,000}{\left(\frac{l}{r}\right)^2}\right) \text{N/mm}^2 \quad ...(3.7)$$

When the slenderness ratio for the column is less than or equal to C_c, then, the failure of column occurs by *inelastic buckling*. The allowable stress in axial compression for $\left(\frac{l}{r}\right) \le C_c$ is given by

$$\sigma_c = \frac{1}{F.S.}\left[1 - 0.5\left(\frac{l}{r}\right)^2 \cdot \left(\frac{1}{C_c^2}\right)^2\right] \cdot f_y \quad ...(3.8)$$

where

$$\text{F.S.} = \frac{5}{3} + \frac{3}{8}\left(\frac{l}{r}\right) \cdot \left(\frac{1}{C_c}\right) - \frac{1}{8}\left(\frac{l}{r}\right)^3 \cdot \left(\frac{1}{C_c}\right)^3 \quad ...(3.9)$$

When $\frac{l}{r} = 0$, F.S.=1.67, and $\frac{l}{r} = C_c$, F.S.=1.92.

The factor of safety of 1.92 remains constant for $\left(\frac{l}{r}\right)$ greater than C_c.

Equations. 3.7 and 3.8 are used for main compression members. For bracing and secondary compression members for which $\left(\frac{l}{r}\right)$ exceed 120, the factor of safety is reduced by multiplying it by a factor ψ equal to $\left[1.6-\left(\frac{l}{r}\right)\left(\frac{1}{200}\right)\right]$. A more conservative value of factor of safety is used (F.S. = 2.12) for the members of the bridge structures since these members are in a hostile atmosphere than those in buildings.

2. Engesser suggested the **tangent modulus formula** for the buckling load of an axially load column.

$$P_{cr} = \left(\frac{\pi^2 E_t I}{l^2}\right) \qquad \text{...(3.10)}$$

or

$$\sigma_{cr} = \left(\frac{P_{cr}}{A} = \frac{\pi^2 E_t}{\left(\frac{l}{r}\right)^2}\right) \qquad \text{...(3.11)}$$

where E_t = Tangent modulus of the elasticity.

Shanley and Von–Karman defined the tangent modulus, load as the smallest value of the axial load at which the bifurcation of the equilibrium position can occur regardless of whether the transition to the bent position-requires an increase of the axial load. Beedle and Tall pointed out that Eq. 3.11 cannot be applied to a steel column if the stress-strain relationship is determined from a small coupon from the section. The buckling strength of the column is reduced due to the existence of residual stress in the cross-section, since there is early localized yielding in the certain portions of the cross-section where the compressive residual stresses prevails. The basic equation for the critical strength of a column containing residual stresses is given by

$$P_{cr} = \left(\frac{\pi^2 E \cdot I_e}{l^2}\right) \qquad \text{...(3.12)}$$

or

$$\sigma_{cr} = \left(\frac{P_{cr}}{A}\right) = \left(\frac{\pi^2 E \cdot \frac{I_e}{I}}{\left(\frac{l}{r}\right)^2}\right) \qquad \text{...(3.13)}$$

3. Beedle and Tall suggested the following relationships for the critical stresses for the buckling about the strong axis and the buckling about the weak axis respectively.

For $\qquad \frac{l}{r} < \pi\left(\frac{E}{\sigma_p}\right)^{1/2}$

$$\sigma_{cr(xx)} = \left[f_y - \frac{f}{\pi^2 E}(\sigma_y - \sigma_P)\cdot\left(\frac{l}{r}\right)^2 \right] \quad ...(3.14)$$

$$\sigma_{cr(yy)} = \left[f_y - \frac{(\sigma_y - \sigma_p)}{\pi}\left(\frac{\sigma_p}{E}\right)^{1/2}\cdot\left(\frac{l}{r}\right) \right] \quad ...(3.15)$$

For $\frac{l}{r} \leq \pi\left(\frac{E}{\sigma_p}\right)^{1/2}$

$$\left[\sigma_{cr(xx)} = \sigma_{cr(yy)} = \frac{\pi^2 E}{\left(\frac{l}{r^2}\right)} \right] \quad ...(3.16)$$

The difference between $(\sigma_y - \sigma_p)$ gives the magnitude of *residual stress*, where σ_p= stress at the *limit of proportionality*.

The effect of stresses was recognised as an important factor in calculating the strength of a centrally loaded column in a report published by Column Research Council.

The residual stresses develop in the hot rolled structural steel sections due to differential cooling. The residual stresses result from cooling after hot rolling, welding or fabrication operations such as plane cutting, cold bending or combining. The flange tips and interior web parts being thinner and more exposed, always cool more quickly than the other parts. The junctions of flange and web are the thickest and most protected parts and always cool last. *Residual tensile stresses* develop in those parts of the section that cool last as the metal tends to contract but it is restrained by colder metal. These residual tensile stresses produce *residual compressive stresses* in the adjacent metal which has cooled earlier.

The welding also produces residual stresses as the hot metal and flame cutting in the vicinity is restrained from contraction by the cooler-to-cold surrounding metal.

Because of high heat penetration due to welding in built-up column, the residual stresses tend to be large. In most cases, the tensile residual stresses at the weld reach the value of yield stress. The distribution of residual stress in a heavy shape (the thickness of thinnest element of which is more than 38 mm) differs from that in a small shape in two major respects. The magnitude of residual stresses is very large and they vary considerably along the thickness.

Nearly all steel members contain both tensile and compressive residual stresses.

4. **AREA (American Railway Engineering Association)** recommends the following for the determination of allowable stress in axial compression.

For $\frac{l}{r} \not> 140$ (Rivet ends)

$$\sigma_c = 15000 - \frac{1}{4}\left(\frac{l}{r}\right)^2 \text{ ...(psi units)} \quad \text{...(3.17)}$$

For $\frac{l}{r} \not> 140$ (Pin ends)

$$\sigma_c = 15000 - \frac{1}{3}\cdot\left(\frac{l}{r}\right)^2 \text{ ...(psi units)} \quad \text{...(3.18)}$$

5. **AASHTO (American Association of State Highway and Transport Officials)** *also recommends the above-mentioned expressions.* In bridges, the compression members may have one end riveted and the other end pin-connected. For example, the end post of a riveted bridge truss is sometimes pin-connected at its lower end. Bridge specifications do not have any provisions for this condition. In the absence of the specific instructions, the allowable stress in axial compression for the member pinned at one end and riveted at the other end may be found from the following :

$$\sigma_c = 15000 - \frac{7}{24}\cdot\left(\frac{l}{r}\right)^2 \text{ ...(psi unit)} \quad \text{...(3.19)}$$

Equations 3.17, 3.18 and 3.19 are the forms of *Johnson's parabolic formula*

$$\sigma_c = f - k\left(\frac{l}{r}\right)^2 \quad \text{...(3.20)}$$

where, σ_c = Permissible stress in axial compression
f = Crushing stress at failure
k = Constant

AISC specifications recommend Johnson's parabolic formula for $\left(\frac{l}{r}\right) \not> 120$

$$\sigma_c = 17000 - 0.485\left(\frac{l}{r}\right)^2 \text{ (psi units)} \quad \text{...(3.21)}$$

The allowable stress in axial compression may be found by a *straight line formula*

$$\sigma_c = f - k\left(\frac{l}{r}\right) \quad \text{...(3.22)}$$

AREA specifications recommend this straight line formula for $\left(\frac{l}{r}\right) \not> 120$

$$\sigma_c = 18000 - 70\cdot\left(\frac{l}{r}\right) \text{ (psi units)} \quad \text{...(3.23)}$$

6. **Rankine-Gorden formula or simply known as Rankine formula** is adopted for $\left(\frac{l}{r}\right)$ greater than 120 and less than 200

$$\sigma_c = \left(\frac{f}{1 + a \cdot \left(\frac{l}{r}\right)^2} \right) \qquad \text{...(3.24)}$$

where a = constant. Its value depends on the material.

It is a semi-empirical formula. AISC recommended this formula in 1949 for $\frac{l}{r} > 120$ and < 200

$$\sigma_c = \left[\frac{18000}{1 + \frac{1}{18000} \cdot \left(\frac{l}{r}\right)^2} \right] \text{...(psi units)} \qquad \text{...(3.25)}$$

Rankine–Gorden formula, straight line formula and Johnson parabolic formula were extensively used in the past. The curves given by Rankine–Gordon formula and Johnson parabolic formula when plotted are seen to be the same curve. Johnsons parabolic formula is simple and it is more generally used.

7. British Standard Specifications for steel and girder bridges recommended *Perry–Roberston formula* for the determination of allowable stress in axial compression

$$\sigma_c = \frac{1}{K}\left[\frac{f_y + (\eta+1)C_0}{2} - \left(\frac{f_y + (\eta+1)C_0^2}{2} - f_y C_0 \right)^{1/2} \right] \qquad \text{...(3.26)}$$

where, K = Factor of safety, 1.7, f_y = Yield stress.

Professor Perry used the value of η as $0.0015 \left(\frac{l}{r}\right)$. Later on Roberston examined the experimental data and suggested the following.

$$\eta = 0.003 \left(\frac{l}{r}\right)$$

The struts are considered as secondary members. These members do not carry primary loads. These members are used to brace other structural members or the structures as a whole. AISC specification 1963 recommended that for the struts having slenderness ratio $\left(\frac{l}{r}\right)$ greater than or equal to 120 and less than 200, the allowable stress is increased by factor, K.

$$K = \left[\frac{1}{1.6 - \frac{1}{200}\left(\frac{l}{r}\right)} \right] \qquad \text{...(3.27)}$$

The factor K for end-conditions is adopted as unity.

In the above column formulae, the columns were considered perfectly straight and uniform. The load was considered perfectly axial. In actual practice, the ideal column and loading do not exist. The load supported by the column is eccentric with respect to the longitudinal axis of the column. The initial crookedness and other imperfections of column and imperfectness of the service load distribution give small eccentricity. The secant formula accounts for the initial crookedness of the column and imperfectness of the axial loading. Indian Standard Institution (ISI) adopted the secant formula for the determination of allowable stress in axial compression in 1962. The secant formula was considered as design formula in IS : 800–1962 as per Indian Standard Specification.

8. Secant Formula. Consider an initially straight column. Suppose a load P is applied to it at a distance e from the centre line of the column. The deflected shape of the column would be as shown in Fig. 3.7.

The bending moment at any point on the deflected shape of the column

$$M = -P \cdot y \qquad \text{...(i)}$$

The negative sign is adopted since the bending moment is sagging. The differential equation to the elastic curve is as under:

$$EI \frac{d^2y}{dx^2} = M \qquad \text{...(ii)}$$

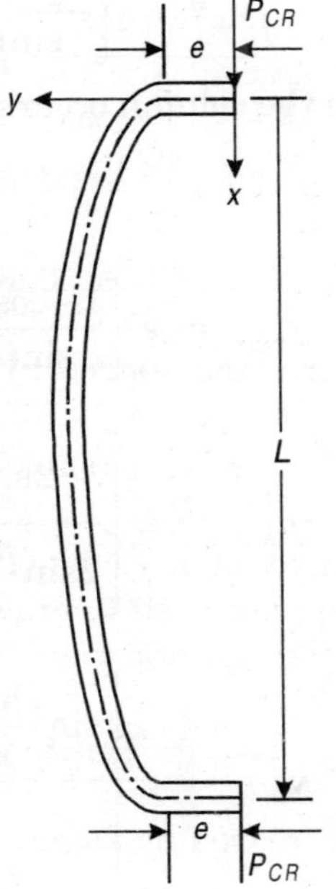

Fig. 3.7

$$EI\frac{d^2y}{dx^2} = -P.y \qquad \text{...(iii)}$$

or
$$\left(\frac{d^2y}{dx^2}\right)+\left(\frac{P}{EI}\right)\cdot y = 0$$

or
$$\left(\frac{d^2y}{dx^2}\right)+n^2 . y = 0 \qquad \text{...(iv)}$$

The general solution of this differential equation is

$$y = A . \sin(nx) + B . \cos(nx) \qquad \text{...(v)}$$

where $n = \left(\frac{P}{EI}\right)^{1/2}$

The constants of this equation can be found by applying the boundary values.

At $x = 0, y = e$ and $x = l, y = e$

$\therefore$ $A = \left(\frac{1-\cos(nl)}{\sin(nl)}\right)\cdot e$ and $B = e$

The expression (v) reduces to

$$y = e . \left[\frac{1-\cos(nl)}{\sin(nl)}\cdot\sin(nx)+\cos(nx)\right] \qquad \text{...(vi)}$$

The maximum deflection in the column occurs at centre i.e., when x is equal to $\frac{l}{2}$

$$y_{max} = e\cdot\left[\frac{1-\cos(nl)}{\sin(nl)}\cdot\sin\frac{(nl)}{2}+\cos\frac{(nl)}{2}\right]$$

$$y_{max} = e\cdot\left[\frac{2\sin^2\frac{nl}{2}}{2\sin\frac{nl}{2}\cos\frac{nl}{2}}\sin\frac{nl}{2}+\cos\frac{nl}{2}\right]$$

$$y_{max} = e\cdot\left[\frac{\sin^2\frac{nl}{2}+\cos^2\left(\frac{nl}{2}\right)}{\cos\frac{nl}{2}}\right]$$

$$\left[\because \sin^2\left(\frac{nl}{2}\right)+\cos^2\left(\frac{nl}{2}\right)=1\right]$$

$$y_{max} = e \cdot \sec\left(\frac{nl}{2}\right) \quad \text{...(vii)}$$

The maximum bending moment in the column

$$M_{max} = P \,.\, e \sec\left(\frac{nl}{2}\right) \quad \text{... (viii)}$$

The maximum stress in the column

$$\sigma_{max} = \left(\frac{P}{A} + \frac{M_{max}}{1} \cdot c\right) \quad \text{...(ix)}$$

where c = Distance to the extreme fibre from centre line of column

$$\sigma_{max} = \left[\frac{P}{A} + \frac{P}{A} \cdot \left(\frac{ec}{r^2}\right) \sec\frac{nl}{2}\right]$$

where $I = Ar^2$, $n = \left(\frac{P}{EI}\right)^{1/2}$, and $= \sigma_c$

Then, $$\sigma_{max} = \frac{P}{A}\left[1 + \frac{ec}{r^2} \sec\frac{l}{2r}\left(\frac{P}{EA}\right)^{1/2}\right]$$

$$\sigma_c = \left(\frac{f_{max}}{1 + \frac{ec}{r^2} \sec\left[\frac{l}{2r}\left(\frac{P}{EA}\right)^{1/2}\right]}\right) \quad \text{...(x)}$$

This expression is known as secant formula.

Let m be the factor of safety, and therefore, $\sigma_c = m \,.\, \sigma_a$

where σ_a = Allowable average stress in axial compression, as it was recommended in IS : 800–1962.

The maximum stress f_{max} reached can be yield stress, f_y.

Then

$$\sigma_a = \left(\frac{\left(\frac{f_y}{m}\right)}{1 + \frac{ec}{r^2} \sec\frac{l}{r}\left(\frac{mf_a}{4E}\right)^{1/2}}\right) \quad \text{...(xi)}$$

Indian Standard Code of Practice for general construction steel IS : 800–1984 has dropped this secant formula for calculating the allowable average stress in axial compression. In secant formula, the value of $\left(\frac{ec}{r^2}\right)$ was taken as 0.20.

The formula takes into account any initial crookedness of the column and imperfectness of axial loading.

Therefore, for

$$\left(\frac{\mathbf{l}}{\mathbf{r}}\right) = \mathbf{0\ to\ 160}$$

$$\sigma_a = \left(\frac{\left(\dfrac{f_y}{m}\right)}{1 + 0.20 \sec\left[\dfrac{l}{r}\left(\dfrac{mf_a}{4E}\right)\right]^{1/2}} \right) \qquad \text{...(3.28)}$$

where f_y = Guaranteed minimum yield stress

m = Factor of safety taken as 1.68 on the yield point

$\left(\frac{l}{r}\right)$ = Slenderness ratio

l = Effective length of the member

r = Appropriate radius of gyration

E = Young's modulus of elasticity, 20,47,000 kg/cm^2

For values of

$$\left(\frac{\mathbf{l}}{\mathbf{r}}\right) = \mathbf{160\ and\ above}$$

$$\sigma'_a = \sigma_a \cdot \left(1.2 - \frac{l}{800r}\right) \qquad \text{...(3.29)}$$

where σ'_a = Allowable average axial stress in compression as it was recommended in IS : 800–1962

for $\left(\frac{l}{r}\right) \geq 160$.

The secant formula for axial compression as it was recommended to determine permissible stress in IS : 800–1962 has been dropped in IS : 800–1984.

Professor Ashok K. Jain of University of Roorkee made the comparison of permissible compressive stresses as given by different national standards in his paper titled as 'compressive stresses in columns' published in International Journal of Structures, Vol. 4, No. 2, April 1984. The disadvantages of secant formula were pointed out. The secant formula provides a conservative estimate of the average column stress at maximum load for an eccentrically loaded steel column. The secant formula is not convenient for hand calculations. It needs number of iterations to calculate the permissible axial stress as this term appears on both the sides of the equation.

The secant formula does not show an yield plateau for the columns of slenderness ratio upto 30. The secant formula gives very conservative values for columns of slenderness ratio upto 90 as compare to those given by AISC, BS 449, IS 802 and AS : 1250. Further, this secant formula gives higher values of permissible stresses for the columns having slenderness ratios beyond 90.

Instead of the secant formula, Merchant–Rankine formula has been specified in IS : 800–1984.

3.8 DESIGN FORMULA

Indian Standard Code of Practice for general construction in steel IS : 800–1984 has recommended Merchant–Rankine formula as below

$$\left(\frac{1}{\sigma_{ac}}\cdot\frac{1}{F\cdot S.}\right)^n = \left(\frac{1}{f_{ac}}\right)^n + \left(\frac{1}{f_y}\right)^n \quad \text{...(i)}$$

or

$$\sigma_{ac} = \left(\frac{f_{cc}\cdot f_y}{F.S.\left[(f_{cc})^n + (f_y)^n\right]^{1/n}}\right) \quad \text{...(ii)}$$

The direct stress in compression on the gross-sectional area of axially loaded compression member was found with a factor of safety of $\frac{10}{6}$. It should *not exceed 0.60* f_y nor the permissible stress σ_{ac}, calculated using the following formula.

$$\sigma_{ac} = 0.6\cdot\left(\frac{f_{cc}\cdot f_y}{\left[(f_{cc})^n + (f_y)^n\right]^{1/n}}\right) \quad \text{...(iii)}$$

where, σ_{ac} = Permissible stress in axial compression in N/mm² (MPa)

f_y = Yield stress in steel in N/mm² (MPa)

f_{cc} = Elastic critical stress in compression, and $\left(f_{cc} = \frac{\pi^2 E}{\left(\frac{l}{r}\right)^2}\right)$

E = Young's modulus of elasticity of steel 2×10^5 N/mm² (MPa)

n = Imperfection factor. It is assumed as 1.4.

The values of permissible stresses for axial compression, σ_{ac} for some of the Indian standard structural steels (having different values of yield stresses varying from 220 N/mm² upto 540 N/mm²) are given in Tables 3.3 (a) and (b), for convenience as per IS : 800–1984.

Table 3.2 (a) *Permissibles stress* $s_{ac,}$ *N/mm² (Mpa) in axial compression for steel with various yield stress* , f_y *(slendemess ratio,* $l = \frac{l}{r}$ *)*

$f_y \rightarrow$ / $\lambda \downarrow$	220	230	240	250	260	280	300	320	340
10	132	138	144	150	156	168	180	192	204
20	131	137	142	148	154	166	177	189	201
30	128	134	140	145	151	162	172	183	194
40	124	129	134	139	145	154	164	174	183
50	118	123	127	132	136	145	153	161	168
60	111	115	118	122	126	133	139	146	152
70	102	10	109	112	115	120	125	130	135
80	93	96	98	101	103	107	111	115	118
90	85	87	88	90	92	95	98	101	103
100	76	78	79	80	82	84	86	88	90
110	68	69	71	72	73	74	76	77	79
120	61	62	63	64	64	66	67	67	69
130	55	55	56	57	57	58	59	60	61
140	49	50	50	51	51	52	53	53	54
150	44	45	45	45	46	46	47	47	48
160	40	41	41	41	42	42	42	42	43
170	36	36	37	37	36	37	38	38	38
180	33	33	33	33	33	34	34	34	34
190	30	30	30	30	30	30	31	31	31
200	27	27	28	28	28	28	28	28	28
210	25	25	25	25	25	25	26	26	26
220	23	23	23	23	23	23	23	24	24
230	21	21	21	21	21	21	22	21	22
240	20	20	20	20	20	20	20	20	20
250	18	18	18	18	18	18	18	18	18

Table 3.2 (b) *Permissibles stress* $s_{ac,}$ *N/mm² (Mpa) in axial compression for steel with various yield stress* , f_y *(slendemess ratio,* $l = \frac{l}{r}$ *)*

f_y	340	360	380	400	420	450	480	510	540
10	204	215	227	239	251	269	287	305	323
20	201	212	224	235	246	263	280	297	314
30	194	204	215	225	236	251	266	280	295
40	183	192	201	210	218	231	243	255	267
50	168	176	183	190	197	207	216	225	233

Contd.

Table 3.2 (b) Contd.

f_y	340	360	380	400	420	450	480	510	540
60	152	158	163	168	173	180	187	193	199
70	135	139	142	147	150	155	160	164	168
80	118	121	124	127	129	133	136	139	141
90	103	105	108	109	111	114	116	118	119
100	90	92	93	94	96	97	99	100	101
110	79	80	81	82	83	84	85	86	87
120	69	70	71	71	72	73	73	74	75
130	61	61	62	62	63	63	64	64	65
140	54	54	54	55	55	56	56	56	57
150	48	48	48	49	49	49	49	50	50
160	43	43	43	43	43	44	44	44	44
170	38	38	39	39	39	39	39	39	39
180	34	35	35	35	35	35	35	35	35
190	31	31	31	31	32	32	32	32	32
200	28	28	28	28	28	28	28	28	28
210	26	26	26	26	26	26	26	26	26
220	24	24	24	24	24	24	24	24	24
230	22	22	22	22	22	22	22	22	22
240	20	20	20	20	20	20	20	20	20
250	18	19	19	19	19	19	19	19	19

Professor D. Allen suggested the use of this formula for the design of columns in his paper titled as 'Merchant–Rankine Approach to Member Stability' a technical note published in Journal of Structural Division, ASCE, Vol. 104. No. ST 12, December 1978. Depending on the degree of residual stresses, the value of parameter takes the value of 1, 1.4 or 2. The value of n is taken as 14 for riveted and normal welded members. Its value is adopted as 1 where the high residual stresses are expected.

A constant stress for the members of very small slenderness ratio is not indicated by Merchant–Rankine formula. A cutoff point below which the permissible stress is constant, is specified by other standards to include this feature. A correction factor in the Merchant–Rankine expression was suggested by Maquoi and J. Rondal, in their discussions on 'Merchant–Rankine Approach to Member Stability' by D. Allen published in Vol. 105, No. ST 11, November, 1979 as below

$$\sigma_{ac} = \frac{1}{F.S.} \cdot \left(\frac{f_{cc} \cdot f_y}{\left[(f_y)^n + (f_{cc})^n \left(1 - 0.15^{2n}\right) \right]^{\frac{1}{n}}} \right) \quad \text{... (iv)}$$

This expression (iv) includes the constant yield plateau for the columns of slendemess ratio less than about 30. It gives about one to four per cent difference in values given by the expressions (iii) and (iv). IS : 800–1984 does not specify the expression (iv).

Ashok K. Jain in his paper referred above has mentioned in the conclusion that Merchant–Rankine formula is very simple. It has lot of potential for use in different situations. It is more convenient for use in case the multiple column curves are to be developed. The value of imperfection factor is simply changed. Only three parameters, namely, yield stress, critical stress and imperfection factor appear in Merchant–Rankine formula. A computer-program may easily be written for use of Merchant–Rankine formula. The values of allowable stresses for columns of slenderness ratio less than 90 obtained by Merchant–Rankine formula are higher than those found by the secant formula. However, these higher values closely agree with those determined by other national standards.

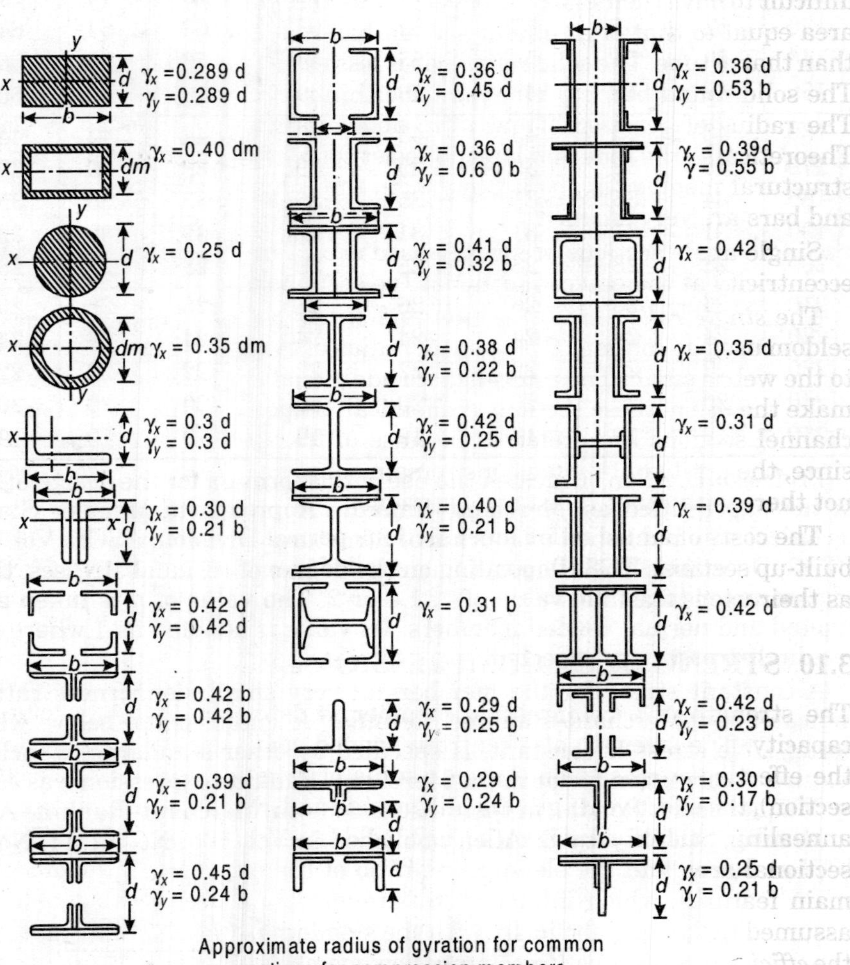

Fig. 3.8 *Common sections for compression members*

(*Reference* : J.A.L. Waddell 'Bridge-engineering' John Wiley and Sons, Inc. New York 1925).

3.9 COMMON SECTIONS OF COMPRESSION MEMBERS

The common sections used for compression members are shown in Fig. 3.8 with their approximate radii of gyration.

A column or a compression member may be made of many different sections to support a given load. Few sections satisfy practical requirement in a given case. A *tubular section* is most efficient and economical for the column free to buckle in any direction. The radius of gyration *r* for the tubular section in all the directions remains same. The tubular section has high local buckling Strength. The tubular sections are suitable for medium loads. However, it is difficult to have their end connections. *A solid round bar* having a cross-sectional area equal to that of a tubular section has radius of gyration, *r* much smaller than that of tube. The solid round bar is less economical than the tubular section. The solid round bar is better than the thin rectangular section or a flat strip. The radius of gyration of flat strip about its narrow direction is very small. Theoretically, the rods and bars do resist some compression. When the length of structural member is about 3 m, then, the compressive strengths of the rods and bars are very small.

Single angle sections are rarely used except in light roof trusses, because of eccentricity at the end connections. Tee sections are often used in roof trusses.

The *single rolled steel I-section* and *single rolled steel channel section* are seldom used as column. The value of radius of gyration *r*, about the axis parallel to the web is small. The intermediate additional supports in the weak direction make the use of these sections economical. Sometimes the use of I-sections and channel sections are preferred because of the method of rolling at the mills, since, the out-to-out dimensions remain same for a given depth. This failure is not there with other rolled steel sections.

The costs of single rolled steel sections per unit weight are less than those of built-up sections. Therefore the single rolled steel sections are preferred so long as their use is feasible.

3.10 STRENGTH OF COMPRESSION MEMBERS

The strength of a compression member is defined as its safe load carrying capacity. The strength of a centrally loaded straight steel column depends on the effective cross-sectional area, radius of gyration (viz., shape of the cross-section), the effective length, the magnitude and distribution of residual stresses, annealing, out of straightness and cold straightening. The effective cross-sectional area and the slenderness ratio of the compression members are the main features, which influence its strength. In case, the allowable stress is assumed to vary parabolically with the slenderness ratio, it may be proved that the *efficiency of a shape of a compression*: member is related to A/r^2. The *efficiency of a shape* is defined as the ratio of the allowable load for a given slenderness ratio to that for slenderness ratio equal to zero. The safe load carrying capacity of compression member of known sectional area may be determined as follows :

Step 1. From the actual length of the compression member and the support conditions of the member, which are known, the effective length of the member is computed.

Step 2. From the radius of gyration about various axes of the section given in section tables, the minimum radius of gyration (r_{min}) is taken. r_{min} for a built-up section is calculated.

Step 3. The maximum slenderness ratio (l/r_{min}) is determined for the compression member.

Step 4. The allowable working stress (σ_{ac}) in the direction of compression is found corresponding to the maximum slenderness ratio of the column from IS : 800–1984.

Step 5. The effective sectional area (A) of the member is noted from structural steel section tables. For the built-up members it can be calculated.

Step 6. The safe load carrying capacity of the member is determined from

$$P = (\sigma_{ac} . A)$$

where P is the safe load

3.11 ANGLE STRUTS

The compression members consisting of single sections are of two types :

1. Continuous members; 2. Discontinuous members.

3.11.1 Continuous Members

The compression members (consisting of single or double angles) which are continuous over a number of joints, are known as *continuous members.* The top chord members of truss girders and principal rafters of roof trusses are continuous members.

The *effective length* of such compression members is adopted between 0.7 and 1.0 times the distance between the centres of intersections, depending upon degree of end restraint provided. When the members of trusses buckle in the plane perpendicular to the plane of the truss, the effective length shall be taken as 1.0 times the distance between the points of restraint.

The *working stresses* for such compression members is adopted from IS : 800–1984 corresponding to the slenderness ratio of the member, and yield stress for steel.

3.11.2 Discontinuous Members

The compression members which are not continuous over a number of joints, i.e., which extend between two adjacent joints only, are known as discontinuous members. The discontinuous members may consists of

(i) Single angle strut, or (ii) Double angle strut.

When an angle strut is connected to a gusset plate or to any structural member by one leg, the load transmitted through the strut, is eccentric on the section of the strut. As a result of this, bending stress is developed along with direct stress. While designing or determining strength of an angle strut, the bending stress developed because of eccentricity of loading is accounted for as follows :

(A) Single Angle Strut. (i) When single angle discontinuous strut is connected to a gusset plate with one rivet as shown in Fig. 3.9 (a) its effective length is adopted as centre to centre of intersection at each end and the allowable working stress corresponding to the slenderness ratio of the member is reduced to 80 per cent. However, the slenderness ratio of such single angle strut should not exceed 180.

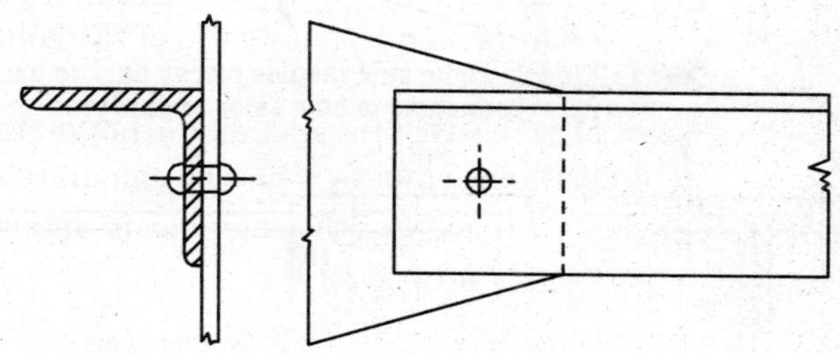

(a) Single angle strut connected with one rivet

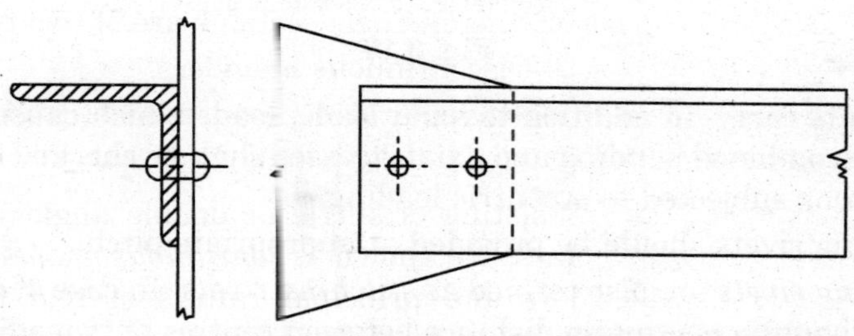

(b) Single angle strut connected two or more rivets

Fig. 3.9 *Single angle strut*

(ii) When a single angle discontinuous strut is connected with two or more number of rivets or welding as shown in Fig. 3.9 (b) its effective length is adopted as 0.85 times the length of strut centre to centre of intersection of each end, and allowable working stress corresponding to the slenderness ratio of the member is *not reduced.*

(B) Double Angle Strut. (i) A double angle discontinuous strut with angles placed back to back and connected to both sides of a gusset or any rolled steel section by not less than two rivets or bolts or in line along the angles at each end or by equivalent in welding as shown in Fig. 3.10 (a), can be regarded as an axially loaded strut. Its effective length is adopted as 0.85 times the distance between intersections, depending on the degree of restraint provided and in the plane perpendicular to that of the gusset, the effective length *T* shall be taken as equal to the distance between centres of intersections.

The tacking rivets should be provided at appropriate pitch.

(ii) The double angles, back to back connected to one side of a gusset plate or a section by one or more rivets or bolts of welds as shown in Fig. 3.10 (b), these are designed as single angle discontinuous strut connected by single rivet or bolt.

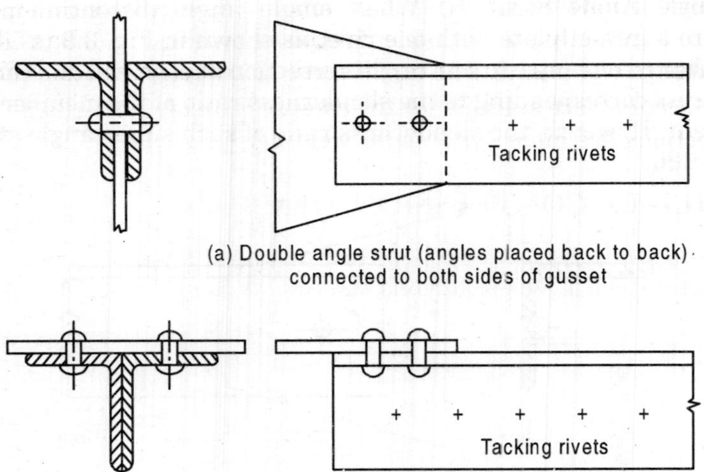

(a) Double angle strut (angles placed back to back) connected to both sides of gusset

(b) Double angle strut (angles placed back to back) connected to one side of gusset

Fig. 3.10

If the struts carry, in addition to axial loads, loads which cause transverse bending, the combined bending and axial stresses shall be checked as described for the columns subjected to eccentric loading.

The tacking rivets should be provided at appropriate pitch.

The *tacking rivets* are also termed as *stitching rivets.* In case of compression members, when the maximum distance between centres of two adjacent rivets exceeds 12 t to 200 mm whichever is less, than tacking rivets are used. The tacking rivets are not subjected to calculated stress. The tacking rivets are provided throughout the length of a compression member composed of two components back to back. The two components of a member act together as one piece by providing tacking rivets at a pitch in line not exceeding 600 mm, and such that minimum slenderness ratio of each member between the connections is not greater than 40 or 0.6 times the maximum slenderness ratio of the strut as a whole, whichever is less.

In case where plates are used, the tacking rivets are provided at a pitch in line not exceeding 32 times the thickness of outside plate or 300 mm, whichever is less. Where the plates are exposed to weather the pitch in line shall not exceed 16 times the thickness of the outside plate or 200 mm, whichever is less. In both cases, the lines of rivets shall not be apart at a distance greater than these pitches.

The *single angle sections* are used for the compression members for small trusses and bracing. The equal angle sections are more desirable usually. The unequal angle sections are also used. The minimum radius of gyration about one of the principal axis is adopted for calculating the slenderness ratios. The minimum radius of gyration of the single angle section is much less than the other sections of same cross-sectional area. Therefore, the single angle sections are not suitable for the compression member of long lengths. The single angle sections are commonly used in the single plane trusses (i.e., the trusses having gusset plates in one plane). The angle sections simplify the end connections.

The tee-sections are suitable for the compression members for small trusses. The tee-sections are more suitable for welding.

3.12 BUILT–UP COMPRESSION MEMBERS

The built-up compression members are needed when the single rolled steel sections are not sufficient to furnish the required cross-sectional area.

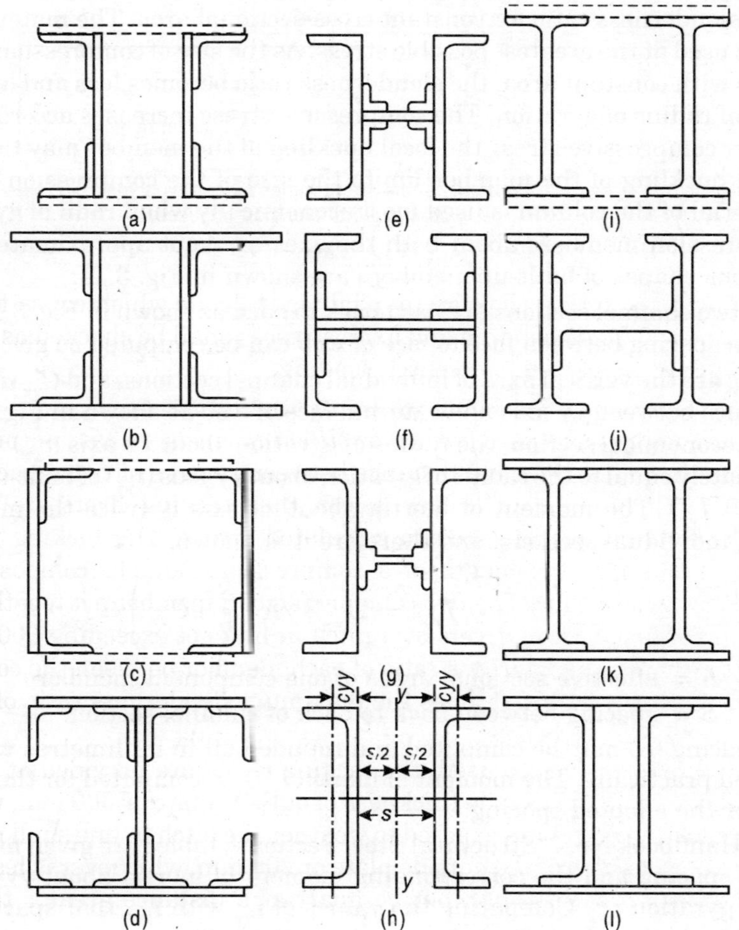

Fig. 3.11 *Built-up compression members*

A built-up compression member may consist of two or more rolled structural steel sections connected together effectively and acts as one compression member. The built-up compression members are given effective column cross-sections.

The built-up sections are used for one or more of the reasons mentioned below.

(i) The built-up sections provide large cross-sectional area which cannot be furnished by single rolled steel sections.

(ii) The built-up sections provide special shape and depth. The special shape and depth facilitate connections between the different members.

(iii) The built-up sections provide sufficient large radius of gyration of more desirable ratio of the radii of gyration in two different directions. In the single rolled steel section, the ratio of radii of gyration in two directions cannot be altered.

The element of built-up members are placed at farthest possible distance from centroid of the section. It gives the largest radius of gyration and the smallest slenderness ratio for constant cross-sectional area. The material of the column is used at the greatest possible stress. As the size of compression member increases with constant area, the slenderness ratio becomes less and less due to increase of radius of gyration. The compressive stress increases accordingly. At the higher compressive stress the local buckling of the member may take place. The local buckling of the member limits the size of the compression member. The material of the column is used most economically when radii of gyration of the compression members about both the axes are kept approximately equal. The various shapes of built-up members are shown in Fig. 3.11.

When two channel sections are used back to back as shown in Fig. 3.11 (h) the maximum spacing between face to face of web can be computed as given below.

yy-axes are the vertical axes of individual channel sections, and C_{yy} represents the distance between *yy*-axis and external face of web as shown in Fig. 3.11 (h). For most economical section, the radius of gyration about *xx*-axis (r_{xx}) should be approximately equal to the radius of gyration about *yy*-axis (r_{yy}) of built-up section (i.e., $I_{xx} \simeq I_{yy}$). The moment of inertia about *xx*-axis is twice the moment of inertia of individual sections, and therefore it is known.

$$I_{YY} = 2I_{yy} + 2A\left(\frac{S}{2} + C_{yy}\right) \simeq I_{xx}$$

where, A = effective sectional areas of one component member

S = spacing between back to back of channel section.

The spacing (S) may be computed and rounded off in millimetres which can be adopted practically. The moment of inertia I_{yy} is recomputed for the built-up section for the adopted spacing.

In ISI Handbook No. 1 'Structural Steel Sections', tables are given for various values of spacing and the corresponding moment of inertia about *yy*-axis and radius of gyration r_{yy}. Comparing the values of I_{yy} with I_{xx}, that spacing (S) is adopted which gives $I_{yy} = I_{xx}$.

The primary object of built-up section is to furnish large radius of gyration and to support heavy loads. However, the arrangement of different sections must allow case of fabrication, connection and maintenance (e.g., painting). The lattice material used should be small.

The two-angle sections placed back to back are most frequently used in the roof trusses. The built-up sections consisting of such sections are particularly economical when single gusset plates are used. The angle sections are tied by rivets at suitable spacings with filler plates between the angles. The unequal angle sections are preferred for such built-up section, in order to obtain radii of gyration about two centroidal axes approximately equal.

Four angle sections are used in a built-up section to maintain the overall depth of the section. In one case, two pairs of four angles are formed, each pair is made of two angles. Two pairs are connected by lacing, when the load is small. Two pairs are connected with a solid plate forming a built-up I-section when the load is large. The long legs of angle sections are kept outstanding. It gives higher value of radius of gyration. The four angle sections are also used to form a box section. It provides large value of radius of gyration. Such a built-up section is economical to carry small load and when the length of the member is long. Such a built-up section needs large amount of lacing. A section built-up of four angles is used for compression members having large length (e.g., masts, crane boom etc.). It provides equal stiffness in both the directions.

A built-up section consisting of *two channels back to back* is occasionally used. Such a built-up section has small radius of gyration about y-axis. A built-up section consisting of *two channel sections placed at a distance apart* is more frequently used. The flanges of channels are kept outward. It gives large value of radius of gyration. When the large value of radius of gyration is not necessary, then, the flanges are kept inward. It reduces the amount of lacing. A built-up section is also made of two channel sections with a solid plate on one side or on both the sidess. The built-up section with solid plate on one side permits the access for inspection and painting. It is not possible, when solid plates are used on both the sides. Two channel sections with solid plates constituting built-up section, provide large cross-sectional area. The bracing between the parts is also stiff.

The built-up sections are also made of solid plates and angle sections to have very large cross-sectional area. When the channels of maximum size do not provide sufficient area, then, such built-up sections are used. Such built-up sections are used in the bridges.

It is to note that the cost of built-up section per unit weight is high because of fabrication and additional material needed for lacing the compression members. The single rolled steel sections are more economical for the equal area and equal slenderness ratio, if the normal connections may be designed for the member.

3.13 THICKNESS OF ELEMENTS IN COMPRESSION MEMBERS

When any plates element is subjected to direct compression bending, or shear stresses or to combination of these stresses, the plate element may buckle locally before the entire member fails In *local buckling,* waves (single wave or series of waves) or wrinkles are formed in the elements of compression members.

ISI in IS : 800–1962 recommended the thickness of element in compression members in terms of outstanding lengths of compression members. These thicknesses have been recommended to avoid the failure of elements owing to *local buckling*.

The outstanding length of compression member in case of plated beam is measured from the free edge to the first row of rivet, as shown in Fig. 3.12 (a).

In case of channels, angles and stem of tee-section, it is measured as nominal width of sections as shown in Fig. 3.12 (b) and in case of flanges of beams and tee-section, it is measured as half the nominal width, as shown in Fig. 3.12 (c).

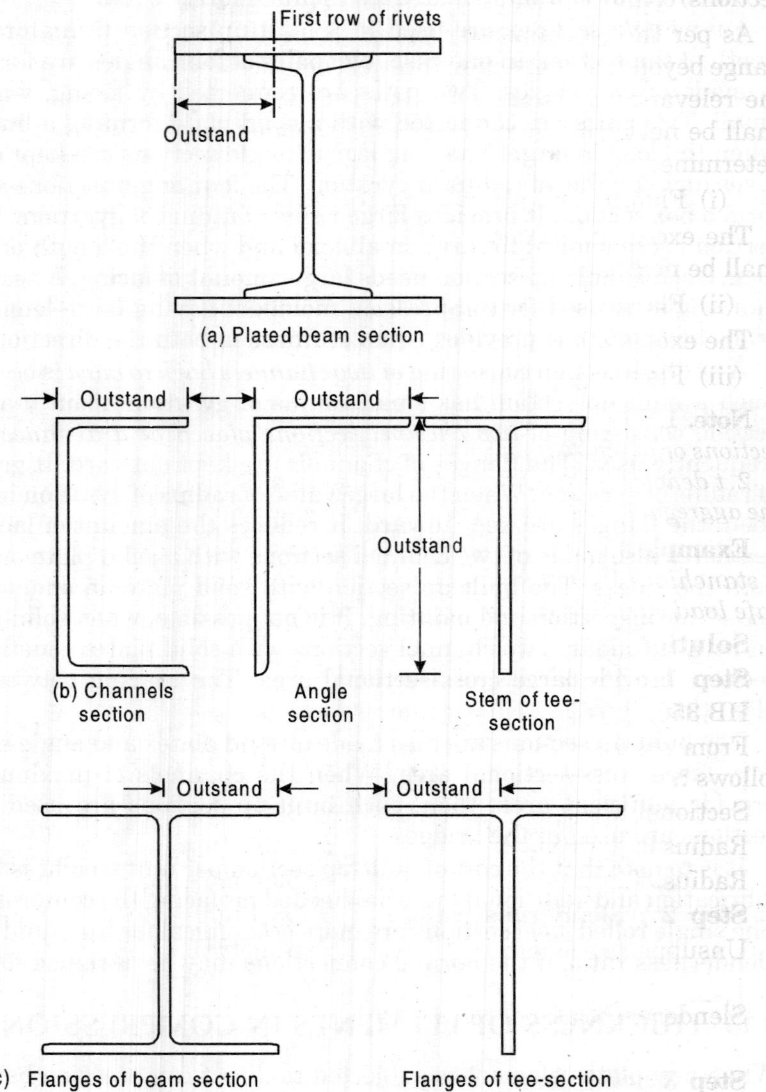

Fig. 3.12 *Out standing lengths compression members*

The thickness is taken as 1/16 of outstand of any member in compression for unstiffened outstand and 1/20 of outstand of any member in compression for stiffened outstand. The thickness is taken as 1/90 of outstand for unsupported width of unstiffened plate forming any part of a compression member, measured between adjacent line of rivets. But for the purpose of calculating effective sectional area and radius of gyration only 50 times the thickness of plate will be taken instead of 90 times the thickness in case single plate is used. However, it is total thickness of two or more plates effectively tacked together.

All hot rolled steel sections as per IS : 808–1964 (beams, channels and angle sections) conform to outstand requirements as per IS : 800–1962 described above.

As per the recommendations in IS : 800–1984, the projection of a plate or flange beyond its connection to a web, or other line of support or the like, exceeds the relevant values given in (i), (ii) and (iii) below, the area of excess flange shall be neglected when the effective geometrical properties of the section are determined :

(i) Flanges and plates in compression with unstiffened edges.

The excess projection if exceeds $256t/(f_y)^{1/2}$ subjected to a maximum of $16t$ shall be neglected.

(ii) Flanges and plates in compression with stiffened edges.

The excess projection if exceeds 20 t to the innermost face of the stiffening.

(iii) Flanges and plates in tension the excess projection if exceeds 20 t.

Note. 1. *The stiffened flanges shall include flanges composed of channels or I-sections or of plates with continuously stiffened edges.*

2. *t denotes the thickness of the flange of a section or of a plate in compression, or the aggregate thickness of the plates, if connected together appropriately.*

Example 3.1 *A rolled steel beam section HB 350 @ 0.674 kN/metre is used as a stanchion. If the unsupported length of the stanchion is 4 metres, determine safe load carrying capacity of the section.*

Solution

Step 1 : Properties of I-section

HB 350, @ 0.674 kN/m section is used as a stanchion.

From steel section tables, the geometrical properties of the section are as follows :

Sectional area A = 8591 mm^2

Radius of gyration r_{xx} = 149.3 mm

Radius of gyration r_{yy} = 53.4 mm

Step 2 : Slendernes ratio, r_{min} = 53.4 mm

Unsupported length l = 4 metres

Slenderness ratio of the stanchion $\dfrac{l}{r_{min}} = \left(\dfrac{4 \times 1000}{53.4}\right) = 75$

Step 3 : Safe load

From IS : 800–1984 for $\dfrac{l}{r}$ = 75.0, and the steel having yield stress, f_y = 260 N/mm^2, allowable working stress in compression

$$\sigma_{ac} = 109 \text{ N/mm}^2 \text{ (MPa)}.$$

The safe load carrying capacity of the stanchion

$$P = (\sigma_{ac} - A) = \left(\frac{109 \times 8591}{1000}\right) = \mathbf{936.42\ kN}$$

Example 3.1 (a) *In Example 3.1, in case standard column section SC 25°, @ 85.6 kg/m is used as a column determine the safe load-carrying capacity of the section.*

Solution

Step 1 : Properties of I-section

From IS : 808–1964, the geometrical properties of the section are as follows.

Sectional area $A = 109 \times 10^2$ mm^2

Radius of gyration $r_a = 107$ mm

Radius of gyration $r_{yy} = 54.6$ mm

Step 2 : Slenderness ratio

Minimum radius of gyration

$$r_{min} = 54.6 \text{ mm}$$

Unsupported length of the stanchion

$$l = 4000 \text{ mm}$$

Slenderness ratio

$$\left(\frac{l}{r_{min}}\right) = \left(\frac{4000}{54.6}\right) = 73.26$$

Step 3 : Safe load

From IS : 800–1984 for $\left(\frac{l}{r}\right) = 73.26$ and the steel having yield sress, $f_y = 260$ N/mm^2, allowable working stress in compression

$$\sigma_{ac} = \left(115 - \frac{12}{10} \times 3.26\right) = 111.088 \text{ N/mm}^2$$

The safe load carrying capacity of the stanchion

$$P = (\sigma_{ac} \,.\, A) = \left(\frac{111.088 \times 109 \times 100}{1000}\right)$$

$$= \mathbf{1210.86 \text{ kN}}$$

Example 3.2 *A single angle discontinuous strut ISA 150 mm× 150 mm × 12 mm (ISA 150 150, @ 0.272 kN/m) with single riveted connection is 3.5 metres long. Calculate safe load carrying capacity of the section.*

Solution

Step 1 : Properties of angle section

ISA 150 mm × 150 mm × 12 mm (ISA 150 150, @ 0.272 kN/m) is used as discontinuous strut. From steel section tables

Sectional area $A = 3459$ mm^2

$$r_{yy} = r_{xx} = 46.1 \text{ mm}$$

$$r_{uu} = 58.3 \text{ mm}, r_{vv} = 29.3 \text{ mm}$$

Step 2 : Slenderness ratio

Effective length of strut $l = 3.5$ m

$$r_{min} = 29.3 \text{ mm}$$

Minimum radius of gyration

Slenderness ratio of strut $\frac{l}{r_{min}} = \left(\frac{3.5\times1000}{29.3}\right) = 119.5$

Step 3 : Safe load

From IS : 800–1984, the allowable working stress in compression for $\frac{l}{r_{min}} =$ 119.5 and the steel having yield stress 260 N/mm^2

$$\sigma_{ac} = 64.45 \text{ N/mm}^2$$

For single angle discontinuous strut with single riveted connection, allowable working stress

$$0.80\ \sigma_{ac} = (0.8 \times 64.45) = 51.56 \text{ N/mm}^2$$

The safe load carrying capacity

$$P = \left(\frac{51.56 \times 3459}{1000}\right) = \mathbf{178.346\ kN}$$

Example 3.2 (a) *In case in Example 3.2, a discontinuous strut 150 × 150 × 15 angle section is used, calculate the safe load carrying capacity of the section*

Solution

Step 1 : Properties of angle section

Angle section 150 × 150 × 15 is used as a discontinuous strut. From IS : 808–1984 cross-section area. $A = 43 \times 100$ mm^2

Radius of gyration $r_{xx} = r_{yy} = 45.7$ mm

$$r_{uu} = 57.6 \text{ mm}$$

$$r_{xv} = 29.3 \text{ mm}$$

Step 2 : Slenderness ratio

Minimum radius of gyration

$$r_{min} = 2.93 \text{ mm}$$

Effective length of strut

$$l = 3.5 \times 1000 \text{ mm}$$

Slenderness ratio of strut

$$\left(\frac{l}{r_{min}}\right) = \left(\frac{3500}{29.3}\right) = 119.5$$

Step 3 : Safe load

From IS: 800–1984, the allowable working stress in compression for $\left(\frac{l}{r}\right) = 119.5$ and steel having yield stress 260 N/mm^2

$$\sigma_{ac} = 64.45 \text{ N/mm}^2$$

For single angle discontinuous strut with single riveted connection, allowable working stress

$$0.80\ \sigma_{ac} = (0.80 \times 64.45) = 51.56 \text{ N/mm}^2$$

The safe load carrying capacity

$$P = \left(\frac{51.56 \times 4300}{1000}\right) = \mathbf{221.708\ kN}$$

Example 3.3 *In Example 3.2, if single angle discontinuous strut is connected with more than two rivets in line along the angle at each end, determine safe load carrying capacity of the section.*

Solution

Step 1 : Properties of angle section

Discontinuous strut ISA 150 mm × 150 mm × 12 mm (ISA 150 150, @ 0.272 kN/m) is used with double riveted connections.

Minimum radius of gyration

$$r_{min} = 29.3 \text{ mm}$$

Sectional area,

$$A = 3459 \text{ mm}$$

Length of strut between centre to centre of intersection

$$L = 3.50 \text{ m}$$

Step 2 : Slenderness ratio

Effective length of discontinuous strut double riveted

$$0.85 \times L = (0.85 \times 3.50) = 2.975 \text{ m}$$

Slenderness ratio

$$\frac{l}{r_{min}} = \left(\frac{2.975 \times 1000}{29.3}\right) = 101.5$$

Step 3 : Safe load

Allowable working stress, for $\frac{l}{r_{min}} = 101.5$ and the steel having yield stress as 260 N/mm²

$$\sigma_{ac} = 71.65 \text{ N/mm}^2 \text{ (MPa)}$$

Allowable working stress for discontinuous strut double riveted is not reduced.

Safe load carrying capacity

$$P = \left(\frac{71.65 \times 3459}{1000}\right) = \mathbf{247.84\ kN}$$

Example 3.4 *A double angle discontinuous strut ISA 125 mm × 95 mm × 10 mm (ISA 12595, @ 0.165 kN/m) long legs back to back is connected to both the sides of a gusset plate 10 mm thick with 2 rivets. The length of strut between centre to centre of intersections is 4 metres. Determine the safe load carrying capacity of the section.*

Solution

Step 1 : Properties of angle section

The double angle discontinuous strut 2 ISA 125 mm × 95 mm × 10 mm (ISA 12595, @ 0.165 kN/m) is shown in Fig. 3.13, Assume the tacking rivets are used along the length.

From steel section tables (*properties of two angles back to back*) Sectional area,

$$A = 4204 \text{ mm}^2$$

$$r_{xx} = 39.4 \text{ mm}$$

Angles are 10 mm apart

$$r_{yy} = 40.1 \text{ mm}$$

Step 2 : Slenderness ratio

Minimum radius of gyration

$\therefore$ $$r_{min} = 39.4 \text{ mm}$$

Effective length

$$l = 0.85\,L = (0.85 \times 4) = 3.40 \text{ m}$$

Slenderness ratio

$$\left(\frac{3.40 \times 1000}{39.4}\right) = 86.4$$

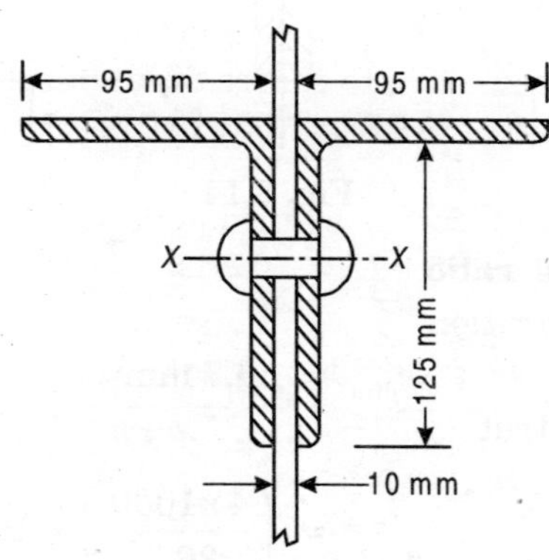

Fig. 3.13

Step 3 : Safe load

From IS : 800–1984 safe allowable stress in compression for $\frac{l}{r_{min}} = 86.4$ and the steel having yield stress as 260 N/mm^2

$$\sigma_{ac} = 95.96 \text{ N/mm}^2$$

Safe load carrying capacity

$$P = \left(\frac{95.96 \times 4204}{1000}\right) = \mathbf{403.416 \text{ kN}}$$

Example 3.5 *In Example 3.4 if double discontinuous strut is connected to one side of a gusset, determine safe load carrying capacity of the strut.*

Solution

Step 1 : Properties of angle sections

A double angle discontinuous strut 2 ISA 125 mm × 95 mm × 10 mm (ISA 12595, @ 0.165 kN/m) connected to one side of a gusset is as shown in Fig. 3.14. Assume that tacking rivets are used along its length.

Effective length of strut whether single riveted or double riveted

$$l = L = 4 \text{ m}$$

From steel section tables.

Section area $A = 4204 \text{ mm}^2$ $r_{xx} = 39.4$ mm

Distance back to back of angles is zero

$$r_{yy} = 36.7 \text{ mm}$$

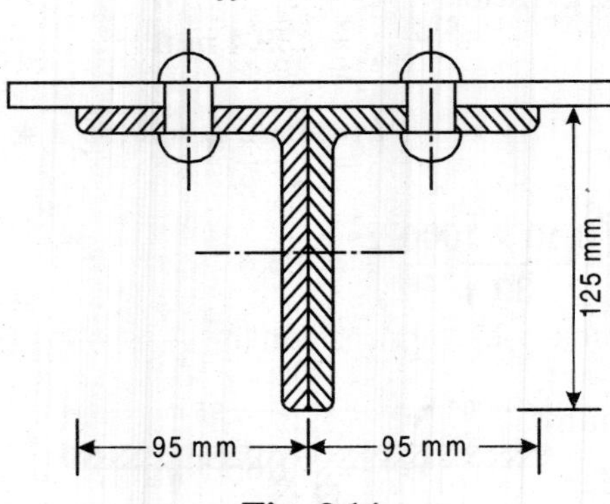

Fig. 3.14

Step 2 : Slenderness ratio

Minimum radius of gyration

$\therefore$ $r_{min} = 36.7$ mm

Slenderness ratio of strut

$$\frac{l}{r_{min}} = \left(\frac{4 \times 1000}{36.7}\right) = 109$$

Step 3 : Safe load

From IS : 800–1984, allowable working stress in compression for $\frac{l}{r_{min}} = 109$

and the steel having yield stress as 260 N/mm²

$$\sigma_{ac} = 73.9 \text{ N/mm}^2$$

For above strut $\sigma_{ac} = 0.8 \times 73.9$ N/mm²

Safe load carrying capacity

$$P = \left(\frac{0.8 \times 73.9 \times 4204}{1000}\right) = \mathbf{248.54\ kN}$$

Example 3.6. *In Example 3.4 double angle strut is continuous and connected with a gusset plate with single rivet, determine safe load carrying capacity of strut.*

Solution

Step 1 : Properties of angle sections

Assume that tacking rivets are used along the length.

The double angle continuous strut 2 ISA 125 mm × 95 mm × 10 mm (ISA 12595, @ 0.165 kN/m) is singly riveted as shown in Fig. 3.13.

$$r_{xx} = 39.4 \text{ mm},$$
$$r_{yy} = 40.1 \text{ mm}$$
$$\text{Sectional area} = 4204 \text{ mm}^2$$

Length of strut between centre of intersection is 4 m.

Step 2 : Slenderness ratio

$$\text{Effective length} = l = L = 4 \text{ m}$$
$$r_{\min} = 39.4 \text{ mm}$$

Slenderness ratio $\dfrac{l}{r_{\min}} = \left(\dfrac{4 \times 1000}{39.4}\right) = 101.5$

Step 3 : Safe load

Allowable working stress in compression for $\dfrac{l}{r_{\min}} = 101.5$ and the steel having yield stress as 260 N/mm^2

$$\sigma_{ac} = 71.65 \text{ N/mm}^2$$

Safe load carrying capacity of strut

$$P = \left(\frac{71.65 \times 4204}{1000}\right) = \mathbf{301.22 \text{ kN}}$$

Example 3.7 *A built-up column consists of three rolled steel beam sections WB 450 @ 0.794 kN/m, connected effectively to act as one column as shown in Fig. 3.15. Determine the safe load carrying capacity of built-up section, if unsupport length of column is 4.25 m.*

Solution

Step 1 : Properties of built-up section

The built-up column consists of 3 WB 450 @ 0.794 kN/m

From steel section tables

Area of section = (3 × 101.15 × 100) = 30345 mm^2

Moment of inertia of built-up section about xx-axis

$$I_{xx} = [2 \times 35057.6 + 1706.7] \times 10^4 = 71821 \times 10^4 \text{ mm}^4$$

Moment of inertia of built-up section about yy-axis

$$I_{yy} = [35057.6 \times 104 + 2 \times 1706.7 \times 10^4 + 2Ah^2]$$
$$= 38471.0 \times 10^4 + 2Ah^2$$

where A = sectional area of one I-section

$$I_{yy} = [38471.0 + 2 \times 101.15\,(22.5 + \tfrac{1}{2} \times 0.92)^2] \times 10^4 \text{ mm}^4$$
$$= 145115.79 \times 10^4 \text{ mm}^4$$

Step 2 : Slenderness ratios

I_{xx} is less than I_{yy}. Therefore, r_{xx} is minimum

$$r_{\min} = \left(\frac{71821.9 \times 10^4}{30345}\right)^{1/2} = 154 \text{ mm}$$

Unsupported length of column

$$l = L = 4.25 \text{ m}$$

Slenderness ratio of column

$$\frac{l}{r_{min}} = \left(\frac{4.25 \times 1000}{154.0}\right) = 27.6$$

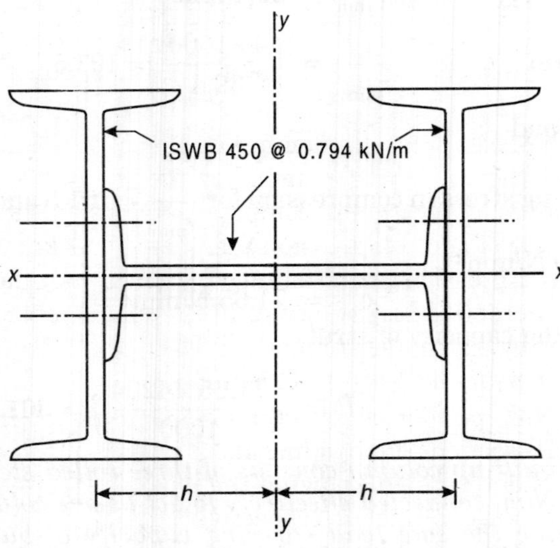

Fig. 3.15

Step 3 : Safe load

From IS : 800–1984, allowable working stress in compression for steel having yield stress as 260 N/mm²

$$\sigma_{ac} = 151.72 \text{ N/mm}^2$$

Safe load carrying capacity of built-up column

$$P = \left(\frac{151.72 \times 303.45}{1000}\right) = \mathbf{46.0394 \text{ kN}}$$

3.14 DESIGN OF AXIALLY LOADED COMPRESSION MEMBER

When a column or compression member is designed, for given load, actual length of the member and its support conditions, the cross-sectional shape of the member is determined.

The cross-sectional shape of axially loaded compression member depends largely on whether the compression member is long or short and whether it carries a small load or a large load. It is difficult to decide, whether a column is short or long. It is arbitrarily decided.

When the slenderness ratio of a column is less than 60, then, it may be considered as a short column. When the slenderness ratio is between 60 and

180, the columns may be considered as long column. Following are the length and load categories arbitrarily made for design of compression members :

1. Short compression members with small loads.
2. Short compression members with large loads.
3. Long compression members with small loads.
4. Long compression members with intermediate load.

The strength of axially loaded compression member depends upon slenderness ratio $\left(\frac{l}{r_{\min}}\right)$. For the design of axially loaded compression member load to be carried, the length of compression member and end conditions are known. The effective length of the compression member for the given end conditions is computed. The radius of gyration of compression member is not known as the cross-sectional shape of the compression member is not known. The allowable working stress in compression can be found when the slenderness ratio is known. There is no direct method of designing a compression member. The compression member is designed by trial and error method. The design of compression member is also done by using safe load tables, if available.

ISI Handbook No.1 provides tables for *safe concentric loads* on rolled steel column sections (HB-sections) for bending about xx-axis and yy-axis. The effective length of column is determined knowing the end conditions. The value of safe concentric loads corresponding to respective effective lengths are given for various sizes of HB-sections. A column section having safe axial load equal to or slightly greater than the required load on the column is selected.

Design procedure. Following are the usual steps in design of compression members.

Step 1. The slenderness ratio for the compression member and the value of yield stress for the steel are assumed.

For the rolled steel beam section compression members, the slenderness ratio varies from 70 to 90. For struts, the slenderness ratio varies from 110 to 130. For compression members carrying large loads, the slenderness ratio is about 40.

Step 2. The effective sectional area (A) required for compression member is determined

$$A = \left(\frac{P}{\sigma_{ac}}\right)$$

where, P = load to be carried by the member.

Step 3. From the steel section tables, section for the compression member of the required area is selected.

The section for the compression member is selected such that it has the largest possible radius of gyration for the required sectional area. It should also be most economical section.

Step 4. Knowing the geometrical properties of the section, slenderness ratio is computed and allowable axial stress in compression is found from IS : 800–1984 for the quality of steel assumed.

Step 5. The safe load carrying capacity of the compression member is determined.

The section selected for the compression member is revised in case the safe load carrying capacity of the compression member is less than or much larger than the load to be carried by it.

Example 3.8 *Design a rolled steel beam section column to carry an axial load 1100 kN. The column is 4 metre long and adequately restrained in position but not in direction at both ends.*

Solution

Design : The slenderness ratio for the column and the value of yield stress for the steel to be used may be assumed as 80 and 260 N/mm^2 respectively.

Step 1 : Selection of trial section

Allowable stress as per IS : 800–1984

$$\sigma_{ac} = 103 \text{ N/mm}^2$$

Effective sectional area required

$$\left(\frac{1100 \times 1000}{103}\right) = 10679.61 \text{ mm}^2$$

Effective length of the column is 4 m

Step 2 : Properties of trial section

From steel section tables, try HB 450, @ 0.872 kN/m section

Sectional area $A = 11114 \text{ mm}^2$

$$r_{xx} = 187.8 \text{ mm}, \quad r_{yy} = 51.8 \text{ mm}$$

Step 3 : Slenderness ratio

$$\therefore \quad r_{min} = 51.8 \text{ mm}$$

Slenderness ratio $\dfrac{l}{r_{min}} = \left(\dfrac{4 \times 1000}{51.8}\right) = 77.2$

Step 4 : Check for safe load

From IS : 800–1984, allowable axial stress in compression, for the steel having yield stress as 260 N/mm^2

$$\sigma_{ac} = 106.36 \text{ N/mm}^2$$

Safe load carrying capacity of the column

$$P = \left(\frac{106.36 \times 11114}{1000}\right) = \mathbf{1182.08 \text{ kN}}$$

The column section lighter in weight than this is not suitable. Hence the design is satisfactory.

Provide HB 450, @ 0.872 kN/m section for the column.

Alternatively :

Step 2 : Properties of trial section

From IS : 808–1984, try SC 250, @ 85.6 kg/m (standard column section)

$$A = 109 \times 100 \text{ mm}^2,$$

$$r_{xx} = 107.0 \text{ mm} \quad r_{yy} = 54.6 \text{ mm}$$

Step 3 : Slenderness ratio

Effective length of column is 4000 mm. Minimum radius of gyration, $r_{\min}$ is 54.6 mm slenderness ratio

$$\left(\frac{l}{r_{\min}}\right) = \left(\frac{4000}{54.6}\right) = 73.26$$

Step 4 : Check for safe load

From IS: 800–1984, allowable stress in axial compression, for the steel having yield stress as 260 N/mm^2

$$\sigma_{ac} = \left(115 - \frac{12}{10} \times 3.26\right) = 111.088 \text{ N/mm}^2$$

Safe load carrying capacity of the column

$$P = \left(\frac{111.088 \times 109 \times 100}{1000}\right) = \mathbf{1210.86\ kN}$$

Example 3.9 *Design a single angle discontinuous strut to carry 110 kN load. The length of the strut between centre to centre of intersections is 3.25 metres.*

Solution

Design :

Step 1 : Selection of trial section

Assuming that the angle strut is connected to the gusset plate with two or more than two rivets.

Effective length of strut

$$l = 0.85\,L = (0.85 \times 3.25 \times 1000)$$
$$= 2762.5 \text{ mm}$$

The slenderness ratio for the single angle discontinuous strut and value of yield stress for the steel may be assumed as 130 and 260 N/mm^2, respectively. Therefore,

Allowable stress in compression for strut

$$\sigma_{ac} = 57 \text{ N/mm}^2$$

Effective sectional area required

$$\left(\frac{110 \times 1000}{57}\right) = 1929.82 \text{ mm}^2$$

The equal angle section is suitable for single angle strut. It has maximum value for minimum radius of gyration.

Step 2 : Properties of trial section

From steel section tables, try ISA 110 mm × 110 mm × 10 mm (ISA 110110 @ 0.165 kN/m).

From steel section tables, try ISA 110 mm × 110 mm × 10 mm (ISA 110110 @ 0.165 kN/m).

Sectional area $A = 2106 \text{ mm}^2$

$r_{xx} = r_{yy} = 33.6 \text{ mm}$

$r_{min} = 42.5 \text{ mm}, \; r_{yy} = 21.4 \text{ mm}$

Step 3 : Slenderness ratio

Slenderness ratio $= \left(\dfrac{2762.5}{21.4}\right) = 129.2$

Step 4 : Safe load

From IS : 800–1984, allowable working stress in compression for the steel having yield stress as 260 N/mm²

$\sigma_{ac} = 57.56 \text{ N/mm}^2$

Safe load carrying capacity

$$P = \left(\frac{57.56 \times 2106}{1000}\right) = \mathbf{121.52 \text{ kN}}$$

The angle section lighter in weight than this is not suitable. Hence the design is satisfactory.

Step 4 : Check for width of outstanding leg

Width of outstanding leg to thickness ratio

$$\left(\frac{\text{Nominal width}}{\text{Thickness}}\right) = \frac{110}{10} = 11 < 16$$

Hence, satisfactory.

Provide ISA 110 mm × 110 mm × 10 mm (ISA 110110, @ 0.165 kN/m) for discontinuous strut.

Alternatively :

Step 2 : Properties of trial section

From IS : 808–1984, try angle section 120 × 120 × 10 (@ 182 kg/m)

Sectional area, $A = 2320 \text{ mm}^2$

$r_{xx} = r_{yy} = 36.7 \text{ mm}$

$r_{uu} = 46.3 \text{ mm}, \quad r_{yy} = 23.6 \text{ mm}$

Step 3 : Slenderness ratio

Effective length of strut is 2762.5 mm

Minimum radius of gyration

$r_{min} = 23.6 \text{ mm}$

Slenderness ratio

$$\left(\frac{l}{r_{min}}\right) = \left(\frac{2762.5}{36.6}\right) = 117.55$$

Step 4 : Safe load carrying capacity

From IS : 800–1984, allowable stress in axial compression for the steel having yield stress as 260 N/mm²

$$\sigma_{ac} = \left(73 - \frac{9}{10} \times 7.55\right) = 66.205 \text{ N/mm}^2$$

Safe load carrying capacity

$$P = \left(\frac{66.205 \times 2320}{1000}\right) = 153.596 \text{ N/mm}^2$$

The angle section lighter in weight than this trial section is not suitable. Hence, the design is satisfactory.

Example 3.10 *Design a double angle discontinuous strut to carry 150 kN load. The length of strut between centre to centre of intersections is 4.00 metre.*

Solution

Design :

Step 1 : Selection of trial section

Assuming that the strut is connected to both sides of gusset 10 mm thick by two or more than two rivets

Length of strut $L = 4.00$ m

Effective length $l = 0.85\,L = (0.85 \times 4) = 3.40$ m

The slenderness ratio of a double angle discontinuous strut and the value of yield stress for steel may be assumed as 120 and 260 N/mm², respectively. Therefore,

Allowable working stress in compression

$$\sigma_{ac} = 64 \text{ N/mm}^2$$

Effective sectional area required

$$\left(\frac{150 \times 1000}{64}\right) = 2343.75 \text{ mm}^2$$

Step 2 : Properties of trial section

From steel section tables (*properties of two angles back to back*) try 2 ISA 100 mm × 65 mm × 8 mm (2 ISA 10065, @ 0.099 kN/m)

Sectional area $A = 2514 \text{ mm}^2$

$$r_{xx} = 31.6 \text{ mm}$$

For angles having 10 mm distance back to back and long legs vertical

$$r_{yy} = 27.5 \text{ mm}$$

Step 3 : Slenderness ratio

$$r_{min} = 27.5 \text{ mm}$$

Step 4 : Safe load

$$\left(\frac{l}{r_{min}}\right) = \left(\frac{3.40 \times 1000}{27.5}\right) = 123.6$$

From IS : 800–1984, allowable working stress in compression for the steel having yield stress as 260 N/mm²

$$\sigma_{ac} = 61.48 \text{ N/mm}^2$$

Safe load carrying capacity of the strut

$$P = \left(\frac{61.48 \times 2541}{1000}\right) = \mathbf{154.50\ kN}$$

Hence, the design is satisfactory.

Provide 2 ISA 100 mm × 65 mm × 8 mm for the strut. Provide tacking rivets 18 mm in diameter at 500 mm spacing.

Example 3.11. *A column 5 metre long is to support a load 4500 kN. The ends of the column are effectively held in position and direction. Design the column if rolled steel beams and 18 mm plates are only available.*

Solution

Design :

Step 1 : Selection of trial section

Length of column = 5 m

Effective length of column is (0.65 × 5) = 3.25 m

In order to support large load, the slenderness ratio for the built-up column and the value of yield stress for the steel may be assumed as 40 and 260 N/mm², respectively.

Allowable working stress in compression

$$\sigma_{ac} = 145 \text{ N/mm}^2$$

Effective sectional area required

$$\left(\frac{4500 \times 1000}{145}\right) = 31034.48 \text{ mm}^2$$

Step 2 : Properties of trial section

From steel section tables, try HB 450 @ 0.925 kN/m

$$r_{xx} = 185.0 \text{ mm}, r_{yy} = 50.8 \text{ mm}$$

For the columns carrying large loads

$$r_{xx} = r_{yy}$$

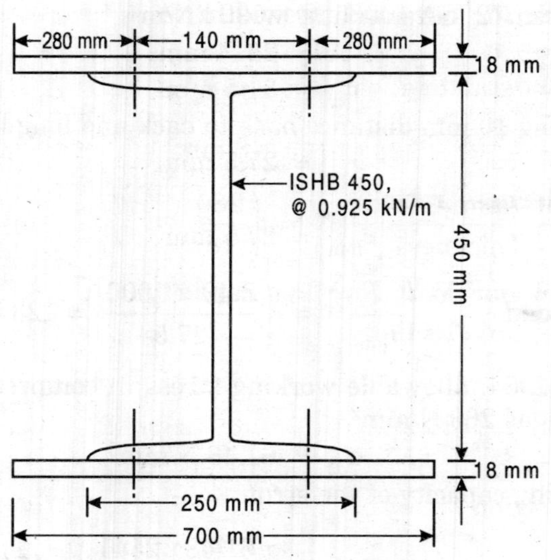

Fig. 3.16

Step 3 : Slenderness ratio

r_{min} for the plated built-up column may be estimated as 185 mm.

Therefore the slenderness ratio for the plated column shall be

$$\frac{l}{r_{min}} = \left(\frac{3.25 \times 1000}{185}\right) = 17.567$$

Step 4 : Area of plates

From IS : 800–1984, allowable working stress in compression for steel having yield stress as 260 N/mm²

$$\sigma_{ac} = 154.486 \text{ N/mm}^2$$

Sectional area required

$$\left(\frac{4500 \times 1000}{154.486}\right) = 29128.74 \text{ mm}^2$$

Area of HB 450, @ 0.925 kN/m is 11789 mm²

Area to be provided by two cover plates

$$= 17339.74 \text{ mm}^2$$

Area to be provided by one plate

$$= 8669.87 \text{ mm}^2$$

Plates available = 18 mm

Width required $= \left(\frac{8669.87}{18}\right) = 481.66$ mm

Provided 700 mm width of cover plate.

Step 5 : Check for outstanding width; thickness ratio for cover plate

$$\text{Outstanding width} = \frac{1}{2}(700 - 140) = \frac{560}{2} = 280 \text{ mm}$$

Thickness = 18 mm

$$\left(\frac{\text{Outstanding width}}{\text{Thickness}}\right) = \left(\frac{280}{18}\right) = 15.55 < 16. \text{ Satisfactory.}$$

Step 6 : Check for load carrying capacity

1. **Properties of section**

I_{yy} of plates $= 2 \times \frac{1}{12} \times 1.8 \times 70^3 \times 10^4 \text{ mm}^4$

$= 102900 \times 10^4 \text{ mm}^4$

I_{yy} of H-section

$= 3045 \times 10^4 \text{ mm}^4$

I_{yy} of compound section

$= 105945 \times 10^4 \text{ mm}^4$

Area of section $= 117.89 + 2 \times 1.8 \times 100 \text{ mm}^2$

$$= 369.89 \times 100 \text{ mm}^2$$

$$r_{yy} = \left(\frac{105945 \times 10^4}{369.9 \times 100}\right)^{\frac{1}{2}} = 169.2 \text{ mm}$$

$$r_{xx} = 185 \text{ mm (for I-section alone)}$$

2. Slenderness ratio of column section

$$\left(\frac{1}{r_{min}}\right) = \left(\frac{3.25 \times 1000}{169.2}\right) = 19.21$$

3. Safe load

From IS : 800–1984, allowable working stress in compression for steel having yield stress as 260 N/mm^2

$$\sigma_{ac} = 154.158 \text{ N/mm}^2$$

Safe load carrying capacity

$$\left(\frac{154.158 \times 36989}{1000}\right) = \mathbf{5702.15 \text{ kN}}$$

Hence, the design is satisfactory.

Provide HB 450, @ 0.925 kN/m with two plates 700 mm × 18 mm. One plate is connected with each flange of I-section.

3.15 LACED AND BATTENED COLUMNS

In built-up columns when rolled steel sections are not connected by plates (viz. load sharing elements) suitable lateral system is needed to connect different load carrying elements of column. The lateral system holds the load carrying elements of the built-up column in their relative positions and it does not share the load. The object of providing lateral system is to carry the transverse shear force which occurs when the column deflects. The following lateral systems are used:

1. Lacing. 2. Batten plates. 3. Perforated plates.

1. ***Lacing.*** The lacing is also termed as *'latticing'* and it is most commonly used. The rolled steel flats, angles and channels are used for lacing. The rolled steel sections or tubes of equivalent length may also be used instead of flats. The lacing is of two types :

1. Single lacing 2. Double lacing.

1. *Single lacing.* In single lacing as shown in Fig. 3.17 (a), the lacing flats are placed to form a single-triangular system. The single lacing on opposite sides of the main components shall preferably be in the same direction so that one may be shadow of the other, instead of being mutually opposed in direction.

2. *Double lacing.* In double lacing as shown in Fig. 3.14 (b), the lacing flats are used to intersect one another and connected at the points of intersections by rivets.

The compression members comprised of two main components laced and tied should, where practicable, have a radius of gyration about the axis perpendicular to the plane of the lacing not less than the radius of gyration at right angles to that axis (i.e., r_{yy}, should not be less than r_{xx}, see Fig. 3.17).

As far as it is practicable, the lacing system should not be varied throughout the length of strut. The tie plates are attached at the top and bottom of a laced column.

The single laced system and double laced system on opposite sides of the main components should not be combined with cross-members (except the plates) perpendicular to the longitudinal axis of the strut unless all forces resulting from deformation of the strut are calculated and accounted for in the lacing and its fastening.

When the components of built-up column are connected by a lateral system, the effect of shearing forces to the deflection of the column is more. The reduction in buckling strength due to the shear deflection is more than that of solid built-up columns. The buckling strength of laced or battened column is given by

$$P = \left(\frac{\pi^2 E_t \cdot I}{(KL)^2}\right)$$

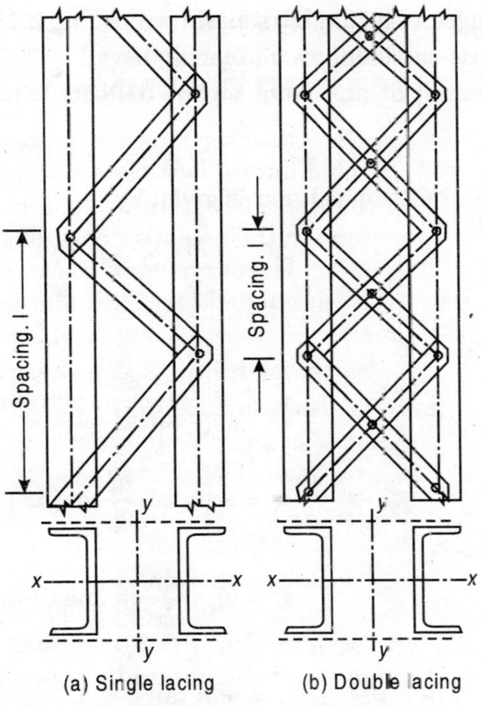

(a) Single lacing (b) Double lacing

Fig. 3.17 *Lacing of compression members*

where E_t = Tangent modulus

I = Moment of inertia of the column section, $(I_0 + 2I_1)$

I_1 = Moment of inertia of each component of column about their own axes

$I_0 = \frac{1}{2}Ah^2$

A = Cross-sectional area of each component of column

h = Distance between yy-axis of components of column

For single lacing the factor K is given by

$$K = \left(1 + \frac{\pi^2 E_t I_0}{L^2} \cdot \frac{d^2}{2E_c \cdot h^2 \cdot A_d}\right)^{\frac{1}{2}} \quad ...(3.30)$$

and *for double lacing*

$$K = \left(1 + \frac{\pi^2 E_t I_0}{L^2} \cdot \frac{d^2}{2E_c \cdot h^2 \cdot A_d}\right)^{\frac{1}{2}} \quad ...(3.31)$$

where c = Panel length, as shown in Fig. 3.18

n = Number of panels

d = Length of lacing bars as shown in Fig. 3.18

A_d = Cross-sectional area of lacing bars.

The bending moment at any point on the deflected shape of the column at buckling

$$M_a = -P_c.y$$

The shear force in built-up column is given by

$$V = \frac{dM_2}{dx} = -P_c \cdot \frac{dy}{dx}$$

From Sec. 3.7,

$$y = e\left[\frac{1-\cos nl}{\sin nl} \cdot \sin(nx) + \cos(nx)\right]$$

$$\frac{dy}{dx} = e\left[n \tan\frac{nl}{2}\cos(nx) - n\sin(nx)\right]$$

At $x = 0$, $\left(\frac{dy}{dx}\right)$ is maximum.

$$\therefore \quad \left(\frac{dy}{dx}\right)_{max} = e.n \tan\frac{nl}{2}$$

and $$n = \left(\frac{P_c}{EI}\right)^{\frac{1}{2}}$$

Therefore,

$$V = -P_c \cdot e\left(\frac{P_c}{EI}\right)^{\frac{1}{2}} \cdot \tan\frac{1}{2}\left(\frac{P_c}{EI}\right)^{\frac{1}{2}}$$

$$= -P_c \cdot \frac{e}{r}\left(\frac{P_c}{EI}\right)^{\frac{1}{2}} \cdot \tan\frac{1}{2r}\left(\frac{P_c}{EI}\right)^{\frac{1}{2}} \qquad \text{... (3.32)}$$

This expression for shear force is lengthy. IS : 800–1984 recommended that lacing bar resists a transverse shear S equal to 2.5 per cent of the axial load in the member.

For battened column, the factor K is given by

$$K = \left[1+\frac{\pi^2 I_0}{24 l_1}\cdot\left(\frac{c}{L}\right)^2 + \frac{\pi E_t I_0}{L^2}\cdot\frac{c \cdot h}{12 E I_b}\right]^{\frac{1}{2}}$$

where, I_b = Moment of inertia of batten plate about the centroidal axis of the cross-section of column perpendicular to plane of buckling.

For simplification (E_t/E) is considered as unity.

Therefore,

$$K = \left[1+\frac{\pi^2 I_0}{24 l_1}\cdot\left(\frac{c}{L}\right)^2 + \frac{\pi^2}{12}\frac{I_0}{I_b}\frac{c \cdot h}{L^2}\right]^{\frac{1}{2}}$$

The second term of the square root depends on flexibility of the component moments of built-up column, and last term depends upon flexural rigidity of batten plates. The last term is small as compared with the sum of first two terms for a properly designed column, and hence neglected. The expression for K becomes as under :

$$K = \left[1+\frac{\pi^2 I_0}{24 l_1}\cdot\left(\frac{c}{L}\right)^2\right]^{\frac{1}{2}} \qquad \text{... (3.33)}$$

The design of lacing or batten plates of built-up column is done in such a way that the lacing or batten plates do not fail before the load carrying capacity of whole column is reached. A laced or battened column is considered to collapse if any element of the column begins to yield locally prior to the buckling load P_{cr} is reached. Therefore, IS : 800–1984 specified rules for design of these elements of built-up column to avoid premature failure of the column.

When perforated plates are used as lateral system then, the factor K for the column is given by

$$K = \left[1+\frac{\pi^2}{12\left(\frac{L}{r}\right)^2}\cdot\left(\frac{a}{r^2}\right)^{\frac{1}{2}}\right] \qquad ...(3.34)$$

where, a = Spacing of perforations
r = Radius of gyration of over-all section of the column
L = Length of the column between supports
r_x = Radius of gyration of main segment about its centroidal axis.

Design of lacing. Following are the usual steps required for the design of lacing as per IS : 800–1984.

Step 1. For lacing the components of built-up column, adopt single or double lacing. The angle of inclination of lacing bars with the longitudinal axis of the component member shall not be less than 40° and not more than 70° to the axis of the member.

Step 2. The spacing of lacing (l) as shown in Fig. 3.17 is computed. The maximum spacing should be such that the following condition is fulfilled :

(*a*) **Compression members composed of two I-sections.** When the compression members are composed of two I-sections, these shall be connected together by riveting so that.

Minimum slenderness ratio $\lambda = (l/r)$ of component of column between consecutive connections $\not>$ 0.7 times most unfavourable slenderness ratio of the ratio of the column as a whole.

or $\not>$ 50, whichever is less

where, 'l' is the distance between the centres of connection of the lattice bars to each component.

(b) **Compression members composed of two channels (back to back).** When the compression members are composed of two angles, two channels or two tees, back to back in contact or separated by a small distance, these shall be connected together by riveting so that

Slenderness ratio of component of column $\not>$ 0.6 times most unfavourable slenderness ratio of the column as a whole

$\not>$ 40

Step 3. The lacing bars resist a total transverse shear 'V' at any point in the length of the member where V equal to 2.5 per cent of the axial load in the member.

The transverse shear V is equally distributed in all transverse systems of parallel planes. Any other shear due to end moment or transverse loading is respectively accounted for.

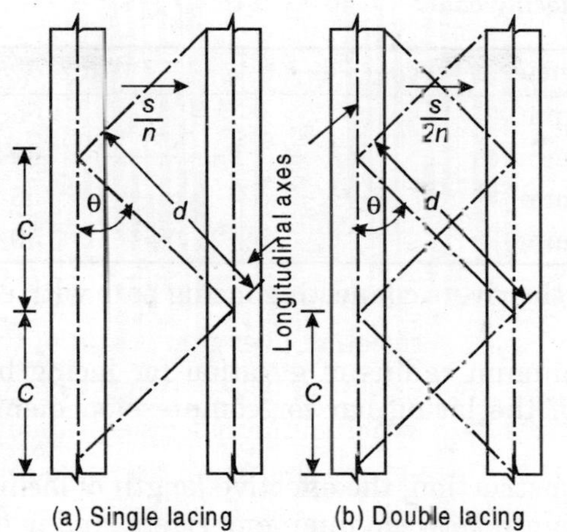

Fig. 3.18 *Transverse shear force in lacing*

Step 4. The force in the lacing bar is computed as given below : In single lacing as shown in Fig. 3.18 (a) transverse shear in each lacing is $\frac{V}{n}$

$$\text{Force in each lacing, } F = \left(\frac{V}{n} \times \frac{d}{n}\right)$$

or
$$F = \left(\frac{V}{n}\right) . \operatorname{cosec} \theta$$

where, θ = Inclination of lacing with the longitudinal axis of member

n = Number of transverse systems in parallel planes

h = Horizontal distance between rivet lines.

In the double lacing as shown in Fig. 3.18 (b) the transverse shear is to be resisted by two lacing bars.

$$\text{Force in each lacing} = \frac{1}{2} \cdot \left(\frac{V}{n} \cdot \operatorname{coses} \theta\right)$$

Step 5. The section for all lacing elements is adopted as below :

Let t be the thickness of flat lacing :

(i) In the single lacing, $t \nless \frac{1}{40}$th (one-fortieth) length between inner end rivets.

(ii) In the double lacing $t \nless \frac{1}{60}$th (one-sixtieth) length between inner end rivets.

IS : 800–1984 specifies the *width of lacing* flats as given in Table 3.4.

Table 3.4 *Width of lacing bars*

Nominal diameter of rivets	*Width of lacing bars*
22 mm	65 mm
20 mm	60 mm
18 mm	55 mm
16 mm	50 mm

The diameter of the rivets connecting lacing bars with components of column is assumed.

Step 6. The minimum radius of gyration for lacing bar is computed. The slendemess ratio of the lacing bar for compression member shall not exceed 145.

In the riveted construction, the effective length of lacing bar in single lacing is adopted as the length between inner end rivets, and in double lacing as 0.7 of this length. In the welded construction, this effective length is taken as 0.7 times the distance between the inner ends of welds connecting the lacing bars to the members.

Step 7. The rivet value for the rivet adopted is computed and should be greater than load coming to the rivet (twice the transverse shear in each lacing.)

Step 8. The lacing bar designed is checked for compressive strength and tensile strength. These should be more than the respective forces in each lacing bar.

The end tie plates are provided at the ends of lacing system in the laced compression members. These tie plates are also provided at the points where the lacing system is interrupted.

The lacing bars are attached to the main members by rivets or welding. The riveting or welding of lacing bars to the main members shall be sufficient to transmit the load in the bars. Where the welded lacing bars overlap the main members, the amount of lap measured along either edge of the lacing bar shall be not less than four times the thickness of the bar or the members, whichever is less. The welding should be sufficient to transmit the load in the bar and shall, in any case, be provided along each side of the bar for the full length of lap.

Where the lacing bars are fitted between the main members, these shall be connected to each member by fillet welds on each side of the bar or by full penetration butt welds. The lacing bars shall be so placed as to be generally opposite the flange or stiffening elements of the main members.

Example 3.12. *A steel column 12 m long carries an axial load of 1000 kN. The column is hinged at both the ends. Design an economical built-up section with double lacing. Design the lacing also.*

Solution

Design :

Step 1 : Design of built-up column

1. Selection of trial section

The length of built-up column, L is $(12 \times 1000) = 12000$ mm

Effective length of built-up column

$$l = L = (12 \times 1000) = 12000 \text{ mm} \quad \text{...(ii)}$$

Axial load carried by the column

$$P = 1000 \text{ kN} \quad \text{...(iii)}$$

The slenderness ratio for the laced built-up column and the value of yield stress for steel may be assumed as 80 and 260 N/mm^2, respectively.

Allowable stress in axial compression for the column

$$\sigma_{ac} = 103 \text{ N/mm}^2 \quad \text{...(iv)}$$

Effective sectional area required

$$A = \left(\frac{1000 \times 1000}{103}\right) = 9708.74 \quad \text{...(v)}$$

2. Properties of trial section

From ISI Handbook No. 1 double channels laced used as columns, try 2 MC 350, @ 0.421 kN/m

$$A = 2 \times 5366 = 10732 \text{ mm}^2 \quad \text{...(vi)}$$

$$I_{xx} = 20016 \times 10^4 \text{ mm}^4 \quad \text{...(vii)}$$

$$I_{xx} = 136.6 \text{ mm} \quad \text{...(viii)}$$

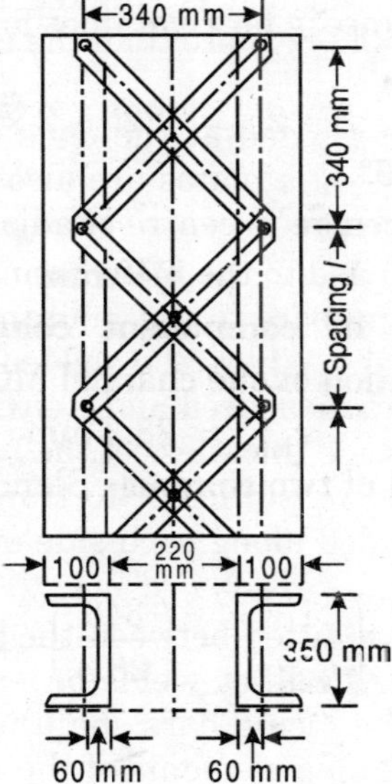

Fig. 3.16 *Laced column*

3. Spacing of channels

For the built-up column, the spacing between back to back of the channels should be such that $I_{yy} \simeq I_{xx}$.

From steel section tables adopt spacing of the channels

$$S = 200 \text{ mm} \quad \text{...(ix)}$$

$$I_{yy} = 20246.8 \times 10^4 \text{ mm}^4 \quad \text{...(x)}$$
$$r_{yy} = 137.4 \text{ mm} \quad \text{...(xi)}$$

4. **Maximum slenderness ratio**

$$\left(\frac{1}{r_{min}}\right) = \left(\frac{12000}{136.6}\right) = 87.848 \quad \text{...(xii)}$$

5. **Safe load**

From IS : 800–1984, maximum permissible stress in axial compression for the steel having yield stress as 260 N/mm^2

$$\sigma_{ac} = 94.367 \text{ N/mm} \quad \text{...(xiii)}$$

Safe load carrying capacity of the section

$$P = \left(\frac{94.367 \times 10732}{1000}\right) = 1012.75 \text{ kN}$$

>1000 kN. Hence, the design is satisfactory.

Step 2 : Design of lacing

1. **Spacing of lacing bar**

The double lacing is provided at inclination of 45° with the longitudinal axis of the column member as shown in Fig. 3.19. The spacing of channels S is equal to 220 mm. The gauge distance, g for rivet lines for MC 350, @ 0.421 kN/m is 60 mm.

$$\therefore \quad S + 2g = (220 + 2 \times 60) = 340 \text{ mm} \quad \text{...(xiv)}$$

Inclination of lacing is 45°.

Spacing of the lacing is centre to centre of adjacent rivets

$$= 340 \text{ mm} \quad \text{...(xv)}$$

2. **Slenderness ratio of component column**

Minimum radius of gyration of one channel MC 350, @ 0.421 kN/m

$$r_{min} = 28.3 \text{ mm} \quad \text{...(xvi)}$$

The column is composed of two channels Slenderness ratio of component of the column

$$\left(\frac{1}{r_{min}}\right) = \left(\frac{340}{28.3}\right) = 12.014$$

< 40 and $< 0.6 \left(\frac{1}{r_{min}}\right)$ of column as a whole (i.e., 0.6×87.8)

22 mm nominal diameter power driven rivets are used for the connection of lacing. Width of lacing flat

$$= 65 \text{ mm} \quad \text{...(xvii)}$$

Length of lacing $= 340\sqrt{2}$ mm ...(xviii)

Thickness of lacing $= \frac{1}{60} \times 340\sqrt{2} = 8.1$ mm ... (xix)

Adopt lacing flat 65 F 10.

Sectional area of flat

$$65 \times 10 = 65 \text{ mm}^2 \quad \text{...(xx)}$$

Minimum radius of gyration

$$\left[\frac{1}{12}\frac{bt^3}{b \cdot t}\right]^{\frac{1}{2}} = \frac{10}{(12)^{\frac{1}{2}}} = 2.88 \text{ mm} \quad \text{...(xvi)}$$

Step 3 : Check

1. Force in lacing bars

Transverse shear to be resisted is 2.5 per cent of axial load

$$V = \left(\frac{2.5 \times 1000}{100}\right) = 25 \text{ kN} \quad \text{...(xxii)}$$

Force in each lacing of double lacing

$$\left(\frac{1}{2} \times \frac{V}{n} \times \text{coses}\,\theta\right) = \left(\frac{1}{2} \times \frac{2.5}{2} \times \sqrt{2}\right) = 8.84 \text{ kN} \quad \text{...(xxiii)}$$

2. Compressive strength of lacing bars

The effective length of lacing bar in double lacing is 0.7 of its length

$$= 0.7 \times 340\sqrt{2} \text{ mm} \quad \text{...(xxiv)}$$

Slenderness ratio for the lacing bar

$$\left(\frac{0.7 \times 340\sqrt{2}}{2.88}\right) = 117 < 145$$

Hence the design is satisfactory.

Form IS : 800–1984, allowable stress in axial compression for the steel having yield stress as 260 N/mm²

$$\sigma_{ac} = 57.7 \text{ N/mm}^2$$

Compressive strength of the lacing bar

$$\left(\frac{57.7 \times 65 \times 10}{1000}\right) = 37.51 \text{ kN} > \text{Force in each flat} \quad \text{...(xxv)}$$

3. Tensile strength of the lacing flat

$$F_t = \left(\frac{(b-d) \times t \times 0.6 f_y}{1000}\right) \text{kN}$$

$$F_t = \left(\frac{(65 - 23.5) \times 10 \times 0.6 \times 260}{1000}\right)$$

$$= 64.74 \text{ kN}$$

$$> \text{Force in the lacing flat.}$$

4. Rivet value

Strength of power driven rivet in single shear

$$\left(\frac{\pi}{4}\times(23.5)^2\times\frac{100}{1000}\right) = 43.35 \text{ kN}$$

The thickness of flange of channel is 135 mm. It is less than that of flat. Therefore, strength of power driven shop rivet in bearing

$$\left(\frac{23.5\times13.5\times300}{1000}\right) = 95.175 \text{ kN}$$

Rivet value, $R = 43.35$ kN

5. Load on the rivet form both the sides

$(2 \times 8.84 \times \cos 45°) = 12.37 \text{ kN} <$ (Rivet value).

Hence, the design is satisfactory.

Provide 2 MC 350, @ 0.421 kN/m with lacing flat 65 F 10 and 22 mm diameter power driven rivets. The tie plates are provided at the ends of lacing system.

Example 3.13 *A building column is made of four ISA 100 mm × 100 mm × 12 mm (ISA 100 100, @ 0.177 kN/m) angles with their backs 350 mm apart as shown in Fig. 3.17. The lacing of column consists of 60 mm × 10 mm flat bars arranged in a single laced system and inclined to the axis of the column at an angle of 45°. The effective length of the column is 8 metres. Using the data given below and using any rational procedure such as recommended in Indian Standard Code of Practice IS : 800, find the safe load on the column. Also check the lacing system for safety against (a) local buckling of column angles, (b) strength of lacing bare in tension and compression against shear load, (c) strength of lacing rivets. Assume the shear load on the column as 2.5 per cent of the axial load. Rivets used have 20 mm nominal diameter (hole diameter = 21.5 mm).*

Properties of ISA 100 mm × 100 mm × 12 mm (ISA 100 100, @ 0.177 kN/m); angle area = 2259 mm², $l_x = I_y = 270 \times 10^4$ mm⁴, minimum radius of gyration = 19.4 mm; distance of centroid from back of angle = 29.2 mm; distance of rivet hole from back of angle = 60 mm.

Adopt permissible stresse in axial compression as per IS : 800.

Solution

Step 1: Properties of ISA 100 mm × 100 mm × 12 mm (ISA 100 100,@ 0.177 kN/m)

$$\text{Area} = 2259 \text{ mm}^2$$

$$I_{xx} = I_{yy} = 207 \times 10^4 \text{ mm}^4$$

$$C_{xx} = C_{yy} = 29.2 \text{ m}, r_{min} = 19.4 \text{ mm}$$

Step 2: Slenderness ratio. Moment of inertia of the whole column section about xx-axis

$$I_{xx} = [(4 \times 207) + 4 \times 22.59\,(17.5 - 2.92)^2] \times 10^4 \text{ mm}^4$$

$$= 20038 \times 10^4 \text{ mm}^4$$

Moment of inertia of the whole column section about yy-axis

$$I_{yy} = 20038 \times 10^4 \text{ mm}^4$$

Cross-sectional area of the whole column section

$$A = (4 \times 22.59 \times 100) = 9036 \text{ mm}^2$$

Radius of gyration of the whole column section

$$r_{xx} = r_{yy} = \left(\frac{20038 \times 10^4}{9036}\right) = 148.9 \text{ mm}$$

The effective length of the column, l is 8 m

Maximum slenderness ratio of the column

$$\left(\frac{1}{r_{min}}\right) = \left(\frac{8 \times 1000}{148.9}\right) = 53.8$$

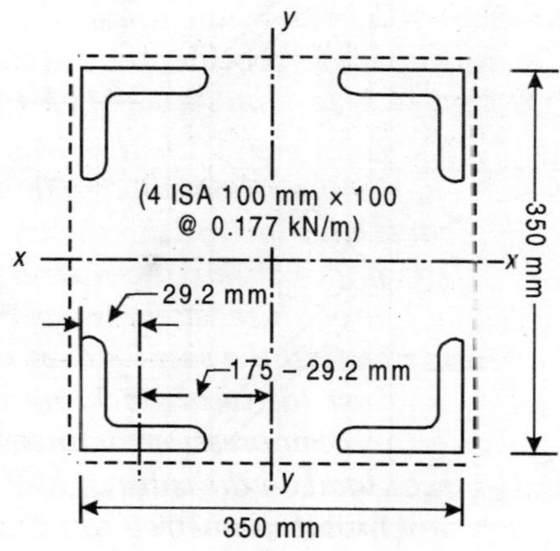

Fig. 3.20

Step 3 : Safe load

From IS : 800–1984, allowable stress in axial compression for the steel having yield stress as 260 N/mm^2.

$$\sigma_{ac} = 132.2 \text{ N/mm}^2$$

Safe load on the column

$$P = \sigma_{ac} \times A = \frac{132.2 \times 9036}{1000} = \mathbf{1194.56 \text{ kN}}$$

Angles have been laced in single laced system as shown in Fig. 3.21. The size of lacing flats is 60 mm × 10 mm.

Step 4 : Check for local buckling of column angles

Spacing of lacing = 460 mm

For angles r_{min} = 19.4 mm

The compression member is composed of angles

For the component of the column

$$\left(\frac{l}{r_{\min}}\right) = \left(\frac{460}{19.4}\right) = 23.7 \not> 40$$

and $\not>$ (0.6 × 53.8)

$$= 32.28 \text{ for column as a whole}$$

Hence, the local buckling of the angles does not occur as per IS : 800–1984.

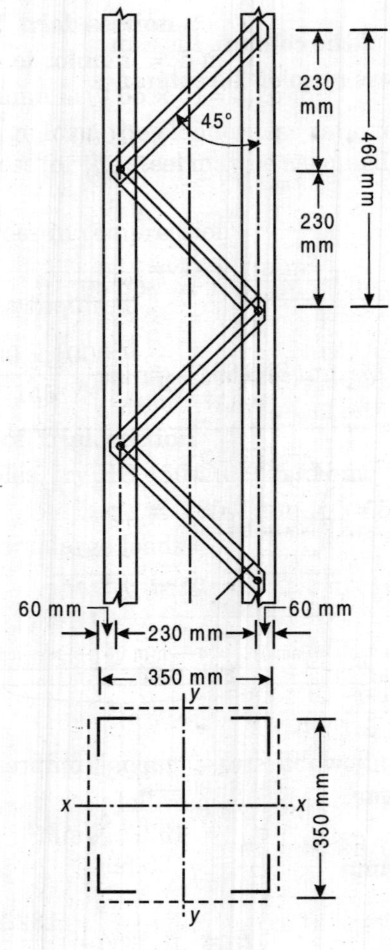

Fig. 3.21

Step 5 : Check for local buckling of lacing flats

Width of lacing flat is 60 mm

Thickness of lacing flat is 10 mm

$$r_{\min} = \left(\frac{t}{\sqrt{12}}\right) = \left(\frac{10}{\sqrt{12}}\right) = 2.89 \text{ mm}$$

Length of lacing flat = $230\sqrt{2}$ = 325 mm

$$\left(\frac{t}{r_{min}} = \frac{325}{2.89}\right) = 112.7 \not> 145$$

Hence satisfactory.

Strength of lacing flat in compression. Slenderness ratio of the lacing flat

$$\left(\frac{l}{r_{min}}\right) = 112.7$$

From IS : 800–1984, the allowable stress in axial compression, for the steel having yield stress as 260 N/mm^2

$$\sigma_{ac} = 70.57 \text{ N/mm}^2$$

Compressive strength of the lacing flat

$$P_1 = \frac{70.57 \times 60 \times 10}{1000} = \mathbf{42.34\ kN}$$

Step 6 : Strength of lacing flat in tension

Rivet diameter = 20 mm

Gross diameter = 21.5 mm

Net area of the lacing flat = (60 – 21.5) × 10 = 38.5 mm^2

Tensile strength of the lacing flat

$$P_2 = \left(\frac{0.6 \times 260 \times 385}{1000}\right) = \mathbf{60.06\ kN}$$

Strength of rivet in single shear

$$\left(\frac{\pi}{4} \times (21.5)^2 \times \frac{100}{1000}\right) = 36.287 \text{ kN}$$

Strength of rivet in bearing

$$\left(\frac{21.5 \times 10 \times 300}{1000}\right) = 64.5 \text{ kN}$$

Rivet value, R = 36.287 kN

Transverse shear resisted in two parallel planes of lacing

$$\left(\frac{2.5 \times 1194.56}{100}\right) = 29.86 \text{ kN}$$

Transverse shear resisted in one plane of lacing

$$\frac{1}{2} \times 29.86 = 14.932 \text{ kN}$$

Force in the, rivet due to transverse shear is 29.86 kN

< Rivet value. Hence, the design is safe.

Force in the lacing flat

$$F_1 = \left(\frac{1}{2} \times 29.86 \times \text{cosec } 45°\right) = 21.117 \text{ kN}$$

It is less than compressive strength and tensile strength of the flat.

Example 3.14. *Design a square column consisting of four angles to support an axial load of 1600 kN. The effective length of column is 10 m.*

Solution

Design :

Step 1 : Design of built-up column

1. Selection of trial section

The effective length of built-up column is 10 m (i.e., 1000 mm). The axial load supported by the column, P is 1600 kN. The square column consists of four angles as shown in Fig. 3.22. The approximate radii of gyration for the built-column from Fig. 3.8.

$$r_x = r_y = r = 0.42\, h$$

2. Slenderness ratio

Assuming the slenderness ratio for the built-up column as 40

$$\left(\because \frac{L}{r} = 40\right)$$

$$\therefore \quad 0.42\, h = \left(\frac{10000}{40}\right), h = 595.24 \text{ mm}$$

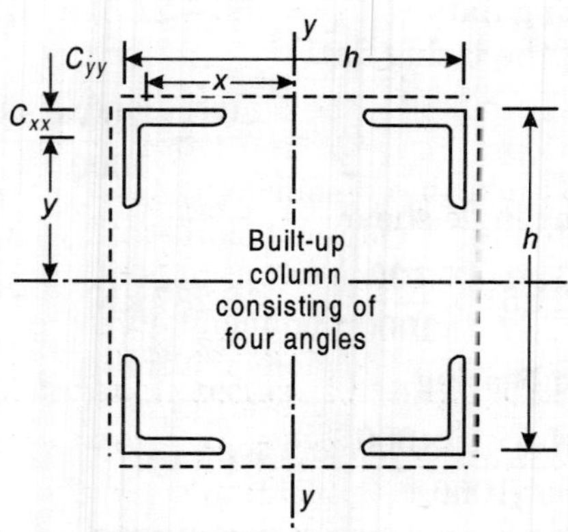

Fig. 3.22

Adopt the size of square column as 600 mm × 600 mm. The slenderness ratio of the built-up column with

$$r_x = r_y = r_{\min} = 0.42 \times 600 = 252 \text{ mm}$$

$$\frac{L}{r} = \left(\frac{10000}{252}\right) = 39.68$$

The permissible stress in axial compression from IS : 800–1984 for the steel having yield stress as 260 N/mm²

$$\sigma_{ac} = 145.192 \text{ N/mm}^2$$

The area required for the built-up column

$$A = \frac{1600 \times 1000}{145.192} = 11019.89 \text{ mm}^2$$

3. Properties of trial section

From ISI Handbook No. 1, try

4 ISA 150 mm × 150 mm × 12 mm (4 ISA 150 150, @ 0.272 kN/m). The cross-sectional area of four angles is 13836 mm^2. The moment of inertia about the reference axes

$$I_{xx} = I_{yy} = 4(735.4 + 34.59 \times 25.2) \times 10^4 \text{ mm}^4$$

$$= 95468.41 \times 10^4 \text{ mm}^4$$

The actual radii of gyration of built-up column

$$r_{xx} = r_{yy} = \left(\frac{95648.41 \times 10^4}{13836}\right)^{\frac{1}{2}} = 262.67 \text{ mm}$$

$$> 252 \text{ mm (approx. value)}$$

Hence, it is satisfactory. Provide 4 ISA 150 mm × 150 mm × 12 mm.

Step 2 : Design of lacing

The double lacing may be provided at an inclination of 45° with the longitudinal axis of the built-up column. It may be designed as it has been done in Example 3.12.

3.16 BATTEN PLATES

The batten plates are also called as the plates, and these are also used in lateral system. The angle sections, channels and I-sections are also used as battens. The components of built-up column sharing the load are connected together by batten plates. The battening of columns shall not be done where the columns are subjected in the plane of battens to eccentric loading.

In the case the battened compression members are subjected, in the plane of battens, to eccentric loading, applied moments or lateral forces shall be designed according to the exact theory of elastic stability or empirically from the verification tests, so that, these have a load factor of not less than 1.7 in the actual structure. It is note that if the column section is subjected to eccentricity or other moments about *yy*-axis (axis perpendicular to the plane of battens), the batons arid the column section should be specially designed for such moments.

The design includes design of built (main) column and design of batten plates. The batten plates are designed as follows.

3.16.1 Design of Batten Plates

Following are the usual steps in design of batten plates. The specifications for the design of plates as per IS : 800–1984 have also been given. The batten plates used for connecting the component members of a built-up column have been shown in Fig. 3.23. The batten shall be placed opposite to each other at each end of the member and at points where the member is stayed in its length and shall, as far as practicable be spaced and proportioned uniformly throughout.

The number of battens shall be such that the member is divided into not less than three bays within its actual length from centre to centre of connection.

Step 1. A battened column is designed with its effective length ten per cent in excess of the usual standard cases.

Step 2. The compression members composed of two main components battened should preferably have their two main components of the same cross-section and symmetrically disposed about their *xx*-axis. The compression members composed of two main components battened, should, where practicable, have a radius of gyration about the axis perpendicular to the plane of batten not less than radius of gyration at right angles to that axis (i.e., $r_{yy} \nless r_{xx}$).

Step 3. The spacing of battens is centre to centre distance between adjacent battens, as shown in Fig. 3.23.

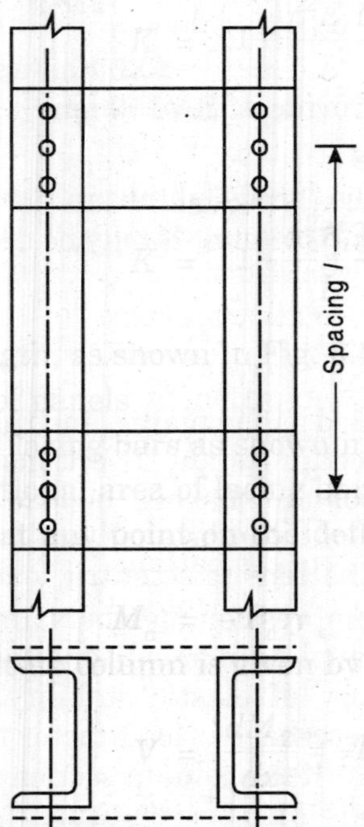

Fig. 3.23 *Battened column*

The spacing of battens is kept equal to length of lesser main component of built-up column. The lesser main component of built-up column is the portion of column between consecutive connections centre to centre of end fastenings. It is computed as below :

(a) **Compression member is compressed of two I-sections**

(i) In the battened compression members not specifically checked for shear stress and bending moment as specified above, the spacing of battens centre to centre of end fastenings shall be such that the slenderness ratio 'λ' of the lesser main component over that distance shall be not greater than 50 or greater than 0.7 times the slenderness ratio of the main member as a whole about its *xx*-axis (axis parallel to the battens).

(b) Compression member is composed of two channels (back to back)

When the compression members composed of two angles, channels, or tees back to back in contact or separated by a small distance shall be connected together by riveting, bolting or welding so that the ratio of slenderness of each member between the connections is not greater than 40 or greater than 0.6 times the most unfavourable ratio of slenderness of the strut as a whole whichever is less.

The minimum radius of gyration of the component member of column is known. The slenderness ratio of the lesser main component is equated to the minimum value as the case may be, length of the lesser main component is computed. The spacing of batten plates is taken equal to this length. The number of batten plates should be such that it divides the column longitudinally in at least three parts.

In no case shall the ends of the strut be connected together with less than two rivets or bolts or their equivalent in welding, and there shall be not less than two additional connections spaced equidistant in the length of strut. Where the members are separated back-to-back, the rivets or bolts through these connections shall pass through solid washers or packings, and where the legs of the connected angles or tables of the connected these are 125 mm wide or over, or where the webs of channels are 150 mm wide or over, not less than two rivets or bolts shall be used in each connection, one on line of each gauge mark.

The battens shall be designed to carry the bending moments and shears arising from the transverse shear force '*V*' of 2.5 per cent of the total axial force on the whole compression member, at any point in the length of the member divided equally between parallel planes of the battens. The main members shall also be checked for the same shear force and bending moments as for the battens.

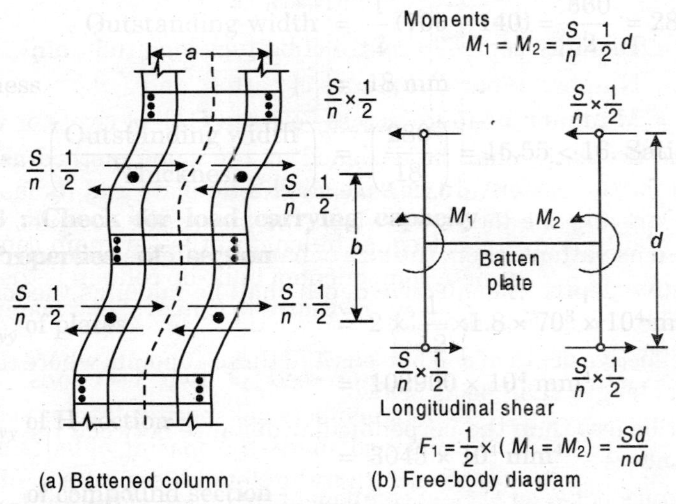

Fig. 3.24

Step 4. The batten plates are designed to resist simultaneously the moments and longitudinal force arising due to transverse shear force. The expression for which may be derived as under :

A battened column carrying an axial load may deform as shown in Fig. 3.24. The transverse shear, S at any point in the length of member is taken as 2.5 per cent of the total axial force on the whole compression member. It is divided equally between n parallel planes of the battens. It may be assumed that there are points of contraflexures, where the curvatures of the column change. Thus, a transverse shear force $\left(\frac{S}{n}\cdot\frac{1}{2}\right)$ shall be acting as shown in Fig. 3.24 (a). Let S be equal to V and n be equal to N.

Consider the free-body diagram of the portion of battened column as shown in Fig. 3.24 (b). The **bending moments** arising at each end of the batten plate due to transverse shear

$$M_1 = M_2 = \left(\frac{S}{n}\cdot\frac{1}{2}\right)\times d = \left(\frac{S\cdot d}{2n}\right)$$

$$\therefore \quad M_1 = M_2 = \left(\frac{V.C}{2n}\right) \qquad (C = d)$$

The **longitudinal shear** arising due to these moments

$$V_1 = \frac{1}{a}(M_1 + M_2) = \left(\frac{S.d}{na}\right)$$

$$\therefore \quad V_1 = \left(\frac{V\times C}{N\times S}\right) \qquad (a = S)$$

where C = Distance centre to centre of battens longitudinally

N = Number of parallel planes of batten, and

S = Minimum distance across between C.G. of rivets or welding

Step 5. The tie plates shall be designed by the same method as the other sections used for the battens. In no case shall a tie plate and its fastenings be incapable of carrying the forces for which the lacing are designed. When the plates are used as battens, then, the size of batten plates is computed as below :

1. Effective depth. The effective depth shall be taken as the longitudinal distance between end rivets or welds.

(a) The effective depth of end battens and those at points where the member is stayed in its length shall be as below :

It shall not be less than the perpendicular distance between the centroids of the main members.

($\nless$ the distance between C.G. of component members of column).

(b) The effective depth of intermediate battens shall be not less than three quarters of this distance.

(≮ $\frac{3}{4}$ ithe distance between C.G. of component members of column.)

But in no case the effective depth of any batten shall be not less than twice the width of one member in the plane of the battens.

(≮ twice the width of one component member of column.)

The total depth is found by adding edge distance on both the sides.

2. Thickness. Thickness of battens (t) should not be less than 1/50th (one-fiftieth) the distance between innermost connecting rivets.

(This clause does not apply for angles, channels, I-sections, used as batten their legs or flanges perpendicular to the main (member) and sufficient to resist longitudinal shear (F_1) and moment (M) arising from transverse shear. However, it should be ensured that the ends of the compression members are tied to achieve adequate rigidity.

Step 6 : The rivets are provided to connect batten plates with components of column to resist shear force (F_1) and moment (M) computed above.

The rivets, bolts or welds in these connections shall be sufficient to carry the shear force and moments, if any, specified for the battened struts. The diameter of rivets shall not be less than 16 mm for members upto and including 10 mm thickness ; 20 mm diameter for members upto and including 16 mm thick; and 22 mm diameter for members over 16 mm thick.

The compressive members connected by such riveting, bolting or welding shall not be subjected to transverse loading in plane perpendicular to the washer riveted, bolted or welded surfaces.

Example 3.15 *Design a built-up column consisting of two channels connected by battens to carry an axial load of 800 kN. The effective length of the column is 6 metres.*

Solution

Design :

Step 1 : Design of built-up column

1. Section of trial section

Effective length of the column is 6 m.

The effective length of battened column is increased by 10 per cent.

Therefore, the effective length of battened column shall be 6.6 mm.

The maximum slenderness ratio for the built-up battened column and the value of yield stress for the steel may be assumed at 80 and 260 N/mm^2, respectively. Permissible stress in axial compression.

$$\sigma_{ac} = 103 \text{ N/mm}^2$$

Effective sectional area required

$$A = \left(\frac{800 \times 1000}{103}\right) = 7766.99 \text{ mm}^2$$

From steel section tables (Double channels battened to be used as columns), try 2 LC 300 @ 0.331 kN/m

Effective sectional area of two channels

$$A = 8422 \text{ mm}^2$$

$$I_{xx} = 12095.8 \times 10^4 \text{ mm}^4, r_{xx} = 119.8 \text{ mm}$$

I_{yy} should be approximately equal to I_{xx} and $r_{yy}, \nless r_{xx}$.

From steel section tables, the spacing between back to back of channels *(S)* is adopted as 200 mm.

$$r_{yy} = 128.7 \text{ mm}$$

2 A. Slenderness ratio of the battened column

$$\therefore \quad r_{min} = 119.8 \text{ mm}$$

$$\frac{l}{r_{min}} = \left(\frac{6.6 \times 1000}{119.8}\right) = 55.2$$

From IS : 800–1984, allowable working stress in compression for the steel having yield stress as 260 N/mm²

$$\sigma_{ac} = 130.8 \text{ N/mm}^2$$

3 A. Safe load carrying capacity

$$\left(\frac{130.8 \times 8422}{1000}\right) = 1101.60 \text{ kN} > 800 \text{ kN}$$

It is not satisfactory.

Try 2 MC 230, @ 0.304 kN/m

Effective sectional area of two channels,

$$A = 7734 \text{ mm}^2$$

$$I_{xx} = 7633.6 \times 10^4 \text{ mm}^4$$

$$r_{xx} = 99.4 \text{ mm}, c_{yy} = 23.0 \text{ mm}$$

Adopt the spacing between back to back of channels as 160 mm

$$I_{yy} = 8643.2 \times 10^4 \text{ mm}^4$$

$$r_{yy} = 105.7 \text{ mm} \nless r_{xx}$$

$$\therefore \quad r_{min} = 99.4 \text{ mm}$$

2 B. Slenderness ratio of battened column

$$\frac{l}{r_{min}} = \left(\frac{6.6 \times 1000}{99.4}\right) = 66.4$$

From IS : 800–1984, allowable working stress in compression for the steel having yield stress as 260 N/mm²

$$\sigma_{ac} = 107.96 \text{ N/mm}^2$$

3 B. Safe load carrying capacity

$$\left(\frac{107.96 \times 7734}{1000}\right) = 834.96 \text{ kN}$$

Hence, the design is satisfactory.

Step 2 : Design of battens

1. Spacing of batten plates

The battened column is composed of two channels back to back separated by small distance.

The maximum slenderness ratios of column about *xx*-axis is as follows :

$$\left(\frac{l}{r_{\min}}\right) = 66.4$$

Minimum radius of gyration of one channel

$$r_{\min} = 23.8 \text{ mm}$$

The column is composed of two channels.

Slenderness ratio of component member of column (slenderness ratio of each member between the connections)

$$\left(\frac{l}{r_{\min}}\right) < 40$$

$$\left(\frac{l}{r_{\min}}\right) = (0.6\ \frac{l}{r_{\min}} \text{ of column as a whole } 0.6 \times 66.4 = 39.84)$$

$$\therefore \quad \left(\frac{l}{23.8}\right) = 39.84$$

$$\therefore \quad l = 39.84 \times 23.8 = 948.192 \text{ mm}$$

Adopt the spacing of battens as 900 mm

Transverse shear due to axial load

$$V = \left(\frac{2.5}{100}\times 800\right) = 20 \text{ kN}$$

Longitudinal shear arising from transverse shear

$$V = \frac{VC}{NS} = \left(\frac{20\times 900}{2\times 205}\right) = 36 \text{ kN}$$

Moment arising from transverse shear

$$M = \frac{VC}{2N} = \left(\frac{20.0\times 900}{2\times 2}\right) = 4500 \text{ mm-kN}$$

2. Size of batten plates

2 (A) Effective depth of end batten

= the distance between C.G. of component members

= (160 + 2 × 23.0) = 206 mm

$\not<$ Twice the width of one component member of the column

Assume 22 mm nominal diameter of rivets connection

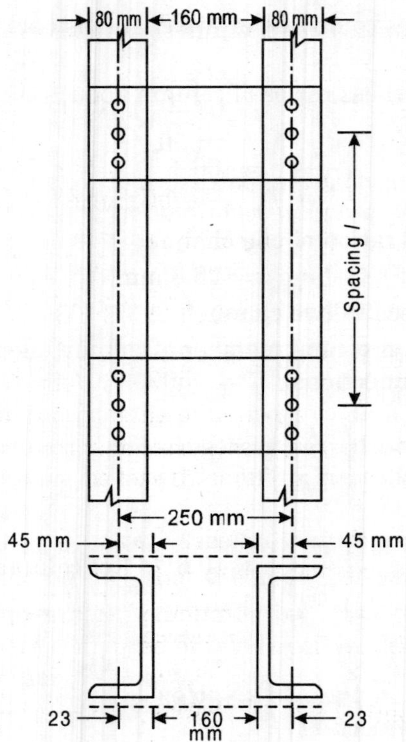

Fig. 3.22 *Battened column*

Overall depth of end batten

$$= 206 + \text{Edge distances on both sides}$$

$$= 206 + 2 \times (35) = 276 \text{ mm}$$

Effective depth of intermediate batten

$$= \frac{3}{4} \text{ distance between C.G. of component member}$$

$$= \left(\frac{3}{4} \times 206\right) = 154.5 \text{ mm}$$

< twice the width of one component member of the column (i.e., 2 × 80 = 160 mm).

2 (B) Effective depth of the intermediate battens

$$= (2 \times 80) = 160 \text{ mm}$$

Overall depth of intermediate battens

$$[1600 + 2 \times (35)] = 230 \text{ mm}$$

Thickness of batten plates

$$t \times \frac{1}{50}\text{th} \times \text{distance between innermost connecting line of rivets}$$

$$t = \left(\frac{1}{50}\times 250\right) = 5 \text{ mm}$$

Adopt 6 mm thickness of batten plates as a provision against corrosion. If the depth of intermediate batten plate required for resisting moment is d mm, then

$$\text{Moment of resistances} = \left(\frac{1}{6} t \times d^2 \times \sigma_{bc}\right) = M$$

$$\frac{1}{6}\times\left(\frac{6 \times d^2 \times 0.66 \times 260}{1000}\right) = 45000, \quad \therefore d = 161.94 \text{ mm}$$

Adopt the size of end batten plates as 300 mm × 6 mm and the size of intermediate plates as 230 mm × 6 mm. The actual depth of batten plate is more than that needed.

3. Check for longitudinal shear

Shear strength of intermediate plate

$$230 \times 6 \times \left(\frac{0.4\times 260}{1000}\right) = 143.52 \text{ kN}$$

> longitudinal shear in plate 36 kN. Hence the design is satisfactory.

4. Design of connection

Shear strength of power driven rivets

$$\left(\frac{\pi}{4}(23.5)^2 \times \frac{100}{1000}\right) = 43.35 \text{ kN}$$

Bearing strength of power driven rivets

$$\left(6\times 23.5\times\frac{300}{1000}\right) = 42.30 \text{ kN}$$

Rivet value, R = 42.30 kN

Adopt 3 rivets in a vertical row as shown in Fig. 3.25

Shear force in each rivet

$$\left(\frac{36}{3}\right) = 12 \text{ kN}$$

$$\text{Force due to moment} = \left(\frac{My}{\Sigma y^2}\right) = \left(\frac{4500\times 80}{2\times 80\times 80}\right) = 28.125 \text{ kN}$$

Resultant force in rivets

$$(12^2 + 28.125^2)^{1/2} = 30.578 \text{ kN} < \text{Rivet value.}$$

Hence the design is satisfactory.

Provide 2 MC 250, @ 0.304 kN/m with batten plates as shown in Fig. 3.25.

3.17 PERFORATED COVER PLATES

The *perforated cover plates* may be used instead of lacing, as a lateral system with the columns. The perforated cover plates are particularly suitable for a built-up box section consisting of four angle sections. The interior of the column remains accessible for painting. Bureau of Indian Standards has not given specifications for the design of perforated plates. Guide to design criteria for metal compression members published by the Column Research Council, U.S.A. has given the following specifications for the design of perforated cover plates.

(1) The perforation of plate should be of ovaloid shape (i.e., two semi-circles connected with straight sides) or they may be elliptical or circular. The long axis of the perforations for the first two cases should be in the direction of the axis.

(2) The spacing of the perforations (i.e., the centre to centre distance between the perforations) should not be less than 1.5 times the length of perforation, C.

(3) The clear distance between perforations should be not less than the distance between the nearest lines of longitudinal fasteners [i.e., $(a - c) \geq d$].

(4) The net section of the column (defined as the section at the perforations) should be used in computing the axial rigidity, EA, and the column moments of inertia about x-axis and y-axis.

The net area of perforated plate is used as the part of cross-sectional area of the column. It is the advantage of the perforated plates as compared with the lacing.

(5) If the slenderness ratio $\dfrac{c}{r_f}$ (length of perforation divided by radius of gyration of flange, see Fig. 3.26 is 20 or less, and also not greater than one-third of the slenderness ratio $\dfrac{l}{r_x}$, the appropriate specification for column stress, applied to the column net area can be used to find the allowable load.

(6) For columns built-up of plates, the net area of each web at the perforation should be sufficient to resist $\dfrac{1}{n}$ times the transverse shear force, where n is the number of perforated plates. Perforated plates designed in accordance with provision (3) need not be checked for shear introduced as a specified percentage of the column load.

(7) The transverse distance from the edge of a perforation to the nearest line of longitudinal fastener, divided by the plafe thickness, i.e., the b/t ratio of plate adjacent to a perforation should, conform to minimum specification requirements for plates in main compression members.

A member with staggered perforations should be treated as if the perforations were opposite to each other. The perforated cover plates should be detailed in such a way that the axial compressive stresses do not cause the failure by local buckling before the strength of column is reached.

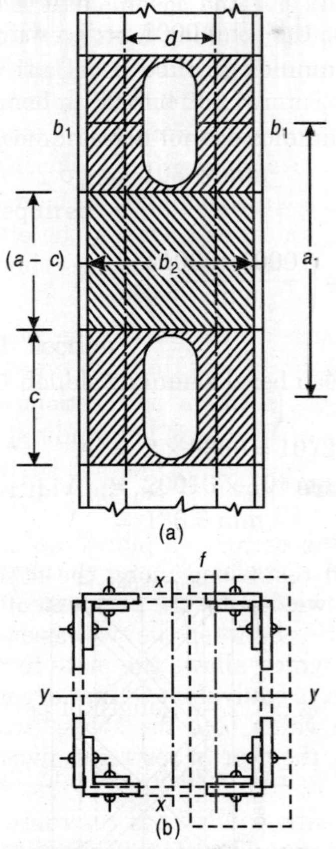

Fig. 3.26 *Perforated cover plates*

The perforated cover plates are designed by the criterion obtained by equating the plate buckling stress $(f_{cr})_{plane}$ and the column strength $(f_{cr})_{column}$. The plate buckling stress is given by

$$(f_{cr})_{plate} = kE \cdot \left(\frac{t}{b}\right)^2 \quad ...(3.35)$$

where, E = Effective modulus of elasticity corresponding to the critical stress

$\frac{t}{b}$ = Thickness to width ratio of plate

k = Coefficient.

The coefficient k depends upon the type of loading, the conditions of edge support and the proportion of plate.

The length of column depends upon the slenderness ratio, yield stress, f_y, and effective modulus of elasticity, E. Therefore, the proportion of perforated plate may be expressed as a function of $\frac{t}{r}$. E and f_y. The function is of complex form. The following expressions give the design criteria for all types of steel. These functions approximate to the complex function with sufficient accuracy.

For $\frac{l}{r} < 60$

$$\frac{b}{\left(t \cdot \sqrt{k}\right)} = 19.2 \qquad \text{...(3.36)}$$

For $\frac{l}{r} > 60$

$$\frac{b}{\left(t \cdot \sqrt{k}\right)} = 0.32 \left(\frac{l}{r}\right) \qquad \text{...(3.37)}$$

When the values of k are known, then, the width thickness ratio $\left(\frac{b}{t}\right)$ may be found. Two portions of the perforated plates are of interest. The portion b_1 is free along one edge and restrained along the other edge. The other edge of portion b_1 is riveted or welded to the segment of the column. The restraint action along the edge varies between the two cases. In one case, it is equivalent to a longitudinal hinge, which allows the plate to rotate about the axis of the edge. In the other case, it is equivalent to that of continuous clamp, which does not allow rotation. The value of k for above two cases are 0.38 and 0.15, respectively. In practice, the type of restraint gives the lower value out of two values.

The portion of plate with width b_2 is restraint along both the edges. The restraint of edges is also equivalent to two above mentioned cases. The values of k for which vary from 3.6 to 6.3. In practice, the type of restraint gives the lower value.

3.18 ECCENTRICALLY LOADED COLUMNS

When a load is applied at an eccentric distance from the centroid of the column section, as shown in Fig. 3.27, the column is called eccentrically loaded.

A column may carry an eccentric load in addition to axial load. The eccentrically loaded column is subjected to bending stress besides the axial compression. The strength of the column is very much reduced.

When the column is subjected to axial loading only, then, the cross-sectional area, A_a required is given by

$$A_a = \left(\frac{P}{\sigma_{ac}}\right) \qquad \text{...(i)}$$

where, σ_{ac} = Allowable working stress in compression on the member subjected to axial load only.

When the column is subjected to bending only then cross-sectional area required is given by

$$A_b = \left(\frac{P \cdot e \cdot y}{\sigma_{bc} \cdot r^2}\right)$$

Fig. 3.27 *Eccentrically loaded column*

where, σ_{bc} = Maximum allowable bending compressive stress on the extreme fibre

y = Distance to the extreme fibre of the section measured from the neutral axis

$A_b \cdot r^2$ = Moment of inertia of the section about the appropriate axis

r = Radius of gyration of the section about the appropriate axis

$P.e$ = Moment due to the eccentricity, e of the load P, and

$$\left(\because \frac{P \cdot e}{I} = \frac{\sigma_{bc \cdot cal}}{y}\right), \therefore \left(\sigma_{bc \cdot cal} = \frac{P \cdot e \cdot y}{I} = \frac{P \cdot e \cdot y}{A_b \cdot r^2}\right)$$

When the column is subjected to both axial load and bending, then, the total area required

$$A = (A_a + A_b) = \left(\frac{P}{\sigma_{ac}} + \frac{P \cdot e \cdot y}{\sigma_{bc} \cdot y^2}\right) \quad \text{...(iii)}$$

Dividing both the sides of expression (iii) by A and writing the expression as below

$$\left[\left(\frac{\frac{P}{A}}{\sigma_{ac}}\right) + \left(\frac{\frac{P \cdot e \cdot y}{Ar^2}}{\sigma_{bc}}\right)\right] = 1$$

$$\left(\frac{\sigma_{ac.cal}}{\sigma_{ac}} + \frac{\sigma_{ac.cal}}{\sigma_{bc}}\right) = 1 \qquad \text{... (3.38)}$$

where, $\sigma_{\text{ac.cal}} = \left(\dfrac{P}{A}\right)$ (calculated average axial compressive stress)

$\sigma_{\text{bc.cal}} = \left(\dfrac{P \cdot e \cdot y}{Ar^2}\right)$ (calculated bending stress in the extreme fibre)

Equation 3.38 gives the fundamental interaction formula for the compression member subjected to the axial load and bending. It is a straight line when plotted graphically, $\left(\dfrac{\sigma_{ac.cal}}{\sigma_{ac}}\right)$ along y-axis and $\left(\dfrac{\sigma_{ac.cal}}{\sigma_{bc}}\right)$ along x-axis. Bureau of Indian Standards adopted this interaction formula. The column subjected to both bending and axial compression shall be so proportioned that the quantity $\left[\left(\dfrac{\sigma_{ac.cal}}{\sigma_{ac}}\right) + \left(\dfrac{\sigma_{ac.cal}}{\sigma_{bc}}\right)\right]$ should not exceed only.

When the bending occurs about both the axes of the member, then the interaction formula may be written as below :

$$\left(\frac{\sigma_{ac.cal}}{\sigma_{ac}} + \frac{\sigma_{acx.cal}}{\sigma_{bcx}} + \frac{\sigma_{acy.cal}}{\sigma_{bcy}}\right) \leq 1.0 \qquad \text{...(iv)}$$

The value of σ_{cbx} and σ_{cby} to be used in the above formulae shall be lesser of the values of the maximum permissible stresses σ_{bc} for bending about the appropriate axis.

At a support and using the values σ_{bcx} and σ_{bcy} at the support.

$$\left(\frac{\sigma_{ac.cal}}{0.60 f_y} + \frac{\sigma_{acx.cal}}{\sigma_{bcx}} + \frac{\sigma_{acy.cal}}{\sigma_{bcy}}\right) \leq 1.0$$

IS and BS codes give the equivalent stress formula for combined bending, bearing and shear stress as under. It is based on von Mises–Hencky's theory

$$\sigma_{e.cal} = (\sigma^2_{bt \cdot cal} + 3\tau^2_{vm \cdot cal})^{1/2}$$

or

$$\sigma_{e.cal} = (\sigma^2_{bc \cdot cal} + 3\tau_{vm.cal})^{1/2}$$

Where a bearing stress is combined with tensile stress or compressive stress bending and shear stresses under the most unfavourable condition of loading the equivalent stress $\sigma_{e.cal}$ obtained from the following formulae, shall not exceed

$$\sigma_e = 0.9 f_y$$

$$\sigma_{e.cal} = [\sigma^2_{bt.cal} + \sigma_{bt.cal}\,\sigma_{p.cal} + 3\tau_{vm.cal}]^{1/2}$$

or

$$\sigma_{e.cal} = [\sigma^2_{bc.cal} + \sigma^2_{p.cal}\,\sigma_{be.cal}\,\sigma_{p.cal} + 3\tau_{vm.cal}]^{1/2}$$

where $\sigma_{ac.cal}$ = calculated average axial compressive stress
$\sigma_{bc.cal}$ = calculated bending compressive stress in the extreme fibre
$\sigma_{bt.cal}$ = calculated bending tensile stress in the extreme fibre
$\tau_{vm.cal}$ = calculated shear stress

$\sigma_{bc.cal}$; $\sigma_{bc.cal}$; $\tau_{vm.cal}$ and $\sigma_{p.cal}$ are the numerical values of the co-existing bending (compressive or tensile stress) shear and bearing stresses. When the bending occurs about both the axes of the member, $\sigma_{bt.cal}$ and $\sigma_{bc.cal}$ shall be taken as the sum of the two calculated fibre stresses. σ_e is the maximum permissible equivalent stress. ($\sigma_e = 0.9f_y$).

The compression flange of the column subjected to bending is laterally unrestrained. As such the allowable bending compressive stress is reduced. The column section is adopted depending upon the effective length of column. Allowable bending compressive stress for compression flange of a section unrestrained, has been discussed in Chapter 6, in design of laterally unrestrained beams.

For design of a column, the eccentricity of loading shall be taken as the distance from the assumed point of application of load to the centroid of the column. For the purpose of determining the stress in a stanchion or column section, the beam reactions or similar loads shall be assumed to be applied 100 mm from the face of the section or at the centre of the bearing whichever dimension gives the greater eccentricity except in the following two cases.

(i) In the case of cap connections, the load shall be assumed to be applied at the face of the column shaft or stanchion section ; or edge of packing if used, towards the span of the beam ; and

(ii) In the case of a roof truss beating on a cap, no eccentricity need be taken for single bearing without connections capable of developing an appreciable moment.

In continuous columns, the bending moment due to eccentricities of loading on the columns at any floor may be taken as

(i) ineffective at the floor levels above and below that floor and

(ii) divided equally between the column's length above and below that floor level, provided that the moment of inertia of either column section, divided by its effective length does not exceed 1.5 times the corresponding value of the other column. In case where this ratio is exceeded, the bending moment shall be divided in proportion to the moment of inertia of the column sections divided by their respective effective length.

The load was assumed to be applied as given in Table 3.5 as per IS : 800–1962. The connections are designed as simple connections to transmit shear only.

Table 3.5 *Assumed eccentricity of loads in columns*

Type of connections	*Assumed point of application of load*
(i) Stiffened bracket	Mid point of stiffened application of load
(ii) Unstiffened bracket	Outer face of vertical leg of bracket
(iii) Clats to the web	Face of strut

The type of connections have been shown in Figs. 3.25 (a), (b) and (c) respectively.

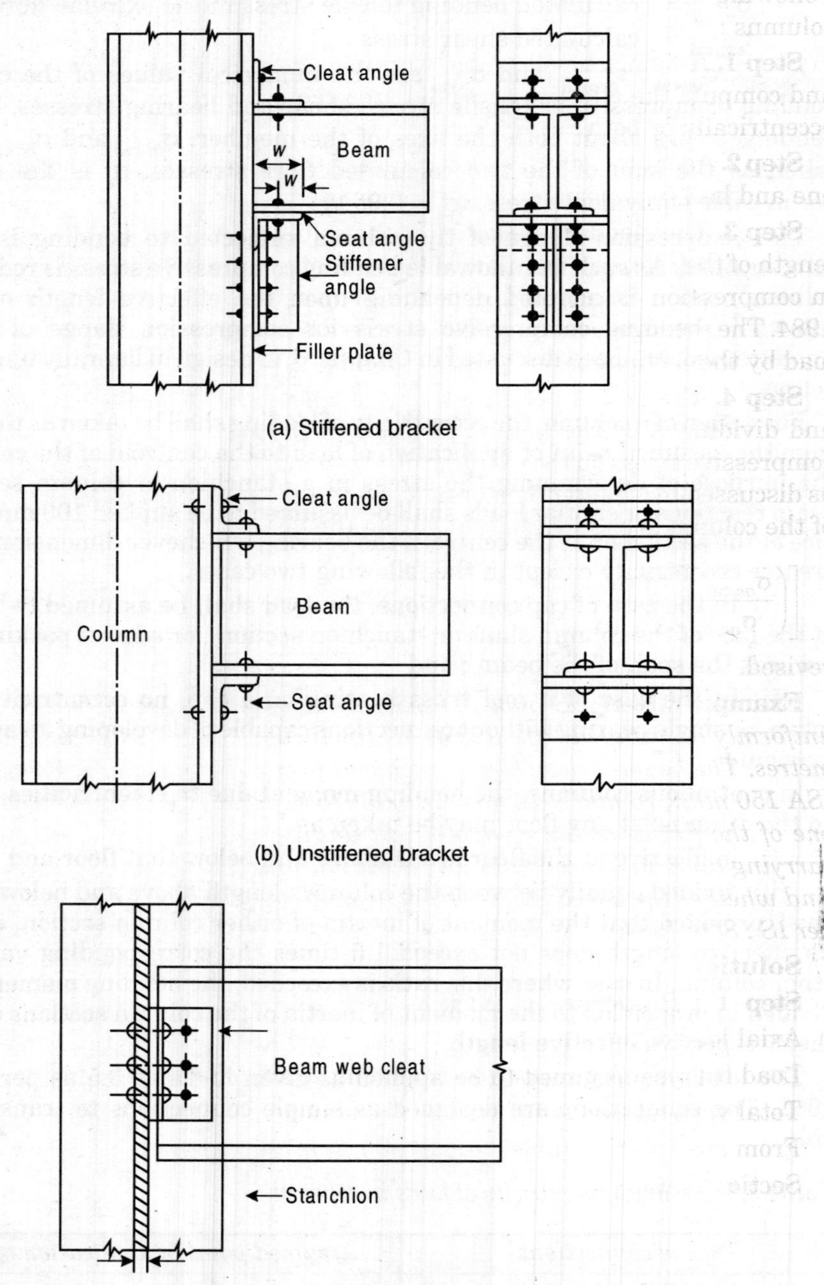

(a) Stiffened bracket

(b) Unstiffened bracket

(c) Cleats to wed of beam types of beam and column connections

Fig. 3.28

3.19 DESIGN OF ECCENTRICALLY LOADED COLUMNS

Following are the usual steps required for the design of eccentrically loaded columns :

Step 1. Assume the allowable working stress in compression for the column and compute the effective sectional area required for the load acting axially or eccentrically or both.

Step 2. Adopt the effective sectional area of eccentrically loaded column about one and half to two times the area computed above.

Step 3. The slenderness ratio for the column is computed from the effective length of the column and minimum radius of gyration. Allowable working stress in compression corresponding to the slenderness ratio is found from IS : 800–1984. The average axial compressive stress is calculated by dividing total vertical load by the sectional area of the column.

Step 4. The actual bending stress is calculated by computing the moment and dividing it by the section modulus of the section. The allowable bending compressive stress on the extreme fibre is noted from IS : 800–1984 or calculated as discussed in Chapter 6, in design of laterally unrestrained beams. The design of the column is checked by computing the quantity.

$\left[\left(\frac{\sigma_{ac\cdot cal}}{\sigma_{ac}}\right)\right]+\left(\frac{\sigma_{bc\cdot cal}}{\sigma_{bc}}\right)$ which should not exceed unity, otherwise design is revised.

Example 3.16 *A LB 500, @ 0.750 kN/m simply supported beam carries a uniformly distributed load of 280 kN (inclusive of self-weight) over a span of 4 metres. The beam is attached at each end to a stanchion of similar section by a ISA 150 mm× 75 mm × 12 mm (ISA 15075 @ 0.201 kN/m) angle iron riveted to one of the flange of the stanchion. In addition to the above, the stanchion is carrying an axial load of 200 kN. If the effective length of the stanchion is 4 m, find whether the stanchion is strong enough to carry the loads. Adopt stress as per IS : 800–1984.*

Solution

Step 1 : Properties of section

Axial load of the stanchion = 200 kN

Load from simply supported beam on the stanchion is 140 kN

Total vertical load supported by the stanchion is 340 kN

From steel section tables for LB 500, @ 0.750 kN/m

Sectional area $A = 9550\ \text{mm}^2$

$r_{xx} = 201.\ 0\ \text{mm}$

$r_{yy} = 33.\ 4\ \text{mm}$

$Z_{xx} = 1543\text{–}2 \times 10^3\ \text{mm}^3$

Step 2 : Slenderness ratio

Effective length of the stanchion is 4 m

Minimum radius of gyration, r_{min} = 33.4 mm

Slenderness ratio

$$\frac{l}{r_{min}} = \left(\frac{4 \times 1000}{33 \cdot 4}\right) = 119.8$$

Step 3: Direct compressive stresses

From IS : 800–1984 allowable working stress in compression for the steel having yield stress as 260 N/mm²

$$\sigma_{ac} = 64.18 \text{ N/mm}^2$$

Average axial compressive stress

$$\sigma_{ac.cal} = \left(\frac{340 \times 1000}{9550}\right) = 35.568 \text{ N/mm}^2$$

Step 4: Bending stresses

Assuming the beam connection with the flange of the stanchion as stiffened bracket connection, the point of application of load from the beam is at mid-point of stiffened seating.

Width of stiffened seating = 75 mm

Eccentricity of the load from the axis of column

$$\left(250 + \frac{75}{2}\right) = 287.5 \text{ mm}$$

Moment due to eccentric load

$$M = (287.5 \times 140) = 40250 \text{ mm-kN}$$

Actual bending stress

$$\sigma_{bc.cal} = \left(\frac{M}{Z_{xx}}\right) = \left(\frac{40250 \times 1000}{1543 \cdot 2 \times 10^3}\right) = 26.082 \text{ N/mm}^2$$

From IS : 800–1984 allowable bending compressive stress

$$\sigma_{bc} = 0.66 f_y = (0.66 \times 260) = 171.6 \text{ N/mm}^2$$

Step 5: Check for interaction expression

For the stanchion subject to the axial compressive stress and bending stress

$$\left(\frac{\sigma_{ac \cdot cal}}{\sigma_{ac}} + \frac{\sigma_{bc \cdot cal}}{\sigma_{bc}}\right) \leq 1$$

Substituting these values,

$$\left(\frac{35 \cdot 068}{64 \cdot 18} + \frac{26 \cdot 082}{171 \cdot 6}\right) = 0\ 706 < 1.$$

Hence, the stanchion is safe

Example 3.17 *A column effectively held in position but not in direction at their end is 4 metres long and carries an axial load of 700 kN and an end moment of 35000 mm-kN. Design the column if only rolled steel beam sections are available. Adopt stress as per IS : 800–1984.*

Solution

Design :

Step 1 : Selection of trial section

The slenderness ratio for the column and the value of yield stress for steel may be assumed as 80 and 260 N/mm^2, respectively. Permissible stress in axial compression.

$$\sigma_{bc} = 103 \text{ N/mm}^2$$

Effective sectional area required for axial load

$$A = \left(\frac{700 \times 1000}{103}\right) = 6796.116 \text{ mm}^2$$

Assume effective sectional area required for moment in addition to the axial load

$$1.5 \times 6796.116 = 10194.174 \text{ mm}^2$$

Step 2 (A) : Properties of trial section

From steel section tables, try MB 500, @ 0.869 kN/m

Effective sectional area

$$A = 11074 \text{ mm}^2$$

$$r_{xx} = 202.1 \text{ mm}$$

$$r_{yy} = 35.2 \text{ mm}$$

$$Z_{xx} = 18087 \times 10^3 \text{ mm}^3$$

Step 3 (A) : Slenderness ratio

Effective length of columns is 4 m

$$r_{\min} = 35.2 \text{ mm}$$

Maximum slenderness ratio

$$\left(\frac{l}{r_{\min}}\right) = \left(\frac{4 \times 1000}{35 \cdot 2}\right) = 113.5$$

Step 4 (A) : Check for interaction expression

Average axial stress in compression

$$\sigma_{ac.\text{cal}} = \left(\frac{700 \times 1000}{11074}\right) = 63.21 \text{ N/mm}$$

Allowable working stress in compression for

$$\left(\frac{l}{r_{\min}}\right) = 113.5$$

and the steel having yield stress as 260 N/mm^2

$$\sigma_{ac} = 69.85 \text{ N/mm}^2$$

End moment = 35000 kN-mm

Bending compressive stress

$$\sigma_{bc.\text{cal}} = \left(\frac{3500 \times 1000}{1808 \cdot 7 \times 10^3}\right) = 19.351 \text{ N/mm}^2$$

From IS : 800–1984 allowable bending compressive stress is $0.66\ f_y$

$$\sigma_{bc} = 171.6 \text{ N/mm}^2$$

$$\left(\frac{\sigma_{ac\cdot cal}}{\sigma_{ac}} + \frac{\sigma_{bc\cdot cal}}{\sigma_{bc}}\right) = \left(\frac{63\cdot21}{69\cdot85} + \frac{19\cdot351}{171\cdot6}\right)$$

$$= 1.0177 > 1 \text{ not satisfactory.}$$

Hence redesign is necessary.

Step 2 (B) : Properties of second trial section

Try WB 500, @ 0.952 kN/m

$$A = 12122 \text{ mm}^2$$

$$r_{xx} = 207.7 \text{ mm}$$

$$r_{yy} = 49.6 \text{ mm}$$

$$Z_{xx} = 2091.6 \times 10^3 \text{ mm}^3,$$

$$r_{min} = 49.6 \text{ mm}$$

Average axial stress in compression

$$\sigma_{ac.cal} = \left(\frac{700\times1000}{12122}\right) = 57.746 \text{ N/mm}^2$$

Step 3 (B) : Slenderness ratio

$$\left(\frac{l}{r_{min}}\right) = \left(\frac{4\cdot0\times1000}{49\cdot6}\right) = 80.645$$

Step 4 (B) : Check for interaction expression

From IS : 800–1984 allowable working stress in compression

$$\sigma_{ac} = 102.29 \text{ N/mm}^2$$

For $\left(\frac{l}{r_{min}}\right) = 80.645$

From IS : 800–1984 allowable bending compressive stress for effective length of 4.0 m

$$\sigma_{bc} = (0.66 \times 260) = 171.6 \text{ N/mm}^2$$

$$\sigma_{bc.cal} = \left(\frac{3500\times1000}{2091\cdot6\times10^3}\right) = 16.734 \text{ N/mm}^2$$

Hence, the design is satisfactory. Provide WB 500, @ 0.952 kN/m.

Alternatively

Step 2 : Properties of trial section

From IS: 808–1984, try Medius weight beam section MB 550, @ 104 kg/m cross-sectional area

$$A = 13200 \text{ mm}^2$$

$$r_{xx} = 222 \text{ mm}$$

$$r_{yy} = 37.3 \text{ mm}$$

$$Z_{xx} = 2360 \times 10^3 \text{ mm}^3$$

Step 3 : Slenderness ratio

Effective length of column is 4000 mm

Minimum radius of gyration

$$r_{min} = 37.3 \text{ mm}$$

Maximum slenderness ratio

$$\left(\frac{l}{r_{min}}\right) = \left(\frac{4000}{37 \cdot 3}\right) = 107.24$$

Step 4 : Check for interaction expression

Average axial stress in compression

$$\sigma_{ac.cal} = \left(\frac{700 \times 1000}{13200}\right) = 53.03 \text{ N/mm}^2$$

Allowable working stress in axial compression for slenderness ratio, 107.24 and the steel having yielding stress as 260 N/mm^2

$$\sigma_{ac} = \left(82 - \frac{9}{10} \times 7 \cdot 24\right) = 75.484 \text{ N/mm}^2$$

$$\text{End-moment} = 35000 \text{ kN-mm}$$

Bending compressive stress

$$\sigma_{bc.cal} = \left(\frac{3500 \times 1000}{2360 \times 1000}\right) = 14.831 \text{ N/mm}^2$$

From IS: 800–1984 allowable bending compressive stress is $0.66 f_y$

$$\sigma_{bc} = (0.66 \times 260) = 171.6 \text{ N/mm}^2$$

$$\left(\frac{\sigma_{ac \cdot cal}}{\sigma_{ac}} + \frac{\sigma_{bc \cdot cal}}{\sigma_{bc}}\right) = \left(\frac{53 \cdot 03}{75 \cdot 484} + \frac{14 \cdot 831}{171 \cdot 6}\right)$$

$$= (0.7025 + 0.086)$$

$$= 0.78896 < 1.00$$

Hence, the design is satisfactory. Provide MB 550, @ 104 kg/m (Medium weight beam section).

Example 3.18 *A column is subjected to an axial load of 500 kN. The beam connected to the flange of a column, has an eccentric load of 100 kN and the beam connected to the web of a column has an eccentric load 50 kN. If the effective length of the column is 4 metres and only rolled steel H-sections are available, design the column. Adopt stresses as per IS : 800–1984.*

Solution

Design :

Step 1 : Selection of trial section

Total vertical load acting on the column

$$(500 + 50 + 100) = 650 \text{ kN}$$

The slenderness ratio of the column and the value of yield stress for the steel may be assumed as 80 and 260 N/mm^2, respectively.

Therefore, permissible stress in axial compression

$$\sigma_{ac} = 103 \text{ N/mm}^2$$

Effective sectional area required

$$A = \left(\frac{650 \times 1000}{103}\right) = 6310.68 \text{ mm}^2$$

Assume effective sectional area required

$$1.5 \times 6310.682 = 9466.023 \text{ mm}^2$$

From steel section tables, try HB 400, @ 0.822 kN/m

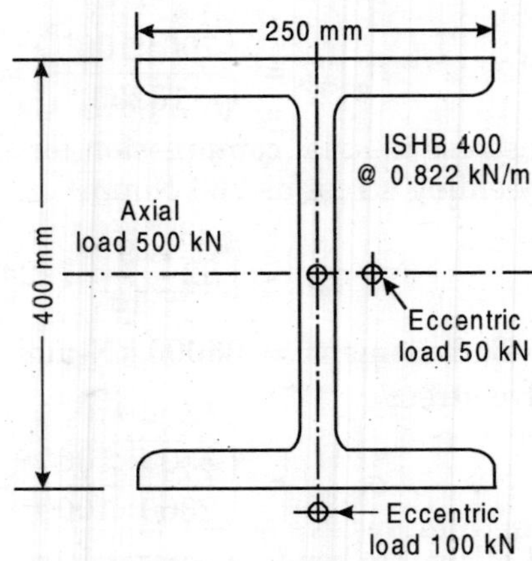

Fig. 3.29

Step 2 : Properties of trial sectional

Effective sectional area and other properties are as follows :

$A = 10466 \text{ mm}^2$		$I_{yy} = 2783.0 \times 10^4 \text{ mm}^4$
$r_{xx} = 166.1 \text{ mm}$		$r_{yy} = 51.6 \text{ mm}$
$Z_{xx} = 1444.2 \times 10^3 \text{ mm}^3$		$Z_{yy} = 221.3 \times 10^3 \text{ mm}^3$
$t_w = 10.6 \text{ mm}$		$t_f = 12.7 \text{ mm}$

Average compressive stress

$$\sigma_{ac.\text{cal}} = \left(\frac{650 \times 1000}{10466}\right) = 62.11 \text{ N/mm}^2$$

Step 3 : Slenderness ratio

Effective length of column is 4 m and $r_{\min} = 51.6$ mm

Slenderness ratio of the column

$$\left(\frac{l}{r_{\min}}\right) = \left(\frac{4 \times 1000}{51 \cdot 6}\right) = 77.5$$

Step 4 : Check for interaction expression

From IS : 800–1984 allowable working stress in compression for the steel having yield stress as 260 N/mm²

$$\sigma_{ac} = 106 \text{ N/mm}^2$$

Eccentric load about xx-axis is 100 kN

This load can be assumed to be acting at an eccentric distance of half the depth of section plus distance upto mid-point of stiffened seating. (Size of stiffened angle may be assumed as ISA 150 mm × 75 mm × 10 mm (ISA 15075, @ 0.169 kN/m) with long leg vertical.)

Moment in plane of yy-axis

$$= (200 + 37.5) \times 100 = 23750 \text{ mm-kN}$$

Eccentric load about yy-axis is 50 kN .

Minimum eccentricity may be assumed as 100 mm

Moment in plane of xx-axis

$$= 50 \times 100 = 5000 \text{ mm-kN}$$

Actual bending stress in plane of yy-axis (i.e., bending is about xx-axis)

$$\sigma_{bc_1\cdot cal} = \left(\frac{23750 \times 1000}{1444 \cdot 2 \times 10^3}\right) = 16.445 \text{ N/mm}^2$$

Actual bending stress in plane of xx-axis (i.e., bending is about yy-axis)

$$\sigma_{bc_2\cdot cal} = \left(\frac{5500 \times 1050}{221 \cdot 3 \times 10^3}\right) = 22.594 \text{ N/mm}^2$$

∴ Allowable bending compressive stress is

$$\sigma_{bc} = 0.66 f_y = (0.66 \times 260) = 171.6 \text{ N/mm}^2$$

There is no possibility of lateral bending in case the bending of column is about xy-axis as shown in Fig. 3.29.

$$\left(\frac{\sigma_{ac\cdot cal}}{\sigma_{ac}} + \frac{\sigma_{bc1\cdot cal} + \sigma_{bc2\cdot cal}}{\sigma_{bc}}\right) = \left(\frac{62 \cdot 11}{106} + \frac{16 \cdot 445 + 27 \cdot 514}{171 \cdot 6}\right)$$

$$= 0.9813 < 1.$$

Hence design is satisfactory. Provide HB 400. @ 0 .822 kN/m as shown in Fig. 3.29.

3.20 MODIFIED AND ADJUSTED INTERACTION FORMULAE

The interaction formula discussed in Sec. 3.18, Eq. 3.36 is the fundamental interaction formula for the column subjected to axial load and bending. The formula takes into account for the additional stress caused by column buckling under an axial load by reducing the value of allowable axial stress, a_{ac} as the maximum slenderness ratio $X = \left(\frac{l}{r}\right)$ increases. The formula does not take into account for the additional stresses for the lateral deflection caused by induced end moment.

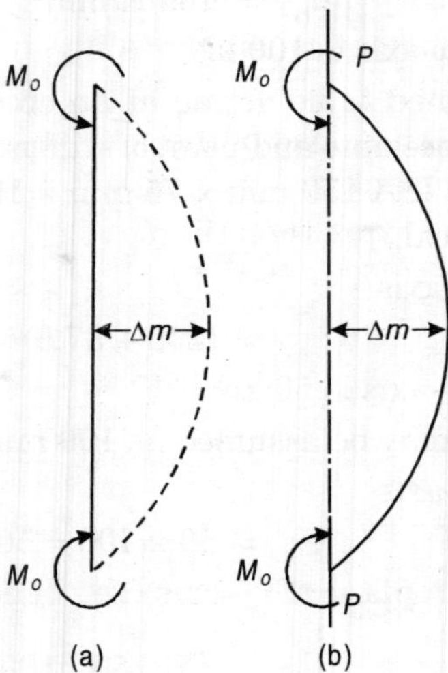

Fig. 3.30 *Column carrying equal end moments*

Consider the sequential column loading as shown in Fig. 3.30. A column is subjected to equal end moments as shown in Fig. 3.30 (a). The dotted line shows the deflected shape of the column on exaggerated scale. The maximum deflection and moment at the centre of column are Δm and M_0 respectively. An axial load P is applied to more deflected column. The column is subjected to additional moment at the centre equal to $P.\Delta m$. The column is subjected to more deflection, due to this, which causes more moment and so on. Final bending stress at the centre of column is given by

$$\sigma_{bc.cal} = \left(\frac{M_0 \cdot y}{I} + \frac{\Delta m P \cdot y}{I}\right) \quad \text{...(i)}$$

It is seen that when the column is subjected to equal end moments at the ends, the lateral deflection depends on the slenderness ratio of the column in the direction in which the bending occurs. A large lateral deflection occurs when the slenderness ratio is large. The bending stresses from the deflection increase with the increasing value of the axial load. AISC specification recommends a method, which simplifies the design procedure. The method is based upon the application of interaction formula. The interaction formula is modified as necessary. The strength of column subjected to bending combined with compressive axial load may be expressed conveniently by interaction formula in terms of the ratios $\frac{P}{P_y}$ and $\frac{M_0}{M_y}$.

$$\left[\frac{P}{P_y} + \frac{M_0}{M_y \frac{(1-P)}{P_e}}\right] = 1.0 \quad \text{...(ii)}$$

where P_y = Axial load causing yielding if it alone occurred

M_y = Bending moment causing yielding, if it occurred in the absence of axial load

P_e = Elastic buckling load of the column for buckling in the plane of applied moment.

The various terms in the expression (ii) may be interpreted as follows :

$$P = \sigma_{ac.cal} A \qquad P_y = \sigma_{ac} A$$

$$M_0 = \left(\frac{\sigma_{ac.cal} I}{y}\right) \qquad M_y = \left(\frac{\sigma_{bc}. I}{y}\right)$$

$$F_c' A = \left(\frac{\pi^2 EI}{L^2}\right)$$

The expression (ii) reduces to

$$\left[\frac{\sigma_{ac\cdot cal} A}{\sigma_{ac} A} + \frac{1}{\left(1 - \dfrac{\sigma_{ac\cdot cal} A}{F_c' A}\right)} \cdot \frac{\sigma_{ac\cdot cal} \dfrac{I}{y}}{\sigma_{bc} \dfrac{I}{y}}\right] = 1.0 \quad \text{... (iii)}$$

or

$$\left[\frac{\sigma_{ac\cdot cal}}{\sigma_{ac}} + \frac{1}{\left(1 - \dfrac{\sigma_{ac\cdot cal}}{F_c'}\right)} \cdot \frac{\sigma_{ac\cdot cal}}{\sigma_{bc}}\right] = 1.0 \quad \text{... (3.39)}$$

where $\dfrac{1}{\left(1 - \dfrac{\sigma_{ac\cdot cal}}{F_e}\right)}$ = **Amplification factor.** It is also known as magnification factor ...(3.40)

The stress F_c' is defined as the limiting Euler stress (divided by a factor of safety). The stress F_e' may be found as follows. From Eq. 3.7.

$$F_c' = \left[\frac{\pi^2 E}{\left(\dfrac{l}{r}\right)^2 \cdot (F \cdot S)}\right] = \left(\frac{9 \cdot 87 \times 2 \cdot 047 \times 10^5}{\left(\dfrac{l}{r}\right)^2 \times 1 \cdot 92}\right) \text{N/mm}^2$$

$$F_e' = \left(\frac{10{,}52{,}000}{\left(\dfrac{l}{r}\right)^2}\right) \text{N/mm}^2 \quad \text{...(iv)}$$

It is to note that the slenderness ratio is measured with respect to the axis about which bending takes place. When the magnitude of stress F_e is very large and/or the magnitude of axial stress f_c is very small, the amplification factor may be neglected. Equation 3.39 gives the *modified interaction formula for the equal column end moments* causing a single curvature deflection.

The amplification factor given by Eq. 3.40 is based upon equal column end moments. The modified interaction formula is further modified depending upon any other combination of end conditions. It introduces a *reduction factor,* C_m. When the interaction formula is further modified, then, it is known as *adjusted interaction formula.* AISC 1963 specification recommends the adjusted interaction formula as below :

$$\left[\frac{\sigma_{ac\cdot cal}}{\sigma_{ac}} + \frac{C_m}{\dfrac{(1-\sigma_{ac\cdot cal})}{F_c'}} \cdot \frac{\sigma_{bc\cdot cal}}{\sigma_{bc}}\right] = 1.0 \qquad \text{...(3.41a)}$$

When the load is eccentric with respect to x-axis, and also with respect to y-axis, then, Eq. 3.41 (a) may be written as

$$\left[\frac{\sigma_{ac\cdot cal}}{\sigma_{ac}} + \frac{C_{mx}\sigma_{bcx\cdot cal}}{\left\{1-\dfrac{\sigma_{ac\cdot cal}}{F_{ex}'}\right\}\cdot\sigma_{bcx}} + \frac{C_{my}\cdot\sigma_{my\cdot cal}}{\left\{1-\dfrac{\sigma_{ac\cdot cal}}{F_{ey}'}\right\}\cdot\sigma_{bcy}}\right] \le 1.0 \qquad \text{...(3.41b)}$$

$$\therefore \qquad F_{ex}' = 0.60\, f_{ce.x}\,;\, F_{ey}' = 0.60\, f_{ce.y}$$

Therefore,

$$\left[\frac{\sigma_{ac\cdot cal}}{\sigma_{ac}} + \frac{C_{mx}\sigma_{bcx\cdot cal}}{\left\{1-\dfrac{\sigma_{ac\cdot cal}}{0\cdot 60 f_{bcx}}\right\}\cdot\sigma_{bcx}} + \frac{C_{my}\cdot\sigma_{my\cdot cal}}{\left\{1-\dfrac{\sigma_{ac\cdot cal}}{0\cdot 60 f_{ecy}}\right\}\cdot\sigma_{bey}}\right] \le 1.0$$

$$\text{where } F_{cc}' = \left(\frac{\pi^2 E}{\left(\dfrac{l}{r}\right)^2}\right) \qquad \text{...(3.41c)}$$

As per IS : 800–1984, the structural members subjected to axial compression and bending about both the axes shall be proportioned to satisfy Eq. 3.41 (c) as above.

The adjusted interaction formula is satisfied for all points on the column between lateral supports so long as $\left(\dfrac{\sigma_{ac\cdot cal}}{\sigma_{ac}}\right)$ is equal to or less than 0.15.

$$\left[\frac{\sigma_{ac\cdot cal}}{\sigma_{ac}} + \frac{\sigma_{bcx.cal}}{\sigma_{bcx}} + \frac{\sigma_{bcy\cdot cal}}{\sigma_{bcy}}\right] \le 1.0 \qquad \text{...(3.41d)}$$

The value of *reduction factor,* C_m depends on the relative size and the direction of column end moments. The AISC 1963 specification recommends the following for the value of reduction factor C_m ($M_1 < M_2$) (as per IS : 800–1984 also)

where, M_1 = Smallest end moment and

M_2 = Largest end moment.

(i) For a member in frames where side sway is not prevented

$$C_m = 0.85$$

(ii) For members in frames where side sway is prevented and not subjected to transverse loading between their supports in the plane of bending

$$C_m = 0.6 - 0.4\left(\frac{M_1}{M_2}\right) \geq 0.4$$

or

$$C_m = 0.6 - 0.4\beta \geq 0.4$$

It is note that the β is the ratio of smaller to the larger moments at the ends of that portion of the unbraced member in the plane of bending under consideration.

It is to further note that $\beta = \left(\frac{M_1}{M_2}\right)$ is negative for single curvature bending and positive for reversed-curvature bending.

(iii) For the members in frames where the sidesway is prevented in the plane of loading and subjected to transverse loading between their supports, the value of C_m may be determined by rational analysis. In the absence of rational analysis, IS : 800–1984 recommends the following values.

(a) For members whose ends are restrained against rotation

$$C_m = 0.85$$

(b) For members whose ends are unrestrained against rotation

$$C_m = 1.00$$

When the end moments act in the same direction, then, these give single curve deflection and the ratio $\left(\frac{M_1}{M_2}\right)$ is *negative.* When the end moments acts in the opposite directions these give double curve deflection and the ratio $\left(\frac{M_1}{M_2}\right)$ is positive.

The AISC specification recommends the values of reduction factor, C_m for the following end conditions which are frequently met in the steel building as shown in Fig. 3.31.

It is to note that the value of C_m should not be less than 0.4. The end conditions of columns carrying transverse loads in addition to axial load is shown in Fig. 3.31 (d). In Fig. 3.31 (d) the bending stress is found at the point using larger moment *M*.

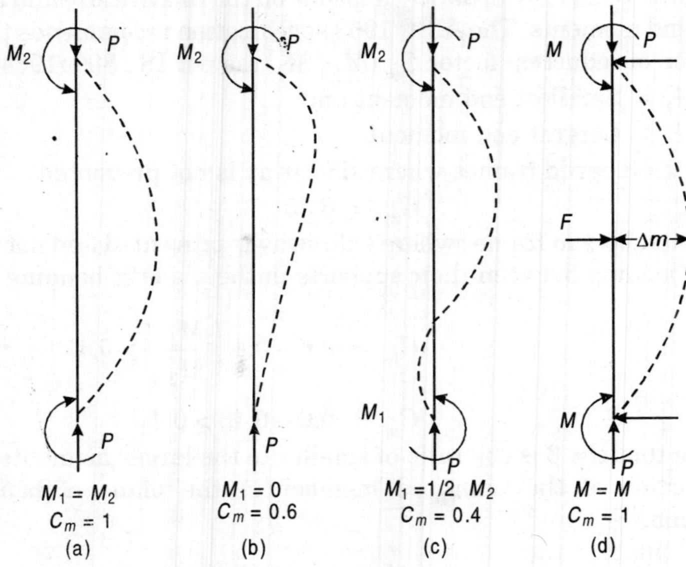

Fig. 3.31

The effective length of column KL is larger than the actual length of a column, in case, there is no bracing against sideway buckling. The value of C_m may be larger than or equal to 0.85 in such cases. The straight line interaction formula may be directly used when $\left(\frac{\sigma_{ac\cdot cal}}{\sigma_{ac}}\right)$ is less than or equal to 0.15.

The modified and adjusted interaction formula have been discussed above for the central section of the column. It is also possible (depending upon the slenderness ratio of a column unbraced in the plane of bending) that the combined stress determined at one end may exceed that at all the points where the lateral deflection is created by the end moments, even when the bending stress at these points have been modified. The AISC specification recommends that the straight line interaction formula may be used with $f_c = 0.6f_y$.

$$\left(\frac{\sigma_{ac\cdot cal}}{0\cdot 6f_y} + \frac{\sigma_{bcx.cal}}{\sigma_{bc}}\right) \leq 1.0 \qquad \text{...(3.43a)}$$

When the load is eccentric with x-axis and also with y-axis the Eq. 3.43 (a) may be written as

$$\left[\frac{\sigma_{ac\cdot cal}}{0\cdot 6f_y} + \frac{\sigma_{bcx.cal}}{\sigma_{bcx}} + \frac{\sigma_{bcy\cdot cal}}{\sigma_{bcy}}\right] \leq 1.0 \qquad \text{...(3.43b)}$$

Therefore, the following interaction formula may be used as per the recommendations of AISC 1963 specifications

(i) For $\left(\frac{\sigma_{ac \cdot cal}}{\sigma_{ac}}\right) \leq 0.15$

$$\left(\frac{\sigma_{ac \cdot cal}}{\sigma_{ac}} + \frac{\sigma_{bcy.cal}}{\sigma_{bc}}\right) \leq 1.0 \quad ...(3.44)$$

(ii) For $\left(\frac{\sigma_{ac \cdot cal}}{\sigma_{ac}}\right) > 0.15$

$$\left[\frac{\sigma_{ac \cdot cal}}{\sigma_{ac}} + \frac{C_m}{\frac{(1-\sigma_{ac \cdot cal})}{F_c'}} \cdot \frac{\sigma_{bc}}{\sigma_{bc}}\right] \leq 1.0 \quad ...(3.45)$$

Equation 3.43 should be satisfied in addition to the above formulae at points of lateral support in the plane of bending.

Instead of end moments, in case transverse loads are acting on the beam column in between end supports, then the interaction formula in different forms shall be used.

3.21 EQUIVALENT AXIAL LOAD METHOD OF DESIGN OF ECCENTRICALLY LOADED COLUMNS

The columns are frequently subjected to eccentric loads in addition to the axial loads. Therefore, the columns are subjected to direct compressive stresses and bending stress (i.e., to combined stress). The method of design of eccentrically loaded columns is to arrive at a trial section and then to check the trial section for the combined stress. The trial section of an eccentrically loaded column may be found conveniently by converting the eccentric load to an *equivalent axial load*. The equivalent axial load is also termed as *equivalent concentric load. The equivalent axial load is the load of sufficient magnitude to produce a stress equal to the maximum stress produced by the eccentric load.* The magnitude of equivalent axial load found is greater than the eccentric load. Consider a column subjected to an eccentric load, P_E as shown in Fig. 3.32.

The eccentricity of load P_E is e_x with respect to x-axis. The column is subjected to direct compressive stress and bending stress. The maximum compressive stress at the extreme fibre of the column is given by

$$f_{c.1} = \left(\frac{P_E}{A} + \frac{P_E \cdot e_x}{Z_{xx}}\right)$$

or

$$f_{c.1} = \frac{P_E}{A} \cdot \left(1 - \frac{A \cdot e_x}{Z_{xx}}\right) \quad ...(i)$$

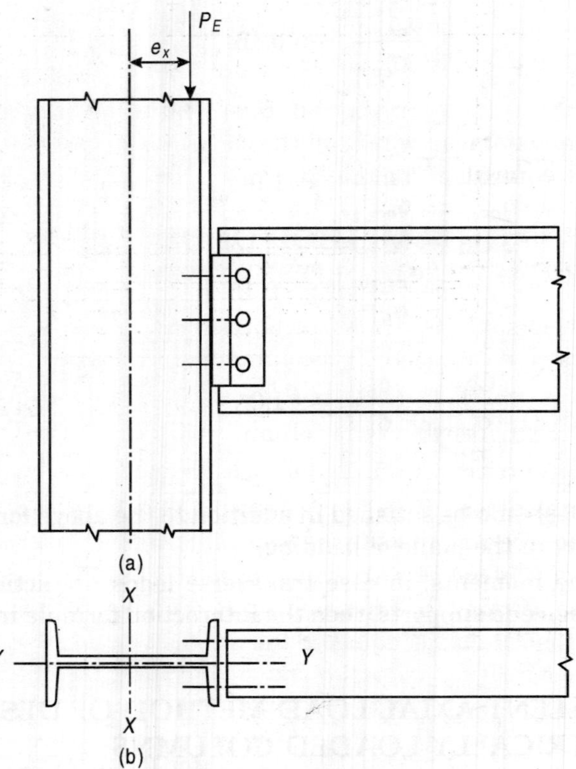

Fig. 3. 32

Let P_{equiv} be the equivalent axial load. The equivalent axial load P_{equiv} is equivalent to the eccentric load, P_E. The equivalent axial load produces the average compressive stress equal to the maximum compressive stress at the extreme fibre of the column. The average compressive stress due to equivalent axial load is given by

$$f_{c.2} = \left(\frac{P_{\text{equiv}}}{A}\right) \quad \text{...(ii)}$$

Since

$$f_{c.2} = f_{c.1}$$

$$\left(\frac{P_{\text{equiv}}}{A}\right) = \frac{P_E}{A}\left(1+\frac{A \cdot e_x}{Z_{xx}}\right)$$

$$P_{\text{equiv}} = P_E\left(1+\frac{A \cdot e_x}{Z_{xx}}\right) \quad \text{...(3.46)}$$

Equation 3.46 gives the equivalent axial load. When the eccentric load is having eccentricity, e_x with respect to *xx*-axis, and eccentricity, e_y with respect *yy*-axis, then, the equivalent load is given by

$$P_{\text{equiv.}} = P_E\left(1+\frac{A\cdot e_x}{Z_{xx}}+\frac{A\cdot e_y}{Z_{yy}}\right) \quad \text{... (3.47)}$$

When the column is also carrying an axial load in addition to an eccentric load, then the total axial load acting on the column may be found as the sum of axial load and the equivalent axial load.

3.22 BENDING FACTOR METHOD OF DESIGN OF ECCENTRICALLY LOADED COLUMN

When the column is subjected to an eccentric load, then, it is subjected to a moment M, due to eccentric load. The column moment is given by

$$M_{xx} = P_E \cdot e_x$$

where P_E = Eccentric load on the column

e_x = Eccentricity of the column with respect of xx-axis.

The equivalent axial load $P_{\text{equiv.}}$ may be found in this method by multiplying the moment due to eccentric load by the *bending factor. The bending factor may be obtained by dividing the cross-sectional area of the column. A, by the section modulus, Z of the section.* It is denoted by B. The compressive stress at the extreme fibre of the column is equal to the average compressive stress due to equivalent axial load, P_{equiv}. Therefore,

$$\left(\frac{P_{\text{equiv}}}{A}\right) = \left(\frac{P_E}{A}+\frac{M_{xx}}{Z_{xx}}\right)$$

or
$$P_{\text{equiv}} = \left(P_E + M_{xx}\frac{A}{Z_{xx}}\right) \quad \text{...(i)}$$

or
$$P_{\text{equiv}} = (P_E + M_{xx} \cdot B_x) \quad \text{...(3.48)}$$

where, $B_x = \left(\frac{A}{Z_{xx}}\right)$, bending factor with respect to axis of x.

When the column is subjected to eccentric load having eccentricity e_x, with respect to x-axis and eccentricity, e_y with respect to y-axis then, the column is subjected to moments in two directions. Therefore,

$$\left(\frac{P_{\text{equiv}}}{A}\right) = \left(\frac{P_E}{A}+\frac{M_{xx}}{Z_{xx}}+\frac{M_{yy}}{Z_{yy}}\right)$$

or
$$P_{\text{equiv}} = \left(P_E + M_{xx}\frac{A}{Z_{xx}} + M_{yy}\frac{A}{Z_{yy}}\right)$$

or
$$P_{\text{equiv}} = (P_E + M_{xx}\cdot B_{xx} + M_{yy}\cdot B_y) \quad \text{...(3.49)}$$

where $M_{xx} = P_E \cdot e_x$,
$M_{yy} = P_E \cdot e_y$ and
B_x = Bending factor with respect to x-axis
B_y = Bending factor with respect to y-axis.

The column may be subjected to moment instead of the eccentric load. It is seen that the moments are reduced to the equivalent axial load. The column may be carrying axial load also in addition to the eccentric load. *The equivalent axial load may be found by having the sum of axial load (if any) and eccentric load, and the products of the bending moments due to eccentric loads and the approximate bending factor.*

It is to note that equivalent axial loads found in Eq. 3.46 and Eq. 3.48 are same, and that found in Eq. 3.47 and Eq. 3.49 are also same.

The usual method to determine the approximate total equivalent load is based upon the requirement of the stability check (i.e., the modified interaction formula). The total equivalent axial load is equal to the sum of total axial load which is used in determining the average axial stress, f_c', and the equivalent axial load which accounts for the effect of eccentricity. The effect of the eccentricity in the stability check is represented in the bending stress ratio $\left(\dfrac{f_{bc}'}{f_{bc}}\right)$

The bending stress ratio $\left(\dfrac{f_{bc}'}{f_{bc}}\right)$ is modified by multiplying by the reduction factor, C_m. In the similar manner, the equivalent axial load is multiplied by the reduction factor C_m, prior to adding its effect to the axial load. Hence, the total equivalent axial load is given by

$$P_{\text{equiv}} = P_E + B \cdot C_m \cdot (P_E \cdot e) \qquad ...(3.50)$$

The trial section obtained for the total equivalent load is checked by Eq. 3.44 or Eq. 3.45. It is to note that when the moments at the ends of a column are either equal to zero, value of C_m is *unity* and Eq. 3.45 assumes the same form as Eq. 3.48 (since $P_E \cdot e = M$).

Example 3.19 *A column HB 300, @ 0.630 kN/m is subjected to an eccentric load. The eccentricity of the load from x-axis of the section is 28 mm. The eccentricity of the load from y-axis of the section is 54 mm. The effective length of column is 4 m. Determine the maximum and safe magnitude of the eccentric-load for the section. Use equivalent axial load method.*

Solution

Step 1 : Properties of section

From ISI Handbook No. 1 for HB 300, @ 0.630 kN/m

$$Z_{xx} = 863.3 \times 10^3 \text{ mm}^3$$
$$r_{xx} = 127.0 \text{ mm}$$
$$Z_{yy} = 178.4 \times 10^3 \text{ mm}^3$$

$$r_{yy} = 52.9 \text{ mm}$$
$$A = 8025 \text{ mm}^2$$

Step 2: Slenderness ratio

Minimum radius of gyration of the section

$$r_{\min} = 52.9 \text{ mm}$$

Effective length of the column $l = 4$ m

Maximum slenderness ratio of the section

$$\left(\frac{l}{r_{\min}}\right) = \left(\frac{4\times1000}{52\cdot9}\right) = 75.7$$

Step 3: Equivalent load

From IS: 800–1984, for the steel having yield stress as 260 N/mm^2 and the maximum allowable stress in axial compression for $\left(\frac{l}{r_{\min}}\right) = 75.7$ is $\sigma_{ac} = 108.16$ N/mm^2.

Maximum safe axial load, which can be carried by the section

$$p_{\text{equiv}} = \sigma_{ac} \times A = \left(\frac{108\cdot16\times8025}{1000}\right) = 867.984 \text{ kN} \quad \text{...(i)}$$

Let P_E be maximum and safe eccentric load for the section. From Eq. 3 47

$$p_{\text{equiv}} = P_E\left(1+\frac{A\cdot e_x}{Z_{xx}}+\frac{A\cdot e_y}{Z_{yy}}\right) \quad \text{...(ii)}$$

$$e_x = 28 \text{ mm}$$
$$e_y = 54 \text{ mm}$$

Substituting the numerical values

$$867.984 = P_E\left(1+\frac{8025\times28}{863\cdot3\times10^3}+\frac{8025\times54}{178.4\times10^3}\right)$$

$$\therefore \quad P_E = \left(\frac{867\cdot984}{1+0\cdot26+2\cdot43}\right) = \left(\frac{867\cdot984}{3\cdot69}\right) = \mathbf{235.23 \text{ kN}}$$

The maximum and safe eccentric load which may be carried by ISHB 300, @ 0.630 kN/m at the given eccentricities is 235.23 kN.

Alternatively

Step 1 : Bending factors

Bending factor for HB 300, @ 630 N/m with respect to x-axis

$$B_x = \left(\frac{A}{Z_{xx}}\right) = \left(\frac{8025}{863\cdot3\times10^3}\right)$$
$$= 0.0093 \text{ mm}^{-1} \quad \text{...(iii)}$$

Bending factor HB 300, @ 0.630 kN/m with respect to y-axis

$$B_x = \left(\frac{A}{Z_{xx}}\right) = \left(\frac{8025}{178.4\times10^3}\right) = 0.045 \text{ mm}^{-1} \quad ...(iv)$$

$$M_{xx} = P_E.e_x = 28P_E \quad ...(v)$$

$$M_{yy} = P_E.e_y = 54P_E \quad ...(vi)$$

Step 2 : Equivalent load

From Eq. 3.49, the equivalent axial load is given by

$$P_{\text{equiv}} = (P_E + M_{xx}B_x + M_{yy}\,.\,B_y)$$

or $$P_{\text{equiv}} = (P_E + 28P_E \times 0.0093 + 53\,P_E \times 0.045)$$

or $$P_E = \left(\frac{P_{\text{equiv}}}{1+0\cdot26+2\cdot43}\right)$$

$$= \left(\frac{867\cdot984}{3\cdot69}\right) = \mathbf{235.23\ kN.}$$

The maximum safe eccentric load found for the given section for given eccentricities by using bending factors if also **235.23 kN**.

Example 3.20 *A column carries an axial load of 400 kN and an eccentric load of 150 kN at 100 mm from x-axis. The actual length of column is 4 m. The column is hinged at both the ends. Design the column by equivalent axial load method.*

Solution

Design :

Step 1: Selection of trial section

Axial load carried by the column, P = 400 kN ....(i)

Eccentric load carried by the column, P_E = 150 kN ...(ii)

Eccentricity of the column from x-axis, e_x = 100 mm ...(iii)

Bending moment

$$M_{xx} = P_F.e_x = (150 \times 100) = 15000 \text{ mm-kN} \quad ...(iv)$$

Equivalent axial load

$$P_{\text{equiv}} = (P + P_E + M_{xx}\,.\,B_x) \quad ...(vi)$$

Since, the column section is not known, the exact bending factor, B_x is also not known. From ISI Handbook No. 1, HB 200, @ 0.400 kN/m is selected tentatively. The bending factor for this column section

$$B_x = \left(\frac{A}{Z_{xx}}\right) = \left(\frac{5094}{372\cdot2\times10^3}\right) = 0.0137 \text{ mm}^{-1} \quad ...(via)$$

$$M_{xx}B_x = (15000 \times 0.0137) = 205.5 \text{ kN} \quad ...(vii)$$

$$P_{\text{equiv}} = (400 + 150 + 205.5) = 755.5 \text{ kN} \quad ...(viii)$$

The slenderness ratio for the column and the value of yield stress are assumed as 80 and 260 N/mm², respectively. Therefore,

Allowable axial stress in compression as 103 N/mm², cross-sectional area required

$$A_{reqd.} = \left(\frac{755.5 \times 1000}{103}\right) = 7334.95 \text{ mm}^2 \quad \text{... (ix)}$$

From ISI Handbook No. l, select the column section HB 300, @ 0.630 kN/m

$$A = 8025 \text{ mm}^2$$
$$r_{xx} = 127.0 \text{ mm}$$
$$Z_{xx} = 863.30 \times 10^3 \text{ mm}^3$$
$$r_{yy} = 52.9 \text{ mm}$$

Minimum radius of gyration of the section

$$r_{min} = 52{\cdot}9 \text{ mm}$$

Step 2 : Revised equivalent load

Revised brending factor

$$B_x = \left(\frac{A}{Z_{xx}}\right) = \left(\frac{8025}{863 \cdot 3 \times 10^3}\right)$$
$$= 0.0093 \text{ mm}^{-1} \quad \text{...(x)}$$

$$M_{xx} \cdot B_x = (1500 \times 0.0093) = 139.5 \text{ kN} \quad \text{...(xi)}$$

Revised equivalent axial load

$$P_{equiv} = (400 + 150 + 139.5) = 689.5 \text{ kN}$$

Area required,

$$A = \left(\frac{689 \cdot 5 \times 1000}{103}\right) = 6694.17 \text{ mm}^2 \quad \text{...(xiii)}$$

Area provided by HB 300, @ 0.630 kN/m is greater than area required.

Effective length of column

$$l = 4 \text{ m} \quad \text{... (xiii)}$$

Maximum slenderness ratio for the column

$$\left(\frac{l}{r_{min}}\right) = \left(\frac{4 \times 1000}{52 \cdot 9}\right) = 75.7 \quad \text{...(xiv)}$$

From IS : 800–1984, the maximum allowable stress in axial compression for the steel having yield stress as 260 N/mm² and

$$\left(\frac{l}{r_{min}}\right) = 75.7 \text{ is } \sigma_{ac} = 108.16 \text{ N/mm}^2$$

Check :

The column section may be checked by interaction formula. The average axial stress

$$\sigma_{ac \cdot cal} = \left(\frac{\text{Total vertical load}}{\text{Cross-sectional area}}\right) = \left(\frac{(400+150) \times 1000}{8025}\right) = 68.536 \text{ N/mm}^2$$

Actual bending stress

$$\sigma_{ac\cdot cal} = \left(\frac{M_{xx}}{I_{xx}}\right)y = \left(\frac{15000\times1000\times150}{12950\cdot2\times10^4}\right) = 17.374\text{N/mm}^2$$

Allowable bending stress in compression from IS : 800–1984

$$0{\cdot}60f_y = (0.60 \times 260) = 156 \text{ N/mm}^2$$

$$\left(\frac{\sigma_{ac\cdot cal}}{\sigma_{ac}} + \frac{\sigma_{bc.cal}}{\sigma_{bc}}\right) = \left(\frac{68\cdot536}{108\cdot16} + \frac{17\cdot374}{156}\right) = 0.745 < 1.00$$

Hence, the design is satisfactory. Provide HB 300, @ 0.630 kN/m column section.

Example 3.21 *Verify that the column designed in Example 3.20, satisfy the AISC requirement.*

Solution :

Step 1 : Properties of section

From ISI Handbook No. 1 for the column section HB 800 @ 0.630 kN/m designed in Example 3.20.

$A = 8025 \text{ mm}^2$ $\quad r_{xx} = 127 \text{ mm}$

$Z_{xx} = 863.3 \times 10^3 \text{ mm}^3$ $\quad r_{yy} = 52.9 \text{ mm}$

Step 2 : Check by interaction expression

Average axial stress in compression

$$\sigma_{ac.cal} = \left(\frac{(400+150)\times1000}{8025}\right)$$

$$= 68.536 \text{ N/mm}^2 \qquad \text{...(i)}$$

Allowable axial stress in compression

$$\sigma_{ac.cal} = 108.16 \text{ N/mm}^2 \qquad \text{...(ii)}$$

$$\left(\frac{\sigma_{ac\cdot cal}}{\sigma_{ac}}\right) = \left(\frac{68\cdot536}{108\cdot16}\right) = 0.634 > 0.15 \qquad \text{...(iii)}$$

From Eqn. 3.45

$$\left[\frac{\sigma_{ac\cdot cal}}{\sigma_{ac}} + \frac{C_m}{\frac{(1-\sigma_{ac\cdot cal})}{F_c'}}\cdot\frac{\sigma_{bc\cdot cal}}{\sigma_{bc}}\right] \le 1.00 \qquad \text{...(iv)}$$

$$\left(\frac{\sigma_{ac\cdot cal}}{\sigma_{ac}}\right) = \left(\frac{17\cdot374}{156}\right) = 0.1114 \qquad \text{...(v)}$$

The column is hinged at both ends. The column end moments are $M_1 = M_2 = 0$. Therefore,

$$C_m = 1 \qquad \text{...(vi)}$$

$$F_c' = \left(\frac{1052000}{(l/r)^2}\right) \text{N/mm}^2 \quad \text{... (vii)}$$

The column has eccentricity about x-axis. The bending takes place about x-axis .

$$\left(\frac{l}{r_{xx}}\right) = \left(\frac{4\times1000}{127}\right) = 31.6 \quad \text{...(viii)}$$

$$F_e' = \left(\frac{1052000}{31\cdot6\times31\cdot6}\right) = 1052 \text{ N/mm}^2 \quad \text{...(ix)}$$

$$\left(\frac{\sigma_{ac\cdot cal}}{F_e'}\right) = \left(\frac{68\cdot536}{1052}\right) = 0.065 \quad \text{...(x)}$$

$$\left(1-\frac{\sigma_{ac\cdot cal}}{F_e}\right) = (1-0.065) = 0.935 \quad \text{...(xi)}$$

$$\left[\frac{C_m}{\left(1-\frac{\sigma_{ac\cdot cal}}{F_c'}\right)}\right] = \left(\frac{1}{0\cdot935}\right) = 1.07 \quad \text{...(xii)}$$

$$\therefore \left[\frac{\sigma_{ac.cal}}{\sigma_{ac}} + \frac{C_m}{\frac{\left(1-\sigma_{ac\cdot cal}\right)}{F_e'}}\cdot\frac{\sigma_{bc}}{\sigma_{bc}}\right] \le 1.00$$

$$(0.634 + 1.07 \times 0.1114) = 0.753 < 1.00$$

when $\left(\frac{\sigma_{ac\cdot cal}}{\sigma_{ac}}\right) > 0.15$, a point braced in the plane of bending, Eq. 3.43 should also be satisfied. Therefore,

$$\left(\frac{\sigma_{ac\cdot cal}}{0\cdot6f_y} + \frac{\sigma_{bcx.cal}}{\sigma_{bc}}\right) = \left(\frac{68\cdot536}{0\cdot6\times260} + 0\cdot1114\right) = 0.551 < 1.00$$

Hence, the design is satisfactory. The yield stress f_y, steel has been adopted as 260 N/mm^2.

3.23 COLUMN SPLICE

A joint in the length of a column provided, when necessary, is known as *column splice*. It is also described as *column joint*. The rolled steel sections are manufactured upto a certain maximum length. When length of the column required is more than the length of rolled steel sections manufactured, in that case, column splice becomes necessary. Secondly, when the columns is used in multi-storey buildings then the sectional area for the column in the upper storey is less than that for the columns in upper storey and lower storey separately. In practice, ends of the column are cut by the following two methods:

1. Ends of column cut by ordinary method.
2. Ends of column cut and milled.

When end of the column are cut by ordinary method, the load is transferred to the lower column through rivets. When the ends of the columns are cut and milled, then column has complete bearing over the whole area and ends of the column are flush ends. Different types of riveted column splices have been shown in Fig. 3.33.

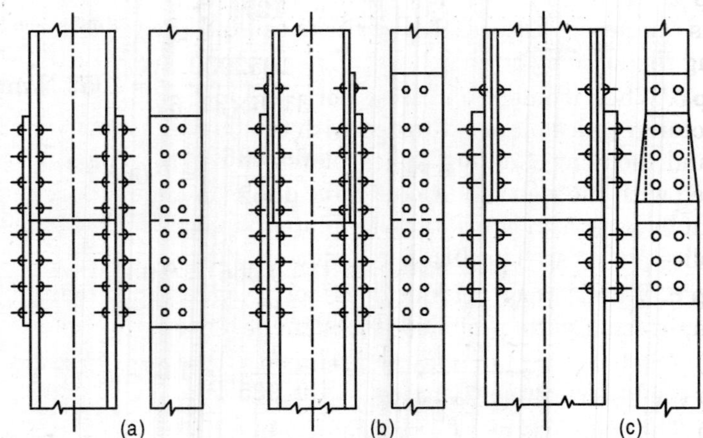

Fig. 3.33 *Column splices*

The column splice, as shown in Fig. 3.33 (a) is used when the depth of columns sections are equal. The column splice, as shown in Fig. 3.33 (b), is used; when the depth of upper column is smaller than the lower column, then filler plates are placed in between column splice plates and column. When the depth of the section of upper column is much smaller than the lower column and it does not provide full bearing, bearing plate is placed as shown in Fig. 3.33 (c).

3.24 DESIGN OF COLUMN SPLICE

Following are the usual steps in the design of column splice. The specifications for the design as per IS : 800–1984 have also been given :

Step 1. (i) When the ends of the column are cut by ordinary methods, or not faced for complete bearing, the column splice is designed for full axial load and other forces to which the joint is subjected.

(ii) When the ends of the column are faced for bearing over the whole area, the joint is designed to hold the connected members accurately in place and to resist tension (if any) and bending if present. The ends of compression members faced for bearing shall invariably be machined to ensure perfect contact of surfaces in bearing. The rivets essential to hold the connected members may be designed for 50 per cent axial load (as advised in IS : 800–1950) and other force acting on the column.

(iii) When possible, splices shall be proportioned and arranged so that the centroidal axis of the splice coincides as nearly as possible with the centroidal axes of members joined in order to avoid eccentricity ; but where eccentricity is present in the joint the resulting stress shall be provided for.

Step 2. The column splice plates may be assumed to act as short column of zero slenderness ratio. Allowable working stress in compression in the splice plates may be found for the steel to be used ; and the sectional area of splice plates required is computed.

Step 3. The width of the splice plates is kept equal to the width of flange of the column and thickness is computed.

Step 4. The nominal diameter of rivets used in the joint is assumed and rivet value is computed. The number of rivets essential to resist force is computed by dividing the force by rivet value.

Step 5. When moment and shear force are also acting in addition to the axial load, column splice plates attached with the flanges are assumed to resist axial force and moment. Column splice plates to resist maximum shear force are provided with the web. Additional force due to moment is computed by dividing moment by lever arm between splice plates attached with the flanges. The rivets and splice plates are checked for tension due to bending.

Step 6. When upper column does not provide full bearing area over lower column, bearing plate provided is designed as below :

(i) The bearing plate may be assumed as short beam to transmit the axial load to the lower column section.

(ii) The axial load of the column is assumed to be taken by flanges, neglecting the load taken by the web. The load transmitted from flanges of upper column andreaction from flanges of lower column are equal and form a couple as shown in Fig. 3.34.

(iii) Moment due to couple, M is $\frac{P}{2} \times a$

(iv) Moment of resistance of plate, is $\frac{1}{6} bt^2 \times \sigma_{bs}$

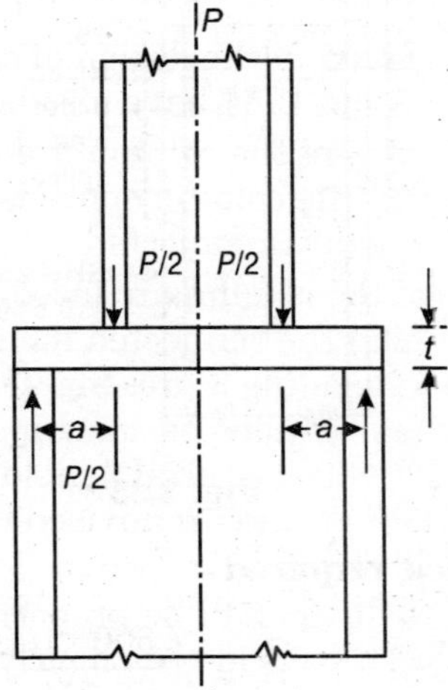

Fig. 3.34

where, b = width of bearing plate and it is equal to the width of flange of column

t = thickness of bearing plate

σ_{bs} = allowable bending stress in slab or bearing plates.

By equating moment of resistance and moment due to couple, the thickness of bearing plate is given by

$$t = \left(\frac{6M}{b \times \sigma_{bs}}\right)^{1/2}$$

Example 3.22 *A column section HB 250, @ 0.510 kN/m carries an axial load of 600 kN. The ends of the column are cut by ordinary method. Design the column splice.*

Solution

Design :

Step 1 : Allowable stress in axial compression

The slenderness ratio for the splice plates (being of small length) and the value of yield stress for the steel to be used may be assumed as zero and 260 N/mm^2, respectively.

Therefore, the allowable stress in axial compression may be adopted as $0.6 f_y$ = 156 N/mm^2

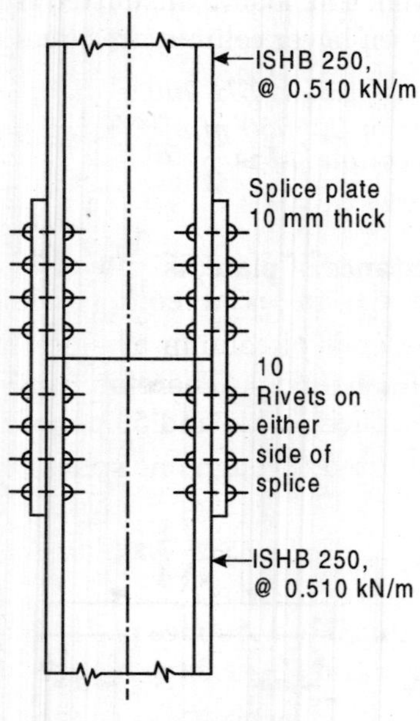

Fig. 3.35

Step 2 : Area of splice required

$$A = \left(\frac{600 \times 1000}{156}\right) = 3846.154 \text{ mm}^2$$

Area of one splice plate required = 1923.07 mm^2

Width the splice plate is kept equal to the width of flange of column = 250 mm.

Required thickness of splice,

$$t = \left(\frac{1923.07}{250}\right) = 7.692 \text{ mm}$$

Adopt thickness of splice plate = 10 mm

Step 3 : Rivet value

Provide 22 mm nominal diameter of rivet. The strength in single shear for power driven rivets

$$\left(\frac{\pi}{4} \times \frac{(23 \cdot 5)^2 \times 100}{1000}\right) = 43.35 \text{ kN}$$

Thickness of flange of column is 9.7 mm.

Strength in bearing for power driven rivets

$$9 \cdot 7 \times \left(\frac{23 \cdot 5 \times 300}{1000}\right) = 68.385 \text{ kN}$$

$$\text{Rivet value } R = 43.35 \text{ kN}$$

Number of rivets required is $\left(\frac{600}{43.35} = 13.84\right)$

Provide 16 rivets, 8 rivets are provided on one side and 8 rivets are provided on the other as shown in Fig. 3.35.

Example 3.23 *A column section BH 400, @ 0.822 kN/m is carrying an axial of 500 kN and a moment of 225000 mm-kN and shear force 45 kN. Design a column splice.*

Solution

Design :

Step 1: Allowable stress in axial compression

It is assumed that the ends of column are faced for complete bearing over whole area (milled or flush ends). Therefore, 50 per cent of axial load is transmitted through the splice plates and 50 per cent by direct bearing.

Axial load to be transmitted through one splice plate.

$$= \left(\frac{1}{4} \times 500\right) = 125 \text{ kN}$$

It is assumed that the thickness of splice plates is 6 mm and the moment is to be resisted fully by flanges and shear force by web only

Lever arm

$$(400 + 6) = 406 \text{ mm}$$

Force to be resisted by splice plate because of moment

$$\left(\frac{22500}{406}\right) = 55.42 \text{ kN}$$

Total load

$$(125 + 55.42) = 180.42 \text{ kN}$$

The slenderness ratio of the splice plates and the value of yield stress for steel to be used are assumed as zero and 260 N/mm² respectively.

The allowable working stress in compression is splice plate

$$0.6 f_y = 156 \text{ N/mm}^2$$

Step 2: Area of one splice plate required

$$\left(\frac{180.42 \times 1000}{156}\right) = 1156.54 \text{ mm}^2$$

Provide width of splice plate equal to 250 mm

∴ Thickness of splice plate required

$$\left(\frac{1156.54}{250}\right) = 4.626 \text{ m}$$

Adopt thickness of splice plate as 6 mm

Step 3 : Rivet value

Provide 22 mm diameter power driven rivets

Strength of rivet in single shear

$$\left(\frac{\pi}{4} \times \frac{(23.5)^2 \times 100}{1000}\right) = 43.35 \text{ kN}$$

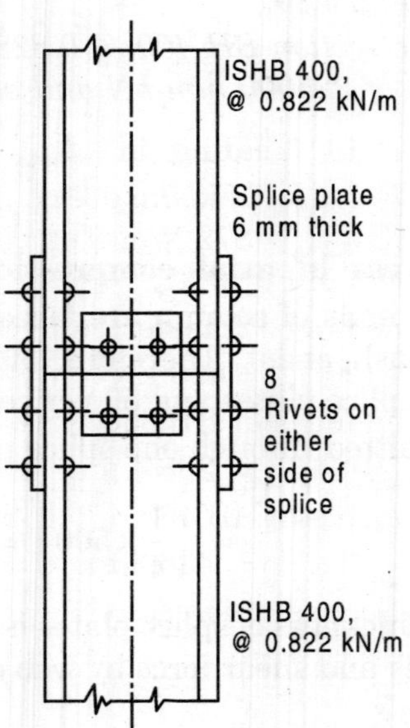

Fig. 3.36

Strength of rivet in bearing

$$\left(\frac{23.5 \times 6 \times 300}{1000}\right) = 42.30 \text{ kN}$$

Rivet value $R = 42.30$ kN

Number of rivets required on one side is $\left(\frac{180.42}{42.30} = 4.26\right)$

Provide 8 rivets.

The tension due to moment in splice plate 55.42 kN is less than axial compression in splice plate 125 kN.

Splice plate for S.F. is designed for maximum shear

Maximum shear force in web = 1.5 × 45 = 67.5 kN

Maximum allowable shear stress

$$0.45f_y = (0.45 \times 260)$$
$$= 117 \text{ N/mm}^2$$

Area of plates required $= \left(\frac{675 \times 1000}{117}\right) = 576.92 \text{ mm}^2$.

Number of rivers required $= \left(\frac{45}{43.35}\right) = 1.038$

Provide two rivets. Provide splice plates 6 mm thick to accommodate two rivets and to give above areas as shown in Fig. 3.36.

Example 3.24 *A column section HB 350, @ 0 674 kN/m is carrying an axial load of 900 kN. It is to be supported over a column section HB 450, @ 0.872 kN/m. Design the column splicing.*

Solution

Design :

Step 1: Allowable stress in bending in compression

The column section HB 350, @ 0.674 kN/m does not provide full bearing area over lower column section HB 450, @ 872 N/m. The bearing plate 450 mm × 250 mm × t mm is placed as shown in Fig. 3.37.

The column load is assumed to be transferred by the flanges.

The load on each flange is (900/2) = 450 kN

Distance between C.G. of flanges in HB 350, @ 0674 kN/m

$$(350–116) = 338.4 \text{ mm}$$

Distance between C.G. flanges in HB 450, @ 0.872 kN/m

$$(450–13\ 7) = 436.3 \text{ mm}$$

Distance between lines of action of forces

$$\frac{1}{2}(436.3 - 338.4) = \frac{1}{2} \times 97.9 = 48.95 \text{ mm}$$

Moment due to couple

$$(450 \times 48.95) = 22000 \text{ mm-kN}$$

Moment of resistance of bearing plate

$$\left(\frac{1}{6}bt^2 \times \sigma_{bc}\right) = 22000$$

Allowable stress in bending in compression

$$\sigma_{bc} = (0.66 \times 250) \text{ N/mm}^2$$

Step 2 : Thickness of splice plate

or $$\left(\frac{1}{6} \times 250 \times t^2 \times 0.66 \times 260\right) = 22000 \times 1000$$

$$\therefore \quad t = 55 \text{ mm}$$

Adopt thickness of 56 mm (as manufactured) and provide filler plate of 50 mm thick on either side.

Splice plate. The column ends are faced for complete bearing over whole area. The slenderness ratio of splice plates and the value of yield stress for steel to be used are assumed as zero and 260 N/mm². Therefore allowable working stress in compression in splice plates $0.6 f_y = 156$ N/mm².

Splice plates are designed for 50 per cent load, viz. 450 kN

Area of splice plates required

$$\left(\frac{450 \times 1000}{156}\right) = 2884.62 \text{ mm}^2$$

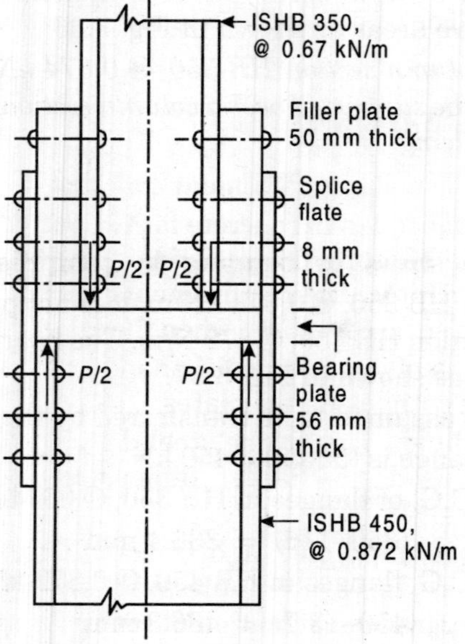

Fig. 3.37

Area of one splice plate required

$$= 1442.31 \text{ mm}^2$$

The width of splice plate is kept equal to the width of flange of column section = 250 mm

Thickness of splice plate is $\left(\frac{1445.31}{250} = 5.769 \text{ mm}\right)$

Provide 8 mm thickness for splice plate.

Step 3 : Rivet connections. Use 22 mm diameter power driven rivets. Strength of rivet in single shear

$$\left(\frac{\pi}{4}(23.5)^2 \times \frac{100}{1000}\right) = 43.35 \text{ kN}$$

Strength of rivet in bearing

$$\left(\frac{23.5 \times 8 \times 300}{1000}\right) = 56.4 \text{ kN}$$

Number of rivets required is $\left(\dfrac{450}{43.35}\right) = 10.38$

In HB 350, @ 0.674 kN/m, rivets are placed through filler plates as shown in Fig. 3.37.

Thickness of filler plate is 44 mm. It is excess than 6 mm

Rivets are increased by 25% for each 200 mm i.e., 27.5%

$$= 10.38 \times 1.275 = 13.23$$

Provide 16 rivets. The additional rivets are provided in each flange in filler plates.

3.25 ENCASED COLUMNS

When a steel column is encased in cement concrete, it is called *encased.*

A steel column is embedded in concrete for two purposes. It may be necessary to make a steel frame building fire resisting. The columns may be encased in cement concrete for architectural requirements. As per IS : 800–1984 a member may be designed as an encased column when the following conditions are fulfilled:

(a) The member is of symmetrical I-shape or a single I-beam or channels back to back, with or without flange plates.

(b) The overall dimensions of the steel do not exceed 750 mm × 450 mm over plating where used, the larger dimensions being measured parallel to web.

(c) The column is unpainted and is solidly encased in ordinary dense concrete with 20 mm aggregate (unless solidity can be obtained with a larger aggregate) and grade designation of concrete minimum M 15.

(d) The minimum width of solid casing is equal to (b_0 + 100 mm) where b_0 is the width of the steel flanges in millimetres.

(e) The surface and edges of the steel column have a concrete cover of not less than 50 mm.

(f) The casing is effectively reinforced with steel wires. The wires shall be at least 5 mm of diameter and the reinforcement shall be in the form of stirrups or bending at not more than 150 mm pitch so arranged as to pass through the centre of the covering of the edges and outer faces of the flanges and supported by longitudinal spacing bars not less than four in number.

(g) The steel cores in encased columns shall be accurately machined at splices and provisions shall be made for the alignment of the column. At the column base the provision shall be made to transfer the load to the footing at safe unit stresses in accordance with IS : 456–1978.

The encased columns are designed as follows :

The steel column section shall be considered as carrying the entire load but allowance may be made for stiffened effect of the concrete. This allowance should be made by assuming the radius of gyration r of the column section about the axis in the plane of its web to be 0.2 (b_0+ 100) mm, where b_0 is the width of the steel flanges in mm. The radius of gyration about its other axis shall be taken as that of the uncased section.

The axial load on the cased column shall not exceed 2 times that which would be permitted on the uncased section, nor shall the slenderness ratio of the uncased section for its length centre to centre of connection exceed 250.

In computing the allowable axial load on the cased strut, the concrete shall be taken as assisting in carrying the load over its rectangular cross-section, any cover in excess of 75 mm from the overall dimension of section of the cased strut being ignored. The allowable compressive load, P in case of encased columns shall be determined as follows:

$$P = (A_{sc} \cdot \sigma_{sc} + A_c \cdot \sigma_c)$$

where, A_{sc}, A_c = cross-sectional area of steel and concrete, and

σ_{sc}, σ_c = permissible stress in steel and concrete

It is to note that above clause does not apply to steel struts of overall dimensions greater than 1000 mm × 500 mm, the dimension of 1000 mm being measured parallel to the web or to box sections.

Design procedure. Following are the steps in design of cased columns :

Step 1. The slenderness ratio for column and the value of yield stress for steel to be used are assumed as about 70 and 260 N/mm^2 (or any other value). And the allowable axial compressive stress, σ_{ac}, is noted from IS : 800–1984.

Step 2. The effective sectional area A, required for the compression member is determined

$$A = \left(\frac{P}{\sigma_{ac}}\right)$$

where, P = load to be carried by the member.

Step 3. From IS Handbook No. 1, a section for the compression member of the required area is selected.

The section for the compression member is selected such that it has the largest possible radius of gyration for the required sectional area. It should also be most economical section.

Step 4. The geometrical properties of section are noted. The effective length of column is found as per end conditions. The slenderness ratio of uncased column is found. The allowable axial compressive stress for uncased column is found from IS : 800–1984.

Step 5. The safe load carrying capacity of the column is found.

Step 6. For the cased column, radius of gyration about *yy*-axis is found. *yy*-axis is kept parallel to plane of web,

$$r_{yy} = 0.2\,(b_0+100)\text{ mm}$$

where, b_0 = width of flange in mm.

Step 7. The slenderness ratio of cased column is determined. The allowable axial compressive stress is determined from IS : 800–1984.

Step 8. The load carrying capacity of the cased column is determined. The load carrying capacity of the column should be equal to or slightly greater than the load on the column. The design may be revised if the load carrying capacity is less than or much greater than load on the column.

Step 9. Checks for slenderness ratio and axial load on the cased column are provided.

Provided sufficient longitudinal steel bars to support stirrups. Provide stirrups of 5 mm diameter at a pitch of 150 mm. The cement concrete is provided to provide a cover of 50 mm on all sides.

Example 3.25 *Design a cased column to carry an axial load of 1100 kN. The column is 4 m long and adequately restrained in position but not in direction at both the ends.*

Solution

Design :

Step 1: Selection of trial section

The slenderness ratio for the column and the value of yield stress for the steel to be used are assumed as 70 and 260 N/m^2, respectively. Therefore,

Allowable axial compressive stress =115 N/mm^2

Effective length of column, l = 4 m

Effective sectional area required

$$A = \left(\frac{1100 \times 1000}{115}\right)$$

$$= 9565.22\text{ mm}^2$$

From ISI Handbook No. 1, try HB 450, @ 0.872 kN/m

Sectional area, $A = 11114\text{ mm}^2$

$$r_{xx} = 187.8\text{ mm},\ r_{yy} = 51.8\text{ mm},$$

$$\therefore \quad r_{min} = 51.8\text{ mm}$$

Step 2 : Maximum slenderness ratio of uncased column

$$\left(\frac{l}{r_{min}}\right) = \left(\frac{4 \times 1000}{51.8}\right) = 77.2$$

From IS : 800–1984, allowable axial compressive stress

$$= 106.36\text{ N/mm}^2$$

Step 3 : Safe load carrying capacity of uncased column

$$\left(\frac{106.36 \times 11114}{1000}\right) = 1182.08\text{ kN}$$

The column is encased in cement concrete. The cement concrete provides a 50 mm cover on all sides. 5 mm diameter stirrups are provided at 100 mm pitch (≯ 150 mm). The stirrups are kept in position by 4 longitudinal bars 10 mm in diameter as shown in Fig. 3.38.

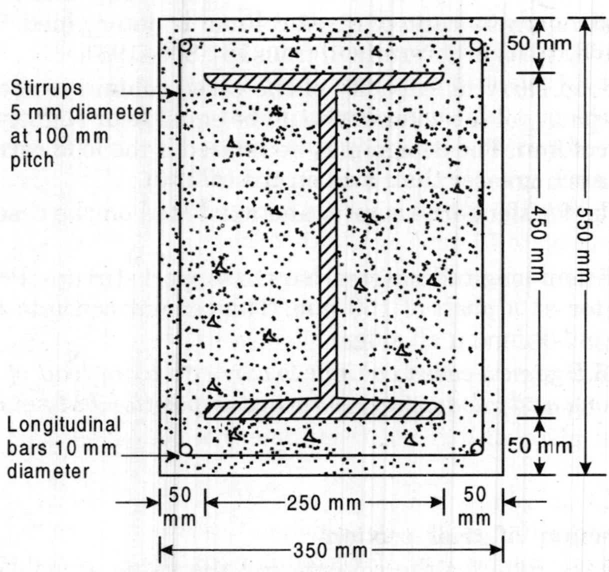

Fig. 3.38 *Cased column*

Step 4 : Radius of gyration for cased column

$$r_{yy} = 0.2\,(b_0 + 100)\text{ mm},\ b_0 = 250\text{ mm}$$

$$r_{yy} = 0.2\,(250 + 1000) = 70\text{ mm}$$

Step 5 : Maximum slenderness ratio of cased column

$$\left(\frac{l}{r_{yy}} = \frac{4 \times 1000}{70}\right) = 57.14$$

From IS : 800–1984, allowable axial compressive stress in steel

$$= 128.86\text{ N/mm}^2$$

Area of concrete = Total area less area of steel core

$$(350 \times 550 - 11114)\text{ mm}^2 = (192500 - 11114) = 181386\text{ mm}^2$$

Let the grade concrete is M 15 for which α_{ac} is 4 N/mm².

Step 6: Safe load carrying capacity of the cased column as per IS : 800–1984

$$P = \left(\frac{181386 \times 4}{1000} + \frac{128.86 \times 11114}{1000}\right)\text{kN}$$

$$P = (725.54 + 1432.15)$$

$$= \mathbf{2157.69\ kN}$$

Check : Slenderness ratio of uncased column is 77.2 (< 250).

PROBLEMS

3.1. A rolled steel beam HB 300, @ 0.588 kN/m is used as a column. The column is fixed in position but not in direction at both ends. Determine the safe load carrying capacity of the column if the length of column is 4.50 m.

3.2. A single angle discontinuous strut ISA 130 mm × 130 mm × 10 mm (ISA 130 130, @ 0.197 kN/m) with single riveted connection is 2.5 m long. Calculate safe load carrying capacity of section.

3.3. In Problem 3.2, if single discontinuous strut is double riveted, determine safe load carrying capacity of the section.

3.4 A double angle discontinuous strut ISA 150 mm × 75 mm × 10 mm (ISA 150 75, @ 0.165 kN/m) long legs back to back, is connected to both sides of a gusset plate 10 mm thick with 2 rivets.

The length of strut between centre to centre of intersections is 3.50 m. Determine the safe load carrying capacity of the section.

3.5. In Problems 3.4, if double angle discontinuous strut is connected to one side of a gusset, determine safe load carrying capacity of the strut.

3.6. In Problem 3.4, if double angle strut is continuous and connected with the gusset plate with single rivet; determine safe load carrying capacity of strut.

3.7. A built-up column consists of two MC 400, @ 0.494 kN/m and two plates 500 mm × 10 mm. The clear distance between back to back of channels is 200 mm. One plate is connected to each flange side. Determine the safe load carrying, capacity of built-up section if the effective length of column is 5 m.

3.8. Design a rolled steel beam section column to carry an axial load of 1120 kN.

The column is 6.20 m long and adequately restrained in position and direction at both ends.

3.9. Design a single angle discontinuous strut to carry 80 kN load. The length of the strut between centre of intersection is 2 .75 m.

3.10. Design a double angle discontinuous strut to carry 120 kN load. The length of strut between centre to centre of intersection is 3.80 m.

3.11. A built-up column consists of two channels MC 250, @ 0.304 kN/m placed back to back with 147 mm gap between them. Determine the safe load on the column when (i) channels are effectively laced together, (ii) channels are effectively battened together. Take effectively length of the column to be 6000 mm.

3.12. Design a built-up column to support 1200 kN axial load. The length of column is 18.5 m. The column is fixed at both ends. Design (i) single lacing, (ii) double lacing for the built-up column.

3.13. Design a built-up column of effective length 5 m to carry an axial load of 1000 kN. Also, design a suitable lacing system for the above column.

3.14. An axially loaded built-up strut is made up of two channels back to back with a gap of 200 mm. The column is 3 metres long and is pinned at both ends. The individual channels of the column are MC 300, @ 0.858 kN/m.

Design a suitable lacing for the column. The properties of the channels are

Area		=	4564 mm^2
Depth of section		=	300 mm
Width of flange		=	90 mm
Thickness of flange		=	13.6 mm
Thickness of web		=	7.6 mm
Centre of gravity	C_{yy}	=	23.6 mm
I_{xx} = 6362.6 × 10^4 mm^4	I_{yy}	=	310.8 × 10^4 mm^4
r_{xx} = 118.1 mm,	r_{yy}	=	26.1 mm

3.15. Design a mild steel built–up column to carry an axial load of 2000 kN with an effective height of 6.10 m. Use 4 m.s. angle sections laced together to form the column. Design also a suitable system of lacing bars.

3.16. A column 400 mm × 400 mm consists of 4 angles of ISA 80 mm × 80 mm × 10 mm (ISA 80 80, @ 0118 kN/m) as shown in Fig. P3.16. The column is 8 m long and is hinged at both ends. Find the maximum safe load for the column and design a suitable lacing system for it. Draw a neat sketch of the column showing connection details (Properties of each angle are: Area = 1505 mm^2, C.G. line $C_{xx} = C_{yy}$ = 23.6 mm, $I_{xx} = I_{yy}$ = 87.7 × 10^4 mm^4, r_{min} = 15.5 mm, rivet gauge distance = 45 mm.

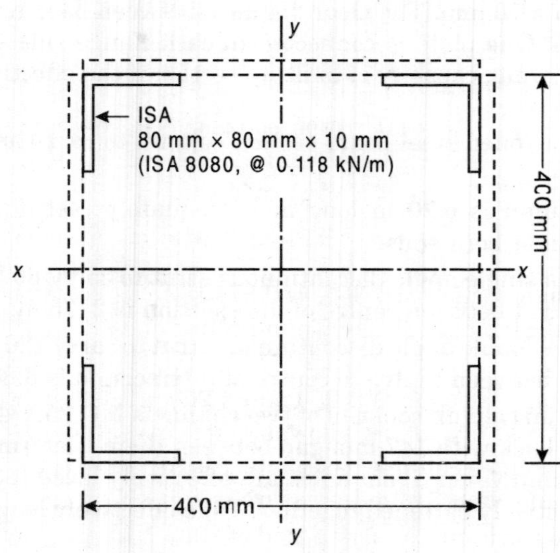

Fig. P 3.16

3.17. Design a built-up column to carry an axial load of 3600 kN. The effective length of column is 5 m. Provide two channel sections and plates.

3.18. Design a steel column 3 metre long carrying an axial load of 500 kN. Assume both the ends to be pin-jointed. Only channel sections are available.

3.19. Design a battened column to carry an axial load of 1100 kN. The effective length of column is 7.00 m.

3.20. The ground floor column of a multistorey building is 5 metres high and is subjected to an axial load of 1000 kN. The column height of the floor next to the ground floor is 4 metres. The beams connected to the flanges of the column have an eccentric load of 10 kN. The beams connected to the web of the column have an eccentric load of 6 kN. Design and sketch the section of the column. Adopt bending stress as per IS : 800–1984.

3.21. The cross-section of a 6 m long, pin ended column consists of 4 ISA 100 mm × 100 mm × 10 mm (4 ISA 100 100, @ 0.149 kN/m) suitably connected with lacing bars. The angles face inwards and the outside dimensions of the cross-section are 350 mm × 350 mm as shown in Fig. P 3.21.

(i) Determine the safe axial compressive load for the column.

(ii) Also design the lacing bars and their connection to angles.

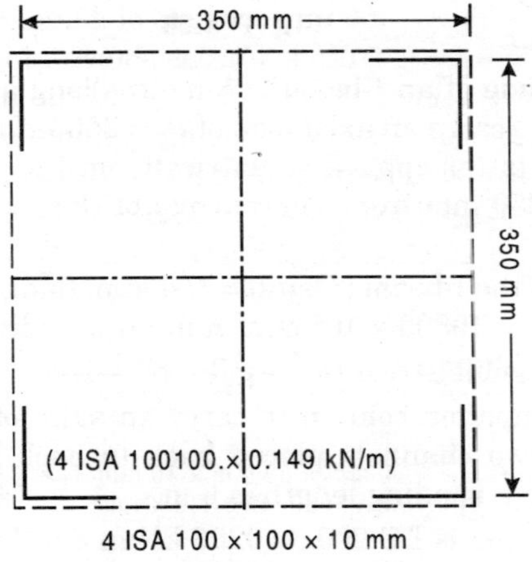

Fig. P 3.21

3.22. A column section HB 350, @ 0.674 kN/m is carrying an axial load of 400 kN and a moment of 18000 mm-kN and shear force 36 kN. Design a column splice.

3.23. A column of section WB 250 mm × 200 mm @ 0.409 kN/m carries an axial load of 800 kN. If the allowable bearing stress in concrete base in 4 N/mm^2, design a suitable base plate. Also design a suitable grillage foundation if the allowable ground pressure is 100 kN/m^2.

3.24. Design a column to carry a load of 800 kN at an eccentric distance of 6 mm from the centroidal axes of the column along the web. The effective length of column is 6.5 m.

3.25. A column of section WB 300, @ 0.481 kN/m (A = 6133 mm^2 ; b_f = 200 mm ; t_f = 100 mm ; I_x = 9821.6 × 10^4 mm^4, I_y = 9901.1 × 10^4 mm^4) along with one plate of 300 mm × 10 mm on each flange as shown in Fig. P3.25 is used as a column over an effective length of 5 metres. Determine the safe load the column can carry and also design a suitable riveted connection between flange and the flange plates. Use 20 mm ϕ rivets. Assume reasonable values of any other data required.

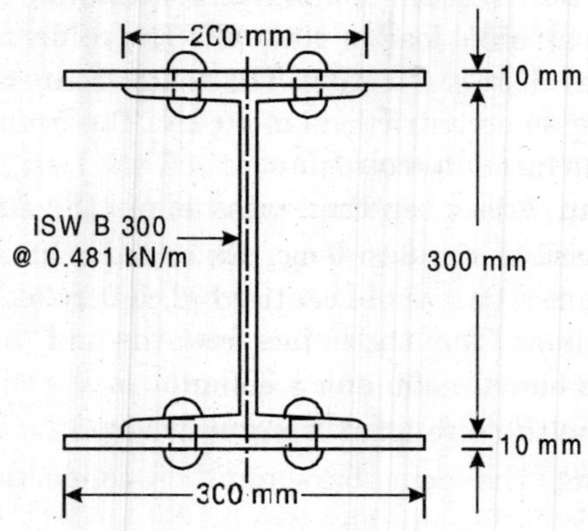

Fig. P 3.25

3.26. A column made of an I-beam is 8 metres long and has hinged ends. The column has to carry an axial load of $P_1 = 250$ kN and a bracket load of $P_2 =$ 100 kN, the latter applied eccentrically on the y-axis of the column at a distance of 250 mm from the centroid of the section. Check the safety of the column.

Properties of the I-beam ; Depth = 350 mm, flange width = 250 mm, area = 9221 mm^2, I_x = 19803 × 10^4 mm^4 minimum radius of gyration = 52.2 mm. Adopt permissible stress (as per IS : 800–1984).

3.27. Design a compound column to carry an axial of 500 kN. The column is built-up of two channels placed back to back and laced together. The equivalent free bending length is 6 m.

Details of channels LC 200, @ 0206 kN/m and MC 200, @ 0221 kN/m are given. Use permissible stress in compression as per IS : 800–1984.

3.28. A steel column to take a central load of 1600 kN is to be built-up of four equal angles forming a square 500 mm × 500 mm. The height of the column is to be 6 m with hinged ends. Design

(a) a suitable column section

(b) a lacing system, and

(c) a steel base plate for the column if it is to rest on a concrete block which can satisfy take a load of 100 kN/m^2.

Give a dimensional sketch for the design. Use permissible as per IS : 800–1984. The distance of the centre of gravity x for the angle is given in Fig. P 3.28.

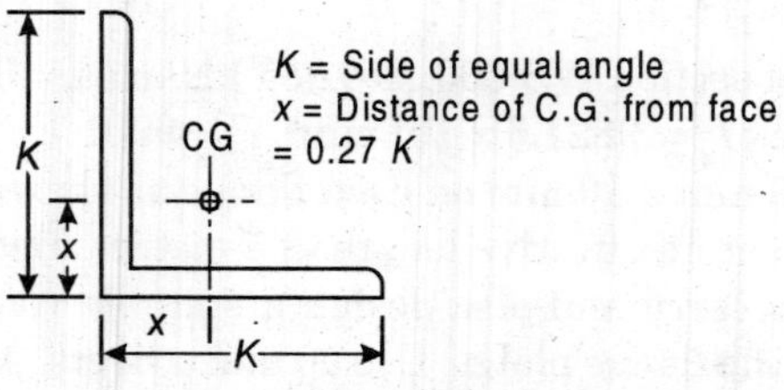

Fig. P 3.28

3.29. The compression member of a bridge is composed of two channel sections, the width over the back of the channel sections being 400 mm. The member carries a load of 100 kN, the length being 10 metres. Using MC 350 channels, design

(a) the compression member, and

(b) the lacing system together with details of connections,

Adopt permissible stresses as per IS : 800–1984.

Consider the location of the centroid of channel section at approximately 0.24 times the width of the flange from the back surface of the web.

3.30. Select a suitable H-section for a column carrying an axial load of 350 kN and a moment 25000 mm-kN about major axis. The effective length of column is 6 m.

3.31. A mid steel column 6 m high has its ends firmly built-in. The column is built-up with two channels MC 300 placed back to back with 180 mm gap between them. The channels are effectively placed together. Determine the safe load on the column. Design also the lacing for it.

Chapter 4

Design of Column Bases and Column Footings

4.1 INTRODUCTION

The columns are supported on the column bases. The column bases transmit the column load to the concrete or masonry foundation blocks. The column load is spread over large area on concrete or masonry blocks. The intensity of bearing pressure on concrete or masonry is kept within the maximum allowable bearing pressure. The safety of the structure depends upon stability of foundation. The column bases should be designed with utmost care and skill. In the column bases, intensity of pressure on concrete block is assumed to be uniform as shown in Fig. 4.1. The column bases shall be of adequate strength, stiffness and area to spread the load upon the concrete, masonry, other foundation or other supports without exceeding the allowable stress on such foundation under any combination of the load and bending moments. The column bases are of two types :

1. Slab bases ; and
2. Gusseted bases.

The column footings are designed to sustain the applied loads, moments and forces and the induced reactions. The column load is spread over large area, so that the intensity of bearing pressure between the column footing and soil does not exceed the safe bearing capacity of the soil. It is ensured that any settlement which may occur shall be as nearly uniform as possible and limited to an accepted small amount. The column load is first transmitted to the column footing through the column base. It is then spread over the soil through the column footing. The column footings are of two types :

1. Independent footings; and
2. Combined footings.

The designs of column base (slab and gusseted bases) and the column footings (independent and combined footing) have been described in this chapter.

4.2 SLAB BASE

The slab base as shown in Fig. 4.1 consists of cleat angles and base plate. The column end is faced for bearing over the whole area.

The gussets (gusset plates and gusset angles) are not provided with the column with the slab bases. The sufficient fastenings are used to retain the parts securely in plate and to resist all moments and forces, other than the direct compression. The forces and moments arising during transit, unloading and erection are also considered.

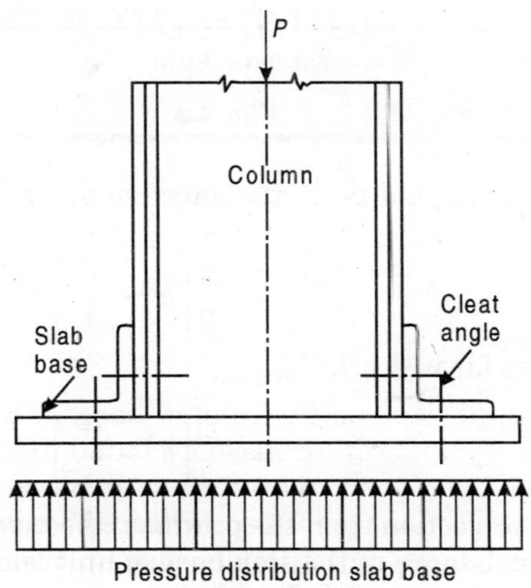

Fig. 4.1 *Slab base*

When the slab alone distributes the load uniformly the minimum thickness of a rectangular slab is derived as below :

The column is carrying an axial load *P*. Consider the load distributed over area $h \times w$ and under the slab; over the $L \times D$ as shown in Fig. 4.2.

Let t = Thickness of the slab

w = Pressure or loading on the underside of the base

a = Greater projection beyond column

b = Lesser projection beyond column

σ_{bs} = Allowable bending stress in the slab bases for all steels, it shall be assumed as 185 N/mm^2 (MPa)

Consider a strip of unit width.

Along the *xx*-axis,

$$M_{xx} = \left(\frac{wa^2}{2}\right)$$

Along the *yy*-axis,

$$M_{yy} = \left(\frac{wb^2}{2}\right)$$

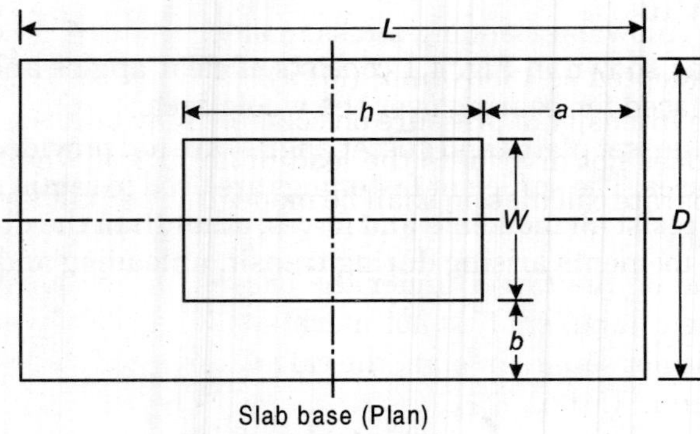

Fig. 4.2

If Poisson ratio is adopted as $\frac{1}{4}$, the effective moment for width D

$$= \frac{w}{2}\left(a^2 - \frac{b^2}{4}\right)$$

Effective moment for width L

$$= \frac{w}{2}\left(b^2 - \frac{a^2}{4}\right)$$

***A* is the greater projection from the column. Effective moment for width *D* is more. Moment of resistance of the slab base of unit width**

$$\text{M.R.} = \left(\frac{1}{6} \times 1 \times t^2 \times \sigma_{bs}\right)$$

$$\left(\frac{1}{6} \times 1 \times t^2 \times \sigma_{bs}\right) = \frac{w}{2}\left(a^2 - \frac{b^2}{4}\right)$$

Thickness of the slab base

$$t = \left[\frac{3w}{\sigma_{bs}}\left(a^2 - \frac{b^2}{4}\right)\right]^{1/2} \quad \text{(as per IS : 800–1984)} \qquad \text{...(4.1)}$$

For solid round steel column, where the load is distributed over the whole area, the minimum thickness of square base as per IS : 800–1984 is given by

$$t = 10\left[\frac{90W}{16\sigma_{bs}}\left(\frac{B}{B - d_0}\right)\right]^{1/2} \qquad \text{...(4.2)}$$

where t = Thickness of plate in mm

W = Total axial load in kN

B = Length of the side of base of cap in mm

d_0 = Diameter of the reduced end (if any) of the column in mm
σ_{bs} = Allowable bending stress in steel
σ_{bs} is adopted as 185 N/mm^2 (MPa).

The allowable intensity of pressure on concrete may be assumed as 4 N/mm^2. When the slab does not distribute the load uniformly or where the slab is not rectangular, separate calculation shall be made to show that stresses are within the specified limits.

When the load on the cap or under the base is not uniformly distributed or where end of the column shaft is not machined with the cap or base, or where the cap or base is not square in plan, the calculations are made on the allowable stress of 185 N/mm^2 (MPa).

The cap or base plate shall not be less than 150 (d_0 + 75) mm in length or diameter.

The area of the shoulder (the annular bearing area) shall be sufficient to limit the stress in bearing, for the whole of the load communicated to the slab to the maximum value $0.75f_y$ and resistance to any bending communicated to the shaft by the slab shall be taken as assisted by bearing pressures developed against the reduced end of the shaft in conjunction with the shoulder.

The bases for bearing upon concrete or masonry need not be machined on the underside provided the reduced end of the shaft terminate short of the surface of the slab, and in all cases the area of the reduced end shall be neglected in calculating the bearing pressure from the base.

In cases where the cap or base is fillet welded direct to the end of the column without boring and shouldering, the contact surfaces shall be machined to give a perfect bearing and the welding shall be sufficient to transit the forces specified above. Where the full length *T*-butt welds are provided no machining of contact surfaces shall be required.

Example 4.1 *A column section HB 250, @ 0.510 kN/m carries an axial load of 600 kN. Design a slab base for the column. The allowable bearing pressure on concrete is 4 N/mm^2. The allowable bending stress in the slab base is 185 N/mm^2 (MPa).*

Solution

Design : Step 1 : Area of slab base required

Axial load of column = 600 kN.

It is assumed uniformly distributed under the slab

Area of the slab base required

$$\left(\frac{600 \times 1000}{4}\right) = 15 \times 10^4 \text{ mm}^2$$

The length and width of slab base are proportioned so that projections on either side beyond the column are approximately equal.

Size of column section HB 250, @ 0.510 kN/mm

= 250 mm × 25 mm

Area of slab base = $(250 + 2a)(250 + 2b)$ mm^2

Step 2 : Projections of base plate

Let the projections a and b be equal

Area of slab

$$(250 + 2a)^2 = 15 \times 10^4, \therefore a = 68.45 \text{ mm}$$

$$\text{Provide projections } a = b = 70 \text{ mm}$$

$$\text{Provide slab base} = (250 + 2 \times 70)\,(250 + 2 \times 70)$$

$$= 390 \text{ mm} \times 390 \text{ mm}$$

Area of slab base provided

$$(390 \times 390) = 152100 \text{ mm}^2$$

Intensity of pressure from concrete under the slab

$$w = \left(\frac{600 \times 1000}{1521000}\right) = 3.945 \text{ N/mm}^2$$

Step 3: Thickness of slab base

$$\text{Thickness of slab base} = \left[\frac{3 \times 3.95}{185}\left(70^2 - \frac{70^2}{4}\right)\right]^{1/2} = 15.34 \text{ mm}$$

Provide 16 mm thick slab base. The fastenings are provided to keep the column in position.

Example 4.1 (a) *A column section SC 250, @ 85.6 carries an axial load of 600 kN. Design a slab base for the column. The allowable bearing pressure on concrete is 4 N/mm². The allowable bending stress in slab base is 185 N/mm². (MPa)*

Solution

Design :

Step 1 : Area of slab base required

Axial load of column is 600 kN. It is assumed uniformly distributed under the slab.

Area of slab base required

$$\left(\frac{600 \times 1000}{4}\right) = 15 \times 10^4 \text{ mm}^2$$

The length and width of slab base are proportioned so that the projections on either side beyond the column are approximately equal.

Size of column section SC 250, @ 85.6 kg/m

$$= (250 \text{ mm} \times 250 \text{ mm})$$

Area of slab base

$$= (250 + 2a)\,(250 + 2b) \text{ mm}^2$$

Step 2: Projections of base plate

Let the projections a and b be equal

Area of slab

$$(250 + 2a)^2 = 15 \times 10^4, \therefore a = 68.45 \text{ mm}$$

Provide projections $a = b = 70$ mm

Provide slab base $= (250 + 2 \times 70)(250 + 2 \times 70)$

$= (390 \text{ mm} \times 390 \text{ mm})$

Area of slab base provided

$(390 \times 390) = 152100 \text{ mm}^2$

$$w = \left(\frac{600 \times 1000}{152100}\right) = 3.945 \text{ N/mm}^2$$

Step 3: Thickness of slab base

Thickness of slab base

$$= \left[\frac{3 \times 3.95}{185}\left(702 - \frac{702}{4}\right)\right]^{1/2} = 15.34 \text{ mm}$$

Provide 16 mm thick slab base. The fastenings are provided to keep the column in position.

The length and width of slab base are proportioned so that the projections on either side beyond the column are approximately equal.

4.3 GUSSETED BASE

In the gusseted base, gusset plates and gusset angles are used on either side of the column. In addition to the gusset plates, cleat angles are used to connect the column with the slab base. The gusseted base plate has been shown in Fig. 4.3. The columns with gusseted bases, the gusset plates, angle cleats, stiffeners, etc., in combination with the bearing area of the shaft should be sufficient to take the loads, bending moments and reactions to the base plate without exceeding the specified stresses. All the bearing surfaces are machined to ensure perfect contact.

When the ends of the columns and gusset plates are not faced for complete bearing, the fastenings connecting them to the base plate shall be sufficient to transmit all the forces to which the base plate is subjected.

Following are the usual steps for the design of gusseted base.

Step 1. The load carried by the column, and intensity of bearing pressure on concrete or masonry block are known. The area required by the gusseted base plate is computed by dividing the load by allowable intensity of pressure in concrete.

Step 2. The gusset materials used in the gusseted base are assumed. The thickness of gusset plate is assumed as 16 mm. The size of gusset angle is assumed such that its vertical leg can accommodate two rivets in one vertical line and corresponding to that leg length other leg is assumed in which one rivet can be provided. The thickness of cleat angle is kept approximately equal to the thickness of gusset plate (i.e., ISA 150 mm × 115 mm × 15 mm is assumed to be used as cleat angle).

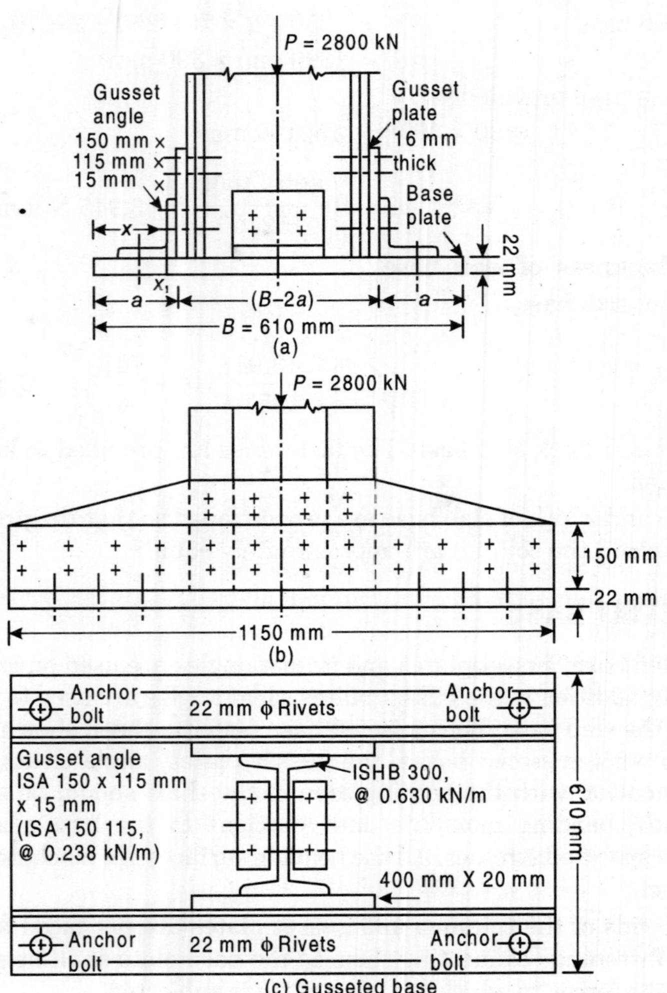

Fig. 4.3 *Gusseted base plate*

Step 3. The depth of column section, thickness of gusset plate and length of leg of angle being known, the width of gusset plate for these distances as shown in Fig. 4.3 (a) is known.

Step 4. The length of gusset plate is computed by dividing area required for gusset plate by the width of the gusset plate.

Step 5. The fastenings of the gusseted base and the end of column faced for complete bearing over whole area shall be sufficient to take loads, bending moments and reactions to the base plate without exceeding specified stresses. (As advised in IS : 800–1956 fastenings having flush ends may be designed for 50 percent axial load and for other forces).

Where ends of the column section and gusset plates are faced for complete bearing, the fastenings connecting them to the base plate shall be sufficient to transmit all the forces to which the base plate is subjected.

A 22 mm nominal diameter is assumed for rivets to be used and rivet value and thus number of rivets to transmit the forces as specified above are computed. The horizontal shear component is equal to the vertical shear. The number of rivets connecting the gusset plate and gusset angle is adopted equal to the number of rivets connecting the gusset plate and the column as shown in Fig. 4.3 (b).

Step 6. The thickness of base plate is computed by equating moment of resistance with the moment due to intensity of pressure at the underside of base plate at two critical sections *X* and *Y* shown in Fig. 4.3 (a).

The thickness whichever is more and safe at both the sections is adopted. The thickness computed is reduced by the thickness of the gusset angle in case the thickness at *XX* section governs the value. It is usual to make the base plates at least as thick as the gussets or as angles.

In the above computation, base plate is assumed to have supports only at gusset plates, and bending of plate takes place in one plane only. But the bending of central portion actually occurs in two planes mutually at right angles. It is also having some support from web of column. The thickness of the plate is made one-half of that calculated. The computation of thickness with the said assumption is on safer side. The thickness of base plate determined at the section *XX* or that determined at the section *YY* whichever is maximum is adopted as the thickness of base plate

When the end of the column is connected directly to the base plate by means of full penetration butt welds the connection should be deemed to transmit to the base all the forces and moments to which the column is subjected.

Example 4.2 *A column section HB 300, @ 0.630 kN/m with one cover plate 400 mm × 20 mm on either side is carrying an axial load of 2800 kN inclusive of self-weight of base and column. Design a gusseted base. The allowable bending pressure in concrete is 4 N/mm². The allowable bending stress in base plate is 185 N/mm² (MPa).*

Solution

Design :

Step 1: Area of base plate required

Axial load on the column = 2800 kN

Area required for gusseted base = $\left(\dfrac{2800}{4}\times 1000\right) = 70 \times 10^4 \text{ mm}^2$

Assume gusset plate 16 mm thick and gusset angles ISA 150 mm × 115 mm × 15 mm. (ISA 150115, @ 0295 kN/m)

Step 2: Width of gusseted base (in the direction parallel to the web of column section)

= [35.00 + 2 × 2.00 + 2 × 1.60 + 2 × 11.5] × 10 = 602 mm

Adopt width of gusseted base = 610 mm

Step 3 : Length of gusseted base

$$= \left(\frac{70\times10^4}{610}\right) = 1147.54 \text{ mm}$$

Adopt length of gusseted base = 1150 mm

Area of gusseted base provided = 610 × 1150 = 701500 mm^2

Intensity of pressure between plate and concrete

$$\left(\frac{2800\times1000}{701500}\right) = 3.99 \text{ N/mm}^2$$

Assume ends of column section, cover plates, gusset plates and angles faced for complete bearing over whole area.

Load transmitted through fastening of both faces of column section is 50 per cent of axial load.

Step 4 : Connection of gusset angle and plate

$$= 1400 \text{ kN}$$

Use 22 mm diameter power driven rivets.

The column section and cover plates attached with the flanges act as one unit. The rivets connecting gusset plate and column section are in single shear.

Strength of rivets in single shear

$$\left(\frac{\pi}{4}\times\frac{(23.5)^2\times100}{1000}\right) = 43.35 \text{ kN}$$

Strength of rivet in bearing

$$\left(\frac{23.5\times16\times300}{1000}\right) = 112.8 \text{ kN}$$

Rivet value R = 43.35 kN

Number of rivets required

$$\left(\frac{1400}{43.35}\right) = 32.29$$

Provide 40 rivets on both faces

i.e., 20 rivets on each face in four vertical rows.

The number of rivets that connecting gusset angles with gusset plate is equal to that connecting gusset plate with column.

Step 5 : Thickness of base plate.

Consider cantilever portion upto critical section XX. The length of cantilever portion

$$x = (115 - 15) + 4 = 104 \text{ mm}$$

Consider width of 10 mm

Bending moment at the critical section XX

$$\left(\frac{3.99\times104^2}{2\times1000}\right) = 216 \text{ mm-kN}$$

Let t_1 be the thickness of base plate inclusive of thickness of cleat angle.

Moment of resistance of strip

$$\text{M.R.} = \left(\frac{1}{6}\times10\times t_1^2\times185\right)\text{mm-N}$$

$$t_1^2 = \left(\frac{216\times6\times1000}{185\times10}\right) \therefore t_1 = 26.47 \text{ mm}$$

Thickness of base plate

$$= (26.47-15) = 11.47 \text{ mm}$$

Consider central portion of unit width strip.

Length $= (B - 2a) = (610 - 2 \times 115 - 2 \times 4) = 372$ mm

Bending moment at critical section YY

$$\left(\frac{w(B-2a)^2}{8}-\frac{wa^2}{2}\right) = \left(\frac{3.99\times372^2}{8}-\frac{3.99\times(104-15)^2}{2}\right)$$

$$= 407.5 \text{ mm-kN}$$

Let t_2 be the thickness of plate

Moment of resistance of strip

$$\text{M.R.} = \left(\frac{1}{6}\times10\times t_2^2\times185\right)\text{N-mm}$$

$$t_2^2 = \left(\frac{407.5\times6\times1000}{185\times10}\right)\text{mm}^2 \therefore t = 36.35 \text{ mm}$$

As central portion of plate in fact has bending in two directions and also supported at web; thickness of plate is made about one half

$$\therefore \text{Thickness of plate} = \frac{1}{2}\times 36.35 = 18.177 \text{ mm} > 16 \text{ mm}$$

Adopt 22 mm plate thickness (manufactured). The base plate should be at least as thick as gusset angles. Provide 2 ISA 150 mm × 115 mm × 15 mm (2 ISA 150115, @ 0.295 kN/m)(cleat angles) and connect with the web of column. The design of gusseted base has been shown in Fig. 4.3.

4.4 COLUMN BASES SUBJECTED TO MOMENT

A column is carrying an axial load 'P' and moment 'M' as shown in Fig. 4.4.

The intensity of bearing pressure between column base and concrete (footing) because of axial load 'P' has been assumed to be uniform as shown in Fig. 4.4. The intensity of pressure is given by

$$P = (P/BL)$$

where, B = Width of the column base

L = Length of the column base

The stress developed between column base and concrete (footing) because of moment 'M' as shown in Fig. 4.4 (b) is given by

$$f_b = \pm \frac{M}{Z}$$

where, Z = Section modulus of column base, $\left(\frac{1}{6}BL^2\right)$

Depending upon the direction of moment, stress due to moment acts upward on one side and downward on other side. The combined stress due to axial load 'P' and moment 'M' is given by

$$(P \pm f_b) = \left(\frac{P}{BL} \pm \frac{6M}{BL^2}\right)$$

The combined stress varies linearly. When intensity of bearing pressure due to axial load is greater than the stress developed due to moment, i.e.,

$$p > f_b$$

or $$\left(\frac{P}{BL} > \frac{6M}{BL^2}\right) \quad \left(\frac{M}{P} < \frac{L}{6}\right)$$

then, the intensity of bearing pressure between column base and concrete is compressive throughout the length of column base as shown in Fig. 4.4 (c).

When the intensity of bearing pressure due to axial load is equal to the stress developed due to moment, i.e.,

$$p = f_b, \quad \left(\frac{P}{BL} = \frac{6M}{BL^2}\right)$$

or $$M = \left(\frac{PL}{6}\right) \quad \text{or} \quad \left(\frac{M}{P} = \frac{L}{6}\right)$$

then, intensity of bearing pressure between column base and concrete is compressive, throughout the length of column base, and it is zero at one edge, as shown in Fig. 4.4(d).

When the intensity of bearing pressure due to axial load is less than the stress developed due to moment, i.e.,

$$p < f_b, \quad \left(\frac{P}{BL} < \frac{6M}{BL^2}\right)$$

or $$M > \left(\frac{PL}{6}\right) \quad \text{or} \quad \left(\frac{M}{P} > \frac{L}{6}\right)$$

then, intensity of bearing pressure between column base and concrete is maximum at one edge and it is zero at a point (o) within the base. The portion 'mo'

will not remain in contact with the base and the intensity of bearing pressure, will, therefore, be zero for all points between 'm' and 'o'. The intensity of bearing pressure acting upward (compressive) has been shown in Fig. 4.4 (e). The resultant passes at a distance greater than $\frac{L}{6}$ from centre of base (i.e., outside the middle third portion). It acts at the C.G. of the triangle at a distance y from the edge, and is equal to and coincides with the line of action of total vertical load acting downward. Length of base of triangle is $3y$. The ordinate of triangle (p_1) represents maximum intensity of bearing pressure. This should not exceed allowable bearing pressure of concrete (footing).

Area of base giving upward pressure

$$= 3y \times B$$

Total upward pressure (resultant)

$$R = \frac{1}{2} \times p_1 \times 3y \times B \qquad \text{(But } R = P\text{)}$$

$$\therefore \quad P = \frac{1}{2} \times p_1 \times 3y \times B \qquad \text{...(4.3)}$$

and taking moment about centre of the base

$$R\left(\frac{L}{2} - y\right) = M, \left(\frac{L}{2} - y\right) = \frac{M}{R}$$

$$\left(\frac{L}{2} - y = \frac{M}{P}\right), \; y = \left(\frac{L}{2} - \frac{M}{P}\right) \qquad \text{...(4.4)}$$

Length of column base is determined by trial and error and the distance y can be computed from Eq. 4.4. Width of column base can be found from Eq. 4.3. Thickness of the base plate is found by equating the moment of resistance of base plate with the moment due to intensity of bearing pressure.

The above analysis is for the base plate *fixed to the column,* and anchor bolts do not resist any moment. Anchor bolts are provided to keep the column base in position. When anchor bolts also resist overturning moment, then resultant of upward pressure

$$R = P + P_1 \qquad \text{...(4.5)}$$

where, P_1 = Tension induced in the anchor bolt in one side due to overturning moment.

Taking moment about C.G. of triangle in Fig. 4.4 (e),

$$P_1 (a - y) + P\left(\frac{L}{2} - y\right) = M \qquad \text{...(4.6)}$$

where a = distance of anchor bolt from edge under maximum pressure.

The position of anchor bolt is known and hence distance a is also known. Value of P_1 is substituted from Eq. 4.5 ; and the distance y is found.

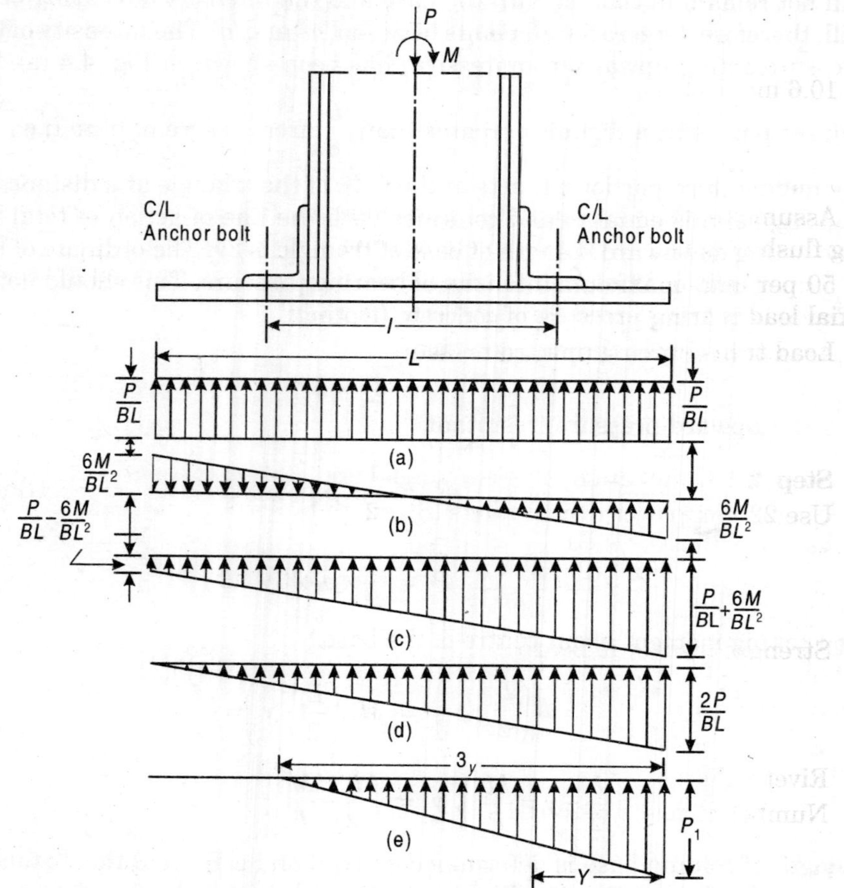

Fig. 4.4 *Distribution of stresses in column carrying load and moment*

The length and width of column base are determined by trial and error method.

Example 4.3 *Design a suitable base for column carrying an axial load of 300 kN and a moment on 40000 mm-kN in the plane of the web. The action of column is HB 300, @ 0.588 kN/m. The allowable bearing pressure on the footing is 4 N/mm^2.*

Solution

Design :

Step 1: Force/Load transmitted

Provide a gusseted base plate

Provide gusset plate 16 mm thick, and gusset angles 2 ISA 150 mm × 115 mm × 15 mm, (2ISA 150115, @ 0.295 kN/m)

Assume, moment is transferred to the gusset plate by flanges only.

$$\text{Force due to moment} = \left(\frac{\text{Moment}}{\text{Lever arm between flanges}}\right)$$

$$F = \left(\frac{40,000}{(300-10.6)}\right)\text{kN}$$

10.6 mm is thickness of flange of HB 300, @ 0.588 kN/mm

$$F = \left(\frac{40,000}{298.4}\right) = 138.217 \text{ kN}$$

Assume that the gusset plates, gusset angles and end of the column are having flush end, i.e., having complete bearing over whole area.

50 per cent of axial load is transferred through bearing and 50 per cent of axial load is transferred through both the faces.

Load transferred through one face

$$\left(\frac{1}{4}\times 300 + 138.217\right) = 213.217 \text{ kN}$$

Step 2 : Connection of cleat angle and base plate

Use 22 mm power driven rivets. Strength of rivet in angle shear

$$\left(\frac{\pi}{4}\times\frac{(23.5)^2\times 100}{1000}\right) = 43.35 \text{ kN}$$

Strength of rivet in bearing

$$\left(\frac{23.5\times 16\times 300}{1000}\right) = 112.8 \text{ kN}$$

Rivet value = 43.35 kN

Number of rivets required

$$\left(\frac{213.217}{43.35}\right) = 4.918$$

Provide 6 rivets on each face in two vertical rows, and 4 rivets are provided in horizontal on either side of centre line to connect the cleat angle and the gusset plate.

Step 3: Base plate. When anchor bolts are not resisting the over-turning moment.

Length of base (dimensions parallel to web)

$$L' = [(3000 + 2 \times 16 + 2 \times 11.5)] - \times 10$$

$$= 562 \text{ mm}$$

Provide length of the base plate equal to 570 mm

From Eq. 4.4

$$\left(\frac{L}{2} - y\right) = \frac{M}{P}, \left(\frac{570}{2} - y\right) = \left(\frac{4,00,00}{300}\right)$$

$$\therefore \quad y = 151.667 \text{ mm}, 3y = 455 \text{ mm}$$

Allowable bearing pressure on footing is 4 N/mm^2

Then from Eq. (4.3)

$$\left(\frac{1}{2}p_1 \times 3y \times B\right) = P$$

or $$\left(\frac{1}{2} \times 4 \times 455 \times B\right) = 300 \times 1000, \quad \therefore \; B = 329.67 \text{ mm}$$

Provide width of base plate B as 400 mm

Actual maximum intensity of pressure p_1', then

$$\frac{1}{2}p_1' \times 455 \times 400 = 3.0 \times 1000$$

$$p_1 = 3.296 \text{ N/mm}^2$$

Length of cantilever portion

$$(115 - 15) + 4 = 104 \text{ mm}$$

The pressure intensity P_1 reduces to zero in a length 455.1 mm from right side. The intensity of pressure at a section XX at a distance 104 mm

$$p = 3.296 \times \left(\frac{455 - 105}{455}\right) = 2.543 \text{ N/mm}^2.$$

The bending moment at section XX, in cantilever portion of 1 mm strip

$$= \frac{1}{2} \times 2.543 \times 104^2 + \frac{1}{2} \times 104\,(3.296 - 2.543) \times \frac{2}{3} \times 104$$

$$= (13752.544 - 2714.816) = 16467.36 \text{ mm-N}$$

Moment of resistance of 1 mm strip

$$\left(\frac{1}{6} \times t^2 \times 0.66 \times 260\right) = 16467.36, \quad \therefore \; t = 23{\cdot}995 \text{ mm}$$

Step 4: The thickness of base plate :

= t–thickness of gusset angle

= (23.995 –15) = 8.995 mm < thickness of gusset plate

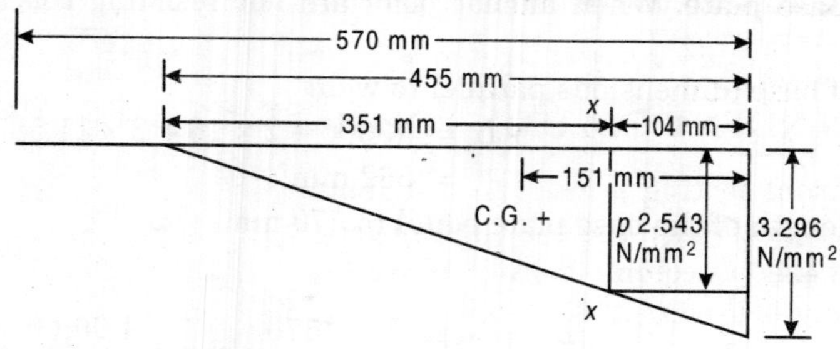

Fig. 4.5 *Stress distribution diagram*

Provide 570 mm × 400 mm × 16 mm thick base plate.

It is fixed to the column. Provide two anchor bolts of 20 mm diameter each.

Example 4.4 *Design available base in Ex. 4.3, when anchor bolts also resist overturning moment.*

Solution

Design :

Step 1 : Forces transmitted

Design of riveted connections remains same. Provide a gusseted base plate of 570 mm × 400 mm size. Provide two anchor bolts in gusset angle on each side at a distance 54 mm from the edge of the base plate upto centre line of anchor bolt.

From Eq. (4.5)

$$(P + P_1) = R$$

or $$(300000 + P_1) = \left(\frac{1}{2} p_1 \times 3y \times B\right)$$

or $$P_1 = \left(\frac{1}{2} \times 4 \times 3y \times 400 - 300 \times 1000\right)$$

or $$P_1 = (2400\,y - 300000) \text{ N}$$

From Eq. (4.6)

$$P_1(a - y) + P\left(\frac{L}{2} - y\right) = M$$

Substitute values of P_1, p, a L, and M

or $$(2400y - 300000)\,(570 - 54 - y) + 30{,}0000\left(\frac{570}{2} - y\right)$$

$$2.4y^2 - 1238\,y + 1093 \times 100 = 0$$

$$y = \frac{1238 \pm (1238^2 - 4 \times 2.4 \times 1093 \times 100)^{1/2}}{2 \times 2.4} = 113.02 \text{ mm}$$

$P_1 = (2400 \times 113.02 - 300000) = -28752 \text{ N} = -28.752 \text{ kN}$

This shows, that the vertical load is larger than that due to overturning.

Two bolts each of 20 mm diameter is provided to maintain the position of the column.

4.5 INDEPENDENT COLUMN FOOTING

The independent footing is used to distribute the load over large area from individual column. When the bearing capacity of the soil is not poor, then concrete block is used as column footing. When the bearing of the soil is poor, then grillage footing is used.

4.5.1 Grillage Footing

The steel grillage is a frame-work of rolled steel beams forming foundation in the soil of poor bearing capacity. In the grillage footing steel beams are used in one tier or more depending upon spread of the base. In case, two or more tiers are used, each tier is placed perpendicular to the tier above it. The beams are embedded in cement concrete and are protected from corrosion. As far as possible, square base is provided in the independent footing.

The column load P is distributed through a column base of length a to the grillage beam of length L as shown in Fig. 4.6.

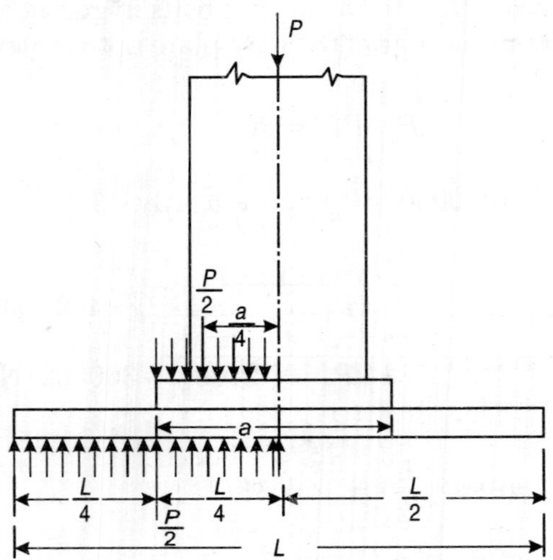

Fig. 4.6 *Shear force and bending moment in steel grillage beams*

It is assumed that each tier distribute the load uniformly to the lower tier and intensity of bearing pressure is also uniform. A load $\frac{P}{2}$ at a distance $\frac{a}{4}$, from the centre line acts downward, and resisting force $\frac{P}{2}$ at a distance $\frac{L}{4}$ from centre line acts upward.

The maximum bending moment occurs at the centre of beam. Maximum bending moment

$$M = \frac{P}{2}\left(\frac{L}{4} - \frac{a}{4}\right) = \frac{P}{8}(L - a) \qquad \text{... (4.7)}$$

The maximum shear force occurs at the edge of the base plate.

Maximum shear force,

$$F = \frac{P}{L}\left(\frac{L-a}{2}\right) = \frac{1}{2}\frac{P}{L}(L - a) \qquad \text{...(4.8)}$$

The grillage beams are kept unpainted and solidly encased in ordinary dense concrete with 100 mm aggregate.

The pipe separators or equivalent are used to keep the grillage beams properly spaced and the beams are spaced apart so that the distance between edges of adjacent flanges is not less than 75 mm.

The thickness of the concrete cover on the upper flange, at the ends, and the outer edges of the sides of the outermost beam is not less than 100 mm.

The concrete is properly compacted solid around all beams.

The distribution of load, soil pressure, shear force and bending moment diagrams are shown in Fig 4.7.

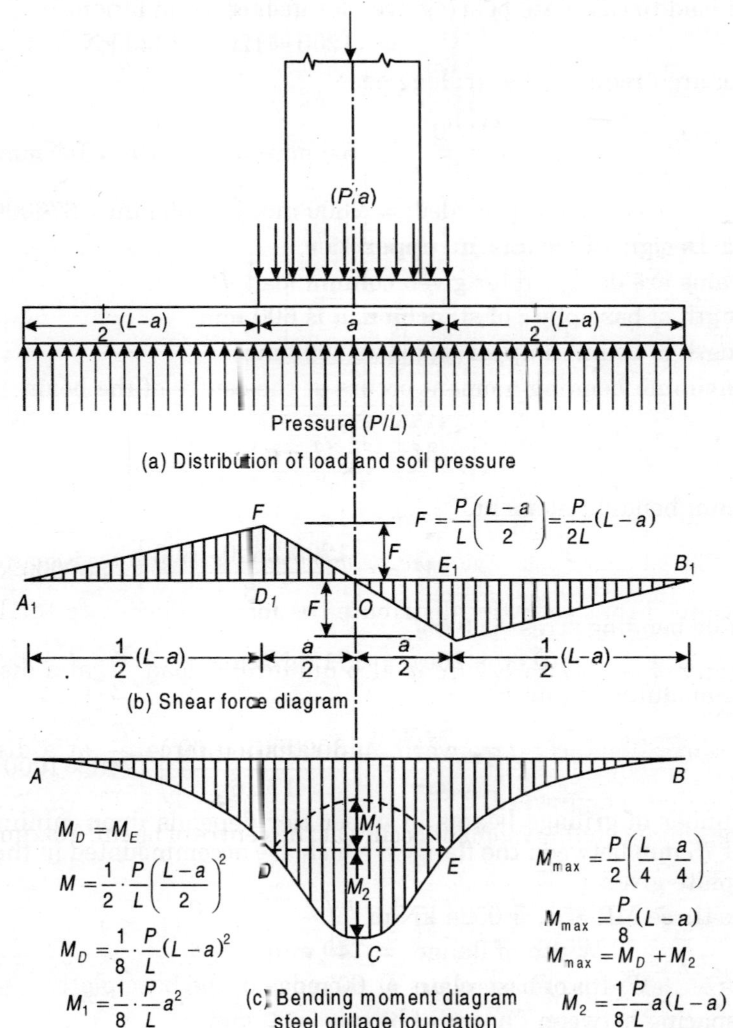

Fig. 4.7

Example. 4.5 *Design two tier grillage foundation to carry a stanchion, designed to carry 1200 kN. The base plate of the stanchion, is 600 mm square in size. The bearing pressure of earth is limited to a value of 200 kN per square metre. Take the allowable bending stress and the allowable average shear stress as per IS : 800–1984.*

Solution

Design :

Step 1 : Area of footing required

Direct axial load on stanchion P = 1200 kN

Add 10 per cent of direct axial load for self-weight of the grillage foundation.

Design load to calculate bearing area for grillage foundation

$$= (1200 + 120) = 1320 \text{ kN}$$

Bearing area required for grillage base

$$\left(\frac{1320}{20}\right) = 6.6 \text{ mm}^2 = 6.6 \times 10^3 \times 10^3 \text{ mm}^2$$

$$\text{Area provided} = 2600 \text{ mm} \times 2600 \text{ mm} = 6760000 \text{ mm}^2$$

Step 2: Design of beams in upper tier

The beams are designed for given column load, P.

The length of base plate of stanchion, a is 600 mm

The length of beam in upper tier, L is 2600 mm

The maximum bending moment occurs at the centre of the beam

$$M = \frac{P}{8}(L-a)$$

Maximum bending moment,

$$M = \left[\frac{1200(2600-600)}{8}\right] = 300000 \text{ kN-mm}$$

Allowable bending stress is $0.66f_y$

$$(0.66 \times 260) = 171.6 \text{ N/mm}^2$$

Section modulus required

$$Z = \frac{M}{f_b} = \left(\frac{300000 \times 1000}{171.6}\right) 1748 \times 1000 \text{ mm}^3$$

The number of grillage beams in upper tier depends upon minimum clear spacing of 75 mm between the flanges, which are accommodated in the width of the base plate given.

Provide three MB 350, @ 0524 kN/m

$$\text{Width of flange} = 140 \text{ mm}$$

$$\text{Width of base plate} = 60 \text{ mm}$$

Clear spacing between flanges of beams is 90 mm

Section modulus provided

$$(3 \times 778.9 \times 1000) = 2336.7 \times 1000 \text{ mm}^3$$

Step 3: Check for shear force

The maximum shear force in the beam is at the edge of the base plate.

Maximum shear force at the edge

$$= \frac{1}{2}\frac{P}{L}(L-a)\frac{1}{n}$$

where, n is the number of beams in upper tier

Maximum shear force

$$\left[\frac{1}{2}\times\frac{1200}{2600}(2600-600)\times\frac{1}{3}\right] = 153.846 \text{ kN}$$

From steel tables thickness of web of beam section

$$t_w = 8.1 \text{ mm}$$

The depth of beam section h is 350 mm

Average shear stress

$$f_s = \left(\frac{153.846\times1000}{8.1\times350}\right) = 54.267 \text{ N/mm}^2$$

It is less than allowable average shear stress

$(0.4 \times f_y) = (0.4 \times 260) = 104 \text{ N/mm}^2$ as per IS : 800–1984.

Step 4: Check for crippling stress

The crippling of web may occur in beams in upper tier (Discussed in Chapter 6)

$$\text{Crippling stress} = \left(\frac{P}{(a+2\sqrt{3h_2})t_w \cdot n}\right)$$

where, a = length of base plate

n = number of beams

h_2 = depth of the root of fillet from top of flange.

From steel section table, h_2 = 31 mm

$$\text{Crippling stress} = \left(\frac{1200\times1000}{10\times(60+2\sqrt{3}\times3.1)\times8.1\times3}\right)\text{N/mm}^2$$

$$= 69.727 \text{ N/mm}^2$$

It is less than allowable bearing stress

$$\sigma_p = 0.75 f_y = (0.75 \times 260) = 195 \text{ N/mm}^2$$

Step 5: Design of beam in lower tier

The beams are designed for given column load P.

Maximum bending moment

$$M = \frac{P}{8}(L-a) = \left(\frac{1000}{8}\times(2600-600\right) = 300000 \text{ mm-kN}$$

Section modulus required

$$Z = \left(\frac{300000}{171.6}\times1000\right) = 1748 \times1000 \text{ mm}^3$$

The number of beams depends upon length of beams in upper tier and minimum clear spacing of 75 mm.

Provide 9 number of beams LB 250, @ 6 279 kN/m in lower tier.

Section modulus of one beam is 297.4×10^3 mm^3

$$\text{Thickness of web} = 8.2 \text{ mm}$$

$$\text{Width of flange} = 125 \text{ mm}$$

Section modulus provided

$$(297.4 \times 10^3 \times 9) = 2676.6 \times 10^3 \text{ mm}$$

Clear distance between flanges

$$\frac{1}{8}\times(2600 - 8\times125) = 208 \text{ mm} > 75 \text{ mm}$$

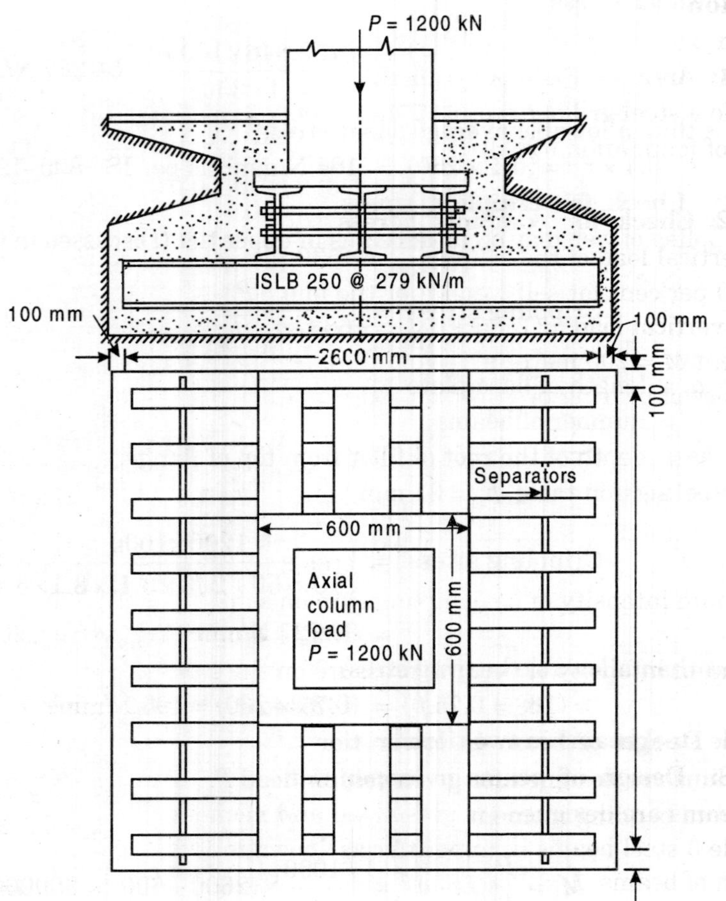

Fig. 4.8 *Steel grillage foundation*

Step 6: Check for shear force: Shear force

$$\frac{1}{2}\frac{P}{L}(L-a)\times\frac{1}{n} = \left(\frac{1}{2}\times\frac{1200}{2600}\times\frac{(2600-600)}{9}\right) = 51.28 \text{ kN}$$

$$\text{Average shear stress} = \left(\frac{51.28\times1000}{8.2\times250}\right) = 25 \text{ N/mm}^2$$

It is less than the allowable average shear stress 104 N/mm^2 as mentioned above.

The web crippling does not occur in lower tier.

Design of grillage foundation has been shown in Fig. 4.8.

Example 4.6 *Design a foundation footing for a column carrying 1530 kN of vertical load along with a moment in the same vertical plane equal to 200 m-kN. The bearing capacity of the soil is 200 kN/m^2. The width of foundation should not exceed 2 metres.*

Solution

Design :

Step 1: Area of footing required

Provide a steel grillage footing 2.0 m wide × 5.0 m long.

Area of foundation footing

$$A = 2.0 \times 5.0 = 10.0 \text{ m}^2$$

Step 2: Check for combined stress

The vertical load of the column, P is 1530 kN

Add 10 per cent for self-weight for the foundation.

Total vertical load = (1530 +153) = 1683 kN

Moment M = 200 metre-kN

Intensity of bearing pressure on soil

$$f_b = \left(\frac{P}{A} \pm \frac{M}{Z}\right) = \left(\frac{1683}{10} \pm \frac{200}{\frac{1}{6} \times 2 \times 5^2}\right)$$

$$= (168.3 \pm 24) \text{ kN/m}^2$$

Maximum intensity of bearing pressure on soil

$$= 192.3 \text{ kN/m}^2 < \text{Bearing capacity of the soil}$$

Minimum intensity of bearing pressure on the soil

$$= (168.3 - 24.0) = 143.3 \text{ kN/m}^2 \text{ compressive}$$

Provide steel beams in two tiers.

Step 3: Design of beams in upper tier

The beams are designed for given load and moment.

Provide 3 steel beams in upper tier as shown in Fig. 4.8

Length of beams L = 2 m

Vertical load on each beam is $\left(\frac{1530}{3}\right)$ = 510 kN

In addition to this, two outer beams take a additional load because of moment. Assume size of base plate of 800 mm × 600 mm and arrange the beams as shown in Fig. 4.9.

Additional load on each beam

$$\left(\frac{200 \times 1000}{600}\right) = 333.3 \text{ kN}$$

Maximum vertical load on outer beam

$$(510 + 333.3) = 843.3 \text{ kN}$$

Maximum bending moment

$$M = \frac{W(L-a)}{8} = \left(\frac{843 \cdot 3(2000-600)}{8}\right) = 147577 \text{ mm-kN}$$

Section modulus required for one beam

$$Z = \frac{M}{\sigma_b} = \left(\frac{147577 \times 1000}{171.6}\right) = 860 \times 10^3 \text{ mm}^3$$

Provide MB 400, @ 0.616 kN/m

Section modulus = 1022.9×10^4 mm^3

Thickness of web = 8.9 mm

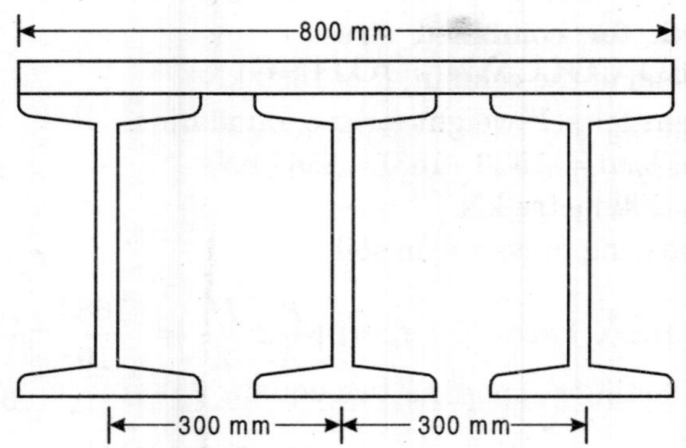

Fig. 4.9 *Steel grillage beams in upper tier*

Step 4: Check for shear force

Maximum shear force

$$\frac{W(L-a)}{2L} = \left(\frac{843.3 \times (2000-600)}{2 \times 2000}\right) \text{kN}$$

$$\left(\frac{843.3 \times 1400}{4000}\right) = 295.1 \text{ kN}$$

$$\text{Average shear stress} = \left(\frac{295.1 \times 1400}{400 \times 8.9}\right) = 82.893 \text{ N/mm}^2$$

It is less than allowable average shear stress 104 N/mm^2

$= (0.4 \times 260)$ N/mm^2 as per IS : 800–1984.

Step 5 : Design of beam in lower tier

The beams are designed for given load and moment

Provide 8 beams in lower tier

Length of beams = 5 m

Moment due to vertical load

$$\left(\frac{1530(5000-800)}{8}\right) = \left(\frac{1530\times4200}{8}\right) = 8032.5 \times 100 \text{ mm-kN}$$

Total moment = 100 × (8032.5 + 2000) = 10032.5 × 100 mm-kN

Moment shared by each beam

$$\left(\frac{10032.5\times100}{8}\right) = 1254.06 \times 100 \text{ mm-kN}$$

Section modulus required for one beam

$$\left(\frac{1254\cdot06\times1000\times100}{171\cdot6}\right) = 730.804 \times 13^3 \text{ mm}^3$$

Provide LB 350, @ 0.495 kN/m

Section modulus = 751.9×10^3 mm³.

4.6 COMBINED COLUMN FOOTING

The combined footing is used to distribute loads from two or more columns over large area. The steel grillage footing used as combined footing to support two column loads, has been described here. Two column loads may be equal and may be unequal.

4.6.1 Steel Grillage Footing to Support two Equal Column Loads

A steel grillage footing supporting two equal column loads has been shown in plan in Fig. 4.10.

The size of the column bases, and centre to centre distance between columns are known. The rectangular shape for the base of the footing is used. The projections of beams in lower tier are kept such that bending moments under the columns and centre of span of beams are approximately equal. It gives economical design. Length of base of footing is thus known. The width of base of footing is computed from area of base required to spread the two column loads. In case

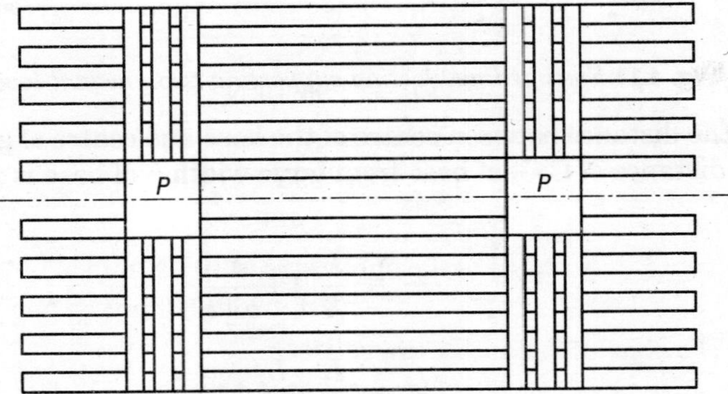

Fig. 4.10 *Steel grillage footing supporting two equal loads*

width of base of footing is restricted comparative more projections are provided in beams in lower tier, and beams are designed for maximum bending moment.

4.6.2 Steel Grillage Footing to Support two Unequal Column Loads

A steel grillage footing supporting two unequal column loads P_1 and P_2 has been shown in plan in Fig. 4.11.

The line of action of the resultant of the two unequal column loads is made to coincide with the centre of gravity of base of the footing. The trapezoidal shape is used for the base of the footing. The centre to centre distance between the two columns K is known. The projections of beams in lower tier m and n are assumed, such that, as far as possible, bending moments under the columns and between two columns are approximately equal. However, it is by trial to fix up projections most satisfactory. In case, there are restrictions for projections, the projections are fixed accordingly. The length of the base of the footing L is known, knowing the centre to centre distance of columns K and fixing the projections of beams in lower tier. The area of base of footing, A, required to spread the two column loads is known, knowing the allowable bearing pressure of soil. The widths of sides of trapezoidal base, b and c as shown in Fig. 4.11 are determined as given below.

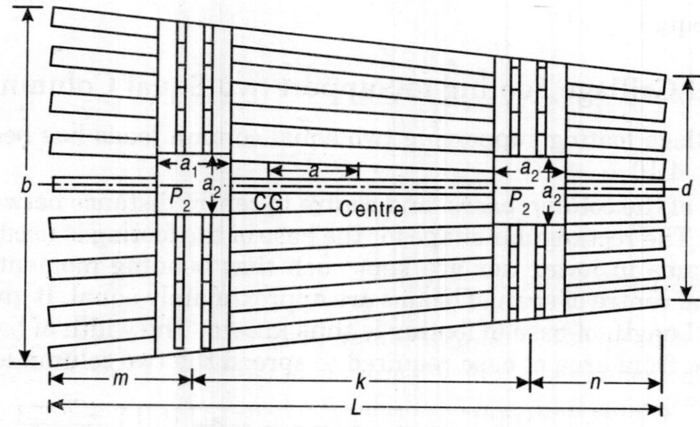

Fig. 4.11 *Steel grillage footing supporting two unequal loads*

Let d be the distance between centre of the base and centre of gravity of base (C.G.). The distance of C.G. of base from large width b of base is given by

$$\bar{x} = \frac{L}{3}\left(1+\frac{c}{b+c}\right)$$

then

$$d = \frac{L}{2}-\frac{L}{3}\left(1+\frac{c}{b+c}\right)$$

or $$6d = 3L - 2L\left(1+\frac{c}{b+c}\right)$$

or $$\frac{6d}{L} = \left(3-2\frac{2c}{b+c}\right), \frac{6d}{L} = \left(1-\frac{2c}{b+c}\right)$$

or $$\frac{6d}{L} = \left(\frac{b+c-2c}{b+c}\right), \frac{6d}{L} = \frac{(b-c)}{(b+c)}$$

Multiply both sides by $\frac{A}{L}$

$$\frac{6d}{L}\times\frac{A}{L} = \frac{(b-c)}{(b+c)}\times\frac{A}{L}, \text{ (but } (b+c)\,.\,L = 2A)$$

$$\frac{6d}{L}\times\frac{A}{L} = \frac{(b-c)A}{2A}$$

or $$\frac{6A\cdot d}{L^2} = \frac{(b-c)}{2} \quad \text{...(i)}$$

and $$\frac{A}{L} = \frac{(b+c)}{2} \quad \text{...(ii)}$$

then from (i) and (ii) $$b = \left(\frac{A}{L}+\frac{6A\cdot d}{L^2}\right) \quad \text{...(4.9)}$$

and $$c = \left(\frac{A}{L}-\frac{6A\cdot d}{L^2}\right) \quad \text{...(4.10)}$$

or b and c are given by $\frac{A}{L}\left(1+\frac{6\cdot d}{L}\right)$

where plus sign gives b, and minus sign gives c.

In case the combined grillage footing is of two tiers, then maximum bending moments for beams in upper tier under the heavy column load P_1 and the other column load P_2 are given below.

Maximum bending moment for beams in upper tier under heavy column load:

$$M_1 = \frac{P_1(L_1-a_1)}{8}$$

where, L_1 = length of the beam under the heavy column

a_1 = Width of the base plate under the heavy column along the length of beams

Maximum bending moment for beams in upper tier under other column load:

$$M_2 = \frac{P_2(L_2-a_2)}{8}$$

where, L_2 = Length of the beam under the other column

a_2 = Width of the base plate under the other column along the length of beams

The beams are designed for maximum bending moments and checked for shear.

For the design of beams in lower tier maximum bending moments are determined at three sections as below :

Let x be the distance from large end of trapezoidal base upto corresponding section as shown in Fig. 4.12. Then, bending moment due to intensity of pressure of soil w acting upward is

From Fig. 4.12,

$$\frac{y}{(L-x)} = \left(\frac{b-c}{2}\right)\cdot\frac{1}{L}$$

$$y = \left(\frac{L-x}{L}\right)\left(\frac{b-c}{2}\right)$$

$$2y = (L-x)\left(\frac{b-c}{L}\right)$$

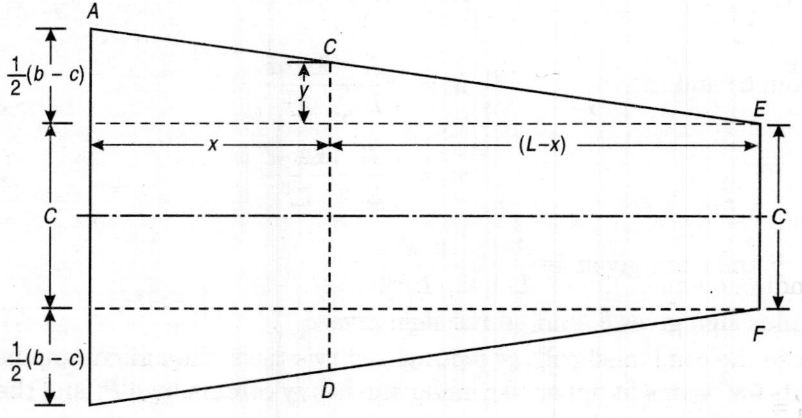

Fig. 4.12

Therefore,

$$CD = (c + 2y) = C + (L-x)\left(\frac{b-c}{L}\right)$$

$\therefore$
$$CD = b - x\left(\frac{b-c}{L}\right) \quad \text{... (ii)}$$

Area of trapezium $ABCD$

$$= \left(\frac{AB+CD}{2}\right)^2 \cdot x$$

$$= \frac{1}{2}\left[b + b - x\left(\frac{b-c}{L}\right)\right]\cdot x$$

$$= \frac{x}{2}\left[2b - \frac{x}{L}(b-c)\right]\cdot x \qquad \text{...(iii)}$$

Total load = (intensity) × area

$$= w\cdot\frac{x}{2}\left[2b - \frac{x}{L}(b-c)\right] \qquad \text{...(iv)}$$

Distance of centroid from right side

$$= \frac{\frac{x}{3}\left[2b + \frac{x}{L}(b-c)\right]}{\left[b + b - \frac{x}{L}(b-c)\right]} \qquad \text{...(v)}$$

Moment = Load × Distance of centroid

$$\frac{wx}{2}\left[2b - \frac{x}{L}(b-c)\right]\times\frac{x}{3}\frac{\left[3b - \frac{x}{L}(b-c)\right]}{\left[2b - \frac{x}{L}(b-c)\right]}$$

$$= \frac{wx^2}{2}\cdot\left[b - \frac{x}{3}\left(\frac{b-c}{L}\right)\right] \qquad \text{...(vi)}$$

$$= \frac{wx^2}{2}\left[b - \frac{x}{3}\frac{(b-c)}{L}\right] \qquad \text{...(4.11a)}$$

Bending moment under the heavy column load

$$M_1' = \frac{wx^2}{2}\left[b - \frac{x}{3}\left(\frac{b-c}{L}\right)\right] - \frac{P_1}{a_1}\times\frac{\left[x - \left(m - \frac{a_1}{2}\right)\right]^2}{3} \qquad \text{...(4.11b)}$$

Bending moment between the two columns

$$M_2' = \frac{wx^2}{2}\left[b - \frac{x}{3}\left(\frac{b-c}{L}\right)\right] - P_1(x-m) \qquad \text{...(4.12)}$$

Bending moment under the other column load

$$M_3' = \frac{wx^2}{2}\left[b - \frac{x}{3}\left(\frac{b-c}{L}\right)\right] - P_1(x-m)\frac{P_2}{a_2}\frac{\left[x - \left(m + K - \frac{a^2}{2}\right)\right]}{2} \qquad \text{...(4.13)}$$

M_1', M_2' and M_3' are differentiated with respect to x, and are equated to zero separately. The values of x are found for maximum bending moments in the above expressions. The maximum bending moments from the above expressions are determined by substituting values of x. The maximum of maximum bending moment is found and it is used for the design of beams in lower tier.

In the trapezoidal shape of the base of the footing, the beams in lower tier are arranged at close spacing at smaller end than that at the other end.

The rectangular shape is also adopted for combined footing to support two unequal column loads. The line of action of the resultant of two column loads is made to coincide with centre of gravity of the base. The cantilever projection near the other column. The length of the base is kept such that bending moments under the heavy column load and between two columns (though these are of opposite sign) are approximately equal.

4.7 APPROXIMATE METHOD OF DESIGN OF COMBINED FOOTING

The exact method of design of combined footing for two columns has been discussed in Sec. 4.5. The approximate method of design of combined footings is discussed here. The first point will be kept in view in dealing the combined grillage footing is that the line of action of the resultant of the two column loads is made to coincide with the centre of gravity of the base of footing.

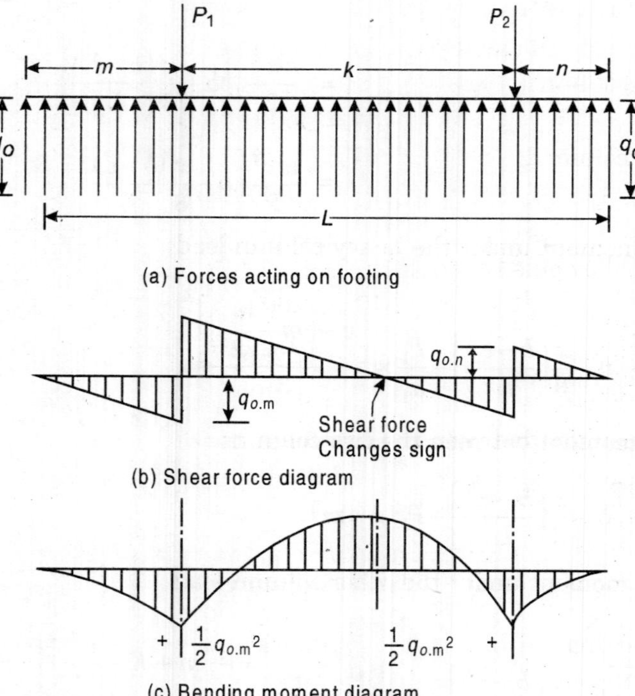

Fig. 4.13 *Design of combined footing by approximate method*

The size of column bases and centre to centre distance between columns are found by the least projection required from a consideration of the connection of a column base to the grillage. The beams in the lower tier in the grillage may be considered to be inverted beams loaded with the uniformly distributed load equal to the intensity of pressure under the grillage footing and supported at two points (i.e., at the columns). Thus, the beams have overhanging portions at both the ends. The beams are subjected to the cantilever moment at each end and the span moment in between the two columns. The cantilever moments and the span moments are of opposite sign. The cantilever moments are assumed to be maximum at the centre line of the columns. The maximum span moment is assumed to occur at the point, where the shear force changes sign. The maximum cantilever moments and the maximum span moments are determined. The beams are designed for the absolute maximum bending moment. The shear force and the bending moment diagrams are drawn. The beams are suitably spaced such that the clear distance between the flanges is not less than 75 mm.

The beams designed are checked for shear, web buckling and web crippling.

Example 4.7 *Design a combined foundation for two columns carrying an axial load of 1000 kN and 1500 kN respectively and placed 4 metre apart centre to centre. The foundation can bear 200 kN/m² of stress safely.*

Solution

Design :

Step 1 : Area of base required

Provide a steel grillage for combined foundation

Load from column 1,

$$P_1 = 1500 \text{ kN}$$

Load from column 2,

$$P_2 = 1000 \text{ kN}$$

Total load $P = 2500$ kN

Maximum allowable soil pressure

$$= 200 \text{ kN/m}^2$$

Area of base required $= \left(\dfrac{2500}{200}\right) = 12.5 \text{ m}^2$

Step 2 : Length and width of grillage base

The distance of line of action of resultant of two columns loads from 1500 kN load

$$\bar{x}_1 = \left(\frac{1000 \times 4}{2500}\right) = 1.60 \text{ m}$$

Provide rectangular base of the foundation and cantilever projection of 1.72 m near column 1.

Centre of gravity of base is kept coinciding with the line of action of resultant of column loads.

Length of column base

$$2(1.60 + 1.72) = (2 \times 3.32) = 6.64 \text{ m}$$

Width of grillage base

$$\left(\frac{12 \cdot 10}{6 \cdot 64}\right) = 1.88 \text{ m}$$

Provide grillage beams in two tiers.

Step 3: Design of beam in upper tier

$$\text{Length of beams} = 1.88 \text{ m} = 1880 \text{ mm}$$

Assume size of base plate under column 1 as 650 mm × 650 mm

Maximum bending moment for beams under column 1

$$\frac{1500}{8}(1880 - 650) = 2305 \times 100 \text{ mm-kN}$$

Total section modulus required

$$Z = \left(\frac{2305 \times 1000 \times 100}{171 \cdot 6}\right) = 1343.24 \times 1000 \text{ m}^3$$

Provide 3 LB 300, @ 0.377 kN/m.Total section modulus provided

$$(3 \times 488.9 \times 10^3) = 1466.7 \times 10^3 \text{ mm}^3$$

Width of flange of LB 300, @ 0.377 kN/m

$$= 150 \text{ mm}$$

Clear distance between flanges

$$= 100 \text{ mm} > 75 \text{ mm}$$

Assume size of base plate under column 2

$$= 600 \text{ mm} \times 600 \text{ mm}$$

Maximum bending moment for beams under column 2

$$\frac{1000}{8}(1880 - 600) = 1538 \times 100 \text{ mm-kN}$$

Total section modulus required

$$\left(\frac{1538 \times 1000 \times 100}{171 \cdot 6}\right) = 896.27 \times 10^3 \text{ mm}^3$$

Provide 3 LB 250, @ 0.279 kN/m

Width of flange = 125 mm

Clear distance between flanges

$$= 112.5 \text{ mm}$$

Total section modulus provided

$$3 \times 297.4 \times 10^3 = 898.2 \times 10^3 \text{ mm}^3$$

The beams under both the sections are checked for shear and web crippling.

Step 4: Check for S.F.

S.F. for beams under column 1

$$\left(\frac{1500}{2\times1880\times3}(1880-650)\right) = 163.8 \text{ kN}$$

From steel section tables for LB 300, @ 0.377 kN/mm²

$$h = 300 \text{ mm},$$
$$t_w = 9.4 \text{ mm},$$
$$h_2 = 27.45 \text{ mm}$$

Shear stress, $\tau_{vf} = \left(\frac{163\cdot8\times1000}{300\times9\cdot4}\right) \text{ N/mm}^2$

$$58.085 \text{ N/mm}^2 < (0.4 \times 260 = 104 \text{ N/mm}^2)$$

S.F. for beams under column 2

$$\left(\frac{1000}{2\times1880\times3}(1880-600)\right) = 113.4 \text{ kN}$$

From steel tables for LB 250, @ 0.279 kN/m

$$h = 250 \text{ mm},$$
$$t_w = 6.1\text{mm},$$
$$h_2 = 23.7 \text{ mm}$$

Shear stress, $\tau_{vf} = \left(\frac{113\cdot4\times1000}{250\times6\cdot1}\right) \text{ N/mm}^2$

$$= 74.36 \text{ N/mm}^2 < 104 \text{ N/mm}^2. \text{ Safe.}$$

Step 5 : Check for web crippling

Beams under column 1

Bearing stress, $\sigma_p = \left(\frac{1500\times1000}{(650+2\sqrt{3}\times27\cdot45)\times9\cdot4\times3}\right) \text{N/mm}^2$

$$71.389 \text{ N/mm}^2 < (0.75 \times 260 = 195 \text{ N/mm}^2). \text{ Safe.}$$

Beams under column 2

Bearing stress, $\sigma_p = \left(\frac{1500\times1000}{(650+2\sqrt{3}\times23\cdot7)\times6\cdot1\times3}\right) \text{ N/mm}^2$

$$80.113 \text{ N/mm}^2 < 195 \text{ N/mm}^2. \text{ Safe.}$$

Step 6 : Design of beams in lower tier

Intensity of soil pressure acting upward

$$= 200 \text{ kN/m}^2$$

The column loads acts downward as uniformly distributed load through the column bases.

The beams in lower tier act as inverted overhanging beams. Let x be the distance from left end of base. The bending moment under column 1 (see Fig.4.14)

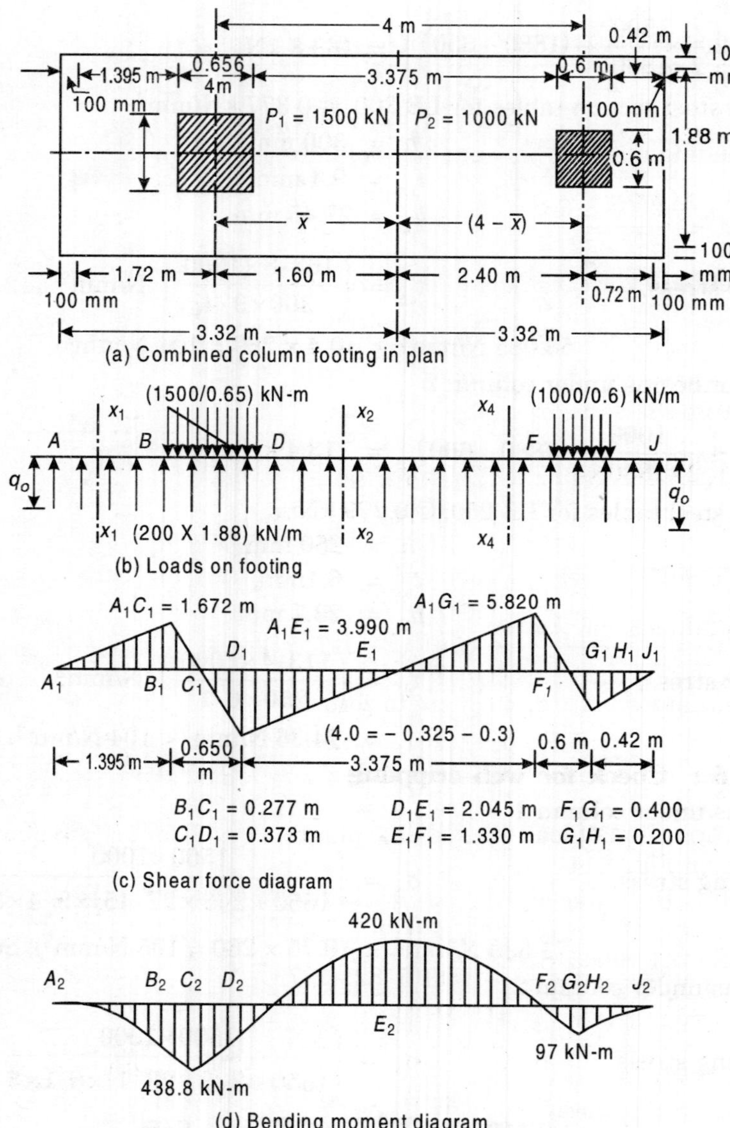

Fig 4.14

$$M_1' = \left(\frac{200 \times 1 \cdot 88 \times x^2}{2} - \frac{1500}{0 \cdot 65} \frac{(x - 1 \cdot 395)^2}{2} \right) \text{kN-m}$$

Differentiate w.r.t. x and equate to zero

$$\left(\frac{dM_1}{dx}\right) = 0 \qquad x = 1.672 \text{ m}$$

Substitute the value of x

$$M_1 = \left(\frac{200\times1\cdot88\times1\cdot672^2}{2} - \frac{1500(1\cdot672-1\cdot395)^2}{0\cdot65\times2}\right)\text{kN-m}$$

$$M_1 = 438.8 \text{ kN-m}$$

The bending moment between two columns

$$M_2' = \left(\frac{200\times1\cdot88\times x^2}{2} - 1500(x-1\cdot72)\right)\text{kN-m}$$

Differentiate w.r.t. x and equate to zero

$$\left(\frac{dM_2'}{dx}\right) = 0 \qquad \therefore x = 3.99 \text{ m}$$

Substitute the value of x

$$M_2' = \left(\frac{200\times1\cdot88\times3\cdot99^2}{2} - 1500(3\cdot99-1\cdot72)\right)\text{kN-m}$$

$$= 420 \text{ kN-m (negative)}$$

Bending moment under column 2

$$M_3' = \left(\frac{200\times1\cdot88\times x^2}{2} - 1500\left(x-1.7^2\right) - \frac{1000}{0\cdot60}\frac{(x-5\cdot42)^2}{2}\right)$$

Differentiate w.r.t. x and equate to zero

$$\left(\frac{dM_3'}{dx}\right) = 0 \qquad \therefore x = 5.82 \text{ m}$$

The distribution of load, shear force diagram and bending moment diagram are shown in Fig. 4.14

Substitute the value of x

$$M_3' = \left(\frac{200\times1\cdot88\times5\cdot82^2}{2} - 1500(5\cdot82-1\cdot72) - \frac{1000}{0\cdot60}\frac{(5\cdot82-5\cdot42)^2}{2}\right)$$

$$= 97 \text{ kN-m.}$$

Alternatively,

Step 1: Shear force

The positions of maximum bending moments may be located by writing the expressions for shear forces and equating them to zero. The shear force under the column 1 (see. Fig. 4.11).

$$F_1 = 200\times1.88x - \frac{1500}{0\cdot65}(x-1\cdot395) = 0, \quad \therefore x = 1.672 \text{ m}$$

The shear force between the two columns

$$F_2 = 200 \times 1.88\,x - 1500 = 0, \quad \therefore x = 3.99 \text{ m}$$

The shear force under the column 2

$$F_2 = \left(200 \times 1 \cdot 88x - 1500 - \frac{1000}{0 \cdot 60}(x - 5 \cdot 42)\right) = 0, \qquad \therefore x = 5.82 \text{ m}$$

Step 2 : Bending moments

The values of bending moment at these locations have been found as above.

Maximum of maximum bending moments occur under column 1

$$M_1' = 438.8 \text{ kN-m}$$

Step 3 : Selection of section

Total section modulus required

$$\left(\frac{438 \cdot 8 \times 1000 \times 1000}{171 \cdot 6}\right) = 2557.11 \times 10^3 \text{ mm}^3$$

Provide 8 LB 275, @ 0.330 kN/m. Total section modulus provided

$$(8 \times 392.4 \times 10^3) = 1319.2 \times 10^3 \text{ mm}^3$$

Width of flange of LB 275, @ 0.330 kN/m

$$= 125 \text{ mm}$$

Clear distance between flanges

$$\left(\frac{1}{7}\right)(1880 - 8 \times 140) \text{ mm} = 108.57 \text{ mm} > 75 \text{ mm}$$

The web crippling does not occur in lower tier. The beams in lower tier may check for shear and these would be found safe.

The design of combined footing for two columns is shown in Fig. 4.15.

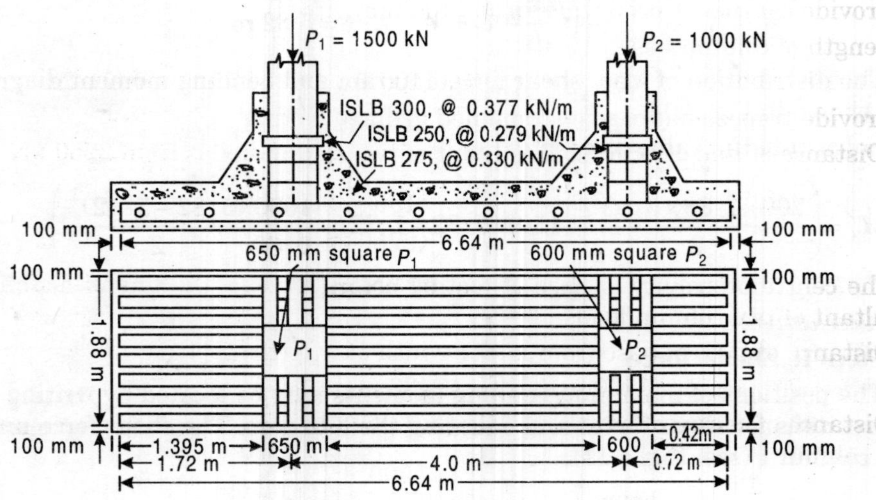

Fig. 4.15 *Rectangular combined footing*

Example 4.8 *Design a combined steel grillage foundation for two columns carrying loads of 2250 kN and 1800 kN spaced 3 m apart. The available space by the side of 1800 kN column is only 1400 mm. The maximum allowable soil pressure is 195 kN/m². Sketch the arrangement of upper and lower tiers with concrete encasing the joints.*

Solution

Design :

Step 1 : Area of base required

The steel grillage foundation is to be designed to support two column loads.

Column load $P_1 = 2250$ kN

Column load $P_2 = 1800$ kN

Total load $= 4050$ kN

Maximum allowable soil pressure

$= 195$ kN/m²

Bearing area of base required

$$A = \left(\frac{4050}{195}\right) = 20.769 \text{ m}^2$$

$$A = (20.769 \times 1000 \times 1000) \text{ mm}^2$$

Provide bearing area of base

Step 2 : Length and width of base

$$A = (21 \times 1000 \times 1000) \text{ mm}^2$$

Distance between centre to centre of column, K is 3 m

Available space by the side of 1800 kN column load is 1.40 m

$\therefore$ Cantilever projection, n as 1400 mm

Provide cantilever projection, m is 1100 mm

Length of column base

$$L = (1100 + 3000 + 1400) = 5500 \text{ mm}$$

Provide trapezoidal base or combined grillage footing

Distance of line of action of resultant of two column loads from 2250 kN

$$\bar{x} = \left(\frac{1800 \times 3000}{(2250 + 1800)}\right) = 1333 \text{ mm}$$

The centre of gravity of base of footing coincides with the line of action of resultant of two column loads.

Distance of C.G. of base from larger width of trapezoidal base

$$(1100 + 1333) = 2433 \text{ mm}$$

Distance of centre of base

$$\frac{L}{2} = \left(\frac{5500}{2}\right) = 2750 \text{ mm}$$

Distance between centre of base and C.G. of base

$$d = (2750 - 2433) = 317 \text{ mm}$$

Larger width of trapezoidal base is given by

$$b = \frac{A}{L} = \left(1+\frac{6d}{L}\right) = \left(\frac{21000000}{5500}\right)\left(1+\frac{6\times317}{5500}\right)$$

$$= 5140 \text{ mm}$$

Smaller width of trapezoidal base is given by

$$c = \frac{A}{L} = \left(1-\frac{6d}{L}\right) = \left(\frac{21000000}{5500}\right)\left(1-\frac{6\times317}{5500}\right)$$

$$= 2500 \text{ mm}$$

Let y_1 be the width of base at the centre of column supporting 2250 kN

$$y_1 = \left(c+(b-c)\frac{K+n}{L}\right)$$

$$= \left[2500+(5140-2500)\times\frac{3000+1400}{5500}\right]$$

$$= 4610 \text{ mm}$$

Let y_2 be the width of base at the centre of column supporting 1800 kN

$$y_2 = \left(c+(b-c)\frac{n}{L}\right)$$

$$= \left[2500+(5140-2500)\times\frac{1400}{5500}\right]$$

$$= 3172 \text{ mm}$$

Step 3 : Design of beams in upper tier

Beams under column supporting 2250 kN

Length of central beam, y_1 is 460 mm

Assume size of base plate under 1250 kN column load as 800 mm × 800 mm.

$\therefore \quad a_1 = 800$ mm

Maximum bending moment

$$M_1 = \frac{P_1}{8}(y_1-a_1) = \frac{2250}{8}(4610-800) = 10.71\times10^5 \text{ kN-mm}$$

Permissible bending stress for beams embedded in concrete

$$\sigma_b = (066 \times 260) = 171.6 \text{ N/mm}^2$$

Section modulus required

$$Z_1 = \left(\frac{M_1}{\sigma_b}\right) = \left(\frac{10.71\times10^5\times10^3}{171.6}\right)$$

$$= 6241 \times 10^3 \text{ mm}^3$$

From ISI Handbook No. 1, provide 3 LB 600, @ 0.995 kN/m

Section modulus of each beam section

$$= 2428.2 \times 10^3 \text{ mm}^3$$

Width of flange of beam, b_f is 210 mm

Total section modulus provided

$$3 \times 2428.9 \times 10^3 = 7286.7 \times 10^3 \text{ mm}^3$$

Clear distance between flanges

$$\frac{1}{2}(800 - 3 \times 210) = 85 \text{ mm} > 75 \text{ mm}$$

Step 4 : Beams under column supporting 1800 kN

Length of central beam, y_2 is 3172 mm.

Assume size of base plate under column supporting 1800 kN as 800 mm × 800 mm.

Maximum bending moment

$$M_3 \frac{P_2}{8}(y_2 - a_2) = \frac{1800}{8} \times (3170 - 800)$$

$$= 5.34 \times 10^5 \text{ kN-mm}$$

Section modulus required

$$Z_y = \left(\frac{M_2}{\sigma_b}\right) = \left(\frac{5.34 \times 10^5 \times 10^3}{171.6}\right)$$

$$= 3111.89 \times 10^3 \text{ mm}^3$$

From ISI Handbook No. 1, provide 3 LB 450, @ 0653 kN/m

Section modulus of three beams section

$$(3 \times 12238 \times 10^3) \text{ mm}^3 = 3671.4 \times 10^3 \text{ mm}^3$$

Width of flange of beam, f_y is 170 mm

Clear distance between flanges

$$\frac{1}{2}(800 - 3 \times 170) = 145 \text{ mm} > 75 \text{ mm}$$

The beams under both the columns may be checked in shear and web crippling and will be found safe.

Step 5 : Design of beam in lower tier

Bearing area of base provided

$$= 21000000 \text{ mm}^2$$

$$= 21 \text{ m}^2$$

$$\text{Total load} = 4050 \text{ kN}$$

Intensity of soil pressure acting upward

$$w = \left(\frac{4050}{21}\right) = 192.5 \text{ kN/m}^2$$

Let x be the distance from larger width of base.

The column loads acts downward as uniformly distributed load through the base plates. However, bending moment under column supporting 2250 kN load

$$M_1' = \frac{wx^2}{2}\left[b - \frac{x}{3}\cdot\frac{b-c}{L}\right] - \frac{2250}{0\cdot80}\frac{(x-0\cdot7)^2}{2}$$

$$= \frac{192\cdot5\times x^2}{2}\left[51\cdot4 - \frac{h}{3}\times\frac{5\cdot14-250}{5\cdot50}\right] - \frac{2250}{0\cdot80}\frac{(x-0\cdot7)^2}{2}$$

Equating $\left(\frac{dM_1'}{dx}\right) = 0$ for M_1' to be maximum, we get $x = 1.0$ m

$$\therefore \quad M_1' = \frac{192\cdot5\times1}{2}\left[51\cdot4 - \frac{1}{3}\times\frac{2\cdot64}{5\cdot50}\right] - \frac{2250}{0\cdot8}\times\frac{(1-0\cdot7)^2}{2}$$

$= +\ 373.5$ kN-m (Hogging)

Bending moment between two column loads

$$M_2' = \left(\frac{wx^2}{2}\right)\left[b - \frac{x}{3}\cdot\frac{b-c}{L}\right] - 2250(x-1\cdot10)$$

$$= \frac{192\cdot5\times x^2}{2}\left[5\cdot14 - \frac{x}{3}\times\frac{5.14-2.50}{5\cdot50}\right] - 2250(x-1\cdot10)$$

Equating $\left(\frac{dM_2'}{dx}\right) = 0$ for M_2' to be maximum, we get $x = 2.60$ m

$$M_2' = \frac{192\cdot5\times2\cdot6^2}{2}\left[5\cdot14 - \frac{260}{3}\times\frac{2\cdot60}{5\cdot60}\right] - 2250(2\cdot6-1\cdot10)$$

$= -275$ kN-m (Sagging)

Bending moment under column supporting 1800 kN

$$M_3' = \left(\frac{wx^2}{2}\right)\left[b - \frac{x}{3}\cdot\frac{b-c}{L}\right] - 2250(x-1\cdot10) - \frac{1800}{0\cdot80}\frac{(x-3\cdot70)^2}{2}$$

$$\frac{192\cdot5x^2}{2}\left[5\cdot14 - \frac{x}{3}\times\frac{5\cdot14-2\cdot50}{5\cdot50}\right] - 2250(x-1\cdot10) - \frac{1800}{0\cdot80}\frac{(x-3\cdot70)^2}{2}$$

Equating $\left(\frac{dM_3'}{dx}\right) = 0$ for M_3' to be maximum, we get $x = 4.15$ m

$$\therefore \quad M_3' = \frac{192\times41\cdot5^2}{2}\left[5\cdot14 - \frac{4\cdot15}{3}\times\frac{2\cdot64}{5\cdot50}\right]$$

$$- 2250(4 \cdot 15 - 1 \cdot 10) - \frac{1800}{0 \cdot 80} \frac{(4 \cdot 15 - 3 \cdot 70)^2}{2}$$

$$= +272.5 \text{ kN-m (Hogging)}$$

The distribution of load, shear force diagram, and bending moment diagram are shown in Fig. 4.16.

The maximum of maximum bending moment is 275 kN-m (Sagging).

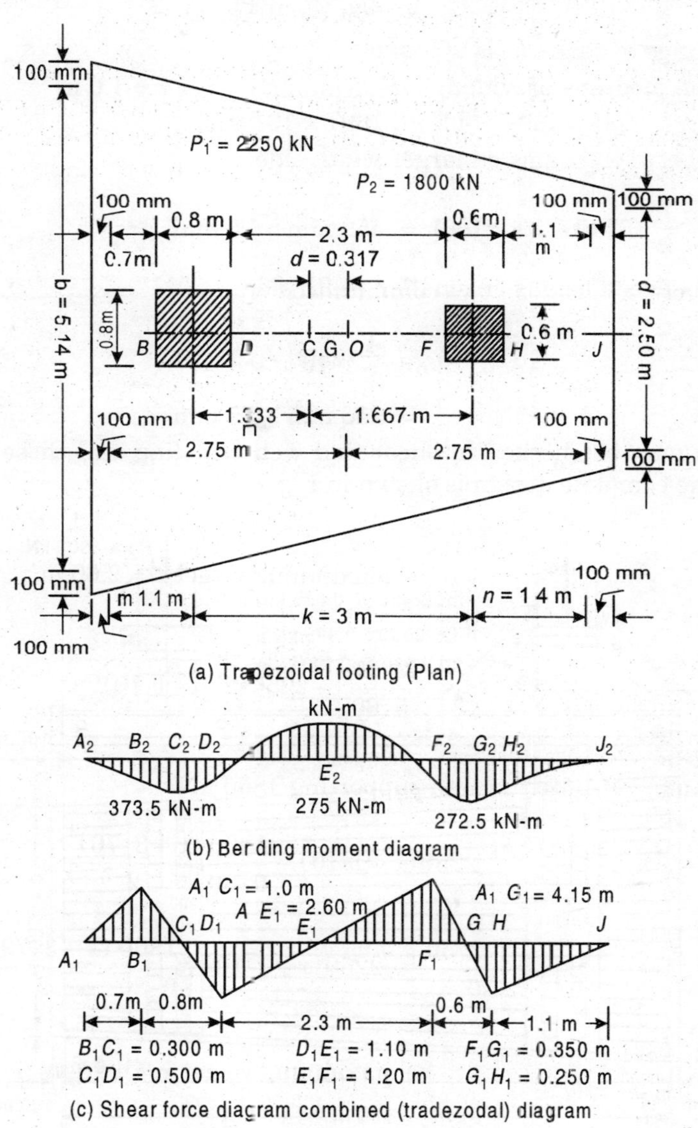

Fig. 4.16

Step 6 : Selection of beam section

The section modulus required

$$Z = \left(\frac{275 \times 10^3 \times 10^3}{171 \cdot 6}\right)$$

$$= 1602.56 \times 10^3 \text{ mm}^3$$

Provide 10 ISLB 200, @ 0.198 kN/m. The section modulus of each beam

$$= 169.7 \times 10^3 \text{ mm}^3$$

Width of flange of beam, b_y is 100 mm

Total section modulus provided

$$10 \times 169.7 \times 10^3 = 1697 \times 10^3 \text{ mm}^3$$

Clearance between beams on larger width side

$$\frac{1}{9}(5140 - 10 \times 100) = 459 \text{ mm} > 75 \text{ mm}$$

Clearance between beams on smaller width side

$$= \frac{1}{9}(2500 - 10 \times 100)$$

$$= 166 \text{ mm} > 75 \text{ mm}$$

The beams may be checked in shear and web crippling and these will be found safe. The complete design is shown in Fig, 4.17.

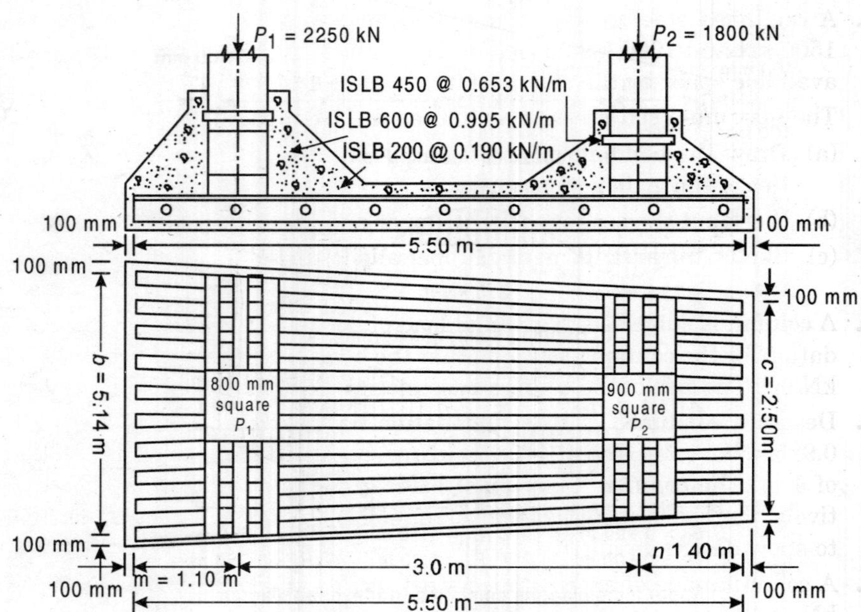

Fig. 4.17 *Trapezodial combined footing*

PROBLEMS

4.1. A steel column HB 300 @ 0.630 kN/m resting on a steel base plate is supported on concrete foundation. The column carries a central load of 3000 kN. Design the base plate and gusseted connection. The safe bearing capacity of concrete is 4 N/mm^2.

4.2. Design the base plate and its anchor bolts for a 300 mm × 140 mm @ 0.442 kN/m I-scction steel stanchion under an axial load of 350 kN and a moment of 4250 kN-mm in the plane of web.

4.3. Design a suitable riveted base plate connection for WB 200 column subjected to an axial load for 300 kN and a moment of 450 kN-mm at the bottom in the plane of web.

4.4. Design a suitable base for a column carrying an axial load of 240 kN and a moment of 25000 kN-mm in the plane of the web. The section of the column is HB 250, @ 0.547 kN/m.

4.5. Design a grillage foundation for a compound column consisting of HB 400, @ 0.822 kN/m with flange plate 300 mm × 18 mm one on each flange and carrying a load of 2300 kN. Draw a neat drawing.

4.6. Design a grillage foundation for a column carrying an axial load of 1500 kN inclusive of self-weight. The bearing capacity of soil is 150 kN/m^2. The column base plate resting on the grillage is 600 mm × 600 mm.

4.7. Design a two tier grillage foundation to carry a stanchion designed to carry 1040 kN. The base plate of the stanchion is 700 mm square. The bearing of earth is limited to a value of 150 kN/m^2.

4.8. A combined steel grillage foundation for two columns carrying loads of 1500 kN and 1000 kN spaced at 6100 mm centres is to be designed. The available space by the side of 1000 kN column is only 460 mm.

The maximum soil pressure is limited to 110 kN/m^2.

(a) Draw the bending moment and shear force diagrams for the upper tier of the grillage foundation.

(b) Work out the sections for the upper and lower tiers of the grillage.

(c) Sketch the arrangement of upper and lower tiers with concrete encasing.

4.9. A column is subjected to an axial load of 2000 kN. Design a grillage foundation for the column assuming that the bearing capacity of the soil is 100 kN/m^2. The size of the column base plate is 800 mm × 600 mm.

4.10. Design a combined grillage foundation for the two columns HB 450, @ 0.925 kN/m and HB 400, @ 0 774 kN/m spaced at centre to centre distance of 4 m. The columns carry axial loads of 2000 kN and 1500 kN respectively. The safe bearing capacity of the soil is 150 kN/m^2. Give neat sketches to show design details.

4.11. A column made of an HB rolled steel section carries an axial load of 400 kN and a bending moment of 36000 kN-mm at the bottom end acting about the major principal axis of the column section. The base plate is connected to the column flange by two 150 mm × 75 mm × 12 mm (2 ISA 15075, @ 0.201 kN/m) angles with 20 mm rivets, the 75 mm legs being outstanding from the column flanges. The end of the I-section is faced for complete bearing on the base plate. Using the data given, find (a) the required horizontal dimensions of the base plate, (b) the required thickness of the base plate and (c) the number of 20 mm rivets necessary to connect each angle to the column flange as governed by the 3600 kN-mm moment.

Properties of sections :

ISHB 250 column section : depth = 250 mm, flange width = 250 mm. flange thickness = 9.7 mm, area = 6971 mm^2.

Rivets : Nominal diameter 20 mm, diameter of rivet holes = 21 .5 mm.

Gauge distance of the rivet holes :

Column flange = 140 mm, 150 mm leg of angles distance from back to first row = 55 mm, from first row to second row = 65 mm.

Adopt *permissible stresses* as per IS : 800–1984.

4.12. It is required to design a two tier grillage foundation for a steel column carrying a load of 1000 kN. The dimensions of the base plate are about 800 mm × 850 mm. The footings is to be supported on concrete base which in turn rests on solid rock. The allowable pressure on the concrete base is 4 N/mm^2. Design the grillage using I-section girder built-up of suitable rectangular plates. Plates 12 mm and 20 mm thick and of any width and length may be assumed to be available. Assume standard working stresses as per IS : 800–1984.

Sketch the arrangement of the built-up beams forming the grillage.

4.13. Design a two tier grillage foundation footing for a HB 350 @ 0.674 kN/m column supporting a load of 1000 kN. The footing is to be supported on soil which has a bearing capacity of 100 kN/m^2.

4.14. Design a grillage foundation for a column consisting of HB 450 section with flange plate 400 mm × 40 mm each carrying an axial load of 4500 kN. The concrete can take a safe bearing pressure of 4 N/mm^2. Sketch the arrangement.

4.15. A column consisting of HB 200, @ 0.373 kN/m with cover plates 300 mm × 20 mm (one on each flange) is fixed at the base and free at the top. It carries a load of P at an eccentricity of 200 mm from the centroidal axis of member giving rise to bending moment about the major axis of the member. If the height of column is 4 m, design a suitable *attached base* for the column. Permissible bearing pressure on concrete is 4 N/mm^2.

Chapter 5

Design of Tension Members

5.1 INTRODUCTION

A *tension member* is defined as a structural member subjected to tensile force in the direction parallel to its longitudinal axis. A tension member is also called as a *tie member* or simply a *tie.* The term tie is commonly used for tension members in the roof trusses. When a tension member is subjected to axial tensile force, then the distribution of stress over the cross-section is uniform. The complete net area of the member is effectively used at the maximum permissible uniform stress. Therefore, a tensile member subjected to axial tensile is said to be efficient and economical member. In the building design, the interest in the use of tension members is increasing. The tension members are used as hangers for floors and cables for roof.

The various members used as tension members are not perfectly straight. The initial crookedness and imperfections of the member result in the small eccentricity. The distribution of stress in the eccentrically loaded tension member is not uniform. The member is subjected to combined stress, (i.e., direct tension and bending). Besides this, a tension member may be subjected to axial tension and bending.

5.2 TYPE OF TENSION MEMBERS

The types of structure and method of connection decide the type of tension member. The compact tension members are generally used. As far as possible, a large portion of the tension member is connected directly to the gusset plates or splice plates. This minimizes stress concentration in the members. The various types of tension members used may be grouped into following four groups:

(i) Wires and cables
(ii) Rods and bars

(iii) Single structural shapes and plates
(iv) Built-up members

5.2.1 Wires and Cables

The wire ropes are used for hoists, derricks, rigging slings, guy wires and hangers for the suspension bridges. The untwisted parallel wires are used for the manufacture of main cables of the suspension bridges. The wire ropes have the advantages of flexibility and strength. The wire ropes require special fittings for proper end connections. The wire ropes are used as guy wires or guy lines with the steel towers. The wire ropes are given initial tension. The initial tension prestresses the frame. The effectiveness of frame in resisting the external load is increased when the frames are prestressed.

5.2.2 Rods and Cables

The square and round bars as shown in Fig. 5.1 (b) and (c) are quite often used for the small tension members. The round bars are threaded at the ends and are used with nuts. The area of cross-section at the ends and are used with nuts. The area of cross-section at the root of the thread is the net area which resists

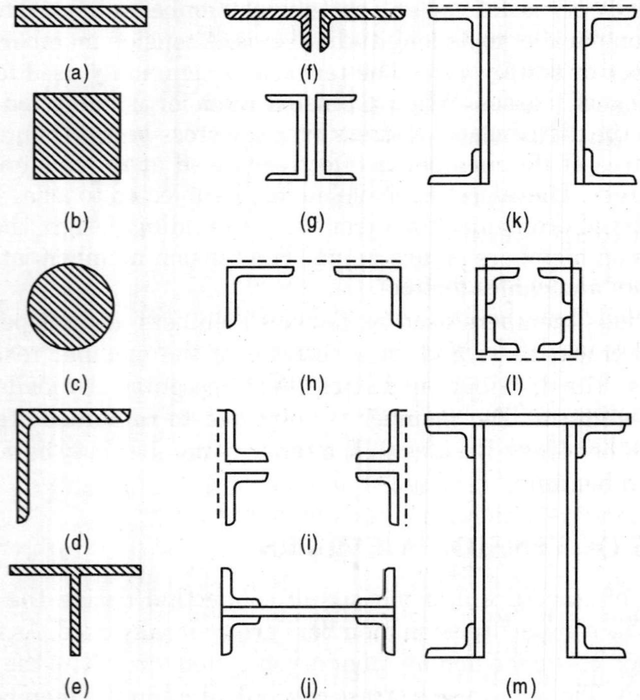

Fig. 5.1 *Various forms of tension members*

the tension. The round bars with threaded ends are used with pin-connections at the ends with standard clevises. Sometimes, the bars are bent to form loops at the ends instead of threads. The round bars with clevises and loop rods are made in two portions. These two portions are joined with turn buckles. The tension members are tightened with these turn buckles. Such members are commonly used for lateral bracing. The ends of rectangular bars or plates are enlarged by forging and then, these are bored to form eye bars. The eye bars are used with pin-connections. The rods and bars have disadvantages of inadequate stiffness resulting in noticeable sag under the self-weight.

5.2.3 Single-structural Shapes and Plates

The single structural shapes, viz., angle sections and tee-sections as shown in Fig. 5.1 (d) and (e) are used as tension members. The angle sections are considerably more rigid than the wire ropes, rods and bars. When the length of tension member is too long, then, the single angle section is also too flexible. The equal angle sections are commonly used when the angles are used as single structural members. The single angle members are also used for bracing. These members can resist small compression if reversal of stress takes place. The single angle sections have the disadvantages of eccentricity in both the planes in the riveted connections.

Sometimes, the single channel sections are also effectively employed as the tension members. The channel sections have eccentricity only in one direction. The single channel sections have high rigidity in the direction of web and low rigidity in the direction of flange.

Occasionally, the I-sections are also used as tension members. I-sections have more rigidity. The single I-sections are more economical than the built-up sections.

5.2.4 Built-up Members

Two or more than two members are used to form built-up members. When the single rolled steel sections cannot furnish the required area, then, the built-up sections are used. The built-up sections may be made more rigid and more stiff than the single structural shapes. The moment of inertia of the built-up sections may be increased. The built-up sections can resist compression if the reversal of stresses takes place.

The double angle sections of unequal legs shown in Fig. 5.1 (f) are extensively used as tension members in the roof trusses. The angle sections are placed back to back on two sides of a gusset plate. When both the angle sections are attached on the same sides of the gusset, then, built-up section has eccentricity in one plane. The built-up section is subjected to tension and bending simultaneously. The two angles sections may be arranged in the star shape (i.e., the angles are placed diagonal opposite to each other with legs on outer sides). The star shape angle sections may be connected by batten plates. The batten plates are alternately placed in two perpendicular directions. The star arrangement provides a symmetrical and concentric connection. Two angle sections as shown in Fig. 5.1 (h) are used in the two plane trusses, where two parallel gussets are used at each connection. Two angle sections as shown in Fig. 5.1 (h) have the advantage of adjusting the distance between them. Four angle sections as shown in Fig. 5. 1 (i) are also used in the two plane trusses. The angles are connected to two parallel gussets. Four angle sections connected by plates as shown in Fig. 5.1 (j) are used as tension members in the bridge truss girders.

A built-up section may be made of two channels placed back to back with a gusset plate in between them. Such sections are used for medium loads in single plane truss. In two plane trusses, two channels are arranged at a distance with their flange turned inward. It simplifies the transverse connections. It also minimizes lacing. The flanges of two channels are occasionally turned outward in order to have greater lateral rigidity.

The heavy built-up tension members in the bridge truss girders are made of angles and plates. Such members can resist compression if the reversal of stresses occurs.

5.3 NET SECTIONAL AREA

The *net sectional area* is used for tension member. The net sectional area of a tension member is the gross-sectional area of the member less the maximum deduction for holes. The projected area for each hole is a rectangle, one side of which is the thickness of the member and the other side is the diameter of the hole.

In chain riveting, the net sectional area of a member at a section is the gross-sectional area minus the area to be obtained from the product of thickness of the plate and the sum of the diameters of all rivet and bolt holes in that section.

In zig-zag riveting, the rivets used to connect the tension member are staggered as shown in Fig. 5.2.

The failure of plate may occur on a zig-zag section passing through all the three rivets instead of a straight right angle section passing through first two

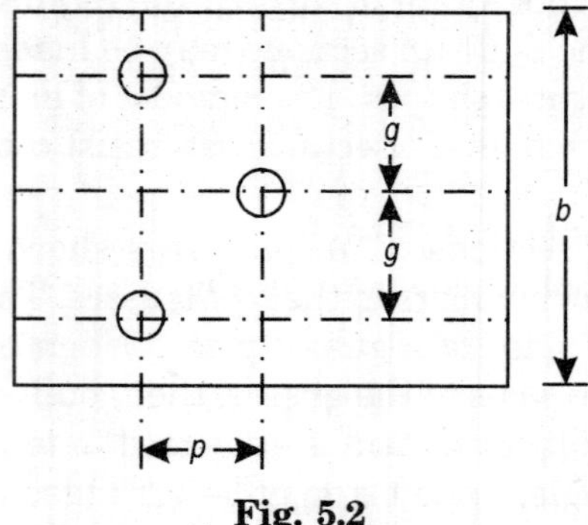

Fig. 5.2

rivets. In case all the three rivets are in one line along a right angle section then, the deduction is made for all the three rivet holes in determining the net area. When one or more than one rivet hole is off the line, the failure of the plate depends upon the staggered pitch, p the gauge distance g, and the diameter of the hole d. When the gauge distance g, is large as compared to the staggered pitch, p then, the failure may occur in zig-zag line. When the gauge distance in small, as compared to the staggered pitch, p then, the failure may occur along a straight right angle section passing through rivet holes in Fig. 5.2. When the staggered pitch, p and the gauge, g are fixed, then, the zig-zag failure become more likely as the diameter of hole, d increases. From the theoretical investigations and laboratory tests it is found that the net area of the plate for Fig. 5.2 may be found as below:

$$A_{net} = \left[A_{gross} - 2dt - d\cdot t\left(1 - 2\times\frac{p^2}{4g\cdot d}\right)\right] \quad \text{...(i)}$$

or

$$A_{net} = \left[bt - 2dt - d\cdot t\left(1 - 2\times\frac{p^2}{4g\cdot d}\right)\right] \quad \text{...(ii)}$$

or

$$A_{net} = \left[b - \left(3d - 2\times\frac{p^2}{4g\cdot d}\right)\right] \quad \text{...(iii)}$$

In general,

$$A_{net} = \left[b - \left(nd - n'\frac{p^2}{4g}\right)\right] \quad \text{...(5.1 a)}$$

where b = width of plate

t = thickness of plate

p = staggered pitch i.e., the distance between any two consecutive rivets in zig-zag chain measured parallel to the direction of stress in the number as shown in Fig. 5.2.

g = gauge space, that is, the distance between the same two consecutive rivets in a zig-zag chain at right angles to the direction of stress in the number as shown in Fig. 5.2. It is transverse spacing between parallel adjacent lines of fasteners.

n = number of rivet holes in zig-zag line.

n' = number of gauge spaces.

d = gross diameter of rivet.

Net sectional area of plate as shown in Fig. 5.2

$$A_{net} = \left[b - \left(3d - 2\frac{p^2}{4g\cdot d}\right)\right]\cdot t \quad \text{...(iv)}$$

Equation 5.1 is known as **Steinman's** formula.

In case, the rivets are connected in different staggered pitches, P_1, p_2, etc. and the gauge distances g_1, g_2, then, the deductions shall be as below :

$$\text{Deduction} = \left[(\text{Sum of section areas of holes}) - \left(\frac{p_1^2}{4g_1} + \frac{p_2^2}{4g_2}\right)t\right]$$

In case, the staggered rivets are used in the two legs of an angle section, then, the net area can also be determined by Eq. 5.1. Then, the riveted connections are in two different planes, viz., in non-planer sections, such as angles with the holes in both the legs. The gross width of angle is the sum of lengths of two legs less the thickness of angle. The gauge distance for the rivets in two legs is equal to the sum of gauge distance for the rivets in each legs less the thickness. It is to note that the standard rivet gauge distances for legs of angles are given is ISI Handbook No. 1.

The effective net area at approximately the root of the thread of a threaded tension member may be found from the following equation. This is as per AISC specification.

$$A_{e.net} = \frac{\pi}{4}\left(D - \frac{0.9743}{n}\right)^2 \text{ mm}^2 \qquad ...(5.1\text{ b})$$

where D = nominal outside diameter of threads in mm

n = number of threads per mm

5.4 NET EFFECTIVE SECTION FOR ANGLES AND TEES IN TENSION

When an angle section is used as tension member, it is connected with the gusset plate by one leg. When a tee-section is used as tension member, it is connected with the gusset plate by the flange. In such cases the rivets connecting tension member and the gusset plate do not lie on the line of action of the load.

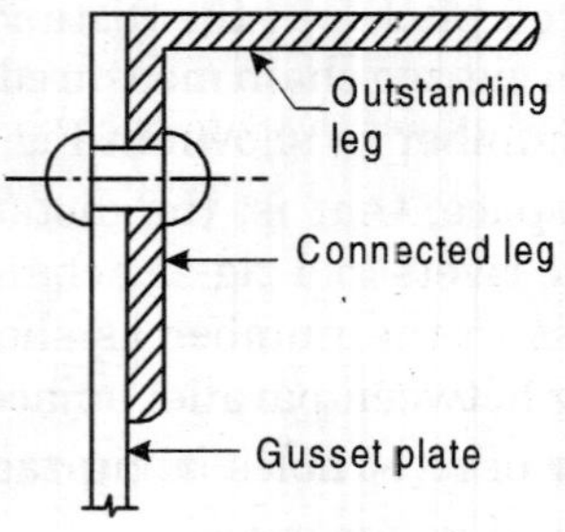

Fig. 5.3

This results in an eccentricity moment on the connection. The distribution of the stress becomes non-uniform because of eccentric connection. The net sectional area of an angle or a tee member is reduced to account for this non-uniform distribution of stress. The reduced net sectional area of such member is called *net effective area.* The net effective section for angles and tees in tension are adopted as specified in IS: 800–1984 as below :

(i) In the case of single angle in tension connected by one leg only, the net effective section of the angle is taken as

$$A_n = (A_1 + A_2 \cdot k) \qquad ...(5.2)$$

where, A_n = net effective sectional area

A_1 = effective cross-sectional area of the connected leg

A_2 = gross cross-sectional area of the unconnected (outstanding) leg, and

$$k = \left(\frac{3A_1}{3A_1 + A_2}\right) \quad ...(5.3)$$

where the leg angles are used, the effective sectional area of the whole of the angle member shall be considered.

(ii) In the case of a pair of angle back to back (or a single tee) in tension connected by only one leg of each angle (or by the flange of a tee) to the same side of a gusset, the net effective area is taken as

$$A_n = (A_1 + A_2 . k) \quad ...(5.4)$$

where, A_1 = effective sectional area of the connected legs (or flange of the tee)

A_2 = gross cross-sectional area of the unconnected (outstanding) legs (or web of the tee), and

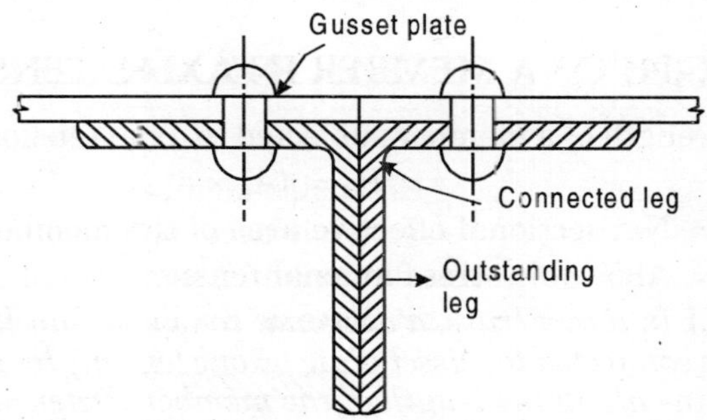

Fig. 5.4 *Pair of angles (back to back) connected by one leg of each angle to same side of gusset plate*

$$k = \left(\frac{5A_1}{5A_1 + A_2}\right) \quad ...(5.5)$$

The tacking rivets are used with the appropriate spacing to connect the angles along their length.

(iii) For double angles or tees carrying direct tension placed back to back and connected to each side of a gusset as shown in Fig. 5.5 or to each side of a rolled section, sectional area to determine the mean tensile stress shall be the full gross area less the deductions for holes, provided the tacking rivets are used with the appropriate spacing to connect the angles along their length.

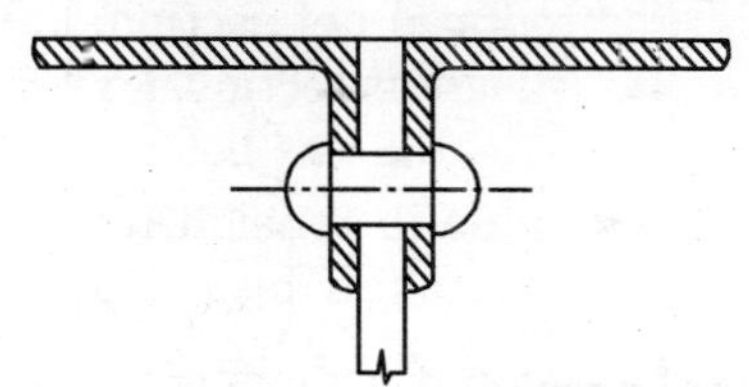

Fig. 5.5 *Two angles connected to each side of gusste plate*

The area of the leg shall be taken as the product of the thickness by the length from the outer corner minus half thickness, and the area of the leg of a tee as the product of the thickness by the depth minus the thickness of the table.

When the angles are back to back and the tacking rivets are not used to connect the angles along their length, then, each angle shall be designed as single angle connected through one leg only as described above.

When two tee sections are placed back to back but the tacking rivets are not used to connect them along the length, then each tee-section shall be designed as single tee-section connected to one side of a gusset as specified above.

5. 5 ALLOWABLE STRESS IN AXIAL TENSION

The permissible stress in axial tension, σ_{at} on the net effective area of the section shall not exceed.

$$\sigma_{at} = 0.6 f_y$$

where, f_y is the minimum yield stress of steel is N/mm^2 (MPa).

5. 6 STRENGTH OF A MEMBER IN AXIAL TENSION

The tensile strength of a member subjected to axial tension P_t is found as below:

$$P_t = (A_n \times \sigma_{at})$$

where, A_n = Net sectional effective area of the member

σ_{at} = Allowable stress in axial tension.

Example 5.1 *In a roof truss, a diagonal consists of an ISA 60 mm × 60 mm × 8 mm and it is connected to gusset plate by one leg only by 18 mm diameter rivets in one chain line along the length of the member. Determine tensile strength of the member.*

Solution

Step 1 : Net area of section

ISA 60 mm × 60 mm × 8 mm (ISA 6060,-@ 0.070 kN/m) is used as a tension member.

Diameter for rivet hole = (18 +1.5) =19.5 mm

The rivets are used in one chain line along the length of the member. Net sectional area of connected leg

$$A_1 = \left(60 - 19{\cdot}5 - \frac{1}{2} \times 8\right) \times 8 = 292 \text{ mm}^2$$

Area of outstanding (unconnected) leg,

$$A_2 = \left(60 - \frac{8}{2}\right) \times 8 = 448 \text{ mm}^2$$

$$k = \left(\frac{3A_1}{3A_1 + A_2}\right) = \left(\frac{3 \times 292}{3 \times 292 + 448}\right) = 0.6616$$

Net effective sectional area

$$A_n = (292 + 0.6616 \times 448) = 588.41 \text{ mm}^2$$

Step 2 : Tensile strength of section

It is assumed that value of yield stress for steel shall be 260 N/mm². The permissible stress in axial tension

$$\sigma_{at} = 0.6 \times 260 = 156 \text{ N/mm}^2$$

Tensile strength of the angle section

$$P_1 = \left(\frac{588 \cdot 41 \times 156}{1000}\right) = \mathbf{91.792\ kN.}$$

Example 5.2 *A double angle tie ISA 150 mm × 75 mm × 10 mm (ISA 15075, @ 0.169 kN/m) (short legs back to back) of a roof truss is connected to the same side of a gusset, with rivets 18 mm in diameter, such that each angle is reduced in section by one rivet hole only. Determine the tensile strength of the member. Stitch rivets have been provided at suitable spacing.*

Solution

Step 1 : Net area of section

Tie member consists of 2 ISA 150 mm × 75 mm × 10 mm (2 ISA 15075, @ 0.169 kN/m)

Diameter of rivet hole

$$(18 + 1.5) = 19.5 \text{ mm}$$

The long legs of angles have been connected to the same side of gusset and the short legs have been kept back to back.

Net sectional area of the connected legs.

$$A_1 = 2\,(150 - 19.5 - 5) \times 10 = 2510 \text{ mm}^2$$

Area of outstanding (unconnected) legs,

$$A_2 = 2 \times \left(75 - \frac{1}{2} \times 10\right) \times 10 = 1400 \text{ mm}^2$$

$$k = \left(\frac{5A_1}{5A_1 + A_2}\right) = \left(\frac{5 \times 2510}{5 \times 2510 + 1400}\right) = 0.8996$$

Net effective area of the angle

Step 2 : Tensile strength of section

$$A_n = (A_1 + A_2 \,.\, k) = 2510 + 1400 \times 0.8996 = 3769.493 \text{ mm}^2$$

It is assumed that the value of yield stress for the steel used is 260 N/mm². The permissible stress in axial tension

$$\sigma_{at} = (0.6 \times 260) = 156 \text{ N/mm}^2$$

Tensile strength of the member in case the angles are connected with tacking rivets along their length at suitable spacing

Example 5.3 *In Example 5.2 if the angles are connected to each side of a gusset, determine the tensile strength of the member.*

Solution

Step 1 : Net area of section

From steel section tables, for 2 ISA 150 mm × 75 mm × 10 mm (2 ISA 15075, @ 0.169 kN/m)

Cross-sectional area

$$2 \times 2156 = 4312 \text{ mm}^2$$

Deductions for rivet holes

$$2 \times 19.5 \times 10 = 390 \text{ mm}^2$$

Net effective sectional area

$$(4312 - 390) = 3922 \text{ mm}^2$$

Step 2 : Tensile strength of section

It is assumed that the value of yield stress for the steel used is 260 N/mm². The permissible stress in axial tension

$$\sigma_{at} = (0.6 \times 260) = 156 \text{ N/mm}^2$$

Tensile strength of the member in case the angles are connected with the tacking rivets along their length at suitable spacing

$$P_t = \left(\frac{3922 \times 156}{1000}\right) = \mathbf{611.832 \text{ kN}}$$

Example 5.4 *Long leg of an ISA 150 mm × 75 mm (ISA 15075) is connected to a gusset plate by 20 mm diameter rivets in two rows. The gauge space is 55 mm and the staggered pitch is 40 mm. Determine the thickness of the angle which would be sufficient to transmit a pull of 250 kN. Allowable tensile stress =* $0.6 f_y$.

Solution

Step 1: Net area of section

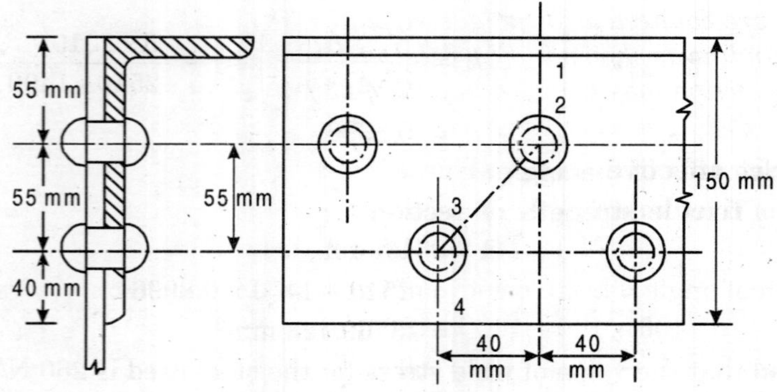

Fig. 5.6

In the zig-zag chain line of rivet holes 1–2–3–4, Fig. 5.6 the net effective width

$$b_1 = b - \left(nd - n'\frac{p^2}{4g}\right)$$

$$= 150 - \left(2 \times 21 \cdot 5 - \frac{1 \times 40^2}{4 \times 55}\right) = 114.273 \text{ mm}$$

Let t be the thickness of the angle

Net effective sectional area of the connected leg

$$A_1 = \left(114 \cdot 273 - \frac{t}{2}\right) \times t \text{ mm}^2$$

Area of the outstanding leg

$$A_2 = \left(75 - \frac{t}{2}\right) \times t \text{ mm}^2$$

Neglecting $\left(\frac{t}{2}\right)$ within the parenthesis

$$\therefore \quad A_1 = 114.273\,t \text{ and } A_2 = 75t$$

$$k = \left(\frac{3A_1}{3A_1 + A_2}\right) = \left(\frac{3 \times 114 \cdot 273t}{3 \times 114 \cdot 273t + 75t}\right) = 0.82$$

Net effective sectional area

$$A_n = (A_1 + A_2 \,.\, k) = (114.273t + 0.82 \times 75 \times t)$$
$$= 175.8\,t \text{ mm}^2$$

Step 2 : Thickness of section

Pull transmitted by the angle

$$= 175.8\,t \times 156 \text{ N}$$
$$175.8\,t \times 156 = 250 \times 1000$$
$$t = 9.12 \text{ mm}$$

Thickness of the angle section (required) is **10 mm**.

Example 5.5 *Both legs of an ISA 100 mm ×100 mm × 10 mm (ISA 100,100 @ 0.149 kN/m) are connected to the gusset plates by 20 mm diameter rivets in staggered chain line as shown in Fig. 5.7. Determine the staggered rivet pitch so that the angle section may transmit a pull of 230 kN.*

Solution :

Step 1 : Net effective sectional area

Diameter of rivet hole

$$30 + 1.5 = 31.5 \text{ mm}$$

Gross width of angle = Sum of width of two legs less thickness of legs

$$(100 + 100 - 10) = 190 \text{ mm}$$

Gauge space, $g = 60 + 60 - 10 = 110$ mm

Net effective width

$$= 190 - \left(2 \times 21 \cdot 5 - \frac{p^2}{4 \times 110}\right) \text{mm}$$

$$\left(190 - 430 + \frac{p^2}{440}\right) = \left(147 \cdot 0 + \frac{p^2}{440}\right) \text{mm}$$

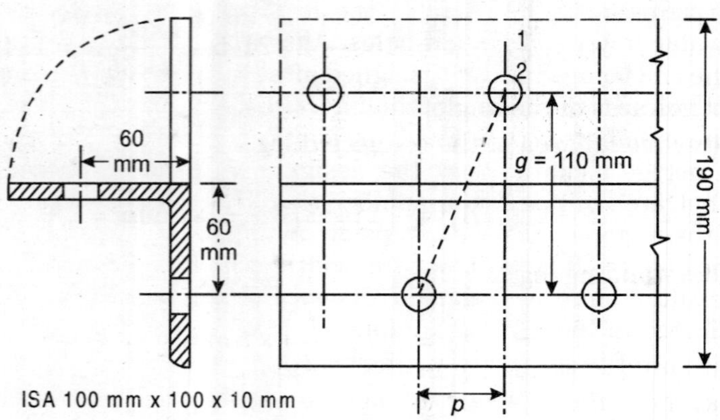

Fig. 5.7

$$\text{Net effective sectional area} = \left(147 \cdot 0 + \frac{p^2}{440}\right) \times 10 \text{ mm}^2$$

Step 2: Staggered pitch of rivets

Pull transmitted by the angle

$$= \left(147.0 + \frac{p^2}{440}\right) \times 156 \text{ N}$$

$$\left(147.0 + \frac{p^2}{440}\right) \times 156 = 230 \times 1000, \qquad \therefore p = 764.22 \text{ mm}$$

Adopt staggered pitch of the rivets equal to **760 mm**

5.7 DESIGN OF TENSION MEMBERS SUBJECTED TO AXIAL TENSION

A tension member subjected to axial tension is designed on the basis of its net sectional area. The end connections of a tension member should be such that the deduction for rivet and bolt holes from the gross-sectional area is minimum. The rivets connecting the tension member with the gusset plate or other sections should be arranged on the line of action of load as far as possible.

The following are the usual steps for the design of tension members. The specifications for the design of tension members as per IS : 800–1984 have also been given.

Step 1. The axial pull to be transmitted by the member and allowable stress in axial tension are known for the steel with yield stress f_y. This determined sectional area of the member required.

$$\text{Net sectional area} = \left(\frac{\text{Axial pull}}{\text{Allowable stress in axial tension}}\right)$$

Step 2. The gross-sectional area of the member is determined by making suitable allowable for rivet and bolt holes. Allowance for holes cannot be made correctly as the end connection are designed after the design of tension member. Allowance for holes in the net area should not be so small that it may require redesign of the member or may result in long connections. This allowance for holes should not be as large that the design may become uneconomical. In general, the net area of a tension member should not be less than 80% of the gross area of the member. In case, a tension member is made of thin plates, the deductions for hole shall be less, but the width of plate needed should be more (which will result in more secondary stresses). The allowance for different cases has been indicated in the solved examples that follow.

Step 3. The suitable section from steel section tables is selected.

Step 4. The end connections of tension member are designed and again the net area of the member is calculated. The net area actually provided is compared with the required net area. If the net area provided is less or much larger than the net area required, the member is redesigned.

If the tension member is normally acting as a tie but subjected to possible reversal of stress due to loads other than wind or seismic forces, the member shall have slenderness ratio not greater than 180.

A member normally acting as a tie truss or a bracing system but subjected to possible reversal of stress resulting from the action of wind or earthquake forces shall have slenderness ratio, λ not more than 350.

Note. *In ISI Structural Engineer's Handbook No. 1, tables are directly given for single angles and double angles used as ties. The sections can be selected directly from these tables.*

The gross cross-sectional area is determined after calculating net sectional area by making suitable allowance for rivet and bolt holes. The suitable allowance for area for rivet and bolt holes may be made according to the following guidelines.

When a tension member consists of single angle section, then, the allowance may be made for one *rivet hole.* When the tension member consists for two angle sections, then the allowance may be made *as one hole for each angle.* When a tension member consists of two channel sections then the allowance may be made *as two holes from each web or one hole from each flange whichever is more.* When a tension member consists of four angles then the allowance may be made as *one hole for each angle.* When a tension member is made of four angles and a plate as a web, then the allowance may be made as *one hole for each angle and two holes for the web.* When the tension member comprises of two channels laced then the allowance may be made *as two hole from each channel web or one hole from each flange, which is more.* When a built-up tension member consists of four angles with or without plates, then, the allowance may be made *as two rivet holes for each angle and one rivet hole* for *every 150 mm width of the plate.*

5.8 DESIGN OF TENSION MEMBERS SUBJECTED TO BOTH BENDING AND AXIAL TENSION

In actual structures, the tension member are seldom subjected to pure axial load. The tension members may be subjected to bending stresses along with direct tensile stress because of various reasons (e.g., the eccentricity in the connections, the member itself may not be straight, resulting in eccentric load at the section the member may not be vertical and it may be subjected to bending due to self-weight). The member may be bending due to wind, vibrations or earthquake forces. The combined stress at any section may be determined, if the axial load and bending moment due to combined effects of eccentricity and transverse forces are known. The combined stress at any section is given by

$$f = \left(\frac{P}{A} + \frac{M_{xx}}{I_{xx}} \cdot y\right) \quad \text{...(i)}$$

where, M_{xx} = Bending moment above xx-axis
y = Distance to the extreme fibre from the neutral axis (xx-axis)
I_{xx} = Moment of inertia of the section about xx-axis
P = Axial tensile load
A = Net cross-sectional area of the section

When the section is subjected to bending about the both the axes, then the combined stress is given by

$$f = \left(\frac{P}{A} + \frac{M_{xx}}{l_{xx}} \cdot y + \frac{M_{yy}}{l_{yy}} \cdot x\right) \quad \text{... (ii)}$$

where, M_{yy} = Bending moment about yy-axis
I_{xx} = Moment of inertia about yy-axis
x = Distance to the extreme fibre from the neutral axis (yy-axis)

When the member is subjected to pure axial tension load P, then the net area required is given by

$$A_a = \left(\frac{P}{\sigma_{at}}\right) \quad \text{...(iii)}$$

where, σ_{at} = Maximum allowable stress in axial tension ($0.6\, f_y$)
f_y = Minimum yield stress of the steel.

When the member is subjected to pure bending then the bending stress is given by

$$\sigma_{bt} = \left(\frac{M \cdot y}{I} = \frac{M \cdot y}{A_b \cdot r^2}\right) \quad \text{...(iv)}$$

The required cross-sectional area is given by

$$A_b = \left(\frac{\dfrac{M \cdot y}{r^2}}{\sigma_{bt}}\right) \quad \text{..(v)}$$

where, I = Moment of inertia of the section about the axis about which bending takes place

r = Radius of gyration of the section

σ_{bt} = Maximum allowable bending stress in tension

When the member is subjected to axial tensile load and bending then the net cross-sectional area required is given by

$$A = (A_a + A_b) = \left(\frac{P}{\sigma_{at}} + \frac{\frac{M \cdot y}{r^2}}{\sigma_{bt}} \right)$$

or
$$\left(\frac{\frac{P}{A}}{\sigma_{at}} + \frac{\frac{M \cdot y}{Ar^2}}{\sigma_{bt}} \right) = 1.0$$

or
$$\left(\frac{\sigma_{at \cdot cal}}{\sigma_{at}} + \frac{\sigma_{bt \cdot cal}}{\sigma_{bt}} \right) \le 1.00 \quad \text{...(vi)}$$

where, $\sigma_{at.cal} = \frac{P}{A}$ = Actual axial tensile stress

$$\sigma_{bt.cal} = \left(\frac{M \cdot y}{Ar^2} = \frac{M \cdot y}{I} \right)$$

= Actual bending tehsile stress in the extreme fibre

When the tension member is subjected to both axial tension and bending then it is so proportioned so that

$$\left(\frac{\sigma_{at \cdot cal}}{0.60 f_y} + \frac{\sigma_{bt \cdot cal}}{0.66 f_y} \right) \le 1.00 \quad \text{...(vii)}$$

When the bending occurs about both the axes of members actual bending tensile stresses $\sigma_{btx.cal}$ and σ_{bry} are calculated.

The member is so proportioned that

$$\left(\frac{\sigma_{at \cdot cal}}{0.6 f_y} + \frac{\sigma_{btx \cdot wcal}}{\sigma_{bt.x}} + \frac{\sigma_{bty.cal}}{\sigma_{br.y}} \right) < 1.00 \quad \text{...(viii)}$$

where, $\sigma_{at.cal}$ = Calculated average axial tensile stress

$\sigma_{bt.cal}$ = Calculated bending tensile stress in the extreme fibre

σ_{at} = Permissible bending tensile stress in the member subjected to axial load

σ_{bt} = Permissible bending tensile stress in extreme fibre.

AISC specification recommends that the member should be so proportioned that

$\left(\frac{\sigma_t}{0.6 f_y} + \frac{\sigma_{bt}}{\sigma_{bt}} \right)$ should not exceed unity

where f_y = yield stress of the steel.

The trial section of a tension member subjected to axial tension and bending may be found by determining the equivalent axial tensile load. *The equivalent axial tensile load is the load, which produces the average axial tensile stress in the section equivalent to the maximum combined stress at the extreme fibre of the section.* The equivalent axial load P_{equiv} is given by

$$P_{equiv} = \left(P + \frac{M}{I} \times A \cdot y\right) \quad \text{...(vii)}$$

or

$$P_{equiv} = \left(P + \frac{M \cdot A}{\left(\frac{I}{y}\right)}\right) \quad \text{...(viii)}$$

or

$$P_{equiv} = P + M\left(\frac{A}{Z}\right) \quad \text{...(ix)}$$

or

$$P_{equiv} = P + M.(B) \quad \text{...(5.6)}$$

where, B = (A/Z) = Bending factor

= Ratio of net area of the section to the section modulus of the area

When the member is subjected to bending about both the axes then the equivalent axial tensile load is given by

$$P_{equiv} = (P + M_{xx} B_x + M_{yy}.B_y \quad \text{...(5.7)}$$

where, B_x = Bending factor of the section about the strong axis (i.e., xx-axis)

B_y = Bending factor of the section about the weak axis (i.e., yy-axis)

Example 5.6 *Design a single angle section for a tension member of a roof truss to carry a pull of 100 kN. The member is subjected to possible reversal of stress due to action of wind. The length of the member from centre to centre of intersection is 3.50 metres.*

Solution

Design :

Step 1 : Selection of trial section

The steel having yield stress f_y as 260 N/mm² shall be used. The permissible bending stress for the member subjected to axial tension

$$\sigma_{at} = (0.60 \times 260) = 156 \text{ N/mm}^2$$

Net effective sectional area required

$$\left(\frac{100 \times 1000}{156}\right) = 641.03 \text{ mm}^2$$

In case of single angle section the gross area is assumed to 40 to 50 per cent in excess of the net area required.

Gross area = $(1.40 \times 641.03) = 897.44$ mm²

From steel section tables try ISA 65 mm × 65 mm × 10 mm

Step 2 : Net effective sectional area

Gross area = 1200 mm²

r_{min} = 12.5 mm

Net effective sectional area

$$A_n = (A_1 + A_2 \,.\, k)$$

Assume 18 mm diameter rivet used for end connections. Diameter of rivet whole

$$(18 + 1.5) = 19.5 \text{ mm}$$

Net sectional area of connected leg,

$$A_x = (6{\cdot}5 - 19.5 - 0.5 \times 10) \times 10 = 405 \text{ mm}^2$$

Area of outstanding leg,

$$A_2 = \left(65 - \frac{10}{2}\right) \times 10 = 60 \text{ mm}^2$$

$$k = \left(\frac{3A_1}{2A_1 + A_2}\right) = \left(\frac{3 \times 405}{3 \times 405 + 600}\right) = 0.669$$

Net effective sectional area provided

$$A_n = (405 + 0.669 \times 600) = 806.65 \text{ mm}^2$$

$$> \text{Net sectional area required } (641.03 \text{ mm}^2)$$

Step 3 : Check for slenderness ratio

Assume two or more than two rivets are used for end connections.

$$\text{Slenderness ratio } \left(\frac{l}{r_{\min}}\right) = \left(\frac{0.85 \times 3.50 \times 1000}{12.5}\right) = 238 < 350$$

Hence design is satisfactory.

Provide ISA 65 mm × 65 mm × 10 mm (ISA 6565 @ 0 094 kN/m) as tension member.

Example 5.7 *Design a tension member consisting of a pair of angles (back to back) and connected by the short legs to the same side of gusset plate. The member is to carry a pull of 250 kN.*

Solution

Design :

Step 1 : Selection of trial section

It is assumed that the steel having yield stress, f_y as 260 N/mm² shall be used.

The permissible stress in axial tension

$$\sigma_{at} = (0.6 \times 260) = 156 \text{ N/mm}^2$$

Pull to be carried by tension member = 250 kN

$$\text{Net area required} = \left(\frac{250 \times 1000}{156}\right) = 1602.56 \text{ mm}^2$$

Assume gross area 40 per cent in excess of the net area required.

$$\text{Gross area} = (1.40 \times 1602.56) = 2243.59 \text{ mm}^2$$

From steel section tables, try 2 ISA 90 mm × 60 mm × 10 mm

Step 2 : Net effective sectional area

Gross area = 2802 mm^2

Use 20 mm diameter rivets. Diameter of rivet hole

= 21.5 mm

Net sectional area of connected leg,

$$A_1 = 2 \times \left(60 - 21.5 - \frac{10}{2}\right) \times 10 = 690 \text{ mm}^2$$

Area of the outstanding leg,

$$A_2 = 2 \times \left(90 - \frac{10}{2}\right) \times 10 = 1700 \text{ mm}^2$$

$$k_1 = \left(\frac{5A_1}{5A_1 + A_2}\right) = \left(\frac{5 \times 699}{5 \times 690 \times 1700}\right) = 0.6727$$

Net effective sectional area

$$(A_1 + A_2 . k) = (699 + 1700 \times 0.6727)$$
$$= 1842.696 \text{ mm}^2$$
$$> \text{Net area required } (1602.56 \text{ mm}^2)$$

Hence, the design is satisfactory.

Provide, 2 ISA 90 mm × 60 mm × 10 mm (2 ISA 9060, @ 0.110 kN/m) section and stitch rivets at pitch of 1.0 metre.

Example 5.8 *If in Example 5.7 the angles are connected to each side of the gusset plate, design the section.*

Solution

Design :

Step 1 : Selection of trial section

$$\text{Net area required} = \left(\frac{250 \times 1000}{156}\right) = 1602.56 \text{ mm}^2$$

The angles are connected to both the sides of the gusset plate.

Use 20 mm diameter rivets for end connections. Diameter of rivet hole

= 21.5 mm

Assume one rivet hole in each angle and 6 mm thickness of angle.

Gross area = (1667 + 2 × 2.15 × 6) = 1925 mm^2

From steel section tables

Try 2 ISA 100 mm × 75 mm × 6 mm (2 ISA 10075, @ 0.080 kN/m)

Step 2 : Net area

Gross area = 2028 mm^2

Net area provided = (2028 – 258) = 1770 mm^2

Hence the design is satisfactory. Provide 2 ISA 100 mm × 75 mm × 6 mm (2 ISA 10075 @ 0.080 kN/m).

5.9 TENSION MEMBER SPLICE

When a joint is to be provided in a tension member, then splice plates are used. Splice plates and rivets are designed for the pull required to be transmitted by the tension member. If the tension members are of unequal thickness, then, the packings are used to have surfaces of tension members in one level. The rivets or bolts carrying calculated shear stress through a packing greater than 6 mm thick shall be increased above the number required by normal calculations by 2.5 per cent for each 20 mm thickness of packing. For double shear connections packed on both sides, the number of additional rivets or bolts required shall be determined from the thickness of the thicker packing. The additional rivets or bolts should preferably be placed in an extension of the packing.

Example 5.9 *In a truss girder of a bridge, a diagonal consists of mild steel flat 400 F16 and it carries a pull of 750 kN. Design a suitable splice for the member.*

Solution

Design :

Step 1 : Design of splice

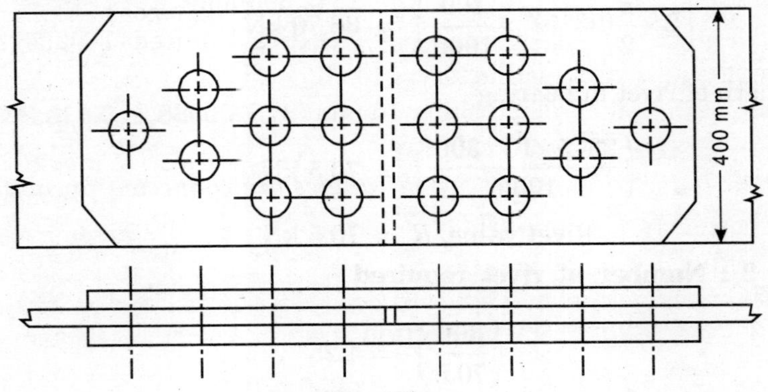

Fig. 5.8

Provide two 400 ISF 8 as splice plates to join the flats.

Step 2 : Rivet value

Provide 22 mm normal diameter, power driven rivets

Gross diameter of rivet = 23.5 mm

Strength of rivet in double shear

$$\left(2\times\frac{\pi}{2}(23\cdot5)^2\times\frac{100}{1000}\right) = 86.70 \text{ kN}$$

Strength of rivet in bearing

$$\left(23\cdot5\times16\times\frac{300}{1000}\right) = 112.8 \text{ kN}$$

Rivet value, R = 86.70 kN

Step 3 : Number of rivets required

$$\left(\frac{750}{86 \cdot 70}\right) = 8.65$$

Provide 9 rivets in diamond group of riveting as shown in Fig. 5.8.

The strength of splice plates is equal to the strength of main member as the total thickness of splice plates is equal to the thickness of the main member.

Example 5.10 A *bridge truss diagonal carries an axial pull of 300 kN. Two mild steel flats 250 ISF 10 and 260 ISF 18 of the diagonal member are to be joined together. Design a suitable splice.*

Solution

Design :

Step 1 : Rivet value

Provide 22 mm nominal diameter power driven rivets.

$$\text{Gross diameter of rivet} = 23.5 \text{ mm}$$

Strength of rivet in double shear

$$\left(2 \times \frac{\pi}{2}(23 \cdot 5)^2 \times \frac{100}{1000}\right) = 86.70 \text{ kN}$$

Strength of rivet in bearing

$$\left(\frac{23 \cdot 5 \times 10 \times 300}{1000}\right) = 70.5 \text{ kN}$$

$$\text{Rivet value, } R = 70.5 \text{ kN}$$

Step 2 : Number of rivet required

$$\left(\frac{300}{70.5}\right) = 4.25$$

Provide 6 rivets.

Thickness of packing required

$$(18 - 10) = 8 \text{ mm} > 6 \text{ mm}$$

Provide additional rivets @ 2.5 per cent for each 2.0 mm thickness of packing.

Additional percentage of rivets

$$\left(\frac{2 \cdot 5 \times 8}{2}\right) = 10 \text{ per cent.}$$

Number of additional rivets

$$\left(\frac{6 \times 10}{100}\right) = 0.61\ \Omega\ 1 \text{ rivet}$$

Provide additional rivet in the extension of packing as shown in Fig. 5.9.

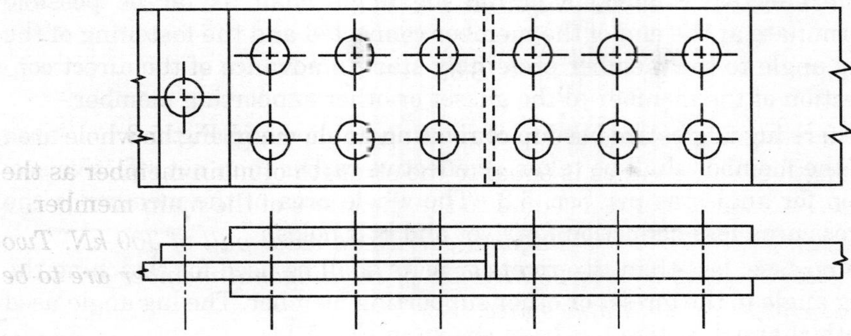

Fig. 5.9

Provide splice plate 6 mm thick. The strength of splice at the weakest section

$$= \left(2(250 - 2 \times 23 \cdot 50) \times 6 \times \frac{0 \cdot 6 \times 260}{1000}\right) \text{kN}$$

= 380.02 kN > Design load. Hence, the design is satisfactory.

5.10 LUG ANGLES

Length of the end connections of angle or channel section to a gusset plate is reduced by using lug angles. Lug angle used with single angle member is shown in Fig. 5.10. Rivets connecting lug angle to the gusset plates are less effective

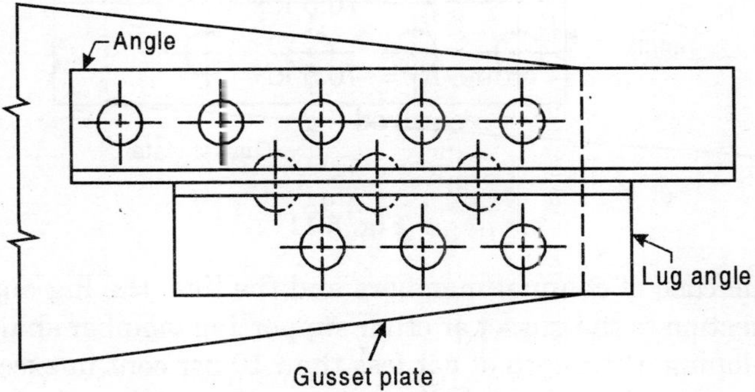

Fig. 5.10

because of possible deformation of outstanding leg of the lug angle. As such, rivets are provided in excess (as per specification) of the number of rivets required by computation.

Specifications. The following specifications for the design of lug angle may be followed:

1. In the case of angle members, the lug angles and their connections to the gusset or other supporting member shall be capable of developing strength not less than 20 per cent in excess of the force in the outstanding leg of the angle, and the attachment of the lug angle to the angle member shall be capable of developing 40 per cent in excess of that force.

2. The effective connection of the lug angle shall as far as possible terminate at the end of the member connected and the fastening of the lug angle to the member preferably start in advance of the direct connection of the member to the gusset or other supporting member.
3. Where lug angles are used to connect an angle member, the whole area of the member shall be taken as effective rather than net effective section for angles as per Sec. 5.3. (The whole area of the member is the gross area less deduction for rivet and bolt holes).
4. In no case, less than two rivets or bolts shall be used for attaching the lug angle to the gusset or other supporting member. The lug angle used with channel section has been shown in Fig. 5.11.

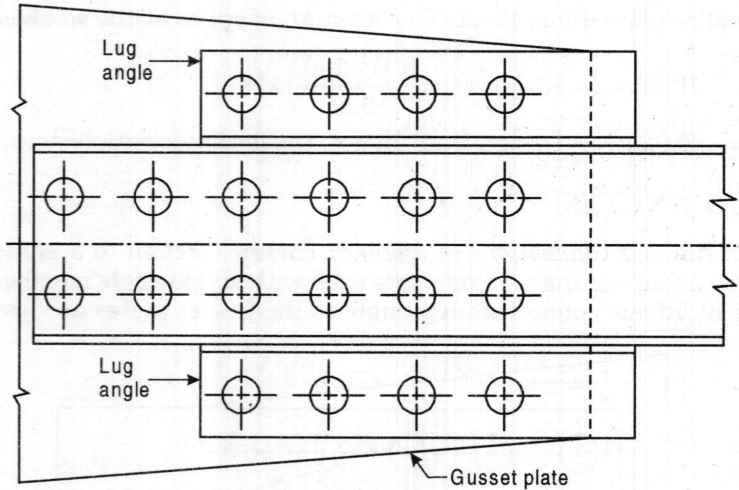

Fig. 5.11

5. In the case of channel members and the like, the lug angles and their connection to the gusset or other supporting member shall be capable of developing a strength of not less than 10 per cent in excess of the force not accounted for by the direct connection of the member and the attachment of the lug angles to the member shall be capable of developing 20 per cent in excess of that force.

Example 5.11 *Design a riveted end connection for the full strength of ISA 100 mm × 116 mm × 10 mm. Provide a lug angle in the connection. Adopt stress as per IS : 800 –1984.*

Solution

Design :

Step 1 : Strength of angle section

Provide 22 mm nominal diameter power driven rivets.

Gross diameter of rivet = 23.5 mm

From steel section tables, gross area of ISA 110 mm × 110 mm × 10 mm 10 110, @ 0.165 kN/m) = 2106 mm²

Assume, the angle section will be reduced by one rivet hole only. Net area of angle section

$$(2106 - 23.5 \times 10) = 1871 \text{ mm}^2$$

Tensile strength of the angle section

$$\left(\frac{1871 \times 0 \cdot 6 \times 260}{1000}\right) = 291.876 \text{ kN}$$

Strength of outstanding leg of the main angle

$$\left(\frac{(110-10)}{(110-10)+110} \times 221 \cdot 876\right) = \left(\frac{100}{210} \times 291 \cdot 876\right) = 138.988 \text{ kN}$$

Strength of connected leg of the main angle

$$= 152.888 \text{ kN}$$

Provide a gusset plate 12 mm thick

Step 2 : Strength of rivet

Strength of rivet in angle in single shear

$$\left(\frac{\pi}{4} \times (23 \cdot 5)^2 \times \frac{100}{1000}\right) = 43.35 \text{ kN}$$

Strength of rivet in bearing on 10 mm thickness

$$\left(\frac{23.5 \times 10 \times 300}{1000}\right) = 70.50 \text{ kN}$$

Assume size of lug angle ISA 110 mm × 110 mm × 8 mm

Strength of rivet in bearing on 8 mm thickness

$$\left(\frac{23 \cdot 5 \times 8 \times 300}{1000}\right) = 56.4 \text{ kN}$$

Step 3 : Number of rivets joining connected leg of main member with the gusset plate

$$\left(\frac{152.888}{43 \cdot 35}\right) = 3.525 \simeq 4$$

Number of rivets joining connected leg of the lug angle with gusset

$$\left(\frac{1 \cdot 2 \times 138 \cdot 988}{43 \cdot 35}\right) = 3.847 \simeq 4$$

Number of rivets joining outstanding legs of lug and main angles

$$\left(\frac{1 \cdot 4 \times 138 \cdot 988}{43 \cdot 35}\right) = 4.488 = 5$$

The design has been shown in Fig. 5.12.

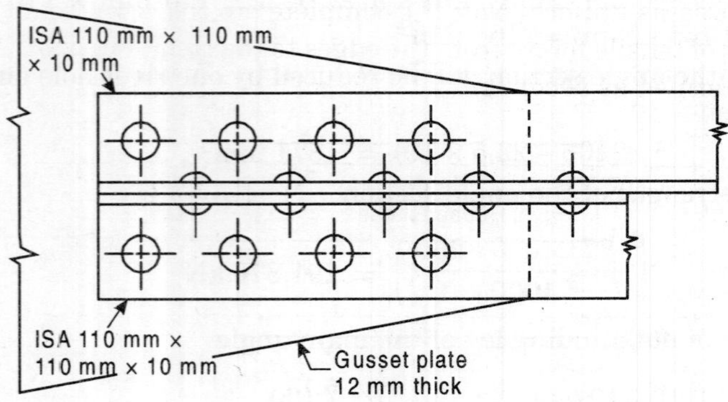

Fig. 5.12

Step 4: Selection of lug angle section

The lug angle should develop strength 20 per cent in excess of the force in the outstanding leg of the angle.

Required strength of lug angle

$$(1.20 \times 138.988) = 166.78 \text{ kN}$$

Net area required for lug angle

$$\left(\frac{166 \cdot 78 \times 1000}{156}\right) = 1069.138 \text{ mm}^2$$

From steel section tables (single angle used as ties)

Provide ISA 110 mm × 110 mm × 8 mm (ISA 110110 @ 0.165 kN/m) as lug angle.

5.11 SHEAR LAG

In case a tension member is built-up of plates and thickness of plates are small, then, the deduction for areas of rivet holes (number of holes × diameter of rivet × thickness of plate) will be less and widths of the plates needed shall be large. The large widths of plates results in more secondary stresses. In the design of riveted connections for the wide plates in tension, certain precautions are necessary, in the absence of which, the strength of such plates becomes less due to effect of *shear lag*. The effect of shear leg may be seen by applying two tensile forces P_1 and P_1 at the edges of a wide rubber sheet as shown in Fig. 5.13. The stretch at the edges AA_1 and CC_1 is large as compared to that at the middle line BB_1. Except at A and A_1 and C and C_1, the stresses in longitudinal direction, at all the points on edges ABC and $A_1B_1C_1$ are zero. The small elements at D and D_1), on the edges ABC and $A_1B_1C_1$, respectively, deform as shown. It shows that the shear stresses are developed in the material of the sheet. These shear stresses transmit the tensile forces from the edges towards the middle line in a certain length approximately equal to the width of sheet. The stresses in tension in this length shall be not equal. However, the stress in tension beyond this length may be treated as uniform over the complete cross-section of the rubber sheet, the transfer of tensile forces from the edges to the complete width through shear stresses is termed as *shear lag*.

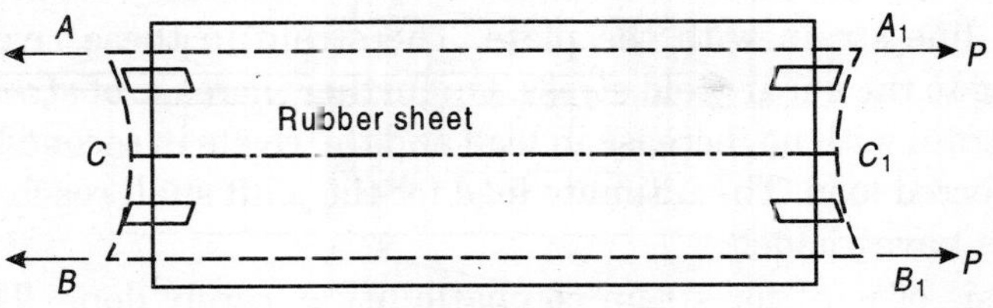

Fig. 5.13

In case, a tension member (angle section or channel section) is connected to the gusset plates, it is necessary to keep length of the joint sufficiently long (viz. equal to the length of transfer or length of shear lag), so that the stresses in the connected member at the beginning remain uniform over its complete cross-section. If the length of joint is smaller, then, there will be stress concentration near the edges of member and its effectiveness will be reduced as shown in Fig. 5.14. The axial force in the tension member shown lags the transfer to the gusset plate, because of its distribution across the member cross-section. If the length of joint is too large, the portion of axial tensile force transmitted by each

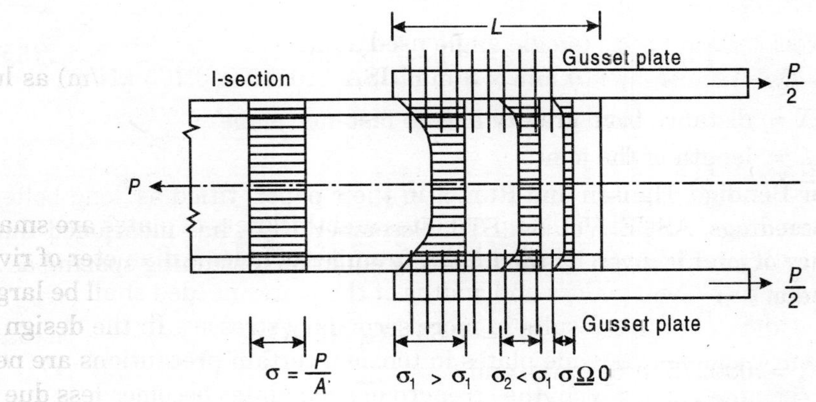

Fig. 5.14

rivet or bolt shall not be equal. In case, N is number of rivets or bolts used in the connection, the rivets in first line carry more than P/N and the rivets in the last line will carry practically zero force. Due to high concentration of stress, the web plate may fall in tearing and tearing of plate progresses across the section. In case, the web plate is adequate in the net section, the tearing of plate does not occur, but it stretches *(PL/AE)* depending upon the value of tensile strain. The rivets in first line will either undergo compatible shear strain or will shear-off if the strain and resulting forward displacement of the hole is too large. The failure of rivets in second line will also fail and so on. As such, it produces a progressive joint failure. This process is called as *unbuttoning.* It is to note that when the axial force in tension to be transmitted is large, this process is almost instantaneous.

If the length of joint is short, then all rivets or bolts carry load. The rivets or bolts in first line strain with the plate. The strain in these rivets increases corresponding to the shear yield stress. On further increase of strain, the rivets continue to strain with no increase in load and the rivets in second line will take up the transferred load. The ultimate load for the joint shall reach when all the rivets or bolts have yielded.

The analysis of joint for strain compatibility is rarely done. The factors of safety used together with the property of steel ductility are such that except for long joint, only the rivets or bolts in first line in a connection (if any) may yield or reach close to the yield.

This is to note that the factor of safety for the connection is kept higher than that used for the members. This is so because, a failure of member (if it occurs) should be earlier than the failure of joint. A joint failure used to be catastrophic, whereas, the failure of member permits time for safety measures to be undertaken.

Professors Munsc and Chesson, in their paper titled as, 'Riveted and Bolted Joints : Net Section Design', Proceedings, ASCE, ST1, February 1963 have given the following equation for shear lag efficiency.

$$\eta = \left(1 - \frac{\bar{x}}{L}\right) \times 100 \qquad \text{...(5.8)}$$

where X = distance from gravity axis to fastener plane

L = length of the joint.

Professor Bendigo, Hansen and Rumpf in their paper, titled as 'long bolted joints', Proceedings, ASCE. Vol. 84, ST6, December 1963, has mentioned that the efficiency of joint is given by the following equation (assuming spacing as 3 *times* diameter)

$$\eta = 0.85 - C_1 \,.\, (L - C_2) \qquad \text{...(5.9)}$$

where C_1 = 000275 in SI units, and

C_2 = 406 mm > $(L \geq C)$

Equation 5.9 indicates that the connections with joint lengths upto 406 mm have an efficiency of 85 per cent. The effect of shear lag may be accounted by calculating the effective area

$$A_{\text{eff}} = A_{\text{net}} \left(1 - \frac{\bar{x}}{L}\right) \qquad \text{...(5.10)}$$

For a single plane connection, x is the centroidal co-ordinate of the entire cross-section, while for a symmetrical double plane connection, it is the distance from each connector plane to the centroid of the half cross-section as shown in Fig. 5.15.

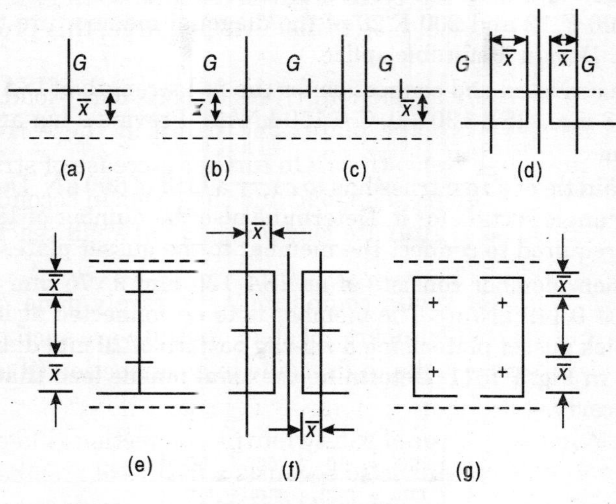

Fig. 5.15

PROBLEMS

5.1 In a roof truss, a diagonal consists of ISA 80 mm × 80 mm × 10 mm (ISA 8080, @ 0.118 kN/m) and is connected to a gusset plate by one leg only by 20 mm diameter power driven rivet in one line along the length of the member. Determine tensile strength of the member.

5.2 A double angle tie ISA 125 mm × 95 mm × 10 mm (viz., ISA 125, 95, @ 0.165 kN/m) (short leg back to back) of a roof truss is connected to the same side of a gusset with rivets 20 mm in diameter, such that each angle is reduced in section by one rivet hole only. Determine the tensile strength of the member. Stitch rivets have been provided at suitable spacing.

5.3 In Problem 5.2, if angles are connected to each side of a gusset, determine the tensile strength of the member.

5.4 Long leg of an ISA 125 mm × 95 mm (ISA 12595) is connected to a gusset plate by 22 mm diameter rivets in two rows. The gauge distances in long legs of angle are 45 mm and 55 mm and staggered pitch is 45 mm. Determine the thickness of the angle which would be sufficient to transmit a pull of 180 kN.

5.5 Design a single angle section for a tension member of a roof truss to carry a pull 100 kN. The member is subjected to possible reversal of stress due to action of wind. The length of the member from centre to centre of intersection is 3.50 m.

5.6 Design a tension member consisting of a pair of angles and connected by their short legs to the same side of a gusset plate. The member is to carry a pull of 180 kN.

5.7 If in Problem 5.6, the angles are connected to each side of the gusset plate, then, design the section.

5.8 A bridge truss diagonal carries an axial of pull of 240 kN. Two mild steel flats 200 F 12 and 200 F 20 of the diagonal member are to be joined together. Design a suitable splice.

5.9 Design a riveted end connection for the full strength of ISA 130 mm × 130 mm × 8 mm. (ISA 130130, @ 0.159 kN/m). Provide a lug angle in the connection.

5.10 The main tie of a roof truss has to carry a load of 800 kN. Design a suitable double angle section for it. Determine also the number of 18 mm diameter rivets required to connect the member to the gusset plates at ends.

5.11 A tension member consists of an ISA 125 mm × 75 mm × 10 mm (ISA 12575,@ 0.149 kN/m). The member is to be connected at its ends to a 12 mm thick gusset plate using a zig-zag pattern of 20 mm diameter rivets as shown in Fig. P 5.11. Determine the axial tensile load that the angle can safely carry.

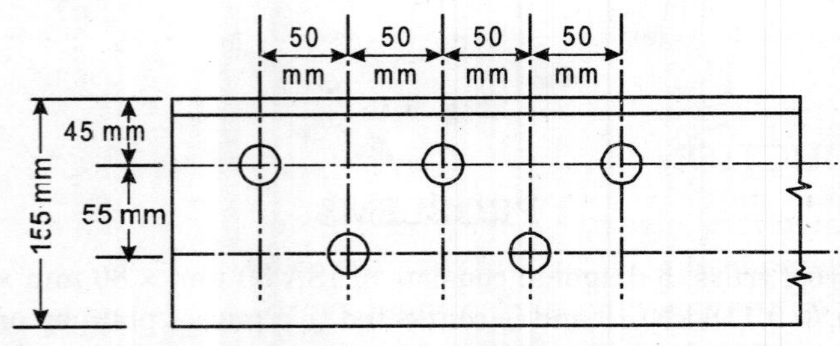

Fig. P 5.11

Chapter 6

Design of Beams

6.1 INTRODUCTION

A *beam* is defined as a structural member subjected to transverse loads. The plane of transverse load is parallel to the plan of symmetry of the cross-section of the beam and it passes through the shear centre, so that the simple binding occurs. The transverse loads produce bending moments and shear forces in the beam at all the sections of the beam.

The term *joist* is used for beams of light sections. Joists support floor construction; they do not support other beams. The term *subsidiary beam* or *secondary beam* is also used for the beam supporting floor construction. *Main beams* are supporting joists for subsidiary beams: these are called *floor beams* in buildings. The term girder is most commonly used in buildings. Any major beam in a structure is known as a girder

In the roof trusses, horizontal beams spanning between the adjacent trusses are known as *purlins*. The beams resting on the purlins are known as *common rafter* or *simply rafters*. In the buildings the beams spanning over the doors, windows and other opening in the walls are known as *lintels*. The beam at the outside wall of a building, supporting is share of the floor and also the wall up to the floor above it are known as *spandrel beams*. The beams framed to two beams at right angle to it, and usually supporting joists on one side of it; used at opening such as stair wells are known as *headers*. The beam supporting the headers is termed as *trimmers*. The beam supporting the stair steps are called as *stringers*. In the bridge floors, the longitudinal beams supported by the floor beams are also called as *stringers*. In the mill buildings, the horizontal beams spanning between the wall columns, and supporting wall covering are called as *girts*.

The beams are also called *simply supported*; *overhanging cantilever*, *fixed* and *continuous* depending upon nature of supports and end conditions.

The rolled steel I-sections, channel, sections, angle sections, tee-sections, flat sections and bars as shown in the Fig. 6.1 are the regular sections, which are used as beams. The *rolled steel I-sections* as shown in the Fig. 6.1(a) are most commonly used as the beams, and as such these sections are also termed as beam sections. The rolled steel I-sections are symmetrical sections. In these sections, more material is placed near top and bottom faces, i.e., in the flanges as compared to the web portion. The rolled steel I-sections provide large moment of inertia about xx-axis with less cross-sectional area. The rolled steel I-sections provide large moments of compared to the other sections and such as these are most efficient and economical beams sections. The *rolled steel wide flange beams* as shown in Fig. 6.1 (b) provide additional desirable features. As the name indicates, the flanges of the sections are wide. These sections provide greater lateral stability and facilitate the connections of flanges to other members. I-sections and wide flanges beam sections have excellent strength. The *rolled steel channel sections* as shown in Fig. 6.1(c) are used as purlins and other small structural members. The channel sections have reasonably good lateral strength and poor lateral stability. The channel sections are unsymmetrical sections about yy-axis. When the channel sections are loaded and supported by vertical forces passing through the centroid of the channel, then the channel sections bend and twist if these are laterally unsupported, except for the special case, wherein the loads act normal to the plane of web, causing bending in the weakest direction. The *rolled steel angle sections* as shown in Fig. 6.1 (d) are also used as purlins

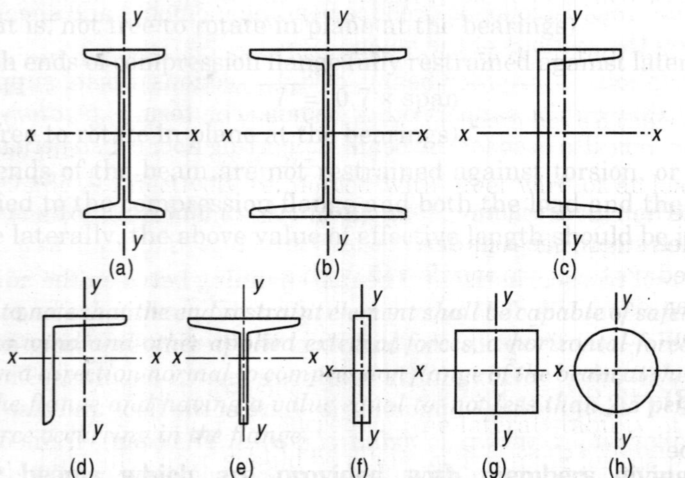

Fig. 6.1 *Rolled steel sections used as beams*

and so other small structural members. The angle sections act as unsymmetrical sections about both xx-axis and yy-axis. The *rolled steel tee-sections* as shown in Fig. 6.1 (e) are used as beams in the rectangular water tanks. The angles and tee-sections are used for light loads. The rolled steel flats and bars as shown in 6.1 (f), (g) and (h) are very rarely used. These sections are weak in resisting bending. Most commonly the beams are loaded in the direction perpendicular to xx-axis, so that the bending of beams occurs about strong (xx-axis) and xx-axis becomes neutral axis. The beams are very rarely loaded in the direction perpendicular to yy-axis. In such cases, yy-axis becomes neutral axis.

In general, in cases of bending of the beam about one (single) axis the load is considered to be applied through the shear centre of the beam sections. In case, the loading passing through the shear centre, the sections may be analyzed for simple and binding and shear. The shear centre for the beam sections is at the centre of area, and this load position produces simple bending about either axis.

When the load does not pass through the shear centre as in channels, angles and some built-up sections, a torisonal moment is produced along with the bending moment and both are considered to avoid over stressing of the member. For such sections, a special load device may be used so that the load passes through shear centre of the section and the torisonal may be avoided.

In addition to the above, *expanded* or *castellated beams* as shown in Fig. 6.2 (b) are used. The castellated beams are light and these are economically used

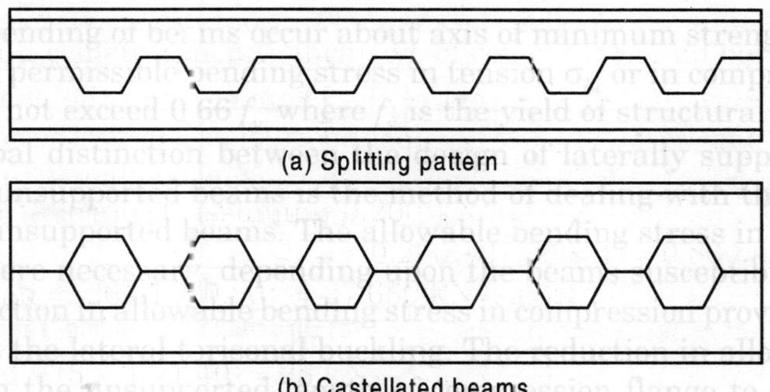

(a) Splitting pattern

(b) Castellated beams

Fig. 6.2

for the light construction. The castellated beams are made by splitting the web of rolled steel I-sections in a predetermined pattern as shown in Fig. 6.2 (a). The splitted portions are rejoined in such a manner as to produce a regular pattern of opening in the web. The modulii of sections of castellated beams increase without the increase of material and weight.

6.2 BENDING STRESS

The bending stress is also termed as *flexural stress*. When beams are loaded, they bend and bending stress are set up all the sections. The established theory of bending is expressed in the following formula:

$$\left(\frac{M}{I} = \frac{\sigma}{y} = \frac{E}{R}\right) \quad ...(6.1)$$

where, M = Bending moment

I = Moment of inertia

σ_b = Bending stress at any point

y = Distance from the neutral axis to the point under consideration
R = Radius of curvature of the beam.

Equation 6.1 holds good when the plane of bending coincides with the plan of symmetry of the beam section. The bending of beam occurs in the principal plane of the beam section. The simple bending of beam occurs i.e., the bending is produced by the application of pure couples at the ends of the beam. In such bending the deflection of beam does not occur due to shear. In Eq. 6.1, further, it assumed that the vertical section of the beam plane before bending remain plane after bending. Then stress in any given fiber is proportional to its strain i.e., Hooke's law holds good. For the material of beam, the value of E is same for the complete beam.

When the load is acting downward in a simply supported beam, then the distribution of bending stress, for any sections of a beam is as shown in Fig. 6.3.

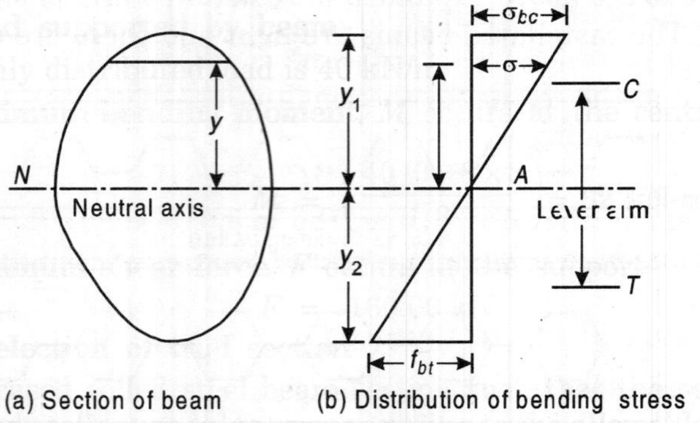

(a) Section of beam (b) Distribution of bending stress

Fig. 6.3

The bending stress varies linearly. The bending stress is zero at the neutral axis. When the load is acting downward, the bending stress is *compressive* above the neutral axis of the section and tensile below it and these are denoted by $\sigma_{bc.cal}$ and $\sigma_{bt.cal}$, respectively. The bending stress is maximum at the extreme fibre.

$$\sigma_{b.max} = \left(\frac{M}{I} \times y_{max}\right) = \left(\frac{M}{\dfrac{I}{y_{max}}}\right) \quad ...(6.2)$$

or

$$\sigma_{b.max} = \left(\frac{M}{Z}\right) \quad ...(6.3)$$

where, Z is the **section modulus** ($Z = I/y_{max}$) is the distance from the neutral axis to the entrance fibre, and σ_{max} is the maximum bending stress.

The maximum bending stress in the beam section (if compressive) should be less than the allowable bending compressive stress and (if tensile); should be less than the allowable bending tensile stress. When the section of beam is symmetrical about the neutral axis then the value of y_{max} is equal to half the depth of section, and the maximum bending stress in compression and in the tension at the extreme fibres are equal. When the beam section is not symmetrical about the neutral axis, then there are two distance y_1 and y_2 to the two extreme fibres from the neutral axis. The bending stress at the extreme top and bottom are not equal. Then, the values of $z_1 = \left(\frac{I}{y_1}\right)$ and $z_2 = \left(\frac{I}{y_2}\right)$ both should be calculated and compared with the section modulus, Z of the beam section provided.

The total compressive force C above the neutral axis is equal to the tensile force T, for the beam is equilibrium. These two forces act in the opposite direction and form a couple. This couple resists the bending moment and this moment is known as **moment of resistance** M_r.

The moment of resistance of a beam section is the moment of the couple which is set up at the section by the longitudinal forces C and T created in the beam due to the bending.

$$M_r = (C \times \text{Lever arm}) = (T \times \text{Lever arm})$$

For the beam in the equilibrium, the moment of resistance M_r would be equal to the maximum bending moment M at any section ($M_r = M$).

6.2.1 Allowable Stress in Bending

The allowable bending stress in the design of rolled steel beam sections considerably depends on the geometrical properties of the section and the lateral support. In case of flange width/flange thickness $\left(\frac{1}{2}\frac{b_f}{t_f}\right)$ and the depth of section/thickness of web $\left(\frac{h}{t_w}\right)$ ratio are not adequate, the elements of beam section will tend to buckle at low compressive stresses which will be due to bending combined with axial loads. If the compression flange is not laterally supported (i.e., supports at intervals or uniformly) along the compression zone, it will either buckle in plane or out-of plane coupled with twisting.

The rolled steel sections are produced with adequate $\left(\frac{1}{2}\frac{b_f}{t_f}\right)$ and $\left(\frac{h}{t_w}\right)$ ratios such that the buckling of the flange or web does not occur. The designers may provide supports at intervals or uniformly along the compression flange such that its bucking is avoided.

The calculated bending compressive stress $\sigma_{bc.cal}$ and bending tensile stress $\sigma_{bt.cal}$ in the extreme fibres should not exceed the maximum permissible bending stress in compression (σ_{bc}) or in the tension σ_{bt} as below.

$$\sigma_{bc} \text{ or } \sigma_{bt} = 0.66 f_y \qquad \text{...(i)}$$

The structural steel used in general construction may have yield stress as 220, 230, 240, 250, 260, 280, 300, 340, 360, 380, 400, 420, 450, 480, 510 or 540 N/mm^2 (MPa). The structural steels having these values of yield stress are also used in flexural members.

The maximum permissible bending compressive stress in beams and channel with equal flanges has been given separately in IS: 800–1984.

For an I-beam or channel with equal flanges bending about the axis of the maximum strength (*xx*-axis), the maximum bending compressive stress on the extreme fibre, calculated on the effective section shall not exceed the value of maximum permissible bending compressive stress, σ_{bc}. The maximum permissible values of stress are given directly in Table 6.1 (a) or 6.1 (b), Table 6.1 (c) or 6.1 (d) and Table 6.1 (e) or 6.1 (f), as appropriate steels with yield stress, f_y = 250 N/mm^2 (Mpa), and 400 N/mm^2 (Mpa), respectively as per IS: 800–1984.

It is to note that Table 6.1 (a), (c) and (e) are applicable for $\left(\frac{T}{t_w}\right)$ ratio greater than 2 (two) and $\left(\frac{d_1}{t_w}\right)$ ratio greater than 85,75 and 67, respectively. Whereas Table 6.1 (b), (d) and (e) are used for $\left(\frac{T}{t_w}\right)$ ratio less than 2 (two) and $\left(\frac{d_1}{t_w}\right)$ ratio less than 85,75 and 67, respectively. The Tables 6.1 (a) to (f) have been prepared by deriving the given values in accordance with the expression recommended by IS: 800–1984.

For the steels having yield stresses other than these included in Tables 6.1 (a) to (f) may be obtained from such recommended expressions by IS: 800–1984.

Following symbols have been used in Tables 6.1 (a) to (f)

D = overall depth of the beam

d_1 = Depth of the beam

(i) For the web of a beam without horizontal stiffeners, the clear distance between the flanges, neglecting fillets or the clear distance between the inner loss of the flanges as appropriate.

(ii) For the web of a beam with horizontal stiffeners, the clear distance between the horizontal stiffener and the tension flange, neglecting fillets or the clear distance between the inner toes of the flanges as appropriate.

l = Effective length of the compression flange.

r_y = Radius of gyration of the section about its axis of minimum strength (*yy*-axis)

T = Mean thickness of the compression flange. It is equal to the area of horizontal portion of flange divided by the width; and

t_w = Web thickness.

For the rolled steel sections, the **mean thickness** *is that which one is given in ISI Handbook No. 1.*

It is to note that for all the rolled steel beam sections given in ISI Handbook No.1, the depth of web/thickness of web ratio is less than 67. At the same time mean thickness of compression flange/thickness of web ratio is less than 2.0 (two) and it is at the most equal to 2.0 (two) for only WB 600, @ 1.451 kN/m section in which t_f = 23.6 mm t_w = 11.8 mm.

As such, the maximum permissible bending stress σ_{bc} in equal flange I-beam is to be adopted from Table 6.1 (b), Table 6.1 (d) and Table 6.1 (f) depending upon the value of yield stresses f_y for the structural steel to be used.

In case of compound girders with the curtailed flanges, D shall be taken as the overall depth of the girder at the point of maximum bending moment, and T shall be taken as the effective thickness of the compression flanges. It shall be calculated as follows:

$T = k_1 \times$ mean thickness of the horizontal portion of the compression flange at the point of maximum bending moment.

The values coefficient k_1 are given in Table 6.2. It allows for the reduction in thickness or breadth of flange between points of effective lateral restraints. Its value depends upon ψ the ratio of total area of both flanges at the point of least bending moment to the corresponding area at the point of greatest bending moment between such points of restraint).

Table 6.1 (a) (i) *Maximum Permissible Bending Stress,* σ_{bc} *N/mm² (MPa). In Equal Flange Beams or Channels with* f_y *=250 N/mm² (Mpa).* $\left(\frac{T}{t_w} > 2.0 \text{ and } \frac{d_1}{t_w} > 85\right)$

$D/T \rightarrow$ l/r_y ↓	8	10	12	14	16	18	20	25	30
40	160	160	159	159	158	158	158	158	158
45	159	158	157	157	156	156	156	155	155
50	158	157	156	155	154	154	153	153	152
55	157	155	154	153	152	151	150	149	149
60	156	153	152	150	149	148	148	146	145
65	154	152	150	148	147	145	144	143	145
70	153	150	143	146	144	142	141	139	138
75	152	148	145	143	141	139	138	136	134
80	180	147	143,	141	138,	136	135	132	130
85	149	145	141	138	136	133	132	128	126
90	147	143	139	136	133	130	128	124	122
95	146	141	137	133	130	128	125	121	118
100	145	140	135	131	128	125	122	117	114
110	142	136	131	127	123	119	116	111	107
120	139	133	127	122	118	114	111	104	100

Contd.

Table 6.1 (a) (i) Contd.

$D/T \rightarrow$ / $l/r_y \downarrow$	8	10	12	14	16	18	20	25	30
130	137	130	124	118	113	109	106	99	94
140	134	127	120	114	109	105	101	93	88
150	132	124	117	110	105	100	96	88	83
160	129	121	113	107	101	96	92	84	78
170	177	118	110	104	98	93	88	80	74
180	124	115	107	100	94	89	85	76	70
190	122	113	104	97	91	86	82	73	66
200	120	110	102	94	88	83	78	70	63
210	118	108	99	92	86	80	76	67	60
220	116	105	97	89	83	78	73	64	58
230	113	103	94	87	80	75	70	62	55
240	111	101	92	84	78	73	68	59	53
250	109	99	90	82	76	70	66	57	51
260	107	97	88	80	74	68	64	55	49
270	106	95	86	78	72	66	62	53	47
280	104	93	84	76	70	65	60	51	45
290	102	97	82	74	68	63	58	50	44
300	100	99	80	72	66	61	57	48	48

Table 6.1 (a) (ii) *Maximum permissible bending stress, σ_{bc} N/mm² (MPa). in equal flange I-beams or channels with f = 250 N/mm² (MPa),* $\left(\dfrac{T}{t_w} > 2.0 \text{ and } \dfrac{d_1}{t_w} > 85\right)$

$D/T \rightarrow$ / $l/r_y \downarrow$	20	25	30	35	40	50	60	80	100
40	158	158	158	157	157	157	157	157	157
45	156	155	155	155	155	155	154	154	154
50	153	153	152	152	152	151	151	151	151
55	150	149	149	148	148	148	148	147	147
60	148	146	145	145	144	144	144	143	143
65	144	143	142	141	140	140	139	139	139
70	141	139	138	137	136	135	135	135	134
75	138	136	134	133	134	131	130	130	129
80	135	132	130	128	128	126	126	125	125

Contd.

Table 6.1 (a) (ii) Contd.

$D/T \rightarrow$ l/r_y ↓	20	25	30	35	40	50	60	80	100
85	132	128	126	124	123	122	121	120	120
90	128	124	122	120	119	117	116	115	115
95	125	121	118	116	114	112	111	110	110
100	122	117	114	112	110	108	107	105	105
110	116	111	107	104	102	99	98	96	95
120	111	104	100	97	94	91	90	88	87
130	106	99	94	90	88	84	82	80	79
140	101	93	88	84	81	78	75	73	72
150	96	88	83	79	76	72	69	67	65
160	92	84	78	74	71	66	64	61	60
170	88	80	74	69	66	62	59	56	55
180	85	76	70	65	62	58	55	52	50
190	82	73	66	62	58	54	51	48	46
200	78	70	63	59	55	50	48	44	43
210	76	67	60	“56	52	47	44	41	40
220	73	64	58	53	49	45	42	38	37
230	70	62	55	51	47	42	39	36	34
240	68	59	53	48	45	40	37	31	32
250	66	57	51	46	43	38	35	32	30
260	44	55	49	.44	41	36	35	30	28
270	62	53	47	43	39	35	32	28	26
280	60	51	45	41	38	33	30	27	25
290	58	50	44	39	36	32	29	25	24
300	57	48	42	38	35	30	27	24	22

Table 6.1 (b) (i) *Maximum permissible bending stress, σ_{bc} N/mm² (MPa). in equal flange I-beams or channels with f_y = 250 N/mm²(MPa),* $\left(\frac{T}{t_w} \leq 2.0 \text{ and } \frac{d_1}{t_w} \leq 85\right)$

$D/T \rightarrow$ l/r_y ↓	8	10	12	14	16	18	20	25	30
40	161	161	160	160	160	160	160	159	159
45	161	160	159	159	158	158	158	157	157
50	160	158	158	157	156	156	156	155	155

Contd.

Table 6.1 (b) (i) Contd.

$D/T \rightarrow$ / $l/r_y \downarrow$	8	10	12	14	16	18	20	25	30
55	159	157	156	155	154	154	153	153	152
60	168	156	154	153	152	152	151	150	149
65	156	153	153	151	150	149	148	147	146
70	155	153	151	149	149	147	146	144	143
75	154	152	149	147	146	144	143	141	140
80	153	150	148	145	143	142	140	138	136
85	152	149	146	143	141	139	138	135	133
90	151	147	144	141	139	137	135	131	129
95	150	146	142	139	137	134	132	128	126
100	149	145	141	137	134	132	129	125	122
110	147	142	137	133	130	127	124	119	115
120	144	139	134	129	126	122	119	113	109
130	142	136	131	126	121	118	114	108	103
140	140	133	128	122	118	113	110	103	97
150	138	131	124	119	114	109	105	98	92
160	136	128	121	115	110	106	101	93	87
170	134	126	119	112	107	102	98	89	83
180	131	123	116	109	104	99	94	85	79
190	129	121	113	106	101	95	91	82	75
200	127	118	111	104	98	92	88	79	72
210	125	116	108	101	95	90	85	76	69
220	123	114	106	99	92	87	82	73	66
230	122	112	103	96	90	84	80	70	63
240	120	110	101	94	87	82	77	68	61
250	118	108	99	92	85	80	75	65	59
260	116	106	97	89	83	77	73	63	57
270	114	104	95	87	81	75	71	61	55
280	113	102	93	85	79	73	69	59	53
290	111	100	91	84	77	72	67	58	51
300	109	98	89	82	75	70	65	56	49

Table 6.1 (b) (ii) *Maximum permissible bending stress, σ_{bc} N/mm² (MPa). in equal flange I-beams or channels with f = 250 N/mm² (MPa),* $\left(\dfrac{T}{t_w} \le 2.0 \text{ and } \dfrac{d_1}{t_w} > 85\right)$

$D/T \rightarrow$ l/r_y ↓	20	25	30	35	40	50	60	80	100
40	160	159	159	159	159	159	159	159	159
45	158	157	157	157	157	157	157	157	157
50	156	155	155	155	154	154	154	154	154
55	153	153	152	152	152	151	151	151	151
60	151	150	149	149	149	148	148	148	148
65	148	147	146	146	145	145	144	144	144
70	146	144	143	142	142	141	141	140	140
75	143	141	140	139	138	137	137	136	136
80	140	138	136	135	134	133	132	132	132
85	138	135	133	131	130	129	128	127	127
90	135	131	129	127	126	125	124	123	123
95	132	128	126	124	122	121	120	119	118
100	129	125	122	120	118	116	115	114	113
110	124	119	115	113	111	108	107	105	105
120	119	118	109	106	104	101	99	97	96
130	114	108	103	99	97	94	91	89	88
140	110	103	97	94	91	87	85	82	81
150	105	98	92	88	85	81	78	76	74
160	101	93	87	83	80	75	73	70	68
170	98	89	83	79	75	70	68	64	63
180	94	85	79	74	71	66	63	60	58
190	91	82	75	71	67	62	59	55	54
200	88	79	72	67	63	58	55	51	50
210	125	76	69	64	60	55	52	48	46.
220	82	73	66	61	57	52	49	45	
230	80	70	63	58	55	49	46	42	40
240	77	68	61	56	52	47	43	40	38
250	75	65	59	54	50	44	41	37	35
260	73	63	57	52	48	42	39	35	33
270	71	61	55	50	46	41	37	33	31
280	69	59	53	48	44	39	35	32	30
290	67	58	51	46	42	37	33	31	28
300	65	56	49	45	41	36	32	29	27

Table 6.1 (c) (i) *Maximum permissible bending stress, σ_{bc} N/mm² (MPa). in equal flange I-beams or channels with f_y = 340 N/mm² (MPa),* $\left(\frac{T}{t_w} > 2.0 \text{ or } \frac{d_1}{t_w} > 75\right)$

$D/T \rightarrow$ / $l/r_y \downarrow$	8	10	12	14	16	18	20	25	30
40	215	214	212	212	211	211	210	210	209
45	213	211	209	208	207	206	206	205	204
50	210	208	205	204	203	202	201	199	199
55	208	204	202	200	198	197	196	194	191
60	205	201	198	195	193	191	190	188	186
65	203	198	194	191	188	186	185	181	180
70	200	195	190	186	183	181	179	175	173
75	198	192	186	182	179	176	173	169	166
80	195	188	183	178	174	170	168	163	159
85	193	185	179	174	169	165	162	157	153
90	190	182	175	169	165	161	157	151	147
95	188	179	172	165	160	156	152	145	141
100	185	176	168	162	156	151	147	140	135
110	180	170	162	154	148	143	138	130	124
120	176	165	155	147	141	135	130	121	115
130	171	159	149	141	134	128	122	113	106
140	167	154	144	135	127	121	116	105	98
150	163	150	139	129	122	115	110	99	92
160	158	145	134	124	116	110	104	93	86
170	155	141	129	120	111	105	99	88	80
180	151	137	125	115	107	100	94	83	76
190	147	131	121	111	103	96	90	79	72
200	144	129	117	107	99	92	86	75	68
210	140	125	113	103	95	88	83	72	64
220	137	122	110	100	92	85	79	69	61
230	134	119	107	97	89	82	76	66	58
240	131	116	104	94	86	79	74	63	56
250	128	113	101	91	83	76	71	61	53
260	125	110	98	88	80	74	68	58	51
270	122	107	95	86	78	72	66	56	49
280	120	105	93	83	76	69	64	54	47
290	117	102	90	81	73	67	62	52	46
300	115	100	88	79	71	65	60	51	44

Table 6.1 (c) (ii) *Maximum permissible bending stress, σ_{bc} N/mm² (MPa). in equal flange I-beams or channels with f_y = 340 N/mm² (MPa),* $\left(\frac{T}{t_w} > 2.0 \text{ and } \frac{d_1}{t_w} > 75\right)$

$D/T \rightarrow$ / $l/r_y \downarrow$	20	25	30	35	40	50	60	80	100
40	210	210	209	209	209	209	209	209	209
45	206	205	204	204	204	203	203	203	203
50	201	199	199	198	198	197	197	197	197
55	196	194	193	192	191	191	190	190	190
60	190	188	186	185	185	184	183	183	183
65	185	181	180	178	177	176	176	175	175
70	179	175	173	171	170	169	168	167	167
75	173	169	166	164	163	161	160	159	159
80	168	163	159	157	156	154	153	151	151
85	162	157	153	150	149	146	145	144	143
90	157	151	147	144	142	139	138	136	136
95	152	145	141	137	135	132	131	129	128
100	147	140	135	131	129	126	124	122	121
110	138	130	124	120	117	114	112	109	108
120	130	121	115	110	107	103	101	98	97
130	122	113	106	101	98	93	91	80	97
140	116	105	98	93	90	85	82	79	78
150	110	99	92	87	83	78	75	72	70
160	104	93	86	80	77	72	68	65	63
170	99	88	80	75	71	66	63	59	58
180	94	83	76	70	66	61	58	54	53
190	90	79	72	66	62	57	54	50	48
200	86	75	63	62	58	53	50	46	44
210	83	72	64	59	55	50	46	43	41
220	79	69	61	56	52	47	43	40	38
230	76	66	58	53	49	44	41	37	35
240	74	63	56	51	47	42	38	35	33
250	71	61	53	48	44	39	36	32	31
260	68	58	51	46	42	37	34	31	29
270	66	56	49	44	41	36	32	29	27
280	64	54	47	42	39	34	31	27	25
290	62	52	45	41	37	32	29	26	24
300	60	51	44	39	36	31	28	25	23

Table 6.1 (d) (i) *Maximum permissible bending stress, σ_{bc} N/mm² (MPa), in equal flangeI-beams or channels with $f_y = 340$ N/mm² (MPa), $\left(\frac{T}{t_w} \le 2.0 \text{ and } \frac{d_1}{t_w} \le 75\right)$*

$D/T \rightarrow$ / l/r_y ↓	8	10	12	14	16	18	20	25	30
40	217	216	215	214	214	213	213	213	212
45	215	214	212	211	211	210	210	209	208
50	213	211	209	208	207	206	206	205	204
55	211	209	206	205	203	202	201	200	199
60	209	206	203	201	199	198	197	195	193
65	207	203	200	197	195	193	192	189	188
70	205	201	197	194	191	189	187	184	182
75	203	198	194	190	187	184	182	178	176
80	201	195	190	186	183	180	177	173	170
85	199	193	187	183	179	175	173	168	164
90	197	190	184	179	175	171	167	162	158
95	195	187	181	175	171	167	163	157	153
100	193	185	178	172	167	163	159	152	147
110	188	180	172	165	159	155	150	142	137
120	184	175	166	159	152	147	142	133	127
130	180	170	161	153	146	140	135	125	119
140	177	165	156	147	140	134	128	118	111
150	173	161	151	142	134	128	122	112	104
160	169	157	146	137	129	122	117	106	98
170	166	153	142	132	124	117	111	100	92
180	162	149	137	128	120	113	107	95	87
190	159	145	133	124	115	108	102	91	82
200	155	141	130	120	111	104	98	86	78
210	152	138	126	116	108	100	94	83	74
220	149	135	123	113	104	97	91	79	71
230	146	132	119	109	101	94	88	76	68
240	143	128	116	106	98	91	85	73	65
250	141	126	113	103	95	88	82	70	62
260	138	123	110	100	92	85	79	68	60
270	135	120	108	98	89	82	77	65	58
280	133	117	105	95	87	80	74	63	56
290	130	115	103	93	84	78	72	61	54
300	128	112	100	90	82	76	70	59	52

Table 6.1 (d) (ii) *Maximum permissible bending stress, σ_{bc} N/mm² (MPa), in equal flangeI-beams or channels with f_y = 340 N/mm² (MPa),* $\left(\frac{T}{t_w} \leq 2.0 \text{ and } \frac{d_1}{t_w} > 75\right)$

$D/T \rightarrow$ l/r_y ↓	20	25	30	35	40	50	60	80	100
40	213	213	212	212	212	212	212	212	212
45	210	209	208	208	208	208	208	207	507
50	206	205	204	203	203	203	203	202	202
55	201	200	199	198	198	197	197	197	197
60	197	195	193	193	195	191	191	191	190
65	192	189	188	187	186	185	184	184	184
70	187	184	182	181	180	178	178	177	177
75	182	178	176	174	173	172	171	170	169
80	177	173	170	168	167	165	164	163	162
85	173	168	164	162	160	158	157	156	155
90	167	162	158	156	154	151	150	148	148
95	163	157	153	150	148	145	143	142	141
100	159	152	147	144	142	138	137	135	134
110	150	142	137	133	130	126	124	122	121
120	142	133	127	123	120	116	113	110	139
130	135	125	119	114	110	106	103	100	99
140	128	118	111	105	102	97	94	91	89
150	122	112	104	99	95	89	86	83	81
160	117	106	98	92	88	82	79	75	74
170	111	100	92	86	82	76	73	69	67
180	107	95	87	81	77	71	67	63	61
190	102	91	82	76	72	66	63	59	56
200	98	86	78	72	68	62	58	54	52
210	94	83	74	69	64	58	54	50	48
220	91	89	71	65	61	55	51	47	45
230	88	75	68	62	58	52	48	44	42
240	85	73	65	59	55	59	45	41	39
250	82	70	62	57	52	45	43	38	36
260	79	68	60	54	50	44	40	36	34
270	77	65	58	52	48	42	38	35	32
280	74	63	56	50	46	40	36	32	30
290	72	61	54	48	44	38	35	31	29
300	70	59	52	46	42	37	33	29	27

Table 6.1 (e) (i) *Maximum permissible bending stress, σ_{bc} N/mm² (MPa), in equal flangeI-beams or channels with f_y = 400 N/mm² (MPa),* $\left(\dfrac{T}{t_w} > 2.0 \text{ and } \dfrac{d_1}{t_w} > 67\right)$

$D/T \rightarrow$ / $l/r_y \downarrow$	8	10	12	14	16	18	20	25	30
40	250	248	247	245	245	244	243	243	242
45	247	244	242	240	239	238	237	236	235
50	244	240	237	234	233	231	230	228	227
55	240	235	232	229	226	224	223	221	219
60	236	231	226	223	220	217	216	212	210
65	233	226	221	217	213	210	208	204	202
70	229	222	216	211	207	203	201	196	193
75	226	217	211	205	200	186	193	188	184
80	222	213	206	199	194	190	186	180	175
85	2f9	209	201	194	188	183	179	172	167
90	216	205	196	188	182	177	173	165	160
95	212	201	191	183	177	171	166	158	152
100	209	197	187	178	171	165	160	151	145
110	203	189	178	169	161	155	149	139	133
120	196	182	170	160	152	145	140	129	121
130	191	176	163	153	144	137	131	119	112
140	185	169	156	i46	137	129	123	111	103
150	179	163	150	139	130	122	116	104	96
160	174	158	144	133	124	116	109	97	89
170	169	152	139	127	118	110	104	92	83
180	165	147	134	122	113	105	97	86	78
190	160	143	129	117	108	100	94	82	74
230	156	138	124	113	104	96	90	78	70
210	152	134	120	109	100	92	86	74	66
220	148	130	116	105	96	88	82	71	63
230	144	126	112	101	92	85	79	67	60
240	141	123	109	98	89	82	76	65	57
250	137	119	106	95	86	79	73	67	53
260	134	116	103	92	83	76	70	60	52
270	131	113	100	89	81	74	68	57	50
280	128	110	97	86	78	71	66	55	48
290	125	107	94	84	76	69	64	53	46
300	122	105	92	82	74	67	62	52	45

Table 6.1 (e) (ii) *Maximum permissible bending stress, σ_{bc} N/mm² (MPa), in equal flangeI-beams or channels with f_y = 400 N/mm² (MPa),* $\left(\frac{T}{t_w} > 2.0 \text{ and } \frac{d_1}{t_w} > 67\right)$

$D/T \rightarrow$ / $l/r_y \downarrow$	20	25	30	35	40	50	60	80	100
40	243	243	242	242	242	241	341	241	241
45	237	236.	235	235	234	234	234	233	233
50	230	228	227	227	226	226	225	225	225
55	223	221	219	218	217	216	216	216	215
60	216	212	210	209	208	207	206	206	205
65	208	204	202	200	199	197	197	196	195
70	201	195	193	191	189	188	187	186	185
75	193	188	184	182	180	178	177	175	175
80	186	180	176	173	171	168	167	166	165
85	179	172	167	164	162	159	158	156	155
90	173	165	160	165	154	151	149	147	146
95	166	158	152	149	146	142	140	138	137
100	160	151	145	141	138	135	133	130	129
110	149	139	133	128	125	121	118	115	114
120	140	129	121	116	113	108	106	103	101
130	131	119	112	106	103	98	95	92	90
140	123	111	103	98	94	88	85	82	80
150	116	104	96	90	86	81	77	74	72
160	109	97	89	83	79	74	71	67	65
170	104	92	83	78	73	68	64	61	59
180	97	86	78	72	68	63	59	55	54
190	94	82	74	68	64	58	55	51	49
200	90	78	70	64	60	54	51	47	45
210	86	74	66	60	56	51	47	43	41
220	82	71	63	57	53	47	44	40	38
230	79	67	60	54	50	45	41	37	36
240	76	65	57	52	47	42	39	35	33
250	73	62	54	49	45	40	37	33	31
260	70	60	52	47	43	38	35	31	29
270	68	57	50	45	41	36	33	29	27
280	66	55	48	43	39	34	31	27	26
290	64	53	46	41	38	33	30	26	24
300	62	52	45	40	36	31	28	25	23

Table 6.1 (f) (i) *Maximum permissible bending stress, σ_{bc} N/mm² (MPa), in equal flange I-beams or channels with f_y = 400 N/mm² (MPa),* $\left(\frac{T}{t_w} \le 2.0 \text{ and } \frac{d_1}{t_w} \le 67\right)$

$D/T \rightarrow$ / $l/r_y \downarrow$	8	10	12	14	16	18	20	25	30
40	253	252	250	249	249	248	248	247	247
45	251	248	246	245	244	243	243	242	241
50	248	245	242	240	239	238	237	235	234
55	245	241	238	236	234	232	231	229	227
60	242	237	234	231	228	226	228	222	220
65	239	234	229	225	222	220	218	215	212
70	236	230	225	220	217	214	212	207	205
75	233	226	220	215	211	208	205	200	197
80	230	223	216	210	206	202	199	193	189
85	227	219	212	205	200	196	192	186	181
90	225	215	207	201	195	190	186	179	174
95	222	212	203	196	190	185	180	172	167
100	219	208	199	191	185	179	175	166	160
110	213	202	191	183	176	169	164	154	148
120	208	195	184	175	167	160	154	144	136
130	203	189	177	167	159	152	146	134	126
140	198	183	171	160	152	144	138	126	117
150	193	178	165	154	145	137	131	118	109
160	188	172	159	148	139	131	124	111	102
170	183	167	154	142	133	125	118	105	96
180	179	172	149	137	127	119	112	99	90
190	175	158	144	132	122	114	108	94	85
200	171	153	139	128	118	110	103	90	80
210	167	149	135	123	114	105	99	86	77
220	163	145	131	119	110	102	95	82	73
230	159	141	127	115	106	98	91	79	70
240	156	138	123	112	102	94	88	75	67
250	152	134	120	108	99	91	85	72	64
260	149	131	117	105	96	88	82	70	61
270	146	128	114	102	93	85	79	67	59
280	143	125	111	99	90	83	77	65	57
290	140	122	108	97	88	80	74	63	55
300	137	119	105	94	85	78	72	61	53

Table 6.1 (f) (ii) *Maximum permissible bending stress,* σ_{bc} *N/mm² (MPa), in equal flange I-beams or channels with* f_y = 400 N/mm² (MPa), $\left(\frac{T}{t_w} \leq 2.0 \text{ and } \frac{d_1}{t_w} \leq 67\right)$

$D/T \rightarrow$ $l/r_y \downarrow$	8	10	12	14	16	18	20	25	30
40	248	247	247	246	246	246	246	246	246
45	243	242	241	241	240	240	240	240	239
50	237	235	234	234	233	233	233	232	232
55	231	229	227	227	226	225	225	225	224
60	225	222	220	219	218	217	217	216	216
65	218	215	212	211	210	209	208	207	207
70	212	207	205	203	202	200	199	198	198
75	205	200	197	195	193	191	190	189	188
80	199	193	189	186	185	182	181	180	179
85	192	186	181	178	176	174	172	171	170
90	186	179	174	171	168	165	164	162	161
95	180	172	167	163	161	157	155	153	152
100	175	166	160	156	153	150	148	145	144
110	164	154	148	143	140	135	133	130	129
120	154	144	136	131	127	123	120	117	115
130	146	134	120	141	117	111	108	105	103
140	138	126	117	111	107	102	98	95	93
150	131	118	109	103	99	93	89	86	84
160	124	111	102	96	92	85	82	78	76
170	118	105	96	90	85	79	75	71	69
180	112	99	90	84	79	73	69	65	63
190	108	94	85	79	74	68	64	60	88
200	103	90	80	75	70	63	60	85	83
210	99	86	77	70	66	59	55	51	49
220	95	82	73	67	62	56	52	48	45
230	91	79	70	64	59	53	49	44	42
240	88	75	67	61	56	50	46	42	39
250	85	72	64	58	53	47	43	39	37
260	83	70	61	55	51	45	41	37	34
270	79	67	59	53	49	43	39	35	32
280	77	65	57	51	47	41	37	33	30
290	74	63	55	49	45	39	35	31	29
300	72	61	53	47	43	37	33	29	27

6.3 SHEAR AND BEARING STRESSES

When the beams are subjected to loads, then, these are also required to transmit large shear forces either at supports or at concentrated loads. For simply supported beams, the shear forces is maximum at the supports. The values of shear forces at the concentrated loads also remain large. Due to shear force, the shear stresses are set up along with the bending stresses at all sections of the beams. The shear stress at any point of the cross-section is given by

$$\tau_v = \left(\frac{F \cdot Q}{I \cdot t}\right) \quad ...(6.4)$$

where, τ_v is the shear stress and F is the shear force at cross-section

Q = static moment about the neutral axis of the portion of cross-sectional area beyond the location at which the stress is being determined.

I = Moment of inertia of the section about the neutral axis

t = Thickness of web (width of section at which the stress is being determined).

The distribution of shear stresses for rectangular section beam and I-beam section are shown in Fig. 6.4 (a) and Fig. 6.4. (b), respectively.

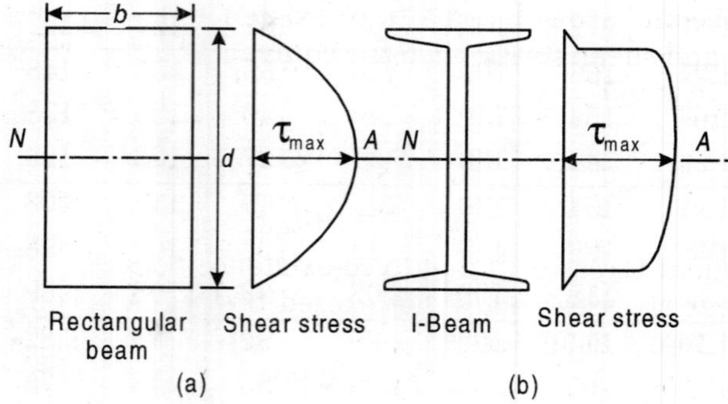

Fig. 6.4

The maximum shear stress occurs at the neutral axis of the section.

The maximum shear stress in a member having regard to the distribution of stresses in conformity with the elastic behaviour of the member in the flexure (bending) should not exceed the value of maximum permissible shear stress, τ_{vm} found as follows:

$$\tau_{vm} = 0.45 f_y \quad ...(i)$$

where, f_y is the yield stress of structural steel to be used.

It is to note that in the case of rolled beams and channels, the design shear is to be found as the average shear.

The average shear stress for rolled beams or channels calculated by dividing the shear force at the cross-section of the web is defined as the depth of the beam or channel multiplied by its web thickness.

Average shear stress for rectangular beam is given by

$$\tau_v = \left(\frac{F}{b \times d}\right) \quad ...(6.5)$$

Average shear stress for I-beam is given by

$$\tau_v = \left(\frac{F}{h \times t_w}\right) \quad ...(6.6)$$

For rolled steel beams and channels, it is assumed that shear force is resisted by web only. The portion of shear resisted by the flanges is neglected.

The average shear stress $\tau_{va.cal}$, in a member calculated on the gross cross-section of web (when web bucking is not a factor) should not exceed in case of unstiffened web of the beam,

$$\tau_{va} = 0.4 f_y \quad ...(ii)$$

The allowable shear stresses as per AISC, AASHTO and AREA specification are as follows:

Specifications	*Allowable shear stress*
AISC	$0.40 f_y$
AASHTO	$0.33 f_y$
AREA	$0.35 f_y$

When the beams are subjected to co-existence bending stresses (tension or compression) and shear stress, then the equivalent stress, $\sigma_{e.cal}$ is obtained from the following formula

$$\sigma_{e.cal} = [\sigma^2_{bt.cal} + 3\tau^2_{vm.cal}]^{1/2} \quad ...(iii)$$

$$\sigma_{e.cal} = [\sigma^2_{bc.cal} + 3\tau^2_{vm.cal}]^{1/2} \quad ...(iv)$$

The equivalent stress $\sigma_{e.ca}$ due to co-existence bending (tension or compression) and shear stresses should not exceed the maximum permissible equivalent stress, σ_e found as under

$$\sigma_e = 0.90 f_y \quad ...(v)$$

When the bearing stress σ_p is combined with tensile or compressive bending and shear stresses under the most unfavourable conditions of loading the equivalent stress $\sigma_{e.cal}$ obtained as below should not exceed

$$\sigma_e = 0.90 f_y \quad ...(vi)$$

$$\sigma_{e.cal} = [\sigma_{bt}^2 + \sigma_{bt\ .cal}^2 + \sigma_{bt\ .cal}^2 + 3\ \tau_{vm\ .cal}^2]^{1/2} \quad ...(vii)$$

$$\sigma_{e.cal} = [\sigma_{bt}^2 + \sigma_{bt\ .cal}^2 + \sigma_{bt\ .cal}^2 + 3\ \tau_{vm\ .cal}^2]^{1/2} \quad ...(viii)$$

$\sigma_{bc.cal}$, $\sigma_{bt\ .cal}$, $\tau_{vm.cal}$ and $\sigma_{p.cal}$ are the numerical values of the co-existence bending (compression or tension), shear and bearing stresses. When the bending occurs about both the axes of the member, $\sigma_{bt\ .cal}$ and $\sigma_{bc.cal}$ should be taken as the sum of the two calculated fibre stresses, σ_e is the maximum permissible equivalent stress.

Bearing stress. The bearing stress in any part of a beam when calculated on the net area of contact should not exceed the value of σ_p calculated as below

$$\sigma_p = 0.75 f_y \quad ...(ix)$$

where σ_p is the maximum permissible bearing stress and f_y is the yield stress.

6.4 EFFECTIVE SPAN AND DEFLECTION LIMITATION

The effective span of a beam shall be taken as the length between the centres of the supports, export in cases where the point of application of the reaction is taken as eccentric to the support, then, it shall be permissible to take the effective span as the length between the assumed points of applications of reaction.

The stiffness of a beam is a major consideration in the selection of a beam section. The allowable deflections of beams depend upon the purpose for which the beams are designed.

The large deflections of beams are undesirable for the following reasons:

(i) When the loads are primarily due to human occupants especially in the case of public meeting places, large deflections result in noticeable to the occupants.

(ii) The large deflections may result in cracking of ceiling plaster, floors or partition walls.

(iii) The large deflection indicates the lack of rigidity. It may cause vibrations and over-stresses under dynamic loads.

(iv) The large deflection may cause the distortions in the connections. The distortions cause secondary stresses.

(v) The large deflection may cause poor drainage, which will lead to ponding of water, and therefore increase the loads.

6.4.1 Limiting Vertical Deflection

The deflection of a member is calculated without considering the impact factor or dynamic effect of the loads causing the deflection. The deflection of a member shall not be such as to impact the strength of efficiency of the structure and lead to damage to finishings.

Generally, the maximum deflection should not exceed (1/325) of the span. This limit may be exceeded in cases where greater deflection would not impair the strength or efficiency of the structure or lead to damage to finishings. The deflection of the beams may be decreased by increasing the depth of beams, decreasing the span, providing greater end restraint or by any other means.

6.4.2 Limiting Horizontal Deflection

At the caps of columns in single storey buildings the horizontal deflection due to lateral force should not ordinarily exceed (1/325) of the actual length l of the column. This limit may be exceeded in cases where the greater deflection would not impair the strength and efficiency of the structure or lead to damage to finishings.

According to AISC specification, the deflection of beams and girders for live load and plastered ceiling should not exceed (1/360) of the span.

6.5 LATERALLY SUPPORTED BEAMS

The laterally supported beams are also called *laterally restrained beams*. When lateral deflection of the compression flange of a beam is prevented by providing effective lateral support, (restraint) the beam is said to be laterally supported. The *effective lateral restraints* is the restrains which produces sufficient resistance in a plane perpendicular to the plane of bending to restrain the compression flange of a beam from lateral buckling to either side at the point of application of the restraint. The concrete slab encasing the top flange (so that the bottom surface of the concrete slab is flush with the bottom of the top flange) is shown in Fig. 6.5 (a). It provides a continuous lateral supports to the top flange of the beam. When other beams frame at frequent intervals into the beam in questions as shown in Fig. 6.5 (b), lateral support is provided at each point of connection but main beam should still be checked between the two supports.

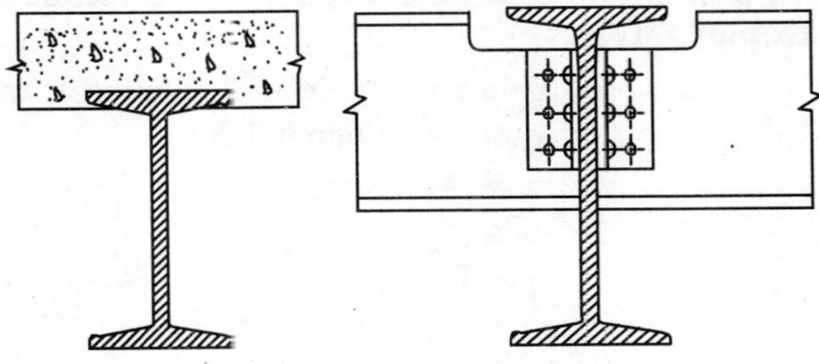

(a) Continuous lateral support (b) Local lateral support

Fig. 6.5 *Adequate lateral support*

In the laterally supported beams, the value of allowable bending compressive stress remains unaltered and the reduction in its value is not made. Bending compressive stress is taken equal to the allowable bending tensile stress, $(\sigma_{bc} = \sigma_{bt} = 0.66 f_y)$. The adequate lateral support is provided to safeguard against the lateral-torsional buckling. In case of doubt for adequate lateral support, the beams should be designed as laterally unsupported. In case the concrete slab holds the top flange (compression flange) of the beam from one side only, then, the lateral support is not credited. The concrete slab simply resting over the top flange of beam without shear connectors also does not provide a lateral support. Sometimes, the plank or bar grating is attached to top flange of beam by means of bolts. When the bolts are firmly fastened, then, they provide adequate lateral support temporarily. Even then, bolts have temporary nature of connections. It is possible that the bolts might be omitted or removed. As such, the top flange should not be considered laterally supported fully. The beams having lateral support from other members may buckle between points of lateral supports. Therefore, the laterally unsupported length of beam is kept short.

Example 6.1 *The effective length of compression flange of simply supported beam MB 500, @ 0.869 kN/m is 8 m. Determine the safe uniformly distributed* *load per metre length which can be placed over the beam having an effective span of 8 metres. Adopt maximum permissible stresses as per IS 800–1984. The ends of beam are restrained against rotation at the bearings.*

Solution

Step 1 : Permissible bending stress

MB 500, @ 0.869 kN/m has been used as simply supported beam. The effective span of beam is 8 m. The effective length of compression flange is also 8 m.

From the steel section table, the section modulus of beam

$$Z = 1808.7 \times 10^3 \text{ mm}^3$$

Means thickness of compression flange,

$$t_f = T = 17.2 \text{ mm}$$

Thickness of web,

$$t_w = 10.2 \text{ mm}$$

It is assumed that the value of yield stress, f_y for the structural steel of MB 500, @ 0.869 kN/mm² (MPa).

Ratio $$\left(\frac{T}{t_w} = \frac{17.2}{10.2}\right) = 1.686 < 2.00$$

Ratio $$\left(\frac{d_t}{t_w}\right) = \left(\frac{h - 2h_2}{t_w}\right) = \left(\frac{500 - 2 \times 37.95}{10.2}\right)$$

$$= \left(\frac{h_1}{t_w}\right) = \left(\frac{424.1}{10.2}\right)$$

$$= 41.578 < 85$$

Ratio $$\left(\frac{D}{T}\right) = \left(\frac{500}{17.0}\right) = 29.07$$

The effective length of compression flange is 8 m

Ratio $$\frac{l}{r_y} = \left(\frac{0.7 \times 8 \times 1000}{35.2}\right) = 159.1$$

From IS: 800–1984, from Table 6.1 (b), the maximum permissible bending stres, for above ratios (by linear interpolation)

$$\sigma_{bc} = 65.121 \text{N/mm}^2 \text{ (MPa)}$$

Step 2: Moment of resistence of beam

$$M_r = (\sigma_{bc} \times Z) = \left(\frac{88.566 \times 1808.7 \times 10^3}{1000 \times 1000}\right)$$

$$= 160.189 \text{ m-kN}$$

MB 500, @ 0.869 kN/m can resist maximum bending moment equal to moment of resistance

∴ Maximum bending moment.

$$M = 160.189 \text{ m-kN}$$

Step 3: Load supported over beam

The effective span of the beam is 8 metres. Let w be the uniformly distributed load per metre length.

The maximum bending moment, M for the beam occurs at the centre,

$$M = \left(\frac{w \cdot l^2}{8}\right)$$

$$\left(\frac{w \times 8 \times 1000}{8}\right) = 160.189$$

$$w = \left(\frac{160.189 \times 8}{8 \times 8 \times 1000} \times 1000\right) = 20.02 \text{ kN/m}$$

The self-weight of the beam is 0.869 kN/m. Therefore, the safe uniformly distrbuted load which can be placed over the team.

$$(20.02 - 0.869) = \mathbf{19.15\ kN.}$$

6.6 DESIGN OF LATERALLY SUPPOTED BEAMS

The design of beams is generally governed by the maximum allowable bending stress and the allowable deflection. Its design is controlled by shear only when the spans are short and loads are heavy. The members are selected such that the sections are symmetrical about the plane of loading and the unsymmetrical bending and torsion are eliminated.

The design of beams deals with proportioning of members, the determination of effective section modulus, maximum deflection and the shear stress. In general, the rolled steel sections have webs of sufficient thickness such that the criterion for design is seldom governed by shear. The following are the usual steps in the design of laterally supported beams:

Step 1. For the design of beams, load to be carried by the beam and effective span of the beam are known. The value of yield stress, f_y for the structural steel to be used is also known. For the rolled steel beams of equal flanges as given in ISI Handbook No. 1, the ratio of mean thickness of the compression flange ($T = t_f$) to the thickness of web used to be less than 2.00. So also, the ratio of the depth of web d_1 to the thickness of web is also smaller than 85. The ends of compression flange of a laterally supported beam remain restrained against lateral bending (i.e., not free to rotate in plan at the bearings).

In the beginning of design, the permissible bending stress in tension, σ_{bt} are in compression, σ_{bc} may be assumed as below:

$$\sigma_{bt} \text{ or } \sigma_{bc} = 0.66 f_y$$

The bending compressive stress, σ_{bc} and the bending tensile stress, σ_{bt} are equal for the laterally supported beam.

Step 2. The maximum bending moment M and the maximum shear force F in the beam are calculated. The required section modulus for the beam is determined,

$$Z = \left(\frac{M}{\sigma_{bc}}\right)$$

Step 3. From the steel section tables, a rolled steel beam section, (which provides more than the required section modulus) is selected. The steel beam section shall have $\left(\frac{D}{T}\right)$ and $\left(\frac{l}{r_y}\right)$ ratios more than 8 and 40, respectively. As such the trial section of beam selected may have modulus sections, Z more than that required. Some of the beam section of different groups (categories, have almost the same value of the section modulus Z. It is necessary to note the weight of beam per metre length and the section modulus, Z. The beam section selected should be such that it has minimum weight and adequate section modulus, Z.

Step 4. The rolled steel beam section is checked for the shear stress. The average and maximum shear stresses should not exceed the allowable average and maximum values of shear stresses.

Step 5. The rolled steel beam is also checked for deflection. The maximum deflection should not exceed the limiting deflection.

ISI Handbook No. 1 provides tables for allowable uniform loads on beams and channels used as flexural members with adequate lateral support for compression flange. The values of allowable uniform loads corresponding to respective effective spans are given for various beams and channel sections. For given span and total uniformly distributed load found, rolled steel beam or channel section may be selected from these tables.

The rolled steel I-sections and wide flange beam sections are most efficient sections. These sections have excellent flexural strength and relatively good lateral strength for their weights.

Example 6.2 *Design a simply supported beam to carry a uniformly distributed load of 44 kN/m. The effective span of beam is 8 metres. The effective length of compression flange of the beam is also 8 m. The ends of beam are not free to rotate at the bearings.*

Solution

Design :

Step 1 : Load supported, bending moment and shear forces

Uniformly distributed load = 44 kN/m

Assume self-weight of beam = 1.0 kN/m

Total uniformly distributed load

$$w = 45 \text{ kN/m}$$

The maximum bending moment, M occurs at the centre

$$M = \left(\frac{w \cdot l^2}{8}\right) = \left(\frac{45 \times 8 \times 8 \times 1000}{8 \times 1000}\right)$$

$$= 360 \text{ m-kN}$$

The maximum shear forces, F occurs at the support

$$F = \left(\frac{w \cdot l}{8}\right) = \left(\frac{45 \times 8}{2}\right) = 180 \text{ kN}$$

Step 2: Permissible bending stress

It is assumed that the value of yield stress, f_y for the structural steel is 250 N/mm^2 (MPa). The ratio $\left(\frac{T}{t_w}\right)$ and $\left(\frac{d_1}{t_w}\right)$ are less than 2.0 and 8.5, respectively.

The maximum permissible stress in compression or tension may be assumed as below:

$$\sigma_{bc} = \sigma_{bt} = (0.66 \times 250)$$
$$= 165 \text{ N/mm}^2$$

Section modulus required

$$Z = \left(\frac{M}{\sigma_{bc}}\right) = \left(\frac{360 \times 1000 \times 1000}{165}\right)$$

$$= 2181.82 \times 10^3 \text{ mm}^3$$

The steel beam section shall have $\left(\frac{D}{T}\right)$ and $\left(\frac{l}{r_y}\right)$ ratios more than 8 and 40, respectively. The trial section of beam selected may have modulus of sections, Z more than that needed, about 25 to 50 per cent more).

Step 3: Trial section modulus

$1.50 \times 2181.82 \times 10^3 \text{ mm}^3 = 3272.73 \times 10^3 \text{ mm}^3$

From steel section tables, try

WB 600, @ 1.337 kN/m

Section modulus, $Z_{xx} = 3540.0 \times 10^3 \text{ mm}^3$

Moment of inertia, $I_{xx} = 106198.5 \times 10^4 \text{ mm}^4$

Thickness of web, $t_w = 11.2$ mm

Thickness of web, $T = t_f = 21.3$ mm

Depth of section, $h = 600$ mm

Step 4 : Check for section modulus

$$\left(\frac{D}{T} = \frac{600}{21.3}\right) = 28.169$$

$$\left(\frac{T}{t_w} = \frac{21.3}{11.2}\right) = 1.901 < 2.0, \text{ also } \left(\frac{d_1}{t_w} < 85\right)$$

The effective length of compression flange of beam is 8 m.

$$\left(\frac{l}{r_y}\right) = \left(\frac{0.7 \times 8 \times 10.0}{52.5}\right) = 106.66$$

From Table 6.1 (b), IS 800–1984, maximum permissible bending stress

$$\sigma_{bc} = 118.68 \text{ N/mm}^2 \text{ (MPa)}$$

Section modulus required

$$\left(\frac{360 \times 1000 \times 1000}{118.68}\right) = 3033.34 \times 10^3 \text{ mm}^3$$

$$< 3540 \times 10^3 \text{ mm}^3 \text{ provided}$$

Further trial may give more economical section.

Step 5: Check for shear force

Average shear stress, $$\tau_{v.cal} = \left(\frac{F}{h \times t_w}\right)$$

$$\tau_{v.cal} = \left(\frac{180 \times 1000}{600 \times 11.2}\right) = 26.78 \text{ N/mm}^2$$

Permissible average shear stress

$$0.4 \times f_y = (0.4 \times 250) = 100 \text{ N/mm}^2$$

$$> \text{Actual average shear stress}$$

Step 6: Check for deflection

Maximum deflection of the beam

$$y_{max} = \left(\frac{5}{384} \times \frac{wl^4}{EI}\right)$$

$$y_{max} = \frac{5}{384} \times \left(\frac{4.5 \times 8 \times 8^3 \times (1000)^3 \times 1000}{2.047 \times 10^5 \times 106198.5 \times 10^4}\right)$$

$$= 24.53 \text{ mm}$$

Allowable deflection

$$\frac{1}{325} \times \text{span} = \left(\frac{1}{325} \times 8000\right)$$

$$= 24.60 \text{ mm}$$

The maximum deflection is less than allowable deflection, hence the beam is safe. Provide WB 600, @ 1.337 kN/m.

Example 6.3 *A class-room is 7 m × 15 m. It is provided with 120 mm thick stone patties over rolled steel beams spaced 3 m centre to centre. A wearing coat of 20 mm thick cement concrete is provided over 160 mm thick lime concrete. The effective length of compression flange would be equal to the effective span of beam. Design rolled steel beams. The ends of beams shall not be free to rotate at the bearings.*

Live load	= 4 kN/m^2
Unit weight of stone patties	= 24 kN/m^3
Unit weight of plain cement concrete	= 24 kN/m^3
Unit weight of lime concrete	= 18 kN/m^3

Solution

Design :

Step 1 : Load supported, bending moment and shear force

The rolled steel beam would carry a load from a strip of 3 m width.

Uniformly distributed load per metre length of beam.

$$\text{Weight of stone patties} = \left(3 \times 1 \times \frac{120}{1000} \times 24\right) = 8.64 \text{ kN}$$

$$\text{Weight of plain cement concrete} \left(3 \times 1 \times \frac{20}{1000} \times 24\right) = 1.44 \text{ kN}$$

$$\text{Weight of lime concrete} \left(3 \times 1 \times \frac{160}{1000} \times 18\right) = 8.64 \text{ kN}$$

Assume self-weight of beam	= 0.60 kN
Live load on the floor	= 3 × 4 = 12.00 kN
Total uniformly distributed load	= 31.32 kN/m
	Say = 31.50 kN/m

Clear span of beams = 7 m

Assume, 300 mm wide bed block would be provided.

The effective span of beam is adopted as the distance centre to centre of bearing or clear span + depth of beam

Adopt effective span = (7000 + 300) mm = 7.30 m

The maximum bending moment, *M* occurs at the centre,

$$M = \left(\frac{w \cdot l^2}{8}\right) = \left(\frac{31.50 \times 7.30 \times 1000}{8 \times 1000}\right)$$

$$= 209.829 \text{ m-kN}$$

The maximum shear, *F* occurs at the support,

$$F = \left(\frac{w \cdot l}{2}\right) = \left(\frac{31.50 \times 730}{2}\right) = 114.975 \text{ kN}$$

Step 2 : Permissible bending stress

It is assumed that the value of yield stress, f_y for the structural steel is 250 N/mm² (MPa). The ratios $\left(\frac{T}{t_w}\right)$ and $\left(\frac{d}{t_w}\right)$ are less han 2.0 and 85, respectively.

The maximum permissible stress may be assumed as below.

$$\sigma_{bc} = \sigma_{bt} = (0.66 \times 250) = 165 \text{ N/mm}^2$$

Step 3 : Section modulus required

$$Z = \left(\frac{M}{\sigma_{bc}}\right) = \left(\frac{209.829 \times 1000 \times 1000}{165}\right)$$

$$= 127.69 \times 10^3 \text{ mm}^3$$

The steel beam section shall have $\left(\frac{D}{T}\right)$ and $\left(\frac{l}{r_y}\right)$ ratios more 8 and 40, respectively. The trial section of beam selected may have modulus of section more than that required (about 25 to 50 per cent more).

Trial section modulus

$$1.50 \times 1271.69 \times 10^3 = 1907.535 \times 10^3 \text{ mm}^3$$

Step 3 : Properties of trial section

From steel section tables, try MB 550, @ 1.037 kN/m

Section modulus provided, $Z_{xx} = 235.8 \times 10^3$ mm³

Moment of inertia, $Z_{xx} = 235.8 \times 10^4$ mm⁴

Thickness of web, $t_w = 11.2$ mm

Depth of section, $h = 600$ mm

Means thickness of flange, $t_f = 19.3$ mm

Step 5 : Check for section modulus

$$\left(\frac{D}{T} = \frac{600}{19.3}\right) = 31.09$$

$$\left(\frac{T}{t_w} = \frac{19.3}{11.2}\right) = 1.723 \text{ and also } \left(\frac{d_1}{t_w} < 85\right)$$

The effective length of compression flange of beams is 7.30 m (equal to effective span of beam)

$$\left(\frac{l}{r_y}\right) = \left(\frac{0.7 \times 7.30 \times 1000}{37.3}\right) = 136.99$$

From Table 6.1 (b), IS: 800–1984, maximum permissible bending stress

$$\sigma_{bc} = 98.086 \text{ N/mm}^2 \text{ (MPa)}$$

Section modulus required

$$= \left(\frac{209.829 \times 1000 \times 1000}{98.086}\right)$$

$= 2139.235 \times 10^3$ mm³
$< 2359.8 \times 10^3$ mm³ provided

Further trial may give more economical section.

Step 6 : Check for shear stress

Average shear stress

$$\tau_{va.cal} = \left(\frac{F}{h \times t_w}\right) = \left(\frac{114.975 \times 1000}{600 \times 11.2}\right)$$

$= 17.11$ N/mm²
$< (0.4 \times 250 = 100$ N/mm²$)$. Hence safe.

Step 7 : Check for deflection

Maximum deflection of beam

$$y_{max} = \left(\frac{5 \times 31.50 \times 7.30 \times (7.30)^3 \times (1000)^3 \times 1000}{384 \times 2.047 \times 10^5 \times 91813 \times 10^4}\right)$$

$= 6.197$ mm

Allowable deflection $= \left(\frac{1}{325} \times 7300\right) = 22.45$ mm

The maximum deflection is less than allowable deflection. Hence, the section of beam adopted is safe. Provide MB 550, @ 1.037 kN/m.

Example 6.4 *The floor of an assembly hall is supported by main beams and secondary beams as shown in Fig. 6.6. The secondary beams would be connected to the web of main beams. The floor consists of 100 mm thick reinforced concrete slab. Design the beams.*

Solution

Design :

The unit weight of reinforced concrete slab is 24 kN/m³, and the live load on the floor is 5 kN/m².

Step 1: (i) Design of secondary beams

Each secondary beam supports load from strip 2 m wide. Uniformly distributed load per metre length of the beam:

1. **Load supported**

   Weight of reinforced concrete slab

$$= \left(2 \times 1 \times \frac{100}{1000} \times 24\right) = 4.80 \text{ kN}$$

Live load on the floor = $(2 \times 1 \times 5) = 10.00$ kN
Assume self-weight of the beam = 0.50 kN
Total uniformly distributed load = 15.30 kN/m say, 15.50 kN/m

2. **Bending moment and shear force**

The effective span of the beam is 5 m

The maximum bending moment, M occurs at the centre,

$$M = \left(\frac{w \cdot l^2}{8}\right) = \left(\frac{15.50 \times 5 \times 5 \times 10.00}{8 \times 100}\right)$$

$$= 48.438 \text{ m-kN}$$

The maximum shear force, F occurs at the support,

$$F = \left(\frac{w \cdot l}{2}\right) = \left(\frac{15.50 \times 5}{2}\right) = 38.75 \text{ kN}$$

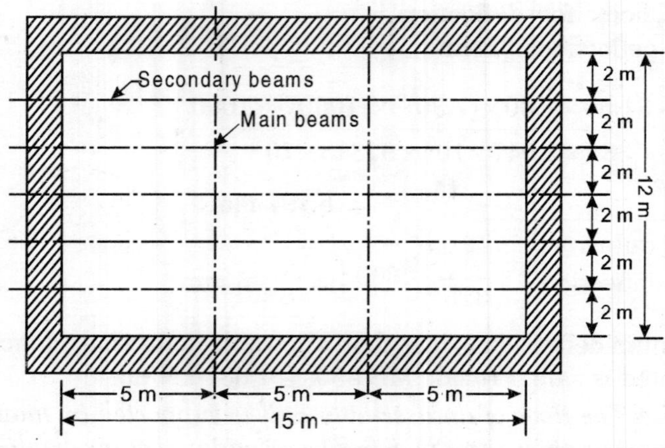

Fig. 6.6

3. **Permissible bending stress**

It is assumed that the value of yield stress f_y for the structural steel is 250 N/mm² (MPa). The ratio of $\left(\frac{T}{t_w}\right)$ and $\left(\frac{d_1}{t_w}\right)$ are less than 2.0 and 85, respectively. The maximum permissible stress in compression or tension may be assumed as below

$$\sigma_{bc} = \sigma_{bt} = (0.66 \times 250)$$

$$= 165 \text{ N/mm}^2$$

4. **Section modulus required**

$$Z = \left(\frac{M}{\sigma_{bc}}\right) = \left(\frac{48.438 \times 1000 \times 1000}{165}\right)$$

$$= 293.56 \times 10^3 \text{ mm}^3$$

The steel beam section shall have $\left(\frac{D}{T}\right)$ and $\left(\frac{l}{r_y}\right)$ ratios more than 8 and 40, respectively. The trial section of beam selected may have modulus of section, Z 1.5 times more than that needed.

Trial section modulus

$$1.50 \times 293.56 \times 10^3 = 440.345 \times 10^3 \text{ mm}^3$$

5. **Check for section modulus**

$$\left(\frac{D}{T} = \frac{300}{9.4}\right) = 31.915$$

$$\left(\frac{T}{t_w} = \frac{9.4}{6.7}\right) = 1.043 < 2.00, \text{ also } \left(\frac{d_1}{t_w} < 85\right)$$

The effective length of compression flange of beam may be assumed equal to effective span,

$$\left(\frac{l}{r_y}\right) = \left(\frac{0.7 \times 5 \times 1000}{28}\right) = 125$$

From Table 6.1 (b), IS: 800–1984, maximum permissible bending stress

$$\sigma_{bc} = 104.66 \text{ N/mm}^2$$

Section modulus required

$$Z = \left(\frac{48.438 \times 1000 \times 1000}{104.66}\right)$$

$$= 462.815 \times 10^3 \text{ mm}^3$$

$$< (488.9 \times 10^3 \text{ mm}^3) \text{ provided.}$$

6. **Properties of trial section**

From steel section tables, try LB 300. @ 0.377 kN/m

Section modulus provided, $Z_{xx} = 488.9 \times 10^3$ mm^3

Moment of inertia, $I_{xx} = 7322.9 \times 10^4$ mm^4

Thickness of web, $t_w = 6.7$ mm

Depth of section, $h = 300$ mm

Means thickness of flange, $t_f = 9.4$ mm

7. **Check for shear force**

$$\text{Average shear stress} = \left(\frac{F}{h.t_w}\right) = \left(\frac{38.75}{300 \times 6.6}\right) = 19.29 \text{ N/mm}^2$$

$< (0.4 \times 250 = 100$ N/mm$^{2)}$. Hence, safe.

8. **Check for deflection**

$$y_{max} = \frac{5}{384} \cdot \left(\frac{wl^4}{EI}\right)$$

$$y_{max} = \left(\frac{5 \times 15.50 \times 5 \times 5^3 \times (1000^3) \times 1000}{384 \times 2.047 \times 10^5 \times 7332.9 \times 10^4}\right)$$

$= 8.4$ mm

Allowable deflection $= \left(\frac{1}{325} \times 5000\right) = 15.39$ mm

The maximum deflection is less than allowable deflection.
Hence, design is satisfactory.

Step : 2 (ii) Design of main beams

1. **Load supported**

Clear span = 12 m

Provide bed blocks = 300 mm wide

The effective span is taken as the distance between centre to centre of bearings. Therefore,

Effective span = 12.30 m

Load transferred from each secondary beam

= (15.50 × 5) = 7.50 kN

Assune self-weight of beam = 2 kN/m

2. **Bending moment**

The maximum bending moment (occurs at centre) due to uniformly distributed load

$$M_1 = \left(\frac{w \cdot l^2}{8}\right) = \left(\frac{2 \times 12.30 \times 12.30 \times 1000}{8 \times 1000}\right)$$

= 37.8225 m-kN

End reaction due to concentrated loads

= (2.5 × 77.500) = 193.75 kN

The maximum bending moment (occurring at centre) due to concentrated loads

M_1 = (193.75 × 6.15 – 77.50 × 4 – 77.50 × 2)

= 726.5625 m-kN

Total bending moment

= (726.5625 + 37.8225)

= 764.384 m-kN

The secondary beams are connected to the web at 2 m centre to centre. The compression flange is assumed to the fully supported against lateral deflection.

3. **Permissible bending stress**

It is assumed that the value of yield stress, f_y for the structural steel is 250 N/mm² (MPa). The ratio $\left(\frac{T}{t_w}\right)$ and $\left(\frac{d_1}{t_w}\right)$ are less than 2.0 and 85, respectively.

The maximum permissible stress in compression or tension may be assumed as under (for lateral laterally supported beam)

$\sigma_{bc} = \sigma_{bt}$ = (0.66 × 250) = 165 N/mm² (MPa)

4. **Section modulus required**

$$Z = \left(\frac{764.3850 \times 10^3 \times 10^3}{165}\right)$$

$$= 4632.6 \times 10^3 \text{ mm}^3$$

From steel section tables (Tale XIV). try WB 500, @ 2.522 kN/m with 25 mm thick plate in each flange, 250 mm wide.

Section modulus provided

$$Z_{xx} = 6916.6 \times 10^3 \text{ mm}^3$$

Moment of inertia of plated beam

$$I_{xx} = 190207.6 \times 10^4 \text{ mm}^4$$

Thickness of web, $t_w = 9.9$ mm

Depth of section, $h = (500 + 2 \times 25) = 550$ mm

Total load on the girder inclusive of its own weight

$$= (15.50 \times 5 \times 5 + 2.522 \times 12.30)$$

$$= 418.52 \text{ kN}$$

5. **Maximum shear force**

$$= 209.26 \text{ kN}$$

6. **Check for average shear stress**

Average shear stress

$$\tau_{va.cal} = \left(\frac{209.26 \times 1000}{500 \times 9.2}\right)$$

$$= 42.27 \text{ N/mm}^2$$

$$< (0.4 \times 250 = 100 \text{ N/mm}^2). \text{ Hence, safe.}$$

7. **Check for moment**

$$\sigma_{bt} \cdot Z = \left(\frac{165 \times 6916.6 \times 103}{1000 \times 1000}\right) \text{kN-m}$$

$$= 1141.14 \text{ kN-m}$$

$$< 764.385 \text{ kN-m. Hence, safe.}$$

Provide MB 500, @ 0.869 kN/m with 25 mm thick and 250 mm wide plate in each flange. This section may be checked for deflection. It will be found safe.

6.7 LATERALLY UNSUPPORTED BEAMS

The *laterally unsupported beams* are also termed as *laterally unrestrained beams*. The symmertrical beams even if loaded in the plane containing principal axis may bend out of the plane of load under certain conditions. The rolled steel beams and channels have moment of inertia about *xx*-axis. I_{xx} (i.e., the other principal axis). The shapes of beas are made economical in the manner. Such shapes have less resistance to bending and torsion in the direction of *yy*-axis (i.e., weak axis). The beams may bend in the weak direction if it is not held or supported in a direction normal to the weaker axis. The beam is weakend further if a horizontal load is applied in the direction normal to *yy*-axis. The bending of beams is accompanied by twisting.This phenomenon is termed as *lateral buckling* or *lateral torisonal buckling*. When the torsional resistance and moment of inertia of any beam for the weak axis are small as compared with the moment of inertia about the strong axis, the beam may buckle laterally wheather the beams are symmetrical or not. The buckling tendency of compression flange increases as the stress increases and this tendency is maximum when the stress attains the value of critical stress. The compression flange buckles side-ways or laterally and it is also accompanied by twisting. The hypothetical buckling of a beam is shown in Fig. 6.7. The tendency of lateral buckling of compression flange also increases as the ratio (I_{xx}/I_{yy}) increases. This tendency further increases when the supported length of compression flange increases.

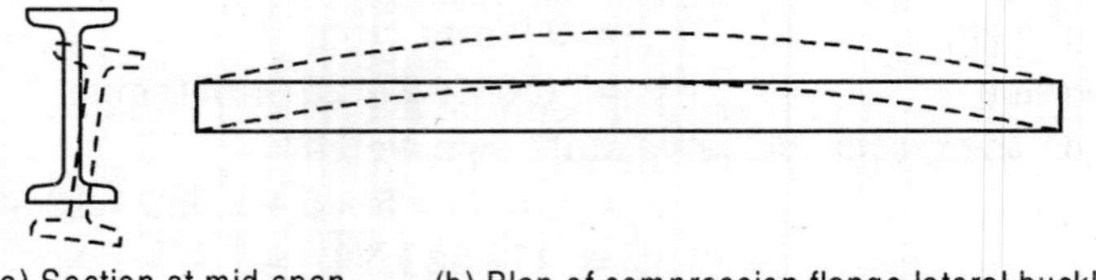

(a) Section at mid-span (b) Plan of compression flange lateral buckling

Fig. 6.7

In order to avoid lateral buckling due to a load in the plane of one of the principal axis of the member, the beam should have enough torsional stiffness to resist torsional forces induced and the beam should be laterally supported in the direction perpendicular to the weaker principal axis. The compression flange of beam tends to buckle if not held in line by attached construction. Sometimes, the special conditions may require that the beams may be loaded in the plane of the web without continuous or intermittent lateral suport at sufficient frequent intervals.

The bending stress in compression for sectional shapes with I_{yy} smaller than I_{xx} (where I_{yy} is moment of inertia of the whole section about the axis lying in the plane of bending. *yy* axis, and I_{xx} is moment of inertia of the whole section about the axis normal to the plane of bending, *xx*-axis) is reduced for lateral buckling in proportion to elastic critical stress, f_{cb} found as follows for sections of single web.

6.7.1 Where the Flanges have Equal Moment of Inertia about yy-Axis

If symmetrical I-beam is subjected to couples *M*, at the ends and beam is restraints against torsion at the ends, the lateral buckling of beam takes place when the value of *M* attains critical value. The critical value of moment *M* is given by

$$M_{cr} = \left(\frac{\pi}{l}\right)\left[El_yC.\left(1+\frac{C_1.\pi^2}{C_2 l^2}\right)\right]^{1/2} \qquad \text{... (6.7)}$$

where, l = Effective length of compression flange
C = Torsional rigidity of beam = GK
G = Shear modulus of elasticity
E = Young's modulus of elasticity
K = Torsional constant of beam
C_1 = Warping rigidity = EC_w

$$C_w = \text{Warping constant} = \left(\frac{1}{24}t_f b^3 h^2 = \frac{1}{2} I_f \cdot h^2\right)$$

$$= \frac{1}{4} I_y \cdot h^2 \text{ (considering } I_f = \frac{1}{2} I_f \text{)}.$$

I_f = Moment of inertia of compression flange about yy-axis of the grider
I = Moment of inertia of beam about yy-axis at the point of maximum bending moment
b = Width of flange of I-beam
t_f = Thickness of flange
h = Distance between the centre of gravity of compression flange ad centre of gravity of tension flange. It may be taken approximately as depth of section minus thickness of flange, $(h - t_f)$.

Substituting the values of C and C_1 in Eq. 6.7

$$M_{cr} = \left(\frac{\pi}{l}\right)\left[1 + \frac{1}{\pi^2} \cdot \frac{4GK \cdot l^2}{EI_y \cdot h^2}\right]^{1/2} \quad \text{...(6.8)}$$

By rearranging the terms and taking the terms $\left(\frac{\pi^2 EI_y \cdot h^2}{4GK}\right)$ common from the parenthesis, Eq. 6.8 is written as follows:

$$M_{cr} = \frac{\pi^2}{2} \frac{EI_y h}{l^2} \cdot \left[1 + \frac{1}{\pi^2} \cdot \frac{4GK \cdot l^2}{EI_y \cdot h^2}\right]^{1/2} \quad \text{...(6.9)}$$

From Eq. 6.8 it may be observed that the critical bending moment depends not so much on the material, since value of E is constant for all steel, but on torsional rigidity, GK and elastic stiffiness, EI_y for the bending in the perpendicular to the load.

Subustituting.

$$G = 0.4\,E,$$

$$K = (0.9\, b_f . b_f^3) \text{ and } I_y = \left(\frac{2 \times t_f \cdot b_f^3}{12}\right)$$

where t_f and b_f are the thickness and width of the flange, respectively.

$$M_{cr} = \frac{\pi^2}{2} \cdot \frac{EI_y h}{l^2} \left[1 + \frac{1}{\pi^2} \times \frac{4 \times 0.4E \times 0.9b_f tf^3}{E \times 2 \times t_f \times b_f^2 / 12} \cdot \frac{l^2}{h^2} \right]^{1/2} \qquad \text{...(i)}$$

$$M_{cr} = \frac{\pi^2}{2} \cdot \frac{EI_y h}{l^2} \left[1 + \frac{8.64}{\pi^2} \times \frac{tf^2}{b_f^2} \cdot \frac{l^2}{h^2} \right]^{1/2} \qquad \text{...(ii)}$$

For an I-section and I-section of wide flanges, the approximate radii of gyrations are given by $r_y = 0.22b$ (viz., $b = 4.545\, r_y$) and $r_y = 0.25\, b$ (viz., $b = 4.000\, r_y$), respectively. As an approximation $b_f \,\hat{=}\, 4.2\, r_y$), then

$$M_{cr} = \frac{\pi^2 \cdot EI_y h}{2l^2} \left[\left(1 + \frac{8.64 t_f^2}{\pi^2 \times 17.64 r_y^2} \cdot \frac{l^2}{h^2} \right) \right]^{1/2} \qquad \text{...(iii)}$$

$$M_{cr} = \frac{\pi^2 \cdot EI_y h}{2l^2} \left[\left(1 + \frac{1}{20} \left(\frac{l \cdot t_f^2}{r_y \cdot h} \right) \right) \right]^{1/2} \qquad \text{...(iv)}$$

The elastic critical for bending for beams with I_y smaller than I_{xx} is given by ($\therefore M_{cr} = f_{cb} \cdot Z_{xx}$)

$$f_{cb} = \frac{\pi^{2E} EI\, h}{2.Z_{xx}\, l^2} \cdot \left[1 + \frac{1}{20} \left(\frac{l \cdot t_f}{r_y \cdot h} \right)^2 \right]^{1/2} \qquad \text{...(v)}$$

The values of modulus of elasticity, E is taken as 2×10^5 N/mm^2 (MPa) and $I_y = A \cdot r_y^2$. Substituing these values in the expression (v)

$$f_{cb} = \frac{9.8526 \times 10^5 A \cdot h}{(l/r_y)^2 Z_{xx}} \left[1 + \frac{1}{20} \left(\frac{l \cdot t_f}{r_y \cdot h} \right)^2 \right]^{1/2} \qquad \text{...(vi)}$$

The value of $\left(\frac{A \cdot h}{Z_{xx}} \right)$ for I- section be taken as 2.688 apporximately.

Then

$$f_{cb} = \frac{26.5 \times 10^5}{\left(\frac{l}{r_y} \right)^2} \left[1 + \frac{1}{20} \left(\frac{l \cdot t_f}{r \cdot h} \right)^2 \right]^{1/2} \qquad \text{...(6.10)}$$

The value of critical stress, C_s is increased by 20 per cent for rolled steel beams, channel and plate girders provided

$$\left(\frac{t_e}{t_w}\right) \ngtr 2 \text{ and } \left(\frac{d_1}{t_w}\right) \ngtr 85$$

where l = Effective length of compression flange

r_y = Radius of gyration about yy-axis of the gross-section of the whole girder at the point of maximum bending moment

h = Overall depth of the girder, at the point of maximum bending moment

t_f = Effective thickness of the compression flange

= k_1 × Mean thickness of the horizontal portion of the compression flange at the point of maximum bending moment.

k_1 makes the allowance for reduction in thickness and breadth of flanges between points of effective lateral restraints and depends on ψ the ratio of the total area of both flange at the point of least bending moment to the corresponding point of greater bending moment between such points of restraints.

The values of k_1 have been given in Table 6.2.

Table 6.2 *Values of k_1 for Beams with curtailed flanges (As per IS: 800–1984)*

ψ	1.0	0.9	0.8	0.7	0.6	0.5	0.4	0.3	0.2	0.1	0.0
k_1	1.0	1.0	1.0	0.9	0.8	0.7	0.6	0.5	0.4	0.3	0.3

Note : *The flanges shall not be reduced in width to give a value y lower than 0.25 where the value of ψ calculated for compression flange alone is smaller than that when both the flanges are combined, this smaller value of ψ shall be used.*

The values of critical bending stress f_{cb} are increased by per cent for all rolled beam and channels and built-up beams for which $\left(\frac{T}{t_w}\right)$ is not greater than 2 and $\left(\frac{d_1}{t_w}\right)$ is not greater than $\left(\frac{1344}{\sqrt{f_y}}\right)$ where, T is the mean thickness of the compression flange and d_1 is the clear depth of the web. The effective thickness of compression flange is equal to k_1 times the mean thickness of the horizontal portion of the compression flange at the point of maximum bending moment. The mean thickness of flange may be found by dividing area of horizontal portion of the compression flange by width of flange.

In order to provide safety against lateral buckling, allowable bending compressive stress σ_{bc} is reduced in proportion to critical bending stress f_{bc} at which lateral buckling of compression flange occurs.

The critical bending stress in compression for section symmertical about the xx-axis may also be found from the following expression.

$$f_{cb} = \frac{\pi}{Z_x \cdot l}\left[\frac{EL_y GK}{\mu}\left(1+\frac{C_1}{C_2}\frac{\pi^2}{l^2}\right)\right]^{1/2} \quad ...(6.11)$$

where, $\mu = \left(\frac{I_x - I_y}{L_z}\right)\left(\frac{I_x - I_y}{I_x}\right)$

I_x = Moment of inertia of whole section about xx-axis

The values C_1 and C are substituted, then

$$f_{cb} = \frac{\pi}{Z_x \cdot l}\left[\frac{EL_yGK}{\mu}\cdot\left(1+\frac{\pi^2}{l^2}\frac{EC_w}{GK}\right)\right]^{1/2} \quad ...(6.12)$$

The values of E, G and C_w mentioned above may be subsitituted. For I-section, Eq. (6.14) reduces to

$$f_{cb} = \frac{4.094\times 10^5}{Z_z}\left[I_yK\left(1+12.3\frac{I_f \cdot h^2}{Kl}\right)\right]^{1/2} \text{N/mm}^2 \quad ...(6.13)$$

For channels, Eq. (6.13) gives conservative values since value of warping constant for channels is more than that for I-sections.

The values of torsional constant 'K' for Indian Standard rolled steel beams and Indian Standard rolled steel channels sections are given in Appendix E in IS: 800–1962. For the sections composed of approximately rectangular sections

$$K = \Sigma\left(\frac{b'\,t'^3}{3}\right)$$

Where, b' and t' are the width and average thickness of the rectangular components of the section, respectively.

For box members, conservative values of 'f_{cb}' are obtained. The value of K is given by

$$K = \left(\frac{4A^2}{\frac{S}{t}}\right) \text{approximately}$$

where, A = Total enclosed area of section

S = Lengthalong the periphery along which t is constant

t = Thickness of box member

t = Effective thickness (in the case of curtailed flanges)

$$K = \frac{2b^2 \cdot d^2t}{(b+d)} \text{ approximately}$$

where, b = Width of box members

d = Depth of box members

For a plate of flat in bending in a plane parallel to its surface, the value of elastic critical stress in bending is given by

$$f_{bc} = \left(\frac{4.094\times 10^5 \cdot t^2}{l \cdot D}\right) \text{N/mm}^2 \quad ...(6.14)$$

where t = Thickness of plate or flat
D = Depth of the plate
t = Effective length of part in compression.

6.7.2 Maximum Permissible Bending Compressive Stress in Beams and Plate Girders

For the beams and plate girders which are likely to bend about the axis of maximum strength (xx-axis) the maximum bending compressive stress on the extreme fibre, determined by using the effective section $\sigma_{bc.val}$ should not exceed the maximum permissible bending compressive stress σ_{bc} in N/mm² (MPa) obtained by using the following formula as recommended by IS: 800–1984.

$$\sigma_{bc} = \left(0.66 \frac{f_{cb} \cdot f_y}{\left[(f_{cb})^n + (f_y)^n\right]^{1/n}}\right) \quad ...(6.15)$$

where, f_{cb} = Elastic critical stress in bending, calculated as described below or an elastic flexural-torsional buckling analysis in N/mm² (MPa)
f_y = Yeild stress of the steel in N/mm² (MPa) and
n = A factor asumed as 1.4

The values of maximum permissible bending compressive stress, σ_{bc} in N/mm² (MPa) calculated from the above formula recommended by IS: 800–1984 for the structural steels (having the value of yield stress, f_y as 220, 230, 240, 250, 260, 280, 300, 320, 340, 360, 380, 400, 420, 510, and 540 N/mm² (MPa) are given in Tables 6.3 (i) and (ii).

Table 6.3 (i) *Values of σ_{bc} calculated from f_{cb} for different value of f_y All units in N/mm² (MPa)*

$f_y \rightarrow$ / $f_{cb} \downarrow$	200	230	240	250	260	280	300	320	340
20	13	13	13	13	13	13	13	13	13
30	19	19	19	19	19	19	19	19	19
40	25	25	25	25	25	25	25	25	26.
50	30	30	31	31	31	31	31	31	31
60	36	36	36	36	36	37	37	37	37
70	41	41	41	41	42	42	42	43	43
80	45	46	46	46	47	47	48	48	48
90	50	50	51	51	51	52	53	53	54
100	54	54	55	55	56	57	57	58	59
110	58	58	59	60	60	61	62	63	64
120	61	62	63	64	64	65	67	67	68
130	65	66	67	67	68	70	71	72	73

Contd.

Table 6.5 (i) Contd.

$f_y \rightarrow$ / $f_{cb} \downarrow$	200	230	240	250	260	280	300	320	340
140	68	69	70	71	72	73	75	76	77
150	71	72	73	74	75	77	79	80	81
160	74	75	77	78	79	81	82	84	85
170	77	78	80	81	82	84	86	88	89
180	79	81	82	84	85	87	89	91	93
190	82	84	85	87	88	90	93	95	97
200	84	86	88	89	91	93	96	98	100
210	86	88	90	92	93	96	99	101	103
220	89	90	92	94	96	99	102	104	106
230	90	93	94	96	98	101	104	107	110
240	92	94	97	88	100	104	107	110	113
250	94	96	99	101	103	106	110	113	115
260	96	98	100	103	105	108	112	115	110
270	97	100	102	104	107	111	114	118	121
280	99	101	104	106	108	113	116	120	123
290	100	103	105	108	110	115	119	122	126
300	102	104	107	110	112	116	121	125	128
310	103	106	108	111	114	118	123	127	130
320	104	107	110	113	115	120	125	129	133
330	105	108	111	114	117	122	126	131	135
340	106	110	113	115	118	123	128	133	137
350	108	111	114	117	120	125	130	134	139
360	109	112	115	118	121	126	131	136	141
370	110	113	116	119	122	128	133	138	143
380	111	114	117	120	123	129	135	140	144
390	111	115	118	121	125	130	136	141	146
400	112	116	119	122	126	132	137	143	148
420	114	118	121	124	128	134	140	146	151
440	115	119	123	126	130	136	142	148	154
460	117	121	124	128	132	138	145	151	157
480	118	122	126	130	133	140	147	153	159
500	119	123	127	131	135	142	149	155	162
520	120	125	129	133	136	144	151	158	164

Contd.

Table 6.3 (i) Contd.

$f_y \rightarrow$ / $f_{cb} \downarrow$	200	230	240	250	260	280	300	320	340
540	121	126	130	134	138	145	153	160	166
560	122	127	131	135	139	147	154	161	168
580	123	128	132	136	140	148	156	163	170
600	124	129	133	137	141	150	157	165	172
620	125	129	134	138	143	151	159	166	174
640	126	130	135	139	144	152	160	168	175
660	126	131	136	140	145	153	161	169	177
680	127	132	136	141	145	154	163	171	178
700	128	132	137	142	146	155	164	172	180
720	128	133	138	143	147	156	165	173	181
740	129	134	139	143	148	157	166	174	182
760	129	134	139	144	149	158	167	175	184
780	130	135	140	145	149	159	168	176	185
800	130	135	140	145	150	159	169	177	186
850	131	137	142	147	152	161	171	180	188
900	132	138	143	148	153	163	172	182	191
950	133	138	144	149	154	164	174	183	193
1000	134	139	145	150	155	165	175	185	195
1050	135	140	145	151	156	167	177	187	196
1100	135	141	146	152	157	168	178	188	198
1150	136	141	147	152	158	168	179	189	199
1200	136	142	147	153	159	169	180	190	200
1300	137	143	149	154	160	171	182	192	203
1400	138	144	149	155	161	172	183	194	205
1500	139	144	150	156	162	173	184	195	206
1600	139	145	151	157	163	174	185	197	208
1700	140	146	151	157	163	175	186	198	209
1800	140	146	152	158	164	176	187	199	210
1900	140	146	152	158	164	176	188	200	211
2000	141	147	153	159	165	177	189	200	212
2200	141	147	154	160	166	178	190	202	213
2400	142	148	154	160	166	179	191	203	215
2600	142	148	154	161	167	179	191	204	216

Contd.

Table 6.3 (i) Contd.

$f_y \rightarrow$ / $f_{cb} \downarrow$	200	230	240	250	260	280	300	320	340
2800	142	149	155	161	167	180	192	204	216
3000	143	149	155	161	168	180	193	205	217
3500	143	149	156	162	168	181	194	206	218
4000	143	150	156	163	169	182	194	207	219
4500	144	150	157	163	169	182	195	208	220
5000	144	150	157	163	170	183	195	208	221
5500	144	151	157	163	170	183	196	208	221
6000	144	151	157	164	170	183	196	209	222

Table 6.3 (ii) *Values of σ_{bc} calculated from f_{cb} for different value of f_y All units in* N/mm² *(MPa)*

$f_y \rightarrow$ / $f_{cb} \downarrow$	340	360	380	400	420	450	480	510	540
20	13	13	13	13	13	13	13	13	13
30	19	19	19	19	19	19	20	20	20
40	26	26	26	26	26	26	26	26	26
50	31	32	32	32	32	32	32	32	32
60	37	37	38	38	38	38	38	38	38
70	43	43	43	44	44	44	44	44	44
80	48	49	49	49	49	50	50	50	50
90	54	54	54	55	55	55	56	56	56
100	59	59	60	60	60	61	61	62	62
110	64	64	65	65	66	66	67	67	67
120	68	69	70	70	71	71	72	72	73
130	73	74	74	75	76	76	77	78	78
140	77	78	79	80	80	81	82	83	84
150	81	82	83	84	85	86	87	88	89
160	85	87	88	89	90	91	92	93	94
170	89	91	92	93	94	95	97	98	99
180	93	94	96	97	98	100	101	102	103
190	97	98	100	102	102	104	106	107	108
200	100	102	103	105	106	108	110	111	113
210	103	105	107	109	110	112	114	116	117
220	106	109	111	112	114	116	118	120	121

Contd.

Table 6.3 (ii) Contd.

$f_y \rightarrow$ / $f_{cb} \downarrow$	200	230	240	250	260	280	300	320	340
230	110	112	114	116	118	120	122	124	126
240	113	115	117	119	121	124	126	128	130
250	115	118	120	122	124	127	130	132	134
260	110	121	123	126	128	131	133	136	138
270	121	124	126	129	131	134	137	139	142
280	123	126	129	132	134	137	140	143	145
290	126	129	132	135	137	141	144	147	149
300	128	131	135	137	140	144	147	150	153
310	130	134	137	140	143	147	150	153	156
320	133	136	140	143	146	150	153	157	160
330	135	138	142	145	148	152	156	160	163
340	137	141	144	148	151	155	159	163	166
350	139	143	147	150	153	158	162	166	169
360	141	145	149	152	156	161	166	169	172
370	143	147	151	155	158	163	168	172	175
380	144	149	153	157	160	166	170	174	178
390	146	151	155	159	163	168	173	177	181
400	148	152	157	161	165	170	175	180	184
420	151	156	160	165	169	175	180	185	189
440	154	159	164	169	173	179	185	190	195
460	157	162	167	172	177	183	189	194	200
480	159	165	170	175	180	187	193	199	204
500	162	168	173	178	183	190	197	203	209
520	164	170	176	181	187	194	201	207	213
540	166	172	178	184	189	197	204	211	217
560	168	175	181	187	192	200	208	215	221
580	170	177	183	189	195	203	211	218	225
600	172	179	185	192	198	206	214	222	229
620	174	181	187	194	200	209	217	225	232
640	175	183	189	196	202	211	220	228	235
660	177	184	191	198	204	214	222	231	238
680	178	186	193	200	207	216	225	234	242
700	180	187	195	202	209	218	228	236	244
720	181	189	196	204	210	220	230	239	247
740	182	190	198	205	212	222	232	241	250
760	184	192	199	207	214	224	234	244	253

Contd.

Table 6.3 (ii) Contd.

$f_y \rightarrow$ / $f_{cb} \downarrow$	200	230	240	250	260	280	300	320	340
780	185	193	201	208	216	226	236	246	255
800	186	194	202	210	217	228	238	248	257
850	188	197	205	213	221	232	243	253	263
900	191	200	208	216	224	236	247	258	268
950	193	202	211	219	227	240	251	262	273
1000	195	204	213	222	230	243	255	266	277
1050	196	206	215	224	233	246	258	270	281
1100	198	207	217	226	235	248	261	273	285
1150	199	209	219	228	237	251	263	276	288
1200	200	210	220	230	239	253	266	279	291
1300	203	213	223	233	243	257	270	284	297
1400	205	215	225	236	246	260	274	288	302
1500	206	217	228	238	248	263	278	292	306
1600	208	219	229	240	250	266	281	295	309
1700	209	220	231	242	252	268	283	298	313
1800	210	221	232	243	254	270	285	301	316
1900	211	222	234	245	256	272	287	303	318
2000	212	226	235	246	257	273	289	305	321
2200	213	225	237	248	259	276	292	309	325
2400	215	226	238	250	261	278	295	312	328
2600	216	227	239	251	263	280	297	314	331
2800	216	228	240	252	264	282	299	316	333
3000	217	229	241	253	265	283	300	318	335
3500	218	231	243	255	267	286	303	321	339
4000	219	232	244	257	269	287	306	324	342
4500	220	233	245	258	270	289	307	326	344
5000	221	233	246	259	271	290	309	327	246
5500	221	234	247	259	272	291	310	328	347
6000	222	234	247	260	273	291	310	329	348

6.7.3 Elastic Critical Stress

The elastic critical stress in bending may be calculated by an **elastic flexural - torsional buckling analysis.** In case, this analysis is not carried out, the elastic critical stress in bending, f_{cb} for the beams and plate girders with the moment of inertia about axis of minimum strength (viz. yy-axis), I_{yy} smaller than that about axis of maximum strength (viz. xx-axis), I_{xx} may be determined from the expression derived as follows:

A paper titled as the basis for design of beams and plate girders in the revised

BS: 153 (British Standards) structural paper 48, proceedings of the Institution of Civil Engineers, London, August 1956 was published by Kerensky, D.A., Flint, A.R., and Brown, W.C. Professor Kerensky gave the expression for critical bending moment for beams and plate girders as under:

$$M_{cr} = \frac{\pi}{l}(EI_y.GK)^{\frac{1}{2}}\left[\left(1+\frac{\pi^2}{l^2}\frac{EI_y\cdot h^2}{4GK}\right)\right]^{\frac{1}{2}} + (2\lambda-1)\frac{\pi h}{2.l}\left(\frac{El_y}{GK}\right)^{\frac{1}{2}}$$

where, λ is the ratio of moment of inertia of compression flange about yy-axis, I_f to the moment of inertia of the whole section about yy-axis I_y. $(2\lambda - 1)$ may be replaced by k_2. For the values of $\lambda = 1$ or 0, k_2 is equal to + 1.0 and – 1.0. Above expression (i) may be written as below [as expression (ii) derived in Eq. 6.10].

$$M_{cr} = \frac{\pi^2 EI_y\cdot h}{2\cdot l^2}\left[1+\frac{1}{20}\left(\frac{lt_y}{r_y\cdot h}\right)^2\right]^{\frac{1}{2}} + k_2\left(\frac{\pi^2}{2}\cdot\frac{EI_y.h}{l^2}\right) \quad \text{...(ii)}$$

$$M_{cr} = \frac{\pi^2 DI'_y\cdot h}{2\cdot l^2}\left[1+\frac{1}{20}\left(\frac{lt_f}{r_y\cdot h}\right)^2\right]^{\frac{1}{2}} + k_2\left(\frac{\pi^2}{2}\cdot\frac{EI_y\cdot h}{l^2}\right) \quad \text{...(iii)}$$

where I_y' is the modified moment of inertia. It is equal to $k_1\cdot I_y'$, (I_y is the moment of inertia of the whole section about yy-axis at the section of maximum bending moment).

Since, $M_{cr} = f_{cb}\,.\,Z_{xx}$, the critical bending stress is given by

$$f_{cb} = \frac{\pi^2 EI'_y\cdot h}{2Z_{xx}\cdot l^2}\left[1+\frac{1}{20}\left(\frac{l\cdot t_f}{r_y\cdot h}\right)^2\right]^{\frac{1}{2}} + k_2\left(\frac{\pi^2 EI'_y h}{2.Z_{xx}\cdot l^2}\right) \quad \text{...(iv)}$$

This expression (iv) may be written as below : (as Eq. 6.10 has been derived).

$$f_{cb} = \frac{26.5\times10^5\cdot k_1}{\left(\frac{l}{r_y}\right)^2}\left[1+\frac{1}{20}\left(\frac{l\cdot t_f}{r_y\cdot h}\right)^2\right]^{\frac{1}{2}} + k_2\cdot\left(\frac{26.5\times10^5\times k_1}{\left(\frac{l}{r_y}\right)^2}\right) \quad \text{...(6.16)}$$

where the moment of inertia of the tension flange about yy-axis exceed that of the compression flange, Eq. 6.16 may be written as below:

$$f_{cb} = \left[\frac{26.5\times10^5\times k_1}{\left(\frac{l}{r_y}\right)^2}\left\{1+\frac{1}{20}\left(\frac{l\cdot t_f}{r_y\cdot D}\right)^2\right\}^{\frac{1}{2}} + k_2\frac{26.5\times10^5\times k_1}{\left(\frac{l}{r_y}\right)^2}\right]\cdot\frac{C_2}{C_1}\ \text{N/mm}^2 \quad \text{...(6.17)}$$

$$f_{cb} = k_1 . [X + k_2 . Y] . \frac{C_2}{C_1} \quad ...(6.18)$$

IS: 800–1984 recommends Eq. 6.18 for calculating elastic critical stress in bending

where $$X = Y\left[1+\frac{1}{20}\left(\frac{l \cdot T}{r_y \cdot D}\right)\right]^{\frac{1}{2}} \text{ N/mm}^2 \text{ (MPa)} \quad ...(6.19)$$

$$Y = \left(\frac{26.5 \times 10^5}{\left(\frac{l}{r_y}\right)^2}\right) \text{ N/mm}^2 \text{ (MPa)} \quad ...(6.20)$$

k_1, l, r_y, D and T symbols as used in Tables 6.1 (a) to (f) have been defined above.

k_1 = a coefficient. It allows for the inequality of the flanges. It depends on ω . ω is the ratio of moment of inertia of the compression flange alone to that of the sum of the moments of inertia of the flanges, each calculated about its own axis parallel to the y-y of the girder, at the point of maximum bending moment. The values of co-efficient, k_2 for different values of ω are given in Table 6.4.

Table 6.4 *Values of k_2 for beams with unequal flanges (As per IS: 800–1984)*

w	1.0	0.9	0.8	0.7	0.6	0.5	0.4	0.3	0.2	0.1	0.0
k_2	0.5	0.4	0.3	0.2	0.1	00	–0.2	–0.4	–0.6	–0.8	–0.10

c_1 = lesser distance to the exterme fibre of the section

c_2 = maximum (greater) distance to the extreme fibre of the section

I_y = moment of inertia of the whole section about the axis lying in the plane of bending (yy-axis); and

I_x = moment of inertia of the whole section about the axis normal to the plane of bending (xx-axis).

(a) For flanges of equal moment of inertia

$$\omega = 0.5, k_2 = 0$$

(b) For tees and angles bending in the plane of web with the xx-axis (neutral axis) nearer the extreme fibre of compression zone.

$$\omega = 1.0, k_2 = 0.5$$

(c) The bending stress in the leg when loaded with the flange or table in compression shall not exceed 0.66 f_y. When loaded with the leg in compression, the permissible bending stress shall be calculated from Eq. 6.17 and Eq. 6.18 with $k_2 = -1.0$ and T as the thickness of leg.

The values of X and Y are given in Table 6.5 for appropriate values of $\left(\dfrac{D}{T}\right)$ and $\left(\dfrac{l}{r_y}\right)$.

The values of elastic critical stress in bending, f_{cb} may be increased by 20 per cent when $\left(\dfrac{T}{t_w}\right)$ ratio is not greater than 2.0 and $\left(\dfrac{d_1}{t_w}\right)$ is not greater than $\dfrac{1344}{(f_y)^{1/2}}$ where d_1 is the depth of web and t_w is the thickness of web.

When the bending of beams occur about axis of minimum strength (*yy*-axis), the maximum permissible bending stress in tension σ_{bc} or in compression σ_{bc} in beams should not exceed $0.66 f_y$, where f_y is the yield of structural steel used.

The principal distinction between the design of laterally supported beams and laterally unsupported beams is the method of dealing with the bending in the laterally unsupported beams. The allowable bending stress in compression is reduced where necessary, depending upon the beams susceptibility of buckling. The reduction in allowable bending stress in compression provide adequate safety against the lateral torisonal buckling. The reduction in allowable stress increases with the unsupported length of compression flange to the width of flange ratio and depth of beam to the thickness of flange ratio.

Table 6.5 (i) *Values of x and y for calculating f_{cb} in N/mm² (MPa)*

$D/T \rightarrow$	X									Y
l/r_y ↓	8	10	12	14	16	18	20	25	30	
40	2484	2222	2066	1965	1897	1849	1814	1759	1728	1656
45	2103	1856	1708	1612	1546	1499	1465	1411	1380	1309
50	1822	1590	1449	1357	1293	1248	1214	1161	1131	1060
55	1607	1389	1254	1166	1105	1061	1028	976	947	876
60	1437	1232	1104	1020	961	918	886	835	806	736
65	1301	1107	985	904	847	806	775	726	697	627
70	1188	1005	889	811	757	717	687	638	610	541
75	1094	920	810	735	682	644	615	567	540	471
80	1014	849	743	672	621	584	556	509	482	414

Contd.

Table 6.5 (i) Contd.

$D/T \rightarrow$	X									Y
l/r_y ↓	8	10	12	14	16	18	20	25	30	
85	945	788	687	618	570	533	506	461	434	367
90	886	735	639	573	526	491	464	420	394	327
95	833	689	597	534	488	454	428	385	360	294
100	787	649	560	499	455	423	398	356	331	265
110	708	582	449	443	402	371	347	307	283	219
120	644	527	451	398	359	330	308	270	247	184
130	591	482	411	361	325	298	277	240	218	157
140	546	444	378	331	297	271	251	217	195	135
150	508	412	350	306	274	249	230	197	177	118
160	474	385	326	284	254	230	212	181	161	104
170	445	360	305	265	236	214	197	167	148	92
180	420	339	286	249	221	200	184	155	137	82
190	397	320	270	235	208	188	172	145	137	73
200	376	304	256	222	197	177	162	136	119	66
210	358	288	243	210	186	168	153	128	112	60
220	341	275	231	200	177	159	145	121	105	55
230	326	262	220	191	169	152	138	115	99	50
240	312	251	211	182	161	145	132	109	94	46
250	299	241	202	175	154	138	126	104	90	42
260	288	231	194	167	148	133	121	99	85	39
270	277	222	186	161	142	127	116	95	82	36
280	267	214	180	155	137	122	111	91	78	34
290	257	207	173	149	132	118	107	88	75	32
300	249	200	167	144	127	114	103	84	72	32

Table 6.5 (ii) *Values of x and y for calculating f_{cb} in N/mm² (MPa)*

$D/T \rightarrow$	X									Y
l/r_y ↓	20	25	30	35	40	50	60	80	100	
40	1814	1759	1728	1709	1697	1683	1675	1667	1662	1656
45	1465	1411	1380	1362	1349	1335	1327	1319	1315	1309
50	1214	1161	1131	1113	1101	1086	1078	1070	1067	1060

Contd.

Table 6.5 (ii) Contd.

D/T→ l/r_y ↓	X									Y
	20	25	30	35	40	50	60	80	100	
55	1028	976	947	929	917	902	894	886	883	876
60	886	835	806	788	776	762	754	746	743	736
65	775	726	697	679	667	653	645	637	634	627
70	687	638	610	592	581	567	559	551	547	541
75	615	567	540	522	511	497	489	481	478	471
80	556	509	482	465	454	440	432	424	421	414
85	506	461	434	417	406	392	385	377	373	367
90	464	420	394	377	366	353	345	337	327	327
95	428	385	360	343	332	319	311	304	300	294
100	398	356	331	314	304	290	283	275	227	265
110	347	307	283	268	257	244	237	229	226	219
120	308	270	247	232	222	209	202	194	191	184
130	277	240	218	204	194	181	174	167	163	157
140	251	217	195	181	172	160	153	145	142	135
150	230	197	177	163	154	142	135	145	124	118
160	212	181	161	148	139	127	121	113	110	104
170	197	167	148	135	126	115	109	102	98	92
180	184	155	137	125	116	105	98	92	88	82
190	172	145	127	115	107	96	90	83	80	73
200	162	136	119	107	99	89	83	76	73	66
210	153	128	112	101	93	82	76	70	66	60
220	145	121	105	94	87	77	71	64	61	55
230	138	115	99	89	82	72	66	60	56	50
240	132	109	94	84	77	67	62	55	52	46
250	126	104	90	80	73	64	58	52	49	42
260	121	99	85	76	69	60	55	48	45	39
270	116	95	82	72	66	57	52	46	42	36
280	111	91	78	69	63	54	49	43	40	34
290	107	88	75	66	60	52	46	41	38	33
300	103	84	72	64	57	49	44	38	35	32

6.8 EFFECTIVE LENGTH OF COMPRESSION FLANGE

In the laterally unsupported of beams, allowable bending compressive stress σ_{ec} is reduced in proportion to critical bending stress f_{bc}. The critical bending stress can be computed from *the effective length of common pression flange 'l'* and other properties of the section.

In the lateral buckling of compression flange, lateral bending of the flange occurs with twisting. The lateral buckling of the compression flange depends on the effective length of the compression flange, which depends on the effective lateral supports. The lateral supports should provide torsional restraints. The torsional restraints can be provided in the following ways:

1. By providing web or flange cleats in the end connection.
2. By providing bearing stiffeners acting in conjuction with bearing of the beam.
3. By providing lateral and frames or any other external support to the end of the compression flange.
4. By fixing the ends of the beam into the walls.

ISI recommends effective length of compression flange in ISI: 800–1984 as follows:

1. **For simply supported beams and girders where each end of the beam is restraint against torsion.**

(i) With ends of compression flange unrestrained (unsupported) against lateral bending

$$l = \text{span}$$

(that is, free to rotate in plane at the bearings)

(ii) With ends of compression flange partially restrained against lateral bending

$$l = 0.85 \times \text{span}$$

(that is, not free to rotate in plane at the bearings)

(iii) With ends of compression flange fully restrained against lateral bending

$$l = 0.7 \times \text{span}$$

(that is, free to rotate in plane at the bearings)

Where the ends of the beam are not restrained against torsion, or where the load is applied to the compression flange and both the load and the flange are free to move laterally, the above value of effective length should be incresed by 20 per cent.

Note. *It is to note that the end restraint element shall be capable of safely resisting, in addition to wind and other applied external forces, a horizontal forces acting at the bearing in a direction normal to compression flange of the beam at the level of the centriod of the flange and having a value equal to, not less than 2.5 per cent of the maximum force occurring in the flange.*

2. **For beams which are provided with members giving effective lateral restraint to the compression flange at interval along the span in addition to the end torsional restraint.**

The effective length 'l' shall be taken as the maximum distance centre to centre of the restraints members.

3. **For cantilever beams builts-in at the support with projecting length 'L'.**

(i) Built-in at the support, free at the end

$$l = 0.85\,L$$

(ii) Built-in at the support, restrained against torsion at the end by continuous construction

$$l = 0.75\,L$$

(iii) Built-in at the support, restrained against lateral bending and torsion at the free end

$$l = 0.5\,L$$

4. **For cantilever beams continuous at the support with projecting length 'L'.**

(i) Unrestained against torsion at the support and free at the end

$$l = 3L$$

(ii) Partially restrained against torsion at the support and free at the end

$$l = 2L$$

(iii) Restained against torsion at the support and free at the end

$$l = L$$

where, L = length of cantilever

If there is a degree of fixidity at the free end, the effective length shall be multiplied by $\left(\frac{0.5}{0.85}\right)$ in 3 (ii) and 3 (iii) above, and by $\left(\frac{0.75}{0.85}\right)$ in 4 (i), 4 (ii) and 4 (iii) above.

Where the beams supported slab construction, the beam shall be deemed to be effective restrained laterally if the frictional or positive connection of the slab to the beam is capable of resisting a lateral force of 2.5 per cent of the maximum force in the compression flange of the beam, considered as distributed uniformly along the flange. Furthermore, the slab construction shall be capable of resisting this lateral force in lateral flexure and shear.

For the beams which are provided with members giving effective lateral restraint of the compression flange at intervals along the span, the effective lateral restraint shall be capable of resisting a force of 2.5 per cent of the maximum force in the compression flange taken as divided equally between the number of points at which the restraint members occur.

In a series of such beams, with soild webs, which are connected together by the same system of restraint members, the sum of the restraining force shall be taken as 2.5 per cent of the maximum flange force in one beam only.

In case of a series of latticed beams, which are connected together by the same system or restraint members, the sum of the restraining forces required shall be taken as 2.5 per cent of the maximum force in the compresssion flange plus 1.25 per cent of the forces of 7.5 per cent.

The maximum slenderness ratio of the compression flange that is, the ratio of the effective length to the appropriate radius should not exceed 300 as in case of column members.

Example 6.5 *MB 550, @ 1.037 kN/m has been used as simply supported beam over a span of 4 metres. The ends of beam are restrained against torsion but not* *against lateral bending. Determine the safe uniformly distributed load per cent metre length which the beam carry.*

Solution

Step 1 : Properties of given section

From ISI Handbook No. 1, for MB 550, @ 1.037 kN/m

Overall depth $D = h = 550$ mm

Means thickness of flange $T = t_f = 19.3$ mm

Radius of gyration, $r_y = 37.3$ mm

Section modulus, $Z_{xx} = 2359.8 \times 10^3$ mm³

Thickness of web, $t_w = 11.2$ mm

$h_2 = 41.25$ mm

Step 2 : Elastic critical stress in bending , f_{cb}

The effective length of the compression flange is 4 m. Slenderness ratio

$$\left(\frac{l}{r_y}\right) = \left(\frac{4\times1000}{37\cdot3}\right) = 107.238$$

Ratio $$\left(\frac{D}{T}\right) = \left(\frac{\text{Overall depth}}{\text{Means thickness of flange}}\right)$$

$$\left(\frac{D}{T}\right) = \left(\frac{550}{19.3}\right) = 28.497$$

From IS: 800–1984, Table 6.5

$X = 303.555$, and $Y = 231.705$

From Eq. 6.18, $$f_{cb} = k_1 \,.\, (X + k_2 \cdot Y)\cdot\frac{c_2}{c_1} \quad \text{...(i)}$$

c_1 and c_2 of the beam and greater distance from the neutral axis to the extreme fibres of the beam section which are equal

$k_1 = 1.0$ $\psi = 1.0$ and k_2 0.0 for $w = 0.5$

Substituting the respective values in the expression (i)

f_{cb} = 1.0 (303.555 + 0.0 × 230.705) × 1.0

= 303.555 N/mm² (MPa)

Let the value of yield stress for the structural steel be 260 N/mm².

Ratio $$\left(\frac{T}{t_w}\right) = \left(\frac{19.3}{11.2}\right) = 1.723 \not> 2$$

Ratio $$\left(\frac{d_1}{t_w}\right) = \left(\frac{550-2\times41\cdot25}{11\cdot2}\right) = 41.74$$

$$\left(\frac{1344}{(f_y)^{\frac{1}{2}}}\right) = \left(\frac{1344}{(260)^{\frac{1}{2}}}\right) = 83.35$$

$$\left(\frac{d_1}{t_w}\right) \not> \left(\frac{1344}{(f_y)^{\frac{1}{2}}}\right)$$

As such, the value of f_{cb} shall be increased by 20 per cent. Therefore,

$$f_{cb} = (1 \cdot 20 \times 303 \cdot 555)$$
$$= 364.266 \text{ N/mm}^2$$

Step 3 : Permissible compressive stress in bending

From Table 6.2, IS: 800–1984,

Maximum permissible bending compressive stress

$$\sigma_{bc} = 121.4 \text{ N/mm}^2 \text{ (MPa)}$$

Step 4 : Moment of resistance of beam section

$$M_r = \left(\frac{121 \cdot 4 \times 2359 \cdot 8 \times 10^3}{1000 \times 1000}\right)$$
$$= 286.48 \text{ m-kN}$$

Step 5 : Uniformly distributed load

Let w be the uniformly distributed load in kN/m

The maximum bending moment, M occurs at the centre

$$M = \left(\frac{w \cdot l^2}{8}\right)$$

$$\left(\frac{w \times 4 \times 4 \times 1000}{8}\right) = (286 \cdot 48 \times 10^3)$$

$$w = 143.24 \text{ kN/m}$$

The self-weight of beam is 1.037 kN/m

The safe uniformly distributed load which can be placed over the beam

$$= (143.24 - 1.037)$$
$$= \mathbf{142.203 \text{ kN}}$$

Example 6.6 *Design a rolled steel I-section for a simply supported beam with a clear span of 6 m. It carries a uniformly distributed load of 50 kN per metre exclusive of self-weight of the girder. The beam is laterally unsupported.*

Solution

Design :

Step 1 : Total load to be supported

The clear span of the simply supported beam is 6 m.

Assuming that 300 mm wide end bed blocks are used

Effective span of the beam l = 6.30 m

Uniformly distributed load carried by the beam exclusive of self-weight

$$= 50 \text{ kN/m}$$

Assuming self-weight of the beam = 1 kN/m

Total uniformly distributed load inclusive of self-weight

$$= 51 \text{ kN/m}$$

Step 2 : Bending moment and shear force

The maximum bending moment occurs at the centre

$$M = \left(\frac{w \cdot l^2}{8}\right) = \left(\frac{51 \cdot 00 \times 6 \cdot 30 \times 6 \cdot 30 \times 1000}{8 \times 1000}\right)$$

$$= 253 \text{ kN-m}$$

The maximum shear force occurs at the ends

$$F = \left(\frac{wl}{2}\right) = \left(\frac{51 \cdot 00 \times 6 \cdot 30}{2}\right) = 160.65 \text{ kN}$$

Step 3 : Permissible stress

Let the value of yield stress f_y for the structural steel to be used be 250 N/mm² (MPa). For the maximum value of f_{cb}, the maximum permissible bending compressive stress should not exceed 0.66 f_y. Therefore, assume

$$\sigma_{bc} = (0 \cdot 66 \times 250) = 165 \text{ N/mm}^2 \text{ (MPa)}$$

Step 4 : Section modulus required

$$Z = \left(\frac{253 \times 1000 \times 1000}{165}\right)$$

$$= 1533.33 \times 10^3 \text{ mm}^3$$

However, the actual value of permissible bending compressive stress, σ_{bc} shall be less than 0.66 f_y. The modulus of section Z needed for the beam section shall be more. The trial section modulus may be adopted 25 percent to 50 percent in excess of above calculated.

Step 5 : Properties of trial section

Trial section modulus = 1.50 × 1533.33 × 10³ = 2299.995 × 10³ mm³

From ISI Handbook No. 1, try WB 550, @ 1.125 kN/m

Overall depth $D = h = 550$ mm

Mean thickness of flange

$$T = t_f = 17.6 \text{ mm}$$

Radius of gyration $r_y = 51.1$ mm

Section modulus

$$Z_{xx} = 2723.9 \times 10^3 \text{ mm}^3$$

Thickness of web,

$$t_w = 10.5 \text{ mm}, \ h_2 = 38.30 \text{ mm},$$

$$I_{xx} = 74906.1 \times 10^4 \text{ mm}^4$$

Step 6 : Check for section modulus

The effective length of compression flange,

$$l = 6.30 \text{ m}$$

Slenderness ratio

$$\left(\frac{6 \cdot 30 \times 1000}{51 \cdot 1}\right) = 123.287$$

Ratio $\left(\frac{D}{T}\right) = \frac{550}{17.6} = 31.25$

From IS: 800–1984, Table 6.5

$$X = 233.799 \text{ and } Y = 175.125$$

From Eq. 6.18,

$$f_{cb} = k_1 \,.\, (X + k_2 .Y). \frac{c_2}{c_1}$$

c_1 and c_2, the smaller and larger distance from the neutral axis to the extreme fibres are equal. Therefore,

$$k_1 = 1.0 \text{ for } \psi = 1.0 \text{ and}$$
$$k_2 = 0.0 \text{ for } \omega = 0.5$$

Substituting the respective values in the expression (i)

$$f_{cb} = 1.0(233.799 + 0.0 \times 175.125 \times) \times 1.0$$
$$f_{cb} = 233.799 \text{ N/mm}^2$$

Ratio $\left(\frac{T}{t_w}\right) = \left(\frac{17.6}{10.5}\right) = 1.676 < 2.0$

$$\left(\frac{d_1}{t_w}\right) = \left(\frac{550 - 2\times 38.50}{10.5}\right) = 45.08$$

$$\left(\frac{1344}{(f_y)^{\frac{1}{2}}}\right) = \left(\frac{1344}{(250)^{\frac{1}{2}}}\right) = 85$$

The value of f_{cb} shall be increased by 20 per cent. Therefore,

$$f_{cb} = (1.20 \times 233.799)$$
$$= 280.559 \text{ N/mm}^2$$

From IS: 800–1984, Table 6.2

$$\sigma_{bc} = 106.112 \text{ N/mm}^2 \text{ (MPa)}$$

Section modulus required

$$\left(\frac{253\times 1000 \times 1000}{106\cdot 112}\right) = 2384.27 \times 10^3 \text{ mm}^2$$

2723.9×10^3 mm^2 provided.

Hence, safe.

Step 7: Check for shear stress

Average shear stress

$$\tau_{va.cal} = \left(\frac{160\cdot 65\times 10^3}{550\times 11\cdot 2}\right)$$
$$= 26.079 \text{ N/mm}^2$$
$$< (0.4 \times 250 = 100 \text{ N/mm}^2). \text{ Hence, safe.}$$

Step 8: Check for deflection

$$y_{max} = \frac{5}{384}\cdot\left(\frac{w\cdot l^4}{EI}\right)$$

$$y_{max} = \frac{5}{384} \times \left(\frac{51 \cdot 00 \times 6 \cdot 3 \times 6 \cdot 30^3 \times 1000^3 \times 1000}{2.047 \times 10^5 \times 74906.1 \times 10^4} \right) \text{ mm}$$

$$= 6.82 \text{ mm}$$

Allowable deflection

$$\left(\frac{1}{325} \times 6300 \right) = 19.4 \text{ mm} > y_{max}. \text{ Hence, safe.}$$

Provide WB 550 @ 1.125 kN/m

6.9 WEB CRIPPLING

The web of rolled steel beam is usually thin. The web is relatively weak when placed under direct compression.

In rolled steel beams, the failure of the web may occur at concentrated loads, and at supports, where it has reaction. The failure of web in direct crushing under concentrated load is known of web crippling or web crimpling. The web crippling may occur if the in-plane compressive stress in the web is too large. It may also occur if $(b + 2k)$ distance (Fig. 6.9) used to deliver load from the beam flange to the web is too narrow or $(b + k)$ distance used to deliver load from the web to the flanges is too small.

The web crippling may also occur if the uniformly distributed load on the flange is too large for web thickness.

The web crippling means the stress concentration due to bottleneck condition at the web toe of the fillet under or over heavy load concentrations. The web crippling is a localized bearing stress caused by the transmission of compression from the comparatively wide flange to the narrow web. The crippling may be brought about by a failure in bearing of the metal at the web toe of the fillet and the resulting tendency of the flange and web to fold over on each other at the plane.

In web crippling, local buckling of the web occurs immediately adjacent to a concentration of stress as shown in Fig. 6.8.

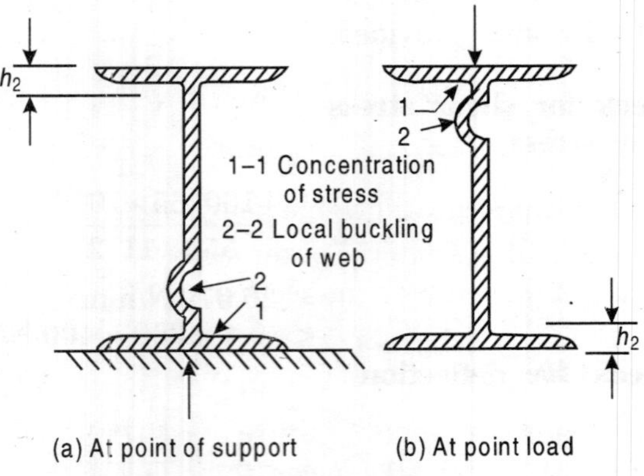

Fig. 6.8 *Web crippling*

For safety against web crippling adequate bearing length along the beam may be provided so that the bearing stress 'σ_p' for beams without stiffeners may not exceed the allowable bearing stress $\sigma_p = 0.75 f_y$ (0.75 times yield stress of IS: 800–1984). As per IS: 800–1962 dispersion of load or reaction through the flange to the web occurs uniformly at an angle of 30 degrees to the horizontal as shown in Fig. 6.9.

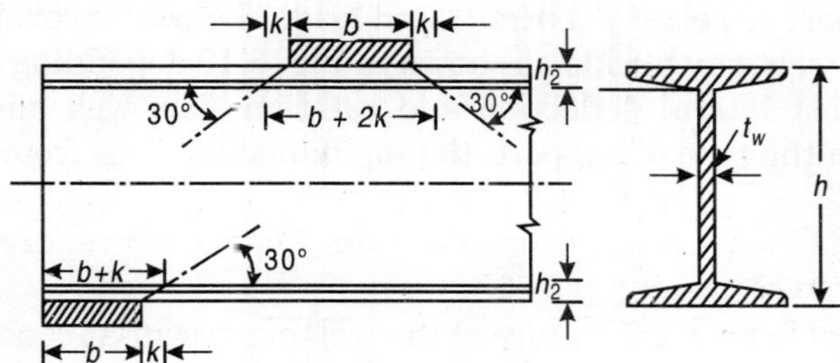

Fig. 6.9 *Bearing length of web*

AISC specification recommend an angle of dispersion of 45° instead of 30 ° as recommended in IS: 800–1984.

The bearing length of the web under concentrated load is given by

$$B = (b_1 + 2k) = \left(b_1 + 2h_2\sqrt{3}\right) \qquad ...(6.17)$$

where b_1 = length of bearing plate

h_2 = depth of root of fillet from the outer surface of flange

Bearing stress $\sigma_p = \left(\dfrac{\text{Concentrated load}}{\left(b_1 + 2h_2\sqrt{3}\cdot t_w\right)}\right)$ N/mm²

This should be less than the allowable bearing stress 0.75 N/mm².

The bearing length of web at the support is given by

$$B = (b + k) = \left(b + h_2\sqrt{3}\right) \qquad ...(6.18)$$

Bearing stress, $\sigma_p = \left(\dfrac{\text{Reaction}}{b + h_2\sqrt{3t_w}}\right)$

This should be less than the allowable bearing stress $\sigma_p = 0.75 f_y$.

This depth of root of fillet from the outer surface of flange 'h_2' for the beam is given in steel section tables under connection details.

6.10 WEB BUCKLING

The *web buckling* means that type of failure of web in which the web, vertically above the bearing plate and the lower flange at the reaction or below a concentrated load (bearing plate and the upper flange) is subjected to the column action and tends to buckle under it. It is out-of plane distortion of web of a beam. It is also called as **vertical web buckling.** It reults due to combination of large value of $\left(\frac{h_1}{t_w}\right)$ ratio and the bending stress. The unbraced length of compression flange may also contribute to the web buckling.

In adding to local crushing or crippling of the web, buckling may also occur above a support or below a concentrated load. The web is considered to act as a column. The web may buckle as shown in Fig. 6.10 depending upon restraint of flanges against lateral deflection and rotation. The web buckling will occur depending on the type of support, the top flange receives from the surrounding construction.

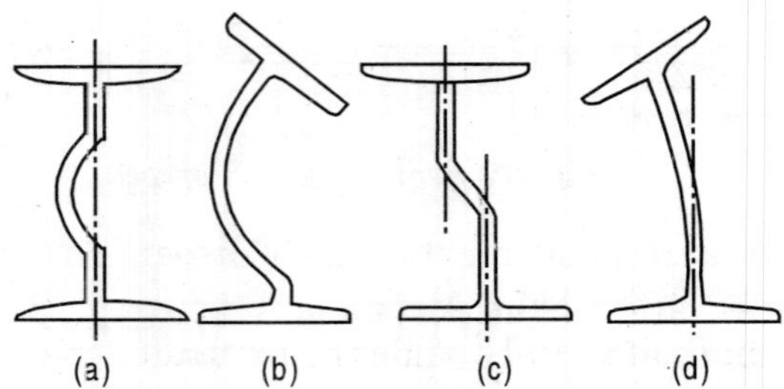

Fig. 6.10 *Web buckling*

The bottom flange is restrained against lateral deflection and rotation in all cases. The restraint conditions for top flange are as below:

1. Restrained agaist lateral deflection and rotation as shown in Fig. 6.10 (a);
2. Restrained agaist lateral deflection and rotation as shown in Fig. 6.10 (b);
3. Restrained agaist lateral deflection and rotation as shown in Fig. 6.10 (c); and
4. Restrained agaist lateral deflection and rotation as shown in Fig. 6.10 (d).

IS: 800–1984 recommends that bearing stiffners should be provided at points of concentrated load and point of support where the concentrated load reaction exceeds the value of ($\sigma_{ac} \cdot t_w \cdot B$)

where, σ_{ac} = Axial stress for column for a slenderness ratio $\left(\frac{d_1\sqrt{3}}{t_w}\right)$

$d_1 = (h - 2 \cdot h_2)$

t_w = Thickness of web

B = The length of the stiff portion of the bearing plus the additional length by dispersion at 45° to the level of the neutral axis, plus the thickness of the scating angle, if any.

(The stiff portion of bearing is that length which cannot deform appreciably in bending and shall not be taken as greater than half the depth of beam for simply supported beams and the full depth of beams continuous over a bearing.)

For concentrated load at any intermediate position on the beam, B is the length of bearing plate under the load plus the depth of beam plus thickness of flange plates if any, e.g., in plated beams (i.e., $B = b + h$) as shown in Fig. 6.11,

d_1 = Clear depth of web between root fillets

where, $d_1 = (h - 2h_2)$

h_2 = Depth of root of fillet from the outer surface of flange.

The selenderness ratio for the portion of web acting as a column, when the two flanges are restrained against lateral displacement and rotation may be found as follows :

$$\text{Slenderness ratio} = \left(\frac{\text{Effective length}}{\text{Minimum radius of gyration}}\right) = \left(\frac{0 \cdot 5d_1}{r_y}\right)$$

$$\left(\frac{\frac{d_1}{2}}{r_y}\right) = \left(\frac{\frac{d_1}{2}}{\left(\frac{I_{yy}}{A}\right)^{1/2}}\right) = \frac{d_1/2}{\left[\frac{1}{12}\frac{B \cdot t_w^3}{B \cdot t_w}\right]^{1/2}} = \left(\frac{d_1}{t_w} \cdot \sqrt{3}\right)$$

The selenderness ratio for the portion of web acting as a column, when the top flange is held in position only while the bottom flange is restrained against lateral displacement and rotation.

$$\text{Slenderness ratio} = \frac{\frac{d_1}{2}}{\left[\left(\frac{1}{12}\right)\frac{B \cdot t_w^3}{B \cdot t_w}\right]^{1/2}} = \left(\frac{d_1}{t_w} \cdot \sqrt{6}\right)$$

Note. *A beam that is safe in web crippling will usually be safe in web buckling, as well.*

The web crippling and web buckling may be prevented by adopting increased thickness of the web, or by adopting the increased length of bearing plates. The web buckling is controlled by limiting either $\left(\frac{d_1}{t_w}\right)$ *ratio. The web buckling may be avoided by stiffening the web. The bearing stiffners may be provided under the heavy loads.*

The diagonal buckling of the web has been discussed in Chapter 8.

Example 6.7 *Check the beam section WB 500, @ 1.451 kN/m against web crippling and web buckling if reaction at the end of beam is 179.6 kN. The length of bearing plate at the support is 120 mm. Design bearing plate. The bearing plate is set in masonry wall.*

Solution

Step 1 : Properties of given section

Assume allowable bearing stress in masonry as 5.5 N/mm^2

End reaction R = 179.6 kN

The length of bearing plates at the support is 120 mm

From the steel section tables for WB 600, @ 1.451 kN/m

Thickness of web t_w = 11.8 mm

Distance between root of fillet and flange of beam

$$h_2 = 46.05 \text{ mm}$$

Depth of beam between root of fillets

$$h_1 = 507.9 \text{ mm}$$

Step 2 : Check for web crippling

Bearing stress at the support, $\sigma_{p.cal}$

$$\left(\frac{\text{End reaction}}{\left(b + h_2\sqrt{3}\right)t_w}\right) = \left(\frac{179 \cdot 6 \times 1000}{\left(120 + 46 \cdot 05\sqrt{3}\right) \times 11.8}\right)$$

$$= 76.3 \text{ N/mm}^2$$

$$< (0.75 \times 260 = 195 \text{ N/mm}^2). \text{ Hence, safe.}$$

Step 3 : Check for web buckling

Reaction should not exceed the value ($\sigma_{ac} \times t_w \times B$)

Slenderness ratio $\left(\frac{h_1}{t_w}\sqrt{3}\right) = \left(\frac{507.9\sqrt{3}}{11 \cdot 8}\right) = 74.55$

Let the value of yield stress, f_y for the steel be 260 N/mm² (MPa)

Axial stress, σ_{bc} = 109.54 N/mm²

The length of stiff portion of bearing

$$B = \left(120 + \frac{600}{2}\right) = 420 \text{ mm}$$

Allowable reaction $= \left(\frac{(109 \cdot 54 \times 11 \cdot 8 \times 420)}{1000}\right) = 542.88$ kN.

> 179.6 kN and reaction. Hence, safe.

Step 4 : Design of bearing plate

End reaction = 179.6 kN

Allowable bearing stress in masonry

= 5.5 N/mm²

Bearing area required

$$\left(\frac{179 \cdot 6 \times 1000}{5 \cdot 5}\right) = 32654.54 \text{ mm}^2$$

Length of bearing plate provided

= 120 mm

Width of bearing plate,

$$B = \left(\frac{32654 \cdot 54}{120}\right) = 272.12 \text{ mm (say 280 mm)}$$

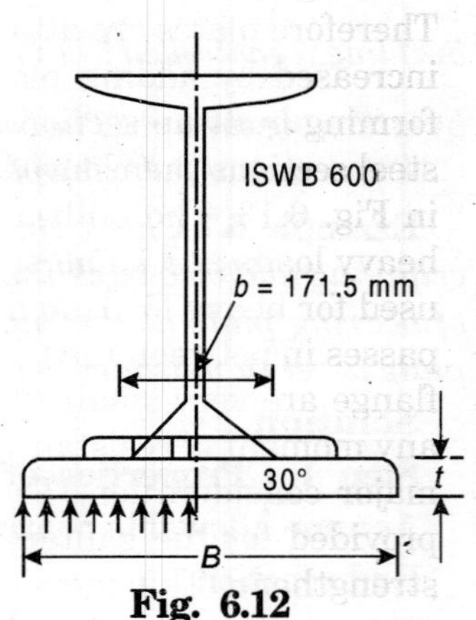

Fig. 6.12

Step 5 : Thickness of bearing plate

The dispersion of reaction occurs uniformly at 30° with the horizontal

$$b = (11.8 + 2 \times 46.05 \times \cot 30^\circ)\text{ mm}$$

$$= (11.8 + 2 \times 46.05 \times \sqrt{3}) = 171.5\text{ mm}$$

The maximum bending moment, M occurs at the centres

$$M = \frac{R}{8}(B-b) = \left(\frac{179\cdot6}{8}\times(280-171\cdot5)\right)$$

$$= 2435.825\text{ mm-kN}$$

The moment resistance of plate section

$$(\sigma_{bc}\cdot Z) = \left(\frac{0\cdot75\times260\times280\times t^2}{1000\times6}\right)\text{kN-mm}$$

$$\left(\frac{0\cdot75\times260\times210\times t^2}{1000\times6}\right) = 2435.825$$

$$t_2 = \left(\frac{2435\cdot825\times6\times1000}{0\cdot75\times260\times280}\right)(\because t = 16.36\text{ mm})$$

Provide base plate 120 mm × 280 mm ×18 mm.

6.11 BUILT-UP BEAM

The built-up beam are also termed as **compound beams or compound girders.**The built-up beam are used when the span, load and corresponding bending moment are of such magnitudes that rolled steel beam section become inadequate to provide required section modulus. The uilt-up beams are also used when rolled steel beams are inadequate for limited depth. In building construction, the depth of beam is limited by the space provided by the architect´s drawings. The beam of small depth do not provide required section modulus. Therefore plates are attached to the beams. The strength of rolled steel beam is increased by adding plates to its flanges. Which is therefore one method of forming built-up section. The other method is to compound a number of rolled steel sections,themselves. The built-up beam section commonly used are shown in Fig. 6.13. The built-up sections shown in Fig. 6.13 (a) and (b) are used for heavy loads and small spans. The built-up section shown in Fig. 6.13 (c) is also used for heavy loads and large spans. In this built-up section, the neutral axis passes in between the two flanges in contact. The bending stresses in these two flange are very small. These two flanges in contact practically do not provide any moment of resistance. Hence, such section is provided where deflection is a major consideration. The built-up section shown in Fig. 6.13 (e) and (f) are provided for the gantery girders. The top flange in these sections have been strengthend. This increases the resistance of the section for the lateral loads.

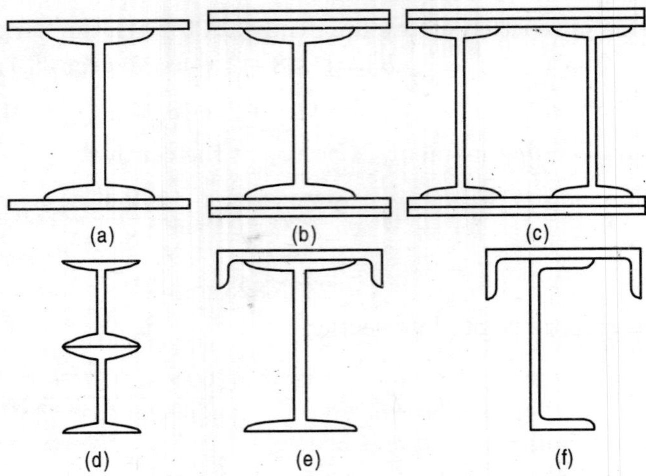

Fig. 6.13 *Built-up beam sections*

In the riveted built-up beams, the area of tension flange is reduced by rivet holes. The deduction is made for rivet holes in tension flange while determining moment of inertia and section modulus of the built-up beam section. In actual

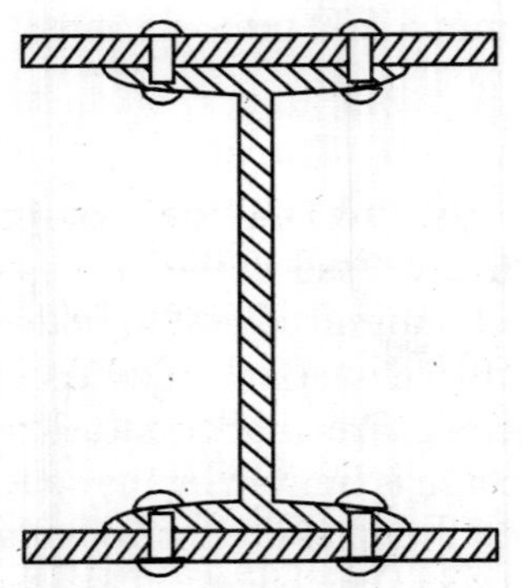

Fig. 6.14

practice, it is usual to make allowance for rivet holes in the both flanges. The rivets connecting flanges plates are staggered so that each flange is reduced by one rivet hole as shown in Fig. 6.14.

The curtailment of cover plates of built-up beams is similar to curtailment of cover plates of plate girder. It is discussed in chapter 8, Design of Plate Girders.

6.12 DESIGN OF BUILT-UP BEAMS

The built-up beams are designed by trial and error method. The built-up section is selected and it is checked for the stresses. The built-up beams with soild webs are proportioned in general on the basis of the moment of inertia of the gross cross-section with neutral axis taken at the centroid of the section.

The following are the usual steps in design of built-up beams.

Step 1. The effective span and load required to be carried by the built-up beam are known. The bending moment M and maximum shear F in the built-up beam are calculated.

Step 2. In the beginning, the value of yield stress f_y, for structural steel to be used is assumed. The permissible bending stress σ_{bc} is assumed as $0.66 f_y$. The required section modulus, Z for the beam section is calculated

$$\text{Required section modulus, } Z = \left(\frac{M}{\sigma_{bc}}\right)$$

where, σ_{bc} is allowable bending compressive stress.

For laterally supported beams, allowable bending compressive stress is equal to the allowable bending tensile stress. For laterally unsupported beams, allowlable bending compressive stress is determined by the condition of lateral restraints and the section of built-up beams.

From ISI Handbook No. 1, a trial section for beam for beam is adopted having the modulus of section, Z about 25 to 50 percent in excess of that required. The geometrical properties of beam section, are noted. The permissible bending compressive stress corresponding to the geometrical dimensions of bean section and restraint conditions is found from IS: 800–1984. As a check, the section modulus, Z needed corresponding to actual permissible stress is found and it is shown that is less than the section modulus of the beam provided.

When the depth of beam is no limited, then the rolled steel beam is selected for the maximum section modulus. The plates are provided for the difference of section modulus i.e., between that required and that provided by the rolled steel beam. When the depth of beam is limited, then, the usual practice is to select from ISI Handbook No. 1, the strongest rolled steel beam that will allow for necessary thickness of cover plates, at top and buttom. The araea of covers plates required is then found by trial and error method. The moment of inertia of trial plated beam section is found, and then checked. It involves a considerable amount of work. The required area of cover plates may be found from the following guidance, which will result in satisfactory selection of the required area of cover plates at the first trial.

Let the moment of inertia of built-up plated beam (symmetrical section) be given by

$$I = \left[I_{\text{beam}} + 2 \cdot A_p \left(\frac{h}{2}\right)^2\right] \quad \text{...(i)}$$

where, I_{beam} is the moment of inertia of the rolled beam section, A_p is the area of cover plates in one flage and h is the distance between the centroids of the top and bottom flanges plates.

The distance between the centroids of the top and bottom flange plates, h may be adopted equal to the depth of rolled steel beam itself, since the thickness of flange plates is small compared with the depth of beam section.

From the expression (i)

$$\left(\frac{I}{\left(\frac{h}{2}\right)}\right) = \left(\frac{I_{beam}}{\left(\frac{h}{2}\right)} + 2A_p.\frac{h}{2}\right) \quad ...(ii)$$

or

$$Z_{reqd} = (Z_{beam} + A_p \,.\, h)$$

or

$$A_p = \left(\frac{Z_{reqd} - Z_{beam}}{h}\right) \quad ...(iii)$$

where,

$$Z_{reqd} = \left(\frac{I}{\left(\frac{h}{2}\right)}\right) \text{ and } Z_{beam} = \left(\frac{I_{beam}}{\left(\frac{h}{2}\right)}\right)$$

The expression (iii) gives the area of cover plates in one flange, when there are not rivet holes in the plates. In the riveted built-up beams, an allowance of about 13 percent of the area of cover plates found is added. The required area of cover plates in one flange is then provided by one or two plates and preferably not more than three plates. The rolled steel beam and cover plates provide the required section modulus. The width of cover plates and thickness are decided with the restrictions of the outstand.

Step 3. From steel section tables, select a suitable section. The outstand of flange plates that is their projection beyond the outer line of connection shall not exceed the relevant values given in IS: 800–1984 in (a), (b) and (c) below. The area of excess flange shall be neglected while calculating the effective geometrical properties of the section.

These values have also been expressed in terms of yield stress of the structural steel as it has been done in recommending the various permissible stresses and other.

(a) Flanges and plates in compression with the unstiffened edges : $\frac{256t}{(f_y)^{1/2}}$

subject to maximum of 16 t.

(b) Flanges and plates in compression with the unstiffened edges : 20 t to the innermost face of the stiffening.

(c) Flanges and plates in tension 20 t.

Step 4. The built-up section is checked for bending stresses in the extreme fibres. The gross moment of inertia of the built-up section is found on the basis of gross cross-section of the built-up section with neutral axis assumed at the centriod of the section. The maximum tensile stress in bending is found from the maximum compressive stress in bending by increasing the maximum compressive stress in bending in the ratio of gross area to the net area of tension flange. The effective sectional area for the compression flange in the gross area with deductions for excessive widths of the plates as discussed for compression members and for open holes occurring in a plane perpendicular to the direction of stress at the section being considered. The effective sectional area for the tension flanges is the gross area with deductions for the rivet holes.

Actual bending compressive stress

$$\sigma_{bc.cal} = \left(\frac{\text{Max Bending Moment}}{\text{Gross Moment of Inertia}}\right) \times \left(\begin{array}{c}\text{Distance of extreme fibre in}\\ \text{compression, from neutral axis}\end{array}\right)$$

$$\sigma_{bc.cal} = \left(\frac{M}{I_{xx}}\right) \cdot y_1$$

The actual bending tensile stress $\sigma_{bt.cal}$ is determined by increasing the actual bending compressive stress in the ratio of gross area of the tension flange to the net area of the tension flange.

$$\sigma_{bt.cal} = \sigma_{bc.cal} \times \left(\frac{\text{Gross area of the tension flange}}{\text{Net area of the tension flange}}\right)$$

The actual bending stresses should not exceed allowable bending stresses.

Step 5. The built-up section is checked for shear force and deflection.

Example 6.8 *Design a simply supported plated rolled steel beam section to carry a uniformly distributed load 50 kN/m inclusive of self-weight of beam. The effective span of beam is 10 metres. The depth of beam should not be more than 500 mm. The compression flange of the beam is laterally supported by floor construction.*

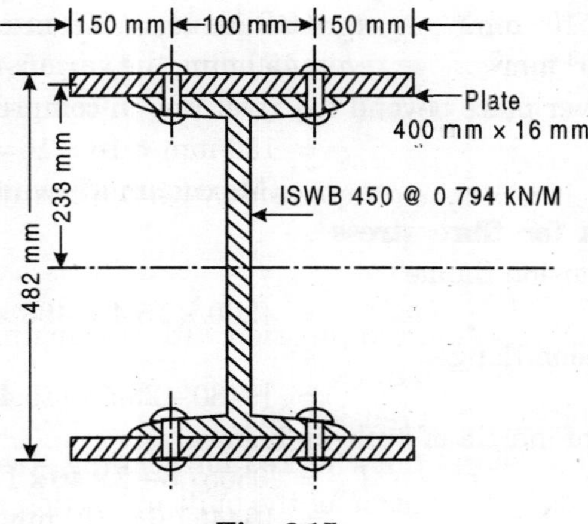

Fig. 6.15

Solution

Design :

Step 1 : Effective span of beam

$$= 10 \text{ m}$$

Step 2: Maximum bending moment,

$$M = \left(\frac{w \cdot l^2}{8}\right)$$

$$= \left(\frac{50 \cdot 00 \times 10 \times 10}{8}\right) = 625 \text{ kN-m}$$

Step 3: Maximum shear force,

$$F = \left(\frac{wl}{2}\right) = \left(\frac{50 \cdot 00 \times 10}{2}\right) = 250 \text{ kN}$$

Step 4 : Permissible bending stress

Let the value of yield stress for the structural steel to be used be 250 N/mm^2. The permissible bending stress in compression may be assumed as

$$(0.66 \times 250) = 165 \text{ N/mm}^2 \text{ (MPa)}$$

Step 5 : Section modulus required

$$Z = \left(\frac{M}{\sigma_{bc}}\right) = \left(\frac{625 \times 1000 \times 1000}{165}\right)$$

$$= 3787.88 \times 10^3 \text{ mm}^3$$

Step 6 : Trial section

From the steel section table, try WB 450, @ 0.794 kN/m with one cover plate 400 mm ×16 mm connected to each flange by 22 mm diameter rivets as shown in Fig. 6.15.

For WB 450, @ 0.794 kN/m

I_{xx} = 35057.6 × 10^4 mm^4, t_w = 9.2 mm, b_f = 200 mm

Z_{xx} = 1558.1×10^3 mm^3, h = 450 mm, t_f = 15.4 mm

Projection of cover plate beyond the rivet line in compression flange

= 150 mm < 16 × 16 = 256 mm

(Maximum allowable outstand)

Step 7 : Check for fibre stress

Gross area of tension flange

$$= (200 \times 15.4 \times 400 \times 16) = 9480 \text{ mm}^2$$

Net area of tension flange

$$= [9480 - 23.5 \times (15.4 + 16)] = 8742.1 \text{ mm}^2$$

Gross moment of inertia of built-up beam

$$I_{xx} = [35057.6 + 2 \times 40 \times 1.6 \times (23.3)^2] \times 10^4 \text{ mm}^4$$

$$= 104457.6 \times 10^4 \text{ mm}^4$$

(Neglecting moment of inertia of plates about their own axes)

Calculated bending tensile stress

$$\sigma_{bc.\text{cal}} = \left(\frac{625 \times 1000 \times 1000 \times 241}{104457 \cdot 6 \times 10^4}\right)$$

$$= 144.197 \text{ N/mm}^2$$

Calculated bending tensile stress

$$\sigma_{bt.cal} = 144.1 \times \left(\frac{9480}{8743}\right) = 156.36 \text{ N/mm}^2$$

The calculated bending stress in compression and that in tension do not exceed

$$(0.66 \times 250) = 165 \text{ N/mm}^2 \text{ (MPa)}$$

Step 8 : Check for shear force

Average shear stress

$$\tau_{bc.cal} = \left(\frac{F}{h \cdot t_w}\right) = \left(\frac{250 \times 1000}{450 \times 9 \cdot 2}\right) = 60.386 \text{ N/mm}^2$$

$$< (0.4 \times 250 = 100 \text{ N/mm}^2. \text{ Hence safe.}$$

Step 9 : Check for deflection

Maximum deflection in the beam

$$y_{max} = \frac{5}{384} \cdot \left(\frac{wl^4}{EL}\right)$$

$$y_{max} = \left(\frac{5 \times 50 \cdot 00 \times 10^3 \times 1000^3 \times 1000}{384 \times 2 \cdot 047 \times 10^5 \times 104457.6 \times 10^4}\right)$$

$$= 30.44 \text{ mm}$$

Allowable deflection

$$\left(\frac{1}{325} \times 10 \times 1000\right) = 30.57 \text{ mm}$$

The actual deflection is less than the allowable deflection. Hence, the section of the beam selected is safe

Provide WB 450 @ 0.794 kN/m and one 400 mm × 16 mm plate on each flange.

Example 6.9 *A simply supported beam is to support a uniformly distributed load 70 kN / m excluding weight of the beam, over a clear span of 8 metres. Design a plated rolled steel beam if MB 500, @ 0.869 kN / m and 10 mm thick plates are only available. The compression flange of the beam is laterally restrained.*

Solution

Design :

Step 1 : Effective span of beam

Clear span of beam = 8

Provide 300 mm wide bed blocks on each side to support the beam

Effective span = Centre to centre of bearing = 8.30 m

Step 2 : Total uniformly distributed load

uniformly distributed load on the beam = 70 kN/m

Assume self-weight of beam = 1.5 N/m

Total uniformly distributed load = 71.50 kN

Step 3 : Bending moment and shear force

The maximum bending moment, M occurs at the centre,

$$M = \left(\frac{w \cdot l^2}{8}\right) = \left(\frac{71.50 \times 8 \cdot 30 \times 8 \cdot 30}{8}\right)$$

$$= 615.7 \text{ m-kN}$$

The maximum shear force, F occurs at the support,

$$F = \left(\frac{71.50 \times 8 \cdot 30}{2}\right) = 296.725 \text{ kN}$$

Let the value of yield stress for the structural steel to be used be 250 N/mm² (MPa).

Step 4 : Permissible bending stress

The maximum permissible bending stress in compression and that in tension are equal for the laterally supported beam and it may be assumed as

$$\sigma_{bc} = (0.66 \times 250) = 165 \text{ N/mm}^2$$

Step 5 : Section modulus required

$$Z = \left(\frac{M}{\sigma_{bc}}\right) = \left(\frac{615 \cdot 7}{165}\right) = 3731.51 \times 10^3 \text{ mm}^3$$

Step 6 : Properties of trial section

From the steel section tables, for MB 500, @ 0.869 kN/m

$$I_{xx} = 45218.3 \times 10^4 \text{ mm}^4,\ t_w = 10.2 \text{ mm}$$

$$t_f = 17.2 \text{ mm}$$

$$Z_{xx} = 1808.7 \times 10^3 \text{ mm}^3,\ h = 500 \text{ mm}$$

$$b_f = 180 \text{ mm},\ r_y = 35.2 \text{ mm}$$

$$h^2 = 37.95 \text{ mm}$$

Provide two cover plates 340 mm × 10 mm on each flange.

Provide 22 mm diameter rivet to connect flange plates.

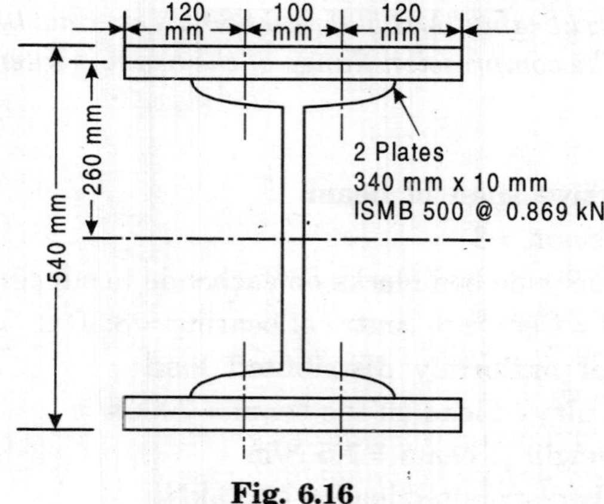

Fig. 6.16

Step 7 : Check for fibre stress

Gross area of tension flange = (180 × 17.2 + 2 × 340)

= 9932 mm^2

Net area of tension flange = [9932 – 23.5 (172 +20)]

= 9057.8 mm^2

Gross moment of inertia of built-up section

$$I_{xx} = [45218.3 + 2 \times 2\times2 \times 34 \times (26)^2] \times 10^4 \text{ mm}^4$$

$$= 137418.3 \times 10^4 \text{ mm}^4$$

Actual bending compressive stress

$$\sigma_{bc.cal} = \left(\frac{615 \cdot 7 \times 10^6 \times 270}{137418 \cdot 3 \times 10^4}\right) = 120.973 \text{ N/mm}^2$$

Actual bending tensile strength

$$\sigma_{bt.cal} = 120 \cdot 973 \times \left(\frac{9932}{9057.8}\right) = 132.648 \text{ N/mm}^2$$

The calculated bending stress in compression and that in tension do not exceed

(0.66 × 250) = 165 N/mm^2 (MPa)

Step 7 : Check for shear force

Average shear stress

$$\tau_{ca.cal} = \left(\frac{F}{h \cdot t_w}\right) = \left(\frac{296 \cdot 725 \times 1000}{500 \times 10 \cdot 2}\right)$$

= 58.18 N/mm^2

< (0.4 × 250 = 100 N/mm^2). Hence, safe.

Step 8 : Check for deflection

Maximum deflection of the beam

$$y_{max} = \frac{5}{384} \times \left(\frac{71 \cdot 50 \times 8 \cdot 30 \times (8 \cdot 30)^3 \times (1000)^3 \times 1000}{2 \cdot 047 \times 10^5 \times 137418 \cdot 3 \times 10^4}\right) \text{mm}$$

= 15.71 mm

$$\text{Allowable deflection} = \left(\frac{1}{325} \times 8300\right) = 25.538 \text{ mm}$$

The maximum deflection is less than the allowable deflection. Hence, the section of beam selected is safe.

Provide MB 500, @ 0.869 kN/m with two cover plates 340 mm × 10 mm on each flange as shown in Fig. 6.16.

Example 6.10 *In Example 6.9, the designed built-up beams restrained against torsion only. Determine the maximum uniformly distributed load per metre length which could be placed over this beam.*

Solution

Step 1 : Properties of given section

Effective length of the beam restrained against torsion,

$$l = \text{Span} = 8 \text{ m}$$

The built-up section of beam consists of BM 500, @ 0.869 kN/m and two cover plates 340 mm × 10 mm connected to each flange as shown in Fig. 6.16.

I_{xx} of MB 500, @ 0.869 kN/m is 1369.8 × 10⁴ mm⁴

$$t_f = 17.2 \text{ mm}$$

$$I_{yy} \text{ of built-up section} = \left(1369.8 + 2\times\frac{1}{12}\times 2\times 34^3\right)\times 10^4 \text{ mm}^4$$

$$= 14471.1 \times 10^4 \text{ mm}^4$$

$$I_{xx} \text{ of built-up section} = 137154.3 \times 10^4 \text{ mm}^4$$

Z_{xx} of built-up section

$$\left(\frac{137154\cdot 3\times 10^4}{270}\right) = 5079.9 \times 10^3 \text{ mm}^3$$

Total cross-sectional area of the built-up beam

$$(11074 + 4 \times 340 \times 10) = 24674 \text{ mm}^2$$

Radius of gyration of the section

$$r_y = \left(\frac{14469\cdot 8\times 10^4}{24674}\right)^{1/2} = 76.579 \text{ mm}$$

Slenderness ratio

$$\left(\frac{l}{r_y}\right) = \left(\frac{8000}{76\cdot 579}\right) = 104.467$$

Overall depth D = 540 mm

Mean thickness of flange = (20 + 17.2) = 37.2 mm

Ratio

$$\left(\frac{D}{T}\right) = \left(\frac{540}{37.2}\right) = 14.516$$

From IS: 800–1984, Table 6.5

$$X = 412.55 \text{ and } Y = 2445.6$$

Step 2 : Critical bending stress

From Eq. 6.18,

$$f_{cb} = k_1(X + k_2Y)\,.\,\frac{c_2}{c_1}$$

c_1 and c_2 the lesser and greater distances from the neutral axis to the extreme fibres of the beam section are equal

$$k_1 = 1.0 \text{ for } \psi = 1.0 \text{ and } k_2 = 0.0\omega = 0.5$$

Substituting the respective value in the expression (i)

$$f_{cb} = 1.0(412.55 + 0.0 \times 244.45)\ 1.0$$

$$= 412.55 \text{ N/mm}^2$$

Let the value of yield stress for the structural steel be 250 N/mm^2.

Ratio $$\left(\frac{T}{t_w}\right) = \left(\frac{37\cdot2}{10\cdot2}\right) = 3.647 > 2.0$$

Ratio $$\left(\frac{d_1}{t_w}\right) = \left(\frac{540-(200+2\times37.95)}{10\cdot2}\right) = 43.54$$

$$\left(\frac{1344}{\sqrt{f_y}}\right) = \left(\frac{1344}{\sqrt{250}}\right) = 85$$

$$\left(\frac{d_1}{t_w} > 85\right). \text{ But } \left(\frac{T}{t_w} > 2.0\right)$$

Therefore $$f_{cb} = 412.55 \text{ N/mm}^2$$

Step 3 : Permissible bending stress

From IS: 800–1984, Table 6.2

$$\sigma_{bc} = 123.255 \text{ N/mm}^2$$

Step 4 : Moment of resistance

$$M_r = (\sigma_{bc}.Z_{xx})$$

$$M_r = \left(\frac{123.255\times5089.56\times10^3}{1000\times1000}\right)$$

$$= 627.314 \text{ kN-m}$$

Maximum bending moment

$$M = M_r = \left(\frac{w\cdot l^2}{8}\right)$$

Step 5: Load which may be supported

$$\left(\frac{w\times8\times8\times1000}{8}\right) = (627.314\times1000)$$

$$w = 78.414 \text{ kN/m}$$

Self weight of beam $= (869 + 1078)$ N/m

$= 1.947$ kN/m

The uniformity distributed load which can be placed over the beam

$= (78.414 - 1.947) =$ **76.467 kN/m.**

6.13 DESIGN OF RIVETS CONNECTING COVER PLATES WITH THE FLANGES OF BEAM

The pitch of rivets connecting cover plates with the flanges of rolled steel beam is determined knowing the shear force at any section of beam.

The shear stress at any point of the cross-section is given by

$$f_x = \left(\frac{F \cdot Q}{I \cdot b}\right)$$

where, F = Shear force at the cross-section of beam.

Q = Static moment about the centroidal axis of the portion of cross-sectional area beyond the location at which the stress is to be found.

I = Moment of inertia of the section about the centroidal axis.

b = Width of section at which the stress is to be found.

The shear force per unit depth of the beam

The shear force per unit depth of the beam

$$F_s = (F \cdot Q/I)$$

The horizontal shear force per unit length of the beam is, equal to the vertical shear force per unit depth of the beam. The horizontal shear force per unit length is found at the level where plates are connected to the flange.

The horizontal shear force per unit length of beam

$$F_s' = \left(\frac{FQ}{I}\right)$$

The horizontal shear force per pitch length

$$= p \times F_s' = p \times \left(\frac{FQ}{I}\right)$$

where, p = pitch of the rivets

The horizontal force per pitch length is resisted by strength of rivet 9 i.e., rivet value R,

$$p.\left(\frac{FQ}{I}\right) = R, \text{ or } p = \left(\frac{R \cdot I}{F \cdot Q}\right)$$

In chain riveting, rivet value R is taken as strength of two rivets. In zig-zag or staggered riveting, staggered pitch of rivet is found and rivet value R is taken as strength of one rivet. The pitch of rivets in chain riveting should not exceed 16 t or 200 mm whichever is less in tension flange and it should not exceed 12 t or 200 mm whichever is less in compression members where t is the thickness of cover plate. When rivets are staggered at equal intervals and the gauge space does not exceed 75 mm, the above limits of pitch may be increased by 50 per cent.

Example 6.11 *In Example 6.9, design rivets connecting flange of rolled steel beam and cover plates.*

Solution

Design :

Step 1 : Static moment

Maximum shear force

$$F = 296.725 \text{ kN}$$

Gross-moment of inertia of the built-up section

$$I_{xx} = 137418.3 \times 10^4 \text{ mm}^4$$

Provide 22 mm diameter power driven rivets in staggered manner. Static moment of cover plates about xx-axis.

$$Q = Ay$$

where, A = Gross-sectional area

y = Distance of C.G. of cross-sectional area from xx-axis

$Q = (340 \times 20 - 23.5 \times 20) \times 260 = 1645.8 \times 10^3$ mm^3

Step 2 : Rivet value

Strength of power driven rivet in single shear

$$\left(\frac{\pi}{4}(23\cdot5)^2 \times \frac{100}{1000}\right) = 43.35 \text{ kN}$$

Strength of power driven rivet in bearing

$$\left(\frac{23\cdot5 \times 17.2 \times 300}{1000}\right) = 121.26 \text{ kN}$$

Rivet value R = 43.35 kN

Step 3 : Pitch of rivets

$$p = \left(\frac{R \cdot I_{xx}}{F \cdot Q}\right)$$

$$\left(\frac{43\cdot35 \times 137418\cdot3 \times 10^4}{29\cdot725 \times 1645 \times 10^3}\right) = 122.04 \text{ mm}$$

Pitch of rivets in one line = (2 × 122.04)
= 244.08 mm

Gauge space is 100 mm

Maximum pitch of rivet in compression flange
= (12 × 10) = 120 mm

Maximum pitch of rivets in tension flange
= (16 × 10) = 160 mm

Provide a pitch of 120 mm in compression flange and 160 mm in tension flange throughout the length of built-up beam.

6.14 LINTELS

The brick masonry or stone masonry over doors, windows and other openings in walls is supported by beams Such beams are known as *steel lintel* or simply as *lintels.* The various types of lintels are shown in Fig. 6.17.

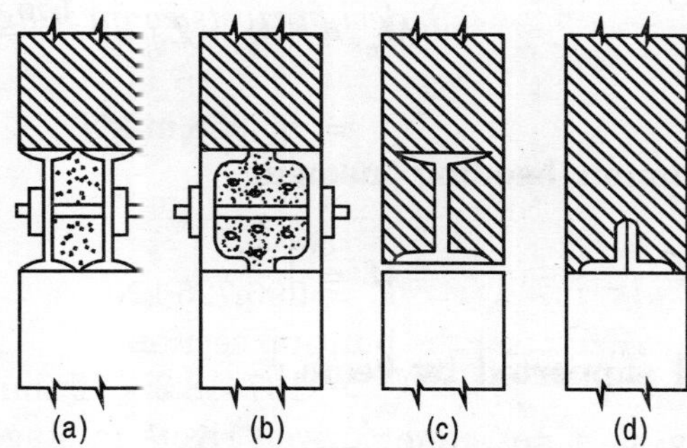

Fig. 6.17 *Steel beams used as lintels*

It is considered that the lintel-supports the masonry contained in an equilateral triangle as shown in Fig. 6.18. The masonry should have well bond and should have an arch action. The masonry should extend at least half the effective span to either side of the opening. The masonry above the lintel should be upto a height of 1.25 times the effective span.Then, the load is considered as uniformly distributed load.

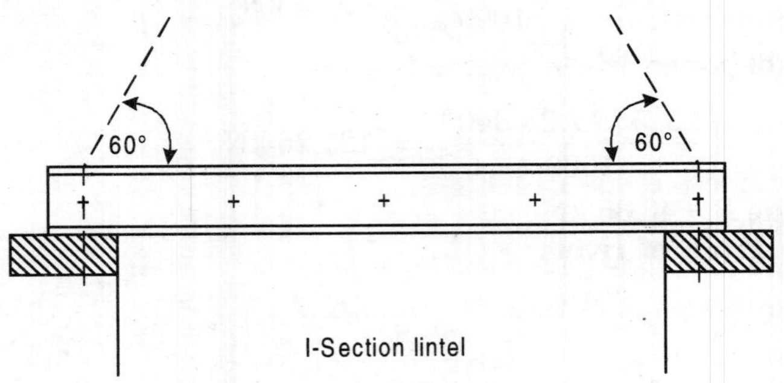

Fig. 6.18

When the height of masonry above the lintel is not sufficient it is considered to support whole masonry about it. The floor or roof loads which are immediately above the opening, are considered as separate load to be supported by the lintels.

Example 6.12 *A lintel consist of two LC @ 0.206 kN/m connected by pipe and bolt separator as shown in Fig. 6.17 (b), The effective span is 3.50 m. Determine the maximum uniformly distributed load for the lintel.*

Solution

Step 1 : Properties of given section

Two LC 200, @ 0.206 kN/m are used as a lintel.

From the steel section tables, the section modulus of two LC 200, @ 0.206 kN/m

$$Z_{xx} = (2 \times 172.6 \times 10^3)$$
$$= 345.2 \times 10^3 \text{ mm}^3$$

Step 2 : Moment of resistance of the lintel

$$M_r = (\sigma_{bc} \times Z) = \left(\frac{165 \times 345 \cdot 2 \times 10^3}{1000}\right)$$
$$= 56.958 \text{ m-kN}$$

Step 3 : Maximum bending moment

$$M = \left(\frac{wl^2}{8}\right)$$

Step 4 : Load supported by beam

$$\therefore \quad \left(\frac{w \times 3 \cdot 5 \times 3 \cdot 50 \times 1000}{8}\right) = 56.958 \times 1000 \times 1000$$

$$w = 37.197 \text{ kN/m}$$

Self-weight of lintel $= 2 \times 206 = 412$ N/m

Load imposed on lintel $= (37.197 - 0.412)$

$= \mathbf{26.785\ kN/m}$

Example 6.13 *A 300 mm thick wall bonded brick well is to be supported over an opening of 2.75 m. An additional load of 15 kN/m is imposed over the wall from the roof. Arch action exists. Design a suitable section for the lintel. Weight of brick masonry = 21.60 kN/m³.*

Solution

Design :

Step 1 : Effective span of lintel

Width of the opening = 2.75 m

Provide 150 mm wide bed blocks to support the lintel.

The effective span of lintel is the distance between centre to centre of bearings

$$l = (2.75 + 0.15) = 2.90 \text{ m}$$

Step 2 : Load supported by beam

Weight of brick masonry in equilateral triangle

$$\left(\frac{1}{2} \times 2 \cdot 90 \times 2 \cdot 90 \times \frac{\sqrt{3}}{2} \times 0 \cdot 30 \times 21 \cdot 60\right) = 23.598 \text{ kN}$$

Additional load imposed over the lintel from roof

$= (2.90 \times 15.00) = 43.50$ kN

Self-weight of lintel $= 2.85$ kN

Total uniformly distributed load

$- (23.598 + 43.50 + 2.85) = 69.948$ kN

Step 3 : Bending moment

The maximum bending moment, M occurs at centre

$$M = \left(\frac{wl^2}{8}\right) = \left(\frac{69 \cdot 948 \times 2 \cdot 90}{8}\right) = 25.356 \text{ kN-m}$$

Step 4 : Shear force

The maximum shear force, F occurs at the centre

$$F = \left(\frac{wl}{2}\right) = \left(\frac{69 \cdot 948}{2}\right) = 34.974 \text{ kN}$$

Step 5 : Section modulus required

$$Z = \left(\frac{25 \cdot 356 \times 1000 \times 1000}{165}\right)$$

$$= 153.67 \times 10^3 \text{ mm}^3$$

Step 6 : Trial section

From steel section tables, try two JC 175, @ 0.112 kN/m

Section modulus provided

$$= (2 \times 82.3 \times 10^3)$$
$$= 164.6 \times 10^3 \text{ mm}^3$$

Thickness of web, $t_w = 3.6$ mm

Depth of section $= 175$ mm

Moment of inertia of section

$$I_{xx} = 719.9 \times 10^4 \text{ mm}^4$$

Step 7: Check for shear force

$$\tau_{va.cal} = \left(\frac{F}{h \cdot t_w}\right) = \left(\frac{34 \cdot 974 \times 1000}{2 \times 175 \times 6}\right)$$
$$= 27.757 \text{ N/mm}^2$$

Step 8 : Check for deflection

Maximum deflection of lintel

$$y_{max} = \left(\frac{5}{384}\right) \times \left(\frac{wl^2}{EI}\right) \qquad (\because wl = 69.948)$$

$$= \left(\frac{5 \times 69 \cdot 948 \times (2 \cdot 90)^3 \times (1000)^3 \times 1000}{384 \times 2 \cdot 047 \times 10^5 \times 2 \times 719 \cdot 9 \times 10^4}\right)$$
$$= 7.537 \text{ mm}$$

Allowable deflection

$$= \left(\frac{1}{325} \times 290\right)$$
$$= 8.923 \text{ mm}$$

The actual deflection is less than the allowable deflection. Hence design is safe. Provide 2 JC, @ 0 .112 kN/m.

6.15 JACK ARCH

Jack-arches are constructed with bricks laid on edge, springing from and supported on the lower flanges of rolled steel beams. The spacing of rolled steel beams shall be generally 1.0 to 1.4 metres, but shall never exceed 20 metres. The beams are designed to support the portion of floor coming over them. Jack arches are arches of small rise. Rise of the arch is kept about $\frac{1}{8}$ th to $\frac{1}{12}$ th of the span. The intermediate beams receive thrust from arches on either side and thrust is balanced. The end beams receive side thrust from the arch. The rods are provided to resist the side thrust of end beams. The tie rods are properly anchored with stout mild steel nut and washers. The details of jack arch floor construction are shown in Fig. 6.19.

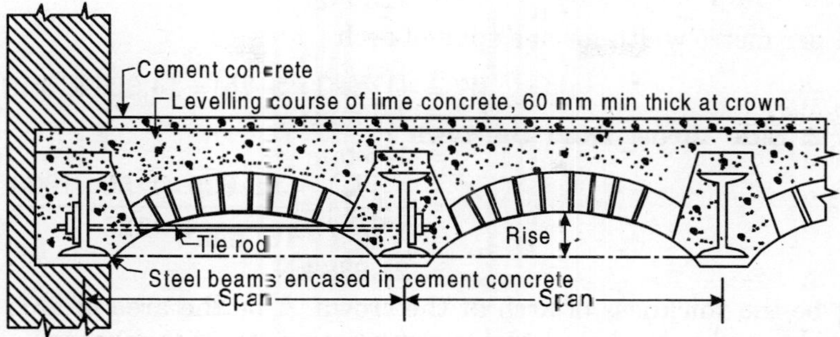

Fig. 6.19 *Jack arch*

Specification. ISI recommends in IS: 2118–1962 the following specifications for the design and construction of jack arch floor or roof:

1. The deflection of the beams shall not exceed $\frac{1}{480}$ of the span.
2. The thickness of arch shall be not less than 100 mm.
3. The tie rods shall be not less than 12 mm in diameter.
4. The tie rods shall be placed generally not less than 75 mm above the bottom of the joist. The spacing of the tie rods shall not exceed 20 times the width of rolled steel beams.

The tie rods are designed to resist side thrust found as follows :

$$T = \left(\frac{WL}{8R}\right)$$

where, F = Side thrust per metre width of arch
W = Live and dead load per metre width of the arch
L = Span of the arch
R = Rise of the arch.

Design of jack arch consists of design of

(i) thickness of arch, (ii) the rods, and; (iii) rolled steel beams.

Example 6.14 *Design a jack arch roof for a room of 6 metres × 4.5 metres. Total uniformly distributed load including self weight of the roof is 9.750 kN/square metre. Allowable compressive stress is masonry is 0.3 N/mm².*

Solution

Design of Jack arch :

Step 1 : Load supported

Provide rolled steel beams on span of 4.5 m, at a regular interval of 1.5 m.

Span of jack arch = 1.5 m

$$\text{Provide rise of the arch} = \left(\frac{1}{12}\times \text{span}\right) = \left(\frac{1}{12}\times 1\cdot 5\times 1000\right)$$

$= 125$ mm

Load per metre width for one span of arch

$$W = (1 \cdot 2 \times 1 \times 9 \cdot 75) = 14.625 \text{ kN/m}$$

Step 2: Side thrust from the arch

$$T = \left(\frac{WL}{8R}\right) = \left(\frac{14 \cdot 625 \times 1 \cdot 50 \times 1000}{8 \times 125}\right)$$

$$= 21.938 \text{ kN}$$

Let t be the thickness of arch at the crown, A be the area of arch ring per metre width and a_c be the allowable compressive stress in masonry

$$A \times \sigma_c = T$$

$$(1000 \times t \times 0.3) = (21.938 \times 1000)$$

$$\therefore \quad t = 73.125 \text{ mm}$$

Provide minimum thickness = 100 mm

Step 3 : Rolled steel beam

Clear span of beam = 4.5 m

Assume bearing of 300 mm on either sides of beam

Effective span of the beam is the distance between centre of bearing

= 4.80 m

Uniformly distributed load w = 4.625 kN/m

Step 4 : Maximum bending moment, M occurs at the centre.

$$M = \left(\frac{wl^2}{8}\right) = \left(\frac{14 \cdot 625 \times 4 \cdot 8 \times 4 \cdot 8}{8}\right) = 42.12 \text{ m-kN}$$

Step 5 : Maximum shear force, F occurs at the support

$$F = \frac{wl}{2} = \left(\frac{14 \cdot 625 \times 4 \cdot 8}{2}\right) = 35.1 \text{ kN}$$

Step 6 : Section modulus required

$$Z = \left(\frac{42 \cdot 12 \times 1000 \times 1000}{165}\right) = 255.273 \times \text{mm}^3$$

Step 7 : Properties of trial section

From steel section Table, try LB 250, @ 0.279 kN/m

Thickness of web $t_w = 6.1$ mm

Depth of section $h = 250$ mm

Width of section $b = 125$ mm

Section modulus $Z_{xx} = 297.4 \times 10^3$ mm^3

Moment of inertia $I_x = 3717.8 \times 10^4$ mm^4

Step 8 : Check for shear

$$\text{Average shear stress} = \left(\frac{F}{h \cdot t_w}\right) = \left(\frac{35 \cdot 10 \times 1000}{251 \times 61}\right) = 23.02 \text{ N/mm}^2$$

$< (0.4 \times 250 = 100$ N/mm^2). Hence, safe.

Step 9 : Check for deflection

Maximum deflection of beam

$$y_{max} = \frac{5}{384} \times \left(\frac{14 \cdot 625 \times 4 \cdot 8 \times (4 \cdot 8)^3 \times (1000)^3 \times 1000}{2 \cdot 047 \times 10^5 \times 3717.8 \times 10^4} \right) \text{mm}$$

$$= 13.283 \text{ mm}$$

Allowable deflection $= \left(\frac{1}{480} \times 4800 \right)$

$$= 10 \text{ mm}$$

LB 250, @ 0.279 kN/m fails in deflection

Increase the moment of inertia of beam

Moment of inertia of beam required

$$\left(\frac{3717 \cdot 8 \times 13 \cdot 283 \times 10^4}{10} \right) = 4860 \times 10^4 \text{ mm}^4$$

Provide LB 275, @ 0.330 kN/m

Step 10 : Design of tie rods

Spacing of the tie rods shall not exceed 20 times width of the beam

(i.e., 20 × 140) = 2800 mm

Provide the rods at 2.0 m. Thrust resisted by tie rod

$$= (21.938 \times 2.0)$$
$$= 43.876 \text{ kN}$$

Allowable tensile stress in tie rods upto 20 mm diameter

$$= (0.6 \times 250) = 150 \text{ N/mm}^2$$

Net area required $= \left(\frac{43 \cdot 876 \times 1000}{150} \right)$

$$= 292.51 \text{ mm}^2$$

Provide tie rods 25 mm in diameter (ISRO 25).

6.16 CRANE GANTRY GIRDER

The overhead travelling cranes are used in factories and workshops to lift the heavy materials, equipments etc., and to carry them from one place to the other. These cranes are either hand operated or electrically operated. The crane consists of bridge spanning the bay of the shop. A trolley or crab is mounted on the bridge. The trolley moves along the bridge. The bridge as a whole moves longitudinally on rails provided at the ends. The rails on either side of the bridge rest on **crane gantry girders.** The rails, supporting the crane and resting over gantry girders are shown in Fig. 6.20 (a) and in partial plan in Fig. 6.20 (b).

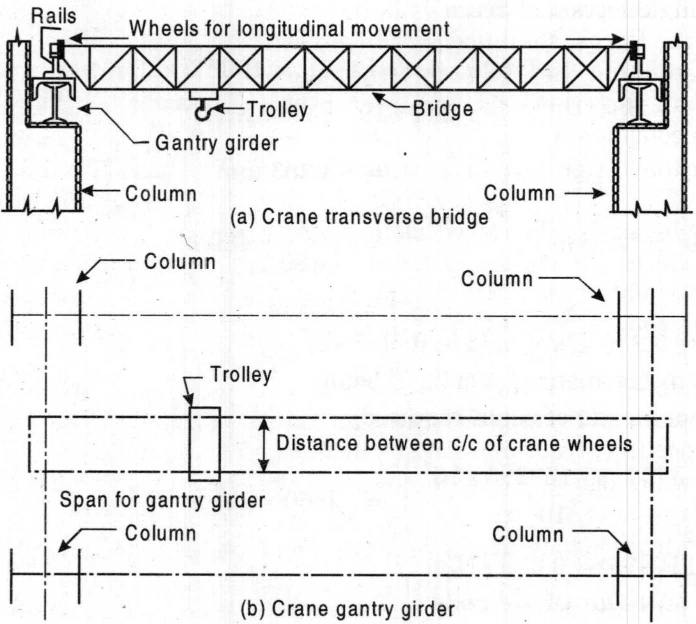

Fig. 6.20

The gantry girders are the girders which support the loads transferred through the travelling (moving) wheels of the cranes.

The effect of cranes to be considered under the imposed loads shall include the following:

1. The vertical loads from the crane.
2. The eccentricity effects induced by the vertical loads, and the impact factors.
3. Lateral (surge) thrust across the crane rail, and
4. Longitudinal horizontal thrust along the crane rail.

The crane loads to be considered shall be as indicated by the customer. In the absence of any specific indications, the load combinations shall be as follows:

(a) The vertical loads with full impact from loaded crane or two cranes in case of tandem operation together with the vertical loads, without impact, from as many loaded cranes as may be positioned for maximum effect along with maximum horizontal thrust (surge) from one crane only or two cranes in case of tandem operation;

(b) For the multibay multicrane gantries the loads specified above, subject to consideration of cranes in maximum of any two bays of the building cross-section;

(c) The longitudinal thrust on a crane track rail shall be considered for the maximum of two loaded cranes on the track; and

(d) The lateral thrust (surge) and the longitudinal thrust acting respectively across and along the crane rail shall not be assumed to act simultaneously. The effect of both the forces, shall however be investigated separately.

While investigating the effect of earthquake forces the resulting effect from dead loads of all the cranes parked in each bay positioned for the maximum effect shall be considered.

The crane runaway girders supporting bumpers shall be checked for bumper impact loads.

The stress developed due to secondary effects such as handling, erection, temperature effects, settlement of foundations shall be appropriately added to the stresses calculated from the combination of loads as recommended by IS: 800–1984. The total stresses thus calculated shall be within the permissible limits.

While considered the simultaneous effects of vertical and horizontal surge loads of cranes, the permissible stresses for the design of crane gantry girders and their supporting structures may be increased by 10 percent.

Where the wind load is the main acting on the structure, no increase in the permissible stresses is allowed.

The gantry girder is to be designed for the following additional loads given in Table 66 as per IS: 875–1984.

The gantry girder is designed on the assumption that either of the horizontal forces transverse to rails or along the rails act at the same time as the vertical loads including the impact load. The horizontal forces act at the rail level. The gantry girder is subjected to bending in vertical plane as well as in horizontal plane along with twisting. The design calculations are simplified by providing a *channel at the top flange* of the girder, and neglecting the bottom flange for transverse load computations. The transverse loads are comparatively small and this simplification in design calculations does not result in serious error. The design example discussed here is for simply supported gantry girder, the compression flange is subjected to horizontal and vertical loads. The horizontal forces act in the plane of the top flange and in the direction perpendicular to it. Therefore, the top flange is strengthened by providing a channel section as shown in Fig. 6.20 to take the horizontal forces in the plane of this flange. The *channel* section provides flange areas to resist bending in horizontal plane due to horizontal forces acting in transverse direction. It increases moment of inertia about *xy*-axis. The flanges of channel section resist the bending in the horizontal plane. The bending of the crane gantry girder occurs about vertical axis as well as about the horizontal axis of the member. The actual bending stresses for bending of the girder in the vertical and horizontal planes are computed. The combined bending compressive stress is taken as the sum of the two calculated fibre stresses. The combined bending compressive stress should be less than or equal to the allowable bending compressive stress.

$$(\sigma_{bc.x.\text{cal}} + \sigma_{bc.y.\text{cal}}) \leq \sigma_{bc}$$

where,

$s_{bc.x.\text{cal}}$ = Actual bending compressive stress in vertical plane

$\sigma_{bc.y.\text{cal}}$ = Actual bending compressive stress in horizontal plane

σ_{bc} = Allowable bending compressive stress.

Table 6.6

Type of load	*Additional load*
1. VERTICAL LOAD	
(a) for electric overhead cranes	25 percent of maximum static wheel loads
(b) for head operated cranes	10 percent of maximum static wheel loads
2. HORIZONTAL FORCES TRANSVERSE TO RAILS	
(a) for electric overhead cranes	10 percent of weight of trolley and weight lifted on the crane
(b) for hand operated cranes	5 percent of weight of trolley and weight lifted on the crane.
3. HORIZONTAL FORCES ALONG THE RAILS	
	5 percent of weight of trolley and weight lifted on the crane

For detailed classification of cranes, the code of practice for design, manufacture, erection and testing (structural portion of cranes and hoists IS: 807–1976) may be referred.

The additional load for overhead travelling cranes has been given in Table 6.6 as the percentage of the maximum static wheel load. The maximum wheel load is the reaction on a wheel due to the total load given by weight of crane plus crab and the lifted load.

The term **'maximum static wheel load'** needs explanation. The minimum approach of crane hook is the distance from the gantry girder upto which the crab or trolley may approach. This extreme position of the crab with respect to the span of crane gives the maximum reaction on one of the gantry girders. This maximum reaction is distributed equally among the crane wheels. The reaction on each wheel is termed as the *maximum static wheel load.*

The allowable bending compressive stresses for bending in horizontal plane is equal to the allowable bending stresses in tension. The allowable bending compressive stress for bending in vertical plane is reduced in proportion of critical stress in bending.

IS: 800–1984 recommends that the allowable stress in axial tension, axial compression, and bending stresses, and allowable stresses for rivets are increased by 10% for the design of gantry girder for the combination of vertical and horizontal loads as discussed above. This increase in allowables stresses is not in addition to that allowed for erection loads with or without wind or seismic forces.

The bridge of a crane as a whole moves longitudinally on rails provided at the ends and mounted on the crane gantry girders. The sections of rails are known as crane rails sections. These sections are designated by the letter ISCR followed by the head width of the rail section in millimeters. The dimensions of crane rail sections are given in Table 6.7 and are shown in Fig. 6 .21 (a) as per IS: 3443–1966 (Indian Standard Specifications for Crane Rail Sections). The sectional properties of the crane rail sections are given in Table 6.8. The dimensions corresponding to Table 6.7 are shown in Fig. 6.21 (b). The dimensions of crane rail section ISCR 140 are shown in Fig. 6.21 (c).

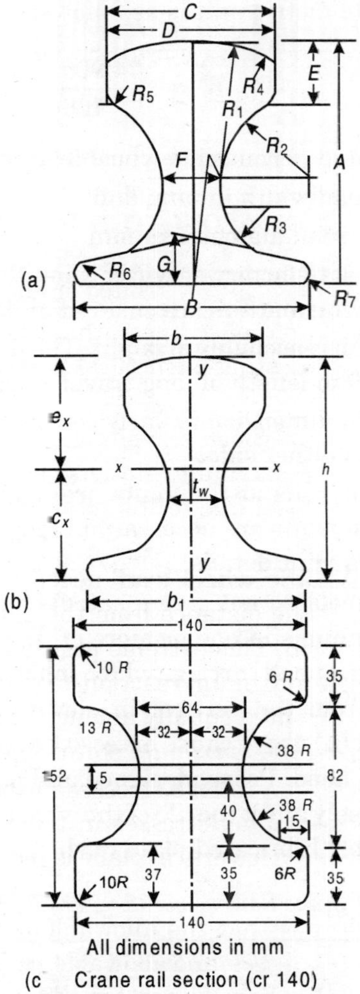

Fig. 6.21 *Crane rail section*

In selecting a crane gantry track rail for a given application on the basis of the crane manufacturer's specified maximum wheel load (and the wheel diameter) and by working to a table of maximum permissible wheel loads as recommended by some authority for a range of rails, it is desirable to remember that the maximum wheel load occurs rarely in service. The maximum wheel load is reached when the load on the hook of the crane is at full rated capacity and when the trolley is at the limit of its traverse motion. These two conditions rarely occur together. Actual wheel loads in service conditions (and with the crane on its long travel motion) normally lie in the range of 50 to 75 percent of the specified possible maximum wheel load. The various formulae are used in determining maximum wheel load. One such formula in use in U.K. for steel mill cranes converted to metric unit is as follows:

$$W = \left(\frac{4.90a \cdot D}{1000}\right)$$

where, W = Maximum permissible wheel load in kN

a = Rail head width in mm, and

D = Wheel tread diameter in mm

The crane gives the satisfactory services when the gantry girder is properly designed and the foilowing matters are ensured in its subsequent maintenance:

(i) The laid track is reasonably straight. The deviation should not be more than 5 mm for every 10 m length of long travel track.

(ii) The track span dimension is fairly constant. The variation should not be more than 10 mm on either side.

(iii) The ends of the rails at the joints are held in close and hard contact except where expansion joints are necessary in long gantries which themselves have definite expansion joint details.

Where rails and wheels do not give a satisfactory life it will generally the consequence of shortcomings in one or more of (i), (ii) and (ii) and/or defect in the crane wheel bearing, or end carriage which make the wheels out of alignment in plan or displaced from the vertical in elevation. These defects and also shortcomings in (i) and (ii) above cause excessive wear of the wheel flanges and also on the sides of rail head. Poorly designed and poorly maintained rail joints will not normally adversely affect the life of the wheels or even the wheel bearings if these later are suitably lubricated plain bush type.

Table 6.7 *Dimensions of Crane Rails as per IS : 3443–1996*

Desig-	*Dimension in mm*														
nation	*A*	*B*	*C*	*D*	*E*	*F*	*G*	*H*	R_1	R_2	R_3	R_3	R_5	R_6	R_7
CR 50	90	90	55	50	25	20	20	9.70	300	26	18	6	6	5	1.5
CR 60	105	105	65.5	60	27.5	24	22	9.82	350	32	20	6	6	5	1.5
CR 80	139	130	87	80	35	32	26	9.75	400	44	26	8	8	6	1.5
CR 100	150	150	108	100	40	38	30	11.25	450	50	30	8	8	8	2.0

However, bed rail joints shorten considerably the life of antifriction bearings in crane-wheels. Also bad rail joints shorten the life of the rails. Quite often poorly designed and poorly maintained rail joints deteriorate to a point where it is deemed desirable to replace along lengths of rails in which there is no defect whatsoever, except at the joints.

Table 6.8 *Sectional properties of crane rail section as per IS : 3443–1966*

Designation	Area	Weight	Head width (b)	Bottom width (b_w)	Overall Height of section (h)	W_{cb} thickness Narrowest Point (t_w)	c_2	e_x	Moment of Inertia		Section modulus		
									I_{xx}	I_{yy}	$Z_{x1} = \left(\frac{I_{xx}}{c_x}\right)$	$Z_{x2} = \left(\frac{I_{xx}}{c_x}\right)$	$Z_y = \left(\frac{I_{yy}}{b_{1/2}}\right)$
(1)	(2)	(3)	(4)	(5)	(6)	(7)	(8)	(9)	(10)	(11)	(12)	(13)	(14)
	mm^2	N/m	mm	mm	mm	mm	mm	mm	cm^4	cm^4	cm^3	cm^3	cm^3
CR 50	3802	298.5	50	90	90	20	432	468	357.54	111.42	82.76	76.40	24.76
CR 60	5099	4003	60	105	105	24	483	567	654.60	195.88	13552	115.45	37.31
CR 80	8113	6389	80	130	130	32	643	657	1547.40	482.39	240.65	235.52	74.21
CR 100	11332	889.6	100	150	150	38	760	740	286473	940.98	376.94	387.12	125.46

6.16.1 Limitation of Vertical Deflection

When the crane structure is subjected to vibration or shock impulses, it may be desirable to maintain reasonable deflection limits such as will produce a stiff structure less apt to vibrate and shake appreciably. For example, the excessive deflection in crane runway support girder will lead to uneven up and down motion of the crane as it proceeds. The impact stresses in such cases would be increased.

In case of crane runway girder, the maximum vertical deflection under dead and imposed loads as per the recommendations of IS: 800–1984, shall not exceed the following values:

(i) Where the cranes are manually operated and to similar loads $\left(\frac{L}{500}\right)$

(ii) Where the cranes are overhead travelling and operated $\left(\frac{L}{750}\right)$ electrically upto 500 kN

(iii) Where the electric overhead travelling cranes operated over $\left(\frac{L}{1000}\right)$ 500 kN

(iv) Other moving loads, such as charging cars, etc. $\left(\frac{L}{600}\right)$

where, L is the span of crane runway girder.

6.16.2 Limitation of Horizontal Deflection

The horizontal deflection at the column cap level of columns supporting crane runway girders in the building shall not exceed as may be specified by the purchaser.

Example 6.15 *Design a simply supported gantry girder to carry one electric over head travelling crane.*

Crane capacity	=	*300 kN*
Weight of crane excluding trolley	=	*190 kN*
Weight of trolley	=	*100 kN*
Minimum approach of crane hook	=	*1.2 metres*
Distance between centres of crane wheel	=	*3.5 metres*
Distance between centres of crane wheel	=	*18 metres*
Span of gantry girder	=	*6 metres*
Weight of rail section	=	*0.300 kN/m*
Height of rail section	=	*75 mm*

Solution

Design :

Step 1 : Maximum wheel load

Weight of trolley + lifted load = (300 + 100) = 400 kN

The weight of crane (excluding trolley) 190 kN acts as uniformly distributed live load as shown in Fig. 6.22.

The vertical reaction on each wheel of crane would be maximum, when trolley is at nearest distance to trolley girder as shown in Fig. 6.22.

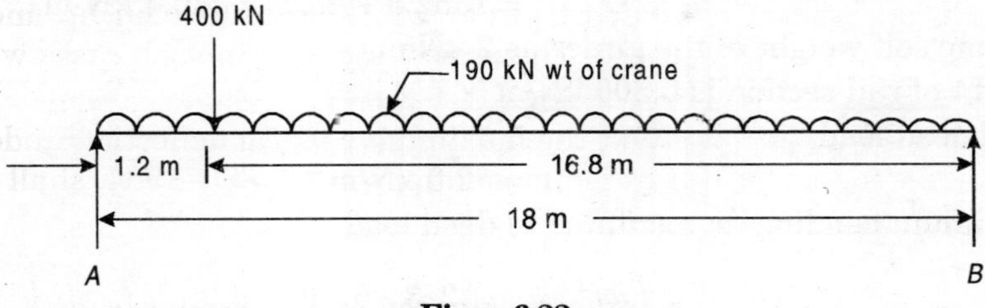

Fig. 6.22

Take moment about B, then reaction at A

$$R_A = \frac{1}{18}\left[400 \times 16 \cdot 8 + 190 \times \frac{18}{2}\right] = 468 \text{ kN}$$

This vertical load at one end of the crane bridge is transferred to the gantry girder through two wheels.

Maximum vertical load on each wheel of crane

$$= \left(\frac{1}{2} \times 468\right) = 234 \text{ kN}$$

Step 2 : Maximum bending moment (due to *D.L.* + *L.L.* + *J.L.*)

The maximum bending moment in the gantry girder under a moving load occurs when the line of action of that load and C.G. of the loads are at equal distance from the centre of span. That is,

$$EC = CF = 0.875 \text{ (Fig. 6.23)}$$

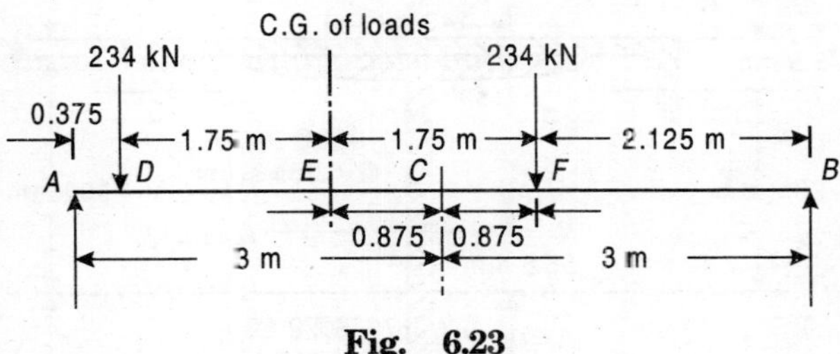

Fig. 6.23

The reaction at the supports A and B are as follows :

$$R_A = 234 \times \frac{1}{6}\left[(6 - 0.375) + 2.125\right]$$

$$= 302.23 \text{ kN}$$

$$R_B = (2 \times 234 - 302 \cdot 23)$$

$$= 165.77 \text{ kN}$$

Maximum bending moment due to moving load

$$M_F = (165.77 \times 2.125) = 352.3 \text{ kN}$$

Add 25 per cent impact moment viz., 88.1 kN-m

(1) **Live load moment**

$$= (352.3 + 88.1) = 440.4 \text{ kN-m}$$

Assume self-weight of the girder as 2 kN/m

Weight of rail section is 0.300 kN/m

Total dead load

$$= 2.3 \text{ kN/m}$$

Maximum bending moment due to dead load

$$\left(\frac{wl^2}{8}\right) = \left(\frac{2{\cdot}3 \times 6 \times 6}{8}\right) = 10.35 \text{ kN-m}$$

(2) **Dead load moment** $= 10.35$ kN-m

(3) **Total vertical moment** $= (440.4 + 10.35) = 450.75$ kN-m

Assume allowable bending compressive stress

$$= (0.66 \times 250) = 165 \text{ N/mm}^2$$

The section modulus required for bending moment is vertical plane (approximately)

$$Z = \left(\frac{450{\cdot}75 \times 1000 \times 1000}{165}\right)$$

$$= 2731.8 \times 10^3 \text{ mm}^3$$

From steel section Tables, try WB 600, @ 1.337 kN/m and LC 300, @ 0.331 kN/m.

The section of the gantry is shown in Fig. 6.24.

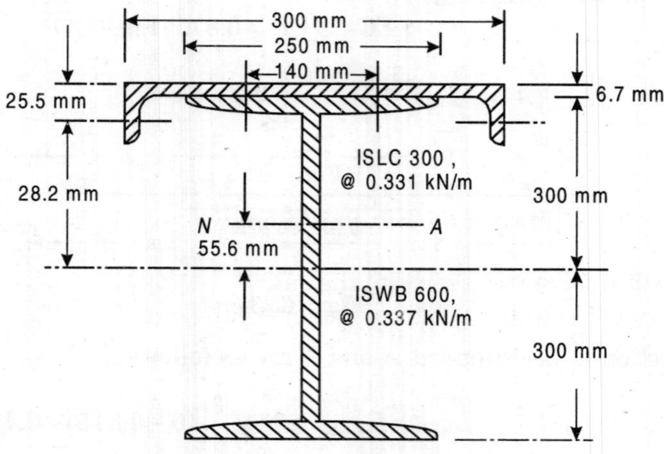

Fig. 6.24 *Gantry girder*

Sectional area of beam section is 17038 mm^2

Section area of channel section is 4211 mm^2

∴ Total section area is 21249 mm^2

Thickness of flange of beam section, t_f is 21.3 mm

Let y be the distance of neutral axis of built-up section from neutral axis of beam section

Moment of inertia of built-up section about xx-axis

$$I_{yy}\text{ (gross)} = [106198.5 + 170.38 \times 5.56^2 + 346 + 42.11 \times (28.12 - 5.56)^2] \times 10^4 \text{ mm}^4$$

$$= 133334.5 \times 10^4 \text{ mm}^4$$

Moment of inertia of built-up section about yy-axis

$$I_{yy}\text{ (gross)} = [4702.5 + 6047.9] \times 10^4$$

$$= 10750.4 \times 10^4 \text{ mm}^4$$

Bending stress due to vertical loading

Actual bending compressive stress for vertical loading

$$\sigma_{bcx.\text{cal}} = \left(\frac{450\cdot 75 \times 1000 \times 1000 \times 251\cdot 1}{133334\cdot 5}\right)$$

$$= 84.8867 \text{ N/mm}^2$$

Actual bending tensile stress for vertical loading

$$\sigma_{bcx.\text{cal}} = \left(\frac{450.75 \times 1000 \times 1000 \times 355.6}{122224.5 \times 10^4}\right)$$

$$= 119.4 \text{ N/mm}^2$$

$$< (1.10 \times 165 = 181.5 \text{ N/mm}^2)$$

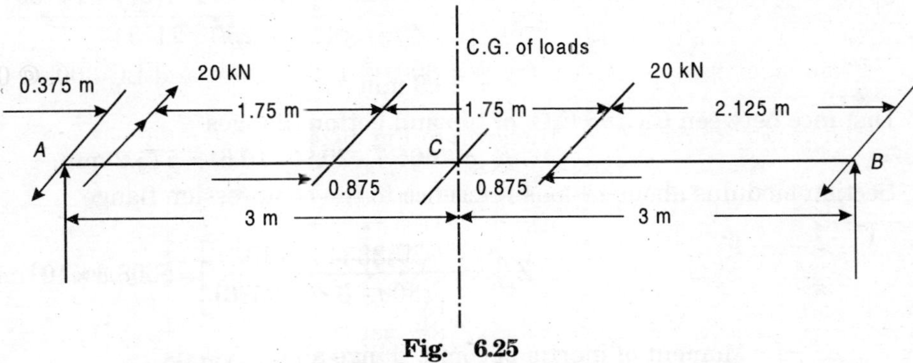

Fig. 6.25

Step 3 : Maximum bending moment due to horizontal (transverse) force

Horizontal force transverse to the rail

= 10 per cent of (weight of trolley + lifted load)

$$= \frac{1}{10} \times (300 + 100) = 40 \text{ kN}$$

Horizontal force transverse to the rail on each wheel or crane

= 20 kN

Horizontal reaction at support A (Figs. 6.23 and 6.25)

$$= \frac{20}{234} \times 302\cdot 33 = 25.83 \text{ kN}$$

Horizontal reaction at support B

$$= \mathbf{14.17\ kN}$$

Horizontal moment

$$14.17 \times 2.125 = 30.1 \text{ kN-m}$$

Step 4 : Bending moment in horizontal plane

$$\text{Horizontal moment} = 30.10 \text{ kN-m}$$

The moment of inertia of compression flange about yy-axis (considering I_{yy} of compression flange of beam section as half of that for beam section)

$$I_{yy} = \left[6047 \cdot 9 + \frac{1}{2} \times 4702 \cdot 5\right] \times 10^4$$

$$= 8399.6 \times 10^4 \text{ mm}^4$$

Bending compressive stress in horizontal plane (Bottom flange is neglected)

$$\sigma_{bc.y.cal} = \left(\frac{30 \cdot 1 \times 1000 \times 1000 \times 150}{8399 \cdot 6 \times 10^4}\right)$$

$$= 53.58 \text{ N/mm}^2$$

Step 5 : Allowable stress in horizontal plane

Let $\bar{y}_1$ be the distance of compression flange from top fibre

$$\bar{y}_1 = \left[\frac{4211 \times 25 \cdot 5 + 250 \times 21 \cdot 3(6 \cdot 7 + 10 \cdot 65)}{4211 + 250 \times 21 \cdot 3}\right]$$

$$= 20.9 \text{ mm}$$

Distance between C.G. to C.G. of top and bottom flanges

$$h = (605.7 - 20.9 - 10.6) = 575.2 \text{ mm}$$

Section modulus about xx-axis reference to the compression flange

$$Z_{xx} = \left[\frac{133334 \cdot 5 \times 10^4}{(300 + 6 \cdot 7 - 55 \cdot 6)}\right] = 5308.8 \times 10^3 \text{ mm}^3$$

$$\omega = \left(\frac{\text{Moment of inertia of comp. flange about } yy\text{-axis}}{\text{Moment of inertia of built-up section about } yy\text{-axis}}\right)$$

$$\omega = \left(\frac{8399 \cdot 6 \times 10^4}{10750 \cdot 4 \times 10^4}\right) = 0.78$$

From IS: 800–1984, $k_1 = 0.28$

Effective length of compression flange = 6000 mm

Radius of gyration of the complete section about yy-axis

$$r_y = \left(\frac{10750 \cdot 4 \times 10^4}{21249}\right) = 79.58 \text{ mm}$$

Slenderness ratio $= \left(\dfrac{6000}{79 \cdot 58}\right) = 75.39$

Overall depth, $D = 606.7$ mm

Mean thickness of flange

$T = (t_f = 21.3 + 6.7) = 28.0$

Ratio $\left(\dfrac{D}{T}\right) = 21.668$

From IS: 800–1984, Table 6.5

$X = 632.02$ and $Y = 503.27$

From Eq. 6.18, the elastic critical stress

$$f_{cb} = k_1 (X + k_2 . Y) \left(\frac{c_2}{c_1}\right)$$

$$= 1.0\,(632.02 + 0.28 \times 503.27) \times \left(\frac{306 \cdot 7}{300}\right)$$

$$= 790.20 \text{ N/mm}^2 \text{ (MPa)}$$

Let the value of yield stress for the structural steel be 250 N/mm²

Ratio $\left(\dfrac{T}{t_w}\right) = \left(\dfrac{28}{11 \cdot 2}\right) = 2.5 > 2.0$

$\therefore f_{cb}$ is not increased by 20 per cent From IS: 800–1984, Table 6.2

$\sigma_{bc} = 145$ N/mm²

Step 6: Check for combined bending compressive stress in extreme fibre

$(\sigma_{bcx.\text{cal}} + \sigma_{bcy.\text{cal}}) = (84.498 + 53.58)$

$137.98 \text{ N/mm}^2 < 1.1 \times 145 = 159.5 \text{ N/mm}^2$

Hence design is safe and satisfactory.

Step 7 : Horizontal (longitudinal) force along the rails

5% of the static wheel load $= \left(\dfrac{1}{20} \times 2 \times 234\right) = 23.4$ kN

Height of rail = 75 mm

Bending moment in the longitudinal direction

$= 23.4 \times (75 + 251.1) = 7630.74$ mm-kN

Stress in longitudinal direction

$$\left(\frac{P}{A} + \frac{M}{Z}\right) = \left(\frac{23 \cdot 4 \times 1000}{21249} + \frac{7630 \cdot 74 \times 1000}{5308 \times 104}\right) \text{N/mm}^2$$

$(1.10 + 14.376) = 2.538$ N/mm² (Very small)

Shear force

Maximum shear force in the gantry girder

$$\left(234 + 234 \times \frac{2 \cdot 5}{6 \cdot 0}\right) = 331 \text{ kN}$$

Add 25% for impact = 82.75 kN

Dead load shear $= \left[\frac{(1337 + 331)6}{2 \times 1000}\right] = 5.61$ kN

Total shear = 419.36 kN

Intensity of horizontal shear stress per mm length

$$f_y = \left(\frac{FQ}{I}\right) \quad (Q = A\bar{y})$$

Consider the portion of web of flange only.

Area = (6.7 × 300) = 2000 mm^2

From *NA*, $\bar{y} = 251 \cdot 1 - \frac{1}{2} \times 6 \cdot 7 = 247.75$ mm

$$\tau_{va} = \left(\frac{419 \cdot 36 \times 2000 \times 247 \cdot 75 \times 1000}{13334 \cdot 5 \times 10^4}\right)$$

$= 155.84$ N/mm^2

Step 8 : Rivet value

Use 22 mm diameter power driven rivets.

Strength of power driven rivets in single shear

$$\left(\frac{\pi}{4} \frac{(23 \cdot 5)^2 \times 100}{1000}\right) = 43.35 \text{ kN}$$

Strength of rivet in bearing

$$\left(23 \cdot 5 \times \frac{6 \cdot 7 \times 300}{1000}\right) = 47.235 \text{ kN}$$

Rivet value, R = 43..35 kN

Pitch of rivets $= \left(\frac{43 \cdot 35 \times 1000}{155 \cdot 84}\right)$

= 278.17 mm

Rivets are provided in two lines

2.*p* = 556.34 mm

Maximum allowable pitch in compression

= (12 × 6.7) = 80.4 mm

Provide rivets at 80 mm pitch throughout the length of gantry girder.

6.17 FILLER JOIST

The steel beams of light sections are termed as joists. The joists placed in plain cement concrete are called *filler joists.* When filler joists are continuous over more than two supports, then, these are called *continuous fillers.* Whenever possible, continuous fillers may extend over three spans. The joists are supported by main steel beams as shown in Fig. 6.26.

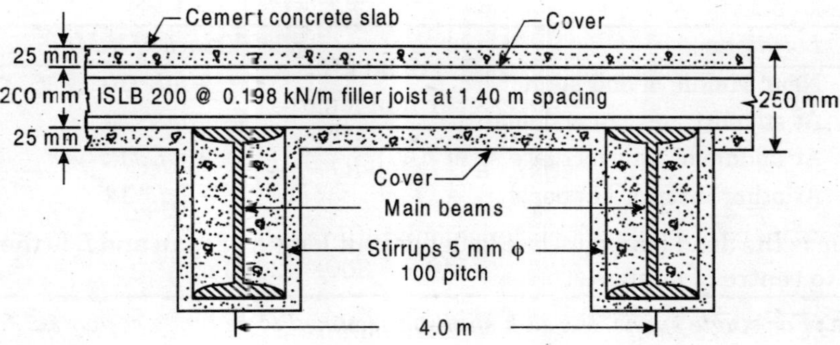

Fig. 6.26 *Continuous fillers*

The joists in continuous fillers are made over supports in such a manner to preserve considerable measure of continuity. The joints are staggered so that full continuity of the fillers occurs in 50 per cent of those crossing any support, such a continuity being evenly displaced.

When filler joists spans between two main steel beams only, then, these are called *discontinuous fillers.* Discontinuous filler joists are shown in Fig. 6.27. These are connected to main beams by means of cleat angles. Overall depth of

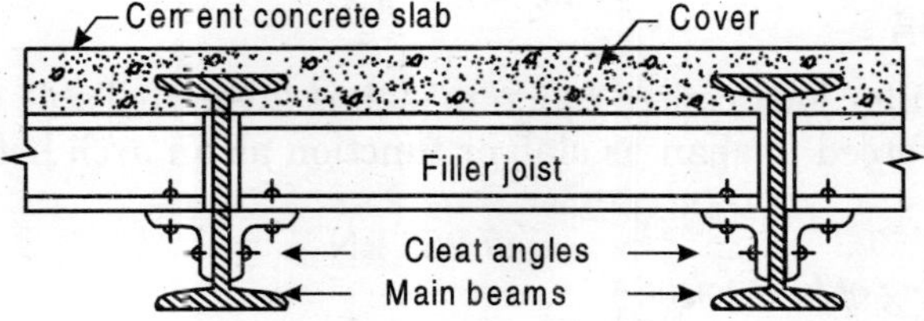

Fig. 6.27 *Discontinuous filler*

construction in case of discontinuous filler joist is small than that of continuous filler joist construction.

The bottom of filler joists remains flush with cement concrete slab.

The design of filler joists is done as per following specifications.

6.17.1. Bending Moment

The bending moments on slabs of which filler joists form part, shall be calculated to satisfy the following :

(a) As slabs spanning continuously over supports and subjected to those combinations of dead and live loads producing the maximum positive and negative moment.

(b) In the case of three or more approximately equal spans of continuous filler joists described above, as slabs designed for uniformly distributed loading, satisfying moment value given in Table 6.9.

Table 6.9 *Bending moment values for filler joists*

	Locations	*Values of B.M.*
1.	Near Middle of end span	$+ wL^2/10$
2.	At support, next to end span	$- wL^2/10$
3.	At middle of intermediate span	$+ wL^2/12$
4.	At other interior supports	$- wL^2/12$

where w is the dead load plus live load per unit length of span and L is the span, centre to centre of supports.

Note: *For single spans, the slab shall be assumed to be freely supports.*

The spans are considered approximately equal when the longest span does not exceed the shortest span by more than 15 percent.

6.17.2 Moment of Resistance

The moment of resistance of the slab shall be calculated from the section properties of the filler joists only except that where the filler joists are completely embedded in a solid concrete slab, the moment of resistance of slab may alternatively be calculated as a composite reinforced concrete section.

6.17.3 Spacing

The spacing of filler joist centre concrete slab is given in Table 6.10 unless the concrete is reinforced to span as slab or function as an arch between the filler joists.

Table 6.10 *Spacing of filler joists*

	Imposed load per m²	*Volume of n*
1.	Upto 2.50 kN	9
2.	Above 2.50 kN upto 5.0 kN	8
3.	Above 5.0 kN upto 7.50 kN	7
4.	Above 7.50 kN upto 10 kN	6
5.	Above 10 kN	5

Where a slab of concrete is designed to function as an arch between the filler joist, the thrust from the arch section shall be taken up by steel tie or by other means so as not to cause appreciable increase of stress in the joists. The concrete shall have a working strength of not less than 15 N/mm^2 at 28 days.

6.17.4 Thickness of Concrete

Where the underside of the concrete is arched between the filler joists, the thickness at the crown shall be not less than 50 mm. The thickness of structural concrete over hollow blocks shall nowhere be less than 30 mm when the filler joists exceed 450 mm centre to centre and nowhere less than 50 mm when the spacing of filler joists is greater than 450 mm centre to centre.

6.17.5 Bending Stress

The bending stress in filler joists other than those designed as part of reinforced concrete slab, shall not exceed (165 + 0.62 t) N/mm^2, where t is an allowance, the value of which is equal to the thickness in mm of the structural concrete cover to the compression flange of the filler joists, provided that the allowance t is applied only in cases where the filler joists are embedded at least flush with underside of their bottom flanges in a solid concrete slab throughout and that any cover less than 25 mm and in excess of 75 mm shall be neglected.

In cases where the underside of the slab is flush with the bottom flanges of the filler joists, the allowance t shall not apply in respect of support moments; nevertheless if the top flange is covered, the allowance may be made in calculating to the resistance sagging moment.

6.17.6 Shear Stress

The shear stress shall be calculated as taken entirely on the filler joists and stress shall not exceed the allowable stress in shear, i.e., 0.4 f_y.

6.17.7 Bearing Stress

The bearing stress shall be calculated as taken entirely on the filler joist and stress shall not exceed the allowable stress in bearing i.e., 0.75 f_y.

6.17.8 Span Depth Ratio

The span of filler joist centre to centre of supports, shall not exceed 35 times the depth from the underside of the joist to the top of the structural concrete or 12 times this depth in the case of cantilever fillers.

6.18 ENCASED BEAMS

When a steel beam is encased in cement concrete throughout the entire length, it is called **encased beam**. The cased beam is shown in Fig. 6.28.

A steel beam is embedded in cement concrete for two purposes. It may be necessary to make a steel frame building fire resisting. The beams may be encased in cement concrete for architectural requirement. As per IS: 800–1984 beams may be designed as encased beams when the following conditions are fulfilled :

(a) The section is of single web and I-shape or of double open channel formed with the web not less than 40 mm apart.

(b) The overall dimensions of the steel section do not exceed 750 mm × 450 mm overplating where used, the larger dimensions being measured parallel to web (as per IS: 800–1962).

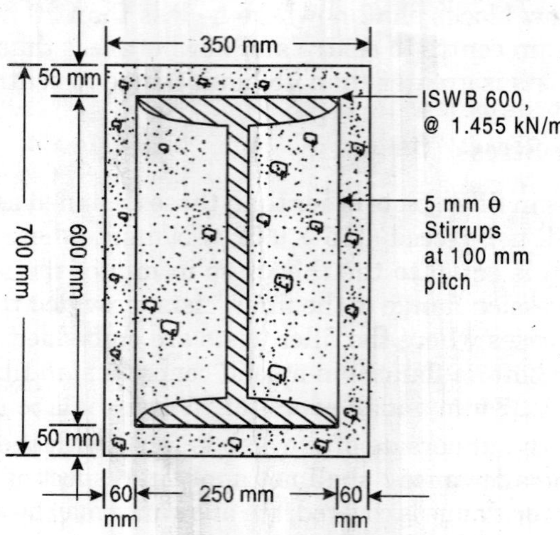

Fig. 6.28 *Cased beam*

(c) The beam is unpainted and it is solidly encased in ordinary dense concrete, with 10 mm aggreagate (unless solidity can be obtained with a large aggregate) and of grade designation M 15, min (IS: 800–1978).

(d) The minimum width of solid casing is equal to $(b_0 + 100)$ mm where b_0 is the width of the steel flanges in mm.

(e) The surface and edges of the flanges of the beam have a concrete cover not less than 50 mm.

(f) The casing is effectively reinforced with steel wire of at least 5 mm diameter and the reinforcement shall be in the form of stirrups or binding at not more than 150 mm pitch, and so arranged as to pass through the centre of the covering to the edges and soffit of the lower flange.

6.18.1 Design of Encased Beam

The steel beam section shall be considered as carrying the entire load but allowance may be made for the effect of concrete on the lateral stability of the compression flange. This allowance should be made by assuming for the purpose of determining the permissible stress in compression, that the equivalent moment of inertia, I_y about yy-axis is equal to Ar_y^2, where A is the steel section and r_y may be taken as $0.2\,(b_0 + 100$ mm). The width of flange b_0, is taken in mm. The other properties required for calculating critical bending stress for Eq. 6.11 may be taken as for uncased section. The permissible bending stress so determined shall not exceed 15 times that permitted for the uncased sections.

It is noted that this does not apply to beams and girders having a depth greater than 1000 mm or a width greater than 500 mm or to box sections.

Example 6.16 *A filler joist is continuous over three equal spans. The distance between centre to centre of main beams is 4 m. The floor consists of 250 mm thick plain cement concrete slab. Unit weight of plain cement concrete is 24 kN/m³. The live load on floor is 5 kN/m². Design filler joist.*

Solution

Design :

Step 1 : Load supported by joist

Distance between centre to centre of main beam = 4 m

Thickness of plain cement concrete slab = 250 mm

Assume, thickness of top cover = 25 mm

Thickness of bottom cover = 25 mm

Therefore, thickness of structural concrete slab = 200 mm

Live load on floor = 5 kN/m²

Therefore, spacing of filler joist centre to centre

n × thickness of structural slab = 7 × 20 = 1400 mm

Dead load

Weight of plain cement concrete inclusive of cover at top and bottom

$$\left(\frac{24 \times 4 \times 1\cdot 400 \times 200}{1000}\right) = 26.88 \text{ kN}$$

Self-weight of filler joist assuming 0.200 kN/m

= (0.200 +27.70 × 4) = 0.800 kN

Total dead load = (26.81 + 0.80) = 27.680 kN (say Ω 27.70)

Live load

Total live load = (5 × 4 ×1.40) = 28 kN

Total dead load plus live load

= (28 + 27.70) = 55.70 kN

Step 2 : Maximum bending moment

The maximum bending moment occurs at middle of end span

$$= \left(\frac{55\cdot 70 \times 4}{10}\right) = 22.28 \text{ kN-m}$$

Allowable bending stress

$(165 + 0.62\, t) = (165 + 0.62 \times 25) = 180.5 \text{ N/mm}^2$

Step 3 : Required section modulus for filler joist

$$Z = \left(\frac{22\cdot 28 \times 10^6}{180\cdot 5}\right) = 126.32 \times 10^3 \text{ mm}^3$$

From ISI Handbook No. 1, try LB 200 @0. 198 kN/m

$$Z_{xx} = 169.7 \times 10^3 \text{ mm}^3$$

Step 4 : Check

Span of filler joist = Centre to centre of main beams (4 m)

Depth from the underside of the joist to the top of the structural concrete

= 225 mm

Span depth ratio $= \left(\frac{4 \times 100}{22 \cdot 5}\right) = 17.75$

$<$ 35. Hence design is satisfactory.

Provide LB 200, @ 0.1908 kN/m filler joist

5 mm in diameter stirrups are provided at 100 mm pitch.

Complete design is shown in Fig. 6.26.

Example 6.17 *An encased beam carries total uniformly distributed load of 40.00 kN/m. The effective span of beam is 8 m. The ends of beams are restricted against torsion, and not restrained against lateral bending. Design the encased beams.*

Solution

Design :

Step 1 : Load supported by beam

Total uniformly distributed load is 40 kN/m

Step 2 : Maximum bending moment, M occurs at the centre

$$M = \left(\frac{40 \cdot 0 \times 8 \times 8}{8}\right) = \mathbf{32\ kN\text{-}m}$$

Step 3 : Maximum shear force, F occurs at the support

$$F = 160.00 \text{ kN}$$

Step 4 : Selection of trial section

Select an uncased rolled steel beam section such that the section of beam provides required section modulus corresponding to the allowable bending compressive stress for the uncased section for the given span.

Select WB 600, @ 1.451 kN/m

$$\sigma_{bc} = (0.66 \times 250) = 165 \text{ N/mm}^2$$

Step 5 : Section modulus required

$$Z = \left(\frac{32 \cdot 00 \times 1000 \times 1000}{165}\right)$$

$$= 193.94 \times 10^3 \text{ mm}^3$$

Step 6 : Properties of trial section

From ISI Handbook No. 1 for WB 600, @ 1.451 kN/m

Sectional area	$A = 18486 \text{ mm}^2$
Depth of section	$= 6000$ mm
Thickness of flange	$t_f = 23.6$ mm
Width of flange	$b = 250$ mm
Section modulus	$Z = 3854.2 \times 10^3 \text{ mm}^3$

Section modulus provided is greater than section modulus required. Hence, safe.

For encased beam, $r_y = 0.2\,(b_0 + 100)$ mm

$= 0.2\,(250 + 100) = 70$ mm

Moment of inertia of cased beam

$$I_y = Ar_y^2 = (18486 \times 70^2)$$
$$= 9020 \times 10^4 \text{ mm}^4$$

Distance between C.G. of flanges

$$= (600 - 23.6) = 576.4 \text{ mm}$$

Step 7: Elastic critical stress in bending

From IS: 800–1984, $K = 249.3 \times 10^4$ mm^4

$$\left(\frac{D}{T}\right) = \left(\frac{600}{23 \cdot 6}\right) = 25.42$$

$$T = t_f = 23.6 \text{ mm}$$

Ratio, $$\left(\frac{T}{t_w}\right) = \left(\frac{23 \cdot 6}{11 \cdot 8}\right) = 2.0$$

Ratio, $$\left(\frac{l}{r_y}\right) = \left(\frac{8000}{70}\right) = 114.28$$

Ratio, $$\left(\frac{d_1}{t_w}\right) = \left(\frac{600 - 2 \times 46 \cdot 05}{11 \cdot 8}\right) = 43$$

$$\left(\frac{1344}{\sqrt{f_y}}\right) = \left(\frac{1344}{\sqrt{250}}\right) = 85, \left(\frac{d_1}{t_w}\right) > \left(\frac{1344}{\sqrt{f_y}}\right)$$

The value of f_{cb} is to be increased by 20 percent From IS: 800–1984, Table 6.5

$$X = 289.18 \text{ and } 7 = 204.2$$

From Eq. 6.18 $$f_{cb} = k_1(X + k_2 T)\left(\frac{c_2}{c_1}\right) \quad \text{...(i)}$$

c_1 and c_2 the smaller and the larger distances from the neutral axis to the extreme fibres are equal

$$k_1 = 1.0 \text{ for } \varphi = 1.0, \text{ and } k_2 = 0.0 \text{ for } \omega = 0{\cdot}5$$

Substituting the respective values in expression (i)

$$f_{cb} = 1.0\,(289.18 + 0.0 \times 204.2)\ 1.00$$
$$= 289.18 \text{ N/mm}^2$$
$$f_{cb} = (1.20 \times 20 \times 289.18)$$
$$= 347.02 \text{ N/mm}^2 \text{ (MPa)}$$

Step 8 : Permissible stress in bending. The permissible stress in bending in compression from Table 6.2

$$\sigma_{fc} = 116.4 \text{ N/mm}^2$$

From IS: 800–1984, allowable bending compressive stress

$$= 165 \text{ N/mm}^2$$
$$\not> (1.5 \times 116.4 = 174.6 \text{ N/mm}^2). \text{ Hence, safe.}$$

Provide WB 600, @ 1.451 kN/m. The beam is cased in concrete. The cement concrete provides a 50 mm cover on all sides. 5 mm diameter stirrups are provided at 100 mm pitch ($\not>$ 150 mm). The section of cased beam is shown in Fig. 6.28. The overall dimensions of steel section are not greater than the allowable dimensions.

PROBLEMS

6.1. A laterally supported beam LB 350, @ 0.495 kN/m has been placed on two supports. Determine the safe uniformly distributed load per metre length which can be placed over the beam for an effective span of 6.30 m.

6.2. Design a beam of 5 metres span carrying a uniformly distributed load of 30 kN per metre including self-weight and a concentrated load of 10 kN at the centre. The beam is simply supported at both ends.

6.3. A simply supported beam of 12 metres span is made of MB 600, @ 1.226 kN/m. The associated properties of the beam are

Depth = 600 mm
Width of flange = 210 mm
Thickness of flange = 20.8 mm
Thickness of web = 12 mm

I_{xx} = 91813 ×10^4 mm^4, I_{yy}= 2651 × 10^4, Z_{xx} = 3060.4 × 10^3 mm^3

Determine that total uniformly distributed load that this beam can carry. The top flange is fully restrained laterally.

6.4. A beam carries a uniformly distributed load 44.00 kN/m. The effective span of beam is 6.40 m. The ends of beam are restrained against torsion and not restrained against lateral bending. Design a suitable section of a rolled steel beam.

6.5. Two LB 600, @ 0.995 kN/m joists at 300 mm with one 36 mm thick plate, 600 mm wise on each flange is used as a simply supported beam on a span of 12 m. Find the maximum uniformly distributed load per metre run which can be allowed on this built-up girder. Neglect rivet holes.

6.6. Design a simply supported plated rolled steel beam section to carry a uniformly distributed load 36.00 kN/m inclusive of self-weight of beam. The effective span of beam is 7.20 m, the depth of beam is limited to 500 mm. The beam is restrained against lateral deflection.

6.7. Design a beam of 6 m span carrying uniformly distributed load of 10 kN/m and two concentrated loads of 80 kN each at 2 m and 4 m, respectively from the left support. Only available rolled steel section is ISMB 350, @ 0.524 kN/m. The compression flange of the beam is laterally supported.

6.8. A beam of 6 m span carries uniformly distributed superimposed load of 75 kN/m and laterally unsupported. (i) Design the central section of the beam using an MB 600 and suitably chosen cover plates attached to both the flanges.

6.9. Design a beam of 5.4 m span carrying a uniformly distributed load of 30 kN/m and two concentrated loads of 50 kN each at third points. The depth of the section is restricted to 350 m. Assume the compression flange to have lateral restraint.

6.10. A freely supported beam spanning 12 m is made up of an I-section MB 500, @ 0.869 kN/m with two pates in each flange of 200 mm × 12 m each. Calculate the maximum uniformly distributed load the beam can carry if the compression flange is fully restrained. Assume the flange connections to be made with 18 mm ϕ rivets. Show calculations for the curtailment of the outermost plate in each flange.

6.11. A plated steel beam of span 12 metres effective is to carry a uniformly distributed load of 20 kN per metre inclusive of its own weight. Design a suitable plated beam including minimum spacing of rivets using any one of the three section listed below. Plate of 10 mm thickness is available in any width. The flange of beam should not be wider than 300 mm. Assume that 22 mm diameter rivets are used for the connection.

MB 500, @ 0.869 kN/m. LB 600, @ 0.995 kN/m

LB 500, @ 07.50 kN/m.

6.12. A simply supported beam of effective span 8 metres is required to carry a.u.d.l. of 50 kN/m including self-weight over the entire span. Check whether the LB 450, @ 0.653 kN/m with the following properties is suitable for the purpose. (Area = 8314 mm²; b_f = 170 mm; t_j = 13.4 mm; t_w = 8.6 mm; I_{xx} = 27566.1 × 10⁴ mm⁴). If not, design a safe section using 150 mm × 10 mm thick plates.

Design a suitable welded connection between top plate and flange of the joint.

Assume that the compression flange is adequately restrained laterally.

6.13. Design a jack arch roof for a hall of size 8 m × 4 m. The self-weight of the roof may be taken as 8.5 kN/m². The permissible compressive stress in the brick masonry is 0.3 N/mm².

6.14. A jack arch roof is to be constructed for a room measuring 5.5 m × 4.2 m. The live loaded acting is 4 kN/m². Design the roof assuming appropriate self-weight for the roof.

6.15. Design a gantry girder to carry an electric overhead travelling crane to suit following data:

Crane capacity	... 200 kN
Wt. of crab alone	... 70 kN
Wt. of crane	... 150 kN
Minimum approach of crane hook	... 1.2 m
Dist. between centres of crane wheels	... 3.5 m
Dist. between cranes of gantry girders	... 15.0 m
Span of gantry girder	... 7 m
Wt. of rail section	... 0.300 kN/m
Height of rail section	... 80 mm

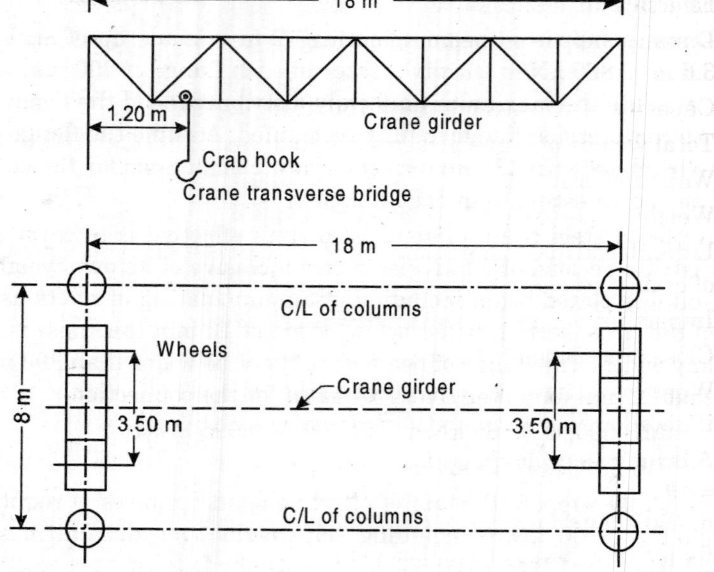

Fig. P 6.16

6.16. A gantry girder is provided for mill building to carry an electric overhead travelling crane with the following data as shown in Fig. P 6.16.

(i) Centre to centre distance between columns (span of gantry girder) = 8 metres

(ii) Centre to centre distance between gantry girders = 18 metres

(iii) Crane capacity = 300 kN

(iv) Self-weight of crane hook = 30 kN

(v) Self weight of crane hook = 30 kN

(vi) Allowable approach of crane hook from the vertical axis of the gantry girder = 1.2 metres

(vii) Distance between centres of wheels, moving on the gantry girder and supporting the crane girder = 3 .5 metres

(viii) Weight of rail section attached on the top of the gantry girder = 0.300 kN/ mètre

Determine :

(a) The maximum B.M. and S.F. produced by the vertical loads

(b) The màximum B.M. and S.F. produced by the lateral loads

(c) The maximum longitudinal force exerted along the rails.

Also design the section of gantry girder.

6.17. A gantry girder is composed of an MC 550 × 150 mm, @ 1.037 kN/m and a channel MC 300 × 90 mm, @ 0 .358 kN/m placed on the top of the joist with its flange down. Calculate the maximum flange and web stresses in the girder section, given the following:

(i) Effective span of gantry girder = 7.3 m

(ii) Effective span of crane girder between centres of gravity of rails = 17 m

(iii) Distance between the pair of carriage wheels moving on each gantry rail = 3.6 m

(iv) Capacity of overhead crane = 150 kN

(v) Total weight of crane girder excluding the crab = 180 kN

(vi) Weight of crab = 50 kN

(vii) Weight of gantry girder inclusive of rail = 1.80 kN/m

(viii) Lateral load due to horizontal surge = 10 percent of lifted weight and weight of crab

(ix) Impact for vertical wheel loads = 25 percent

(x) Closest position of crane hook from the centre of gantry rails = 1 mm

(b) What would have been the effect on the vertical moment and shear diagrams if an allowance for longitudinal load were also taken.

6.18. A beam of simple framed building is made up of LB 550, @ 0.863 kN/m (b_f = 190 mm, t_f = 15 mm, A = 10.97 mm², t_w = 9.9 mm) and carries a total load of 400 kN uniformly distributed over the entire span. It is connected to the flange of column at ends of section WB 600, @ 04.81 kN/m b_f = 200 mm, t_f = 10 mm, A = 6133 mm². t_w = 7.4 mm). Using 20 mm ϕ rivets, design a suitable connection for the joints. Draw a neat sketch of the joint.

6.19. A beam has an effective span of 10 m and carries a uniformly distributed load of 100 kN. Design the cross-section of the beam if its depth is limited to 45 mm.

6.20. The cross-section of a built-up beam consists of an LB 325 with a flange plate of 1200 mm × 10 mm provided at the top as well as at the bottom. It uses 18 mm diameter rivets.

It is simply supported over an effective span of 6 m. Calculate the maximum uniformly distributed load the beam can carry, if the beam is laterally supported throughout its length along the compression flange. Consider bending moment only.

Chapter 7 Design of Beam Connections

7.1 INTRODUCTION

A steel structure is a framework of assembly of structural members. The beams are connected to beams or columns (stanchions). The connections are necessary to attach the members and to allow an orderly flow of the load to the foundation by continuing the transfer of loads from the adjoining members. The connections are designed adequately. The joints are made safe economical and practical. More practical connections are not more economical as fabrication cost greatly influences the economy of both connections and members. The connections must be designed with almost care, so that the failure of joint does not occur. A joint may fail earlier than the failure of a member. When the failure of member occurs, it is likely to allow time for safety to be undertaken. A failure of a joint will be more catastropic. The factor of safety used for the design of connections is kept more than that for the members to be connected. The transfer of loads through the joint is made gradual and the joint is not overloaded. Various factors for example, length of the joint, edge distance, distribution of fasteners, gauge distance, distribution of stress in the fasteners, length of fasteners themselves, factor of safety, etc. influence the design of connections.

The frame connections are classified according to their rotational characteristics as *flexible connections*, *semi-rigid connections* and *rigid connections*.

7.1.1 Flexible Connections

Flexible connections permit large angles of rotation and transmit no negligible moment. Actually a small moment may be developed, but it is ignored in the design. Any joint eccentricity less than about 63 mm is neglected. The deformations and moments in beams with flexible connections are treated as for ideally simply supported beams. The ends of beams are connected to resist shear only. The frameworks with flexible connections are referred as simple frame constructions.

7.1.2 Semi-rigid Connections

These connections allow small end rotation and transmit appreciable moment (moment less than the full moment capacity of connected members). Design of these connections needs an assumption regarding moment capacity. Arbitrarily, the moment capacity is considered as *twenty, thirty,* or *seventy five percent* of member capacity. The frameworks with semi-rigid connections are referred as semi-rigid constructions.

7.1.3 Rigid Connections

Rigid connections do not allow any end rotation and retain a constant relative angle between the connected members under any joint rotation and transmit moment equal to full moment capacity of the members connected. The original angles between the connected members remain unchanged. The connections are fully restrained. The frameworks with rigid connections are referred as rigid frame connections.

The simple beam end connections are of two types :

1. Frame connections and 2. Seated connections.

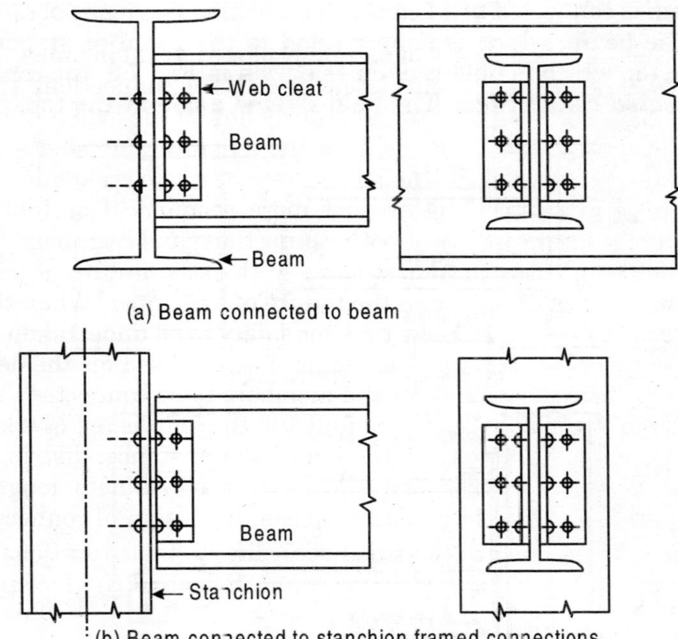

Fig. 7.1 *Framed connections*

7.1.3.1 Framed Connections

When a beam is connected to a beam or a stanchion by mean of two angles riveted to them as shown in Fig. 7.1, the connection is known as *a framed connection*.

In case of framed connections, when the flanges of the beams are to be kept as the same level, the connecting beams are cut as shown in Fig. 7.2.

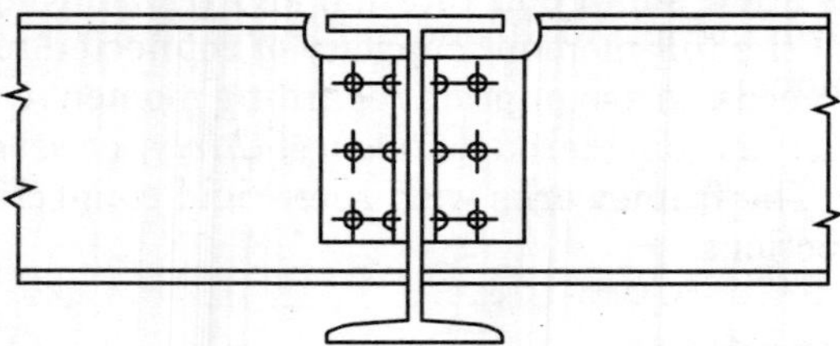

Fig. 7.2 *Flanges of beams at same level (framed connections)*

The framed connections are normally used to connect beams to beams. When the beams are to be connected to webs of stanchions, or to flanges of stanchions, depth of the web or width of the flange of stanchion may be insufficient to accommodate the connecting angles. In such cases, the framed connections are not possible, and therefore, the seated connections are employed.

7.1.3.2 *Seated Connections*

When a beam is connected to a beam or a stanchion by means of an angle at the bottom of the beam, which is shop riveted to the beam or stanchion and an angle at the top which is field riveted as shown in Fig. 7.3, the connections are known as *seated connections*. The field riveted angle at the top, prevents the

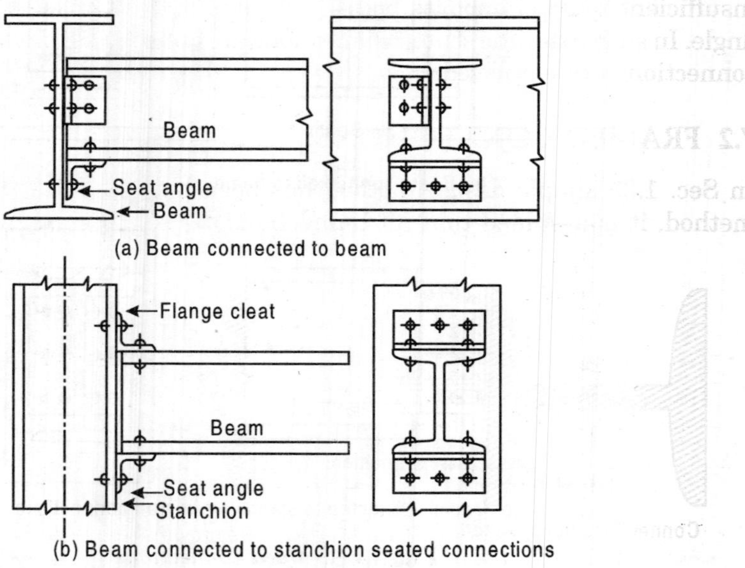

Fig. 7.3

lateral deflection of the beam. The connection as shown in Fig. 7.3 (b) is known as *unstiffened seated connection*. When the end reaction of the beam is large then one or two stiffener angles are provided to support the outstanding leg of the seat angle, as shown in Fig.7.4. The stiffening angles should be tightly fitted under seating angle. Such connections are known as *stiffened seated connections*. The stiffener angles are also used when the vertical leg of seat angle is inadequate to accommodate the required number of rivets. Beams connected to flange of a stanchion are shown in Fig. 7.3 (b) and in Fig. 7.4. Beams can also be connected to webs of stanchions.

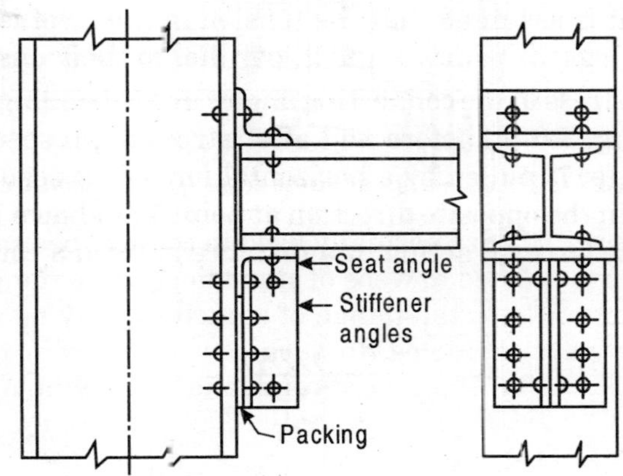

Fig. 7.4 *Stiffened seated connection*

The seated connections are normally used to connect beams to stanchions. When the beams are connected to beams, the depth of beam sections may be insufficient to accommodate both the beams to be connected and the seating angle. In such cases the seated connections are not possible, and therefore framed connections are employed.

7.2 FRAMED CONNECTIONS

In Sec. 1.39 simple design method has been discussed. In the simple design method, it is assumed that all connections of beams to columns are virtually

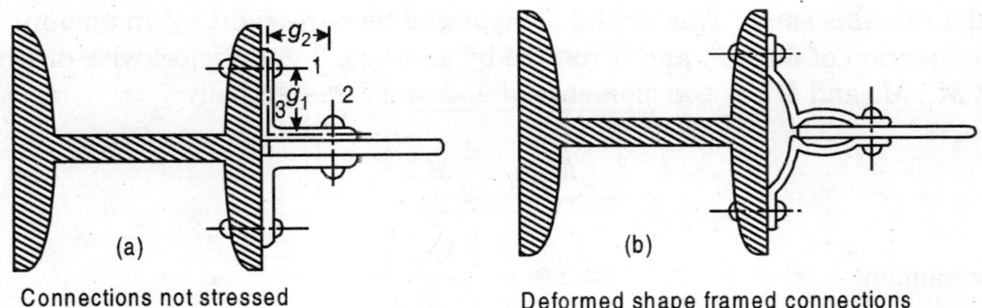

Fig. 7.5

flexible and are proportioned for the reaction shears applied at the appropriate eccentricity. The connections do not offer resistance to rotation of the end of the beam in the vertical plane. If the rotation of end of the beam does not take place fully, moment is induced at the end of the beam and bending stress is caused in the connecting angle. Figure 7.5 shows a section through a *framed connection*. Figure 7.5 (a) shows the connections when these are not stressed. Figure 7.5 (b) shows the deformed shape of connecting angles under severe loading on the tension side of connections. Connecting angles deform more at the top than at the bottom of the connection. This deformation decreases gradually from top towards bottom. It is assumed that the tension in the rivet is sufficient to keep the centre line of legs, at points 1 and 2, parallel to their unstressed positions.

For analysis of stress in the connecting angles due to bending moment, consider Fig. 7.6 (a) and Fig. 7.6 (b) before and after stressing, respectively. The frame shown in Fig. 7.6 (a) is pulled by a horizontal force *P*. It results in a horizontal reaction *P* acting in the opposite direction at point 1 as shown in Fig. 7.6 (b). Two equal and opposite forces *P* acting at a distance g_1 form a couple Pg_1.

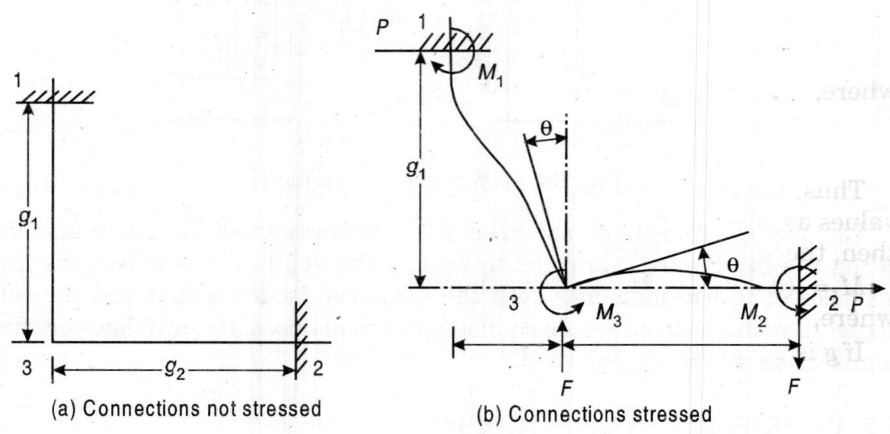

Fig. 7.6

This couple acts in anticlockwise direction. The original direction of frame at 1 and 2 remains same. This joint 3 is displaced towards right by an amount δ, in the direction of force *P*, and it rotates by an angle θ in anticlockwise direction. Let M_1, M_2 and M_3 be the moments at 1, 2 and 3 respectively

$$M_2 = \frac{1}{2}M_3 \quad \text{...(i)}$$

For moment

$$3\text{–}2\ \theta = \left(\frac{M_3 \cdot g_2}{4EI}\right) \quad \text{...(ii)}$$

where, I = Moment of inertia of member 3–2,

g_2 = Length of member 3–2

We have

$$(M_1 + M_3) = P \,.\, g_1$$

$$M_l = (P \,.\, g_1 - M_3) \quad \text{...(iii)}$$

For member 3.1, by area moment method

$$\theta = \left(\frac{M_1 g_1}{2EI} - \frac{M_3 \cdot g_1}{EI}\right)$$

Substitute the value of θ from (iii) and M_1 from (iv)

$$\left(\frac{M_3 \cdot g_2}{4EI}\right) = \frac{g_2}{2EI} \cdot (Pg_1 - M_3) - \left(\frac{M_3 \cdot g_1}{2EI}\right)$$

$$M_3 = Pg_1 \cdot \left(\frac{2g_1}{4g_1 + g_2}\right) \quad ...(7.1)$$

Therefore,
$$M_1 = Pg_1 - Pg_1\left(\frac{2g_1}{4g_1 + g_2}\right)$$

$$M_1 = Pg_1 \cdot \left(\frac{2g_1 + g_2}{4g_1 + g_2}\right) \quad ...(7.2)$$

where, g_1 = Length of member 3–1

I = Moment of inertia of member 3–1, same as moment of inertia of 3–2

Thus, the moment induced at the points 1 and 3 in Fig. 7.5 have the same values as given by Eq. 7.2 and Eq. 7.1, respectively. In case g_1 and g_2 are equal, then, the values of moments induced are given by

$M_1 = (0.6\,P.g_1)$, $M_2 = (0.2P.g_1)$, and $M_3 = (0.4P.g_1)$ ...(7.3)

where, $g_1 = g_2$

If g is gauge distance for leg of angle, then

$$\therefore \quad g_1 = \left(g - \frac{1}{2} \cdot t\right)$$

The difference between the values g_x and g_2 is not significant. The value of bending moment M_1 is maximum in the angle. The connecting angle should be able to resist M_1. The value of pull P should not be greater than maximum allowable tensile force in the rivet.

The tensile force in the rivet connecting the leg 3–2, in Fig. 7.6 (b) is F. This force is given by

$$\therefore F.g_2 = (M_2 + M_3), \qquad F = \left(\frac{3}{2} \cdot \frac{M_3}{g_2}\right) \qquad \left(\because M_2 = \frac{1}{2} M_3\right)$$

Substituting the values of M_2, and taking, $g = g_2$

$$F = 0{\cdot}6\,P, \delta = \left(\frac{1}{2} \cdot \frac{M_1 g_1}{EI} \cdot \frac{2}{3} g_1 \cdot \frac{1}{2} \cdot \frac{M_3 g_1}{EI} \cdot \frac{1}{3} g\right) \quad ...(7.4)$$

Substituting the values of M_1 and M_3 from Eq. 7.3

$$\delta = 0.133\left(\frac{P \cdot g^3}{EI}\right)$$

7.3 UNSTIFFENED SEATED CONNECTIONS

In case of the unstiffened seated connections, seat angle supports a beam. The action of seat angle is approximately similar to that of a cantilever beam. In case, thicker seat angles are used, then, they have a tendency to concentrate the reaction at the toe of outstanding leg. While if thinner seat angle is used, they have a tendency to distribute the reaction over more length. The stiffness of flanges of beam and thickness of web also affect the distribution of reaction. If the connection of beam is not done with the seat angle the bending of seat angle takes place about a critical section. The critical section is at the top of row of rivets. When the beam is loaded, the bottom flange of the beam elongates. It pushes the back of seating angle against the column. As a result of this, the vertical leg of seating angle is relieved from bending moment. The connection of seat angle and beam flange restrains the bending of horizontal leg as a simple cantilever. Therefore, it is difficult to locate the critical section about which bending moment is to be found. Location of critical section is assumed at the edge of the fillet in the horizontal leg. The end reaction is assumed to be uniformly distributed on the outstanding leg of seat angle, which is sufficient to meet the web crippling requirement of beam. The length of beam resisting the web crippling has been found in Sec. 6.9.

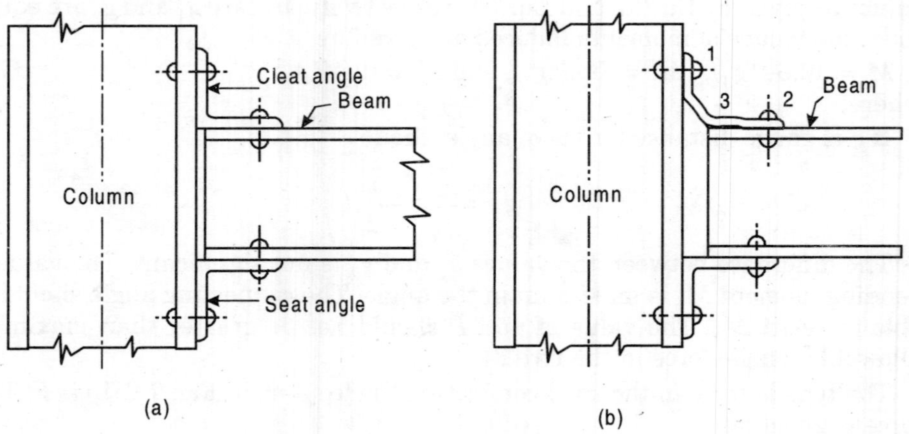

Fig. 7.7

Figure 7.7 (a) shows an unstiffened seated connection before loading. Figure 7.7 (b) shows an unstiffened seated connection after loading. When the rotation of end of beam takes place, then the seat angle is pressed against the column, and cleat angle is deformed as shown in Fig. 7.7 (b). If the rotation of end of beam does not take place fully, moment is induced at the bending stress is caused in the connecting angle.

The analysis of the stress in the connecting angle due to bending is same as that for connecting in framed connection. When the section of connecting angle is considered, it is horizontal plane. The section for the connecting angle (cleat angle) is in the vertical plane. The values of moment M_1, M_2 and M_3 found by Eq. 7.3 hold good at same respective point as in Fig. 7.7 (b).

7.4 STIFFENED SEATED CONNECTIONS

When the end reaction of the beam is large, then one or two stiffener angles are used to support the outstanding leg of the seat angle. The length of outstanding leg of stiffening angle is extended upto the toe of seating angle as far as possible. The beam reaction acts on the outstanding leg of stiffening angle. In case, a single stiffener angle is used, the reaction is eccentric with respect to the rivets connecting the stiffener to the column as shown in Fig. 7.8. The end reaction R, through the stiffener has eccentricity e_{xx} along the xx-axis and e_{yy} along the yy-axis. In addition to direct shear, rivets are subjected to twisting moment, $R \, . \, e_{yy}$ and bending moment, $R. \, e_{zx}$. The twisting moment acts in a plane parallel to YZ- plane, (the plane in which rivets are connected). The rivets are subjected to additional shear due to twisting. The bending moment acts in a plane parallel to XZ plane. The rivets above the neutral axis (the line of rotation) are subjected to tension, and rivets below the neutral axis are subjected to compression against column flange. The rivets connecting stiffener angle to column flange are designed to resist direct shear, twisting moment and bending moment.

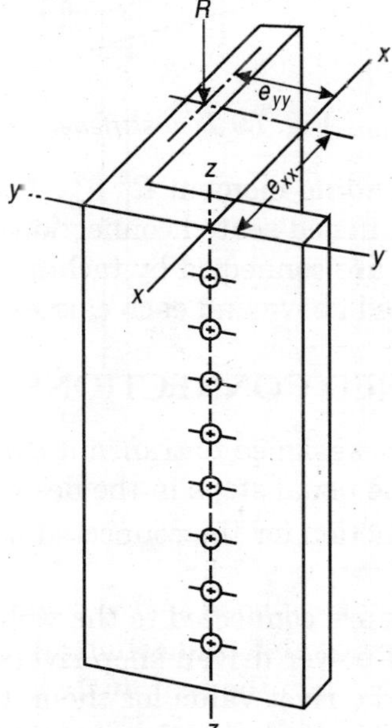

Fig. 7.8 *Single stiffener angle*

In case two stiffener angles are used, their outstanding legs are connected together, by means of tacking rivets. Two stiffener angles act as one unit. The reaction R, transmitted by stiffening angles has eccentricity along xx-axis only as shown in Fig. 7.9. The rivets connecting stiffener angles to the flange of

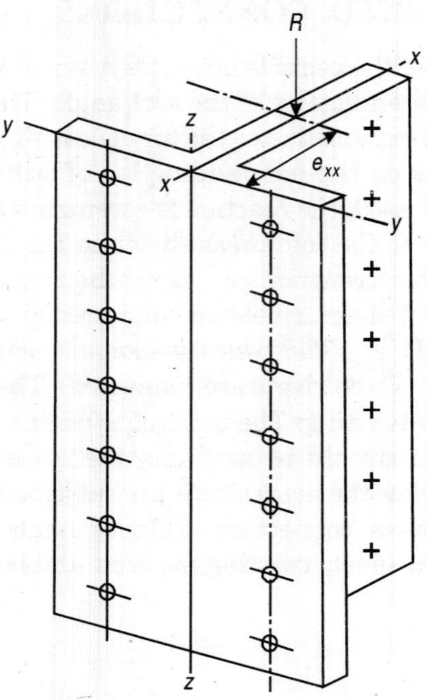

Fig. 7.9 *Two stiffener angles*

column are subjected to bending moment $R \cdot e_{xx}$. The bending moment acts in XZ-plane. The design of stiffened seated connections are discussed in Sec. 7.7. If the two stiffener angles are connected by tacking rivets they act as a single stiffening angle as discussed above and each carries half reaction.

7.5 DESIGN OF FRAMED CONNECTIONS

The framed connections are designed to transmit the end reaction of connected beam. The following are the usual steps in the design of framed connections.

Step 1. Compute end reaction for the connected beam for the loading on the beam.

Step 2. The cleat angles are connected to the web of the connected beam in the workshop by means of power driven shop rivets. Assume the diameter of the rivets and determine the rivet value for them. Compute number of rivets required to transmit the end reaction of the connected beam as under :

$$\text{Number of rivets} = \left(\frac{\text{End reaction of connected beam}}{\text{Rivet value}}\right)$$

The size of cleat angles depends upon whether the rivets can be accommodate in one line or in two lines.

Step 3. The power driven field rivets are used to connect the cleat angles to the supporting beam. Assume the diameter of the rivets and determine rivet value of them. Compute the number of rivets required.

Step 4. The cleat angles are checked for shear stress. The maximum shear-stress should not exceed 0.45 f_y N/mm^2.

Maximum shear stress is found as below.

$$t_{v.max} = \frac{3}{2} \times \left(\frac{F}{2d \times t} \right)$$

where, F = End reaction of the connected beam

d = Depth of cleat angles

t = Thickness of cleat angles.

Example 7.1 *An LB 350 @ 0.495 kN/m transmits an end reaction of 290 kN to the web of an MB 500, @ 0.869 kN/m. Design a framed connection and give a neat sketch. Use 22 mm diameter rivets.*

Solution

Design :

Step 1 : End reaction

End reaction of connecting beam = 290 kN

From steel section tables,

Thickness of web of connecting beam = 7.4 mm

Step 2: Rivet value

Strength of power driven rivets in double shear

$$\left(2 \times \frac{\pi}{4} \times (23.5)^2 \times \frac{100}{1000} \right) = 86.70 \text{ kN}$$

Strength of power driven rivets in bearing

$$\left(23.5 \times 7.4 \times \frac{300}{1000} \right) = 52.17 \text{ kN}$$

$$\text{Rivet value} = 52.17 \text{ kN}$$

Step 3 : Design of connections

$$\left(\frac{290}{52.17} \right) = 5.558. \text{ (Say 8 rivets)}$$

Accommodate the rivets in two rows. Pitch of rivets is 65 mm

Edge distance = 40 mm

Use 2 cleat angles ISA 150 mm × 115 mm × 10 mm (2 ISA 150 115 @ 0.200 kN/m).

Thickness of web of supporting beam, t_w is 10.2 mm

Step 4 : Rivet value

Strength of power driven rivets in single shear

$$\left(\frac{\pi}{4} \times \frac{(23.5)^2 \times 100}{1000} \right) = 43.35 \text{ kN}$$

Strength of power driven rivets in bearing

$$\left(23.5\times10.2\times\frac{300}{1000}\right) = 71.91 \text{ kN}$$

$$\text{Rivet value} = 43.35 \text{ kN}$$

Step 5 : Design of connections

$$\text{Number of rivets} = \left(\frac{290}{43.35}\right) = 6.689. \text{ Say 8 rivets.}$$

Provide 4 rivets for each cleat angle.
Pitch of rivets = 65 mm
Edge rivets = 40 mm

Step 6 : Check for maximum shear stress

Maximum shear stress

$$\left(\frac{3}{2}\times\frac{290\times1000}{2\times10\times275}\right) = 79.09 \text{ N/mm}^2$$

$$< (0.5 \times 250 = 112.5 \text{ N/mm}^2)$$

Hence design is satisfactory.
Design of framed connection is shown in Fig. 7.10.

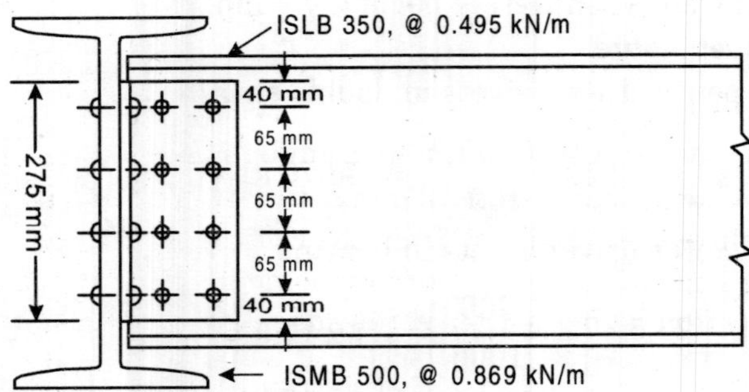

Fig. 7.10 *Beam connected to beam (Framed connection)*

7.6 DESIGN OF UNSTIFFENED SEATED CONNECTIONS

The unstiffened seated connections are used to trasmit end reaction of beam upto 200 kN. The following are the usual steps for the design of unstiffened seated connections:

Step 1. The seat angle is designed to transmit end reaction, F of connected beam. The end reaction, 'F' is considered as uniformly distributed on the outstanding leg of seat angle over a length b as shown in Fig. 7.11. Compute the bearing length b.

The length b is such that web cripping of beam does not occur. The length of beam 'B' resisting the web cripping at the end is found as below :

$$B = \left(\frac{F}{\sigma_{p.tw}}\right), \text{ and also } B = \left(b + \sqrt{3h_2}\right)$$

where, a_p = Allowable bearing stress in web ($0.75f_y$ N/mm²)
t_w = Thickness of web of beam
h_2 = Depth of root of fillet from extreme fibre of flange.

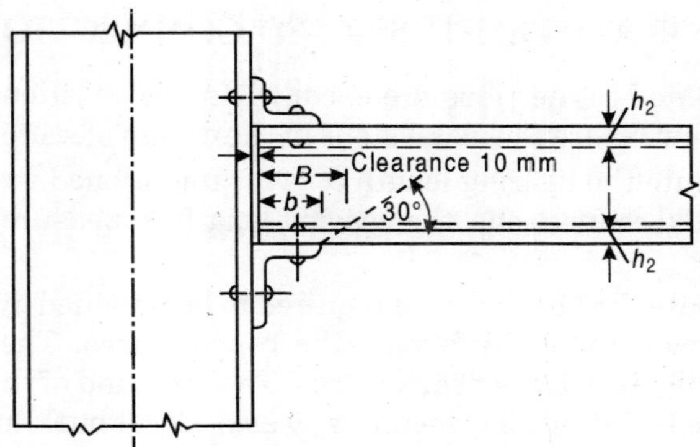

Fig. 7.11 *Beam connected to Stanchion unstiffened seated connections*

The bearing length

$$b = \left(\frac{F}{f_b \cdot t_w} - \sqrt{3h_2}\right) \nless \frac{1}{2}B, \text{i.e.,} \left(\frac{1}{2}\frac{F}{\sigma_p \cdot t_w}\right)$$

Step 2. Select a trial section for seat angle.

Step 3. Provide a clearance of 10 mm between the beam and stanchion. Compute the distance of end reaction from the critical section of the outstanding leg of the seating angle. Calculate bending moment, 'M' at the critical section, XX. The critical section is the section at the edges of the fillet as shown in Fig. 7.12.

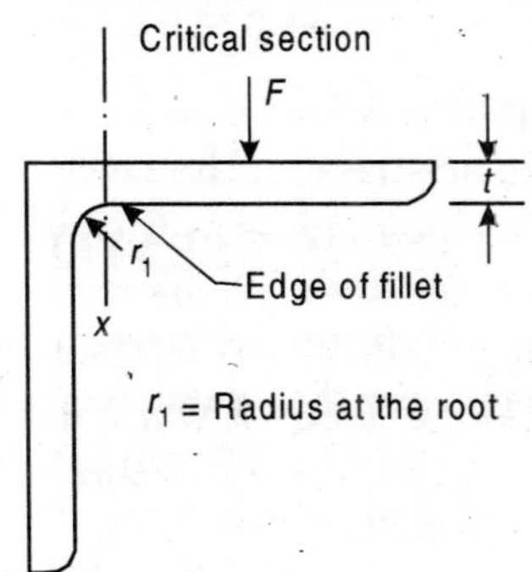

Fig. 7.12 *Critical section for angle section*

Step 4. Assume the length of the seating angle equal to the width of the flange of the beam. Compute moment of resistance of seating angle, which should be greater than bending moment M. Then trial section of seating angle is satisfactory.

Step 5. The seating angle is connected to the stanchion in the workshop by power driven rivets. Assume diameter of rivets and determine the rivet value. Compute the number of rivets required.

Step 6. Connect a cleat angle at the top by two rivets in either leg. The diameter of the rivets should be equal to the diameter of rivets used to connect the seating angle.

7.7 DESIGN OF STIFFENED SEATED CONNECTIONS

The stiffened seated connections are used, when end reaction of beam exceeds 200 kN. The following are the usual steps in the design of stiffened connections :

Step 1. Compute the bearing length b as for unstiffened seated connections. In stiffened seated connections, the bearing length is measured from the end of stiffening leg.

Step 2. Compute the bearing area required to be provided by stiffener angles. Select the stiffener angles to furnish the bearing area. The outstand of the stiffener angles to furnish the bearing area. The outstand of the stiffener angles should not exceed 16 times its thickness to avoid local buckling of the angles.

Step 3. Provide a seating angle depending upon the length of outstanding leg of stiffener angles.

Step 4. The rivets connecting the stiffener angles to the stanchion are subjected to direct shear and tension. It is better to provide two stiffener angles and connect their outstanding legs by stitching rivets. The distance of end reaction is found from the face of the stanchion and bending moment, 'M' is computed. The number of rivets is computed from Eq. 2.12 as below :

$$n = \left(\frac{6M}{p \cdot R}\right)^{1/2}$$

The rivets are checked for shear and tension. The quantity $\left[\frac{\tau_{vf.cal}}{\tau_{vf}} + \frac{\sigma_{tf\cdot cal}}{\sigma_{tf}}\right]$ should be less than or equal to 1.4

where, $\tau_{v}f._{cal}$ = Actual shear stress in the rivet

τ_{vf} = Allowable shear stress in the rivet

$\sigma_{tf.cal}$ = Actual tensile stress in the rivet

σ_{tf} = Allowable tensile stress in the rivet.

Example 7.2 *An MB 500, @ 0.869 kN/m transmits an end reaaction of 130 kN to the flange of stanchion HB 250, @ 0.510 kN / m. Design a unstiffened seated connections. Use 22 mm diameter rivets.*

Solution

Design :

Step 1 : End reaction

End reaction of beam, F = 130 kN

Required bearing length of seat angle

$$b = \left(\frac{F}{\sigma_p \times f_w} - \sqrt{3h_2}\right)$$

From steel section tables, for MB 500, @ 0.868 kN/m

Width of flange = 180 mm

Thickness of web t_w = 10.2 m

h_2 = 37.95 mm

Permissible bearing stress = $0.75 f_y$

σ_p = (0.75 × 250) = 187.5 N/mm²

$$b = \left(\frac{130 \times 1000}{187.5 \times 10.2} - \sqrt{3} \times 37.95\right) = 2.24 \text{ mm}$$

But $$b \nless \left(\frac{1}{2}\frac{F}{\sigma_p \times t_w}\right) \nless \left(\frac{1}{2} \times \frac{130 \times 1000}{187.5 \times 10.2}\right) \text{ mm}$$

$$\nless 33.987 \text{ mm}$$

The bearing length of seating angle

= 33.987 mm

Step 2 : Selection of trial section

Try seating angle ISA

150 mm × 115 mm × 12 mm (ISA 150115 @, 0.238 kN/m)

Radius at roof = 11 mm

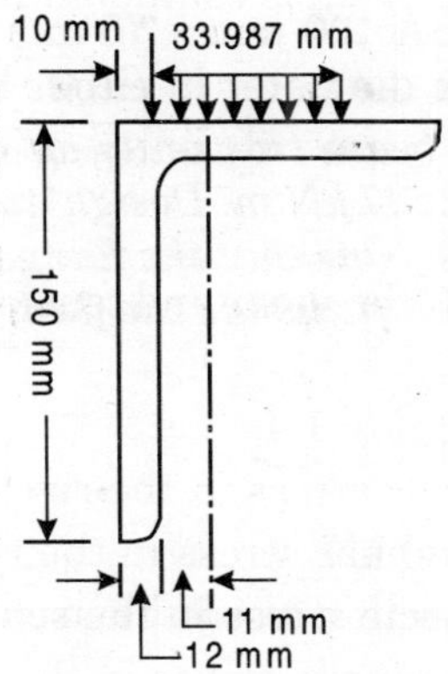

Fig. 7.13 *Bearing length of seat angle section*

The distance of end raction from critical sections

$$\left[\left(10 + \frac{1}{2} \times 33.987) - (12 + 11\right)\right] = 3.994 \text{ mm}$$

Step 3 : Bending moment at the critical section

(3.994 × 130) = 519.155 mm-kN

Length of seating angle

Width of flange of beam = 180 mm

Moment of resistance of seating angle

$$(\sigma_{bc} = 0.66 \times 250 = 165 \text{ N/mm}^2)$$

$$\left(\frac{165 \times 180}{1000} \times \frac{12^3}{6}\right) = 712.8 \text{ mm-kN}$$

Hence seating angle may be provided.

Step 4. Rivet value

Strength of 22 mm power driven rivets in single shear

$$\left(\frac{\pi}{4} \frac{(23.5)^2 \times 100}{1000}\right) = 43.35 \text{ kN}$$

Strength of power driven rivet in bearing

$$\left(\frac{23.5 \times 10.2 \times 300}{1000}\right) = 71.91 \text{ kN}$$

$$\text{Rivet value} = 43.35 \text{ kN}$$

Step 5 : Design of connections

Number of rivets required is $\left(\dfrac{130}{43.35} = 2.999\right)$

Provide 4 rivets in two rows.

Connect top cleat angle ISA 100 mm × 75 mm × 10 mm (ISA 10075 @ 0.130 kN/m with two rivets 22 mm diameter in either leg of the angle).

Example 7.3 *An WB 550 beam transmits an end reaction of 280 kN to the web of a column HB 250, @ 0.547 kN/m. Design and sketch a suitable connection using 22 mm diameter rivets. Note that the flange of the beam will not fit inside the flange of the beam will not fit inside the flange of the column as such.*

Solution

Design :

Step 1: End reaction

End reaction of beam = 280 kN

The stiffened seated connection would be used to connect the beam with the stanchion.

Step 2 : Properties of given section

From the steel section tables, for WB 550, @ 1.125 kN/m

$$\text{Width of flange} = 250 \text{ mm}$$
$$t_w = 10.4 \text{ mm}$$
$$h_2 = 38.30 \text{ mm}$$

For HB 250, @ 0.547 kN/m

$$\text{Width of flange, } b_t = 250 \text{ mm}$$
$$t_w = 8.8 \text{ mm}$$
$$t_f = 9.7 \text{ mm}$$

The flanges of WB 550, @ 1.125 kN/m are cut and fitted as shown in Fig. 7.14.

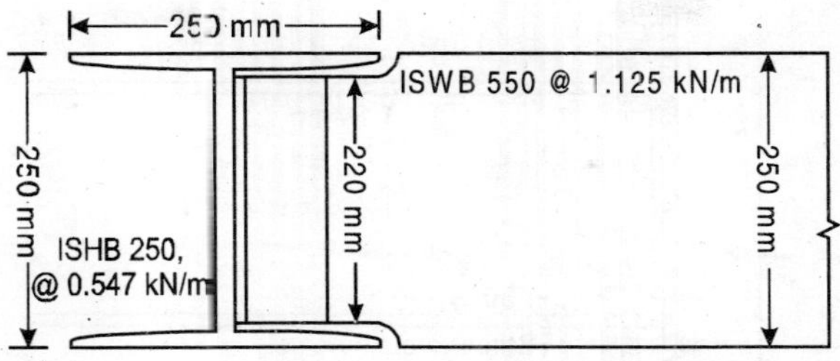

Fig. 7.14 *Beam connected to Stanchion (Stiffened seated connection)*

Step 3 : Permissible bearing stress

$$(0.75 \times 250) = 187.5 \text{ N/mm}^2$$

The bearing length $b = \left(\frac{F}{\sigma_p \times t_w} \sqrt{3h_2}\right)$

$$\left(\frac{280 \times 1000}{187.5 \times 10.5} - \sqrt{3} \times 38.3\right) = 75.885 \text{ mm}$$

But $b \nless \left(\frac{1}{2} \frac{F}{\sigma_p \times t_w}\right) \times \left(\frac{1}{2} \frac{280 \times 1000}{187.5 \times 10.5}\right)$

$$= 71.11 \text{ mm} < 75 \text{ mm}$$

Provide a clearance of 10 mm

Length of outstanding leg of stiffener angle

$$= (75.885 + 10)$$
$$= 85.885 \text{ mm}$$

Provide ISA 100 mm × 75 mm × 12 mm (ISA 10075, @ 0.154 kN/m) for seating angle. Outstanding leg of seating angle is 100 mm.

Step 4 : Bearing area required for stiffener angles

$$\left(\frac{280 \times 1000}{187.5}\right) = 1493.33 \text{ mm}^2$$

Outstanding legs of stiffener angles furnish bearing area.

Provide 2 ISA 90 mm × 90 mm × 10 mm (2 ISA 9090, @ 0.1703 kN/m)

Bearing area furnished

2 (90 –10) × 10 =1600 mm^2 > Bearing area required.

The distance of end reaction of the beam from end of stiffener angle

$$\left(\frac{1}{2} \times 75.885\right) = 37.94 \text{ mm}$$

The distance of end reaction from face of web of stanchion

$$(90 + 12 - 37.3) = 64.7 \text{ mm}$$

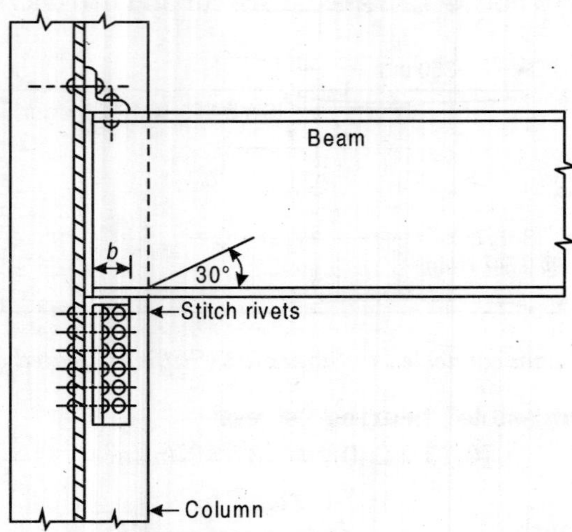

Fig. 7.15 *Beam connected to Stanchion (Stiffened seated connection)*

Step 5 : Bending moment about the face of web

$$\left(\frac{64.7 \times 280}{1000}\right) = 18.116 \text{ m-kN}$$

Step 6 : Rivet value

Strength of 22 mm power driven rivets in single shear

$$\left(\frac{\pi}{4} \times \frac{(23.5)^2 \times 100}{1000}\right) = 43.35 \text{ kN}$$

Strength of rivets in bearing

$$\left(\frac{(23.5) \times 8.8 \times 300}{1000}\right) = 62.04 \text{ kN}$$

Strength of rivets in tension

$$\left(\frac{\pi}{4} \times \frac{(23.5)^2 \times 18}{1000}\right) = 34.68 \text{ kN}$$

$$\text{Rivet value, } R = 34.68 \text{ kN}$$

Step 7: Design of connections

The rivets are provided in two rows at pitch of 70 mm.

Number of rivets required in one row

$$n = \left(\frac{6 \times 18.11 \times 1000}{2 \times 70.0 \times 34.68}\right)^{1/2} = 4.73$$

Provide 5 rivets in each row. Force due to direct shear

$$\frac{280}{10} = 28 \text{ kN}$$

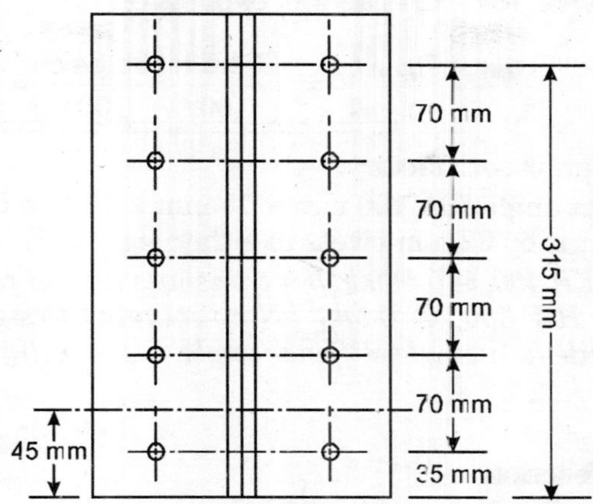

Fig. 7.16

Step 8 : Shear stress

Actual shear stress $\tau_{av.cal} = \left(\dfrac{28\times1000}{\dfrac{\pi}{4}\times(23.5)^2}\right) = 64.59 \text{ N/mm}^2$

Allowable shear stress in power driven rivet

$$\tau_{vf} = 100 \text{ N/mm}^2$$

Step 9 : Moment resisted by rivets in tension

$$M' = \left(\frac{M}{1+\dfrac{2h}{2t}\dfrac{\Sigma y}{\Sigma y^2}}\right)$$

$$\Sigma y = 2[60 + 130 + 200 + 270] = 1320 \text{ mm}$$

$$\Sigma y^2 = 2[(6^2 + 13^2 + 20^2 + 27^2) \times 10^2]$$

$$= 2668 \times 100 \text{ mm}^2$$

$$M' = \left(\frac{18.11}{1+\dfrac{2\times315\times320}{21\times2668\times100}}\right) = 15.785 \text{ m-kN}$$

Step 10 : Tensile force in the top rivets

$$F_t = \left(\frac{15.785\times1000\times270}{2668\times100}\right) = 15.974 \text{ kN}$$

Tensile stress in the rivet,

$$\sigma_{tf.cal} = \left(\frac{15.974\times1000}{\dfrac{\pi}{4}\times(23.5)^2}\right) = 36.847 \text{ N/mm}^2$$

Allowable tensile stress in power driven rivets

$$= 100 \text{ N/mm}^2$$

Step 11 : Check for interaction expression

$$\left[\frac{\tau_{vf.cal}}{\tau_{vf}} + \frac{\sigma_{tf.cal}}{\sigma_{tf}}\right] = \left(\frac{64.59}{100} + \frac{36.847}{100}\right) = 1.0144 \not> 1.4$$

Hence, the design is satisfactory.

Provide top cleat angle ISA 100 mm × 75 mm × 10 mm (ISA 10075, @ 0.130 kN/m) and connect it by 22 mm rivets in either leg.

Example 7.4 *A LB 400, @ 0.569 kN/m transmits an end reaction of 250 kN to the flange of an HB 300, @ 0.588 kN/m column. Design stiffened seated connections. Provide (i) a single stiffener angle, (ii) two stiffener angles.*

Solution

Design :

Step 1: End reaction

The end reaction for the beam is 250 kN

Step 2 : Properties of given section

From ISI Handbook No. 1, LB 400, @ 0.569 kN/m

Width of flange –165 mm

Thickness of web, t_w = 8.0 mm, h_2 = 31.90 mm

For HB 300, @ 0.588 kN/m

Width of flange, b_f = 250 mm

Thickness of web, t_w = 7.6 mm

Thickness of flange, t_f = 10.6 mm

Step 3: Permissible bearing stress

$$\sigma_p = (0.75 \times 250) = 187.5 \text{ N/mm}^2$$

The bending length, b

$$\left(\frac{F}{\sigma_p \cdot t_w} - \sqrt{3}h_2\right) = \left(\frac{250 \times 1000}{187.5 \times 8.0} - \sqrt{3} \times 31.90\right)$$

$$= 111.414 \text{ mm}$$

But,

$$b \nless \left(\frac{1}{2}\frac{F}{\sigma_p \cdot t_v}\right) \nless \left(\frac{1}{2}\frac{250 \times 1000}{187.5 \times 8}\right) \nless 83.33 \text{ mm}$$

Provide a clearance of 10 mm

Length of outstanding leg of stiffener angle

$$= (111.414 + 10)$$

$$= 121.414 \text{ mm}$$

Provide ISA 130 mm × 130 mm × 10 mm ISA 130130, @ 0197 kN/m for seating angle.

Step 4 : Bearing area required for stiffener angle/angles

$$\left(\frac{250 \times 1000}{187.5}\right) = 1333.33 \text{ mm}^2$$

(i) Provide a single stiffener angle ISA 110 mm × 110 mm × 15 mm. (ISA 110 110, @ 0.242 kN/m.

The outstanding leg provides bearing area.

Bearing area provided

$$(110 - 15) \times 15 = 1425 \text{ mm}^2$$

$$> \text{Bearing area required. Hence, safe.}$$

Length of outstanding leg

$$= (110 - 15) = 95 \text{ mm}$$

Thickness of angle, t is 15 mm

Outstanding ratio $\left(\frac{95}{15}\right) = 6.33 < 16$. Hence, safe.

Width of seating angle is 130 mm

Point of application of reaction from face of column

$$\left(\frac{130}{2}\right) = 65.0 \text{ mm}$$

∴ Eccentricity, $e_{xx} = 65.0$ mm.

From ISI Handbook No. 1, the gauge distance for ISA 110 mm × 110 mm × 15 mm for 110 mm leg is 65.0 mm.

Eccentricity, $e_{yy} = 65.0$ mm

Step 5 : Bending moment and twisting moment

$$M_{xx} = R.e_{xx}\left(\frac{250 \times 65.0}{1000}\right)$$

$$= 16.250 \text{ kN-mm}$$

The twisting moment

$$M_{yy} = R.e_{xx} = \left(\frac{250 \times 65.0}{1000}\right)$$

$$= 16.250 \text{ m-kN}$$

Step 6 : Rivet values

Use 22 mm power driven rivets. The strength of rivets in single shear

$$\left(\frac{\pi}{4}\frac{(23.5)^2 \times 100}{1000}\right) = 43.35 \text{ kN}$$

Strength of power driven rivets in bearing

$$\left(\frac{23.5 \times 10.6 \times 300}{1000}\right) = 74.73 \text{ kN}$$

Strength of power driven rivet in tension

$$\left(\frac{\pi}{4} \times \frac{(23.5)^2 \times 100}{1000}\right) = 43.35 \text{ kN}$$

The rivet value for rivets subjected to shear and twisting,

$$R = 43.35 \text{ kN}$$

The rivet value for rivets subjected to shear and bending,

$$R = 43.35 \text{ kN}$$

The rivets are provided in one row at a pitch of 80 mm

Number of rivets $n = \left(\dfrac{6M}{p \cdot R}\right)^{1/2}$

$$\left(\frac{6\times16.25\times1000}{80\times43.35}\right)^{1/2} = 5.30. \text{ Provide 8 rivets.}$$

Step 7 : Forces in rivets

The direct shear force in a rivet $\dfrac{250}{8} = 31.25$ kN

$$\Sigma y^2 = 2 \times [4^2 + 12^2 + 20^2 + 28^2] \times 100$$
$$= 2688 \times 100 \text{ mm}^2$$

The shear due to twisting

$$\left(\frac{M_y}{\Sigma y^2}\right) = \left(\frac{16.25\times1000\times280}{268800}\right) = 16.927 \text{ kN}$$

The resultant force in a rivet

$$(31.25^2 + 16.927^2)^{1/2} = 35.54 \text{ kN}$$

Hence, rivets are safe in direct shear and twisting

Shear stress in rivets

$$\left(\frac{31.25\times1000}{\left(\frac{\pi}{4}\right)\times(23.5)^2}\right) = 72.08 \text{ N/mm}^2$$

Step 8 : Moment resisted by rivets in tension

$$M' = \left(M / \left(1+\frac{2h}{2l}\frac{\Sigma y}{\Sigma y^2}\right)\right)$$

Height of line of rotation above the bottom of stiffener angle

$$\left(\frac{1}{2}\times600\right) = 85.5 \text{ mm}$$

Distance to extreme rivet is 514.5 mm

$\Sigma y = [3.45 + 11.45 + 19.45 + 27.45 + 35.45 + 43.45 + 51.45] \times 10 = 1921.5$ mm

$\Sigma y^2 = 2 \times [3.45^2 + 11.45^2 + 19.45^2 + 27.45^2 + 35.45^2 + 43.45^2 + 51.45^2] \times 100$

$= 7053 \times 100$ mm^2

$$M' = \left(\frac{16.25}{1+\dfrac{2\times600\times1921.5}{21\times7053\times100}}\right) = 14.06 \text{ kN-m}$$

The tensile force in the top rivet,

$$F_t = \left(\frac{14.06\times1000\times514.5}{7053\times100}\right) = 10.275 \text{ kN}$$

The tensile stress in the rivet,

$$\sigma_{tf.cal} = \left(\frac{10.257\times1000}{\frac{\pi}{4}\times(23.5)^2}\right) = 23.66 \text{ N/mm}^2$$

Step 9 : Check for interaction expression
As per IS: 800–1984

$$\left[\frac{\tau_{vf.cal}}{\tau_{vf}} + \frac{\sigma_{tf.cal}}{\sigma_{tf}}\right] = \left[\frac{72.08}{100} + \frac{23.66}{100}\right] = 1.9574 \not> 1.4$$

Hence, the rivets are safe for shear tension.

Complete design of seated connection, using a single stiffener angle is shown in Fig. 7.17.

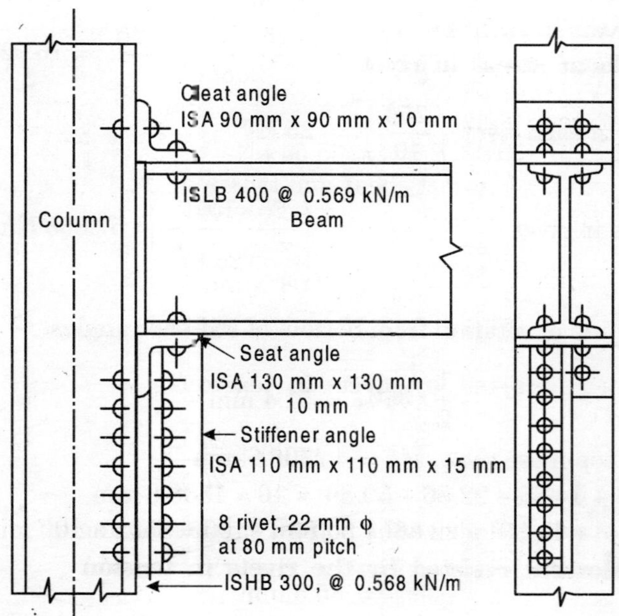

Fig. 7.17 *Beam connected to stanchion (Stiffened seated connections) single angle section*

(ii) **Provide, two stiffener angle** ISA 90 × mm × 90 mm × 10 mm (2 ISA 9090,@ 0.134 kN/m)

Bearing area provided = 2 × (90 – 10) × 10 = 1600 mm²

>1322 mm². (Bearing area required). Hence, safe.

Length of outstanding leg

$$(90 - 10) = 80 \text{ mm}$$

Thickness of angle t is 10 mm

Outstanding ratio $\frac{80}{10} = 8 < 16$. Hence, safe.

Point of application of reaction from face of column

$$\left(\frac{130}{2}\right) = 65 \text{ mm}$$

Eccentricity $e_{xx} = 65.0$ mm

The bending moment

$$\left(\frac{250\times65}{1000}\right) = 16.25 \text{ kN-m}$$

Step 10 : Design of connection

Number of rivets in each row

$$n = \left(\frac{6\times16.25\times1000}{2\times80\times43.35}\right)^{1/2} = 3.75$$

Provide 5 rivets in each row

Step 11 : Shear stress in rivet

Shear force in each rivet $\left(\frac{250}{10}\right) = 25$ kN

$$\text{Shear stress in rivet} = \left(\frac{25\times1000}{\frac{\pi}{4}\times(23.5)^2}\right) = 57.668 \text{ N/mm}^2$$

Distance of line of rotation from bottom of stiffener angles

$$\frac{1}{2}\times360 = 51.4 \text{ mm}$$

Distance to top-most rivet = 308.6 mm

$\Sigma y = 2\ [6.86 + 14.86 + 22.86 + 30.86] \times 10 = 1508.8$ mm

$\Sigma y^2 = 2\ [6.86^2 + 14.86^2 + 22.86^2 + 30.86^2] \times 100 = 3502 \times 100$ mm

Step 12 : Moment resisted by the rivets in tension

$$M' = \left[\frac{M}{1+\frac{2h}{2l}\cdot\frac{\Sigma y}{\Sigma y^2}}\right] = \left[\frac{16.25}{1+\frac{2\times360\times1508.8}{21\times3502\times100}}\right] = 14.159 \text{ kN-m}$$

Tensile force in top-most rivet

$$F_t = \left(\frac{14.159\times1000\times308.6}{350200}\right) = 12.476 \text{ kN}$$

Tensile stress in the rivet

$$F_t = \left(\frac{12.476\times1000}{(\pi,4)\times(23.50)^2}\right) = 28.779 \text{ N/mm}^2$$

Step 13 : Check for interaction expression

As per IS: 800–1984

$$\left[\frac{\tau_{vf.cal}}{\tau_{vf}}+\frac{\sigma_{tf.cal}}{\sigma_{tf}}\right]=\left[\frac{57.66}{100}+\frac{28.779}{100}\right]=0.8644 \not> 1.4$$

Hence the rivets are safe.

The complete design of seated connections using two stiffener angles are shown in Fig. 7.18.

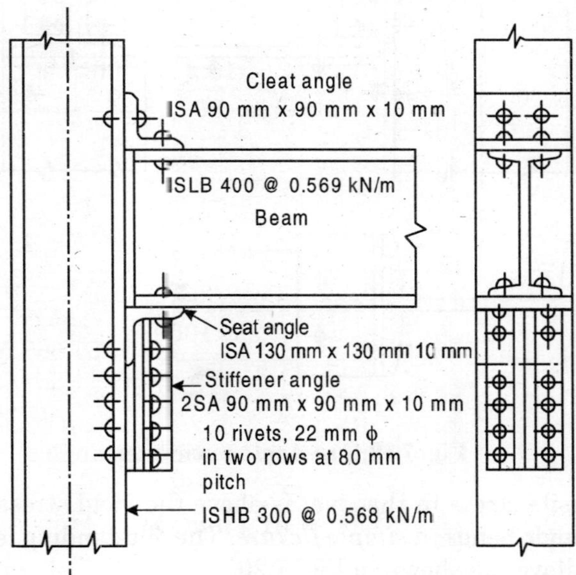

Fig. 7.18 *Beam connected to stanchion (Stiffened seated connections) two angle sections*

7.8 SMALL MOMENT RESISTANT CONNECTIONS

When the values of moments are small, then the clip-angle connections as shown in Fig. 7.19 are used.

The beam is connected to column by four angles. The *web angles* are used to connect the web of beam. One *clip angle* at the top and one *clip angle* at the bottom are used to connect the flanges of beam. These angles are called *flange clip angles.* It is assumed that web connection resists no moment and flange connections resist no shear. When the connections are subjected to clockwise moment then the rivets at 1–1 are subjected to tension and rivets at 2–2 are subjected to shear. When the connections are subjected to anticlockwise moment then, the rivets at 1′–1′ are subjected to tension and rivets at 2′–2′ are subjected to shear. Two rivets are used in one gauge line to connect the vertical leg of the clip angle. If instead of two rivets, four rivets are used in one gauge line, then the distribution of tension is not uniform. If instead of one gauge line rivets are provided in two guage lines in the vertical leg of the clip-angle, then the distribution of tension is not uniform. It is seen that number of rivets limits moment resisting capacity of the connections. Therefore, these types of connections are used for small moments.

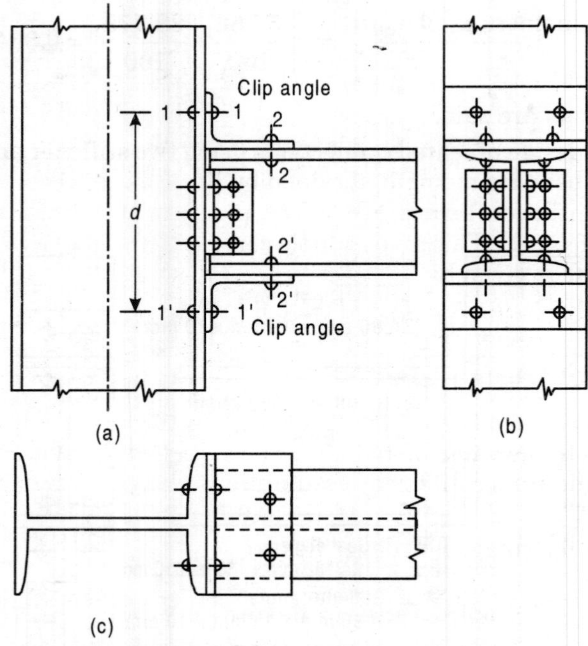

Fig. 7.19 *Clip-Angle connections*

When the tensile stress in the rivet is above the yield stress, then, the outstanding leg of angle bends in *simple flexure.* The outstanding leg of angles acts as a simple cantilever as shown in Fig. 7.20.

If P is pull on the rivets, g is gauge distance for connected leg of angle, and t is the thickness of angle, the bending moment to be resisted by angle is

$$M = P.(g-t) \quad ...(7.5)$$

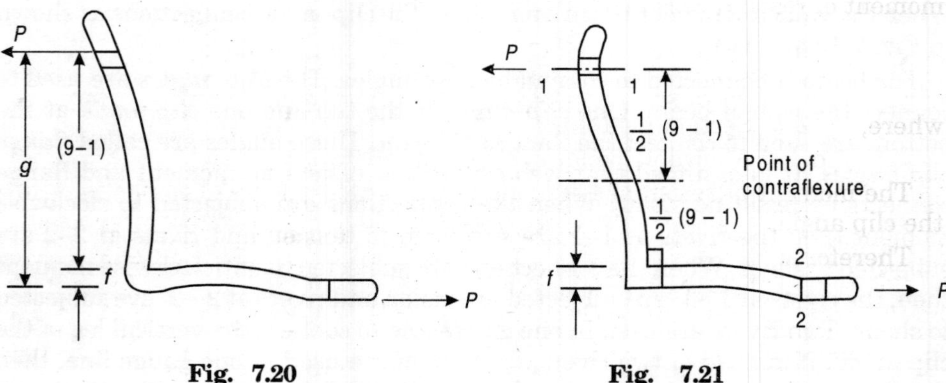

Fig. 7.20 **Fig. 7.21**

When the initial tension in the rivets is sufficient to keep the centre lines of the angle legs at 1–1 and 2–2 parallel to their original position as shown in Fig. 7.21

the angle bends *in double flexure.* The rotation of angle at 1–1 and 2–2 is prevented. There exists a point of contraflexure halfway between the rivet and face of angle.

The bending moment is given by

$$M = P.\left(\frac{g-t}{2}\right)$$

$$M = 0.5\,P.\,(g-t) \qquad \text{...(7.6)}$$

The value of bending moment obtained by Eq. 7.6 is half that obtained of Eq. 7.5 for single cantilever flexure. For safety, the point of contraflexure is assumed at a distance 0.6 $(g - t)$. This gives a little conservative design of clip angles. The bending moment is given by

$$M = 0.6\,P\,.\,(g-t) \qquad \text{...(7.7)}$$

7.7.1 Choice of Design Method

The connections shown in Fig. 7.19 are designed for small moment. There is no clear limitation to the value of moment. The values of moments to be resisted by clip angles, found under different assumptions are given below:

From Eq. 7.3 $\quad M = (0.6\,P.\,g_1) = 0.6\,P\,.\,(g - ½\,t)$

From Eq. 75 for single cantilever flexure

$$M = P.\,(g-t)$$

From Eq. 7.7 for double cantilever flexure

$$M = 0.6\,P\,.\,(g-t)$$

The design of clip angle based on simple cantilever flexure is conservative and neglects the initial tension of rivets. A thick section of clip angle is obtained by this method, which increases the weight and cost of the structure. The values of moments obtained by Eq. 7.3 and Eq. 7.7 both account for the initial tension in the rivet. The difference between two values is small. The value of moment obtained by Eq. 7.3, viz., $M = 0.6\,P\,.\,g_1$ is maximum and it is based on detail analysis. The design of clip angle is done on the basis of this Eq. 7.3.

If, l is the length of connected leg of angle and it is equal to rivet spacing, the moment of resistance is given by

$$M_R = (\sigma_{bt}\,.\,Z) = \left(\frac{1}{6}l\cdot t^2\cdot\sigma_{bt}\right) \qquad \text{...(7.8)}$$

where, $\quad Z$ = Section modulus

σ_{bt} = Allowable bending stress in tension.

The moment of resistance of angle leg is equal to moment to be resisted by the clip angle.

Therefore

$$\left(\frac{1}{6}l\cdot t^2\cdot\sigma_{bt}\right) = M_1$$

$$t = \left(\frac{6M_1}{\sigma_{bt}}\right)^{1/2} \qquad \text{...(7.9)}$$

where, $M_x = 0.6\,P\,.\,g_1 = 0.6\,R.\left(g - \frac{1}{2}t\right)$

The thickness of the angle is determined by Eq. 7.9. The allowable bending stress in tension, $0.66f_y$ used for the design of beam results in over design. The value of bending stress in tension, σ_{bc} in Eq. 7.9 is adopted as 185 N/mm², which is permissible bending stress for slab bases as per IS : 800–1984.

7.9 LARGE MOMENT RESISTANT CONNECTIONS

The maximum allowable value of tension in rivets limits the use of connections shown in Fig. 7.19. In case, the connections are subjected to large moment, *bracket connections* or *split beam connections* are used instead of flange clip angles. The bracket connections and split beam connections considered as heavy connections.

7.9.1 Bracket Connections

The bracket connections shown in Fig. 7.22 are composed of gasset plate and two sets of angles. One set of angles is connected to the column. The other set consisting of two pairs of angles is used to connect the flange of beam. In the bracket connections shown in Fig. 7.22 (a) the angles which are connected to the column are continuous. In the bracket connections shown in Fig. 7.22 (b), the angles which are connected to the column, are discontinuous.

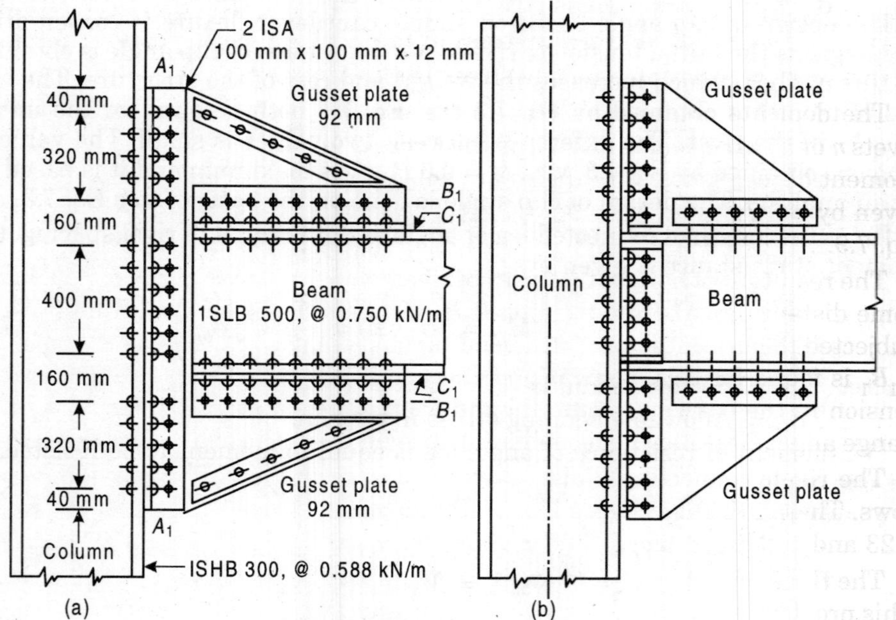

Fig. 7.22 *Bracket connections*

Consider bracket connection shown in Fig. 7.22 (a). Rivets on line A_1A_2 resist moment and shear. The other rivets are in double shear. The bearing strength of the outer rivets is found, taking thickness of bracket plate. (The thickness of the bracket plate is kept equal to the thickness of web of beam). It is assumed that rivets are provided at uniform pitch. The number of rivets is found from Eq. 2.12.

$$n = \left(\frac{6M}{pR}\right)^{1/2}$$

The rivets are not provided opposite to the flanges of beam. Therefore, it is not possible to provide uniform pitch for these rivets. The resultant of force due to direct shear and bending should be less than the rivet value.

The rivets connecting angles to the flanges of column are provided in two rows. The rivets are subjected to direct shear and tension. The rivet value is governed by strength of rivet in tension. The number of rivets is found from Eq. 2.12.

$$n_1 = \left(\frac{6M}{2pR_1}\right)^{1/2}$$

As per IS : 800–1984 these rivets are so proportioned that the quantity.

$\left[\frac{\tau_{vf.cal}}{\tau_{vf}} + \frac{\sigma_{tf.cal}}{\sigma_{tf}}\right]$ does not exceed 1.4

where, $\tau_{vf.cal}$ = Actual shear stress in the rivet

$\sigma_{vf.cal}$ = Actual tensile stress in the rivet

τ_{vf} = Allowable shear stress in the rivet

σ_{tf} = Allowable tensile stress in the rivet.

The depth of angles connected to the column is kept to accommodate the rivets n or n_1 whenever is more. The thickness of angle is determined for required moment of resistance. The angles are subjected to maximum bending moment given by Eq. 7.3 (viz., $M_1 = 0.6\, P \,.\, g_1$). The thickness of angles, t is found from Eq. 7.9.

The resultant of horizontal forces of rivets 4 to 8 in line A_1A_1 Fig. 7.23 acts at some distance apart from the line B_1B_1. Therefore, the rivets in line B_1B_1 are subjected to horizontal force and bending moment. The number of rivets in line B_1B_1 is found for this moment and checked as discussed above. The allowable tension in the extreme rivets if found from allowable bending moment in the flange angle. The allowable bending moment in the flange angle is given by Eq. 7.3.

The rivets connecting flange angles to flanges of beam are provided in two rows. These are subjected to horizontal forces as for rivets in line B_1B_1 in Fig. 7.23 and bending moment due to this horizontal force at level C_1C_1.

The thickness of bracket plate is kept equal to the thickness of web of beam. This provides ease in construction. The bracket plates and web of beam act as a rectangular beam. These must be strong enough to resist bending moment acting at line A_1A_1 in Fig. 7.22. The bending stress at the extreme fibre of this rectangular beam is given by

$$\sigma_b = \left(\frac{M \cdot y}{I} \times \frac{A_{\text{gross}}}{A_{\text{net}}} \right) \qquad \text{...(7.10)}$$

where, I = Gross-moment of inertia of beam formed by bracket plates and web of beam

y = Distance to the extreme fibre of rectangular beam from neutral axis as shown in Fig. 7.23 (b)

A_{gross} = Gross area of flanges of the beam

A_{net} = Net area of flanges of the beam

The strength of bracket plate (gusset plate) may be checked for bending moment in line B_1B_1.

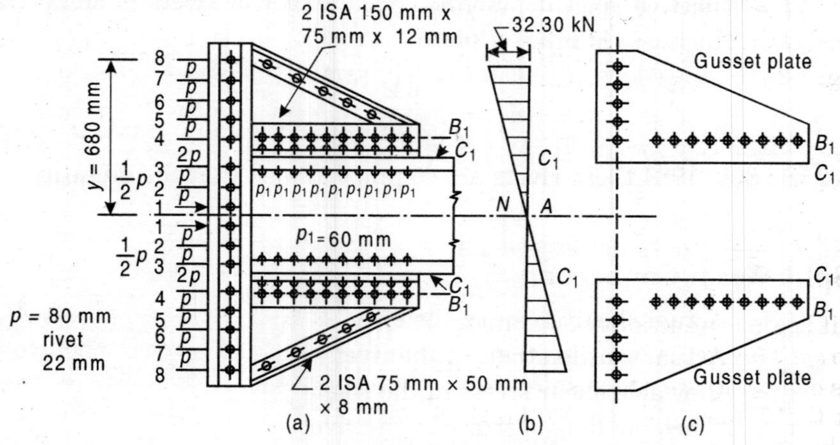

Fig. 7.23

The inclined edge of top or bottom bracket plate will be in tension or compression depending on whether the moments are clockwise and anti-clockwise. When the inclined edge of the beam will be in compression, the lateral buckling of plate may take place. The lateral buckling of plate may be prevented either by providing sufficient thickness of the plate or by providing stiffening angles as shown in Fig. 7.24.

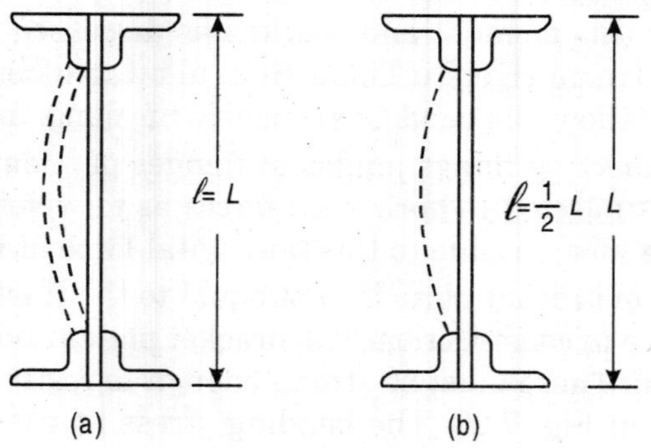

Fig. 7.24

In Fig. 7.24 (a), it is assumed that inclined edge of the bracket plate acts as a column and its edges are free to rotate. The lateral buckling of bracket plate will not occur, in case $\left(\frac{l}{r}\right)$ is less than 60.

$$\left(\frac{l}{\frac{t}{\sqrt{12}}}\right) < 60, \left(\text{For rectangular plate } r = \frac{t}{\sqrt{12}}\right)$$

$$\left(\frac{l}{r}\right) < 17.3 \qquad \text{...(7.11)}$$

where, l = Effective length of inclined edge of plate

t = Thickness of bracket plate.

In Fig. 7.24 (b), the edges are assumed to be restrained against rotation, then

$$\frac{l}{r} < 60, \therefore \left(\frac{l}{\frac{2t}{\sqrt{12}}}\right) < 60, \therefore \frac{l}{t} < 34.6 \qquad \text{...(7.12)}$$

7.9.2 Split Beam Connection

The split beam connection shown in Fig. 7.25 (a) consists of a pair of *web angles* and two *split beams*. A beam-section is cut at the middle of the web. The two cut sections of a beam are called *split beams*. Split beams are used as flange clips to connect the flanges of beam to flange of column. Web clip angles connect the web of beam to the column flange. Tee-sections are also used instead of split beams in these types of connections. In split beam connections, it is assumed that split beams resist moment only and web angles resist shear only.

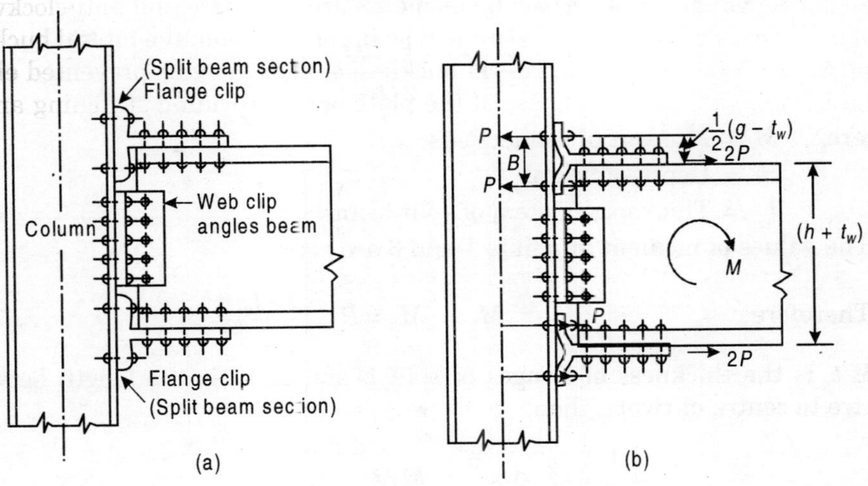

Fig. 7.25 *Split beam connections*

Figure 7.26 (a) and Fig. 7.26 (b) show modified split beam connections. The depth of connection is increased by providing flange clip seat in between flange clips and beam. The value of pull transmitted by flange clip is reduced. Therefore, the number of rivets connecting the split beams to column is reduced.

Figure 7.25 (a) shows split beam connections before loading. Fig. 7.25 (b) shows split beam connections after loading. If the connections are subjected to a clockwise moment, the top of split beam undergoes flexure. Because of symmetry, the stem of split beam does not bend. If one split beam is to transmit a pull equal to $2P$, the rivets connecting split beam to column in each gauge line are subjected to pull equal to P.

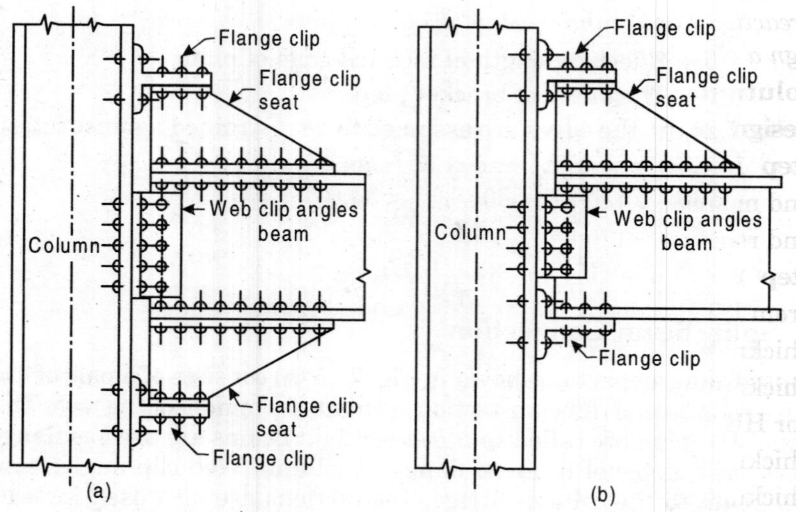

Fig. 7.26 *Modified split beam connections*

$$2P \times (h + t_w) = M$$

$$P = \frac{1}{2}\frac{M}{(h+t_w)} \quad \text{... (7.13)}$$

where, M = Moment of connections

h = Depth of beam

t_w = Thickness of stem of split beam.

The values of moment at points 1 and 3 are equal

Therefore $$M_1 = M_3 = P \cdot \frac{(g-t_w)}{2} \quad \text{...(7.14)}$$

If t_f is the thickness of flanges of split beam, and l is the length between centre to centre of rivets, then

$$\left(\frac{1}{6}\cdot l \cdot t_f^2 \cdot \sigma_{bt}\right) = M_1$$

$$t_f = \left(\frac{6M_1}{l \cdot \sigma_{bc}}\right)^{1/2} \quad ...(7.15)$$

where, σ_{bt} = Allowable bending stress in tension (185 N/mm²)

From Eq. (7.14), $M_1 = 0.5\,P\,.\,(g - t_w)$

The thickness of stem of split beam, t_w is kept sufficient to transmit the pull, 2*P* and it provides bearing strength to rivets greater than the strength of rivets connecting split beams and flange of beam in single shear. The web clip angles are designed to resist end shear.

Example 7.5 *In a framed connection, a LB 400, @ 0.569 kN/m transmits an end reaction 170 kN and a moment 14.0 kN-m to a column HB 300, @ 0.588 kN-m. Design a clip connection.*

Solution

Design :

Step 1: End moment and end reaction

End moment = 140 kN-m

End reaction =170 kN

Step 2 : Properties of given section

From ISI Handbook No. 1, for LB 400, @ 0569 kN/m

Thickness of flange t_f = 12.5 mm

Thickness of web t_w = 8.0 mm

For HB 300, @ 0.588 kN/m

Thickness of flange t_{f1} = 10.6 mm

Thickness of web t_{w1} = 1.6 mm

Step 3 : Rivet values

22 mm power driven rivets are used

Strength of rivets in single shear

$$\left(\frac{\pi}{4}\times(23.5)^2\times\frac{100}{1000}\right) = 43.35 \text{ kN}$$

Strength of power driven rivets in double shear

$$\left(2\times\frac{\pi}{4}\times(23.5)^2\times\frac{100}{1000}\right) = 86.70 \text{ kN}$$

Strength of power driven rivets in bearing (on t_w = 8.00 mm)

$$\left(23.5\times8.0\times\frac{300}{1000}\right) = 56.4 \text{ kN}$$

Strength of power driven rivets in bearing (on t_{f1} = 10.6 mm)

$$3.5\times10\,6\times\frac{300}{1000} = 74.73 \text{ kN}$$

Strength of power driven rivets in tension

$$\left(\frac{\pi}{4}\times(23.5)^2\times\frac{100}{1000}\right) = 43.35 \text{ kN}$$

Rivet values

The rivet value for connecting flange to column is governed by tension

$$R_1 = 43.35 \text{ kN}$$

The rivet value for connecting flange clip angles to beam is governed by single shear

$$R_2 = 43.35 \text{ kN}$$

The rivet value for connecting web angles to column is governed by bearing

$$R_3 = 74.73 \text{ kN}$$

The rivet value for connecting web angles to beam is governed by bearing

$$R_4 = 56.4 \text{ kN}$$

Step 4 : Flange clip angles

Two power driven rivets 22 mm in diameter are used to connect flange clip angles.

Maximum allowable tension for rivets

$$= (2 \times 43.35) = 86.70 \text{ kN}$$

Provide flange ISA 200 mm × 100 mm

The gauge distance for rivets for 100 mm leg from ISI Handbook No. 1 is 60 mm.

Distance between gauge lines of rivets

$$= (400 + 2 \times 60) = 520 \text{ mm}$$

Pull transmitted by two rivets

$$P = \left(\frac{14+1000}{520}\right) = 26.923 \text{ kN}$$

< Maximum allowable tension. Hence, rivets are safe.

The maximum moment induced in clip angles is given by Eq. 7.3

$$M_1 = (0.6 \times g_1) = 0.6\, P.(g - \frac{1}{2}\, t)$$

$$= [0.6 \times 26.923\,(60 - 7.5)]$$

$$= 848.07 \text{ kN-mm}$$

(t is assumed as 15 mm)

From ISI Handbook No. 1, rivet spacing for HB 300, @ 0.585 kN/m column

$$l = 140 \text{ mm}$$

Thickness of clip angles of clip angle is given by Eq. 7.9

$$t = \left(\frac{6\times M_1}{l\times\sigma_{bc}}\right)^{1/2} = \left(\frac{6\times 848.07\times 1000}{140\times 185}\right)^{1/2}$$

$$= 14.02 \text{ mm}$$

Adopt thickness of clip angle 15 mm

Provide flange clip angle ISA 200 mm × 100 mm × 15 mm (ISA 200 100, @ 0.536 kN/m)

Horizontal shear in rivets connecting flange clip angle to beam

$$P_1 = \left(\frac{\text{Moment at the connection}}{\text{Depth of beam}}\right) = \left(\frac{14 \times 1000}{400}\right) = 35 \text{ kN}$$

Number of rivets required to connect flange clip to beam

$$\left(\frac{35}{43.35}\right) = 0.807$$

For the rigid connections, at least 4 rivets are needed. These rivets are provided in two rows.

Web angles

The number of rivets required to connect web angles to column

$$\frac{170}{74.73} = 2.275$$

Four rivets are provided in two vertical rows. The number of rivets required to connect web angles to beam

$$\left(\frac{170}{50.4}\right) = 3.014$$

Four rivets are provided at a pitch of 80 mm, 2 ISA 100 mm × 75 mm × 8 mm are provided. The long leg is connected to the beam. The clip angle connection design, is shown in Fig. 7.27.

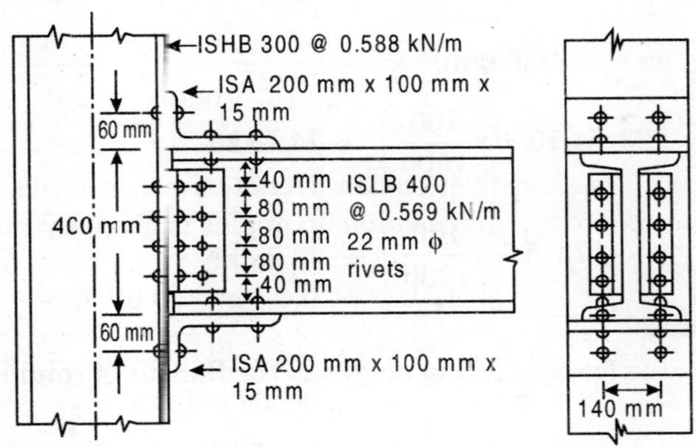

Fig. 7.27 *Clip angle connection*

Example 7.6 *In a framed connection, an LB 500, @ 0.759 kN/m transmits an end reaction 100 kN and an end moment 140 0 kN-m to a column HB 300, @ 0.588 kN/m. Design a bracket connection.*

Solution

Design

Step 1: End moment and end reaction

End moment = 140.0 kN-m

End moment = 100 kN

Step 2: Properties of given section

From ISI Handbook No. 1, for LB 500, @ 0.759 kN/m

Thickness of flange $t_{f1} = 14.1$ mm

Thickness of web $t_{w1} = 9.2$ mm

For HB 300, @ 0.538 kN/m, thickness of flange

$$t_{f1} = 10.6 \text{ mm}$$

Thickness of web $t_{wl} = 7.6$ mm

Thickness of bracket plate is kept equal to the thickness of web of beam = 9.2 mm

Step 3: Rivet values

Use 22 mm power driven rivets.

Stength of rivets

In single shear $\left(\frac{\pi}{4}\times(23\cdot5)^2\times\frac{100}{1000}\right) = 43.35$ kN

In double shear $\left(2\times\frac{\pi}{4}\times(23\cdot5)^2\times\frac{100}{1000}\right) = 86.70$ kN

In bearing (on t_w = 9.2 mm)

$$\left(23\cdot5\times9\cdot2\times\frac{300}{1000}\right) = 64.86 \text{ kN}$$

In bearing (on t_{f1} = 10.6 mm)

$$\left(23\cdot5\times10\cdot6\times\frac{100}{1000}\right) = 74.73 \text{ kN}$$

In tension $\left(\frac{\pi}{4}\times(23\cdot5)^2\times\frac{100}{1000}\right) = 43.35$ kN

Rivet values

The rivet value for connecting angles to the flanges of column is governed by tension,

$$R_1 = 43.35 \text{ kN}$$

The rivet value for connecting angles to bracket plate and web of beam is governed by bearing

$$R_2 = 64.86 \text{ kN}$$

The rivet value for connecting angles to the flanges of beam is also governed by tension

$$R_3 = 43.35 \text{ kN}$$

The rivet value for connecting anges to the bracket plate is governed by bearing,

$$R_4 = 64.86 \text{ kN}$$

Step 4: Rivets

Let number of rivets connecting angles to column in each row at a pitch of 80 mm be n_1. There are two vertical rows of rivets.

From Eq. 2.12

$$n_1 = \left(\frac{6 \times M}{2p \times R_1}\right)^{\frac{1}{2}} = \left(\frac{6 \times 140 \times 1000}{2 \times 80 \times 43 \cdot 35}\right)^{\frac{1}{2}} = 11.0$$

Let number of rivets connecting angles to bracket plate and web of beam at a pitch of 80 mm be n. These rivets are provided in one row. From Eq. 2.12

$$n_1 = \left(\frac{6M}{p \times R_2}\right)^{\frac{1}{2}} = \left(\frac{6 \times 140 \times 1000}{80 \times 64 \cdot 86}\right)^{\frac{1}{2}} = 12.723$$

Sixteen rivets are provided to connect angles to column, and angles to bracket plate and beam in each vertical row as shown in Fig. 7.22 (a).

Step 5: Check for rivets

The rivets connecting angles to column are subjected to direct shear and tension,

Shear for in rivet

$$\frac{100}{(2 \times 16)} = 3.125 \text{ kN}$$

Shear stress in rivet

$$\tau_{vf.cal} = \left(\frac{3 \cdot 125 \times 1000}{\left(\frac{\pi}{4}\right) \times (23 \cdot 5)^2}\right) = 7.208 \text{ N/mm}^2$$

Allowable shear stress in the power driven rivets

$$\tau_{vf} = 100 \text{ N/mm}^2$$

Distance to the extreme rivet from neutral axis, y is 680 mm

$$\Sigma y^2 = 4 \times 8^2 \left[\left(\frac{1}{2}\right)^2 + \left(\frac{3}{2}\right)^2 + \left(\frac{5}{2}\right)^2 + \left(\frac{9}{2}\right)^2 + \left(\frac{11}{2}\right)^2 + \left(\frac{13}{2}\right)^2 + \left(\frac{15}{2}\right)^2 + \left(\frac{17}{2}\right)^2\right] \times 100$$

$$= 64 \times 920 \times 100 = 58880 \times 100 \text{ mm}^2$$

Maximum tension in extreme rivet

$$F_1 = \left(\frac{140 \times 1000 \times 680}{5888000}\right) = 16.168 \text{ kN}$$

Tensile stress in rivet

$$\tau_{vf.cal} = \left(\frac{16 \cdot 168 \times 1000}{\frac{\pi}{4} \times (23 \cdot 5)^2}\right) = 37.296 \text{ N/mm}^2$$

Step 6: Check for interaction expression

$$\left[\frac{\tau_{vf\cdot cal}}{\tau_{vf}} + \frac{\sigma_{vf\cdot cal}}{\tau_{vf}}\right] = \left[\frac{7\cdot 208}{100} + \frac{37\cdot 396}{100}\right] = 0.445 < 1.4 \text{ (Safe)}$$

The rivets connecting angles to web are in direct shear and bending.

$$\text{Shear force in rivet} = \left(\frac{100}{16}\right) = 6.25 \text{ kN}$$

$$\Sigma y_1^2 = \left(\frac{1}{2}\times 58880\times 10\right)$$

$$= 29440 \times 100 \text{ mm}^2$$

Force in the rivet due to bending moment

$$F_b = \left(\frac{140\times 1000\times 680}{29440\times 100}\right) = 32.337 \text{ kN}$$

Resultant force, $R = \left[(F_b)^2 + (F_b)^2\right]^{\frac{1}{2}}$

$$[6.75^2 + 32.337^2]^{1/2} = 32.935 \text{ kN} < \text{Rivet value. Hence, safe.}$$

Angles connecting bracket plate to the column :

Depth of angles $= (2\times 720) = 1440$ mm

Maximum tension in one rivet,

$$P = 16.188 \text{ kN}$$

Use 2 ISA 100 mm × 100 mm

For 100 mm leg, gauge distance is 60 mm

Assume thickness of angle as 12 mm

Step 7: Maximum moment induced in the angle is given by Eq. 7.3

$$M_1 = 0.6\,P \times g_1 = 0.6\,P\,.\left(g - \frac{1}{2}t\right)$$

$$= (0.6 \times 16.168 \times 54)$$

$$= 523.84 \text{ kN-mm}$$

The spacing for rivets for HB 300, @ 588.84 kN/m is 140 mm

Thickness of angle is given by Eq. 7.9

$$\left(\frac{6M}{l\times\sigma_{bc}}\right)^{\frac{1}{2}} = \left(\frac{6\times 523\cdot 84\times 1000}{140\times 185}\right)^{\frac{1}{2}} = 11.016 \text{ mm}$$

Adopt thickness of angles as 12 mm

Provide 2 ISA 100 mm × 100 mm × 12 mm to connect bracket plate to column as shown in Fig. 7.22 (a).

Step 8: Design of riveted connection

Rivets in the line B_1B_1

Force in extreme rivet in line A_1A_1

$$= 32.935 \text{ kN}$$

Force in rivet in line A_1A_1 at the level of B_1B_1

$$= \left(\frac{32 \cdot 935 + 360}{680}\right) = 17.436 \text{ kN}$$

Similarly, force in other rivets can be found. Resultant of horizontal force rivets from 4 to 8

Figure 7.23 (a), $$F = \left(\frac{32 \cdot 935 + 17 \cdot 436}{2}\right) \times 5 = 125.927 \text{ kN}$$

Line of action of horizontal force above B_1B_1

$$\frac{320}{3} \times \left(\frac{2 \times 32 \cdot 935 + 17 \cdot 436}{32 \cdot 935 + 17 \cdot 436}\right) = 176.41 \text{ mm}$$

Moment at the level B_1B_1

$$\left(\frac{125 \cdot 927 \times 176.41}{1000}\right) = 22.215 \text{ kN}$$

Moment at the level C_1C_1

$$\left(\frac{125 \cdot 927[176.41 + (360 \quad 250)]}{1000}\right) = 36.0667 \text{ kN-m}$$

Let n be number of rivets in line B_1B_1 at the pitch of 60 mm

These rivets are in one row

$$n_1 = \left(\frac{6 \times 22 \cdot 215 \times 1000}{60 \times 64 \cdot 86}\right)^{\frac{1}{2}} = 5.85$$

Let n_1 be number of rivets connecting angles to flange of beam at a pitch of 60 mm. These rivets in two rows

$$n_1 = \left(\frac{6 \times 36 \cdot 0667 \times 1000}{2 \times 60 \times 43 \cdot 35}\right)^{\frac{1}{2}} = 6.45$$

Provide 9 rivets in each row, connect angles to bracket plate and angles to beam as shown in Fig. 7.22 (a) and Fig. 7.23 (a).

Step 9: Check for rivets

The rivets in line B_1B_1 are in direct shear and bending

Direct shear in rivets

$$F_{s1} = \left(\frac{125 \cdot 927}{9}\right) = 13.99 \text{ kN}$$

$$\Sigma y_1^2 = 2 \times 6^2 [1^2 + 2^2 + 3^2 + 4^2] \times 100$$
$$= 2160 \times 100 \text{ mm}^2$$

Distance to the extreme rivet y_1 is 240 mm

Force in rivet due to bending

$$F_{b1} = \left(\frac{22 \cdot 215 \times 1000}{21600} \times 240\right) = 24.683 \text{ kN}$$

Resultant force in the rivets

$$\left[F_{s1}^2 + F_{b1}^2\right]^{\frac{1}{2}} = [13.99^2 + 24.683^2]^{1/2} = 28.372 \text{ kN}$$
$$< 64.86 \text{ kN, rivet value. Hence, safe.}$$

The rivets in the C_1C_1 are in direct shear and tension

Direct shear $$F_{s2} = \left(\frac{125 \cdot 927}{2 \times 9}\right) = 13.99 \text{ kN}$$

Shear stress $$\tau_{vf.cal} = \left(\frac{6 \cdot 996 \times 1060}{\left(\frac{\pi}{4}\right) \times (23 \cdot 5)^2}\right) = 16.138 \text{ N/mm}^2$$

$$\Sigma y_2^2 = 4320 \times 100 \text{ mm}^2$$

Tension in rivets

$$F_{t2} = \left(\frac{36 \cdot 0667 \times 1000}{432000} \times 240\right) = 20.037 \text{ kN}$$

Tensile stress in rivets

$$\sigma_{tf.cal} = \left(\frac{20 \cdot 037 \times 1000}{\left(\frac{\pi}{4}\right) \times (23 \cdot 5)^2}\right) = 46.22 \text{ N/mm}^2$$

Step 10: Check for interaction expression

$$\left[\frac{\tau_{vf \cdot cal}}{\sigma_{vf}} + \frac{\sigma_{vf \cdot cal}}{\sigma_{tf}}\right] = \left[\frac{16 \cdot 138}{100} + \frac{46 \cdot 22}{100}\right] = 0.624 < 1.4$$

Hence, the rivets are safe.

Angle connecting bracket plate to beam

Depth of angle = (9 × 60) = 540 m

Tension in extreme rivet is 20.37 kN

Use 2 ISA 150 mm × 75 mm. The gauge distance for 70 mm leg from ISI dbook No. 1 is 40 mm.

Moment induced in angle

$$M_1 = \left(\frac{0 \cdot 6 \times 20 \cdot 037 \times (40 - 6)}{1000}\right) = 0.4088 \text{ kN-m}$$

Rivet space for ISLB 500, @ 0.750 kN/m is 100 mm

Thickness of angle is obtained from Eq. 7.3

$$t = \left(\frac{6\times0\cdot4088\times1000\times1000}{100\times185}\right)^{\frac{1}{2}}$$

$$= 11.514 \text{ mm}$$

Provide 2 ISA 150 mm × 75 mm 12 mm (2 ISA 15075, @ 0.201 kN/m) shown in Fig. 7 .22 (a).

The long leg is connected to bracket plate.

Bracket plate. Thickness of bracket plate is kept equal to the thickness web of beam = 9.2 mm. It will be prepared by matching from 10 mm thickness of plate.

At section A_1A_1. Distance to the extreme fibre of bracket plate

$$= 720 \text{ mm}$$

Moment of inertia $= \left(\frac{1}{12}\times9\cdot2\times(720)^2\right) = 28650 \times 10^4 \text{ mm}^4$

Gross area, A (gross) $= (9.2 \times 720) = 64.9 \times 100 \text{ mm}^2$

Net area, A (net) $= (64.9 \times 100 - 16 \times 23.5 \times 9.2)$

$= 3030.8 \text{ mm}^2$

Tensile stress in extreme fibre of bracket plate from Eq. 6.28

$$\sigma_b = \left(\frac{M\cdot y}{I_{xx}}\frac{A_{\text{gross}}}{A_{\text{net}}}\right) = \left(\frac{140\times10^3\times10^3}{28650\times10^3}\times\frac{6490}{3030\cdot8}\right)$$

= 10.4 N/mm^2 < Allowable stress. Hence, safe.

2 ISA 75 mm × 50 mm × 8 mm (2 ISA 7550, @ 0.074 kN/m) are connected to inclined edge of the bracket plate to avoid lateral buckling.

Example 7.7 *In a framed connection, an LB 600, @ 0 .995 kN/m transmits an end reaction 200 kN and an end moment of 40 kN-m to HB 300 @ 0.588 kN/m column. Design split-beam connections.*

Solution

Design

Step 1: End moment and end shear

End moment = 40 kN-m

End shear = 200 kN

Step 2: Properties of given sections

From ISI Handbook No 1, for LB 600, @ 0.995 kN/m

Thickness of flange t_f = 15.5 mm

Thickness of web t_w = 10.5 mm

Width of flange b = 210 mm

For BH 300, @ 0.588 kN/m, thickness of flange

t_{f1} = 10.6 mm

Thickness of web $t_{w1} = 7.6$ mm

Step 3: Rivet value

Use 22 mm power driven rivets.

Strength of rivets

In single shear $\left(\frac{\pi}{4}\times(23\cdot5)^2\times\frac{100}{1000}\right) = 43.35$ kN

In double shear $\left(2\times\frac{\pi}{4}\times(23\cdot5)^2\times\frac{100}{1000}\right) = 86.70$ kN

In beating (on $t_w = 10.5$ mm)

$$\left(23\cdot5\times10\cdot5\times\frac{300}{1000}\right) = 74.03 \text{ kN}$$

In bearing (on $t_f = 10.6$ mm)

$$\left(23\cdot5\times10\cdot6\times\frac{300}{1000}\right) = 74.73 \text{ kN}$$

In tension $\left(\frac{\pi}{4}\times(23\cdot5)^2\times\frac{100}{1000}\right) = 43.35$ kN

Rivet values. The rivet value for rivets connecting split beams to the flanges of column is governed by tension

$$R_1 = 43.35 \text{ kN}$$

The rivet value for rivet connecting split beam to the flanges of beam is governed by single shear

$$R_2 = 43.35 \text{ kN}$$

The rivet value for rivets connecting web angles to flanges of column is governed by single shear

$$R_3 = 43.35 \text{ kN}$$

The rivet value for rivets connecting web angles to web of beam is governed by beaing

$$R_4 = 74.3 \text{ kN}$$

Split beams. Two rivets 22 mm is diameter are used to connect spit beam to column.

Maximum allowable tension

$$= 2 \times 43.35 = 86.70 \text{ kN}$$

Two rivets are subjected to a pull acting through the centroid of the rivets. The rivets are assumed to be equally stressed.

Assume thickness of split beam as 10 mm

From Eq. 7.13, pull in split beam

$$P = \frac{1}{2}\frac{M}{(h+t_w)} = \left(\frac{1}{2}\times\frac{40\times1000}{(600+10)}\right) = 32.787 \text{ kN}$$

$$< \text{Maximum allowable tension}$$

Assume gauge distance for rivet in split beam as 140 mm

Moment induced in split beam is given by Eq. 7.14.

$$M_1 = \frac{1}{2}P\cdot(g-t_w)$$

$$= \left(\frac{1}{2}\times\frac{32\cdot787(140-10)}{1000}\right) = 2.13 \text{ kN-m}$$

From ISI Handbook No. 1, rivet spance for ISHB 300, @ 0.585 kN/m column, l is 140 mm

The thickness of flange of split beam is given by Eq. 6.33.

$$t_f = \left(\frac{6M_1}{l\times\sigma_{bc}}\right)^{\frac{1}{2}} = \left(\frac{6\times2\cdot13\times1000\times1000}{140\times185}\right)^{\frac{1}{2}}$$

$$= 22.22 \text{ mm}$$

From ISI Handbook No. 1 select WB 600 @ 1.451 kN/m.

This beam section is cut at half the depth. This gives two split beams.

Thickness of flange t_f = 23.0 mm

Thickness of web t_w = 11.8 mm

Length of stem = 300 mm

Minimum gross width of stem = Width of flange of beam = 210 mm

Net width of stem = (210 – 2 × 23.5) = 163.0 mm

The pull transmitted by split beam

$$2P = (2\times32.787) = 65.574 \text{ kN}$$

Strength of split beam in axial tension

$$\left(163\cdot0\times11\cdot9\times\frac{0\cdot6\times250}{1000}\right) = 288.51 \text{ kN} > 2P. \text{ Hence, safe}$$

Horizontal shear at the flanges of beam

$$F_s = \left(\frac{40\times1000}{43\cdot35}\right) = 66.667 \text{ kN}$$

Number of rivets required to connect split beam

$$\left(\frac{66\cdot667}{43\cdot35}\right) = 1.538$$

Four rivets are provided in two rows for rigid frame connections.

Web clip angles and connections

Number of rivets to connect web angles to beam

$$\left(\frac{200}{74\cdot73}\right) = 2.676$$

Four rivets are provided at 80 mm pitch.
Number of rivets required to connect web angles to column

$$\left(\frac{200}{43 \cdot 35}\right) = 4.616$$

Eight rivets are provided in two vertical rows.
2 ISA 100 mm × 75 mm × 10 mm (2 ISA 10075, @ 0.130 kN/m) are provided.
The long leg is connected to the beam.
A complete design of split beam connection is shown in Fig. 7.28.

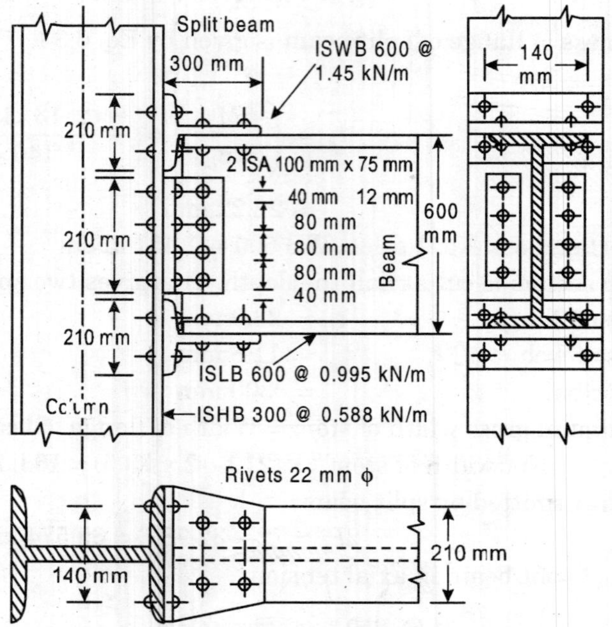

Fig.7.28 *Split beam connection*

7.10 SEMI-RIGID CONNECTIONS

Flexible, semi-rigid and rigid connections of a beam with columns at its are shown in Fig. 7.29 (a), (b) and (c), respectively. In order to simplify the calculations of moments in the members of a frame, in the analysis of structures, the connections are generally idealized as either flexible rigid. With the flexible end connections, a beam shall be simply supported and the maximum bending moment occurs at the centre, $M_c = \left(\frac{wL^2}{8}\right)$ where w is the intensity of uniformly distributed load and L is the span of beam. The beam is then designed for moment, M_c. The design of beam shall be not economical. With the infinitely rigid connections, a beam shall be fixed beam and the moments at the fixed ends shall be $M_A = M_B = \left(\frac{wL^2}{12}\right)$. By having some appropriate choice of connections (say, semi-rigid), the end moments of beam M_A, M_B and moment at the centre of beam, M_c may be made equal and equal to maximum bending moment any where in the beam M = 0.5 times $\left(\frac{wl^2}{8}\right) = \frac{wL^2}{16}$ In that case, the modulus of section for a beam needed with semi-rigid connection shall be 50% of that required for beam with flexible connections, and 75% of that necessary for beam with rigid end connections.

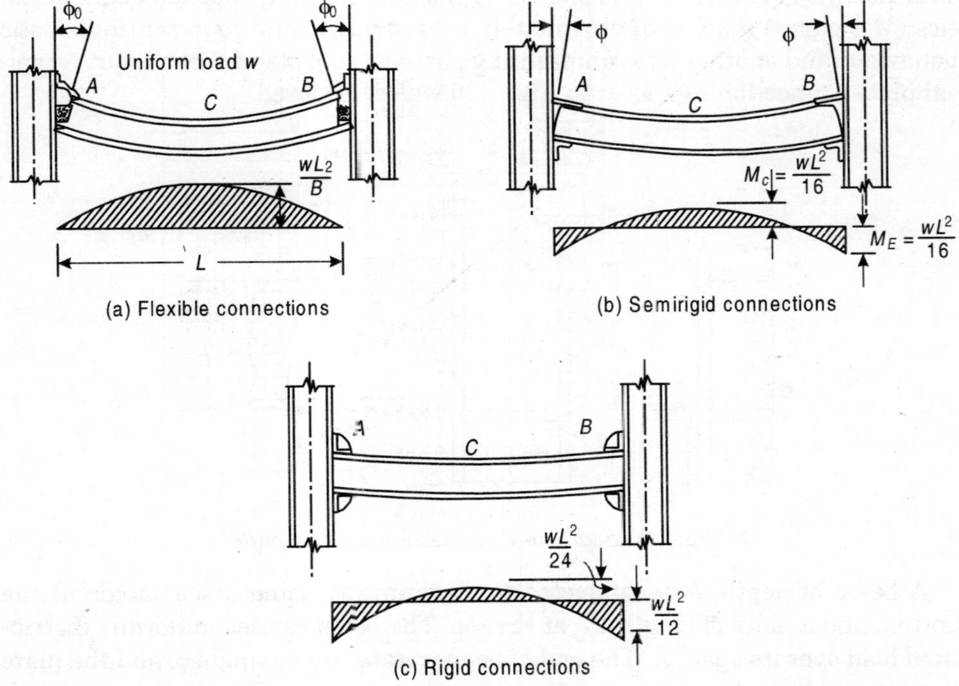

Fig. 7.29 *Flexible semi-rigid and rigid connections*

The rigidity of connections is defined as the ratio of resulting end-moment with semi-rigid connections to the fixed end moment with rigid connections. For above suggested semi-rigid connections, the rigidity of connections becomes $\left[\left(\frac{wL^2}{16}\right)\right.$ divided by $\left.\left(\frac{wL^2}{12}\right)\right]$ equal to 75%. It means, the connections provide 75 percent fixity and the moment developed at the end shall be $\left(0 \cdot 75\frac{wL^2}{12}\right) = \frac{wL^2}{16}$ In other words it permits 25 percent of rotation for simply-supported beam. (**The flexibility of connections** *shall be® unity minus rigidity of connections)*.

When the semi-rigid connections are used in the frame structures, the redistribution of moments occurs and the values of maximum bending moment is reduced. The modulus of section necessary for beam also becomes smaller as compared with other types of connections. However, in practice, all the riveted connections act as semi-rigid connections.

7.11 BEHAVIOUR OF SEMI-RIGID CONNECTIONS

The behaviour of semi-rigid connections may be studied by plotting its moment-rotation characteristics. Professor B.G. Johnston and E. Mount have shown in their paper titled as 'Analysis of building frames with semi-rigid connections, transactions of ASCE 107, pp 993, 1942 that the moment-rotation characteristics (M-ϕ curve) may be approximately by a straight line representing elastic behaviour and another line representing post-yielding plastic behaviour. A typical plate connection as shown in Fig. 7.30 was considered.

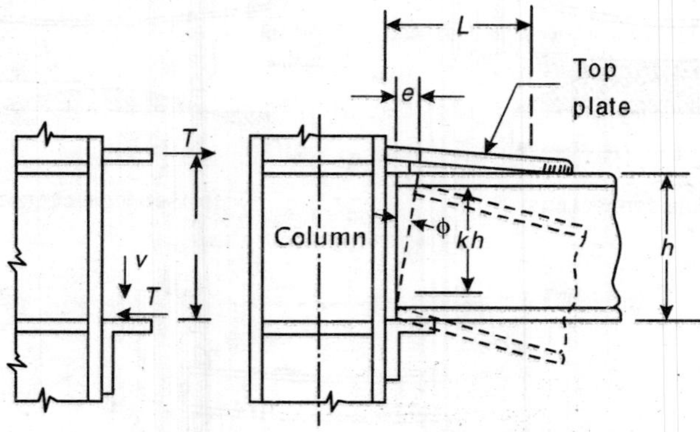

Fig. 7.30 *Semi-rigid connection with to plate*

A beam of depth, h is connected to a column by using a seat angle at the bottom and a plate of length, L_p at the top. The beam carries uniformly distributed load over its span, L. The end of beam rotates by an angle ϕ, and the plate is stretched by e, due to tension T developed in the plate. Due to end rotation, a thrust, C (equal to tension T) also acts against the face of the column. Two forces tension, T and thrust, C are equal in magnitude and act in opposite direction and form a couple, $M = T.h$ at the connection. The end rotation ϕ may be expressed as under:

$$\phi = \frac{e}{(k \cdot h)} \quad \text{...(i)}$$

where, $e = \dfrac{T \cdot L_p}{(A_p \cdot E)}$

A_p = Area of the plate

$k.h$ = Depth from the top plate to the centre of rotation.

The position of line of rotation depends upon the deformation of connection between flange of beam and the support. The value of k varies from 0.5 to 1.00.

The behaviours of all elements of the connections may be assumed elastic and the deformation of support may be neglected. The value of e may be rules substituted in expression (i), then

$$\phi = \left(\frac{T \cdot L_p}{A_p \cdot E \cdot kh}\right) \quad \text{...(iii)}$$

The slope of moment-rotation ($M - \phi$) line, θ as shown in Fig. 7.31 may be stated stated as below:

$$\theta = \left(\frac{\theta}{M}\right)\left(\frac{T \cdot L_p}{A_p \cdot E \cdot kh \cdot T \cdot h}\right) = \left(\frac{T \cdot L_p}{A_p \cdot E \cdot kh^2}\right) \quad \text{...(iv)}$$

When the stress in top plate becomes yield stress, σ_y, the rotation at the end of beam, θ will increase and the moment, M remains constant. The moment-rotation curve (M– φ curve) may be idealised as *bi-linear* (consisting of two straight lines) as shown in Fig. 7.31.

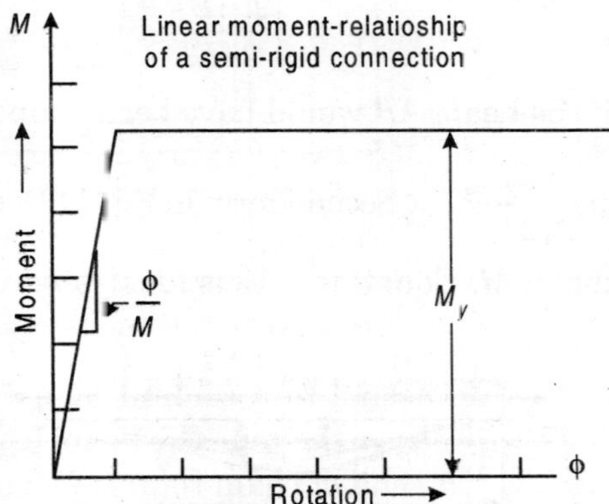

Fig. 7.31 *Linear moment-rotation diagram (semi-rigid connection)*

The value of end moment, M and rotation at the end of beam, φ also depend on the intensity of load, w, the span of beam. L and the geometrical properties of beam section. The value of end moment M for the beam AB, Fig. 7.29 (b) may be written by slope-deflection as follows. In general,

$$M_{AB} = M_{FAB} - 2E.K\,(2\phi_{AB} - \phi_{BA}) \quad \text{...(v)}$$

$$M_{BA} = M_{FBA} - 2E.K\,(2\phi_{BA} - \phi_{AB}) \quad \text{...(vi)}$$

where, M_{AB} and M_{BA} are the end moments and M_{FAB} and M_{FBA} are the fixed end moments for a beam, K is the relative stiffness of the beam (ratio of moment of inertia of the cross-section and length of beam). By adding the expression (v) and (vi)

$$\left(\frac{M_{AB} + M_{BA}}{2}\right) = \left(\frac{M_{FAB} + M_{FBA}}{2} - 2EK\frac{\phi_{AB} + \phi_{BA}}{2}\right) \quad \text{...(vii)}$$

In case, the load is uniformly distributed load on the beam, then

$$M_{AB} = M_{BA} = M \quad \text{...(viii)}$$

$$M_{FAB} = M_{FBA} = M_F \quad \text{...(ix)}$$

The expression (vii) may be rewritten as under:

$$M = \left(M_F - \frac{2EI}{L} \cdot (\phi) \right) \quad \text{...(xi)}$$

or

$$M = \left(\frac{w \cdot L^2}{12} - \frac{2EI}{L} \cdot (\phi) \right) \quad \text{...(7.16)}$$

From Eq. 7.16, the value of end rotation, ϕ may be written as below:

$$\phi = \left(\frac{w \cdot L^2}{24EI} - \frac{M \cdot L}{2EI} \right) \quad \text{...(7.17)}$$

Since,

$$M = T.h$$

$$\phi = \left(\frac{w \cdot L^3}{24EI} - \frac{T \cdot h \cdot L}{2EI} \right) \quad \text{...(7.18)}$$

It is to note that if the beam *AB* would have been simply supported, the slope at end of beam shall be $\left(\frac{L_2}{24} \cdot El \right)$. Second term in Eq. 7.18, viz., $\frac{T \cdot h \cdot L}{(2EI)}$ represents the effect of end moment, *M*. Equation 7.18 is illustrated in Fig. 7.32.

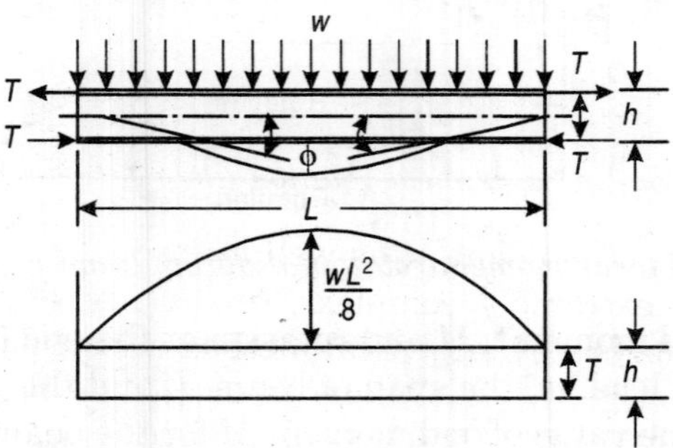

Fig. 7.32

In the analysis and design of semi-rigid connections of a structure, the moments at the ends of members for the known loads are determined and the linear rotation characteristics of a particular connection is found.

The moment-rotation relationship of the connection may be assumed carried. The analysis of frame structure with semi-rigid connection may then be carried out either by graphical method (known as beam-line method) or by deal methods, (namely ; modified slope-deflection method and moment-distribution procedure).

Actually, the moment rotation relationship for the semi-rigid connection is near as shown in Fig. 7.33. It shows *M*-ϕ relationship for semi-rigid connections 1, 2 and 3 and beam lines *QR* and *SV* for two beams *AB* and *CD*, respectively.

The semi-rigid connection-1 is moment resisting connection with connection on web and moment connection by using Tee-sections or bracket type. This type of connection is used when the span moment is larger than the end moment. The semi-rigid connection-2 is moment resisting connection with shear connection with the web and moment connection by seat lip angles. In this type of connection, the difference between the end moment and span moment is smaller. Whereas, the semi-rigid connection-3 is frame type or seat type connection with ordinary cleat angle at the top. In this type the end moments are bigger than the span moment. An arrangement of connection would be economical in case the span moments and end moments are equal. In such a case, a beam is designed for mimimum moment.

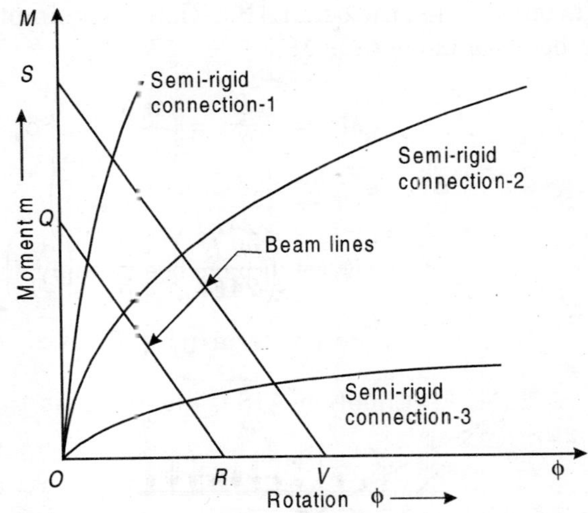

Fig. 7.33 *Non-linear moment-rotation diagram semi-rigid connection*

In case, the sign of moment M as well as its magnitude are considered, the analysis becomes extremely complex. Further, when the lateral forces are considered, the analysis by analytical method becomes impractical.

7.12 BEAM-LINE METHOD

Professor Batho, C and Rowan, H.C. in their paper titled as 'The analysis of the moments in the members of a frame having rigid or semi-rigid connections' published in second report, Steel Structures Research Committee, London, H.M.S.O., 1934 and Professor Batho, C in his paper titled as 'The analysis and design of beams under given end restraints published in final report, Steel Structures Research Committee, London, H.M.S.O. 1936 gave graphical method applicable to single-bay frames in which there is no away or sideways deflection.

The relation between end moment, M and end rotation, ϕ for a beam AB of uniform section, loaded symmetrically with identical semi-rigid connections at its ends, Fig. 7.29 is given by the straight line

$$M = \left(M_F - \frac{2EI \cdot (\phi)}{L}\right) \quad \text{...(i)}$$

where, M_F is the fixed end moment for the beam AB with fixed ends. The straight line represented by the expression (i) is called as *beam-line*. This light line is plotted graphically as QP as shown in Fig. 7.34. Point Q is plotted for the beam AB fixed at its both the ends for which end-moment $M = \frac{wL^2}{12}$ and end-rotation $\phi = \phi_0$ is zero. Point P is plotted for the beam AB simply supported at its both the ends for which end-moment M is zero and end rotation $\phi = \phi_0$. Points Q and P are joined, which represents the beam-line QP the beam AB, having semi-rigid connections at its ends. When the beam AB designed for simply supported ends, then, maximum bending moment is M_c.

$$M_c = \frac{wL^2}{8} = \left(\frac{\sigma_b . I}{y}\right) = \left(\sigma_b . \frac{I}{\frac{h}{2}}\right) \quad \text{...(ii)}$$

and

$$\phi_c = \left(\frac{w \cdot L^2}{24EI}\right) = \frac{2}{3} = \left(\frac{\sigma_b}{E}\right) = \left(\frac{L}{h}\right) \quad \text{...(iii)}$$

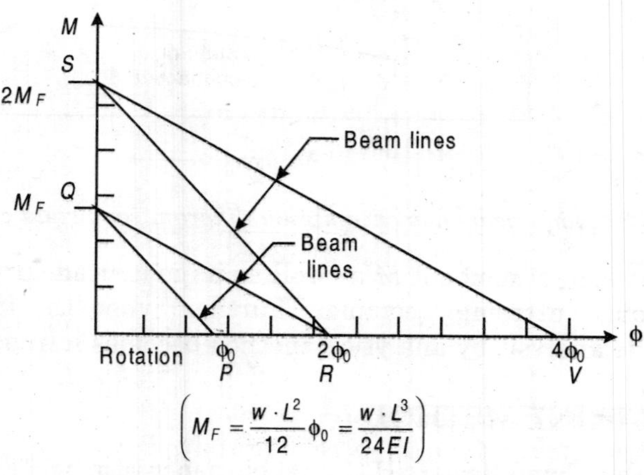

Fig. 7.34 *Moment-rotation diagram (Semi-rigid connections)*

This expression (iii) shows that (ϕ_0 is directly proportional to ratio $\left(\frac{L}{h}\right)$ for simply supported beam for given bending stress, σ_b, but the moment, M_c = $\left(\frac{w \cdot L^2}{8}\right) = \sigma_b\left(\frac{I}{y}\right)$ is not dependent on the ratio $\left(\frac{L}{h}\right)$. The end rotation ϕ_0

$\left(\frac{wL^2}{24El}\right)$ and the fixed end-moment $M = M_P = \left(\frac{w \cdot L^2}{12}\right)$ are directly portional to the intensity of uniformly distributed load, w. The influence of rotation span/depth and load, w on M–ϕ relationship for a beam having semi-rigid connections at its ends is also shown in Fig. 7.34.

The straight line QR represents a beam-line when the load $w_1 = \left(\frac{w}{4}\right)$ and span $L_x = 2L$. From Eqs. 7.16 and 7.17

$$M_{F1} = \left(\frac{w_1 \cdot L_1^2}{12}\right) = \frac{1}{12} \cdot \left(\frac{w}{4}\right). \times (4\,L^2) = M_F \qquad \text{...(iv)}$$

$$\phi_{0.2} = \left(\frac{w_1 \cdot L_1^3}{24EI}\right) = \frac{1}{24} \cdot \left(\frac{w}{4}\right)\left(\frac{8L^3}{EI}\right) = 2\phi_0 \qquad \text{...(v)}$$

The straight line SR represents a beam-line when the load $w_2 = (2w)$ and the span $L^2 = I$. From Eqs. 7.16 and 7.17

$$M_{F2} = \left(\frac{w_2 L_2^2}{12}\right) = \frac{(2w) \cdot L^2}{12} = 2 \times \frac{2 \cdot wL^2}{4} = 2M_F \qquad \text{...(vi)}$$

$$\phi_{0.2} = \left(\frac{w_2 L_2^3}{24EI}\right) = \left(\frac{(2w) \cdot L^3}{24EI}\right) = 2\phi_0 \qquad \text{...(vii)}$$

The straight line *SV* represents the beam-line, when load $w_2 = (0.5\,w)$ and the span $L^3 = L$. From Eqs. 7.16 and 7.17

$$M_{F3} = \left(\frac{w_3 L_3^2}{12}\right) = \frac{1}{2}\left(\frac{(2w) \cdot L^3}{24EI}\right) = 2\phi_0 \times \frac{2 \cdot wL^2}{4} = 2M_F \qquad \text{...(viii)}$$

$$\phi_{0.2} = \left(\frac{w_3 L_3^2}{24EI}\right) = \frac{1}{24} \cdot \left(\frac{w}{2}\right)\frac{8L^3}{24EI} = 4\phi_0 \qquad \text{...(ix)}$$

The beam lines *QP, QR, SR* and *SV* as shown in Fig. 7.34 are distinctly different from the moment rotation *(M–ϕ)* relationship of a semi-rigid connection shown in Fig. 7.31.

For a particular structure, the line may be defined and the end moment and end rotation (viz. M–ϕ) characteristics of the beam and column connections are known, then the corresponding values of end moment, M and end rotation, ϕ may be found graphically as shown in Fig. 7.35.

Figure 7.35 shows beam lines *QR* and *SV* for beams *AB* and *CD,* respectively. It also shows moment and end rotation characteristics *OAB* and *OCB* (assuming the relationship to be linear) for the semi-rigid connections 1 and 2, respectively, having different rigidities. The points of intersection of the beam line *QR* and M–ϕ curve for the semi-rigid connection-1, *OAB,* (viz., point *E*) gives the values of end moment M and end rotation ϕ. These values for the beam AB and the semi-rigid connection-1 remains same. The point of intersection of the beam-line, SV and M–ϕ curve OAB (viz., point F) gives the values of end moment M,

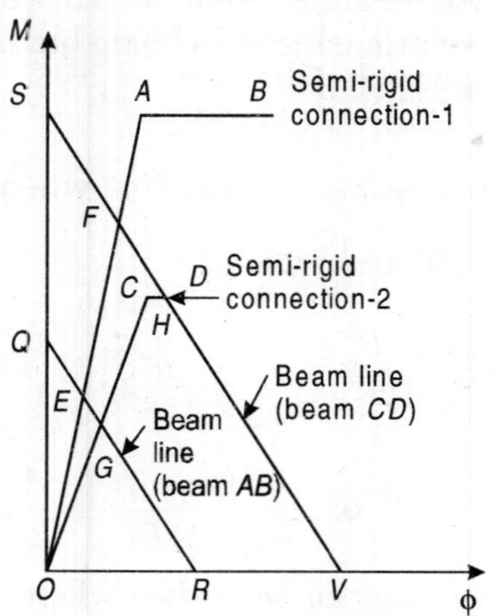

Fig. 7.35 *Beam-lines and moment-rotation characteristics (semi-rigid connections)*

and ϕ end rotation ϕ_1. It is seen that the semi-rigid connection is suitable and adequate for both the beams AB and CD. Whereas, the semi-rigid connection-2 is suitable and adequate only for beam AB. This connection does not have reserve capacity for overload for the beam CD.

7.13 MODIFIED SLOPE-DEFLECTION METHOD

Batho's graphical method (beam line method) is based on the true moment-rotation characteristics curves for the semi-rigid connections. It was not possible to extend his graphical method for the analysis of multi-storey multi-bay frames. The relationship between end moment and end rotation may be made linear by replacing these curves by their chords. Then, the analysis of semi-rigid connections may be carried out by modified slope-deflection method discussed as under. Following sign convention and assumption are made for the analysis.

The clockwise rotations of joints are considered positive. The end-moments applied by the joints to the members are positive when the rotation due to moment is clockwise. The deflection of one end of a member relative to the other is positive when the line joining the ends rotates clockwise.

The frame structures are considered two dimensional and consist of lines of vertical columns, fixed at supports, and connected by lines of horizontal beams. Any change in the section at a column is made at the floor level and the splices are perfectly rigid. The section of each beam is uniform throughout its length. The width of column is taken into account. However, the beams are represented by their neutral axes. The relative rotation of the members at a beam to column connection is proportional to the moment transmitted by the connection, (viz.

the moment-rotation relationships of the connections known and assumed linear i.e., M. = stiffness × rotation). The deformations resulting from shear and direct stresses in the members, and the strut effect in the columns are neglected. In any particular problem, many of these assumptions may be amended to represent practical conditions more closely. It is desirable to keep the general expressions which follow as simple as possible. Therefore, it is assumed that only uniformly distributed vertical loads, w per unit length, act on the structure.

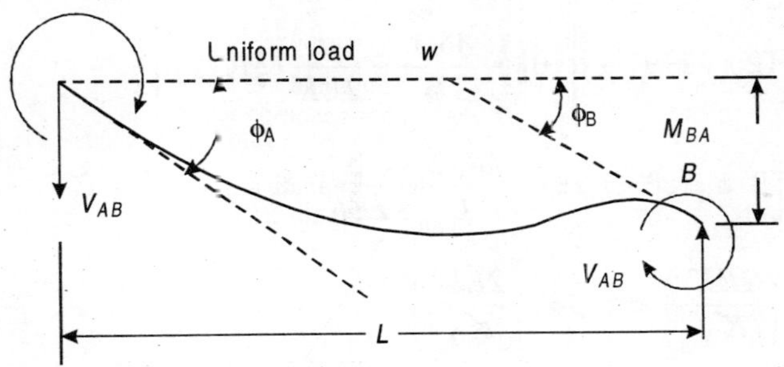

(a) Simply supported beam

(b) Part of a loaded frame with semi-rigid beam connection

Fig. 7.36

Consider a simply supported horizontal beam of length, L and relative stiffness, K with end couples M_{AB} and M_{BA} applied to its ends as shown in Fig. 7.36 (a). The end B of beam AB settles down by a height, δ relative to its end A. Let ϕ_A and ϕ_B be the rotations at the ends. Then

$$M_{AB} = 2EK\left[2_A + \phi_B - \frac{3\delta}{L}\right] + M_{F \cdot AB} \quad \text{...(i)}$$

$$M_{BA} = \left[2_A + 2\phi_B - \frac{3\delta}{L}\right] + M_{F \cdot AB} \quad \text{...(ii)}$$

where $M_{F.AB}$ and $M_{f.BA}$ are the fixed end moments due to loads acting over the beam, taken with correct sign. A part of a loaded frame structure having semi-rigid connections of beams with the columns is shown diagrammetically in Fig. 7.36 (b). The tangents and rotations at the ends of beams and columns are also shown. Total rotations at the joints A and B are $(\theta_A - \phi_A)$ and $(\theta_A - \phi_B)$ respectively. Since, the moments at the ends are proportional to the end rotation.

$$M_{AB} = K_A.(Q_A - \phi_A) \quad \text{...(iii)}$$

$$M_{BA} = K_B.(\theta_B - \phi_B) \quad \text{...(iv)}$$

where K_A and K_B are the rigidity co-efficients of the connections at A and B. Equating the moments in the expressions (i) and (ii) and moments in the expressions (ii) and (iv) and rearranging the terms

$$\phi_A = \left[(2\alpha + 1)\theta_A + \alpha \cdot \theta_B - \frac{3\delta \cdot \alpha}{L} + \frac{\alpha}{2EK} \cdot M_{F \cdot AB}\right] \quad \text{...(v)}$$

$$\phi_B = \left[\beta \cdot \phi_A (2\beta + 1) \cdot \theta_B - \frac{3\delta \cdot \beta}{L} + \frac{\beta}{2EK} \cdot M_{F \cdot BA}\right] \quad \text{...(vi)}$$

where $\alpha = \left(\frac{2EK}{K_A}\right)$ and $\beta = \left(\frac{2EK}{K_B}\right)$

Solving the expression (v) and (vi) for ϕ_A and ϕ_B,

$$\phi_A = \left(\frac{1}{3\alpha\beta + 2\alpha + 2\beta + 1}\right)\left[(2\beta + 1)\phi_A - \alpha \cdot \theta_B + 3\alpha(\beta + 1)\frac{\delta}{L} - \frac{\alpha}{2EK}\{(2\beta + 1)M_{F \cdot AB}\beta \cdot M_{F \cdot BA}\}\right] \quad \text{...(vii)}$$

$$\phi_B = \left(\frac{1}{3\alpha\beta + 2\alpha + 2\beta + 1}\right)\left[(-\beta \cdot \theta_A + 1) \cdot \phi_B - \alpha \cdot \theta_B + 3\beta(\alpha + 1)\frac{\delta}{L} - \frac{\alpha}{2EK}\{-\alpha \cdot M_{F \cdot AB}\beta \cdot (2\alpha + 1)M_{F \cdot BA}\}\right] \quad \text{...(viii)}$$

Values of θ_A and θ_B are substituted in the expressions (i) and (ii) and the terms are rearranged. The end moments M_{AB} and M_{BA} may be found as below:

$$M_{AB} = \left(\frac{(2\beta + 1)M_{F \cdot BA} - \beta \cdot M_{F \cdot BA}}{(3\alpha \cdot \beta + 2\alpha + 2\beta + 1)}\right) + \left(\frac{2EK}{(3\alpha \cdot \beta + 2\alpha + 2\beta + 1)}\right) \times \left[(3\beta + 2)\theta_B - (\beta + 1)\frac{3\delta}{L}\right] \quad \text{...(7.19)}$$

$$M_{BA} = \left(\frac{-\alpha . M_{F \cdot AB}(2\beta + 1)M_{F \cdot BA}}{(3\alpha \cdot \beta + 2\alpha + 2\beta + 1)}\right) + \left(\frac{2EK}{(3\alpha \cdot \beta + 2\alpha + 2\beta + 1)}\right) \times \left[\theta_A + (3\alpha + 2)\theta_B - (\alpha + 1)\frac{3\delta}{L}\right] \quad \text{...(7.20)}$$

In case, there is no settlement of end B, and for uniformly distributed load, w,

$$M_{F.AB} = \left(\frac{-wL^2}{12}\right) \text{ and } M_{F.BA} = \left(\frac{wL^2}{12}\right)$$

$$M_{AB} = \left(\frac{2EK}{(3\alpha \cdot \beta + 2\alpha + 2\beta + 1)}\right)\left[(3\beta + 2)\theta_A + \theta_B \frac{(3\beta + 1)wL^2}{24E \cdot K}\right] \quad ...(7.21)$$

$$M_{BA} = \left(\frac{2EK}{(3\alpha \cdot \beta + 2\alpha + 2\beta + 1)}\right)\left[\theta_A\,(3\alpha + 2)\theta_B + \theta_B \frac{(3 + 1)wL^2}{24E \cdot K}\right] \quad ...(7.22)$$

The shears at the ends of beam are given by

$$V_{AB} = V_{S.AB} + \left(\frac{M_{AB} + M_{BA}}{L}\right) \quad ...(7.23)$$

$$V_{BA} = V_{S.BA} + \left(\frac{M_{AB} + M_{BA}}{L}\right) \quad ...(7.24)$$

where $V_{S.AB}$ and $V_{S.BA}$ are the shears at the ends of beam as a simply supported beam considering the shear to the right of section as positive.

Consider the *equilibrium of any joint,* say B. Figure 7.37 (a). The end shears eccentric with respect to the joint. Let the eccentries of end shears be e_{BA} and e_{BC} as shown in Fig. 7.37 (b) for the equilibrium of joint B.

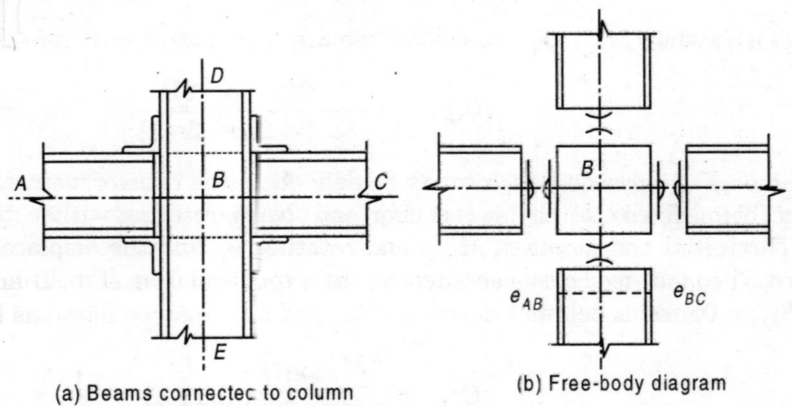

(a) Beams connected to column

(b) Free-body diagram

Fig. 7.37 *Equilibrium of joint*

$$M_{BA} + M_{BC} + M_{BD} + M_{BE} + M_{BE} + V_{BA} \cdot e_{BA} + V_{BC} \cdot e_{BC} = 0$$

$$\text{or } \Sigma M_B + V_{S.BA} \cdot e_{BA} + V_{SBC} \cdot e_{BC} + \frac{e_{BA}}{L_{BA}}(M_{AB} + M_{BA}) + \frac{e_{BC}}{L_{BC}}(M_{BC} + M_{CB}) = 0 \quad ...(7.25)$$

For the equilibrium of a storey

Σ (column moments of a storey) + $P_{H}h = 0$ ...(7.26)

where P_H is the total horizontal shear and h is the storey height.

7.14 MODIFIED MOMENT DISTRIBUTION METHOD

Analysis of the bending frames with semi-rigid beam to column connections may also be done by the modified moment distribution method. The expressions for fixed end moments, stiffness and carry-over factors for the members will be modified because of end rotations in the semi-rigid connections.

The joints are considered first fixed in position. The deformations (viz., joint rotations θ_A and θ_B, and the displacement, δ) shall be zero. The *modified fixed end moments* are now called as *initial end moments*. The initial end moments $M_{I.AB}$ and $M_{I.BA}$ may now be written from Eqs. 7.19 and 7.20 below

$$M_{i.AB} = \left(\frac{(2\beta+1)M_{F.AB} - \beta \cdot M_{F.BA}}{(3\alpha \cdot \beta + 2\alpha + 2\beta + 1)}\right) \quad ...(7.27)$$

$$M_{i.AB} = \left(\frac{-\alpha \cdot M_{F.AB} + (2\alpha+1)M_{F.BA}}{(3\alpha \cdot \beta + 2\alpha + 2\beta + 1)}\right) \quad ...(7.28)$$

The modified stiffness may be defined as the moment required for a unit rotation (viz., 1 radian) at the joint when the far end is kept fixed. Then, M_{IBA}, θ_B and δ shall remain zero, and $\theta_A = 1$ from Eq. 7.19.

$$K'_{AB} = \left(\frac{2EK \cdot (3\beta+2)}{(3\alpha \cdot \beta + 2\alpha + 2\beta + 1)}\right) \quad ...(7.29)$$

Similarly, when M_{IBA}, θ_A and δ shall remain zero, and $\theta_B = 1$ from Eq. 7.20.

$$K'_{AB} = \left(\frac{2EK \cdot (3\beta+2)}{(3\alpha \cdot \beta + 2\alpha + 2\beta + 1)}\right) \quad ...(7.30)$$

The *modified carry over factor* may be defined as the ratio of moment at the far end (being fixed) to that at the near end (being rotating without translation). The initial end moment, M_{iBA}, end rotating, θ_B and the displacement, δ are zero. The *modified carry-over factor* from *A* to *B* and from *B* to *A* ($\theta_B = 8 = 0$, and M_{iAB}, = 0) are as below :

$$C_{AB} = \left(\frac{M_{BA}}{M_{BA}}\right) = \left(\frac{1}{(3\beta+2)}\right) \quad ...(7.31)$$

$$C_{BA} = \left(\frac{M_{AB}}{M_{BA}}\right) = \left(\frac{1}{(3\beta+2)}\right) \quad ...(7.32)$$

The moment developed at fixed ends due to settlement of any support, *d* relative to the other may be determined by considering initial end moments M_{iAB}, and M_{iBA},, and end rotations θ_A and θ_B as zero. Then

$$M_{AB.}\delta = \left(\frac{6(\beta+1)EK}{(3\alpha \cdot \beta + 2\alpha + 2\beta + 1)}\right) \quad ...(7.33)$$

$$M_{BA}\delta = \left(\frac{6(\alpha+1)EK}{(3\alpha\cdot\beta+2\alpha+2\beta+1)}\right) \quad ...(7.34)$$

The analysis may be simplified by having advantage of symmetry or anti-symmetry, if any. For symmetry, the end rotations $\theta_A = -\theta_B$ and $\alpha = \beta$. The initial end moments and displacement, δ shall be zero. From Eq. 7.19

$$K_{AB\,.\,sym} = K_{BA\,.\,sym} = \left(\frac{2EK}{(\alpha+1)}\right) \quad ...(7.35)$$

And for initial end moments

$$M_{F\,.\,AB} = -M_{F\,.\,BA}$$

$$M_{i\;AB\,.\,sym} = M_{i\,.\,BA\,.\,sym} = \left(\frac{M_{F\cdot AB}}{(\alpha+1)}\right) \quad ...(7.36)$$

For *anti-symmetry, the end rotations* $\theta_A = \theta_B = 1$ and $\alpha = \beta$. The initial end moments and displacement, δ shall be zero. From Eq. 7.19

$$K_{AB\,.\,sym} = K_{BA\,.\,sym} = \left(\frac{6EK}{(3\alpha+1)}\right) \quad ...(7.37)$$

And for initial end moments

$$M_{F\,.\,AB} = +M_{F\,.\,BA}$$

$$M_{i\,.\,AB\,.\,asym} = M_{i\,.\,BA\,.\,asym} = \left(\frac{M_{F\cdot AB}}{(3\alpha+1)}\right) \quad ...(7.38)$$

For uniformly distributed load

Since, $\left(-M_{F\cdot AB} = M_{F\cdot BA} = \frac{wL^2}{12}\right)$, the initial end moments from Eq. 7.19 are as follows:

$$M_{i\,.\,AB} = \left(\frac{-(3\alpha+1)}{(3\alpha\cdot\beta+2\alpha+2\beta+1)}\right)\cdot\frac{wL^2}{12} \quad ...(7.39)$$

$$M_{i\,.\,AB} = \left(\frac{(3\alpha+1)}{(3\alpha\cdot\beta+2\alpha+2\beta+1)}\right)\cdot\frac{wL^2}{12} \quad ...(7.40)$$

Example 7.8 *A beam of uniform section 6 m long carries a uniformly distributed load 40 kN/m. The end moments and span moment are equal. Determine the stiffness of end connections and end rotation.*

Solution

The end moments and span moment are equal. It is possible when

$$-M_{AB} = M_{BA} = \left(\frac{wL^2}{16}\right) \quad ...(i)$$

And, the similar semi-rigid connections having equal rigidities are used at both the ends. Therefore,

$$\alpha = \beta = \left(\frac{2EK}{K_A}\right)$$

Fixed end moments

$$-M_{F\cdot AB} = M_{F\cdot BA} = \left(\frac{wL^2}{12}\right) \quad \text{...(ii)}$$

The initial end moment from Eq. 7.27

$$M_{i.AB} = \left(\frac{(2\beta+1)M_{F\cdot AB} - \beta\cdot M_{F\cdot BA}}{(3\alpha\cdot\beta+2\alpha+2\beta+1)}\right) \quad (\because \alpha = \beta)$$

$$-\frac{wL^2}{16} = \left(\frac{-(3\alpha+1)\cdot wL^2}{(3\alpha+1)(\alpha+1)\cdot 12}\right) \quad \therefore \alpha = \frac{1}{3}$$

The stiffness of semi-rigid connection at A

$$K_A = \left(\frac{2EK}{\frac{1}{3}}\right) = \mathbf{6\ E.K}$$

The bending moment diagram shall be as shown in Fig. 7.29 (b). From the area-moment theorem, area of bending moment diagram from end to half the span.

End rotation

$$\theta_A = \frac{1}{EI}\left(\frac{-wL^2}{16}\cdot\frac{L}{2}+\frac{2}{3}\cdot\frac{wL^2}{8}\cdot\frac{L}{2}\right)$$

$$\theta_A = \left(\frac{wL^3}{96EI}\right) = \left(\frac{40\times6^3\times(1000)^3}{96EI}\right) \text{ rads.}$$

$$\theta_A = 90\times(1000)^3\times(EI)^{-1} \text{ rads.} \quad \text{... (iii)}$$

For simply supported span, the end rotation

$$\theta_A = \left(\frac{wL^3}{24EI}\right) = \left(\frac{40\times6\times(1000)^3}{24EI}\right) \text{ rads}$$

$$\theta_A = 360\times(1000)^3\times(EI)^{-1} \text{ rads}$$

From the expressions (iii) and (iv) it is seen that the end rotation for semi-rigid connection is 25 percent of that for simply supported span. The rigidity of end connections is 75 percent.

Example 7.9 *A single bay single storey portal frame carries uniformly distributed load 40 kN/m over its beam 6 m long. The rigidity of semi-rigid connection at either end of beam with the column is 80 percent. Determine the design moments. The columns and beam have uniform cross-section.*

Solution

Step 1: End moment and end rotation

The rigidity of semi-rigid connection at end *B*, (and 50 also of end *C*) of beam *BC* of portal frame *ABCD* as shown in Fig. 7.38 is 80 percent. Therefore, the fixed end moment developed shall be 80 percent and free span rotation at the end shall be 20 percent.

End moment, $(M_{BC} = M_{CB})$

$$M_{BC} = \left(\frac{0\cdot 8\times wL^3}{12}\right) = \left(\frac{0\cdot 8\times 40\times 6\times 6}{12}\right) = 96 \text{ kN-m}$$

End rotation, $(\theta_{BC} = \theta_{CB})$

$$\theta_{BC} = \left(\frac{0\cdot 20\times wL^3}{24EI}\right) = \left(\frac{0\ 20\times 40\times 6^3}{24}\right) 12 \times 6\ (EI)^{-1} \text{ rads.}$$

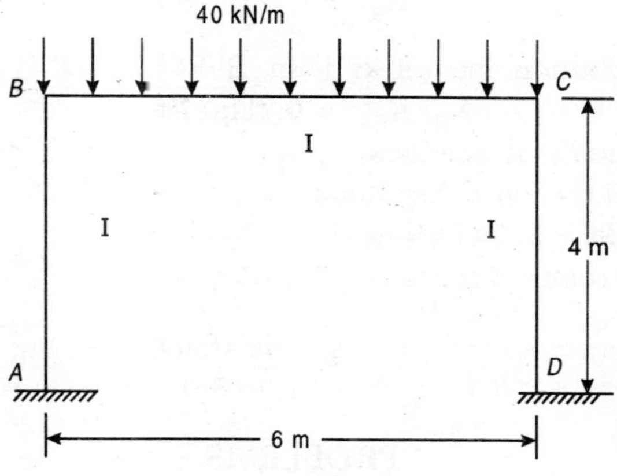

Fig. 7.38

Step 2: Stiffness of members

Modified stiffness of joint $K_B = K_C$

$$K_B = \left(\frac{\text{End moment}}{\text{End rotation}}\right) = \left(\frac{90}{12{\times}6(EI)^{-1}}\right) = \frac{8EI}{6} = \text{rad.}$$

Relative stiffness of beam $K_{BC} = \dfrac{1}{6}$

$$\text{By symmetry } \alpha = \beta = \left(\frac{2EK_{BC}}{K_B}\right) = \left(\frac{2E\cdot\dfrac{I}{6}}{8E\dfrac{I}{6}}\right) = 0.25$$

Fixed end moment $(M_{F.BC} = -M_{F.CAB})$

$$M_{F.BC} = \left(\frac{wl^2}{12}\right) = \left(\frac{40\times 6\times 6}{12}\right) = 120 \text{ kN-m}$$

For symmetry

Initial end moment $(M_{i.BC}, M_{i.CB})$

$$M_{i.BC} = \left(\frac{M_{F.BC}}{1+0\cdot25}\right) = \left(\frac{120}{1+0\cdot25}\right) = 96 \text{ kN-m}$$

Relative stiffness of end B

$$K_{BC.\,sym} = \left(\frac{2E \cdot K_{BC}}{1+\alpha}\right) = \left(\frac{2E \cdot \dfrac{I}{6}}{1+0\cdot25}\right) = 0.267\ EI$$

For column BA fixed at A

$$K_{BA} = \left(\frac{4EI}{4}\right) = EI$$

Step 3: Distribution factors at joint B

$$K_{BC}.K_{BA} = 0.211{:}0.789$$

Step 4: Moments at sections

Moment at B at the top of the column

$M_{BA} = 0.789 \times 96 = 75.744$ kN-m

Moment at the centre of beam

$$= \left(\frac{40\times6\times6}{8}\right) - 75.256 \text{ kN-m}$$

PROBLEMS

7.1. AN MB 200, @ 0.254 kN/m transmits an end reaction of 186 kN to the web of an MB 450, @ 0.794 kN/m. Design a framed connection. Give a neat sketch.

7.2. An LB 325, @ 0.431 kN/m transmits an end reaction of 124 kN to the flange of a column HB 300, @ 0.630 kN/m. Design an unstiffened seated connection.

7.3. An LB 600, @ 0.995 kN/m transmits an end reaction of 286.0 kN to the flange of an HB 400, @ 0.774 kN/m column. Design stiffened seated connections.

Provide (i) a single stiffener angle (ii) double stiffener angles.

7·4. In a framed connection, an LB 325, @ 0.431 kN/m transmits an end reaction 148.0 kN and a moment 21.0 kN-m to a column HB 300, @ 0.530 kN/m. Design a clip angle connection.

7.5. In a framed connection, an MB 450, @ 0.724 kN/m transmits an end reaction 120 kN and an end moment 156.0 kN-m to a column HB 400, @ 0.774 kN/m. Design a bracket connection.

7.6. In a framed connection, an MB 500, @ 0.869 kN/m transmits an end reaction 162 kN and an end moment 36 kN-m to HB 400, @ 0.822 kN/m. Design a split beam connection.

7.7. Design an unstiffened seat connection to support a beam LB 350, @ 0.495 kN/m transmitting end reaction of 100 kN to the flange of column HB 200, @ 0.400 kN/m. Use 16 mm dia. rivets. Show the details in a neat sketch.

Chapter 8

Design of Plate Girder

8.1 INTRODUCTION

When the span and load combination is such that the rolled steel beams become insufficient to furnish the requirement, and built-up beams become uneconomical, then plate girders are used. The built-up beams are used where overall depth is limited. As such the use of beams, built-up beams and plate girder is a step by step approach for the increase in loads and the spans. The object of design is to achieve overall economy, which involves the cost of fabrication in addition to the cost of material. The cost of fabrication is more for built-up beams as compared to beams, and it is still higher for plate girders as compared to both. Attempt is made to provide deep sections for economy as regards material and cost of fabrication.

The *plate girders* are deep structural members subjected to transverse loads. They consist of plates and angles riveted together. Plates and angles form an I-section. They are used in building constructions and also in bridges. The plate girders are economically used for span upto about 30 m in building construction. The depth of plate girder may range upto 5 m or more, 1.5 m to 2.5 m depths are very common. Common sections used for plate girders are shown in Fig. 8.1.

Figure 8.1 (a) shows the simplest form of a plate girder. In case, this simple section cannot furnish sufficient flange material, additional plates are riveted to outstanding legs of angles as shown in Fig. 8.1 (b) and (c). When number of cover plates becomes excess, then the section of plate girder is modified as shown in Fig. 8·1 (d). In case of deck girders of rail-road bridges, sometimes a form shown in Fig. 8.1 (e) is used. Sometimes for long span bridges carrying heavy loads, heavy sections shown in Fig. 8.1 (f) and (g) are used. Section shown in Fig. 8.1 (h) is adopted for crane runway girders.

Sometimes in building construction, the depth of a girder is limited to provide necessary head room. Limitations of head room require the depth of girder much less than the economical depth. It becomes necessary to provide large

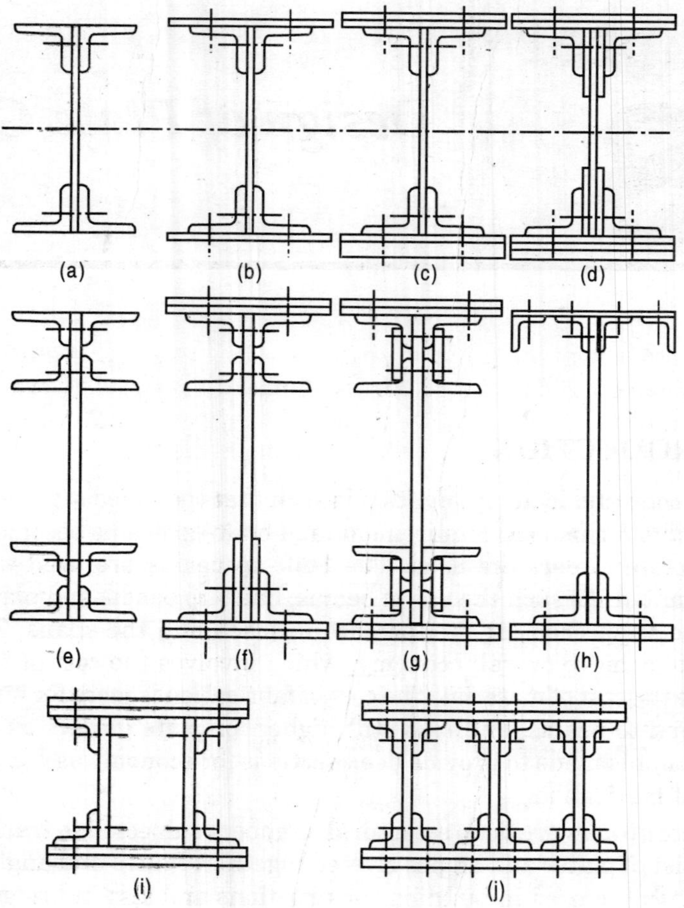

Fig. 8.1 *Common section of plate girder*

web area to resist shear forces. In such cases, sections shown in Fig. 8.1 (i) and (j) are often used in which two or more webs are provided. These are called *box girders.* In addition to shear, webs resist bending moment to a considerable extent.

8.2 ELEMENTS OF PLATE GIRDER

The vertical plate of the plate girder is termed as *web plate.* The angles connected at the top and bottom of the web plate are known *as flange angles.* The horizontal plates connected with the flange angles are known as *flange plates* or *cover plates*. The web plate, flange angles and flange plates are shown in Fig. 8.2. The bearing stiffeners, intermediate stiffeners and horizontal stiffeners used with the plate girder are shown in Fig. 8.3.

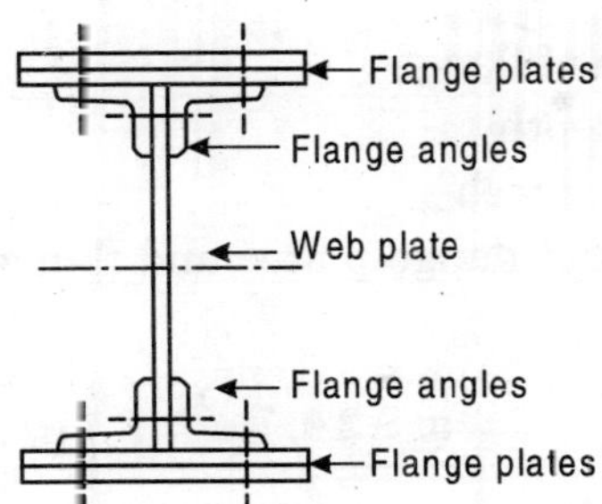

Fig. 8.2 *Elements of plate girder*

A web plate is kept *unstiffened* when the ratio of clear depth to thickness of web is less than 85. It does not require stiffeners. A web plate is called *stiffened*, when the ratio of clear depth to thickness of web is greater than 85 and stiffeners are provided to contribute additional strength to web.

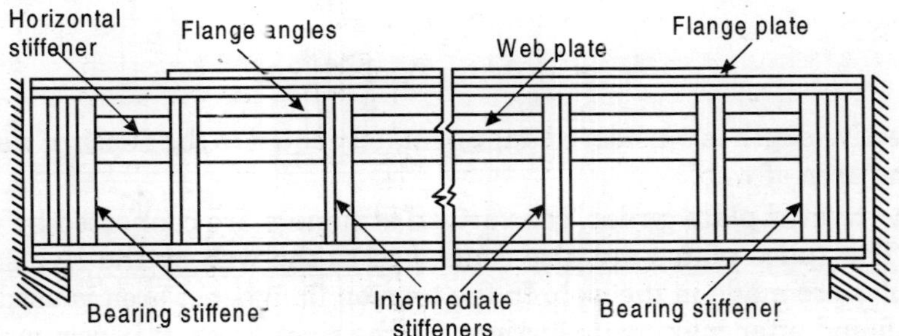

Fig. 8.3 *Elements of a plate girder*

8.3 AREA OF FLANGES OF PLATE GIRDER

When a plate girder is carrying transverse load, its top flange is in compression and bottom flange is in tension. The area of flange comprises of areas of flange angles, flange plates (if flange plates are attached) and part of the web plate. The parts of web plate sandwiched between flange angles also resist bending moment. The part of web plate which is available for flange area is computed as below.

Bending moment, $$M = \sigma_{bc}\left(\frac{I_{xx}}{y_{\max}}\right)$$

where, I_{xx} = Gross moment of inertia of plate girder

σ_{fc} = Allowable bending stress

$y_{\max}$ = Vertical distance of extreme fibre of section from neutral axis

The gross moment of inertia of plate girder

$$I_{xx} = \left[2A_f\left(\frac{d_e}{2}\right)^2 + \frac{t_w \cdot d_e^{\,3}}{12}\right] \text{approximately} \quad ...(i)$$

where, A_f = Gross area of flange
d_e = Effective depth
t_w = Thickness of web

The moment of inertias of flange plates and flange angles about their own axes are neglected. Then

$$M = \left(\frac{\sigma_b \times 2A_f \times d_e^{\,2}}{\frac{d_e}{2\times 4}}\right) + \left(\frac{\sigma_b \times t_w \times d_e^{\,3}}{\frac{d_e}{2\times 12}}\right) \quad (ii)$$

$$M = (A_f \times \sigma_b \times d_e) + (t_w \times d_d) \times \sigma_b \times \frac{d_e}{6} \quad (iii)$$

$$M = (A_f \times \sigma_b \times d_e) + \left(A_w \times \sigma_b \times \frac{d_e}{6}\right) \quad (iv)$$

$$\therefore \quad M = \left(A_f + \frac{1}{6}A_w\right) \times \sigma_b \times d_e \quad (8.1)$$

The effective depth is considered approximately equal to the depth of web, where, A_w is the area of web.

In the riveted plate girder, the vertical stiffeners are connected by using the rivets. The splice plates are also connected to the web by the rivets. As such rivet holes are made in the web. In the tension flange, net area is used. The net area is found after making deductions for the rivet holes. It is assumed that 25 mm diameter holes are made for rivets at 100 mm spacing. Then the net area of web is three-fourth of the gross-area.

Therefore, net area of tension flange

$$A_{f\,\text{net}} = \left(A'_f + \frac{3}{4} \times \frac{1}{6} A_w\right) = \left(A'_f + \frac{1}{8} A_w\right) \quad ...(v)$$

where, A'_f = Net area of flange excluding web.

From this, it is seen that approximately $\frac{1}{6}$ th web area acts as flange area if the gross area is considered, and $\frac{1}{8}$ th web area (on account of reduction in flange area for rivet holes) acts as flange area if the net area is considered. The portion of web which acts as flange, is termed as *web equivalent*.

The effective sectional area of compression flange shall be the gross area with deductions for excessive width of plates as it was specified for compression members and open holes occurring in a plane perpendicular to the direction of stress at the section being considered.

The effective sectional area of tension flanges shall be the gross-area with the deduction for holes where the rivets are also used.

8.4 DEPTH OF PLATE GIRDER

The depth between outer surfaces of flanges is termed as **over-all depth** d_0 or depth of the plate girder. In general, the depth of plate girder is kept $\frac{1}{10}$ to $\frac{1}{12}$ of span. The distance between C.G. of compression flange and C.G. of tension flange is known as **effective depth** d_e of the plate girder. The distance between vertical legs of flange angles at the top and at the bottom is known as **clear depth** d of plate girder, in case horizontal stiffeners are not used. The overall depth, effective depth and clear depth of plate girder are shown in Fig. 8.4. The depth of plate girder is an essential factor for the design of plate girder. Sometimes, structural

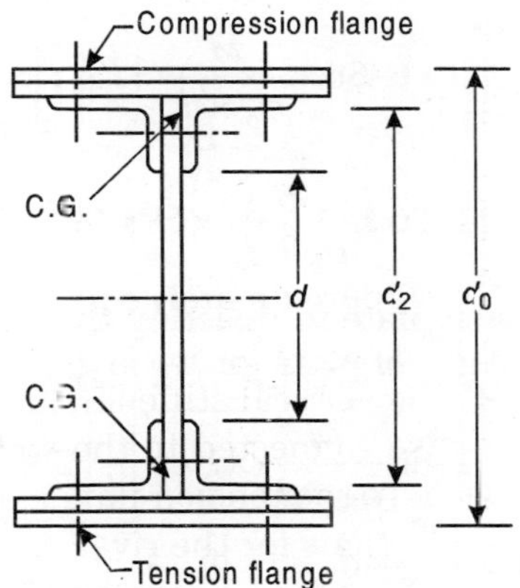

Fig. 8.4 *Different depths of plate girder*

requirements limits the depth of plate girder. When the depth of plate girder is less than 750 mm, then, such girders are known as **shallow plate girders.** The shallow plate girders resemble rolled steel beams. When the depth of plate girder is more than 750 mm or when the depth is atleast eight times the depth of vertical leg of the flange angles, then, such girders are known as **deep plate girders.**

8.5 ECONOMICAL DEPTH OF PLATE GIRDER

The *economical depth* of the plate girder is the depth which gives the minimum weight of plate girder. The economical depth of plate girder is found as below:

Let M = Maximum bending moment

d_e = Effective depth of plate girder (Depth of web plate)

t_w = Thickness of web plate

σ_b = Allowable bending stress

l = Length of plate girder

ρ = Unit weight of steel

Area of flange $A_f = \dfrac{M}{(\sigma_b \cdot d_e)}$

When two or more cover plates are used in plate girder in each flange, then cover plates are curtailed. The depth of plate girder varies as the cover plates are curtailed. The part of total weight of plate girder (comprising of flanges, web plates, splice plates, stiffeners and filler plates) varies with the variation in the depth of plate girder. As such, the average section of flange is approximately reduced to 80 per cent. The weight of stiffeners is approximately 60 per cent of web plate.

The total weight of plate girder

W = Average weight of flanges + weight of web plate inclusive of stiffener

$$W = \left[2 \times 0 \cdot 80 \times \frac{M}{\sigma_b \times d_e} \times l \times \rho\right] + [1 \cdot 6 \times t_w \times d_e \times l \times \rho] \quad \text{...(i)}$$

$$W = \left[1 \cdot 60 \times \frac{M}{\sigma_b \times d_e} \times l \times \rho\right] + [1 \cdot 6 \times t_w \times d_e \times l \times \rho] \quad \text{...(ii)}$$

In case, the portion of web plate in resisting the bending moment is considered for, then, the total weight of plate girder is given by

$$W = \left[1 \cdot 60 \frac{M}{\sigma_b \times d_e} \times l \times \rho\right] - \left(2 \times \frac{1}{8} t_w \times d_e \times l \times \rho\right)$$

$$+ [1.6 \times t_w \times d_e \times l \times \rho] \quad \text{...(iii)}$$

Differentiating the expression (iii) for W with respect to d_e

$$\left(\frac{dW}{dd_e}\right) = -\left[1 \cdot 60 \frac{M}{\sigma_b \times d_e^2} \times l \times \rho\right] + [1.6 \times t_w \times l \times \rho]$$

$$-\left[\frac{1}{4} \times t_w \times d_e \times l \times \rho\right] \quad \text{...(iv)}$$

For the economical depth, $\left[\left(\dfrac{dW}{d(d_e)}\right)\right] = 0$

$$\therefore \quad d_e = 1.1\left(\frac{M}{\sigma_b \times t_w}\right)^{1/2} \quad \text{...(8.2)}$$

In case the portion of web plate in resisting the bending moment is not considered, then the total weight of plate girder is given by the expression (i). Differentiating the expression (ii), and equating to zero, the economical depth of plate girder works out to be $\sqrt{\left(\frac{M}{\sigma_b \times t_w}\right)}$. When the span of plate girder is small, then the flange plates may not be curtailed, and the section is uniform throughout the span. For such plate girder, the economical depth works out to $1.21\left(\frac{M}{\sigma_b \times t_w}\right)^{1/2}$.

The weight of plate girder varies 1 percent for 10 percent variation in the depth of the girder. In practice, economical depth of plate girder is assumed 15 to 20 percent less than that computed.

8.6 SELF-WEIGHT OF PLATE GIRDER

From Sec. 8.5, the total weight of plate girder

$$W_1 = \left[1.60\frac{M}{\sigma_b d_e} \times l \times \rho + 1.60\, t_w \times l \times \rho\right] \quad \text{...(i)}$$

The economical depth of plate girder, from Eq. 8.2

$$d_e = 1.1\left(\frac{M}{\sigma_b \cdot t_w}\right)^{1/2}$$

$$\therefore \quad t_w = 1.21\left(\frac{M}{\sigma_b \cdot d_e^2}\right) \quad \text{...(ii)}$$

Therefore,

$$W_1 = \left[\left(1.60\frac{M}{\sigma_b d_e} \times l \times \rho\right) + \left(1.60 \times 1.21 \times \frac{M}{\sigma_b \cdot d_e^2} \times d_e \times l \times \rho\right)\right.$$

$$\left. - \left(\frac{1}{4} \times 1.21 \times \frac{M}{\sigma_b \cdot d_e^2} \times d_e \times l \times \rho\right)\right] \quad \text{...(iii)}$$

$$W_1 = \left[\left(1.60\frac{M}{\sigma_b d_e} \times l \times \rho\right) + \left(1.35 \times 1.21 \times \frac{M}{\sigma_b \cdot d_e} \times l \times \rho\right)\right] \quad \text{...(iv)}$$

$$W_1 = 3.2\left(\frac{M \cdot l\rho}{(\sigma_b d_e)}\right) \quad \text{...(v)}$$

The expression (v) is for the weight of plate girder with gross-area of the flanges without making deductions for rivet holes. The weight of plate girder is about 10 to 20 percent greater than that given by the expression (v). The unit weight of steel ρ is 78450 N/m³. Therefore, the weight of plate girder is given by

$$W_1 = \left(\frac{1 \cdot 2 \times 3 \cdot 2 \times 78450 \times Ml}{100 \times 100 \times 100\sigma_b \times d_e} \times 100 \times 10\right)$$

or

$$W_1 = \left(\frac{300 \times Mi}{\sigma_b \cdot d_e}\right) \quad \text{...(vi)}$$

where, W_1 = Weight of the plate girder in N
M = Bending moment in N-mm
l = Span of the plate girder in metre
σ_b = Bending stress in N/mm²
d_e = Effective depth of plate girder in mm.

The expression (vi) is useful when the bending moment M and the depth, d_e are known. Therefore, this expression is used to check the self-weight of the plate girder.

For all practical purpose, self-weight of girder is taken as

$$W_1 = \left(\frac{W \cdot l}{300}\right) \quad \text{...(8.3)}$$

where W = Total superimposed load
l = Span of plate girder in metres

The expression for self-weight of plate girder is approximate since it is assumed that web plate resists shear only and bending moment is resisted by flanges.

8.7 IMPACT

The dynamic action of the moving load on a structure is called *impact.*

The impact is caused due to sudden application of load, unbalanced driving wheel of locomotives, uneven track, wheel tyres worn out of truth, etc. The stresses and strains (deformations) due to sudden application of load are twice than those due to gradual application of load simultaneously and the actual increase in deformation is generally less than twice that due to gradually applied load because of damping effect of other construction. Therefore, the effect of impact is considered less than 100 percent. The effect of impact depends upon nature of moving load, speed of moving load, type of structure, loaded length of structure etc. Provision for effect of impact is made by impact allowance or impact load. The effect of impact is generally expressed as the percentage of the moving load. Impact load is determined as a product of *impact factor* and live (moving) load. The impact factor is specified by different authorities separately for different type of moving loads and different type of structures. Indian Railway Board has specified for steel girders for single track span, for metre gauge and broad gauge impact factor as under.

Impact factor $$i = \left(\frac{20}{14 + L}\right) \ngtr 1.00 \quad \text{...(8.4)}$$

where, L is the loaded length of span in metres.

8.8 ASSUMPTIONS FOR DESIGN FOR PLATE GIRDER

The plate girders are designed by trial and revised until the actual bending stresses are less than or equal to the permissible bending stresses. The stiffeners are attached as needed, until the shear stress is less than or equal to the allowable shear stress. The web plate is designed on the basis of economical depth of the web. The initial design of flange size is made by *the flange area method* and it is finally checked by moment of inertia method. The moment carried by the flange is

$$M_f = \left(A_f \cdot \sigma_b' \left(d_0 - t_f\right)\right) \qquad \text{...(i)}$$

and the moment shared by the web is

$$M_w = \left(\sigma_b'' \cdot Z_w\right) = \left(\frac{\sigma_b'' t_w \cdot d_e^{\ 2}}{6}\right) \qquad \text{...(ii)}$$

The total moment resisted by the plate girder

$$M = (M_f + M_w) \qquad \text{...(iii)}$$

However, to simplify, the following assumptions are made for the design of the plate girder:

1. The web plate resists the shear force, and the shear stress is uniformly distributed over whole cross-sectional area of web.
2. The flanges resist the bending moment. The distribution of bending stress in the flanges in uniform.

8.9 ALLOWABLE BENDING STRESS

The maximum bending compressive stress calculated for the plate girders bending about the axis of maximum strength (*xx*-axis), $\sigma_{bc\cdot cal}$ shall not exceed the maximum permissible bending compressive stress, σ_{bc} in N/mm² (MPa) obtained by the following formula

$$\sigma_{bc} = 0.66 \left(\frac{f_{cb} . f_y}{\left[(f_{cb})^n + (f_y)^n\right]^{1/n}}\right)$$

where f_{cb} = Elastic critical stress in bending
f_y = Yield stress in the steel in N/mm² (MPa); and
n = A factor assumed as 1.4.

The expression to calculate the elastic critical stress, f_{cb} has been given in Chapter 6 'Design of Beam'.

The values of σ_{bc} as derived from the above formula for some of the Indian Standard structural steels have also been given in Table 6.2 in Chapter 6, 'Design of Beam'. Same values as given in Table 6.2 are used for the plate girders.

8.10 ALLOWABLE SHEAR STRESS

The allowable average shear stress for unstiffened web of plate girder calculated on the cross-section of the web $\tau_{va.cal}$ shall not exceed, the permissible average shear stress as recommended by IS: 800–1984, τ_{va} as below :

$$\tau_{va} = (0.4 f_y)$$

where, f_y is the yield stress for the structural steel to be used.

The web plate of plate girder may be classified as (i) unstiffened (ii) stiffened for shear, and (iii) stiffened for shear and bending. The unstiffened web plate must be sufficiently thick to resist the shear with an adequate factor of safety against shear buckling. The stiffened web plate is designed to resist the shear with some margin of the respect to shear buckling. The longitudinal stiffeners are used for web stiffened for both shear and bending. The longitudinal stiffeners increase the buckling resistance of the web in the region of the compressive bending stress.

For the stiffened webs, the average shear stress in a member calculated on the cross-section of the web, $\tau_{va \cdot cal}$ shall not exceed the permissible average shear stress as recommended by IS: 800 –1984. The value of permissible average shear stress τ_{va} are given in Tables 8.1 (a), (b) and (c) for appropriate values of yield, stresses 250, 340 and 400 N/mm² (MPa), respectively.

The values τ_{va} for stiffened webs for steel for which the yield stress is not given in Table 8.1 (a), (b) and (c) shall be found by using the following formulae, provided that the average stress t_{va} shall not exceed 0.4 f_y.

1. For the webs where the distance between the vertical stiffeners is less than the clear depth *d*.

$$\tau_{va} = 0.4 f_y \left[1 \cdot 3 - \frac{f_y^{1/2} \cdot \dfrac{c}{t_w}}{4000 \left\{ 1 + \dfrac{1}{2} \left(\dfrac{c}{d} \right)^2 \right\}} \right] \text{N/mm}^2 \qquad \text{...(8.5)}$$

2. For the webs where the distance between the stiffeners is more than, *d*.

$$\tau_{va} = 0.4 f_y \left[1 \cdot 3 - \frac{f_y^{1/2} \cdot \dfrac{c}{t_w}}{4000 \left\{ 1 + \dfrac{1}{2} \left(\dfrac{c}{d} \right) \right\}} \right] \text{N/mm}^2 \qquad \text{...(8.6)}$$

where τ_{va} = maximum permissible average shear stress

c = distance between the vertical stiffeners

t_w = thickness of web.

For the vertically stiffened webs without horizontal stiffeners

d is the clear distance between flange angles, or where there are no flange angles, the clear distance between the flanges ignoring the fillets.

For the vertically stiffened webs with the horizontal stiffeners

d is the clear distance between the tension flange (angles flange plates) and the horizontal stiffeners.

Table 8.1 (a) *Permissible average shear stress τ_{va} in stiffened webs of steel with $\sigma_y = 250$ N/mm² (MPa)*

$\left(\frac{d}{t_w}\right)$	*Stress τ_{va} (MPa) for Different Distance c Between Stiffners*												
	0.3d	*0.4d*	*0.5d*	*0.6d*	*0.7d*	*0.8d*	*0.9d*	*1.0d*	*1.1d*	*1.2d*	*1.3d*	*1.4d*	*1.5d*
90	100	100	100	100	100	100	100	100	100	100	100	100	100
95	100	100	100	100	100	100	100	100	100	100	100	100	99
100	100	100	100	100	100	100	100	100	100	100	99	99	98
105	100	100	100	100	100	100	100	100	100	99	98	97	96
110	100	100	100	100	100	100	100	100	99	98	96	95	94
115	100	100	100	100	100	100	100	100	98	96	95	94	93
120	100	100	100	100	100	100	100	98	96	95	93	92	91
125	100	100	100	100	100	100	98	97	95	93	92	91	90
130	100	100	100	100	100	99	97	96	94	92	90	89	88
135	100	100	100	100	100	98	96	94	92	90	89	87	86
140	100	100	100	100	99	96	95	93	91	89	87	86	85
150	100	100	100	100	97	94	92	90	88	86	84	83	81
160	100	100	100	98	94	92	89	88	85	83	81	80	78
170	100	100	100	96	92	89	87	85	82	80	78	76	75
180	100	100	98	94	90	87	84	82	80	77	75	73	72
190	100	100	97	92	88	84	82						
200	100	100	95	90	86	82	81						
210	100	99	93	88	83	81							
220	100	98	91	86	81	80							
230	100	96	90	84	79					Non-applicable zone.			
240	100	95	88	83	77								
250	100	93	86	82	74								
260	100	92	85	81									
270	99	90	84	81									

Note : Intermediate values may be obtained by linear interpolation.

Table 8.1 (b) *Permissible average shear stress τ_{va} in stiffened webs of steel with f_y = 340 N/mm² (MPa)*

(d/t_w)	Stress τ_{va} (MPa) for Different Distance c Between Stiffners												
	0.3d	0.4d	0.5d	0.6d	0.7d	0.8d	0.9d	1.0d	1.1d	1.2d	1.3d	1.4d	1.5d
75	136	136	136	136	136	136	136	136	136	136	136	136	136
80	136	136	136	136	136	136	136	136	136	136	136	136	136
85	136	136	136	136	136	136	136	136	136	136	136	134	133
90	136	136	136	136	136	136	136	136	136	135	133	132	131
95	136	136	136	136	136	136	136	136	135	133	131	129	128
100	136	136	136	136	136	136	136	135	132	130	128	127	126
105	136	136	136	136	136	136	135	133	130	128	126	124	123
110	136	136	136	136	136	135	133	131	128	126	124	122	120
115	136	136	136	136	136	133	131	129	126	123	121	119	118
120	136	136	136	136	135	131	129	127	124	121	119	117	115
125	136	136	136	136	133	129	127	125	121	119	116	114	113
130	136	136	136	135	131	127	125	122	119	116	114	112	110
135	136	136	136	134	129	126	123	120	117	114	109	108	108
140	136	136	136	132	127	124	121	118	115	112	109	107	105
150	136	136	135	129	124	120	117	114	110	107	104	102	100
160	136	136	132	126	120	116	113	110	106	102	99	97	95
170	136	136	129	123	117	112	109	106	101	98	95	92	90
180	136	135	127	119	113	108	105	102	97	93	90	87	84
190	136	133	124	116	110	105	100						
200	136	130	121	113	106	101	96						
210	136	128	118	110	103	97							
220	136	126	116	107	99	93				Non-applicable zone			
230	135	123	113	103	96								
240	134	121	110	100	92								
250	132	119	107	97	89								
260	130	116	104	94									
270	128	114	102	91									

Note. Intermediate values may be obtained by linear interpolation.

Table 8.1 (c) *Permissible average shear stress τ_{va} in stiffened webs of steel with $f_y = 400$ N/mm² (MPa)*

$\left(\frac{d}{t_w}\right)$	Stress τ_{va} (MPa) for Different Distance c Between Stiffners												
	0.3d	0.4d	0.5d	0.6d	0.7d	0.8d	0.9d	1.0d	1.1d	1.2d	1.3d	1.4d	1.5d
70	160	160	160	160	160	160	160	160	160	160	160	160	160
75	160	160	160	160	160	160	160	160	160	160	160	160	159
85	160	160	160	160	160	160	160	160	160	160	159	157	156
90	160	160	160	160	160	160	160	160	160	158	156	154	152
95	160	160	160	160	160	160	159	157	154	152	149	147	146
100	160	160	160	160	160	160	157	155	151	149	146	144	143
110	160	160	160	160	159	155	152	149	146	143	140	138	136
115	160	160	160	160	156	152	149	147	143	140	137	135	133
120	160	160	160	159	154	150	147	144	140	137	134	132	129
125	160	160	160	157	152	147	144	141	137	134	131	128	126
130	160	160	160	155	150	145	141	139	134	131	128	125	123
135	160	160	160	153	147	143	139	136	132	128	125	122	120
140	160	160	158	151	145	140	136	133	129	125	122	119	116
150	160	160	155	147	141	135	131	128	123	119	115	112	110
160	160	160	151	143	136	130	126	123	117	113	109	106	103
170	160	158	148	139	132	126	121	117	112	107	103	100	97
180	160	155	144	135	127	121	116	112	106	101	97	93	90
190	160	152	140	131	123	116	111						
200	160	149	137	127	118	111	<u>106</u>						
210	160	146	133	123	114	106							
220	157	143	130	119	109	<u>101</u>							
230	155	140	126	114	105			Non-applicable zone.					
240	153	137	123	110	100								
250	151	134	119	106	<u>96</u>								
260	148	131	116	102									
270	146	128	112	98									

Note. Intermediate values may be obtained by linear interpolation.

It is to note that the allowable shear stresses given in Table 8.1 (a), (b) and (c) apply provided any reduction in the web cross-section is due to only rivet holes. Where the large apertures are cut in the web, a special analysis is to be carried out to ensure the maximum permissible average shear stresses laid down in IS: 800 –1984 are not exceeded.

The maximum shear stress having regard to the distribution of stresses in confirmity with the elastic behaviour in bending, shall not exceed the value τ_{vm}, found as under

$$\tau_{vm} = (0.45\, f_y)$$

where, f_y is the yield stress of the structural steel to be used.

8.11 DESIGN OF WEB PLATE

In the design of web plate of plate girder, the economical depth and thickness of web plate are determined. Approximately, the economical depth of web plate is given by the following formula:

Economical depth of web plate

$$= 1.1\left(\frac{M}{\sigma_{bc} \times t_w}\right)^{1/2}$$

where, M = Maximum bending moment

σ_{bc} = Allowable bending stress

t_w = Thickness of web plate.

For the economy, the depth of web plate selected should be such that web plates are manufactured and are available for that depth. The depth of plate girder varies from $\frac{1}{10}$th and $\frac{1}{12}$th the span. The thickness of web plate should be such that maximum shear stress in the web does not exceed maximum allowable shear stress and average shear stress in the web does not exceed average allowable shear stress. The thickness of web plate should also provide necessary bearing area for rivets connecting web plate and flange angles, so that rivets are provided at proper spacing. A minimum thickness of 6 mm is adopted to provide for corrosion. The thickness of web plate is fixed keeping in view the buckling of web in shear, and shear along with bending.

The economical depth of web plate may be approximately determined by the use of *Rawter and Clark formula* also.

Economical depth of web plate

$$= K \cdot \left[\frac{M}{\sigma_{bc}}\right]^{\frac{1}{2}}$$

where, K = A parameter, which is adopted as 4.5 for the riveted plate girder

M = Maximum bending moment in N-mm

σ_{bc} = Allowable bending stress in N/mm^2 (MPa).

8.11.1 Web Buckling

The web plate of simply supported plate girder may be considered as subjected to pure shear close to the support, to pure bending at the centre, and to combined shear and bending at the intermediate positions, when subjected to uniformly distributed load. The web plates used are deep and thin. The web plate may fail by buckling.

Buckling is defined as sudden bending, warping, curling or crumpling of the elements under compressive stresses.

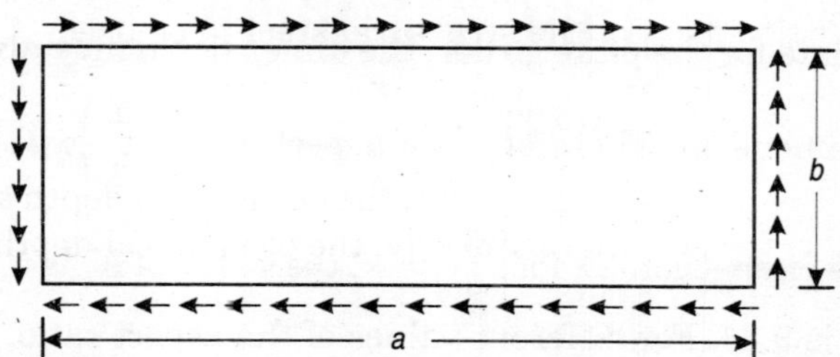

Fig. 8.5 *Uniformly distributed shear along edges*

8.11.1.1 Web Buckling in Pure Shear

When the web plate is subjected to pure shear uniformly distributed along the edge as shown in Fig. 8.5, then, the critical shear stress is given by the expression given below :

$$\tau_c = \left(\frac{\pi^2 \cdot E}{12\left(1-\mu^2\right)\left(\dfrac{b}{t_w}\right)^2} \cdot K_s \right) \quad \text{...(8.7)}$$

where μ = Poisson's ratio

K_s = Constant, for the plate subjected to pure shear

$\left(\dfrac{a}{b}\right)$ = Aspect ratio

a = Length of the plate

b = Width of the plate

t_w = Thickness of the plate

The value of constant, K_s depends upon aspect ratio $\left(\dfrac{a}{b}\right)$ and the support conditions for the edges of plate. The concept of deep plate girders was developed by considering the importance of this aspects ratio.

Plate simply supported along all four edges

The values of constant K_s are given by the following expressions:

$$K_s = \left[5 \cdot 34 + \frac{4 \cdot 00}{\left(\frac{a}{b}\right)^2}\right]\left(\text{for}\left(\frac{a}{b}\right) > 1\right) \quad \text{...(i)}$$

and

$$K_s = \left[4 \cdot 00 + \frac{5 \cdot 34}{\left(\frac{a}{b}\right)^2}\right]\left(\text{for}\left(\frac{a}{b}\right) < 1\right) \quad \text{...(ii)}$$

In the web plate for the plate girder, the dimension a may always be selected as the larger dimension. Therefore, the aspect ratio $\left(\frac{a}{b}\right)$ will be greater than unity. From the expression (i), for $\left(\frac{a}{b}\right) = \infty$, the value of K_s is 5.34. For $\left(\frac{a}{b}\right) = 1$, the value of K_s is 9.34. For different values of the aspect ratio, the values of K_s can be found from the expression (i). The values of K_s for some values of the aspect ratio have been given in Table 8.2

Table 8.2

$\left(\frac{a}{b}\right)$	1.0	1.2	1.4	1.6	1.9	2.0	3.0	5.0	∞
K_s	9.34	8.12	7.38	6.90	6.58	634	5.78	5.36	5.34

As the values of aspect ratio increase from 5.0 to ∞ the change in value of is K_s is insignificant. For unstiffened web, value of K_s can be assumed equal to 5.34 for all practical purposes. For structural steel value of Poisson's ratio is 0.3 and Young's modulus of elasticity E is 2.11×10^5 N/mm². Substituting these values in above expression Eq. (8.7) reduces to

$$\sigma_c = \left(\frac{10 \cdot 05 \times 10^5}{\left(\frac{b}{t_w}\right)^2}\right) \text{N/mm}^2 \quad \text{...(8.8)}$$

The value of *width of web plate may be* substituted equal to *clear depth of web* viz. *d*,

$$\therefore \quad \sigma_c = \left(\frac{10 \cdot 05 \times 10^5}{\left(\frac{d}{t_w}\right)^2}\right) \text{N/mm}^2 \quad \text{...(8.9)}$$

It is seen that the critical shear stress in the web is inversely proportional ratio $\frac{d}{t_w}$. When the value of critical shear stress reaches the value *of yield in shear,* viz. 150 N/mm², then

$$\left(\frac{d}{t_w}\right)^2 = \left(\frac{10 \cdot 05 \times 10^5}{150}\right)$$

$$\left(\frac{d}{t_w}\right)^2 = 6700, \left(\frac{d}{t_w} = 82\right)$$

When the ratio $\frac{d}{t_w}$ is less than 82, the failure of web plate takes place by yielding. When the ratio of $\frac{d}{t_w}$ is greater than 82, the buckling of web takes place. Therefore the allowable shear stress is decreased. As $\frac{d}{t_w}$ ratio increases, value of critical shear stress decreases rapidly. The value of critical shear stress may be increased by decreasing the aspect ratio. When the aspect ratio is decreased, the value of constant K_s increases. The aspect ratio may be decreased reducing the length *a*, by providing transverse stiffeners on the web plate, as shown in Fig. 8.6. The value of constant K_s increases as the length *a* (i.e., the spacing between the transverse stiffeners, *c*) reduces. The value of critical stress increases. When the spacing between the transverse stiffeners *c*, becomes less than the clear depth of web, *d*, then, value of critical stress further increases since, *b* is the smaller dimensions in Eq. 8.7, and it is universely proportional to square of $\left(\frac{b}{t_w}\right)$ ratio. Sub-dividing a simply supported plate of length, *a*, by sufficiently rigid transverse stiffeners of spacing *c*, is shown in Fig. 8.6, smaller panels are formed. *The smaller panels may be considered approximately as simply supported along the four edges.* The buckling strength of web plate is increased, and economical design of web may be done.

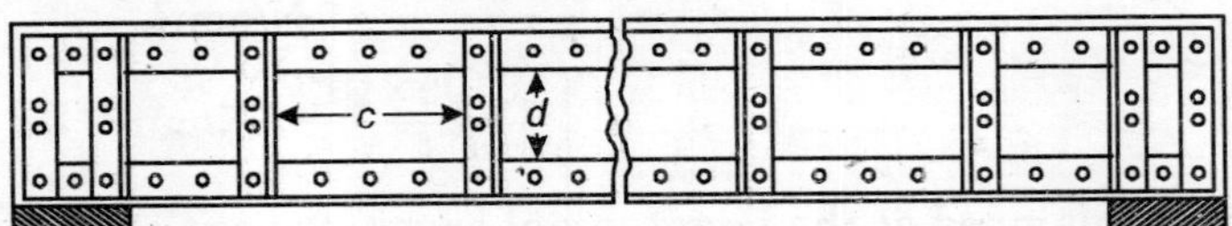

Fig. 8.6 *Panel dimensions of plate girder*

8.11.1.2 *Web Buckling in Pure Bending*

When the web plate is subjected to pure bending, then, the critical bending stress is given by

$$\sigma_c = \left(\frac{\pi^2 \cdot E}{12(1-\mu^2)\left(\frac{b}{t_w}\right)} \cdot K_b\right) \qquad ...(8.11)$$

where, K_b = constant for the plate subjected to pure bending.

For the plates simply supported on all the four edges, the minimum value constant K_b is 23.9 as integer multiples of $\left(\frac{a}{b}\right) = \left(\frac{2}{3}\right)$. It is the common practice to use this minimum value of K_b for plates with $ab > 0.67$. For the plates simply supported on loaded edges and clamped on the other edges, the minimum value of $K_b = 3.96$ as integer multiples of $\left(\frac{a}{b}\right) = 0.47$.

8.11.1.3 *Web Buckling in Combined Shear and Bending*

The web plate of simply supported plate girder is subjected to combined shear and bending at the positions in between the support and the centre. The plate is subjected to non-uniform longitudinal bending stress in compression above neutral axis, along with shear. The bending stress in compression causes the local buckling of the web plate above the neutral axis. The value of constant K is given in Table 8.3 for simply supported plates and for plate fixed or clamped at the longitudinal edges.

Table 8.3

$\left(\frac{a}{b}\right)$ ratio	0.4	0.47	0.5	0.6	0.667	0.7	0.8	1.0	1.5
K simple	29.1	—	25.6	24.1	23.9	—	24.4	25.6	24.1
K fixed	42.7	39.6	39.7	41.8	—	45.8	—	—	—

For simply suppbrted edges, the minimum value of K_b is 23.9. The critical stress for bending is given by

$$\sigma_c = \left[\frac{45 \cdot 4 \times 10^5}{\left(\frac{b}{t_w}\right)^5}\right] \text{ N/mm}^2 \qquad \text{...(8.12)}$$

For the plate clamped at the longitudinal edges, the critical stress for bending is given by

$$\sigma_c = \left[\frac{75 \cdot 2 \times 10^5}{\left(\frac{b}{t_w}\right)^2}\right] \qquad \text{...(8.13)}$$

The critical stress for combined bending and shear lies in between these two values. For combined bending and shear, the web plate may be checked by calculating the equivalent stress, $\sigma_{e.cal}$ due to co-existent bending (tension or compression) and the shear stresses obtained as below. $\sigma_{e.cal}$ shall not exceed the value, $\sigma_e = 0.9 f_y$ (the maximum permissible equivalent stress):

$$\sigma_{e.cal} = \left[\sigma_{bt\cdot cal}^{2} + 3\tau_{vm\cdot cal}^{2}\right]^{\frac{1}{2}} \quad ...(8.14\text{ a})$$

$$\sigma_{e.cal} = \left[\sigma_{bc\cdot cal}^{2} + 3\tau_{vm\cdot cal}^{2}\right]^{\frac{1}{2}} \quad ...(8.14\text{ b})$$

The theoretical values of critical stress for the buckling of web have been determined by small-deflection buckling theory making the following assumptions:

(i) The thickness of web plate is small compared with its surface dimensions.

(ii) The plate is perfectly straight before loading.

(iii) The plate deflections are small compared with the plate thickness.

The small-deflection buckling theory provides a safe but often over conservative design. When the web plate is adequately supported then, it will carry stresses substantially exceeding theoretical buckling stresses.

1. The thickness of the unstiffened web plate shall be not be less than the following:

(i) $$t_{w\,.\,min} = \left(\frac{d_1 \cdot (\tau_{va\cdot cal})^{\frac{1}{2}}}{816}\right) \text{ and}$$

(ii) $$t_{w\,.\,min} = \left(\frac{d_1 \cdot (f_y)^{\frac{1}{2}}}{1344}\right)$$

whichever is more, but

(iii) $$t_{w\,.\,min} \nless \left(\frac{d_1}{85}\right)$$

where, d_1 is the depth of the web, and $\tau_{va.cal}$ is the calculated average shear stress in the web due to shear force.

2. The thickness of the vertically stiffened web plate shall be not be less than the following

(i) $\left(\frac{1}{180}\right)$ of the smaller clear panel dimension and

(ii) $\left(\frac{d_2 \cdot (f_y)^{\frac{1}{2}}}{3200}\right)$ whichever is more in (i) and (ii)

(iii) $\left(t_{w\cdot min} < \frac{d_2}{200}\right)$

3. The thickness of webs stiffened both vertically and horizontally with a horizontal stiffener at a distance from the compression flange equal to 2/5 of the distance from the compression flange to the neutral axis.

(i) $\left(\frac{1}{180}\right)$ of the smaller dimension in each panel and

(ii) $\left(\frac{d_2 \cdot (f_y)^{\frac{1}{2}}}{4000}\right)$ whichever is more in (i) and (ii)

(iii) $\left(t_{w \cdot \min} < \frac{d_2}{250}\right)$

4. **The thickness of webs stiffened additionally with a horizontal stiffener at the neutral axis.**

(i) $\left(\frac{1}{180}\right)$ of the smaller dimension in each panel, and

(ii) $\left(\frac{d_2 \cdot (f_y)^{\frac{1}{2}}}{6400}\right)$ whichever is more in (i) and (ii)

(iii) $\left(t_{w \cdot \min} < \frac{d_2}{400}\right)$

where, d_2 in the above expressions is twice the clear distance from the compression flange angles or plate to the neutral axis.

The minimum thickness of web plates for the different values of yield stress are given in Table 8.4 (i) and (ii) as specified by IS: 800–1984.

Table 8.4 (i) *Minimum thickness of web plate*

	Minimum thickness of web for yield stress f_y N/mm² (in MPa) of								
$f_y \rightarrow$	*220*	*230*	*240*	*250*	*260*	*280*	*300*	*320*	*340*
$\left(\frac{d_1 \cdot (f_y)^{\frac{1}{2}}}{1344}\right)$	$\frac{d_1}{85}$	$\frac{d_1}{85}$	$\frac{d_1}{85}$	$\frac{d_1}{85}$	$\frac{d_1}{83}$	$\frac{d_1}{80}$	$\frac{d_1}{78}$	$\frac{d_1}{75}$	$\frac{d_1}{73}$
$\left(\frac{d_2 \cdot (f_y)^{\frac{1}{2}}}{3200}\right)$	$\frac{d_2}{200}$	$\frac{d_2}{200}$	$\frac{d_2}{200}$	$\frac{d_2}{200}$	$\frac{d_2}{198}$	$\frac{d_2}{191}$	$\frac{d_2}{185}$	$\frac{d_2}{179}$	$\frac{d_2}{174}$
$\left(\frac{d_2 \cdot (f_y)^{\frac{1}{2}}}{4000}\right)$	$\frac{d_2}{250}$	$\frac{d_2}{250}$	$\frac{d_2}{250}$	$\frac{d_2}{250}$	$\frac{d_2}{248}$	$\frac{d_2}{239}$	$\frac{d_2}{231}$	$\frac{d_2}{224}$	$\frac{d_2}{217}$
$\left(\frac{d_2 \cdot (f_y)^{\frac{1}{2}}}{6400}\right)$	$\frac{d_2}{400}$	$\frac{d_2}{400}$	$\frac{d_2}{400}$	$\frac{d_2}{400}$	$\frac{d_2}{396}$	$\frac{d_2}{382}$	$\frac{d_2}{370}$	$\frac{d_2}{358}$	$\frac{d_2}{348}$

Table 8.4 (ii) *Minimum thickness of web plate*

$f_y \rightarrow$	*Minimum thickness of web for yield stress f_y N/mm² (in MPa) of*								
	340	*360*	*380*	*400*	*420*	*450*	*480*	*510*	*540*
$\left(\frac{d_1 \cdot (f_y)^{\frac{1}{2}}}{1344}\right)$	$\frac{d_1}{73}$	$\frac{d_1}{71}$	$\frac{d_1}{69}$	$\frac{d_1}{67}$	$\frac{d_1}{66}$	$\frac{d_1}{63}$	$\frac{d_1}{61}$	$\frac{d_1}{60}$	$\frac{d_1}{58}$
$\left(\frac{d_2 \cdot (f_y)^{\frac{1}{2}}}{3200}\right)$	$\frac{d_2}{174}$	$\frac{d_2}{169}$	$\frac{d_2}{164}$	$\frac{d_2}{160}$	$\frac{d_2}{156}$	$\frac{d_2}{151}$	$\frac{d_2}{146}$	$\frac{d_2}{142}$	$\frac{d_2}{138}$
$\left(\frac{d_2 \cdot (f_y)^{\frac{1}{2}}}{4000}\right)$	$\frac{d_2}{217}$	$\frac{d_2}{211}$	$\frac{d_2}{205}$	$\frac{d_2}{200}$	$\frac{d_2}{195}$	$\frac{d_2}{189}$	$\frac{d_2}{183}$	$\frac{d_2}{177}$	$\frac{d_2}{172}$
$\left(\frac{d_2 \cdot (f_y)^{\frac{1}{2}}}{6400}\right)$	$\frac{d_2}{348}$	$\frac{d_2}{338}$	$\frac{d_2}{328}$	$\frac{d_2}{320}$	$\frac{d_2}{312}$	$\frac{d_2}{302}$	$\frac{d_2}{292}$	$\frac{d_2}{284}$	$\frac{d_2}{276}$

The *effective cross-sectional area of the web plate* in the plate girder shall be calculated as the product of the full depth of the web plate and the thickness of web plate.

For the riveted plate girder, in the exposed situation and which do not have flange plates for their entire length, the top edge of the web plate shall be flush with the angles and the bottom edge of the web plate shall also be flush with the flange angles.

The thickness of web plate selected should be such that web plates are manufactured and are available for that thickness. For the purpose of economy, the thickness of web is kept minimum practicable and stiffeners are provided if required.

8.12 DESIGN OF FLANGES

In the riveted plate girder, the *flange section* consists of flange plates, flange angles and that part of the web which is between the flange angles. The *effective sectional area of compression flange* is the gross area with deductions for excessive width of plates and for open holes, occurring in a plane perpendicular to the direction of stress at the section. The *effective area of tension flange* is the gross area with deduction of holes.

In practice, the gross area of compression flange and tension flange are kept equal. The tension flange of the plate girder is designed on the net area basis and the corresponding gross area is provided for compression flange. After selecting thickness and depth of web plate, flanges are designed by the following methods:

1. Flange area method; 2. Moment of inertia method.

8.12.1 Flange Area Method

The trial section of plate girder is determined by *flange area method.* The flange area method is also termed as *approximate method.* The bending stress in compression flange and in tension flange are assumed uniformly distributed as shown in Fig. 8.7, apart from triangular stress distribution. This assumption is satisfactory, so long as the depth of the flange is small in compression to overall depth of the plate girder.

The forces in the compression flange C and in the tension flange T are equal and act at the C.G., of flanges in opposite direction. They form a couple which resists the bending moment.

$$M = (C \times d_e) \quad \text{...(i)}$$

$$M = (T \times d_e) \quad \text{...(ii)}$$

$$M = (A_f' \times \sigma_b' \times d_e) \quad \text{...(iii)}$$

$$\therefore \quad A_f' = (M / (\sigma_b \times d_e)) \quad \text{...(8.15 a)}$$

where, M = Maximum bending moment

σ_b' = Average bending stress

d_e = Effective depth of plate girder

A_f' = Net area of tension flange.

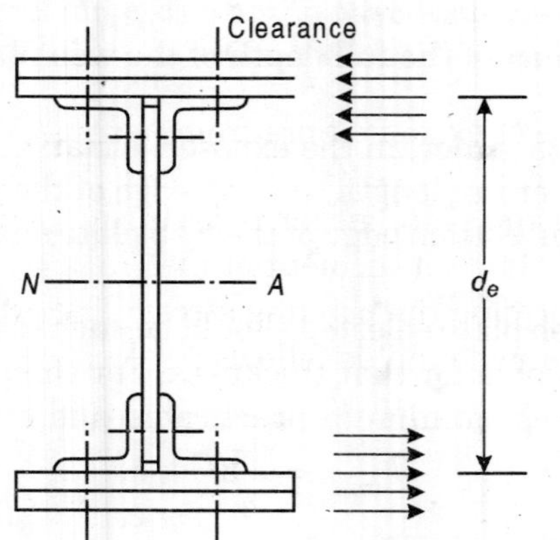

Fig. 8.7 *Distribution of bending stresses over flanges*

In the design of net area of tension flange, a large part of the area of the flange is provided by the flange angles and the number of flange plates is kept to a minimum. Preferably not less than one-third net area of the tension flange is provided by angles. One-eighth area of web plate acts as flange area. The remaining flange area is provided by flange plate. It is also possible that the flange of a plate girder consists of only flange angles and web equivalent and flange plates. The number of flange plates in each flange should not be more than three. When more than three plates are used, then, the length of rivets connecting flange plates and flange angles becomes large. The large abjected to bending in addition to shearing and bearing. As such the rivet grip becomes less effective. The net area of flange provided should be little more than that required.

Corresponding gross area to the net area adopted for compression flange. The unequal angle sections are used for flange angles, and the long legs are kept horizontal. The outstand of flange plate, i.e., their projection beyond the outer line of connections to flange angle is provided as recommended for the compression to flange angles.

In order to transmit load through flanges and not by direct bearing through web plate clearance of 5 mm is kept between cover plates and web plate as shown in Fig. 8.7.

8.12.2 Moment of Inertia Method

The *moment of inertia method* of design of plate girder section is BIS method of design of plate girder section. This is also known as *exact method.* A trial section of plate girder, designed by flange area method is checked by *moment of inertia method.* The moment of resistance of trial section is computed. If it is equal to or greater than the maximum bending moment,then the trial section designed is satisfactory.

Moment of resistance,

$$\text{M.R.} = \sigma_b \times \left(\frac{I_{\text{gross}}}{y_{\max}}\right) \quad \text{...(i)}$$

where, $y_{\max}$ = Distance of extreme fibre from neutral axis $\left(\frac{d}{2}\right)$.

The neutral axis of the plate girder is assumed at the C.G. of gross area of plate girder. However, the net moment of inertia of plate girder may be calculated and used to determine the bending stress. The gross moment of inertia of the plate girder section is found as follows :

$$I_{\text{gross}} = \left[2I_f + 2A_{f1}\cdot\left(\frac{d}{2}\right)^2 + \frac{1}{12}t_w d^3\right] \quad \text{...(ii)}$$

where, A_{f1} = Flange area excluding web equivalent.

It is to note that the difference between the overall depth, the effective depth and the depth of the web is very small. Therefore, the depth of web and the effective depth have been assumed to be equal to the depth of plate girder section. The moment of inertia of flange plates about their own axes, I_f is very small. Hence, I_f is neglected. Therefore, the gross moment of inertia is given by

$$I_{\text{gross}} = \left[\frac{1}{2}A_{f1}\cdot d^2 + \frac{1}{12}t_w d^3\right] \quad \text{...(iii)}$$

or

$$I_{\text{gross}} = \left(A_{f1} + \frac{1}{6}A_w\right)\frac{d^2}{2} \quad \text{...(iv)}$$

The moment of resistance is found by substituting this in the expression (i)

$$\therefore \quad \text{M.R.} = \frac{\sigma_{bc}}{\left(\frac{d}{2}\right)}\left(A_{f1} + \frac{1}{6}A_w\right)\frac{d}{2} \qquad \text{...(v)}$$

$$\text{M.R.} = \sigma_{bc} \cdot \left(A_{f1} + \frac{1}{6}A_w\right) . \, d \qquad \text{...(vi)}$$

When the moment of resistance is equal to or greater than the maximum bending moment, then, the trial section designed is satisfactory. The moment of resistance of the plate girder section may also be written as

$$\text{M.R.} = \sigma_{b1}\left(A_{f1} + \frac{1}{8}A_w\right) . \, d \qquad \text{...(vii)}$$

In case, σ_b' and σ_b'' in the expression (i) and (ii) are taken equal and $(d_0 - t_f)$ is adopted as d_e, then, the bending moment resisted by flanges, then

$$M_f = (M - M_w)$$

Substituting the values of these moments.

The compression and the tension flanges are checked for bending stresses in extreme fibre.

Equating (vi) and (vii)

$$\sigma_{bt} \cdot \left(A_{f1} + \frac{1}{8}A_w\right) . \, d = \sigma_{bc} \cdot \left(A_{f1} + \frac{1}{6}A_w\right) . \, d$$

$$\sigma_{bt} \,.\, A_{\text{net}} = \sigma_{bc} \,.\, A_{\text{gross}}$$

$$\sigma_{bt} = \sigma_{bc} \times \left(\frac{A_{\text{gross}}}{A_{\text{net}}}\right)$$

then

$$\text{M.R.} = \sigma_b' \left(A_f + \frac{1}{6}t_w \cdot d_w\right) \cdot d_e \qquad \text{...(viii)}$$

$$= \sigma_b' \left(A_f + \frac{1}{6}A_w\right) \cdot d_e$$

$$(A_f \times \sigma_b' \times d_e) = \left(M - \sigma_b' \frac{t_w \cdot d_e^2}{6}\right) \qquad \text{...(ix)}$$

The bending stress in the extreme fibre in compression flange is given by

$$\sigma_{bc.\text{cal}} = \left(\frac{M}{I_{\text{gross}}} \times y_{\max}\right) \qquad \text{...(8.15)}$$

The bending stress in the extreme fibre in tension flange is computed by increasing σ_{bc} in proportion of gross area and net area of flanges.

The bending stress in the extreme fibre in tension flange

$$\sigma_{bt.cal} = \sigma_{bc.cal} \times \left(\frac{\text{Gross area of flange}}{\text{Net area flange}}\right) \quad ...(8.16)$$

σ_{bt} should be less than allowable bending stress.

Example 8.1 *A plate girder simply supported at ends is composed of web plate 1000 mm depth × 12 mm thickness, and two flange angles ISA 200 mm × 100 mm × 15 mm (ISA 20010, @ 0.336 kN/m) and two flange plates 500 mm wide × 20 mm thickness in each flange. The effective span of the plate girder is 12 m. The diameter of rivets used for connecting flange plates to flange angles and flange angles to web plate is 22 mm. Determine the maximum uniformly distributed load (inclusive of self weight) which can be carried by the plate girder. Assume maximum allowable stress as per IS : 800 –1984.*

Solution

Step 1 : Moment of inertia of given plate girder section

The gross-moment of inertia of the plate girder I_{xx} is found about xx-axis. xx-axis is assumed to be located at the centre. The moment of inertia of the flange plates about their own axis is neglected.

I_{xx} (gross) = Moment of inertia of web plate + moment of inertia of flange angles about their own axis + Moment of inertia of flange angles about xx-axis + Moment of inertia of flange about xx-axis.

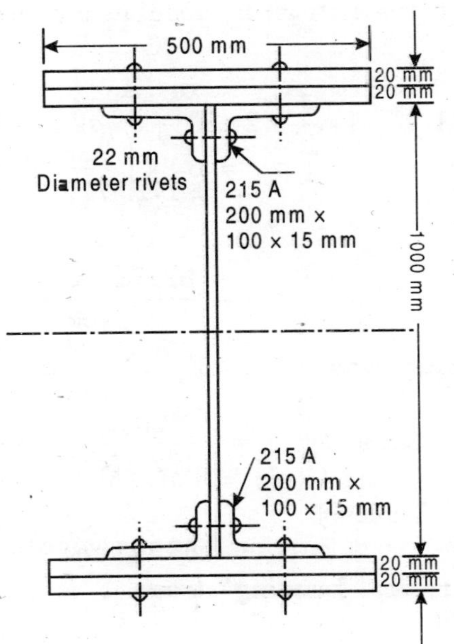

Fig. 8.8

From ISA Handbook No. 1

Moment of inertia of one flange about its own axis

$$= (298.1 \times 10^4)\ \text{mm}^4$$

$$\text{Area} = 4278\ \text{mm}^2$$

Distance of its C.G. is 22.2 mm

Distance between C.G. and *xx*-axis = (500 – 22.2) = 477.8 mm

$$I_{xx}\,(\text{gross}) = \left[\frac{1}{12}\times 1\cdot 2\times 100^3 + 4\times 298.1 + 4\times 42.78\times 47.78^2 + 2\times 50\times 4\times 52^2\right] \times 10^4\ \text{mm}^4$$

$$= 1573992 \times 10^4\ \text{mm}^4$$

$$\text{Net flange area} = 26776\ \text{mm}^2$$

$$\text{Gross flange area} = 30556\ \text{mm}^2$$

Overall depth of the plate girder

$$D = (1000 + 40 + 40) = 1080\ \text{mm}$$

Mean thickness of the compression flange

$$T = (20 + 20 + 15) = 55\ \text{mm}$$

$$\text{Ratio}\left(\frac{D}{T}\right) = \left(\frac{1080}{55}\right) = 19.636$$

Effective length of compression flange

$$l = (12 \times 1000) = 12000\ \text{mm.}$$

Moment of inertia of the plate girder about its axis of minimum strength (*yy*-axis)

$$I_{yy} = \left[\frac{1}{12}\times 100\times 1\cdot 2^3 + 4\times\frac{1}{12}\times 2\times 50^3 + 4\times 42\cdot 78\times 7\cdot 18^2\right] \times 10^4\ \text{mm}^4$$

$$= 99172.377 \times 10^4\ \text{mm}^4$$

$$r_y = \left[\frac{99172\cdot 377\times 10^4}{30556}\right]^{\frac{1}{2}} = 180.155\ \text{mm}$$

Step 2 : Slenderness ratio

$$\text{Ratio,}\left(\frac{l}{r_y}\right) = \left(\frac{12000}{180\cdot 155}\right) = 66.61, \left(\frac{D}{T}\right) = 19.636$$

From IS: 800–1984, Table 6.5

$$X = 186.727 \text{ and } Y = 81.86$$

Step 3 : Elastic critical bending stress

From IS : 800 –1984

$$f_{cb} = \left(k_1\,(X + k_2\,.Y)\,\frac{c_2}{c_1}\right)$$

$$k_1 = 1\,00 \text{ for } \varphi = 1.0 \text{ and } k_2 = 0.0 \text{ for } \omega = 0.5$$

and $\frac{c_2}{c_1}$ = 1.00 (N.A. assumed at centre)

$$f_{cb} = 1.00\,(186.726 + 0.0 \times 81.86) \times 1$$
$$= 186.726 \text{ N/mm}^2$$

Step 4 : Allowable bending stress. From IS : 800 –1984, Table 6.2 for the value of yield stress for the structural steel as 250 N/mm^2.

Bending stress in compression is 86.02 N/mm^2.

Flange Area

Description	*Gross Area (mm)^2*	*Deduction for rivet hole (mm)^2*	*Net area (mm)^2*
Flange angles 2 ISA 200 mm × 100 mm × 15 mm	2 × 4278 = 8556	4 × 235 × 15 = 1410	7146
Flange Plates 2 × 500 mm × 20 mm	2 × 500 × 20 = 20000	4 × 235 × 20 = 1880	18120
Web equivalent	$\frac{1}{6}A_w = \frac{1}{6}\times 1000$ × 12 = 2000	—	$\frac{1}{6}A_w = \frac{1}{6}\times 1000$ × 12 = 1500
Total	**30556**	—	**26766**

From theory of bending

$$\left(\frac{M}{I_{xx}} = \frac{\sigma_{bc}}{y}\right)$$

Step 5 : Moment of Resistance of plate girder section

$$M = \left(\frac{\sigma_{bc} \times I_{xx}}{y}\right) = \left(\frac{86 \times 15373992 \cdot 4 \times 10^4}{540}\right)$$
$$= 2507.25 \text{ kN-m}$$

Step 6 : Load supported by plate girder

Let w be the uniformly distributed load inclusive of self weight of plate girder per unit length

$$\text{Maximum bending moment} = \left(\frac{w \times l^2}{8}\right)$$

Equating bending moment to moment of resistance of plate girder

$$\left(\frac{w \times 12 \times 12}{8}\right) = 2507.25,$$

$$\therefore \quad w = 139.29 \text{ kN/m}$$

The plate girder can carry 139.29 kN/m load inclusive of self weight.

Example 8.2 *A mild steel plate girder simply supported at two ends has an effective span of 20 m. It carries a dead load of 50.00 kN/m and uniformly distributed live load of the same intensity, longer than the span. Design the maximum section of the plate girder. Allow for impact.*

Solution

Design :

Step 1: Impact load

The effective span of the plate girder is 20 m

Impact factor $$i = \left(\frac{20}{14+L}\right)$$

where L = Effective span in metres

$$i = \left(\frac{20}{14+20}\right) = 0.588$$

Intensity of live load = 50.00 kN/m

Impact load = (0.588 × 50.00) = 29.40 kN/m

Step 2: Dead load = 50.00 kN/m

Superimposed load over the plate girder

$$W = 129.40 \text{ kN/m}$$

Self-weight of the plate girder

$$W_1 = \left(\frac{WL}{300}\right) \quad (W_1 = \text{Self-weight of the plate girder})$$

where, W = Superimposed load over the plate girder

L = Effective span of plate girder

$$W_1 = \left(\frac{129.40 \times 20}{300}\right) \times 20 = 175.52 \text{ kN.}$$

Self-wcight of the plate girder per unit length is 8.626 kN/m

Step 3: Total uniformly distributed load

(50.00 + 50.00 + 29.40 + 8.626) = 138.026 kN/m

(Say = 138.10 kN/m)

Step 4: Maximum bending moment

$$M = \left(\frac{wL^2}{8}\right) = \left(\frac{138 \cdot 10 \times 20 \times 20}{8}\right)$$

$$= 6905 \text{ kN-m}$$

Maximum S.F., $$F = \left(\frac{wL}{2}\right) = \left(\frac{138 \cdot 10 \times 20}{2}\right) = 1381 \text{ kN}$$

Step 5: Design of web plate

Let the value of yield stress for the structural steel to be used be 250 N/mm^2. The maximum permissible stress in compression flange may be assumed as

$$\sigma_{bc} = 0.66 \times 250 = 165 \text{ N/mm}^2$$

Assume thickness of web plate, t_w = 12 mm

Economical depth of the plate girder

$$d = 1.1\left(\frac{M}{\sigma_{bc} \times t_w}\right)^{\frac{1}{2}} = 1.1\left(\frac{6905000 \times 1000}{165 \times 12}\right)^{\frac{1}{2}} = 1867.45 \text{ mm}$$

Adopt depth of the web plate as 2000 mm

Step 6 : Design of flanges (Flange area method)

Net flanges area required for tension flange

$$A_{f.\text{net}} = \left(\frac{M}{\sigma_{bc} \times d_e}\right) = \left(\frac{6905000 \times 1000}{165 \times 2000}\right) = 20924.24 \text{ mm}^2$$

Assume 22 mm diameter rivets for connecting flange plates to flange angles and flange angles to web plate.

Net flange area provided in tension flange = 24642 mm²

Gross flange area provided in compression flange

= 28556 mm²

Flange Area

Description	*Gross Area (mm)²*	*Deduction for rivet hole (mm)²*	*Net area (mm)²*
Flange angles 2 ISA 200 mm × 100 mm × 15 mm	 2 × 4278 = 8556	 4 × 23.5 × 15 =1410	 7146
Flange Plates 2 × 500 mm × 16 mm	2 × 500 × 16 = 16000	4 × 23.5 × 16 = 1504	 14496
Equivalent web area	$\frac{1}{6}A_w = \frac{2000 \times 12}{6}$ = 4000	—	$\frac{1}{6}A_w = \frac{2000 \times 12}{6}$ = 3000
Total	**28556**	—	**24642**

The maximum section of the plate girder is shown in Fig. 8.9.

The maximum section selected by the flange area method is checked by moment of inertia method. The neutral axis (xx-axis) of the plate girder is assumed to be located at the centre. The moment of inertia of plates about their own axis is neglected.

Step 7 : To check the designed section by (moment of inertia method)

From ISI Handbook No.1

The moment of inertia of the flange angle about its own axis

= 298.1 × 10⁴ mm⁴

Area A = 42.78 × 100 mm². Distance of C.G. is 22.2 mm

$$I_{xx}(\text{gross}) = \left[\frac{1}{12} \times 1.2 \times 200^3 + 4 \times 298.1 + 4 \times 42.78 \times (100 - 2.22)^2\right.$$

$$\left. + 2 \times 50 \times 2 \times 1.6 \times 101.6^2\right] \times 10^4 \text{ mm}^4$$

$$= 5736642.2 \times 10^4 \text{ mm}^4$$

Step 8 : Stress in compression flange

$$\sigma_{bc.cal} = \left(\frac{M}{I_{xx}} \times y\right) = \left(\frac{6905000 \times 1000 \times 1032}{5736642 \cdot 4 \times 10^4}\right)$$

$$= 124.2 \text{ N/mm}^2$$

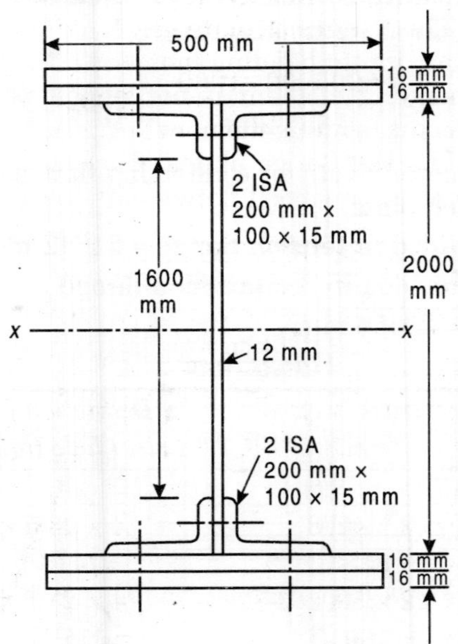

Fig. 8.9

Step 9: Stress in tension flange

$$\sigma_{bt.cal} = 124 \cdot 2 \times \left(\frac{28556}{24642}\right)$$

$$= 143.8 \text{ N/mm}^2 < 165 \text{ N/mm}^2. \text{ Hence, safe.}$$

Step 10: Check for shear stress

$$\tau_{v.cal} = \left(\frac{1381000}{2000 \times 12}\right)$$

$$= 57.6 \text{ N/mm}^2$$

$$< (0\ 4 \times 250) = 100 \text{ N/mm}^2. \text{ Hence, safe.}$$

$$\left(\frac{d}{t_w}\right) = \left(\frac{1800}{12}\right)$$

$$= 150.$$

It requires intermediate stiffeners.

The maximum section of the plate girder shown in Fig. 8.9 is satisfactory for plate girder. It requires the intermediate stiffeners.

8.13 LENGTH OF FLANGE PLATES : (CURTAILMENT OF FLANGE PLATES)

When the plate girder is subjected to loading, then, the maximum bending moment occurs at one section usually. When the plate girder is simply supported at the ends, and subjected to the uniformly distributed load, then, maximum bending moment occurs at the centre. The values of bending moment decrease towards the supports. The flanges of plate girder or the plate girder section is designed to resist the maximum bending moment. The flange area designed to resist the maximum bending moment is not required at other sections. The flange area is reduced as the bending moment decreases. The reduction in flange area is done by curtailing the flange plates. Each flange plate is curtailed or made shorter than the neighbourhood flange plate nearer to the flange angles.

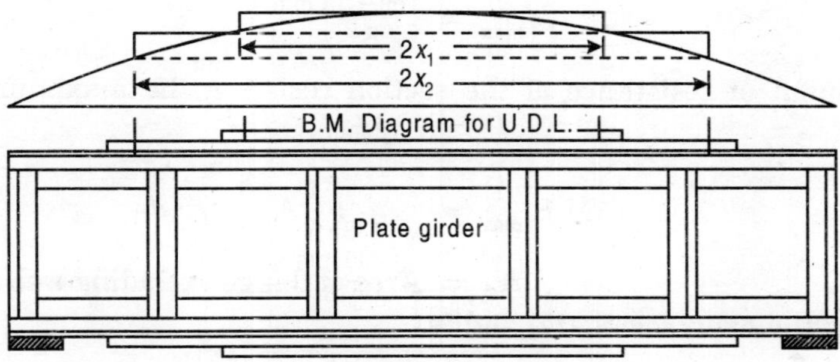

Fig. 8.10 Curtailment of flange plates

The curtailed length of the flange plates suits the variation in the bending moment. It gives economy as regards the material and cost. The actual lengths of flange plates curtailed are found as follows :

When the plate girder is carrying uniformly distributed load, the bending moment diagram has parabolic curve.

Let A_f = Total flange area including web equivalent

$A_1, A_2, \ldots, A_n$ = Area of individual cover plates, counted from outer side

$2x_1, 2x_2 \ldots 2x_n$ = Theoretical length of individual plates as shown in Fig. 8.10

l = Span of plate girder.

The flange area is given by

$$A_f = \frac{M}{(\sigma_{bc} \times d_e)} \qquad \ldots(i)$$

From this, it is seen that flange area varies as the moment. As such, flange area also varies following parabolic curve. The effective depth of plate girder is assumed as constant.

Then
$$\left(\frac{A_1}{A_f}\right) = \left(\frac{4x_1^2}{l^2}\right) \qquad \ldots(ii)$$

$$x_1 = \frac{l}{2}\left(\frac{A_1}{A_f}\right)^{\frac{1}{2}}, x_2 = \frac{l}{2}\left(\frac{A_1 + A_2}{A_f}\right)^{\frac{1}{2}} \quad \text{...(iii)}$$

$$x_n = \frac{l}{2}\left(\frac{A_1 + A_2 + ...A_n}{A_f}\right)^{\frac{1}{2}} \quad \text{...(8.17)}$$

Alternatively. Equation 8.17 may also be found by equating the bending moment at any section to the moment of resistance of the section. The maximum bending moment for the uniformly distributed load of intensity w per unit length over the simply supported plate girder

$$M_{max} = \left(\frac{wL^2}{8}\right) \quad \text{...(iv)}$$

The moment of resistance of the section resisting the maximum bending moment

$$\text{M.R.}_{max} = \left(A_{f1} + \frac{1}{6}A_w\right)\cdot \sigma_{bc} \cdot d \quad \text{...(v)}$$

where, A_{f1} = Area of flange excluding web equivalent

Equating the expressions (iv) and (v)

$$\left(A_{f1} + \frac{1}{6}A_w\right)\cdot \sigma_{bc} \cdot d = \left(\frac{wL^2}{8}\right) \quad \text{...(vi)}$$

In case, one flange plate is curtailed from the top, two flange plates are curtailed from the top and n flange plates are curtailed from the top, then, the expressions similar to the expression (vi) may be written as follows :

$$\left[\left(A_{f1} + \frac{1}{6}A_w\right) - A_1\right]\sigma_{bc} \cdot d = \left(\frac{wL^2}{8} - \frac{w(2x_1)^2}{8}\right) \quad \text{...(viii)}$$

$$\left[\left(A_{f1} + \frac{1}{6}A_w\right) - (A_1 + A_2)\right]\sigma_{bc} \cdot d = \left(\frac{wL^2}{8} - \frac{w(2x_2)^2}{8}\right) \quad \text{...(viii)}$$

$$\left[\left(A_{f1} + \frac{1}{6}A_w\right) - (A_1 + A_2 + A_n)\right]\sigma_{bc} \cdot d = \left(\frac{wL^2}{8} - \frac{w(2x_n)^2}{8}\right) \quad \text{...(ix)}$$

By dividing the expressions (vii), (viii) and (ix), respectively, by the expression (vi)

$$\left(1 - \frac{A_1}{A_{f1} + \frac{1}{6}A_w}\right) = \left\{1 - \frac{(2x_1)^2}{l^2}\right\} \quad \text{...(x)}$$

$$\left(1-\frac{A_1+A_2}{A_{f1}-\frac{1}{6}A_w}\right) = \left\{1-\frac{(2x_2)^2}{l^2}\right\} \quad ...(xi)$$

$$\left(1-\frac{A_1,A_2,....A_n}{A_{f1}+\frac{1}{6}A_w}\right) = \left\{1-\frac{(2x_n)^2}{l^2}\right\} \quad ...(xii)$$

Therefore,

$$x_1 = \frac{l}{2}\left(\frac{A_1}{A_2}\right)^{\frac{1}{2}} \quad ...(xiii)$$

$$x_2 = \frac{l}{2}\left(\frac{A_1+A_2}{A_f}\right)^{\frac{1}{2}} \quad ...(xiv)$$

$$x_3 = \frac{l}{2}\left(\frac{A_1+A_2+....A_n}{A_f}\right)^{\frac{1}{2}} \quad ...(xv)$$

where, $A_f = \left(A_{f1}+\frac{1}{6}A_w\right)$

The computations are done for the curtailment of flange plates in tension flange. Net areas of plates and flanges are used in calculations. The curtailment of plates in compression flange is done at the same positions as that in tension flange, for economy in fabrication.

8.13.1 Graphical Method of Curtailment of Flange Plates

For other system of loading, other than that of uniformly distributed load, analytical expression for the theoretical lengths of cut-off the flange plates becomes complicated. In such a situation, the graphical method of curtailment of flange plates is adopted. The graphical method is suitable for all types of loading including the system of uniformly distributed load. In the graphical method, the bending moment diagram for the given system of loading is drawn to the scale. Let the bending moment diagram for the given system of loading be as shown in Fig. 8.11. The maximum section of the plate girder is designed for the maximum value of the bending moment DD_1.

The moment of resistance of the section resisting the maximum bending moment from expression (v) above.

$$\text{M.R.} = \left(A_{f1}+\frac{1}{6}A_w\right)\cdot\sigma_{bc}\cdot d$$

or

$$\text{M.R.} = \left(A_{fa}+A_1+A_2+....A_n+\frac{1}{6}A_w\right)\sigma_b\cdot d$$

or

$$\text{M.R.} = \left(A_{fa}+\frac{1}{6}A_w+A_1+A_2+....A_n\right)\sigma_b\cdot d \quad ...(xvi)$$

where, A_{fa} = Area of the flange-angles

$A_1, A_2, ..., A_n$ = Area of individual cover plates, counted from outer side

The expression (xvi) gives moment of resistance of the section. The moments of resistance of the section with one cover plate curtailed, with two cover plates curtailed, with n cover plates curtailed are given by the following expressions

$$M_1.R_1 = \left[\left(A_{fa} + \frac{1}{6}A_w\right) + A_2 + ...A_n\right]\sigma_b \cdot d \quad ...(xvii)$$

$$M_2.R_2 = \left[\left(A_{fa} + \frac{1}{6}A_w\right).... + A_n\right]\sigma_b \cdot d \quad ...(xviii)$$

$$M_n.R_n = \left[\left(A_{fa} + \frac{1}{6}A_w\right)\right]\sigma_b \cdot d \quad ...(xix)$$

The moments of resistance of the sections, $M_1.R_1, M_2.R_2, ...M_n.R_n$ are found and then plotted on the same bending moment diagram as shown in Fig. 8.11. The horizontal lines intersecting with bending moment diagram given the theoretical lengths of curtailment of plates, *EF, GH,...,* and *JK.*

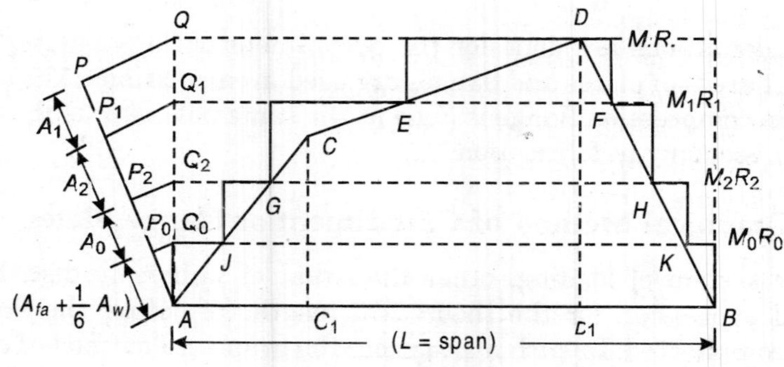

Fig. 8.11

Alternatively, from the expression (xvi) it is seen that the total moment of resistance comprises of those due to flange angles and equivalent of web, i.e., $\left(A_{fa} + \frac{1}{6}A_w\right)\sigma_b d$, due to one cover plate, i.e., $A_1 \ \sigma_b d$, due to second cover plate, i.e., $A_2 \sigma_b d$ due to second cover plate, i.e., $A_n \ \sigma_b d$. It is to note that the moments of resistance of different flange elements are proportional to their respective areas.

A line AQ is drawn perpendicular to AB at A to represent the moment of resistance of the maximum section, ($AP = DD_2$, maximum bending moment). A line AP is drawn at any inclination from A. On this line, AP, AP_n, P_2P_n, P_1P_2 and PP_1 are marked equal to $\left(A_{fa} + \frac{1}{6}A_w\right)$, A_n, A_2 and A_1, respectively to some scale. The points P and Q are joined. Then, from P_1, P_2, P_n lines P_1Q_1, P_2Q_2 and P_nQ_n are drawn parallel to PQ. The horizontal lines Q_tE, Q_2GH and Q_nJK and drawn through Q_1, Q_2 and Q_n, respectively. The intersections of these horizontal lines with the bending moment diagram give the theoretical lengths of cut-off as *EF, GH* and *JK*. In this method, it is not necessary to calculate the moments of resistance. Thus, the theoretical lengths of cut-off of the cover plates are found graphically.

The flange plates are extended beyond the theoretical cut-off points. IS: 800–1984 recommends that the extension shall contain sufficient rivets to develop in the plate, the load calculated for bending moment and girder action (taken to include the curtailed plate at the theoretical cut-off points. When the plate girder is used in the exposed situation, at least one plate of top flange shall extend the full length of girder, unless the top edge of the plate is machined flush with the flange angles. Where two or more flange plates are used on the one flange, then, the tacking rivets shall be provided, if necessary.

The actual lengths of curtailed flange plates shall be theoretically lengths of cut-off plus the extended lengths on both the sides of the plates.

The outstand of flange plates with unstiffened edges, that is, their projection beyond the outer line of connections to flange angles shall not exceed the values specified in the design of compression members.

Example 8.3 *A plate girder simply supported at ends is composed of flange plates, flange angles and web plate as shown in Fig. 8.12. Effective span of plate girder is 20 m. The plate girder is subjected to maximum bending moment of 9000 kN-m due to uniformly distributed load. Determine actual lengths of the flanges if they are curtailed.*

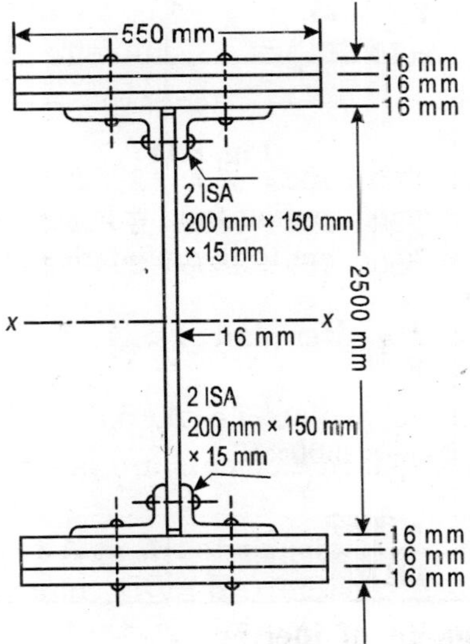

Fig. 8.12

Solution

Design :

Step 1 : Area of flanges

The bending moment diagram due to uniformly distributed load is parabola as shown in Fig. 8.13. The maximum bending moment occurs at the centre of plate girder. From table below :

Net area of tension flange

$$= 377784 \text{ mm}^2$$

Gross area of compression flange = 43117 mm²

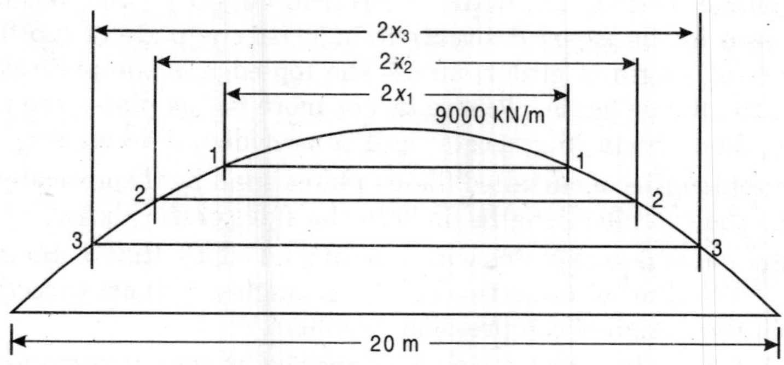

Fig. 8.13 *Bending moment diagram*

Flange Area

Description	*Gross Area* $(mm)^2$	*Deduction for rivet hole* $(mm)^2$	*Netarea* $(mm)^2$
Flange angles 2 ISA 200 mm × 100 mm × 15 mm	2 × 5025 = 10050	4 × 23.5 × 15 = 1410	8640
3 Flange Plates			
550 mm × 16 mm	550 × 160 = 8800	2 × 23.5 × 16 = 752	8048
550 mm × 16 mm	550 × 160 = 8800	2 × 23.5 × 16 = 752	8048
550 mm × 16 mm	550 × 160 = 8800	2 × 23.5 × 16 = 752	8048
Equivalent web area	$\frac{1}{6}A_w$	—	$\frac{1}{6}A_w$ area
	$=\frac{1}{6}\times 2500\times 16$	—	$=\frac{2500\times 12}{8}$
	= 6667		= 5000
Total	**43117**		**37784**

Step 2 : Gross moment of inertia

The neutral axis (*xx*-axis) of the plate girder is assumed to be located at the centre. The moment of inertia of flange plates about their own axis is neglected.

Number of flange plates	Gross moment of inertia				
	Web plate $10^4 \times mm^4$	*Flangee ∠s about own axis $10^4 \times mm^4$*	*Flanges ∠s about xx-axis $10^4 \times mm^4$*	*Flanges about xx-axis $10^4 \times mm^4$*	*I_{xx} (gross) $10^4 \times mm^4$*
Nil	2083333.3	4 × 969.9 = 3879.6	4 × 50.25 × (12.5 – 372) = 2960000		50472129
One	2083333.3	3879.6	2960000	$2 \times 55 \times 1.6 \times 125.8^2$ = 2790000	7837212.9
Two	2083333.3	3879.6	2960000	$2 \times 55 \times 3.2 \times 1256^2$ = 5650000	13487212.9
Three	2083333.3	38796	2960000	$2 \times 55 \times 48 \times 1274^2$ = 8650000	22137212

Step 3: Curtailments of plates in tension flange (Theoretical length of cut-off of plates)

Let the theoretical length of cut-off flange plates be $2x_1$, $2x_2$, and $2x_3$ as shown in Fig. 8.13.

Length
$$x_1 = \frac{1}{2}L\left(\frac{A_1}{A_{fn}}\right)^{\frac{1}{2}}$$

where, A_1 = Net area of the plate curtailed

Af_n = Net flange area

Length
$$x_1 = \frac{20}{2}\left(\frac{8048}{37784}\right)^{\frac{1}{2}} = 4.62 \text{ m}$$

Length
$$x_2 = \frac{1}{2}L\left(\frac{A_1 + A_2}{A_{fn}}\right)^{\frac{1}{2}}$$

$$x_2 = \frac{20}{2} \times \left(\frac{2 \times 8048}{37784}\right)^{\frac{1}{2}} = 6.52 \text{ m}$$

Length
$$x_3 = \frac{L}{2}\left(\frac{A_1 + A_2 + A_n}{A_{fn}}\right)^{\frac{1}{2}}$$

$$x_3 = \frac{20}{2}\left(\frac{3 \times 8048}{37784}\right)^{\frac{1}{2}} = 8.00 \text{ nm}$$

The plates to be extended beyond their theoretical cut-off points. The value of bending moments at the theoretical cut off points are found as follows:

Equation of parabolic bending moment diagram with A as origin, is given by

$$y = \frac{4a}{L^2} x \cdot (L - x)$$

At $x = \frac{L}{2}, y = 9000$ kN-m

$\therefore$ $a = 9000$

$\therefore$ $y = 36000\, x \frac{(L - x)}{L^2}$

Step 4: Distances to sections for theoretical cut-off

Distance of cross-section 1,

from A = (10 – 4.62) = 5.38 m

Distance of cross-section 2,

from A = (10 – 6.52) = 3.48 m

Distance of cross-section 3,

from A = (10 – 8) = 2 m

Step 5: Bending moment at sections for theoretical cut-off

Bending moment at section 1,

$$M_1 = \frac{36000}{400} \times 5 \cdot 38 \times 14 \cdot 62 = 7019 \text{ kN-m}$$

Bending moment at section 2,

$$M_2 = \frac{36000}{400} \times 3 \cdot 48 \times 16 \cdot 52 = 5174 \text{ kN-m}$$

Bending moment at section 3,

$$M_3 = \frac{36000}{400} \times 2 \times 18 = 3240 \text{ kN-m}$$

The extensions of the plates beyond the theoretical cut-off points contain vets to develop in the plate, the load calculated for the bending moments and rder section (taken to include the curtailed plate) at the theoretical cut-off) points. The bepding stresses in compression are found considering the gross moment of inertia of the sections as follows:

Step 6: Bending stress in compression in the top most cover plate

The gross moment of inertia of section cut-off the top-most plate

M_1 = 7079 kN-m

The distance upto the centre of top-most cover plate

$$y_1 = \left(125 + 2 \times 1.6 + \frac{1}{2} \times 1.6\right) = 129 \times 10 \text{ mm}$$

The bending stress in compression flange at section, 1

$$\sigma_{bc.cal.1} = \left(\frac{7079 \times 10^6 \times 1290}{22137212.9 \times 10^4}\right) = 41.25 \text{ N/mm}^2$$

Step 7: Bending stress in compression in the second cover plate

The gross moment of inertia of section including the second plate to be curtailed

$$I_{xx}(\text{gross})_2 = 13487212.9 \times 10^4 \text{ mm}^4$$

The bending moment at the theoretical cut-off of the second plate

$$M_2 = 5174 \text{ kN-m}$$

The distance upto the centre of second cover plate

$$y_2 = \left(125 + 1\cdot6 + \frac{1}{2} \times 1\cdot6\right) = 127.4 \times 10 \text{ mm}$$

The bending stress in compression flange at section, 2

$$\sigma_{bc.\text{cal}.2} = \left(\frac{5174 \times 10^6 \times 1274}{13487212\cdot3 \times 10^4}\right) = 48.87 \text{ N/mm}^2$$

Step 8: Bending stress in the third cover plate

The gross moment of inertia of section including the third cover plate to be curtailed

$$I_{xx}(\text{gross})._3 = 7837212.9 \times 10^4 \text{ mm}^4$$

The bending moment at the theoretical cut-off of the third cover plate

$$M_3 = 3240 \text{ kN-m}$$

The distance upto the centre of third cover plate

$$y_3 = \left(125 + \frac{1}{2} \times 1\cdot6\right) = 125.8 \times 10 \text{ mm}$$

The bending stress in compression flange at section, 3

$$\sigma_{bc.\text{cal}.3} = \left(\frac{3240 \times 10^6 \times 1258}{7837212\cdot9 \times 10^4}\right)$$

$$= 52 \text{ N/mm}^2$$

Step 9: Bending stress in tension upon the centre of respective cover plate at the respective sections

$$\sigma_{bc.\text{cal}.1} = \left(\frac{41\cdot25 \times 43117}{37784}\right)$$

$$= 47.072 \text{ N/mm}^2$$

$$\sigma_{bc.\text{cal}.2} = \left(\frac{48\cdot87 \times (43117 - 8800)}{(37784 - 8048)}\right)$$

$$= 56.396 \text{ N/mm}^2$$

$$\sigma_{bc.cal.3} = \left(\frac{52\times(43117-2\times8800)}{(377.84-2\times8048)}\right)$$

$$= 61.18 \text{ N/mm}^2$$

Step 10 : Tensile force in each cover plate at the respective sections

$$F_1 = \left(47\cdot072\times\frac{8048}{1000}\right) = 378.834 \text{ kN}$$

$$F_2 = \left(48\cdot87\times\frac{8048}{1000}\right) = 393.304 \text{ kN}$$

$$F_3 = \left(61\cdot18\times\frac{8048}{1000}\right) = 492.37 \text{ kN}$$

Step 11 : Rivet value

Use 22 mm diameter power driven rivets. The strength of rivet in single shear

$$\left(\frac{\pi}{4}\times\frac{(23\cdot5)^2\times100}{1000}\right) = 43.35 \text{ kN}$$

Strength of rivet in bearing

$$\left(23\cdot5\times\frac{16\times300}{1000}\right) = 112.8 \text{ kN}$$

Rivet value, R = 43.35 kN

Step 12 : Design of riveted connections

Number of rivets at section, 1–1

$$\left(\frac{378\cdot834}{43\cdot35}\right) = 8.739$$

Number of rivets at section, 2–2

$$\left(\frac{393\cdot304}{43\cdot35}\right) = 9.07$$

Number of rivets at section, 3–3

$$\left(\frac{492\cdot37}{43\cdot35}\right) = 11.36$$

Provide 16 rivets at each section. These rivets are provided in two rows in extended portion of the plate beyond the theoretical sections with 1.5 times leter of rivets as edge distance and 3 times diameter as pitch.

Step 13 : Actual length of curtailment of top-most cover plate

$$\left(2\times4\cdot62+\frac{2}{100}\times8\times3\times23\cdot5\right) = 10.36 \text{ m}$$

Actual length of curtailment of second cover plate

$$\left(2\times 6\cdot 52+\frac{2}{100}\times 8\times 3\times 23\cdot 5\right)= 14.16\text{ m}$$

Actual length of curtailment of third cover plate

$$\left(2\times 8+\frac{2}{100}\times 8\times 3\times 23\cdot 5\right) = 17.12\text{ m}$$

The flange plates in compression flange are curtailed at the same positions, where the flange plates in the tension flange are curtailed. If the plate girder is used in the exposed conditions, one plate in top flange is continued throughout the length of girder.

8.14 CONNECTION OF FLANGE ANGLES TO WEB

The flange angles of plate girder are connected to the web by sufficient rivets to transmit the maximum horizontal shear force resulting from bending moments in the girder, combined with any vertical loads which are directly applied to the flanges.

The design of connection of flange angles to web depends upon horizontal shear. A vertical component of load is added in the horizontal shear. The rivets connecting flange angles to web resist the resultant of vertical component of load and horizontal shear as shown in Fig. 8.14. The pitch of rivets is found as under :

The shear stress at any point of the cross-section is given by

$$\tau_{vf.\text{cal}} = \left(\frac{VQ}{Ib}\right) \qquad ...(8.18)$$

where, $t_{vf.\text{cal}}$ = Calculated shear stress

V = Shear force at the cross-section

Q = Static moment about the neutral axis of the portion of cross-sectional area beyond the location at which stress is determined

l = Moment of inertia of section

b = Width of section, at which stress is determined.

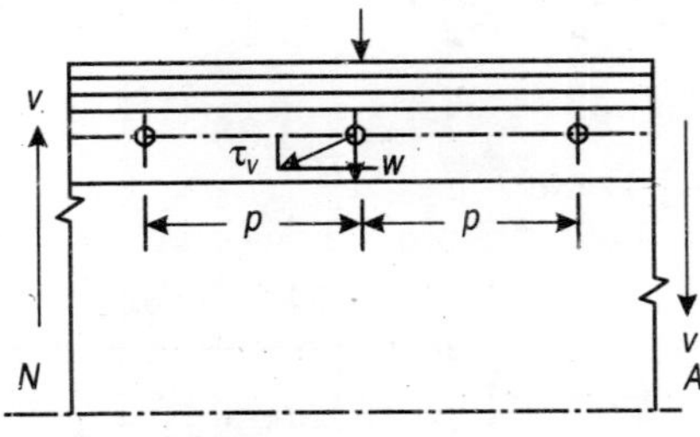

Fig. 8.14

Shear per unit depth

$$\tau_{vf.cal} = \left(\frac{VQ}{Ib} \times b \times 1\right) \quad \text{...(i)}$$

$$\tau_{vf.cal} = \left(\frac{V \cdot Q}{I}\right) \quad \text{...(ii)}$$

Moment of section

$$I = \left(2A_f \times \frac{d_e}{2} \times \frac{d_e}{2}\right) \quad \text{...(iii)}$$

and $$Q = \left(A \cdot \bar{y} = A_f \times \frac{d_e}{2}\right) \quad \text{...(iv)}$$

Shear per unit depth

$$\tau_{vf.cal} = \left[\frac{V \times A_f \times \left(\frac{d_e}{2}\right)}{2A_f \times \left(\frac{d_e}{2}\right) \times \left(\frac{d_e}{2}\right)}\right] = \left(\frac{V}{d_e}\right) \quad \text{...(v)}$$

The shear stress per unit depth is accompanied by horizontal shear per unit length of equal intensity. The horizontal shear per unit length is therefore, given by

$$\tau_{vf.cal} = \frac{V}{d_e} \times \frac{A_{f1}}{\left(A_{f1} + \frac{1}{6} \cdot A \cdot w\right)} \quad \text{...(8.19)}$$

where A_{f1} is the gross area of flange excluding web equivalent. The resultant of horizontal shear per unit length and vertical component of load w is given by

$$= \left[(\tau_{vf.cal})^2 + (w)^2\right]^{\frac{1}{2}} \quad \text{...(vii)}$$

The resultant force per pitch length should be less than or equal the rivet value.

Then, $$p \times \left[(\tau_{vf.cal})^2 + (w)^2\right]^{\frac{1}{2}} = R$$

$$p = \left(\frac{R}{\left[(\tau_{vf.cal})^2 + (w)^2\right]^{\frac{1}{2}}}\right) \quad \text{...(viii)}$$

Substitute the value of $\tau_{vf.cal}$ The pitch of rivets connecting flange angles to web for loaded flange is given by

$$p = \frac{R}{\left[\frac{V}{d_e} \times \frac{A_{f1}}{\left(A_{f1} + \frac{1}{8}A_w\right)^2 (w)^2}\right]^{\frac{1}{2}}} \quad ...(8.20)$$

The pitch of rivets connecting flange angles to web for unloaded flange ($w = 0$) is given by

$$p = \frac{R}{\left(\frac{V}{d_e} \times \frac{A'_{f1}}{\left(A'_{f1} + \frac{1}{8}A_w\right)}\right)} \quad ...(8.21)$$

where, A'_{f1} is net area of tension flange. The bottom flange is in tension and it is unloaded flange.

Where a load is directly applied to the top flange, it is considered as dispersed uniformly at an angle of 30 degrees to the horizontal.

8.15 CONNECTION OF FLANGE PLATES TO FLANGE ANGLES

The flange plates of the plate girder are connected to the flange angles by sufficient rivets.

The design of connection of flange plate to flange angles is designed for the horizontal shear. The horizontal shear per unit length (for tension flange)

$$\tau_{vf.cal} = \frac{V}{d_e} \times \left(\frac{A_1 + A_2 + ... + A_n}{A_{f1} + \frac{1}{8}A_n}\right) \quad ...(i)$$

where, $A_1\, A_2 ..\, .A_n$ = Areas of flange plates, counting from outside.

The pitch of rivets for tension flange, if rivets are to be provided in one straight line, is given by

$$p = \frac{R \times d_e}{V} \times \left(\frac{A'_{f1} + \frac{1}{8}A_w}{A'_1 + A'_2 + ... + A'}\right) \quad ...(8.22a)$$

For the compression flange

$$p = \frac{R \times d_e}{V} \times \left(\frac{A'_{f1} + \frac{1}{6}A_w}{A'_1 + A'_2 + ... + A'}\right) \quad ...(8.22b)$$

But the rivets are provided in two parallel lines, as such rivets are provided at double the pitch computed above.

Example 8.4 *A simply supported plate girder spans 24 m and supports a uniformly distributed load of 3840 kN excluding self-weight. The section of plate girder at supports is shown in Fig. 8.15.*

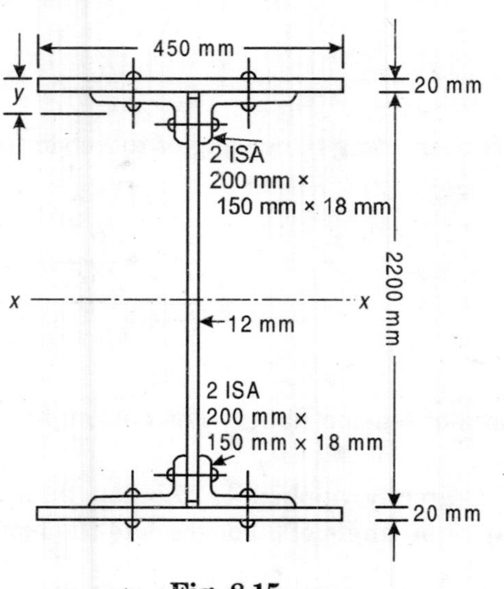

Fig. 8.15

Design the connection of flange angles to web and flange plates to flange angles.

Solution

Design:

Step 1: Total load supported plate girder

Self-weight of plate girder $= \left(\dfrac{3840 \times 24}{300}\right)$ 307 ≏ 310 kN

Total uniformly distributed load $= (3840 + 310) = 4150$ kN

Shear force at support V $= \left(\dfrac{4150}{2}\right) = 2075$ kN

Step 2: Effective depth of plate girder

Let $\bar{y}$ be the distance of C.G. of compression flange from top. From ISI Handbook No. 1, cross-sectional area of one flange angle

$= 59.75 \times 100 \text{ mm}^2$

Distance of its C.G. $= 38.4$ mm

$$\bar{y} = \left(\frac{45 \times 2 \times 1 + 2 \times 59 \cdot 76\,(2 + 3 \cdot 84)}{45 \times 2 + 2 \times 59 \cdot 76}\right)$$

$= 3.75 \times 10$ mm

Effective depth of plate girder at supports

$d_e = (224 - 2 \times 3.75)$

$= 216.50 \times 10 = 2165$ mm

Step 3: Area of flanges

Net area of tension flange = 21622 mm^2

Gross area of compression flange = 25352 mm^2

Flange Area

Description	*Gross Area (mm)^2*	*Deduction for rivet hole (mm)^2*	*Net area (mm)^2*
Flange angles 2 ISA 200 mm × 150 mm × 18 mm	2 × 5976 = 11952	4 ×23.5 ×18 = 1690	10262
Flange Plates 450 mm × 20 mm	450 × 20 = 9000	2 × 23.5 × 20 = 940	8060
Equivalent web area	$\frac{1}{6}A_w$	—	$\frac{1}{8}A_w$ area
	$=\frac{1}{6}\times 2200\times 12$		$=\frac{2200\times 12}{8}$
	= 4400	—	= 3300
Total	**25352**	**—**	**21622**

Step 4 : Connection of flange angles to web in compression flange

Load on compression flange per mm length

$$w = \left(\frac{3840}{24}\times\frac{1000}{1000}\right) = 160 \text{ N/mm}$$

Use 22 mm diameter power driven rivets

Strength of rivet in double shear

$$\left(2\times\frac{\pi}{4}\times\frac{(23\cdot 5)^2\times 100}{1000}\right) = 86.70 \text{ kN}$$

Strength of rivet in bearing

$$\left(\frac{23\cdot 5\times 12\times 300}{1000}\right) = 84.6 \text{ kN}$$

Rivet valued R = 84.6 kN

Let p be the pitch of rivets connecting flange angles to web. From Eq. (8.20)

$$p = \frac{R}{\left[\left(\frac{V}{d_e}\times\frac{A_{f1}}{A_{f1}+\frac{1}{6}A_w}\right)^2 + w^2\right]^{\frac{1}{2}}} \text{ mm}$$

$$p = \frac{84 \cdot 6}{\left[\left(\frac{2075\times1000}{2165 \cdot 0}\times\frac{11952+9000}{25352}\right)^2+(160)^2\right]^{\frac{1}{2}}} = 106.42 \text{ mm}$$

Provide 100 mm pitch for rivets connecting flange to web plate.

Step 5: Connection of flange plates to flange angles in compression flange

Use 22 mm diameter power driven rivets

Strength of rivet in single shear

$$\left(\frac{\pi}{4}\times\frac{(23 \cdot 5)^2\times100}{1000}\right) = 43.35 \text{ kN}$$

Strength of rivet in bearing

$$\left(\frac{23 \cdot 5\times18\times300}{1000}\right) = 126.9 \text{ kN}$$

$$\text{Rivet value } R = 43.35 \text{ kN}$$

Let p be the pitch of the rivets, if rivets arc provided in one row. From Eq. 8.22 (b)

$$p = \frac{R\times d_e}{V}\times\left(\frac{A_{f1}+\frac{1}{6}A_w}{A_1+A_2+\ldots.+A_n}\right)$$

$$p = \frac{43 \cdot 35\times217 \cdot 50}{2075\times1000}\times\left(\frac{20952+4400}{9000}\right)$$

$$= 128.248 \text{ mm}$$

Rivets are used in two rows

$$\therefore \quad 2p = 256.496 \text{ mm}$$

Provide 230 mm pitch for rivets in two rows connecting flange plates to flange angles.

Step 6: Connection of flange angles to web in tension flange

Rivet value, $R = 84.6$ kN

Tension flange is unloaded flange

$$\therefore \quad w = 0$$

Let p be the pitch of the rivets connecting flange angles to web. From Eq. 8.21

$$p = \frac{R}{\left(\frac{V}{d_e}\times\frac{A'_f}{A'_f+\frac{1}{8}A_w}\right)}$$

$$p = \frac{84 \cdot 6}{\left(\frac{2075\times1000}{2165 \cdot 0}\times\frac{10262+8060}{21622}\right)} = 105.75 \text{ mm}$$

Provide 100 mm pitch for rivets connecting flange angles to web plate.

Step 7 : Connection of flange plates to flange angles in tension flange

Use 22 mm diameter power driven rivets.

Rivet value $R = 43.35$ kN

Let p be the pitch or rivets if rivets are provided in one row. From Eq (8.22a)

$$p = \frac{A \times d_e}{V} \times \left(\frac{A'_{f1} + \frac{1}{8} A_w}{A'_1 + A'_2 + ... + A'_n} \right)$$

$$p = \left(\frac{43 \cdot 35 \times 217 \cdot 50}{2075 \cdot 0 \times 1000} \times \frac{21622}{8060} \right) = 121.88 \text{ mm}$$

Rivets are used in 2 rows

$$\therefore \quad 2p = 243.77 \text{ mm}$$

Provide 230 mm pitch for rivets in two rows connecting flange plates of angles.

8.16 DIAGONAL BUCKLING OF WEB

The diagonal buckling of web occurs when the ratio of clear depth to thickness of web $\left(\frac{d}{t_w}\right)$ exceeds 85. It occurs because of diagonal compression as shown in Fig. 8.16. A small element of the web plate is subjected to the bending stress σ_b acting horizontally and shear stress, acting vertically and which one is also

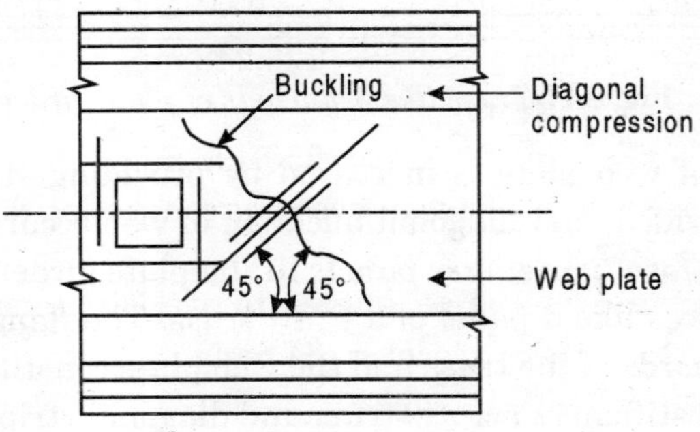

Fig. 8.16 *Diagonal buckling of web*

accompanied by complimentary shear stress, τ_{vf} acting horizontally.The principal stress which small be acting on this element is under :

$$\sigma_1 = \frac{\sigma_b}{2} + \left[\left(\frac{\sigma_b}{2} \right)^2 + (\tau_{vf})^2 \right]^{1/2} \quad ...(i)$$

$$\sigma_b = \frac{\sigma_b}{2} - \left[\left(\frac{\sigma_b}{2} \right)^2 + (\tau_{vf})^2 \right]^{1/2} \quad ...(ii)$$

At the end, the plate girder is simply supported, where, the bending moment and thereby bending stress are zero. Therefore, $\sigma_1 = \tau_{vf}$ and $\sigma_2 = \tau_{vf}$ (numerically). One principal stress is compressive, while the other is tensile. The inclination of the principal stress is given by

$$\tan 2\theta = \left(\frac{-\tau_{vf}}{\left(\frac{\sigma_b}{2}\right)} = \frac{-\tau_{vf}}{\text{zero}} \right) = \infty$$

$$2\theta = 90°, \therefore \theta = 45°$$

Both these stresses σ_1 and σ_2 act along the diagonals inclined 45° with the horizontal line. The compressive stress acting along the diagonal causes buckling of the web plate and phenomenon is known as diagonal buckling. In this phenomenon, the **diagonal compression** and the **diagonal tension** of equal intensities, acting mutually at right angles are set up in the web at the neutral axis at an inclination of 45° with horizontal. *At the neutral axis, the intensity of diagonal compression is equal to that of shear.* Intensity and inclination of diagonal compression and of diagonal tension are changed by bending compressive stress above the neutral axis, and bending tensile stress below the neutral axis.

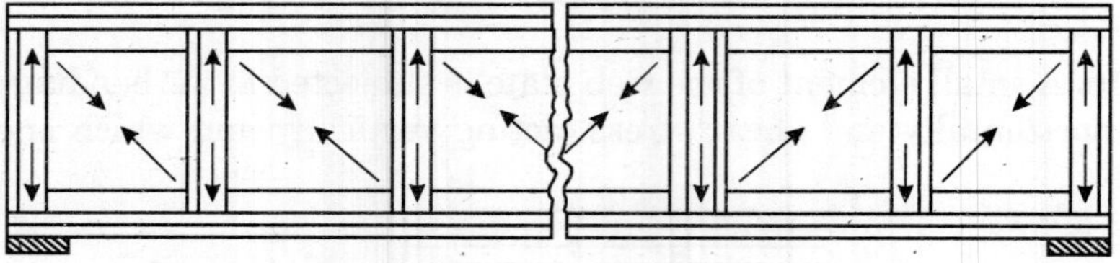

Fig. 8.17 *Truss like behaviour of plate girder*

The strength of web plate is increased by providing stiffeners. When the suffeners are provided, and diagonal buckling of web occurs, then flanges and *vertical stiffeners form imaginary panels* in the plate girder. Each panel of the plate girder behaves like a panel of a Pratt truss. The flanges behave like the top and bottom chords of the truss and the web plate constitutes the web members. The vertical stiffeners act as struts and diagonal strips of web act as ties as shown Fig. 8.17. The *buckling strength of a plate girder* web that has regularly spaced transverse stiffeners is increased by its capacity because of truss like behaviour. The diagonal carrying tension are considered as effective. The contribution of those diagonals carrying compression to the shear resistance of the web is neglected (although each would, in fact, contribute its small buckling load). Therefore, the shear resistance between any two transverse stiffeners of an adequated stiffened web is the sum of the shear buckling strength of the web and the vertical component of the web yield strength in direct diagonal tension. This latter is sometimes known as *tension field strength.*

8.17 BEARING STIFFENERS

The bearing stiffeners are attached with the web plate of the plate girder to avoid local bending failure of the flange, local crippling or buckling and crushing of web (web crippling). The bearing stiffeners are provided under concentrated loads and at the points of supports. The bearing stiffeners strengthen the web and transmit heavy concentrated loads or reactions to the flanges of plate girder. When these are provided at ends, these are also termed as end bearing stiffeners. They should be joggled or crimped. They are provided straight as shown in Fig. 8.18. The bearing stiffener extend the full dept between flange to flange and have a tight bearing against the flange transferring the load. The filler plates of thickness equal to the thickness of flanges angles are inserted between web plate and stiffeners.

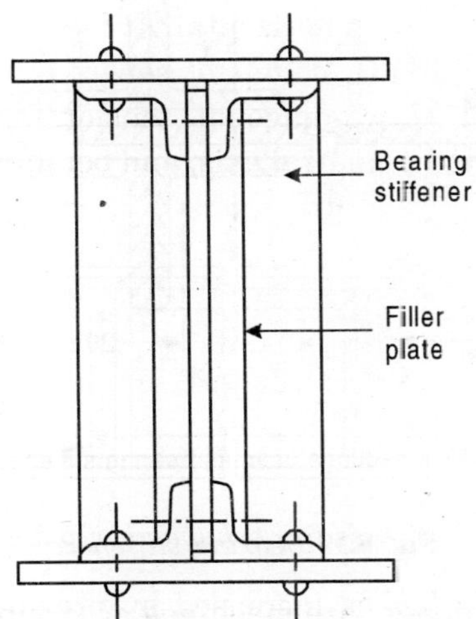

Fig. 8.18 *Bearing stiffener*

The various types of bearing stiffeners used are shown in Fig. 8.19. The angle sections in the end bearing stiffeners are arranged with reference to the point of points of maximum pressure on the flanges. In the fixed bearing (plate bearings and cast-iron shoe bearings), the maximum pressures are likely to occur at the outer and inner edges of the sole plate (the plate attached to the bottom flange of plate girder at the support). At such plates, the arrangements of the angles as shown in Fig. 8.19 (b) and (d) are suitable. In the pin or centre-bearing, the maximum pressure occurs at the centre of bearing. Therefore, the arrangement of the angles as shown in Fig. 8.19 (c) i.e., in the form of cluster of four angles is suitable. Sometimes, two angles (one on either side of the web plate) are attached to the free or outer edge of the web, and four angles in the form of a cluster are arranged at the point of maximum pressure.

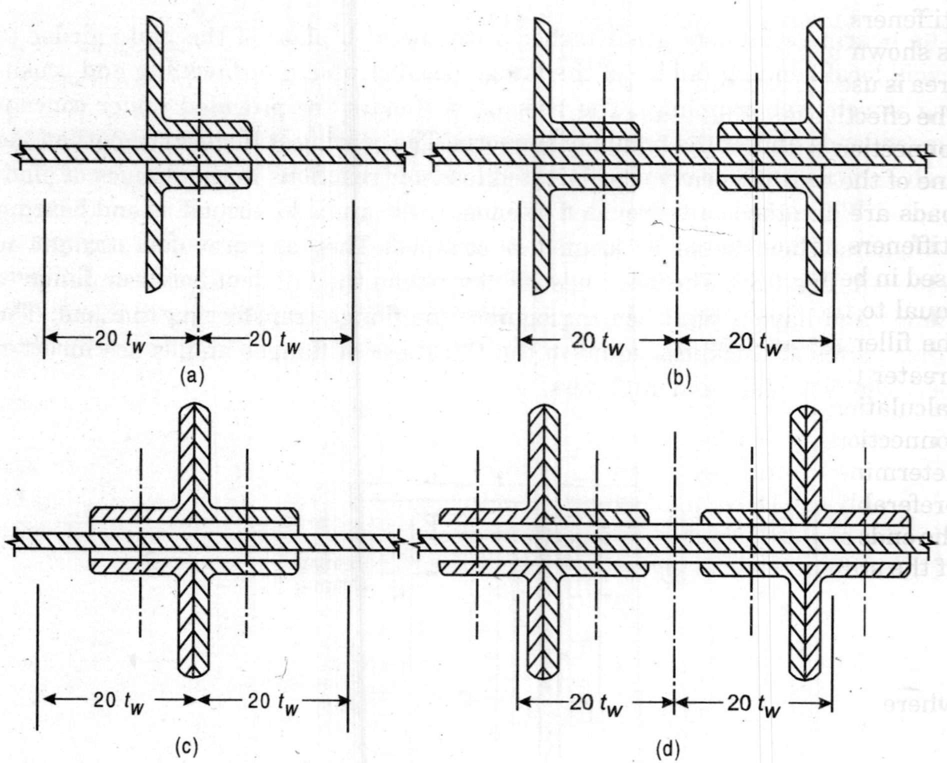

Fig. 8.19 *Bearing stiffeners*

The end of bearing stiffeners are machined or ground, and are fitted lightly between top and bottom flange angles. They are provided symmetrically about the web. The outstanding width of the bearing stiffener is kept such as to extend as nearly as practicable upto the outer edges of flange angles. The bearing stiffeners are initially selected on the basis of bearing area required. The intensity of bearing pressure should not exceed $\sigma_p = 0.75\ f_y$ N/mm². The bearing areas outstanding legs are taken clear of root of the flange angles as shown in Fig. 8.20. The load bearing capacity calculated using the effective area should be greater than or equal to the applied load or the reaction.

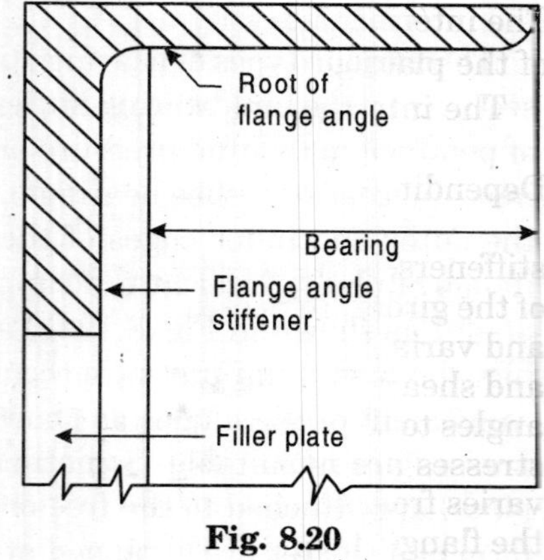

Fig. 8.20

The bearing stiffeners are designed as columns of sectional area consisting of stiffeners together with a length of web in each side of centre line of stiffeners as shown in Fig. 8.19 equal to 20 times the thickness of web, if available. This area is used to determine the radius of gyration and to check the column stresses. The effective length is assumed, 0.7 of length of the stiffeners because of secured connection with the web. The radius of gyration is computed about the centre line of the web. The rivets connecting stiffeners to the web under concentrated loads are designed to transmit these loads. At supports the rivets connecting stiffeners to the web are designed to transmit the reaction. The filler plates are used in between the stiffeners and the web plate. The thickness of filler plate is equal to the thickness of the flange angles. The rivets are connected through the filler plates. The rivets carrying calculated shear stress through packing greater than 6 mm thick are increased above the number required by normal calculations by 2.5 percent for each 2 mm thickness of packing. For double shear connections packed on both sides, the number of additional rivets required are determined from the thickness of thicker packing. The additional rivets are preferably be placed in an extension of the packing. When torsional restraint at the ends of plate girder is provided only by stiffeners, then the moment of inertia of the stiffeners about the centre line of the web shall not be less than

$$\left(\frac{D^3T}{250}\times\frac{R}{W}\right)$$

where D = Overall depth of the girder
T = Maximum thickness of compression flange
R = Reaction on the bearing
W = Total load on girder.

The load carrying capacity of the bearing stiffeners as a column should be greater than or equal to the applied load or the reaction.

8.18 INTERMEDIATE STIFFENERS

The intermediate stiffeners are used for the economical design of the web plate of the plate girder.

The intermediate stiffeners are used to avoid diagonal buckling of the web. Depending upon the ratio of clear depth to the thickness of web $\left(\frac{d}{t_w}\right)$, vertical stiffeners, or vertical and horizontal stiffeners are provided throughout the length of the girder. In Sec. 8.16, the diagonal buckling of web, it is seen that a complex and variable stress pattern is set in the web due to combined bending moment and shear. The inclined compressive stresses are set up in the direction at right angles to the compressive stresses and at the neutral axis, the intensities of the stresses are numerically equal. The direction of inclined compressive stresses varies from 45° at the neutral axis to a considerably flatter angle at the edge of the flange. Due to this, the web plate tends to buckle in one direction and tends to straighten in the right angle direction. The intermediate stiffeners prevent the web plate from buckling under the inclined compressive stress. The vertical intermediate stiffeners divide the web plate into small panels. These panels are supported along the lines of stiffeners. The resistance of web plate to buckling is measurably increased. The intermediate vertical stiffeners also have a second function. When the vertical stiffeners are fitted against the top and bottom flanges, then, they maintain the original 90° angle between the flanges and the web. When the dimensions of web are very large, then the panel dimensions are reduced by providing the horizontal stiffeners on the compression side of the web.

Unless the vertical stiffeners are closely spaced, these are not sufficient in increasing the resistance to buckling in bending. The horizontal stiffeners placed on the compression zone of the web, effectively, increase its resistance to buckling due to bending. A combination of both vertical and horizontal web stiffeners is often used depending upon $\left(\frac{d}{t_w}\right)$ ratio. When the web plate is thoroughly stiffened, it becomes effective upon permissible stresses in shear.

8.18.1 Vertical Stiffeners

The vertical stiffeners are also termed as *transverse stiffeners*. The vertical stiffeners are provided throughout the length of the girder when the thickness of web is less than the limits specified for the minimum thickness of the web plate. They are *joggled* or *crimped* as shown in Fig. 8.21. They may be provided straight. In that case, filler plate of thickness equal to that of flange angles is inserted between the stiffener and web plate. They are fitted tightly between outstanding legs of top and bottom flange angles. When two angles are used for the vertical stiffeners, they are provided on either side of web. When single angle is used for the vertical stiffeners, then they are placed alternately on opposite sides of the web.

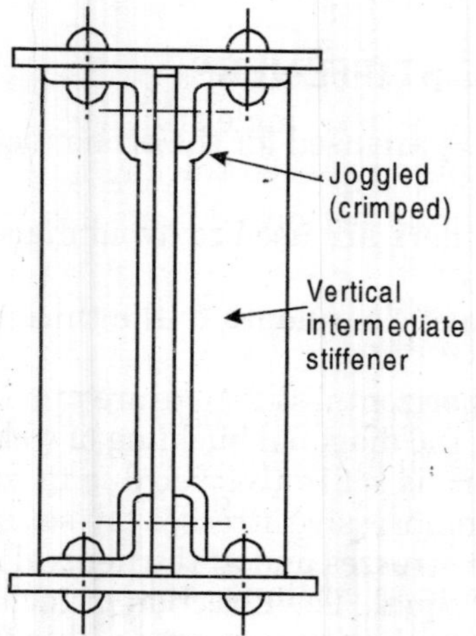

Fig. 8.21 *Intermediate (vertical) stiffeners*

The vertical stiffeners are provided at spacing not greater than 1.5 d and not less than 0.33 d, where, d is the distance between flanges angles or where there are no flange angles, the clear distance between flanges ignoring fillets. When the horizontal stiffeners are provided, d shall be taken as the clear distance between the horizontal stiffener and the tension flange (farthest flange) ignoring fillets. The vertical stiffeners divide the web plate into number of panels of dimensions c and d as shown in Fig. 8.6. The panels are supported along the lines of stiffeners. The greater unsupported clear dimension of web panel should not be greater than 270 times the thickness of web, and the lesser unsupported clear dimension of the same web panel should not be greater than 180 times the thickness of web. These limiting values of panel dimensions have been adopted by keeping the aspect ratio $\frac{c}{t}$ as 1.5 and to avoid the web buckling. The spacing of stiffeners is kept less near the supports, where the shear force is more. The spacing in increased towards the centre since the shear force decreases towards the centre. Actual average shear stress in the web should not exceed the allowable average shear stress in the stiffened web.

The length of outstanding leg of vertical stiffener may be taken equal to 1/30 of the clear depth of girder plus 50 mm. The outstand of stiffeners from the web shall not be more than $\frac{256 \cdot t}{f_y^{1/2}}$ (where, t is the thickness of the section, so long as the outer edge of each stiffener is not continuously stiffened, or 12 t for the flat sections. The length of connected leg of vertical stiffener should be sufficient to accommodate the rivets connecting the stiffener to the web. The moment of inertia I of the stiffener selected should not be less than

$$I \nless \left(\frac{1 \cdot 5d^3 t_w^3}{c^2}\right)$$

where, I = The moment of inertia of a pair of stiffeners about the centre line of web, or of a single stiffener about the face of the web

t_w = The minimum required thickness of the web

c = The maximum permitted clear distance between vertical stiffeners

8.18.2 Horizontal Stiffeners

The horizontal stiffeners are also termed as **longitudinal stiffeners.** When the thickness of web is less than the limits specified (for minimum thickness of web plate), are provided on the web at a distance from compression flange equal to $\frac{2}{5}$th of distance of compression flange to the neutral axis i.e., at $\left(\frac{1}{5}\right)$th × d, where d is the clear depth of the plate girder. The moment of inertia I, of the horizontal stiffener should not be less than $4c_1 . t_w^2$.

where c_1 is the actual distance between the transverse (vertical) stiffeners.

When the thickness of web is less than the limits specified, second horizontal stiffeners in addition to vertical stiffeners and horizontal stiffeners described above are provided at the neutral axis of the girder. The moment of inertia '*I*' of this horizontal stiffeners should not be less than $d.tw^3$. The moment of inertia is computed about the line as described for vertical stiffeners. The outstand of the stiffeners from the web shall not be more than $\frac{256 \cdot t}{f_y^{\frac{1}{2}}}$ (where, t is the thickness of the section) or 12 t for the flat sections.

The requirements for moment of inertia for vertical stiffencrs are based on the theoretical and experimental work. Timoshcnko used a value of 0.3 in the expression of moment of inertia for the vertical stiffeners. Whereas Moore, R.L. proposed a value of 1.33 based on his investigation on effectiveness of the stiffeners on shear resistance of plate girder webs. Moore's this value was found to be adequate by ultimate load test, but it has been raised to 1.5. The expressions for moment of inertia of horizontal stiffeners are based on theoretical work and it is due to Blcich, F.

Connection of intermediate stiffeners to web plate. The intermediate vertical stiffeners and horizontal stiffeners (when they are not subjected to external load) are connected to the web by rivets to withstand shear force not less than

$$\left(\frac{125 t_w^2}{h}\right) \text{ kN/m}$$

where, h is the outstand of stiffener in mm and t_w is the thickness of web in mm.

External Forces on intermediate stiffeners. In case, the vertical loads are acting eccentric or transverse forces are acting on the vertical intermediate stiffeners causing bending moment and shear, the moment of inertia of the stiffeners mentioned above shall be increased as under :

(a) The vertical load acting eccentric with respect to the vertical axis of the web causing bending moment

$$\text{Increase of } I = \left(\frac{150 MD^2}{E \cdot t_w}\right) \times 10^4 \text{ mm}^4$$

(b) The lateral loading of stiffener

$$\text{Increase of } I = \left(\frac{0 \cdot 3VD^2}{E \cdot t_w}\right) \times 10^4 \text{ mm}^4$$

where, M = Applied bending moment in kN-m

D = Overall depth of girder in mm

E = Young's modulus of elasticity, 2.047 × 105 N/mm² (MPa)

t_w = Lateral force to be taken by the stiffener and deemed to be applied at the compression flange of girder in kN.

Example 8.5 *A simply supported plate girder spans 20 m and carries a uniformly distributed load, (including its own weight), of 3000 kN. The section of plate girder at supports is shown in Fig. 8.22. Design end bearing stiffeners. Also design the necessary intermediate stiffeners.*

Solution

Design :

Stepl: Design of bearing stiffener (End reaction)

The uniformly distributed load including own weight of plate girder is 3000 kN.

Support reaction = 1500 kN

Allowable bearing stress (yield stress for steel 250 N/m²)

$$\sigma_p = (0.75 \times 250)$$
$$= 187.5 \text{ N/mm}^2$$

Step 2: Bearing area required

$$\left(\frac{1500 \times 1000}{187 \cdot 5}\right) = 8000 \text{ mm}^2$$

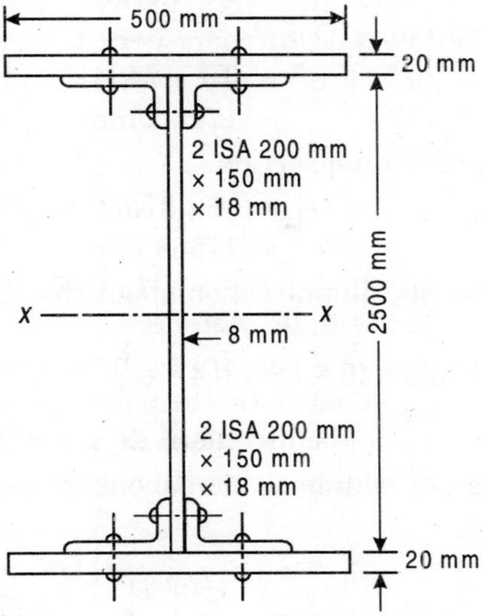

Fig. 8.22

From ISI Handbook No. 1, select 4 ISA 150 mm × 115 mm × 15 mm (4 ISA 150 115, @ 0.394 kN/m)

Radius at root, r_1 for the flange angle is 13.5 mm

Bearing area provided

$$= 4\,(150 - 13.5) \times 15$$
$$= 8190 \text{ mm}^2$$

The bearing provided is grater than bearing are a required. Provide 18 mm thick filler plates, as shown in Fig. 8.23.

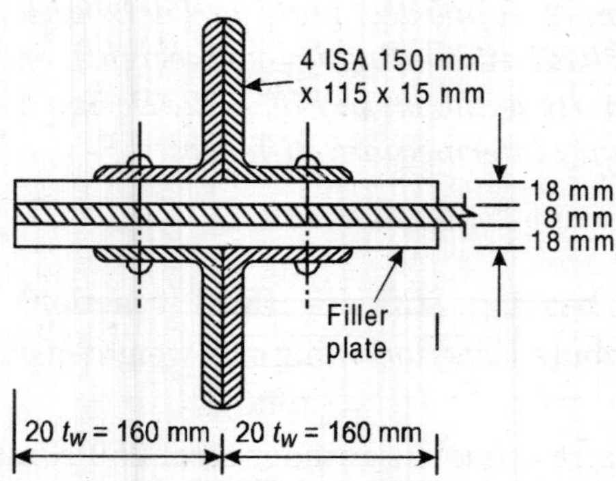

Fig. 8.23 *Bearing stiffener*

Step 3 : Check for load carrying capacity

The bearing stiffener acts as a column

Actual length of bearing stiffener

$$= (250 - 2 \times 18)$$
$$= 2464 \text{ mm}$$

Effective length of column

$$l = 0.7 \times 2464$$
$$= 1724.8 \text{ mm}$$

Cross-sectional area of column section

$$A = [4 \times 37.52 + (40 \times 0.8) \times 0.8] \times 100$$
$$= 17568 \text{ mm}^2$$

The moment of inertia of column section about the centre line of web,

$$I = [4 \times 823.5 + 4 \times 37.52(4.76 + 1.8 + 0.4)^2 + \frac{1}{2} \times 0.8^3]\ 10^4 \text{ mm}^4$$
$$= 10595.36 \times 10^4 \text{ mm}^4$$

The radius of gyration of column section about the center line of web,

$$r = \left(\frac{I}{A}\right)^{\frac{1}{2}} = \left(\frac{10595 \cdot 36 \times 10^4}{17568}\right)^{\frac{1}{2}}$$
$$= 77.66 \text{ mm}$$

Slenderness ratio of column section

$$\left(\frac{l}{r}\right) = \left(\frac{1724 \cdot 8}{77 \cdot 66}\right)$$
$$= 22.23$$

From IS : 800–1984, allowable stress in axial compression, for the steel having yield stress as 250 N/mm^2

$$\sigma = 147.331 \text{ N/mm}^2$$

Load carrying capacity of stiffener

$$\left(\frac{147 \cdot 331 \times 17568}{1000}\right) = 2588.31 \text{ kN} > 1500 \text{ kN}$$

(Support reaction)

Hence design of bearing stiffener is safe.

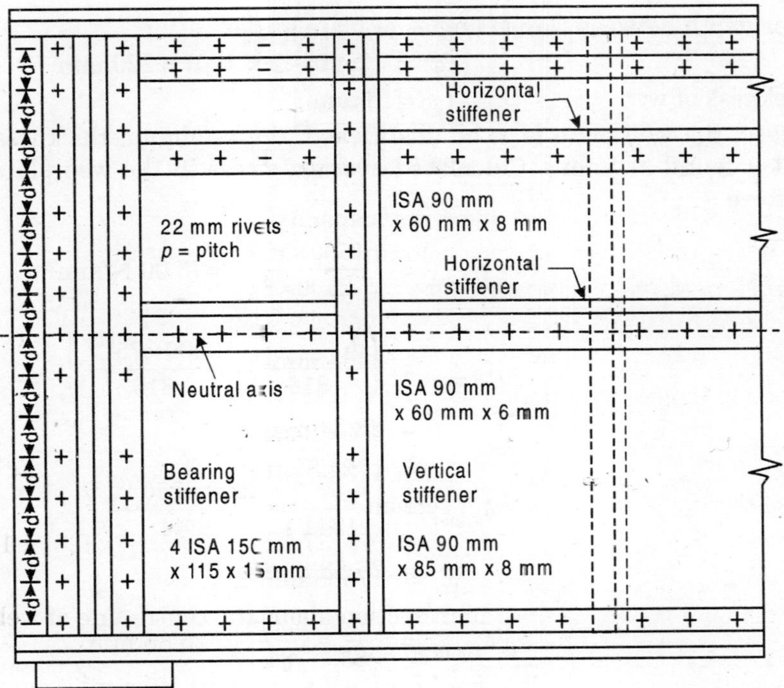

Fig. 8.24 *Details of bearing and vertical stiffeners*

Step 4: Connection of bearing stiffener to web plate

Use 22 mm diameter power driven rivets

Strength of rivet in double shear

$$\left(2 \times \frac{\pi}{4} \times \frac{(23 \cdot 5)^2 \times 100}{1000}\right) = 86.70 \text{ kN}$$

Strength of rivet in bearing

$$\left(23 \cdot 5 \times 8 \times \frac{300}{1000}\right) = 56.4 \text{ kN}$$

Rivet value, $R = 56.4$ kN

Number of rivets required to transmit reaction

$$= \left(\frac{1500}{56 \cdot 4}\right) = 26.59$$

The filler plates are provided on both the sides of web plate.

Thickness of filler plate $= 18$ mm

The filler plate (packings) are properly fitted with the bearing stiffeners. These filler plates are subjected to direct compression only.

Provide 30 rivets in 2 rows at pitch

$$p = 130 \text{ mm}$$

Step 5: Design of Intermediate stiffeners

Clear depth between flange angles of plate girder

$$d = (2500 - 2 \times 150) = 220 \text{ mm}$$

Thickness of web $t_w = 8$ mm

In case, the web plate is to be unstiffened, the minimum thickness of web needed is found as under. Calculated average stress in the web plate due to shear force.

$$\tau_{va.cal} = \left(\frac{1500 \times 100}{2500 \times 8}\right) = 75.00 \text{ N/mm}^2$$

(i) $$t_{w.\min} = \left(\frac{d_1 \cdot \tau_{va \cdot cal}^{1/2}}{816} = \frac{2200 \times 75^{1/2}}{816}\right)$$

$$= 23.34 \text{ mm, or}$$

(ii) $$t_{w.\min} = \left(\frac{d_2 \cdot f_y^{1/2}}{1344} = \frac{2200 \times 250^{1/2}}{1344}\right)$$

$$= 25.88 \text{ mm, or}$$

(iii) $$t_{w.\min} = \left(\frac{d_1}{85} = \frac{2200}{85}\right) = 25.88 \text{ mm}$$

Actual thickness of the web 8 mm is less than the above values of $t_{w.\min}$, as such the vertical stiffeners are required.

In case, the vertical stiffeners are used only, then, the thickness of web required is as under ($d_2 = 2200$ mm).

(i) $$t_{w.\min} = \left(\frac{d_2 \cdot f_y^{1/2}}{3200} = \frac{2200 \times 250^{1/2}}{3200}\right) = 10.870 \text{ mm}$$

(ii) $$t_{w.\min} = \left(\frac{d_2}{200} = \frac{2200}{200}\right) = 11 \text{ mm}$$

Actual thickness of the web 8 mm is less than $t_{w.\min}$, then, both, the vertical and horizontal stiffness become necessary.

Therefore, thickness required shall be as below, ($d_2 = 2200$ mm)

(i) $$t_{w.\min} = \left(\frac{d_2 \cdot f_y^{1/2}}{4000} = \frac{2200 \times 250^{1/2}}{4000}\right) = 8.696 \text{ mm}$$

(ii) $$t_{w.\min} = \left(\frac{d_2}{250} = \frac{2200}{250}\right) = 8.8 \text{ mm}$$

Since, actual thickness of web 8 mm is still less than $t_{w.\min}$, a horizontal stiffener is also necessary at the neutral axis, in which case, the minimum thickness of web needed is as follows :

$$(d_2 = 2200 \text{ mm})$$

(i) $$t_{w.\min} = \left(\frac{d_2 f_y^{1/2}}{6400} = \frac{2200 \times 250^{1/2}}{6400}\right)$$

$$= 5.435 \text{ mm}$$

(ii) $$t_{w.\min} = \left(\frac{d_2}{250}\right) = 5.5 \text{ mm}$$

Therefore, the web of 8 mm thickness has to be stiffened using vertical and horizontal stiffeners at a distance from the compression equal to $\frac{2}{5}$th of the distance from the compression flange to the neutral axis $\left(\frac{2}{5} \cdot 1100 = 440 \text{ mm}\right)$ and also at the neutral axis of the plate girder.

Step 6 : Design of vertical stiffeners

At support, shear force $= 1500$ kN

Actual average shear stress in web plate

$$\tau_{va.cal} = \left(\frac{1500 \times 1000}{2500 \times 8}\right)$$

$$= 75 \text{ N/mm}^2$$

Ratio, $$\frac{d}{t_w} = \left(\frac{2200}{8}\right) = 275$$

The smaller clear panel dimensions for the actual thickness of web

$$= 180 \times 8 = 1440 \text{ mm}$$

The greater clear panel dimension for the actual thickness of web

$$= 270 \times 8 = 2160 \text{ mm}$$

The vertical stiffeners may be provided at spacing smaller than 1440 mm.

Let the spacing of vertical stiffeners be

$$= (0.6 \times d) = 0.6 \times 2200 = 1320 \text{ mm}$$

From IS: 800 –1984, Table 6.6 (a), the permissible average shear stress, τ_{va} in the stiffened web plate of steel with $f_y = 250$ N/mm^2 and 0.6 d spacing and $\frac{d}{t_w}$ ratio

$$\tau_{va} = 81 \text{ N/mm}^2 > (\tau_{va.cal} = 75 \text{ N/mm}^2)$$

Length of outstanding leg of the vertical stiffener

$$\left(\frac{1}{30}\times \text{clear dept of girder} + 50\right)\text{mm}$$

$$\left(\frac{2200}{30}+50\right) = 123.33 \text{ mm}$$

Provide ISA 125 mm × 95 mm × 8 mm (ISA 125 95, @ 0.133 kN-m). The length of outstanding leg of the angle section is 90 mm.

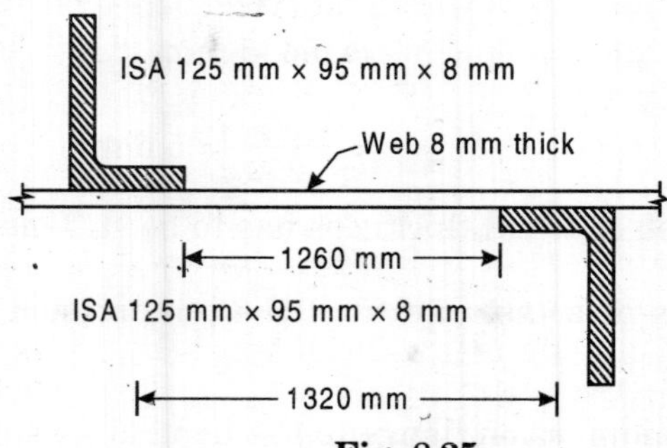

Fig. 8.25

Clear distance between vertical stiffeners

$$c = (1320 - 60) = 1260 \text{ mm}$$

$$\text{Depth of plate girder} = 2500 \text{ mm}$$

Minimum required thickness of web

$$t_{w1} = \left(\frac{1500\times 1000}{2500\times 81}\right) = 7.41 \text{ mm}$$

Required moment of inertia,

$$I = \left(\frac{1\cdot 5d^3 t_{w_1}^3}{c^2}\right) = \left(\frac{1\cdot 50\times 2200^3\times 7\cdot 41^3}{1260^2}\right)$$

$$= 409 \times 10^4 \text{ mm}^4$$

Moment of inertia about the face of web plate provided

$(266 + 3.80^2 \times 1698) \times 10^4 = 511 \times 10^4 \text{ mm}^4$.

Step 7: Connection of vertical stiffener to web plate

$$\text{Shear force} = \left(\frac{125t^2}{h}\right) = \left(\frac{125\times 8^2}{90}\right)$$

$$= 88.89 \text{ kN/m}$$

Use 22 mm diameter power driven field rivets

Strength of power driven rivet in single shear

$$\left(\frac{\pi}{4}\times (23\cdot 5)^2\times \frac{100}{1000}\right) = 43.35 \text{ kN}$$

Strength of rivet in bearing

$$\left(23 \cdot 5 \times 8 \times \frac{300}{1000}\right) = 56.4 \text{ kN}$$

Rivet value, $R = 43.35$ kN

$$\text{Pitch of rivets} = \left(\frac{43 \cdot 35}{88 \cdot 89}\right) = 0.487 \text{ m} = 487 \text{ mm}$$

Provide rivets at a pitch of 200 mm

Provide ISA : 125 mm × 95 mm × 8 mm (ISA 125 95, @ 0.133 kN/m) and 22 mm rivets to connect the stiffener with the web at 200 mm pitch. The vertical stiffeners are shown in Fig. 8.24.

8.19 WEB SPLICE

A joint in the web plate provide to increase the length is known as *web splice*. The plates are manufactured upto a limited length. When the maximum manufactured length of plate is insufficient for full length of plate girder, web splice becomes essential. It also becomes essential when the length of plate girder is too long to handle conveniently during transportation and erection.

As far as possible, web splices may be located at sections where excess flange areas are available. The excess flange areas are available at sections prior to the curtailment of flange plates. Preferably, splices may be located under stiffeners. The stiffeners provide additional strength to the splice. The splices should not be located at the sections, where maximum bending moments occur. In case, only one splice may suffice for full length of the girder, it may be located at such sections.

The web splice are designed to resist the shears and moments at the spliced sections. The splice plates are provided on each side of the web. There are following three types of web splices, which are commonly used :

1. Rational splice
2. Moment splice
3. Shear splice.

8.19.1 Web Splice (Rational Splice)

This type of web splice is shown in Fig. 8.26. The stresses are transmitted directly in this type. This type is most satisfactory than other two types. The splice plates A as shown in Fig. 8.26 are provided between flange angles. A clearance of 6 mm is left between splice plates and flange angles. The splice plates B are provided for portion of web underneath the flange angles. If sufficient excess flange area is available at the splice section, the splice plates B need not be provided. These plates are designed for shear and moment, which would be resisted by the portion of the web, if the web was not spliced. The rivets are provided at uniform spacing in this type. The pitch of rivets connecting splice plates to the web is found as under:

Vertical shear per unit depth

$$\tau_{v.cal} = \left(\frac{V}{d_e}\right) \quad \text{...(i)}$$

The bending stress upto the level of rivets connecting flange angles,

$$\sigma_{b.cal} = \left(\frac{M}{I}\right) \times y_1 \quad \text{...(ii)}$$

where, M = Bending moment at the splice section

I = Moment of inertia of the girder

y_1 = Distance upto the level of rivets connection flange angles from neutral axis.

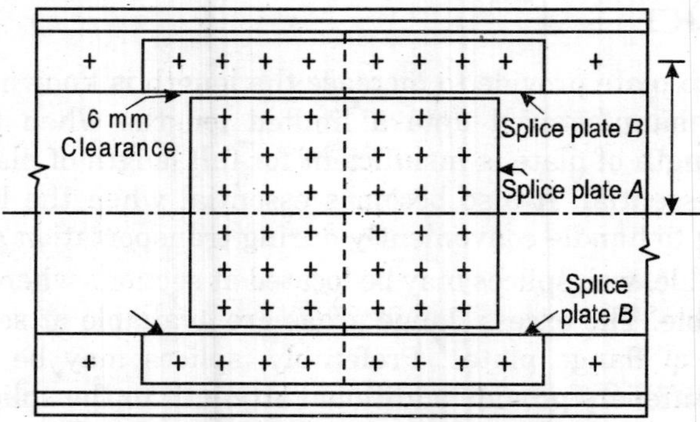

Fig. 8.26 *Web splice*

The horizontal force per unit length due to moment

$$= (\sigma_{bc} \cdot t_w)$$

If the rivets are provided in one vertical row, and p is the pitch of rivets, the resultant of vertical and horizontal forces per pitch should not exceed the rivet value, R.

$$\text{Then} \quad p \times \left[(\sigma_{cal} \times t_w)^2 + \left(\frac{V}{d_e}\right)^2\right]^{1/2} = R$$

$$\therefore \quad p = \frac{R}{\left[(\sigma_{b\cdot cal} \times t_w)^2 + \left(\frac{V}{d_e}\right)^2\right]^{1/2}} \quad \text{...(8.23)}$$

If the computed pitch of rivets is less than minimum pitch, rivets are provided in two or three vertical rows. The rivets are provided at spacing of twice pitch computed above, if rivets are provided in two vertical rows. The rivets are provided at spacing of three times the pitch computed above if rivets are provided in three vertical rows. The thickness of splice plates A is kept equal to half the thickness of web, but not less than 6 mm. The width of splice plates *A* is kept sufficient to accommodate the rivets.

The area of portion of web beneath the flange angles is found. The splice plates *B* provide this area. The rivets connecting splice plates *B* to the flange angle area designed to resist horizontal shear per pitch length between splice plates and angles, and horizontal force in the portion of web beneath flange angles due to moment.

The horizontal shear force per pitch length

$$P_1 = \left(\frac{V}{d_e} \times \frac{A_f}{A_f \frac{1}{6} A_w} \times p\right)$$

The pitch of rivets is assumed

A_f = Gross area of flange excluding web equivalent

p = Pitch of rivets, which one is assumed

The horizontal force in the portion of web beneath flange angles due to moment

$$P_2 = \frac{M \cdot y}{I} \times \left(\begin{matrix}\text{Area of portion of web} \\ \text{beneath flange angles}\end{matrix}\right)$$

where, y is the distance of rivets connecting splice plates *B* to the flange angles from the neutral axis.

If n is the number of rivets required

$$n \,.\, R = n \,.\, P_1 + P_2$$

$$n = \frac{P_2}{(R - R_1)} \qquad \text{...(8.24)}$$

The rivets connecting splice plates *B* to flange angles are provided at close spacing, so that their length is small.

Example 8.6 *The bending moment and shear force at a particular section of a plate girder are 5760 kN-m and 1080 kN respectively. The cross-section of plate girder at that section is as shown in Fig. 8.27. Design the web splice. 22 mm power driven rivets are used.*

Solution

Design :

Step 1: Effective depth of plate girder

Gross flange area = 25517 mm^2

Net flange area = 21688 mm^2

Let $\bar{y}$ be the distance of compression flange from top

$$\bar{y} = \left[\frac{2 \times 50 \cdot 25(1 \cdot 6 + 3 \cdot 27) + 1 \cdot 6 \times 5 \cdot 5 \times 0 \cdot 8}{2 \times 50 \cdot 25 + 1 \cdot 6 \times 55}\right] \times 10 = 32.1 \text{ mm}$$

Effective depth of the plate girder

$$d_e = (253.2 - 2 \times 3.21) \times 10 = 2467.8 \text{ mm}$$

Step 2: Gross moment of inertia of the plate girder about *xx*-axis.

$$I_{xx} = \left[\frac{1}{12}\times1\cdot6\times(250)^3+4\times969.2+4\times50\cdot25\right.$$

$$\left.\times(125-3\cdot72)^2+2\times55\times1\cdot6\times125\cdot8^2\right] \times 10^4\ \text{mm}^4$$

$$I_{xx} = 7829002.1 \times 10^4\ \text{mm}^4$$

Step 3 : Area of flanges

The short legs of flanges are connected with the web plate. From ISI Handbook No. 1, the rivets are provided at 90 mm from top of the flange angles.

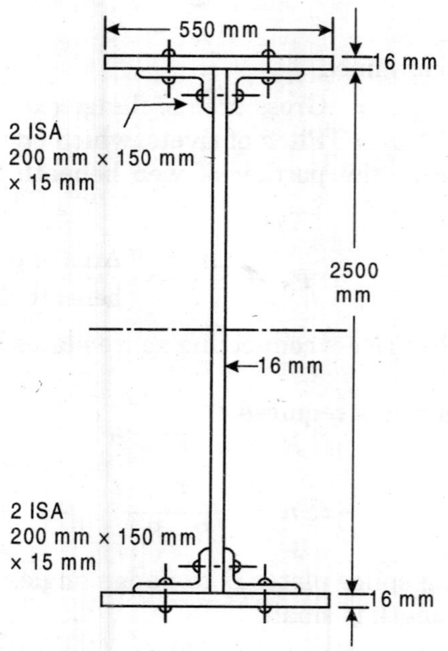

Fig. 8.27

Flange Area

Description	*Gross Area* $(mm)^2$	*Deduction for rivet hole* $(mm)^2$	*Net flange area* $(mm)^2$
Flange angles 2 ISA 200 mm × 150 mm × 15 mm	2 × 5025 = 10050	4 ×23.5 ×15 = 1410	8640
Flange plates 550 mm × 16 mm	550 × 16 = 8800	2 × 23.5 × 16 = 752	8048
Equivalent web area	$\frac{1}{6}A_w = \frac{2500\times16}{6}$ = 6667	—	$\frac{1}{8}A_w = \frac{2500\times16}{8}$ = 5000
Total	**25517**		**21688**

The distance upto the level of rivets connecting flange angles to web

$$(1250 - 90) = 1160 \text{ mm}$$

Step 4: Bending moment

$$M = 5670 \text{ kN-m}$$

Step 5 : Bending stress at the level of rivets connecting flange angles

$$s_{bc.cal} = \frac{M}{I_{xx}} \cdot y_1 = \left(\frac{5760 \times 10^6 \times 1160}{7829002 \cdot 1}\right)$$

$$= 85.3 \text{ N/mm}^2$$

Thickness of web = 16 mm

Step 6: Design of riveted connection

Use 22 mm diameter power driven rivets

Strength of rivets in double shear

$$\left(2 \times \frac{\pi}{4} \times \frac{(23 \cdot 5)^2 \times 100}{1000}\right) = 86.70 \text{ kN}$$

Strength of rivet in bearing

$$\left(\frac{23 \cdot 5 \times 16 \times 300}{1000}\right) = 112.8 \text{ kN}$$

$$\text{Rivet value, } R = 86.70 \text{ kN}$$

Let p be the pitch of rivets, in case rivets are used in one row

$$p = \frac{R}{\left[(\sigma \times t_w)^2 + \left(\frac{V}{d_e}\right)^2\right]^{1/2}}$$

$$p = \frac{86 \cdot 70 \times 1000}{\left[(85 \cdot 2 \times 16)^2 + \left(\frac{1080 \times 1000}{2467 \cdot 8}\right)^2\right]^{1/2}}$$

$$= 60.49 \text{ mm}$$

Provide rivets in rows at 165 mm pitch, using 13 rivets in each row.

Depth of splice plates $A = (2200 - 2 \times 6) = 2188$ mm

Thickness of splice plate $A = \frac{1}{2} t_w = \frac{1}{2} \times 16 = 8$ mm

Width of splice plates A to accommodate rivets = 423.0 mm

This provides distance between adjacent rows of rivets = $3d$ and edge distance = 1.5 d, where d is the gross diameter of rivets used.

Step 7 : Splice plates *B*

Area of web plate under flange angles = (150 × 16) = 2400 mm²

Provide 130 mm wide plates on both sides

$$\text{Thickness of plates} = \left(\frac{2400}{130 \times 2}\right) = 9.23 \text{ mm}$$

Provide two plates 130 mm ×10 mm

The horizontal shear force per pitch length between flange angles and splice plates

$$P_1 = \left(\frac{V}{d_e} \times \frac{A_f}{A_f \frac{1}{6} A_w} \times p\right)$$

Assume pitch of rivets p as 60 mm ($p > 2.5\ d$ minimum pitch)

$$P_1 = \left(\frac{1080 \times 1000}{2467 \cdot 8} \times \frac{18850}{18850 + 6667} \times \frac{60}{1000}\right)$$

$$= 19.40 \text{ kN}$$

The horizontal force in the portion of web beneath the flange angles due to moment

$$P_2 = \frac{M}{I} \cdot y \times (\text{Area of web beneath flange angles})$$

or

$$= \left(\frac{5760 \times 10^6}{7829002 \cdot 1} \times \frac{1160 \times 2400}{10^4 \times 1000}\right) = 204.83 \text{ kN}$$

Number of rivets required with 60 mm pitch, from Eq. 8.24

$$n = \left(\frac{P_2}{R_1 - P_1}\right) = \left(\frac{204.50}{(86 \cdot 70 - 19 \cdot 40)}\right) = 4$$

The length of splice plates is 480 mm

The complete web splice is shown in Eq. 8.28.

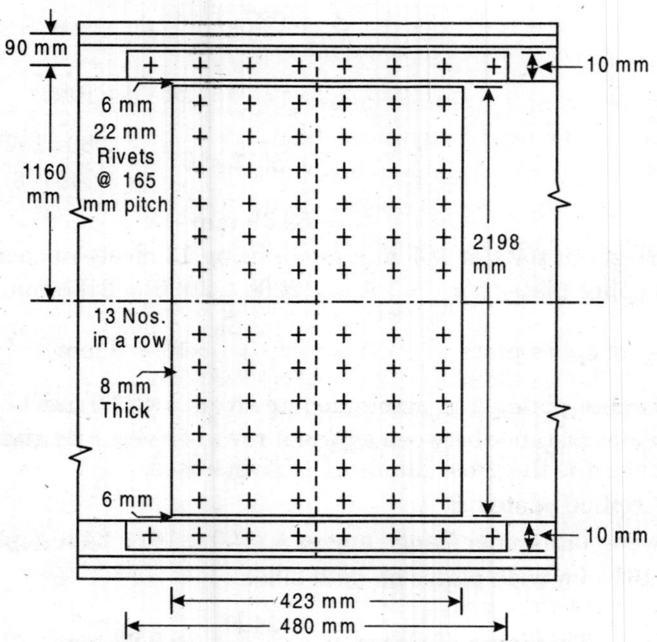

Fig. 8.28 *Web splice (Rational Splice method)*

8.19.2 Web Splice (Moment Splice)

This type of web splice is shown in Fig. 8.29. There are four moment plates, (two on each face), marked as splices A plates and two shear plates (one on each face), marked as splices plate B. It is assumed that moment plates resist moment resisted by web, and shear plates resist shear resisted by web. In fact, each set of plate resist shear as well as moment. But in case of deep girders, shear resisted by splice plates A is small compared with plates B. Similarly the

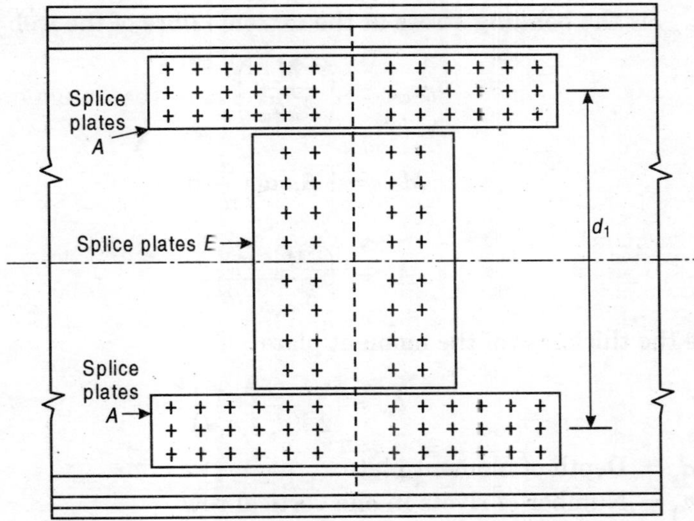

Fig 8.29 *Web splice (Moment splice method)*

moment resisted by splice plates B is small compared with splice plates A. This type of splice may be used for girders about 2 metres deep. A clearance of 6 mm is provided between splice plates A and flange anges, and between splice plates A and splice plates B. The web splice (moment splice) is designed as under :

The moment resisted by the web plate is as under

$$M_w = \left(\frac{I_w}{I} \times M\right) \quad \text{...(8.25)}$$

where, I_w = Gross moment of inertia of the web

I = Gross moment of inertia of the girder

M = Bending moment at the web splice section

8.19.2.1 *Moment Plates*

The moment of resistance of four moment plates (splice plates A) is equal to moment resisted by web.

The moment of resistance of four moment plates is $A_s . \sigma_1 . d_1$

where, A_s = Net area of two moment plates
d_1 = Distance between centre to centre of splice plates A (Moment plates)
σ_1 = Bending stress at the centre of splice plate A, and assumed uniform in these plates.

Therefore
$$M_w = A_s \cdot \sigma_{b1} \cdot d_1 \quad \text{...(i)}$$

$$\sigma_{b1} = \left(\frac{d_1}{d_w} \times \sigma_{bc}\right) \quad \text{...(ii)}$$

where, $\sigma_{bc \cdot cal}$ is the bending stress at the extreme fibre of the web plate

$$\sigma_{bc \cdot cal} = \left(\frac{M}{A_n d_w}\right) \quad \text{...(iii)}$$

$$\therefore \quad M_w = \left(A_s \cdot d_1 \cdot \frac{d_1}{d_w} \sigma_{bc \cdot cal}\right)$$

$$A_s = \left(\frac{M_w \cdot d_w}{\sigma_{bc \cdot cal} \cdot d_1^2}\right) \quad \text{...(8.26)}$$

Let t_s be the thickness of the moment plate

$$t_s = \frac{1}{2} \cdot \frac{A_s}{(d_s - n_1 d)} \quad \text{...(8.27)}$$

where, d_s = Depth of moment plate
n_1 = Number of rivets in one vertical row
d = Diameter of rivets.

The horizontal force in the two moment plates A

$$= (A_1 \times \sigma_{b1})$$

The number of rivets required to connect moment plates A to the web plate on each side of web splice is given by

$$n = \left(A_s \cdot \frac{\sigma_{b1}}{R}\right) \quad \text{...(8.28)}$$

where, R = Rivet value.

8.19.2.2 *Shear Plates*

The shear plates B resist shear at the web splice section. The combined thickness of these plates is designed to resist shear at web splice section. The width of the splice plates B is kept sufficient to accommodate rivets. The number of rivets, required to resist shear

$$n = \left(\frac{V}{R}\right)$$

where, V is the shear at the web splice and R is the rivet value.

These rivets are provided on each side of the web splice.

Example 8.7 *Design the web splice given in Example 8.6.*

Solution

Design :

Step 1: Moment resisted by web plate

Gross moment of inertia of plate girder

$$I_{xx} = 7829002.1 \times 10^4 \text{ mm}^4$$

Gross moment of inertia of web plate

$$I_w = 2083334.3 \times 10^4 \text{ mm}^4$$

The bending moment resisted by the web plate from Eq. 8.25

$$M_w = \left(\frac{I_w}{I} \times M\right) = \left(\frac{2083333\cdot 3}{7829002\cdot 1} \times 5760\right)$$

$$= 1535 \text{ kN-m}$$

The depth of web plate, d_w is 2500 mm

Net flange are $A_{fn} = 21688 \text{ mm}^2$

Step 2: Moment plates

The depth of moment plates is kept sufficient to accommodate rivets in 3 rows. 22 mm diameter rivets are used at 100 mm spacing centre to centre and 50 mm edge distance.

The distance between centre to centre of moment plates,

$$d_1 = (2500 - 2 \times 152 - 2 \times 150 - 2 \times 6)$$

$$= 1888 \text{ mm}$$

The bending stress at the extreme fibre of web plate,

$$\sigma_{bc.cal} = \left(\frac{5760 \times 1000 \times 1000}{21688 \times 2500}\right)$$

$$= 106.23 \text{ N/mm}^2$$

The net area of two moment plates from Eq. (8.22)

$$A_s = \left(\frac{M_w \cdot d_w}{s_{bc\cdot cal} d_1^2}\right) = \left(\frac{1535 \times 10^6 \times 2500}{106\cdot 23 \times 1888^2}\right)$$

$$= 10134.38 \text{ mm}^2$$

The thickness of moment plates as per Eq. (8.23)

$$t_s = \frac{1}{2}\left(\frac{A_s}{d_s - n\cdot d}\right)$$

$$= \frac{1}{2} \times \left(\frac{10134\cdot 38}{(300 - 3 \times 23\cdot 5)}\right)$$

$$= 22.08 \text{ mm}$$

Provide 18 mm thick plates.

Step 3 : Design of riveted connections

Strength of power driven rivets in double shear

$$\left(2\times\frac{\pi}{4}\times\frac{(23\cdot5)^2\times100}{1000}\right) = 86.70 \text{ kN}$$

Strength of rivets bearing

$$\left(23\cdot5\times18\times\frac{300}{1000}\right) = 126.9 \text{ kN}$$

Rivets value, R = 86.70 kN.

Number of rivets required to connect moment plates to web

Bending stress $\sigma_{b1} = \left(106\cdot23\times\frac{1800}{2500}\right)$

$= 80.22$ N/mm^2

From Eq. 8.28, $n = \left(\frac{A_s\times\sigma_{b1}}{R}\right) = \left(\frac{18\times2(300-3\times23\cdot5)}{86.70}\right)\times\frac{80\cdot22}{1000}$

$= 7.64$

Provide 12 rivets on each side of web splice section

The distance between rows of rivets is 70 mm

Edge distance = 35 mm

Width of moment plates = 560.0 mm

Step 3 : Shear plates

6 mm clearance is kept between moment plates and shear plates.

Depth of shear plates = (2500 – 2 × 150 – 2 × 300 – 4 × 16) mm

= 1536 mm

Shear force at web splice = 1080 kN

Allowable shear stress = $0.4f_y$ = 0.4 × 250 = 100 N/mm^2

Thickness of two shear plates

$$\left(\frac{1080\times1000}{1576\times100}\right) = 7.03 \text{ mm}$$

Thickness of one shear plates

= 3.515 mm

Provide 6 mm thick shear plate

Step 4: Design of riveted connection

Number of rivets required

$$\left(\frac{1080}{86.70}\right) = 12.45$$

Provide 14 rivets in two rows at 225 mm pitch.

Width of splice plate = 280.0 mm

The distance between rows of rivets in 70 mm

Edge distance = 35 mm

The complete web splice of the web plate is shown in Fig. 8.30.

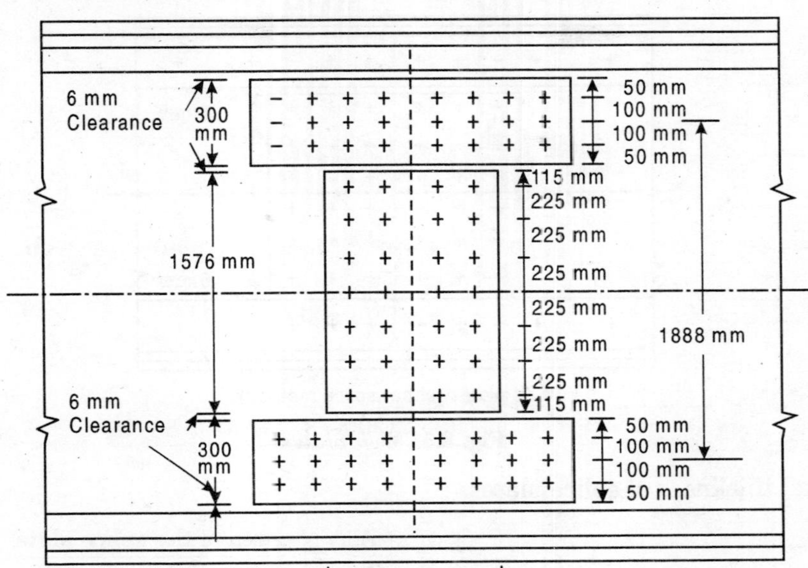

Fig. 8.30 *Web splice (moment splice method)*

8.19.3 Web Splice (Shear Splice)

This type of web splice is shown in Fig. 8.31. The splice plates are provided between flange angles. A clearance of 6 mm is left between splice plates and flange angles. The web splice (shea splice) is designed as under :

The moment of resistance of splice plate is kept equal to moment of resistance of web plates

$$\left(\sigma_{b1} \times \frac{1}{6} \times t_s \times (d)^2\right) = \left(\sigma_{bc} \times \frac{1}{6} \times t_w \times (d_w)^2\right)$$

where, σ_{b1} is the bending stress at the extreme fibre of splice plate and σ_{bc} the bending stress at the extreme fibre of web plate. From the triangular distribution of bending stress,

$$\sigma_{b1} = \left(\sigma_{bc} \cdot \frac{d_e}{d_w}\right)$$

$$\therefore \quad \left(\sigma_{bc} \cdot \frac{d_e}{d_w} \cdot \frac{1}{6} \cdot t_s \cdot (d_s)^2\right) = \left(\sigma_{bc} \cdot \frac{1}{6} \cdot t_w \cdot (d_w)^2\right)$$

Thus

$$A_s = A_w \cdot \left(\frac{d_w}{t_s}\right) \quad \text{...(8.29)}$$

where, A_s and d_s = Area and depth of splice plates

A_w and d_w = Area and depth of web plates

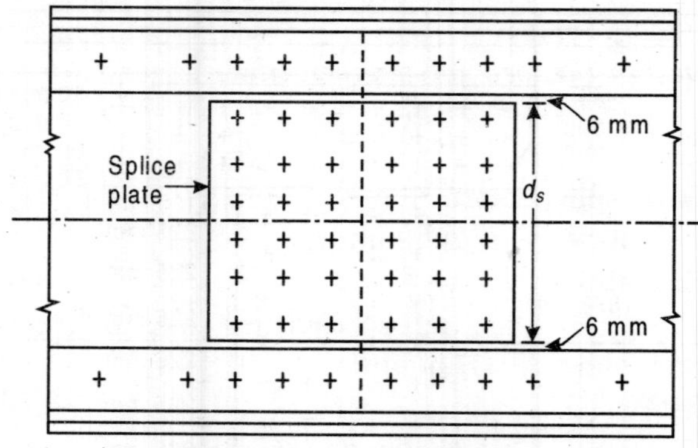

Web plates (shear splice method)

Fig. 8.31 *Web plates*

Total thickness of splice plates

$$t_s = \left(\frac{A_s}{b_s}\right) = \left(\frac{\text{Area of the splice plate}}{\text{Width of the splice plate}}\right)$$

The width of splice plate is kept sufficient to accommodate the rivets. The splice plates are designed to resist shear and moment which would be resisted by web, if web was not spliced

$$M_w = \left(\frac{I_w}{I} \times M\right)$$

where, I_w = Gross moment of inertia of the web

I = Gross moment of inertia of the girder

M = Bending moment at the splice section.

The vertical shear V at the splice section is shown in Fig. 8.32 (a). It acts at an eccentric distance from C.G. of rivets, connecting splice plates to the web. It is considered equivalent to a vertical force V and an eccentric moment $M_e = V \times e$ as shown in Fig. 8.32 (b).

The splice plates resist a total moment M_s

$$M_s = (M_e + M_w)$$

(a) (b)

Fig. 8.32

The rivets connecting splice plates to the web are designed to resist a vertical force V and a moment M_e. This becomes design of eccentric riveted connections, explained in Art. 2.27, Type 1.

Example 8.8 *Design the web splice given in Example 8.6.*

Solution

Design :

Step 1 : Area of splice plate

The depth of web plate, d_w is 2500 mm

Depth of splice plates

$$d_e = (2500 - 2 \times 150 - 2 \times 6) = 2188 \text{ mm}$$

Thickness of web plate, t_w = 1.6 mm

Area of web plate

$$A_w = d_w.t_w$$
$$= (2500 \times 16) = 40000 \text{ mm}^2$$

Area of splice plate required, from Eq. 8.25

$$A_s = A_w\left(\frac{d_w}{d_s}\right)^2$$

$$= 4000 \times \left(\frac{2500}{2188}\right)^2 = 52221 \text{ mm}^2$$

Step 2 : Total thickness of splice plates required

$$t_s = \left(\frac{A_s}{d_s}\right) = \left(\frac{52221}{2188}\right) = 23.887 \text{ mm}$$

Provide two splice plates 12 mm thick one on each side of web plate. The width of the splice plate is kept to accommodate three rows of rivets on either side of splice section.

Step 3 : Width of splice plate

Use 22 mm power driven rivets

$\therefore$ Width of splice plates, $b_s = (6 \times 3d)$

where, d the gross diameter of rivet

$$b_s = (6 \times 3 \times 23.5) = 423.0 \text{ mm}$$

From Example 8.6,

The gross moment of inertia of plate girder, I_{xx} is 7829002.1 × 10^4 mm^4

The gross moment of inertia of web plate 2500 mm deep × 16 mm thick from ISI Handbook No. 1.

$$I_w = 2083333.3 \times 10^4 \text{ mm}^4$$

Bending moment at web splice section

$$M = 5760 \text{ kN-m}$$

Bending moment resisted by the web plate

$$M_w = \left(\frac{I_w}{I} \times M\right) = \left(\frac{2083333 \cdot 9}{782900 \cdot 12}\right) \times 5760 \text{ kN-m}$$

Vertical shear at web splice section,

$$V = 1080 \text{ kN}$$

Eccentricity from C.G. of group of rivets

$$e = (4.5\,d) = (4.5 \times 23.5) = 105.0 \text{ mm}$$

Eccentric moment

$$M_e = \left(\frac{1080 \times 105 \cdot 5}{1000}\right) = 113.4 \text{ kN-m}$$

Splice plates resist a total moment

$$M_s = (M_e + M_w) = (113.4 + 1535)$$
$$= 1648.4 \text{ kN-m}$$

Step 5: Design of riveted connections

Group of rivets resist a moment

$$= 1648.4 \text{ kN-m}$$

and a vertical force $= 1080$ kN

Strength of power driven rivet in double shear

$$\left(2 \times \frac{\pi}{4} \times \frac{(23 \cdot 5)^2 \times 100}{1000}\right) = 86.70 \text{ kN}$$

Strength of power driven rivet in bearing

$$\left(23 \cdot 5 \times 16 \times \frac{300}{1000}\right) = 112.8 \text{ kN}$$

Rivet value, $R = 86.70$ kN

Let n be the number of rivets to be provided in one row. The rivets are provided in 3 rows. Assume pitch of rivets p in each row at 90 mm.

$$n = \left(\frac{6M_s}{(3p) \times R}\right)^{1/2} = \left(\frac{6 \times 1648 \cdot 4 \times 1000}{3 \times 90 \times 86 \cdot 70}\right)^{1/2}$$
$$= 20.55$$

Provide 24 rivets in each row.

Step 6: Check for forces in rivets

The rivets are arranged in a narrow strip. The vertical distance to the extreme rivet, is as below.

$$y_n = 1035 \text{ mm}$$

$$\Sigma_y^2 = 6.\ [4.5^2 + 13.5^2 + 22.5^2 + 31.5^2 + 40.5^2 + 49.5^2$$
$$+ 58.5^2 + 67.5^2 + 76.5^2 + 85.5^2 + 94.5^2 + 103.5^2] \times 100 = 6 \times (46557.20 \times 100) \text{ mm.}$$

The force due to direct shear in each rivets is as under

$$F_1 = \left(\frac{1080}{24 \times 3}\right) = 15 \text{ kN}$$

Force in an extreme rivet resisting the moment

$$F_2 = \left(\frac{M \cdot y_n}{\Sigma y^2}\right) = \left(\frac{1648 \cdot 4 \times 1000 \times 1000 \times 1035}{6 \times 46557 \cdot 20 \times 100 \times 1000}\right)$$

$$= 61 \text{ kN}$$

$$\cos \theta = \frac{70 \cdot 5}{\sqrt{(70 \cdot 5^2 + 103 \cdot 50^2)}} = 0.068$$

Resultant of two forces

$$F = (F_1^2 + F_2^2 + 2F_1F_2 \cos \theta)^{1/2}$$
$$= ((15)^2 + (61)^2 + 2 \times 15 \times 61 \times 0.068)^{1/2}$$
$$= 63.90 \text{ kN} < 867.0 \text{ kN, rivet value.}$$

Hence, the design of rivet is satisfactory.

The complete web splice is shown in Fig. 8.33.

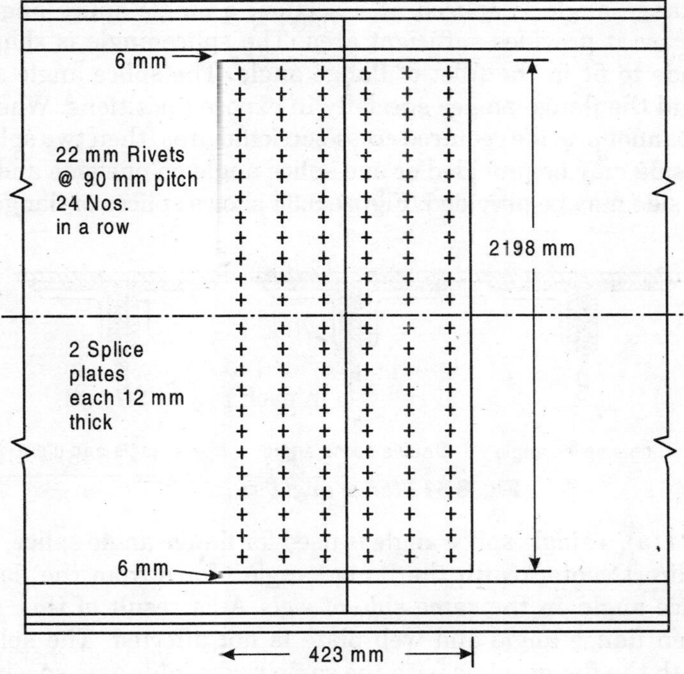

Fig. 8.33 *Web splice (shear splice method)*

8.20 FLANGE SPLICES

A joint in the flange element provided to increase the length of flange angle or flange plate is known as flange splice.

The flange splices should be avoided as far as possible. In general, the flange angles and flange plates can be obtained for full length of the plate girder. In spite of availability of full length of flange angles and flange plates, sometimes, it becomes necessary to make flange splices, for example, the transportation facilities may not permit transportation of plate girder for the entire span as one piece. The flange splices should be located at the section where some excess of flange area is available and not at the points where web splice is done. It should preferably not be located at points of maximum stress. Only one element of the flange, viz., one flange angle or one plate should be spliced at one section. Where splice plates are used, their area shall be not less than 5 percent in excess of the area of flange element spliced. The centre of gravity of splice plates should coincide as nearly as possible with that of the element spliced. The rivets are designed for force 5 percent in excess to force in the element spliced. The strength of splice should be equal to or greater than 50 percent of the strength of splice elements.

8.20.1 Splice of Flange Angles

When one flange angle is spliced at a section, a single splice angle may be provided in case it provides sufficient area. The splice angle is shaped at the heel and made to fit in the fillet of flange angle. The splice angle should not project beyond the flange angles specially in exposed positions. When a single splice angle cannot provide required cross-sectional area, then two splice angles, one on each side may be provided or one splice angle on one side and one plate on the other side may be provided. Figure 8.34 shows splices of flange angles of three types.

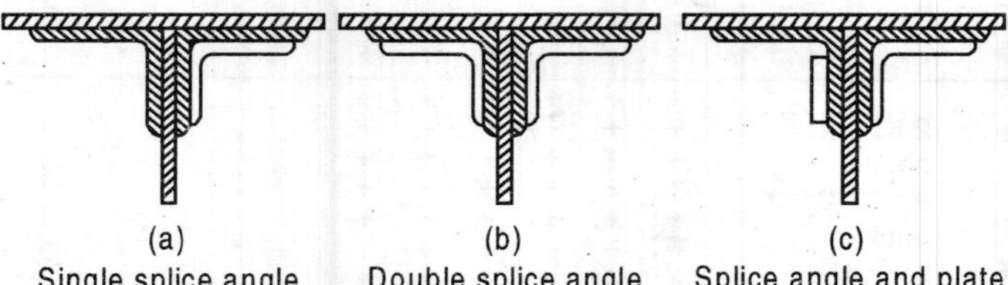

Fig. 8.34 *Flange angle splices*

In Fig. 8.34 (a), a single splice angle is used for flange angle splice, and splice angle is in direct contact with the flange angle. Force from the flange angle shifts to splice angle on the same side of web. As a result of this, horizontal shear between flange angle and web plate is not affected. The splice and is connected with the flange angle with the same rivets and same spacing as used for connecting flange angle with the web. In order to minimize the length of splice angle, the rivets may be provided as less spacing. The length of splice angle is kept sufficient to accommodate rivets required to transmit force from flange to splice angle and back. In Fig. 8.34 (c), a single splice angle and a splice plate is used for flange angle splice. The plate is not in direct contact with the element spliced. The part of the force from flange angle shifts to opposite side of the web through the web and unspliced flange angle. As a result of this, horizontal shear between flange angle and web is affected. It requires additional rivets for connections. The force in the flange angle is assumed to be distributed to the elements of the splice in proportion to their cross-sectional area. The strength to be transmitted by rivets connecting splice plate is equal to force in splice plate plus force due to horizontal shear. Therefore,

$$(n \,.\, R_s = P + \text{Horizontal shear})$$

where, n = Number of rivets required to connect splice plate on each side of splice

P = Force to be carried by splice plate

R_s = Strength of rivet in single shear

The strength of rivets is considered single shear because P is force to be carried in one plane only.

The horizontal shear per unit length

$$\tau_{vf.cal} = \left(\frac{V}{d_e}\right)$$

The horizontal shear per unit length in one plane

$$\frac{1}{2}.\tau_{vf.cal} = \left(\frac{1}{2}\frac{V}{d_e}\right)$$

The horizontal shear per unit length in one plane

$$F_s = \left(\frac{1}{2}\frac{V}{d_s}\times n \cdot p\right)$$

where, p = Pitch of rivets

V = Shear force at the splice section

d_e = Effective depth of the girder

$$\therefore \qquad n \,.\, R_s = \left(P + \frac{1}{2}\frac{V}{d_s}\times n \cdot p\right)$$

$$\therefore \qquad n = \frac{P}{\left(R_s - \frac{1}{2}\frac{V}{d}\cdot p\right)} \qquad \text{...(8.30)}$$

8.20.2 Splice of Flange Plates

In case, it becomes necessary to splice an outer flange plate, a splice plate of same cross-section as the plate is provided. The length of splice plate is kept sufficient to accommodate necessary number of rivets. The strength of rivet is found in single shear. The splice plate is in direct contact. Therefore, the force to be transmitted by splice plate is not affected.

In case, it becomes necessary to splice an inner flange plate, the splice may be located at the theoretical cut-off of the next outer plate. The outer plate is extended. This serves as a splice plate. Splice of an inner plate is also done by providing an extra plate, which is placed outside of all flange plates. In this, splice plate is not in direct contact with the flange plate splices. The area of cross-section of splice plate is kept equal to the cross-sectional area of flange plate cut. The rivets may be designed to take full strength of flange plate cut and the shearing stress due to transmission of flange increment is neglected.

Example 8.9 *A plate girder is composed of a 1600 mm × 12 mm web plate, 4 ISA 150 mm × 150 mm ×18 mm flange angles and two 500 mm × 12 mm thick flange plates. The gross moment of inertia of plate girder is 2355120.2 × 10⁴ mm⁴. The web of bending moment at a section is 2100 kN-m and shear force 1600 kN. One flange angle is spliced at this section. Design flange angle splice. Provide a single splice angle.*

Solution

Design :

Step 1: Design of splice of flange angle

Size of flange angle ISA 150 mm × 150 mm × 18 mm (ISA 150 150, @ 0.399 kN/m)

Use 22 mm diameter power driven rivets for connections

Net area of flange angle (5079 – 2 × 235 × 18) = 4233 mm^2

∴ Add 5 per cent = 212 mm^2

Total area required = 4445 mm^2

From ISA 200 mm × 200 mm × 25 mm (ISA 200 200, @ 0.736 kN/m) cut a splice angle 130 mm × 130 mm × 25 mm. The splice angle is cut and shaped to clear root of flange angle, and it does not project beyond flange angle.

Gross area of splice angle

[13 × 25 × (13 – 2.5) × 2.5] = 58.75 × 100 mm^2

Deduct 2 rivet holes

2 × 23.5 × 25 = 1175 mm^2

Net area provided by splice angle

(5875 –1175) = 4700 > 4445 mm^2

Strength of power driven rivet in single shear

$$\left(\frac{\pi}{4}\times(23\cdot 5)^2\times\frac{100}{1000}\right) = 43.35 \text{ kN}$$

Strength of rivets in bearing

$$\left(23\cdot 5\times 12\times\frac{300}{1000}\right) = 84.6 \text{ kN}$$

Rivet value = 43.35 kN

Net flange area = 16501 mm^2 (See Table on page no. 555)

Gross flange area = 19358 mm^2

The stress at the level of C.G. of flange angles

$$\sigma_b = \left(\frac{2100\times 10^6}{2355120\cdot 2\times 10^4}\right)\times\left(\frac{(812-12-43\cdot 8)\times 19358}{16501}\right) = 78.1 \text{ N/mm}^2$$

Force in flange angle

$$\left(\frac{78\cdot 1}{1000}\times 4445\right) = 347.15 \text{ kN}$$

Number of rivets required

$$\left(\frac{347 \cdot 15}{43 \cdot 35}\right) = 8.008.$$

Step 3 : Area of flanges

Flange Area

Description	*Gross Area* $(mm)^2$	*Deduct for rivet hole* $(mm)^2$	*Net area* $(mm)^2$
Flange angle 2 ISA 150 mm × 150 mm × 18 mm	2 × 5071 = 10158	4 ×23.5 ×18 = 1692	8466
Flange Plate 550 mm × 12 mm	500 × 12 = 6000	2 × 23.5 × 16.5 = 565	5635
Equivalent web area	$\frac{1}{6}\times1600\times12$ = 3200	—	$\frac{1}{8}\times1600\times12$ = 2400
Total	**19358**		**16501**

Provide 10 rivets on each side of flange angle, 5 rivets are provided in each leg at 60 mm spacing. The length of splice angle

$$= 60 \times 10 = 600 \text{ mm}$$

The complete flange splice is shown in Fig. 8.35.

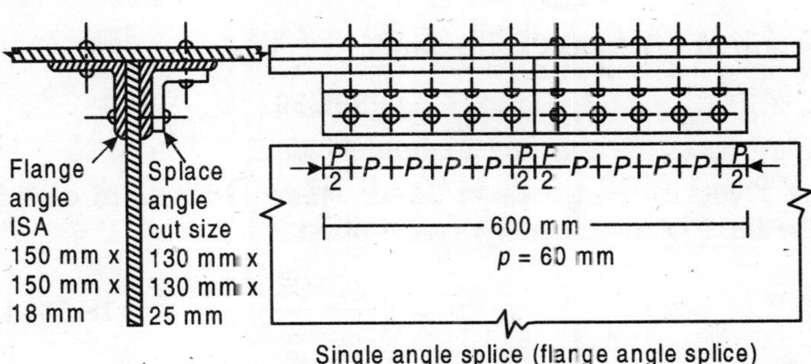

Fig. 8.35 *Flange angle splice*

Example 8.10 *Design flange angle splice in Example 8.9. Provide a splice angle and a rectangular plate.*

Solution

Design :

Step 1: Design of splice angle

From Example 8.9

Net area of flange angle ISI 150 mm × 150 mm × 18 mm (ISA 150 150, @ 0.399 kN/m)

$$= 4233 \text{ mm}^2$$

Add 5 per cent = 212 mm^2

Total area of splice elements required

= 4445 mm^2

Provide ISA 130 mm × 130 mm × 15 mm (ISA 130 130, @ 0.289 kN/m) as splice angle

The splice angle does not project beyond flange angle

New area provided by splice plate

(36.81 – 2 × 2.35 × 1.5) × 100 = 29.76 × 100 mm^2

Step 2 : Design of splice plate

Area required from splice plate

= (4445 – 2986) = 1469 mm^2

Provide 120 mm × 16 mm splice plate

Net area provide by splice plate

= (120 × 16 – 23.5 × 16) = 1544 mm^2

Total area provide by splice elements

= (2976 + 1544) = 4520 mm^2

From Example 8.9, the stress at the level of C.G. of flange angle

σ_b = 78 1 N/mm^2

Force to be transmitted by flange angle

= (78.1 × 4445) = 347.15 kN

Force transmitted by splice angle

$$\left(\frac{347.15 \times 2976}{4520}\right) = 228.56 \text{ kN}$$

Rivets required to connect splice angle

$$n = \left(\frac{228.56}{43.35}\right) = 5.27$$

Provide 6 rivets on each side of flange splice, 3 rivets in each leg at 6 mm spacing. The force transmitted by splice plate

$$p = \left(\frac{347.15 \times 1544}{4520}\right) = 118.584 \text{ kN}$$

Step 3: Rivets required to connect splice plate

The distance of C.G. of compression flange from outside

$$\bar{y} = \left(\frac{50 \times 1.2 \times 0.6 + 2 \times 50.79 \times (1.2 + 4.38)}{50 \times 1.2 + 2 \times 50.79} \times 10\right) = 37.1 \text{ mm}$$

Effective depth of the plate girder

d_e = (1624 – 2 × 37.1) = 1549.8 mm

The number of rivets needed for splice plate, from Eq. 8.26

$$n = \frac{P}{\left(R_s - \frac{1}{2} \times \frac{V}{d_e} \cdot p\right)}$$

$$n = \frac{118 \cdot 584}{\left(43 \cdot 35 - \frac{1}{2} \times \frac{1600 \times 1000}{1549 \cdot 8} \times 600\right)}$$

$= 12.11$ mm

Provide 14 rivets on each side of flange splice. The length of splice plate.

$= 28 \times 1680$ mm

The complete flange splice is shown in Fig. 8.36.

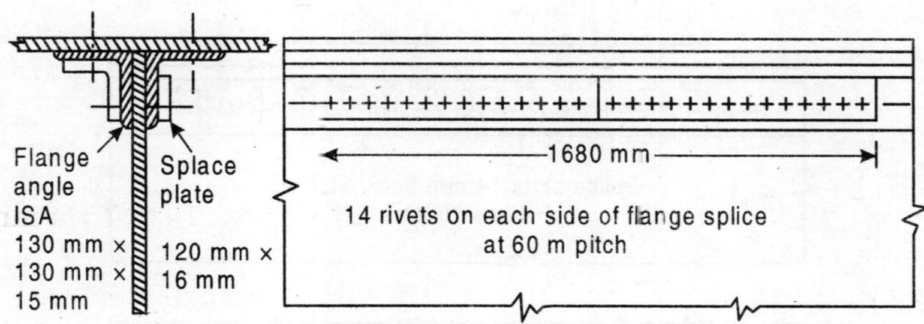

Fig. 8.36 *Flange angle splice*

Example 8.11 *It is required to splice flange plate only in Example 8.9. Design splice for flange plate.*

Solution

Design :

Step 1: Force in flange plate

From Example 8.9, net area of flange plate

$= 5635 \text{ mm}^2$

Add 5 per cent $= 282 \text{ mm}^2$

Net area required by splice element

$= 5197 \text{ mm}^2$

Provide 50 mm × 14 mm splice plate

Net area provided $= (50 \times 1.4 - 2 \times 2.35 \times 1\ 6) \times 100$

$= 6245 \text{ mm}^2 > 5917 \text{ mm}^2$

The stress in spliced flange plate

$$\left(\frac{2100 \times 10^6 \times 19558}{2355120 \cdot 2 \times 10^4 \times 16501}\right) = 84.31 \text{ N/mm}^2$$

Force in flange plate $= \left(\frac{84 \cdot 31 \times 5635}{100}\right) = 475.1$ kN

Force 5 percent in excess of force in flange plate

$= 1.05 \times 475.1 = 498.855$ kN

Step 2 : Design of riveted connections

Rivet value, $R = 43.35$ kN

The number of rivets required

$$\left(\frac{498 \cdot 855}{43 \cdot 35}\right) = 11.50$$

Provide 12 rivets in two rows on each side of flange plate splice at 6 mm spacing. The length of splice plate

$$= 12 \times 60 = 720 \text{ mm}$$

The complete flange plates splice is shown in Fig. 8.37.

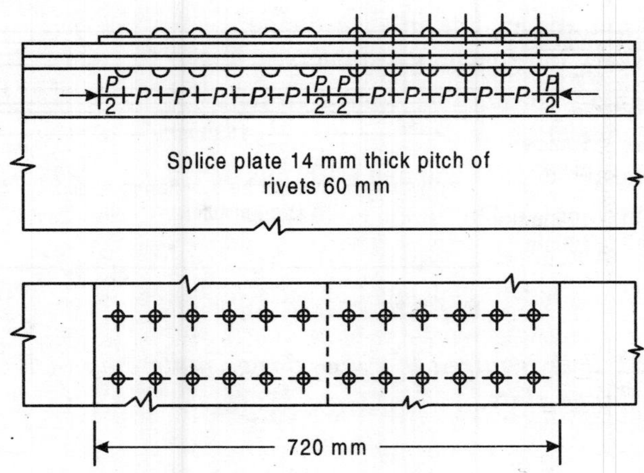

Fig. 8.37 *Flange splice*

PROBLEMS

8.1. A plate girder simply supported at ends is composed of a web plate 1250 mm depth × 16 mm thickness, and, two flange angles ISA 200 mm 100 mm × 15 mm (ISA 200 100, @ 0.336 kN/m wide and two flange plates 550 mm wide × 16 mm thickness in each flange. The effective span of the plate girder is 15 m. The diameter of power driven rivets used for connecting flange plates to flange angles and flange angles to web plate is 22 mm. Determine the maximum uniformly distributed load (inclusive of self weight) which can be carried by the plate girder.

8.2. A plate girder, having an effective span 12.20 m is required to carry a uniformly distributed load of 90 kN/m inclusive of its own weight, ends being simply supported. Compression flange has got adequate lateral support. Design the plate girder and draw :

(i) An elevation showing the plate girder with details

(ii) A plan of the flange showing curtailment of flange plates

(iii) A cross-section showing details of the make-up.

8.3. Design the central section of a plate girder for an effective span of 20 metres if the dead and live loads amount to 30 kN/m and 60 kN/m, respectively. Show the curtailment of flanges on a diagram.

8.4. A plate girder consists of a web plate 1000 mm × 10 mm, flange angles 100 mm × 100 mm × 12 mm (ISA 100 100, @ 0.177 kN-m) and cover plates 300 mm wide by 1.20 mm thick (one each at top and bottom). Design rivets connecting the flanges angles with the web and the cover plates at a section subjected to 730 kN shear.

8.5. A simply supported plate girder spans 13.60 m and carries a uniformly distributed load of 80 kN/m inclusive of self weight of girder. The plate girder consists of 1600 mm deep × 12 mm thick web plate, 2 ISA 200 mm × 100 mm ×15 mm (2 ISA 200 100, @ 0.336 kN/m) flange angles and 500 mm × 12 mm cover plate in each flange. Design end bearing stiffeners, and necessary intermediate stiffeners.

8.6. Design a web splice for a plate girder fro the following data : Web plate : 2000 mm ×12 mm

Flange angles : ISA 200 mm × 100 mm × 15 mm (ISA 200 100, @ 0.336 kN/m)

Net area of tension flange : 24642 mm^2

Gross area of compression flange : 28556 mm^2

Maximum bending moment at section : 6900 kN-m

Corresponding shear force at section : 1380 kN

Flange plates in each flange : 2 × 500 mm × 16 mm.

8.7. A plate girder has an effective span of 16 m and is simply supported at its ends. It carries a uniformly distributed load of 100 kN/m exclusive of its self-weight. Design (i) the central section (ii) the rivets connecting the components, using at least two flange plates on each flange.

Determine the theoretical points of cut-off for the flange plates of the plates girder.

8.8. A riveted plate girder section is designed to carry a uniform load of 50 kN/m excluding the self-weight of the girder. In additions, the girder has to support two concentrated loads of 400 kN each at one-third points. The effective span is 18 m. The following section has been provided.

Web plate 2000 mm × 6 mm

Flange angles ISA 150 mm × 150 mm × 12 mm (ISA 150 150, @ 0.272 kN/m)

Flange plates 2 Nos. 400 mm × 12 mm

Design the bearing stiffeners to be provided at the supports and the vertical intermediate stiffeners.

8.9. A plate girder section consists of the following components

Web plate 1500 mm × 10 mm

Flange angles ISA 150 mm × 150 mm × 10 mm (ISA 150 150, @ 0.228 kN/m)

Flange plates 2 Nos. 500 mm × 10 mm on each flange

If the end reaction produced by the applied loads is 1000 kN design the bearing stiffener for it.

For the plate girder above, design all the necessary intermediate stiffeners.

8.10. It is required to design a suitable mid-span section of built-up plates girder, supporting a uniformly distributed load of 90 kN/m including its own weight.

The girder has a simply supported span of 23 m.

Use 22 mm diameter power driven rivets.

8.11. A plate girder of span 20 m has to carry a uniformly distributed load of 90 kN/m inclusive of its own weight. Design the maximum section of the girder if the web depth is not to exceed 1600 mm, calculate the spacing of 22 mm diameter power driven rivets connecting the flange angles and the flange plates near the ends.

8.12. A simply supported plate girder having a span of 4 metres has to support floor beams that frame at 2 metres centre to centre as showñ in Fig. P.8.12.

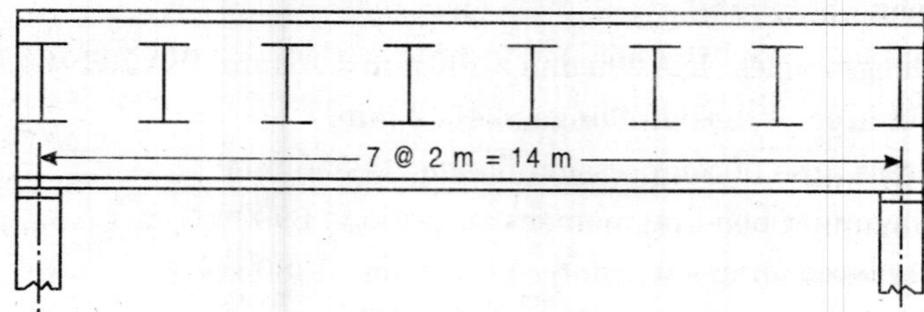

Fig. P.8.12

Each floor beam introduces a concentrated load of 1000 kN on the girder. In addition, the girder has to carry a uniformly distributed load of 18 kN/m including its own weight. The top flange of the girder is to be restrained effectively and the girder is to be provided with vertical stiffeners. Assuming that the depth of web plate is to be 1000 mm and using the data given, find (a) the suitable thickness of the web plate, (b) the section of the flange at the centre of span. Show by a dimensioned sketch the cross-section of the girder at the centre span.

8.13. Evaluate the girder cross-section of a riveted plate girder to carry a super load of 120 kN per metre on a simple span of 2 metres. Furnish typical connections of:

(i) flange plates

(ii) web and flange angle connection.

(**N.B:** Determine the preliminary section and check by rigorous moment of inertia method).

The girder is laterally restrained throughout.

C.G. of equal angles = 0.28 × Leg length from the heel.

8.14. A simply supported plate girder spans 24 m and carries a total uniformly distributed load of 4000 kN inclusive of self-weight of the girder. The plate girder consists of 1600 mm deep × 12 mm thick web plate, 2 ISA 200 mm × 150 mm × 18 mm (2 ISA 200 150, @ 0.469 kN/m) angles and 550 mm × 12 mm cover plate in each flange. Design the riveted connection of flange angles to web and flange plates to flange angles.

8.15. A plate girder section is made up of a web of 200 mm × 12 mm and flange angles 150 mm × 150 mm × 12 mm (ISA 150 ×150, @ 0.279 kN/m) and one cover plate in each flange of 45 mm × 12 mm. The girder is supported at either end on bearing plates 500 mm × 500 mm. If the maximum end reaction is 1640 kN, design end stiffener using a cluster of 4 angles ISA 130 mm × 130 mm × 12 mm. (ISA 130 130, @ 0.234 kN/m).

8.16. It is required to a design a triple plate, i.e., with moment and shear plates web splice for a plate girder where the shear is 480 kN and moment 54,00 kN-m. The size of web plate is 1800 mm × 10 mm. The available clear distance between toes of flange angles is 1520 mm. The total moment of inertia of the girder section is 466 × 10^8 mm^4.

Use 22 mm diameter power driven rivets.

8.17. Design a suitable web splice for a plate girder at a section at which the bending moment is 5000 kN-m and shear force is 1000 kN. The plate girder is built-up from a web plate 2500 mm deep × 16 mm thick and two ISA 200 mm × 150 mm × 15 mm (ISA 200 150, @ 0.394 kN/m) angles and a 600 mm × 16 mm cover plate in each flange. The rivets used are of 22 mm diameter. Draw a sketch of the splice. The long legs of the flange angles are placed horizontally.

Chapter 9

Design of Roof Trusses

9.1 INTRODUCTION

The roof trusses are the frame structures in which separate straight members are so arranged and connected at their ends that the members form triangles. The axes of members meeting at one joint interest at a common point. The riveted joints used for the connections of the members are considered to act as pin-joints. The external loads are applied to the joints of the truss so that the members carry direct forces only. When beams are subjected to bending, then bending stress varies from zero at the neutral axis to maximum towards the extreme fibres. The different fibres in a cross-section of the beams are subjected to different intensity of stress. The entire section of beam is not utilized. Whereas the bending of trusses is quite distinct from the beams. In roof trusses, the entire section of each member is subjected to uniform stress. The strength of member is fully utilized. The bending action in roof trusses is provided by elongation and shortening of the members of the truss. It further results in deflection of truss. The forces in various members are either compressive or tensile. The members carrying compressive forces in a roof truss are called *struts,* and those carrying tensile forces are called *ties*. In a roof truss, members are so arranged that the length of members in compression are small while the lengths of members in tension are long. As far as possible, it is seen that the length of a member is not more than 3 metres. The members of a truss are classified as *main members* and *secondary members.* The *main members* are the structural members which are responsible for carrying and distributing the applied loads and stability of the truss. The *secondary members* are the structural members which are provided for stability and or restraining the main members from buckling or similar modes of failure.

The roof trusses are used at places which require sloping roofs. The sloping roofs are necessary at places where rainfall is more and at places where snow fall occurs. The roof trusses are also used in many single storeyed industrial buildings, workshops, godowns, warehouses, residential buildings and schools, where large column free spaces are required for operational purposes. The roof trusses are suitable for relatively light loads and large spans. The roof trusses are also used to span the distance between walls or end supports when the least dimension of the building becomes large and unsuitable and economical to span with simple beams or joists, columns and girder. The roof trusses have the advantages of permitting a wider variety of roof shapes and greater unobstructed interior floor area at less cost. The trusses are specially advantageous where greater depth is required to satisfy the condition of adequate stiffness.

The primary function of a roof truss is to support the roof covering, external loads carried by roof covering and ceiling load if any. These loads are transmitted as reactions to the walls or to the supporting stanchions. In general, the external loads are applied at the joints of truss. Sometimes, it becomes necessary to apply loads at intermediate points. In such cases, the members are subjected to bending in addition to direct forces.

9.2 TYPES OF TRUSSES

The various types of roof trusses used are shown in Fig. 9.1. The suitable spans for these trusses have also been shown for each type of truss. *King post truss* shown in Fig. 9.1 (a) is a wooden truss. It can also be built of a combination of wood and steel. *Queen post* truss shown in Fig. 9.1 (b) is also a wooden truss. Howe truss had been named after Willium Howe. Howe trusses shown in Fig. 9.1 (c) are made of combination of wood and steel. The vertical members are tension members and made of steel. *Pratt trusses* shown in Fig. 9.1 (d) are made of steel. The Pratt truss had been named after Thomas W. and Caleb Pratt. The diagonal members near the top are highly stressed because of their sharp inclination. These are less economical than fink trusses. The vertical members are compression members and diagonal members are tension members. The modified shape of Pratt roof truss is also used. *Fink roof trusses* shown in Fig. 9.1 (e) are made of steel. They are very economical form of roof trusses. The lengths of compression members are small in these types of trusses. Fink trusses are also called *French roof trusses.* The Fink truss had been named after its originator *Albert Fink.* A Fink truss is constructed by drawing perpendicular to and from the centre of the top-left and right hand chords and extending until they meet the lower chord. The remaining two web members are drawn from these intersections to the peak as shown in Fig. 9.1 (e). The number of panels in half the truss may be increased to *four* by sub-dividing each panel into 2, as shown in Fig. 9.1 (e). This process may be repeated again resulting in *eight* panels as shown in Fig. 9.1 (e). A disadvantage of this type of truss is that the number of panels can be increased only by doubling the previous number. A modification of fink truss that permits greater flexibility in number of panels is the *fan truss. Fan roof trusses* shown in Fig. 9.1 (f) are made of steel. Fan roof trusses are form of fink roof trusses. In fan trusses top chord is divided into small lengths, in order to provide supports for purlins which would not come at joints in trusses shown in Fig. 9.1 (e). The web numbers do not intersect the top chord at right angles.

This feature of fan truss may be used or combined with the fink truss. The number of panels in the combined fink-fan truss may be made *six* instead of *four*, i.e., in between four and eight. The sky light may be mounted on top of most of trusses when desired. For unsymmetrical layouts, the natural light is received by using saw tooth trusses or north light trusses. The roof trusses shown in Fig. 9.1 (g) are used for industrial buildings and drawing rooms. Such type of trusses provide north light. These trusses are called *North light roof trusses.* The trusses shown in Fig. 9.1 (h) are used for roofs. The trusses shown in Fig. 9.1 (j) are used for very large span such as railway sheds and auditoriums.

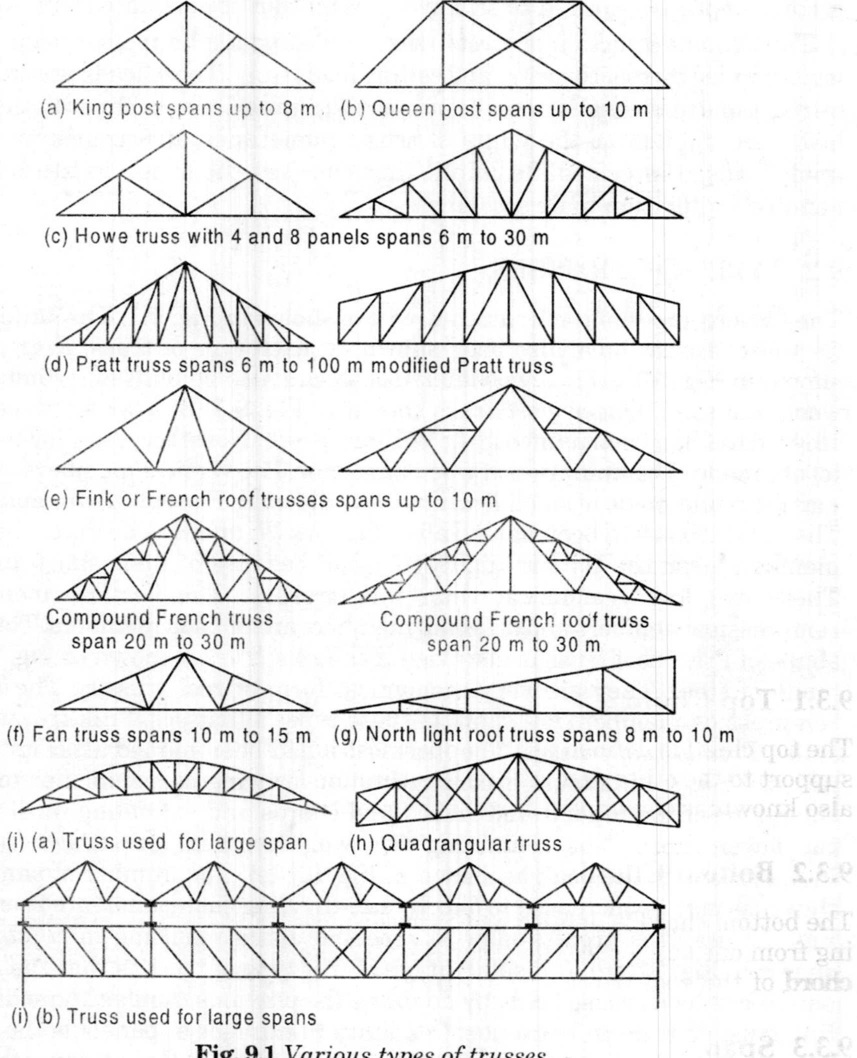

Fig. 9.1 *Various types of trusses*

In general, the lower chord of the truss is kept straight. In large rooms, series of trusses are used one behind the other. In such cases, trusses with straight chords appear to sag. Then, the lower chord is cambered. It is done for better appearance. As a result of this, the depth of truss is reduced. Therefore, the cost of trusses arrangement will be economical and pleasing.

9.3 VARIOUS TERMS USED IN ROOF TRUSSES

The various terms used in roof trusses are given below and are shown in Fig. 9.2.

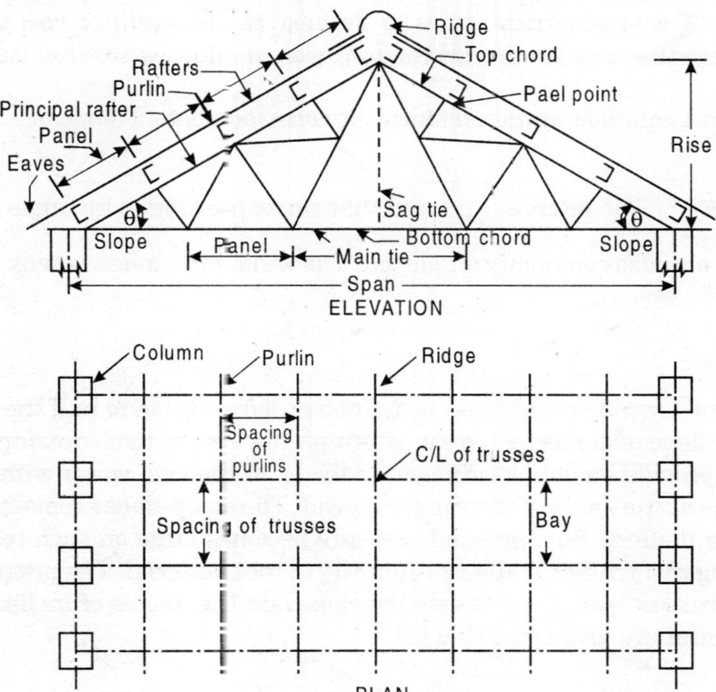

Fig. 9.2 *Elements of roof truss*

9.3.1 Top Chord

The top chord is defined as the uppermost line of members extending from one-support to the other and that passes through the peak of truss. The top chord is also known as the *upper chord* of the roof truss.

9.3.2 Bottom Chord

The bottom chord is defined as the lower most line of members of truss extending from one support to the other. The bottom chord is also known as the **lower chord** of the roof truss.

9.3.3 Span

The span of a roof truss is defined as the distance between centre to centre of supports. The span of a roof truss is decided by the dimensions of area to be kept free of columns.

9.3.4 Rise

The rise of a roof truss is defined as the distance from the highest point to the line joining supports.

9.3.5 Pitch

The pitch of a symmetrical truss is defined as the **ratio of rise to the full span.** Where the roofs are to carry snow loads in addition to wind load, a pitch of $\frac{1}{4}$ is most common and economical. It corresponds to a slope of 1 in 2 or an angle of $26\frac{1}{2}°$. The pitches $\frac{1}{3}, \frac{1}{5}$ and that corresponding to an angle of inclination of 30° are also commonly employed. The wind force is less when the pitch of roof truss is less.

9.3.6 Slope

The slope of a symmetrical truss is defined as ratio of rise to half the span. The minimum slope depends to a great extent on the type of roof covering material. The slope provided must be sufficient to drain off the rain water without allowing leakage at the joints of covering material. The steep slopes remain better for architectural effect. But the wind pressure becomes large on such roofs. It will need strong roof trusses and large quantity of roof material. The pitches of symmetrical trusses used, their respective slope and the angles of inclination with the horizontal are given in Table 9.1.

Table 9.1

Pitch	*Slope*	*Angle of inclination with horizontal*
$\frac{1}{2}$	1	45° 00′ 00"
$\frac{1}{3}$	$\frac{2}{3}$	33° 41′ 00"
$\frac{1}{4}$	$\frac{1}{2}$	20° 34' 00"
$\frac{1}{5}$	$\frac{2}{5}$	21° 48' 00'
$\frac{1}{6}$	$\frac{1}{3}$	18° 26' 00"
$\frac{1}{8}$	$\frac{1}{4}$	14° 02' 00"

The slope of a roof is defined as the tangent of the angle that the plane of the roof makes with horizontal; or the tangent of the angle between the top and bottom chords of the roof trusses provided the bottom chord is horizontal. *The slope of the roof therefore is not equal to the pitch and great care must be exercised to see that two terms are not used synonymously.* The slope of the roof is equal to *twice* the numerical value of pitch in all the cases whether truss is symmetrical or unsymmetrical.

9.3.7 Purlins

The purlins are structural members subjected to transverse loads and rest on the top chords of roof trusses. The purlins support the sheathing that carries roof covering or roof covering directly. The purlins are horizontal beams spanning between the two adjacent trusses.

9.3.8 Sub-purlins

The sub-purlins are secondary system of purlins resting on the rafter. The sub-purlins are spaced to support the tiles or slate coverings.

9.3.9 Rafters

The rafters are beams and rest on the purlins. The rafters support the sheathing. These may support sub-purlins directly. These are called common rafters to distinguish from principal rafter.

9.3.10 Sheathing

The sheathing are covering of boards or reinforced concrete. The sheathings are supported on purlins or rafters. The sheathings provide support for the roof covering.

9.3.11 Panel

The panel is defined as a distance between two adjacent joints in a principal rafter of a roof truss. It is also defined as the distance between the two adjacent purlins.

9.3.12 Bay and Spacing of Purlins

The bay is defined as the distance between the adjacent trusses. The spacing of purlins is defined as the distance between two adjacent panel points, if purlins are placed at panel points only. It depends upon type of roof covering material and on the slope of the truss. The spacing of purlins should be less than or equal to the safe span of roof covering material if these are placed directly over purlins. In case, the roof covering material is supported on common rafters, then, purlins must be placed at any desired spacing. As far as possible, the purlins must be placed at the panel points so that in the top chord of truss, there is no bending moment due to purlin load. Generally, the spacing of purlins varies from 2 m to 3 m.

9.3.13 Spacing of Trusses

The spacing of trusses is decided by local conditions and requirements. The positions of doors, windows and columns, also control the spacing of trusses It is desirable to have a uniform spacing of trusses in one portion or in a wing cf a structure, so that as many trusses as possible may be made identical.

9.3.14 Ridge Line and Eaves

The ridge line or a ridge is a line joining the vertices of the trusses. The bottom edges of an inclined roof surface or a pitched roof is termed as eaves.

9.3.15 Sag Tie

As shown in Fig. 9.2, a tie member is provided to join the peak of truss and middle tie member. The length of middle tie member used to be large. The deflection of this member due to self weight may be large. When a sag tie is provided, it decreases the deflection of middle tie member.

9.3.16 Principal Rafter

The top chord members as shown in Fig. 9.2 are called principal rafters. The principal rafter carries compressive force.

9.4 ECONOMICAL SPACING OF ROOF TRUSSES

For a roof structure, the most economical spacing of trusses depends on many factors. It is not possible to give a simple formula. The economic spacing of trusses depends upon the weight per square metre of floor area of trusses, purlins columns, roof covering etc. It also depends on relative cost of each type of material. In case, the spacing of trusses is large, the cost of purlins will be large. If the spacing of trusses is kept small, the cost of purlins is small, but the cost of trusses will be large. Therefore, the spacing of trusses should be kept such that the overall cost of roof structure as a whole is minimum. The *economic spacing of trusses is defined as the spacing which makes the cost of trusses, purlins, columns, roof covering etc., a minimum.*

Let s, be the spacing of trusses ; t, be the cost of trusses ; p, be the cost of purlins : and r, be the cost of roof covering material.

The cost of trusses, t, is inversely proportional to the spacing of trusses, s. Therefore,

$$t = \left(\frac{c_1}{s}\right) \quad \text{...(i)}$$

The size of purlins depends on the bending moment. The bending moment carried by purlin varies as the square of the span of purlin. Hence, the cost of purlins, p, is directly proportional to the square of spacing of trusses.

Therefore,

$$p = (c_2 \,.\, s^2) \qquad \text{...(ii)}$$

The cost of roof covering materials, *r*, is directly proportional to the spacing of trusses. Therefore,

$$r = (c_3 \,.\, s) \qquad \text{...(iii)}$$

where c_1, c_2 and c_3 are constants

The overall cost of roof structure

$$X = (t + p + r)$$

Therefore,

$$X = \left(\frac{c_1}{s} + c_2 \cdot s^2 + c_3 \cdot s\right) \qquad \text{...(iv)}$$

For, economic spacing of the trusses, the overall cost of these items should be minimum.

Therefore,

$$\left(\frac{dX}{ds}\right) = 0 \qquad \text{...(v)}$$

$$\left(\frac{dX}{ds}\right) = \left(-\frac{c_1}{s} + c_2 \cdot s + c_3\right)$$

$$\left(-\frac{c_1}{s} + 2c_2 \cdot s^2 + c_3 \cdot s\right) = 0,$$

or

$$(-t + 2p + r) = 0$$

$$t = (2p + r) \qquad \text{...(9.1)}$$

Therefore, *for economic spacing of trusses the cost of trusses should be equal to twice the cost of purlins plus cost of roof covering material.* In this expression the factor *s*, spacing of trusses does not occur. Therefore, it is not possible to determine the spacing directly from this expression. This expression provides only a check to verify that the spacing provided is economical or not.

The economical spacing will usually be kept between 3 m to 5 m for spans upto 15 m and 45 m to 7.5 for spams 15 m to 30 m.

9.5 ROOF COVERINGS

The following materials are used as roof covering materials.

9.5.1 Slates

The slates are most durable roof covering material. The slates are not affected by heat, rain or other atmospheric actions. The slates are heavy in weight. The slates are attached to battens by slating nails. The slating nails are made of copper or non-rusting material. A minimum pitch of $\frac{1}{2}$ is required for roof trusses using slates as roof covering materials. The slopes of 27° and 33° are provided for 500 mm × 250 mm size and 400 mm × 200 mm size of slates, respectively.

9.5.2 Tiles

The tiles are used as roofing material as substitute for slates. The tiles are manufactured from asbestos, clay and many compositions. The tiles are available in market in various forms. Tiles are cheap in comparison to slates. The tiles are not durable as slates. A minimum pitch of $\frac{1}{4}$ is required for roof trusses using tiles as roof covering materials.

9.5.3 Lead

The lead-sheets were formerly used as roof covering materials specially on flat roofs. The lead is heavy in weight. The lead is also a costly material. The lead sheets are very durable. But these are not suitable for steep slopes.

9.5.4 Zinc

The zinc is also one of the materials used for roof covering. The zinc sheets are comparatively light in weight and cheap in cost in comparison to lead. Under atmospheric action, the corrosion and decay of zinc sheets may take place. The zinc sheets can be used on steep slopes.

9.5.5 Glass

The glass sheets are manufactured in various sizes and thickness. The glass sheets are mainly used for light from sloping roof.

9.5.6 Galvanized Corrugated Iron Sheets

The galvanized steel sheets are made by galvanizing black sheets, rolled from good quality low-carbon mild steel, free from cracks, pittings, blisters, laminations, twists, scales and other surface defects. The black sheets used for galvanizing are annealed. The galvanizing is carried out by the hot-dip process by first pickling the black sheets, and then dipping them in bath of molten zinc, at a temperature suitable to produce a complete and uniform adhesive coating of zinc. The zinc coating contains not less than 97.5 percent of pure zinc. The galvanizing corrugated sheets used for roof covering should be free from twist or buckle and should have uniform corrugations, true in depth and pitch and parallel to the sides of sheets.

The depth of corrugation is 18 mm (nominal), and pitch of corrugations is 5 mm. The number of corrugations in a galvanised corrugated sheet is 8 or 10. The nominal overall width, of galvanized corrugated sheets measured between the crown of outside corrugations in case of 8 corrugations per sheet, is 750 mm before corrugations and 600 mm after corrugation and that in case of 10 corrugations per sheet, is 900 mm before corrugation and 800 mm after corrugations. The galvanized sheets are manufactured in lengths of 18 m, 2.2 m, 2.5 m, 2.8 m and 3.2 m. The dimensions of galvanized corrugated sheets are given as per IS : 277–1962.

These sheets are used for slopes of roof 1 in 4 and above. These sheets are very light in weight, and can be fixed to purlins by J-bolts. The holes in these sheets should be made on ground. Bitumen washers are used below. G.I. washers to make the joint water-tight. The decay of galvanized sheets starts rapidly at holes due to atmospheric action. The decay of these sheets takes place due to atmospheric action as soon as the galvanizing has worn off. The life of galvanized corrugated sheet is comparatively short.

9.5.7 Asbestos Cement Sheets

A.C. sheets for roof covering are available in two varieties, namely, *'Asbestos cement corrugated sheet'* and *'Asbestos cement Trafford sheet'*. A.C. sheets are good insulators against temperature. A.C. sheets do not decay because of atmospheric action. These are not perishable. The asbestos cement corrugated sheets are shown in Fig. 9.3 (a) and asbestos cement trafford sheets are shown in Fig. 9.3 (b). The trafford sheets are also termed as semi-corrugated sheets.

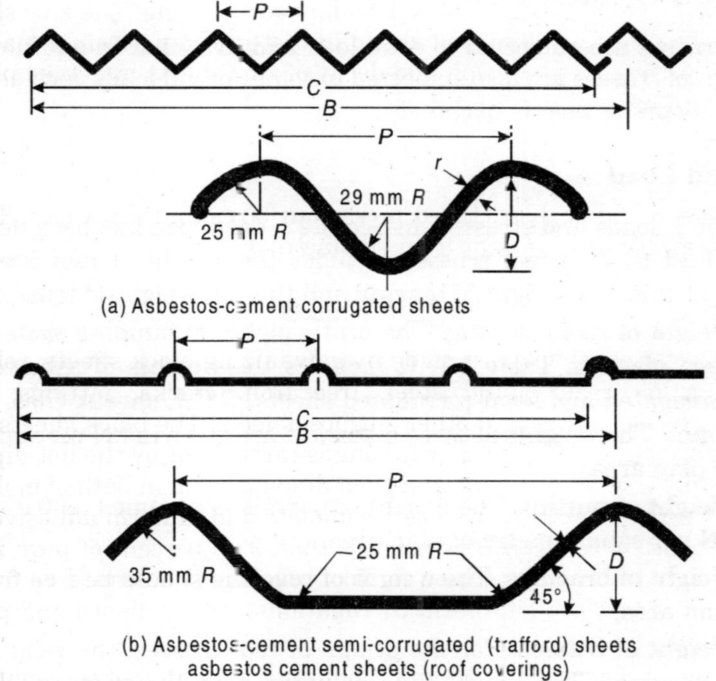

Fig. 9.3 *Asbestos cement sheets*

Both the varieties of A.C, sheets are available in lengths of 1.50 m, 1.75 m, 2 m, 2.25 m, 2.50 m, 2.75 m and 3 m. Both the sheets are manufactured in nominal thickness, *T*, as 6 mm and 7 mm. For corrugated sheets, the pitch *P* is 146 mm, and depth of corrugation *D*, is 48 mm. The overall width *B* of these sheets is 1.50 m and effective width *C* is 1.01 m. For trafford sheets, the pitch *P* is 338 mm and depth of corrugation *D* is 45 mm. The overall width *B* of the trafford sheets is 1.01 m and effective width is 1.014 m. The dimensions of these A.C. sheets are given as per IS : 459–1962.

The asbestos cements sheets are used for 30° slope and above. A.C. sheets are fixed with purlins with J-bolts. The holes are drilled on the top of corrugations. The bitumen washers are placed under G.I. washers in order to make it water-tight.

9.5.8 Corrugated Aluminium Sheets

The corrugated aluminium sheets find extensive application where use of aluminium is justified due to its corrosive resistance, lightness and thermal efficiency. The dimensions of corrugated aluminium sheets may be noted from IS : 1254–1965. The corrugated aluminium sheets are known as general purpose sheets. The thickness of sheet is kept not less than 0.45 mm. The widths available are 650 mm and 800 mm and lengths, 1800, 2400 and 3600 mm.

9.6 LOADS ON ROOF TRUSSES

The roof trusses are subjected to dead load and live load. In addition to these loads, the roof trusses are also subjected to some special loads such as ceiling or suspended floors or heavy machinery.

9.6.1 Dead Load

In Chapter 1; loads and stresses, in Sec. 1.22, dead load has been described in general. Dead load on roof trusses includes the weight of roof covering, the weight of purlins, the weight of bracing and the self weight of trusses.

(i) **Weight of roof covering.** The unit weights of building materials have been given in Table 1.7. It includes the weights of asbestos cement corrugated and semi-corrugated sheets, C.G.I, sheets, tiles, glass and slates. The weights of these materials are given in kN per square metre of plan areas.

(ii) **Weight of purlins.** The weight of purlins is assumed as 0.070 to 0.150 kN per square metre of plan area.

(iii) **Weight of bracings.** The weight of bracings is assumed as 0.015 kN of plan area.

(iv) **Weight of trusses.** For the design of roof trusses, the weight of truss is assumed. The weight of truss varies with the span, and the rise of truss, the spacing of trusses, the type of roof covering material, the geographical situation of the roof structure. It is not possible to have an exact formula which includes all these factors. The self weight of truss is a small part of the total design loads for the roof truss. The self weight of truss may be assumed as 0.90 to 0.150 kN per square metre of plan area. The self weight of truss can also be found by empirical formula given below :

The self weight of truss in kN per square metre of plan area

$$w = \frac{1}{100}\left(\frac{l}{3}+5\right) \text{kN/m}^2 \quad ...(9.2)$$

where, l is the span of truss in metres.

This formula is applicable for roof trusres with C.G.I, sheets and pitch equals to 1 to 4 and for 4 m spacing of trusses. For other spacing of trusses, a proportionate value for self weight is determined.

The weight of trusses w in N/m^2 of area covered, depending upon the design loads q in N/m^2 may be found from the following formulae for the spans mentioned:

$$\text{Span } L = 18 \text{ m}, w = \frac{1}{100}\left(2.2+\frac{q}{12.5}\right) \text{kN/m}^2 \quad ...(i)$$

$$\text{Span } L = 24 \text{ m}, w = \frac{1}{100}\left(2.78+\frac{q}{5.42}\right) \text{kN/m}^2 \quad ...(ii)$$

$$\text{Span } L = 30 \text{ m}, w = \frac{1}{100}\left(4.44+\frac{q}{3.47}\right) \text{kN/m}^2 \quad ...(iii)$$

$$\text{Span } L = 36 \text{ m}, w = \frac{1}{100}\left(5.27+\frac{q}{2.1}\right) \text{kN/m}^2 \quad ...(iv)$$

The weight of steel roof trusses may also be estimated from the following expressions. These expressions include the plan area covered by the steel roof trusses. The weight of riveted steel trusses are found as follows:

For sheeted roof trusses.

$$w = \frac{1}{100}(4.88 + 0.075 \times A) \text{ kN/mm}^2 \quad ...(v)$$

For partly glazed roof trusses

$$w = \frac{1}{100}(4.88 + 0.088 \times A) \text{ kN/m}^2 \quad ...(vi)$$

where A = plane area in m^2

The weight of welded roof trusses area approximated as follows :

For sheeted roof trusses

$$w = \frac{1}{100}(5.37 + 0.053 \times A) \text{ kN/m}^2 \quad ...(vii)$$

For partly glazed roof trusses

$$w = \frac{1}{100}(5.37 + 0.064 \times A) \text{ kN/m}^2$$

9.6.2 Live Loads

In Chapter 1, the loads and stresses : Sec. 1.23 live loads have been described in general. The imposed (live) loads on various types of roofs other than wind load and snow load, as per IS : 875–1987, for roofs with slopes upto and including

10 degrees, is adopted as 1.5 kN per square metre of plan area where access is provided to roof. The minimum live load measured on plan shall be 3.75 kN uniformly distributed over any span of one metre width of the roof slab and 9.0 kN uniformly distributed over the span in the case of all beams. Where the access is not provided, except the maintenance, live load on roofs is adopted as 0.750 kN per square metre of plan area. In ths case, the minimum live load measured on plan shall be 1.9 kN uniformly distributed over any span of one metre width of the roof slab and 4.50 kN uniformly distributed over the span in the case of beams. The live load for sloping roof with slopes greater than 10 degrees is adopted as 0.750 kN per square metre of plan area less 0.020 kN/m^2 for every degree increase in slope over 10 degrees subjected to minimum of 0.400 kN/m^2 per square metre of plan area.

The live loads for curved roofs with slopes at springing greater than 10 degrees shall be $(0.75–52\ \gamma^2) \times$ kN/m^2,

where, $r = \left(\frac{h}{l}\right)$

h = the height of the highest point of the structure measured from springing

l = the chord width of the roof if singly curved and shorter of the two, if doubly curved.

This is subjected to minimum of 0.400 kN/m^2. The live load also includes snow load and wind load.

9.6.3 Snow Loads

The snow loads depends upon geographical situation of roof structure. It depends upon latitude of the place and atmospheric humidity. If a roof is subjected to snow load, it should be designed for the actual load due to snow or for the live load mentioned above whichever is more severe. The actual load due to snow will depend upon the shape of roof and its capacity to retain the snow and each case shall be treated on its own merits. In the absence of any specific information, the bending due to the collection of snow may be assumed to be 25 N/m^2 per mm depth of snow. The possibility of total or partial snow load should be considered, that is, one half of the roof fully loaded with the design snow load and the other half loaded with half the design snow load. In the case of roof with slopes greater than 50°, snow load may be disregarded. Where, however, there are possibilities of formation of snow pockets, these should be taken into accounts.

9.6.4 Wind Loads

The wind load is one of the most important loads that an engineer has to deal with and is also one of that is most difficult to evaluate properly. The magnitude of wind pressure depends on **wind velocity** and the shape of the structure. The magnitude of wind velocity varies with the geographical location of the structure and the height of the structure.

Sir Isaac Newton established the relationship between the wind pressure and the wind velocity and is known as *Newtonian theory*. When a stream of air flows around the object, then, the wind creates a pressure on the nose of object. The pressure developed may be determined from the consideration of energy involved. According to **Bernoullis theorem,** the sum of energies at all points.

Therefore. $(K.E. + P_1. E_1 - P_2. E_2) =$ Constant ...(v)

where, $K.E.$ = Kinetic energy, which a mass M possesses, $\frac{W_u^2}{2g}$

W = Weight of that mass

g = Acceleration due to gravity

u = Linear velocity of the wind

$P_1 E_1$ = Potential energy, $W.h$

h = Height above the datum plant

$P_2 E_2$ = Pressure energy, $\frac{W \cdot p}{w}$

P = Pressure of the fluid (wind)

W = Weight this pressure can lift

w = Unit weight of the fluid (wind)

The expression may be written in terms of the expressions for these energies. Therefore,

$$\left(\frac{W \cdot u^2}{2g} + W \cdot h - \frac{W \cdot p}{w}\right) = \text{Constant} \quad \text{... (vi)}$$

The expression (vi) is divided by W

$$\therefore \quad \left(\frac{v^2}{2g} + h + \frac{p}{w}\right) = \text{Constant} \quad \text{... (vii)}$$

The sum of three terms in the expression (vii) is known as total head. It is represented by H. These terms are known as velocity head, gravity head and pressure head, respectively.

$$H = \left(\frac{v^2}{2g} + h + \frac{p}{w}\right)$$

or

$$H.w = \left(\frac{w \cdot v^2}{2g} + w \cdot h + p\right) \quad \text{...(viii)}$$

The expression (viii) gives total energy

$$E = \left(\frac{w \cdot v^2}{2g} + w \cdot h + p\right) \quad \text{...(ix)}$$

The unit of total energy is now N per sq. metre. The energy per unit volume given by the expression (ix) remains constant. The energy at the nose of object and the stagnation point (the point where wind velocity is zero) are equal

$$\left(p_0 + \frac{w \cdot u_0^2}{2g} = p_s \right)$$

where, p_0 = Pressure at the nose of object

p_s = Pressure at the stagnation point of object

u_0 = Velocity at the nose

From the expression (x)

$$(p_s - p_0) = \left(\frac{w \cdot u_0^2}{2g} \right) \quad \text{...(xi)}$$

When the wind is brought to rest, then it results in rise in pressure $(p_s - p_0)$. It is termed as impact pressure, *(kinetic pressure)* of the air and is represented by q. Hence

$$q = \left(\frac{w \cdot v_0^2}{2g} \right) = \left(\frac{w \cdot v^2}{2g} \right) \quad \text{...(xii)}$$

When the temperature and the barometric pressure change, then, the weight of wind per unit volume also changes. The weight of wind at sea level at a temperature of 17°C is 0.347 N. At 0°C, the weight of wind is 0.365 N. It increases with the decreases of temperature. The value of acceleration due to gravity may be adopted at 9.81 m/sec². The expression (xii) may be written as

$$q = \left(\frac{K}{v^2} \right) \quad \text{...(xiii)}$$

where, K is the constant or a coefficient. It takes into account the unit of wind velocity in km per hour.

The basic wind pressure p is adopted equal to $q \times C_D$ and the value of K is taken as 0.06 in IS : 875–1964. Therefore,

$$p = (0.06 \times v^2) \times 10 \text{ N/m}^2$$

$$= 0.6\, v^2 \text{ N/m}^2 \quad \text{...(xiv)}$$

where, C_D is the shape factor.

The velocity of wind (wind speed) varies with the height above ground elevation. The relationship showing the variation of wind is as follows :

$$\frac{v_1}{v_2} = \left(\frac{H_1}{H_2} \right)^{0.225} \quad \text{...(xv)}$$

where, v_1 is the velocity of wind (wind speed) in km per hour at height H_1 in metres and v_2 is the velocity of wind in km per hour at height H_2 in metres.

The wind velocities (wind speed) are measured by meteorological observatories with the help of instrument known as *anemometers.* These instruments are kept at standard height of 10 m in the different parts of our country and the wind velocities (speeds) are measured at this height.

Basic wind speed: The *basic wind speed,* V_b is the wind speed measured in a 50 year return period. The basic wind speed is based on peak gust velocity averaged over a short interval of time of about 3 (three) seconds and it corresponds to mean heights above groud level in an open terrain (category 2).

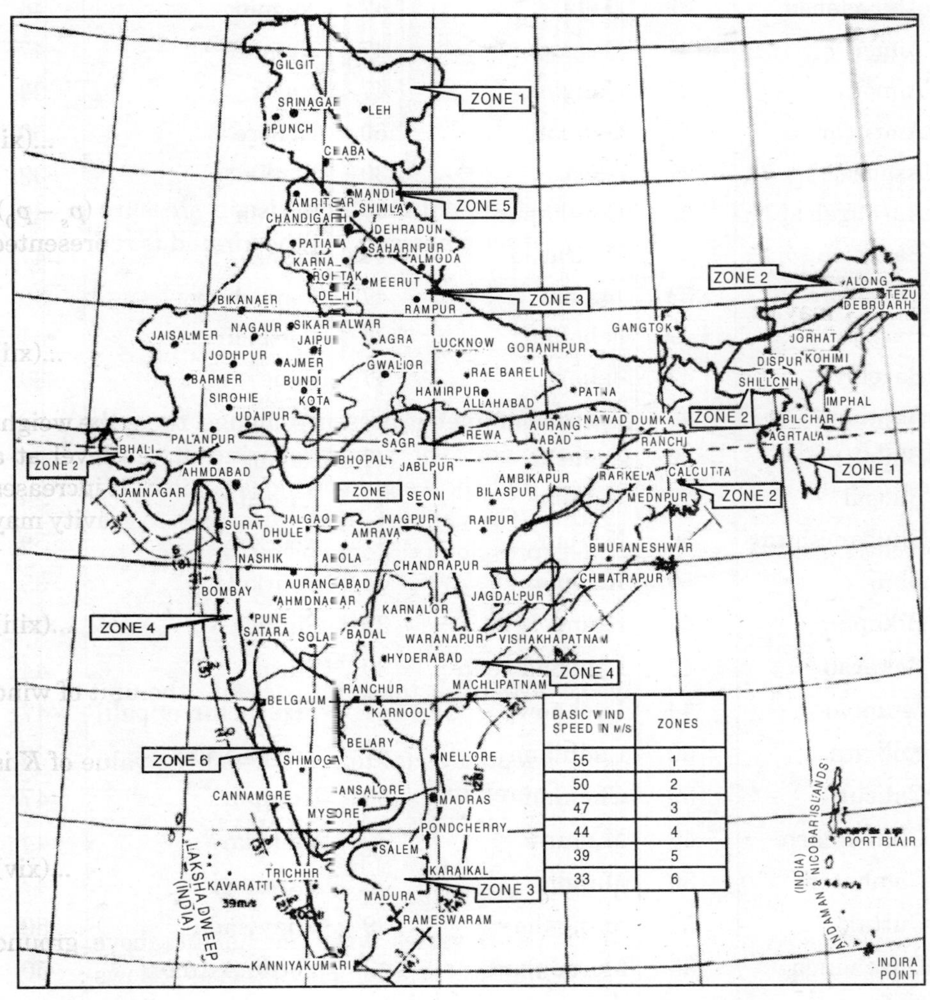

BASIC WIND SPEED IN m/s	ZONES
55	1
50	2
47	3
44	4
39	5
33	6

Fig. 9.4

Figure 9.4 shows basic wind speed map of our country, as applicable to 10 m height above mean ground level for different zones. As per IS : 875 (Part 3 – wind loads) – 1987, six wind zones have been formed which correspond to basic wind speed of 55, 50, 47, 44, 39 and 33 metre per second, respectively, as shown in Fig. 9.4. The basic wind speeds for some important cities/towns are given in Table 9.2.

Table 9.2 *Basic wind speeds in m/sec (As per IS : 875 (Part 3)–1987)*

City/Town	*Speed (m/sec)*	*City/Town*	*Speed (m/sec)*	*City / Town*	*Speed (m/sec)*
Agra	47	Dehradun	47	Mysore	33
Ahmadabad	39	Delhi	47	Nagpur	44
Ajmer	47	Durgapur	47	Nainital	47
Almora	47	Gangtok	47	Nasik	39
Amritsar	47	Gauhati	50	Nellore	50
Asansol	47	Gaya	39	Panjim	39
Aurangabad	39	Gorakhpur	47	Patiala	47
Bahraich	47	Hydrabad	44	Patna	47
Bangolore	33	Imphal	47	Pondicherry	50
Barauni	47	Jabalpur	47	Portblair	44
Bareilly	47	Jaipur	47	Pune	39
Bhatinda	47	Jamshedpur	47	Raipur	39
Bhilai	39	Jhanshi	47	Rajkot	39
Bhopal	39	Jodhpur	47	Ranchi	39
Bhubaneshwar	50	Kanpur	47	Roorkee	39
Bhuj	50	Kohima	44	Rourkela	39
Bikaner	47	Kumool	39	Simla	39
Bokarao	47	Lakshadweep	39	Surat	44
Mumbai	44	Lucknow	47	Tiruchchirrappalli	47
Kolkata	50	Ludhiana	47	Trivendrum	39
Calicut	39	Chennai	50	Udaipur	47
Chandigarh	47	Madurai	39	Vadodara	44
Combatore	39	Mandi	39	Varanasi	47
Cuttak	50	Mangalore	39	Vijaywada	50
Darbhanga	55	Moradabad	47	Visakapatnam	50
Darjeeling	47				

It is to note that the occurrence of a tornado is possible in virtually any part of India. They are particularly more severe in the northern parts of our country. The recorded number of these tornados is too small to assign any frequency. The devastation caused by a tornado is due to exceptionally high winds about its periphery, and the sudden reduction in atmospheric pressure at its centre, resulting in an explosive outward pressure on the elements of the structure. The regional basic wind speeds do not include any specific allowance for tornado. It is the usual practice to allow for the effect of tornados unless special requirements are called for as in the case of important structures, such as, nuclear power reactors and satellite communication towers.

It is to further note that the total number of cyclonic storms that have struck different sections of east and most coasts are included in Fig. 9.4 based on available records for the period from 1877 to 1982. The figures above the lines (between the stations) indicate the total number of severe cyclone storms with or without a core of hurricane winds (speeds above 87 kmph) and the figures in the brackets below the lines indicate the total number of cyclone storms. Their effect on land is already reflected in the basic wind speeds specified in Fig. 9.4. These have been included only as an additional information.

The cyclonic storms gradually reduce in speed as they approach the sea coast. The cyclonic storms generally extend upto 60 km inland after striking the sea coast. Their effect onland is already reflected in basic wind speed specified in Fig. 9.4. The influence of wind speed off the coast upto a distance of about 200 km may be taken as 115 times the value on the nearest coast in the absence of any definite wind data.

9.7 DESIGN WIND SPEED

The basic wind speed, V_b for any site shall be noted from Fig. 9.4 and it is modified to include the effect of risk level, terrain roughness, height and size of the structure, and local topography. The design wind speed, V_z may be mathematically expressed as under. The design wind speed is the wind speed for which the structure is designed.

$$V_z = (k_1 . k_2 . k_3) . V_b \qquad ...(9.3)$$

where,

V_z = design wind speed at any height z in m/sec,

k_1 = risk coefficient (probability factor),

k_2 = terrain, height and structure size factor, and

k_3 = topography factor

Above factors k_1, k_2 and k_3 have been described in IS : 875 (Part 3) –1987, which are as follows: It is to note that the **design wind speed** upto 10 m height from the mean ground level shall be considered constant.

9.7.1 Risk Coefficient (Probability Factor), k_1

The basic wind speed is based on 50 years mean return period. Figure 9.4 gives basic wind speeds for terrain (category 2) as applicable at 10 m ground level. The design life of a structure is based on the **function aspect** and **importance of a structure.** A structure is designed to withstand the **frequency distribution** of the wind over that period. The values of risk coefficients based on probability concept are given in Table 9.3.

Table 9.3 *Risk coefficients for different classes of structures in different wind speed zones (As per IS : 875 (Part 3)–1987)*

Class of structures	*Meam Probable life years*	*k_1 factors for basic wind speeds (m/s)*					
		33	39	44	47	50	55
1. Temporary structures	5	0.82	0.76	0.73	0.71	0.70	0.67
2. Structures presenting a low degree of hazard to life and property	25	0.94	0.92	0.91	0.90	0.90	0.89
3. All general buildings and structures	50	1.0	1.0	1.0	1.0	1.0	1.0
4. Important building and structures	100	1.05	1.06	1.07	1.07	1.08	1.08

Note: *It is to note that the temporary sheds and structures include those used during construction operations (e.g., form work and falsework), structures during construction stages and boundary walls. Isolated towers in wooded areas, farm buildings other than the residential buildings present a low degree of hazard to life and property in the event of failure. Hospitals, communication buildings / towers, power plant structures are important buildings and structures.*

The risk factor k_1 is based on statistical concepts which take into consideration of the degree of reliability needed and period of time in years daring which these will be exposed to wind, (i.e., life of the structures). Whatever wind speed is adopted for the design purposes, there is always a probability (however small) that it may be exceeded in a storm of exceptional voilance; the greater the period of years over which these will be exposed to wind, the greater is the probability.

Higher return periods ranging from 100 years to 1000 years (implying lower risk level) in the association with greater periods of exposure may have to be selected for exceptionally important structures, such as, nuclear power reactors and satellite communication towers.

The expression given below may be used in such cases to estimate k_1 factors for different periods of exposure and chosen probability of exceedance (risk level). The probability level of 0.63 is normally considered sufficient for the design of buildings and structure against wind effects and the values of k_2, corresponding to this risk level are given below:

$$k_1 = \left(\frac{X_{N.P}}{X_{50,\ 0.63}}\right) \qquad \text{...(9.4)}$$

$$k_1 = \frac{\left(A - B\left[\log(-\frac{1}{N}\log(1-P_N))\right]\right)}{(A+4B)} \qquad \text{...(9.5)}$$

where,

N = mean probable design life of a structure in years

P_N = risk level in N consecutive years (probability that the design wind speed is exceeded atleast once in N successive years), normal value is 0.63.

$(X_{50,\,0.63})$ = extreme wind speed for (N = 50 years and P_N = 0.63)

The values of A and B coefficients for different basic wind speed zones.

Values of	*Wind zones of speeds (m/sec)*					
	33	*39*	*44*	*47*	*50*	*55*
A	83.2	84.0	88.0	88.0	88.8	90.8
B	9.2	14.0	18.0	20.5	22.8	27.3

9.7.2 Terrain, Height and Structure Size Factor, k_2

Terrain category means the characteristics of the surface irregularities of an area which arise from natural or constructed features. The terrain categories are numbered in the increasing order of roughness. The effect of obstructions in the surrounding of the structure creates the turbulence and roughness, which also depends upon the direction of wind. The terrain is classified into following *four* categories:

(a) **Category 1 :** The exposed open terrain with few or no obstruction and in which the average height of any object surrounding the structure is less than 15 m above ground surface. It is to note that open sea coasts and flat plains having no trees are included in this category.

(b) **Category 2 :** The open terrain with scattered obstructions having heights usually between 15 m to 10 m above ground surface. It is to note that air fields, open parklands and undeveloped sparsely built-up out skirts of towns and suburbs are included in this category. This category may also include open land adjacent to sea coast due to roughness of large sea waves at high winds.

(c) **Category 3 :** The terrain having numerous closely spaced obstractions (e.g., building structures upto 10 m in height with or without a few isolated tall structures.

It is to note that well wooded areas, and shrubs, towns and industrial areas full or partially developed are included in this category. It is likely that the next higher category than this will not exist in most design situations and that selection of a more severe category will be deliberate. The particular attention must be given to performance of obstructions in areas affected by fully developed tropical cyclones. The vegetation which is likely to be blown down or defoliated cannot be relied upon to maintain the conditions of category 3. Where such situation may exist, either an intermediate category with velocity multipliers midway between the values for category 2 and 3 given or category 2 should be selected having due regard to local conditions.

(d) **Category 4 :** The terrain with numerous large high closed spaced structures. It is to note that large city centres, (generally with obstructions above 25 m and well developed industrial complexes) are included in this category.

Table 9.4 *Values of factors, k_2 (as per IS: 875 (Part 3)–1987)*

Height (m)	*Terrain categories and classes of buildings*											
	(1)			(2)			(3)			(4)		
	A	B	C	A	B	C	A	B	C	A	B	C
10	1.05	1.03	0.99	1.00	0.98	0.93	0.91	0.88	0.82	0.80	0.76	0.67
15	1.09	1.07	1.03	1.05	1.02	0.97	0.97	0.94	0.87	0.80	0.76	0.67
20	1.12	1.10	1.06	1.07	1.05	1.00	1.01	0.98	0.91	0.80	0.76	0.67
30	1.15	1.13	1.09	1.12	1.10	1.04	1.06	1.03	0.96	0.97	0.93	0.83
50	1.20	1.18	1.14	1.17	1.15	1.10	1.12	1.09	1.02	1.10	1.05	0.95
100	1.26	1.24	1.20	1.24	1.22	1.17	1.20	1.17	1.10	1.20	1.15	1.05
150	1.30	1.24	1.20	1.24	1.22	1.17	1.20	1.17	1.10	1.20	1.15	1.05
200	1.32	1.30	1.26	1.30	1.28	1.24	1.27	1.24	1.18	1.27	1.22	1.13
250	1.34	1.32	1.28	1.32	1.31	1.26	1.29	1.26	1.20	1.28	1.24	1.16
300	1.35	1.34	1.30	1.34	1.32	1.28	1.31	1.28	1.22	1.30	1.26	1.17
350	1.37	1.35	1.31	1.36	1.34	1.29	1.32	1.30	1.24	1.31	1.27	1.19
400	1.38	1.36	1.32	1.37	1.35	1.30	1.34	1.31	1.25	1.32	1.28	1.20
450	1.39	1.37	1.33	1.38	1.36	1.31	1.35	1.32	1.26	1.33	1.29	1.21
500	1.40	1.38	1.34	1.39	1.37	1.32	1.36	1.33	1.28	1.34	1.30	1.22

Note. *Intermediate values may be obtained by linear interpolation, if desired. It is permissible to assume constant wind speed between two heights for simplicity. The height mentioned above is the height of the structure.*

The terrain and building sizes classified into following *three* classes which depend on their sizes.

(i) **Class A :** The structures and/or their components such as cladding, glazing, roofing etc. having maximum dimensions (horizontal/vertical) is less than 20 m above ground surface).

(ii) **Class B :** The structures and/or their components such as cladding, glazing, roofing etc. having maximum dimension (horizontal/vertical) between 20 m and 50 m above ground surface.

(iii) **Class C :** The structures and/or their components such as cladding, glazing, roofing etc. having maximum dimension (horizontal/vertical) more than 50 m above ground surface.

Table 9.4 gives multiplying factor, k_2 by which the basic wind speed noted from Fig. 9.4 must be multiplied to get the wind speed at different heights, in each terrain category for different sizes/classes of buildings/structures.

Depending on the direction of wind, the terrain category used in the design of a structure may vary. The velocity profile for a given terrain category does not develop to full height immediately with the commencement of that terrain category. It develops gradually to height h_x which increases with the fetch or upward distance. Fetch is the distance measured along the wind from a boundary at which a change in the type of terrain occurs. The developed height, h_x in metres may be noted from Table 9.5.

Table 9.5 *Fetch length and develope height, h_x (As per IS : 875 (Part 3)–1987)*

Fetch length x km	*Developed height, h_x (m) (Terrain category)*			
	1	2	3	4
0.2	12	20	35	60
0.5	20	30	35	95
1	25	45	80	130
2	35	65	110	190
5	60	100	170	300
10	80	140	250	450
20	120	200	350	500
50	180	300	400	500

For the structures of heights more than the developed height h_x as noted or least rough terrain.

9.7.3 Topography Factor k_3

The basic wind speed, V_b as given in Fig. 9.4 takes into consideration the general level of site above sea level. This does not allow for local topographic features such as hills, valleys, cliffs, escarpments, or ridges which may significantly influence the wind speed in their surrounding. Near the summits of hills or crests of cliffs, escarpments or ridges, the wind gets accelerated, where as near the foot of cliffs, steep escarpments or ridges, the wind gets decelerated. The wind is parallel to the ground in the planes. The wind is assumed to be normal to the face of the face of the structure. The topography factor k_3 is given by

$$k_3 = (1 + C \cdot s) \qquad ...(9.6)$$

For $3° < \phi \leq 17°$

$$C = 12 . \left(\frac{z}{L}\right) \qquad ...(9.7)$$

$$Le = L \qquad ...(9.8)$$

For $\phi > 17°$,

$$C = 0.36 \qquad ...(9.9)$$

$$L = \left(\frac{z}{0.3}\right) \qquad ...(9.10)$$

where ϕ = upward slope of the ground in the wind direction
S = a factor taken from Figs 9.5 and 9.6
L = actual length of the upward wind slope in the direction of wind
z = effective height of the crest (feature)
Le = effective horizontal length

The effect of topography becomes significant at a site when the upward wind slope, ϕ is more than 30°, and below that, the value of k_3 may be taken equal to *unity.* The value of k_3 lies in the range of 1.0 to 1.36 for slopes greater than 3°.

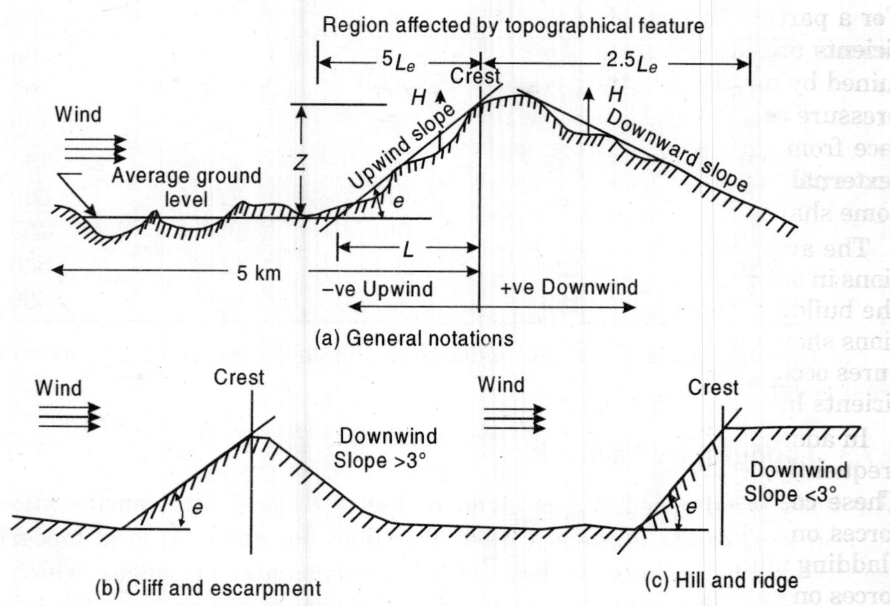

Fig. 9.5 *Topographical dimension*

From the expressions (9.8) and (9.10), it is seen that the values of factors *s* depend upon the effective horizontal length, *Le*. For the design of all the structure within the effected zone (1.5 *Le* on the windward side and 2.5 *Le* on the loeward side) as shown in Fig. 9.5, this coefficient has to be considered.

9.8 DESIGN WIND PRESSURES

The design wind pressure, p_z depends upon the basic wind speed, V_b, the structure above ground level, the terrain categories, the local topography, the aspect ratio (viz., length and breadth of the building/structure, i.e., size of structure), the shape of structure and the solidity ratio or openings the structures.

The design wind pressure at any height above mean ground level shall be obtained from the following expression

$$p_z = 0.6\,(V_z)^2 \text{ N/m}^2 \quad ...(9.11)$$

$$p_z = 0.0006\,(V_z)^2 \text{ kN/m}^2 \quad ...(9.12)$$

where, V_z is the design wind speed in m/sec at height *z*. This coefficient 0.6 (or 0.0006) (in S.I.units) in the above expressions depends on a number of factors and mainly on the atmospheric pressure and temperature.

It is to note that wind pressures and wind forces (loads) acting over the structures shall be calculated for the building as a whole, the individual structurals elements as roofs and walls and the individual cladding units (glazing and their fixing).

9.9 WIND FORCES (LOADS) ON BUILDINGS/STRUCTURES

For a particular surface or part of the surface of a building, the **pressure coefficients** are always given. The wind load acting normal to a surface is determined by multiplying the area of that surface or its appropriate potion by the pressure coefficient, C_p and the design wind pressure at the height of the surface from the ground. The averages values of the these pressure coefficients (**external pressure coefficients** and **internal pressure coefficients**) for some shapes of buildings have been given in subsequent sections.

The average values of pressure coefficients are given for critical wind directions in one or more quadrants. In order to calculate the maximum wind load on the building, the total wind load should be found for each of the critical directions shown from all the quadrants. In case, the considerable variation of pressures occurs over a surface, it has been subdivided and the mean pressure coefficients have been mentioned for each of its several parts.

In addition, the areas of high local suction (negative pressure concentration) frequently occurring near the edges of walls and roofs are separately noted. These coefficients for the local effects should only be used for calculation of forces on these local areas influencing roof sheeting, glass panels, individual cladding units including their fixtures. These should not be used for calculating forces on entire structural elements such as roof walls or structure as a whole.

It is to note that the pressure coefficients have been found from measurements on models in wind tunnels, and the great majority of data available has been used in conditions of relatively smooth flow. In case of rectangular buildings, sufficient field data are available. For these building; turbulent flow has been used. The design of wall glazing and cladding has become a source of major concern in recent years. The damage to glass may be hazardous and cause considerable financial losses.

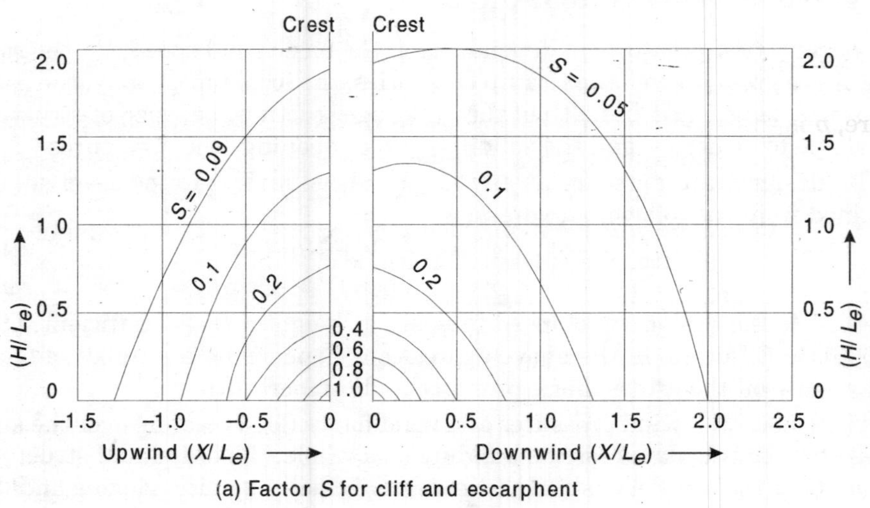

(a) Factor *S* for cliff and escarphent

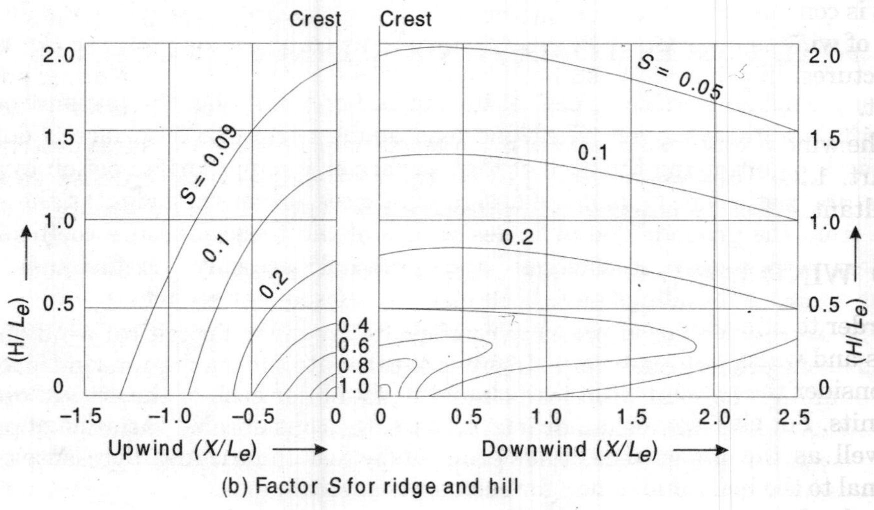

(b) Factor *S* for ridge and hill

Fig. 9.6

In Chapter 1, loads and stresses Sec. 1.24, the wind load has been described. In order to determine the wind load on roofs and other inclined surfaces, the direction of wind is assumed as horizontal. The roof surfaces are also inclined to the horizontal. The normal component and tangential component of wind on inclined roof surfaces may be found. It is to note that these components of horizontal wind pressure cannot be found by resolving. The tangential component of wind pressure is small and it is neglected. The normal component of wind pressure can be found from empirical formula.

In the beginning in 1829, Col. Duchemin, a French Army Officer derived an empirical formula from experiments. This empirical formula bears his name. The Duchemin's formula is

$$p_n = p \cdot \left(\frac{2 \sin \alpha}{1 + \sin^2 \alpha}\right) \quad \text{...(9.5 a)}$$

where, p is the wind pressure on surface in the direction of the wind.

p_n = Component of wind pressure normal to the inclined roof surface.

α = Angle which the inclined roof surface makes with the horizontal direction of wind. The angle α is shown in Fig. 9.6

$$p_v = p_n \cdot \cos \alpha$$

or

$$p_v = p \cdot \left(\frac{2 \sin \alpha + \cos \alpha}{1 + \sin^2 \alpha}\right) \quad \text{...(9.5 b)}$$

and

$$p_h = p_v \cdot \sin \alpha$$

or

$$p_h = p \cdot \left(\frac{2 \sin \alpha \cos \alpha}{1 + \sin^2 \alpha}\right) \quad \text{...(9.5 c)}$$

It is considered that 1.5 kN/m² wind pressure on surface normal to the direction of wind is sufficient for roof structure in an exposed position. For the roof structures, in protected position, 1.00 to 1.25 kN/m² of wind pressure is sufficient.

The wind pressures are expressed in terms of basic wind pressures described in Art. 1.24. The design wind load on the roof surface is determined as the resultant effect of **internal air pressure and external air pressure.**

9.10 WIND LOAD ON INDIVIDUAL MEMBERS

In order to find the wind load on individual structural elements/members (e.g., roofs and walls, and individual cladding units and their fittings, it is essential to consider the pressure difference between the opposite faces of such elements or units. For cladding, it is therefore, necessary to know the **internal pressure** as well as the **external pressure.** The wind load, W_L acting in a direction normal to the individual structural element or cladding unit may be determined from the following expression.

$$W_L = (C_{pe} - C_{pi}) \cdot A \cdot p_d \quad \text{...(9.13)}$$

where,

C_{pe} = external pressure coefficient

C_{pi} = internal pressure coefficient

A = surface area of structural element or cladding unit, and

p_d = design wind pressure

The **external pressure coefficients** and **internal pressure coefficients** are described in the subsequent sections.

It is note that the external pressure coefficient is positive if it is acting outside to inside. However, the internal air pressure is treated positive if it is acting inside to outside.

9.11 EXTERNAL PRESSURE COEFFICIENTS

Average external pressure coefficients shall be considered for walls, pitched roofs, monoslope roofs and canopy roofs, as described hereunder:

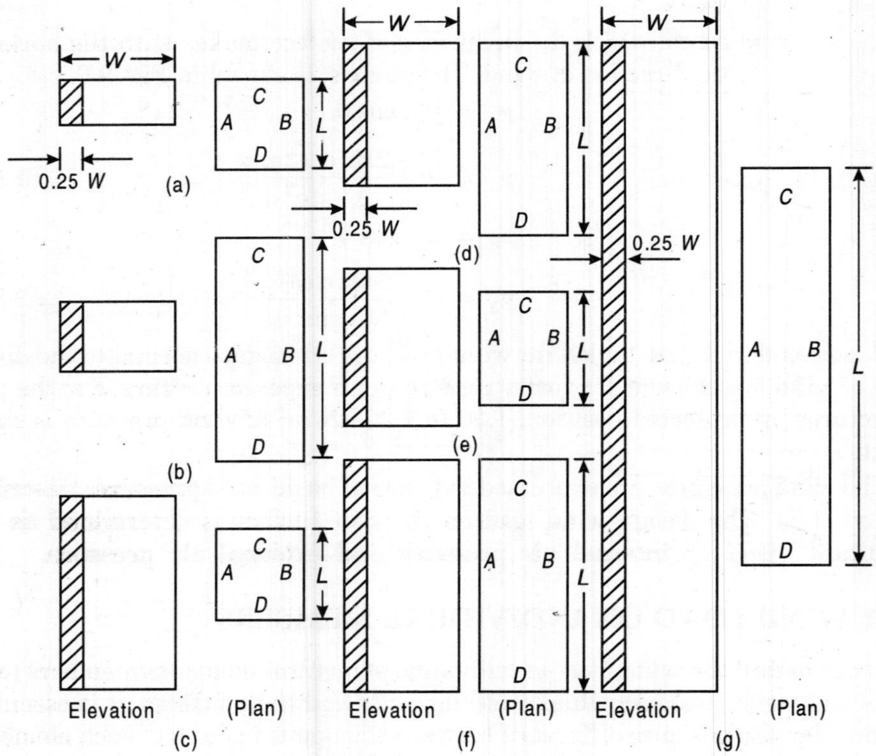

Fig. 9.7 *Walls of rectangular clad buildings*

9.11.1 Walls

The average external pressure coefficients for the walls of clad buildings of rectangular plan as shown in Fig. 9.7, are given in Table 9.6.

9.11.2 Pitched Roofs of Rectangular Clad Buildings

The average external pressure coefficients and pressure concentration coefficients for pitched roofs of rectangular clad building as shown in Fig. 9.8

are noted from Table 9.7, where, the pressure coefficients are not given, the average coefficients shall be used.

It is to note that the pressure concentration is assumed to act outward (suction pressure) at the ridges, eaves, cornices and 90 degree corners of roofs. This pressure concentration is not included with the net external pressure while calculating overall loads.

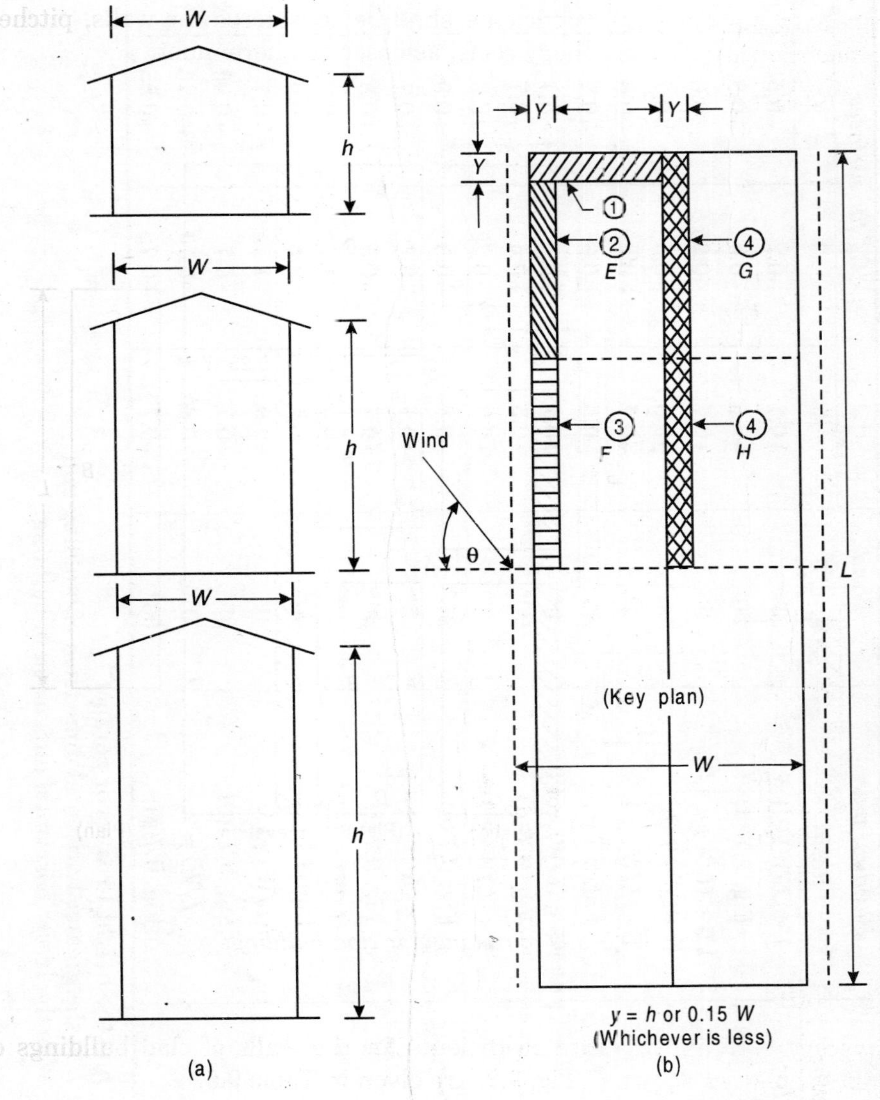

Fig. 9.8 *Pitchd roof rectangular clad buildings*

Table 9.6 *External pressure coefficients.* C_p *for walls of rectangular buildings (as per IS : 875 (Part 3) – 1987)*

Building Height Ratio	*Building plan Ratio*	*Wind Angle, θ*	C_{Pc} *for surfaces* A	B	C	D	*Local,* C_{pe} *shaded area*
$(h/w) \leq 0.5$	$1 < (L/W) \leq 1.5$	0°	+0.7	–0.2	–0.5	–0.5	–0.8
	Fig. 9.7 (a)	90°	–0.5	–0.5	+0.7	–0.2	
	$1.5 < (L/W) < 4.0$	0°	+0.7	–0.25	–0.6	–0.6	–1.0
	Fig. 9.7 (b)	90°	–0.5	–0.5	+0.7	–0.1	
	$1 \leq (L/W) \leq 1.5$	0°	+0.7	–0.25	–0.6	–0.6	–1.1
$0.5\ (h/w) \leq 1.5$	Fig. 9.7 (c)	90°	–0.6	–0.6	+0.7	–0.251	
	$1.5 \leq (L/W) < 4.0$	0°	+0.7	–0.3	–0.7	–0.7	–1.1
	Fig. 9.7 (d)	90°	–0.5	–0.5	+0.7	–0.1	
	$1 < (LIW) \leq 1.5$	0°	+0.8	–0.25	–0.8	–0.8	–1.2
$1.5\ (h/w) < 6$	Fig. 9.7 (e)	90°	–0.8	–0.8	+0.8	–0.25	
	$1.5 \leq (L/W) < 4.0$	0°	+0.7	–0.4	–0.7	–0.7	–1.2
	Fig. 9.7 (f)	90°	–0.5	–0.5	+0.8	–0.1	
	$(L/W) = 1.5$	0°	+0.95	–1.85	–0.9	–0.9	–1.2
	Fig. 9.7 (g)	90°	–0.8	–0.8	+0.9	–0.85	
	$(L/W) = 1.0$	0°	+0.95	–1.25	–0.7	–0.7	–1.25
$(h/w) > \infty$	Fig. 9.7 (g)	90°	–0.7	–0.7	+0.95	–1.25	
	$(L/W) = 2$	0°	+0.85	–0.75	–0.75	0.75	–1.25
	Fig. 9.7(g)	90°	–0.75	–0.75	+0.85	–0.75	

Note. *h* is the height to caves or parapet
L is the greater horizontal dimension of building and
W is the lesser horizontal dimension of building.

Table 9.7 *External pressure coefficients, C_{pe} for pitched roof of rectangular clad buildings (As per IS: 875 (Part 3)–1987)*

Building height ratio	*Angle of Roof*	*Wind Angle, θ*		*Wind Angle, θ*		*Local coefficients*			
	α	*EF*	*GH*	*EG*	*FH*	*1*	*2*	*3*	*4*
	0°	−0.8	−0.4	−0.8	−0.4	−2.0	−2.0	−2.0	
	5°	−0.9	−0.4	−0.8	−0.4	−1.4	−1.2	−1.2	−10
	10°	−1.2	−0.4	−0.8	−0.6	−1.4	−1.4	−1.2	−1.2
(*h*/*w*) ≤ 0.5	20°	−0.4	−0.4	−0.7	−0.6	−1.0			−1.2
	30°	0	−0.4	−0.7	−0.6	−0.8			−1.1
	45°	+0.3	−0.5	−0.7	−0.6				−1.1
	60°	+0.7	−0.6	−0.7	−0.6				−1.1
	0°	−0.8	−0.6	−1.0	−0.6	−2.0	−2.0	−2.0	
	5°	−0.9	−0.4	−0.9	−0.6	−20	−20	−2·0	−1.0
	10°	−1.1	−0.6	−0.8	−0.6	−2.0	−2.0	−1.5	−1.2
0.5 (*h*/*w*) ≤ 1.5	20°	−0.7	−0.5	−0.8	−0.6	−1.5	−1.5	−1.5	−1.0
	30°	−0.2	−0.5	−0.8	−0.8	−1.0			−1.0
	45°	+0.2	−0.5	−0.8	−0.8				
	60°	+0.6	−0.5	−0.8	−0.8				
	0°	−0.7	−0.6	−0.9	−0.7	−2.0	−2.0	−2.0	
	5°	−0.7	−0.6	−0.8	−0.8	−2.0	−2.0	−1.5	−1.0
	10°	−0.7	−0.6	−0.8	−0.8	−2.0	−2.0	−1.5	−1.2
0.5 (*h*/*w*) ≤ 1.5	20°	−08	−0.6	−0.8	−0.8	−1.5	−1.5	−1.5	−1.2
	30°	−1.0	−0.5	−0.8	−0.7	−1.5			
	40°	−0.2	−0.5	−0.8	−0.7	−1.0			
	50°	−0.2	−0.5	−0.8	−0.7				
	60°	−0.5	−0.5	−0.8	−0.7				

Note. *h is the height to eaves or parapet and w is the lesser horizontal dimension of the building. Where the local coefficients are not given the overall coefficients apply. For hipped roofs, the local coefficients for the hip ridge may be conservatively taken as the appropriate ridge value.*

9.11.3 Mono-slope Roofs of Rectangular Clad Buildings

The average pressure coefficients and pressure concentration coefficient for mono-slope (lean – to) roofs of rectangular clad buildings as shown in Fig. 9.9 may be noted from Table 9.8.

Table 9.8 *External pressure coefficients, C_{le} for mono–slope roofs for rectangular clad buildings with (h/w) < 2 (as per IS : 875 (Part 3)–1987)*

Angle of roof α (degrees)	Angle of wind, θ										Local C_{fe}					
	0°		45°		90°		135°		180°							
	H	L	H	L	H&L	H&L	H	L	H	L	H_1	H_2	L_1	L_2	He	Le
5	–1.0	–0.5	–1.0	–0.9	–1.0	–0.5	–0.9	–1.0	–0.5	–1.0	–2.0	–1.5	–2.0	–1.5	–2.0	–2.0
10	–1.0	–0.5	–1.0	–0.8	–1.0	–0.5	–0.8	–1.0	–0.4	–1.0	–2.0	–1.5	–2.0	–1.5	–2.0	–2.0
15	–0.9	–0.5	–1.0	–0.7	–1.0	–0.5	–0.6	–1.0	–0.3	–1.0	–1.8	–0.9	–1.8	–1.4	–2.0	–2.0
20	–0.8	–0.5	–1.0	–0.6	–0.9	–0.5	–0.5	–1.0	–0.2	–1.0	–1.8	–0.8	–1.8	–1.4	–2.0	–2.0
25	–0.7	–0.5	–1.0	–0.6	–0.8	–0.5	–0.3	–0.9	–0.1	–0.9	–1.8	–0.7	–0.9	–0.9	–2.0	–2.0
30	–0.5	–0.5	–1.0	–0.6	–0.8	–0.5	–0.1	–0.6	0	–0.6	–1.8	–0.5	–0.5	–0.5	–2.0	–2.0

Note. *These values apply to length (w/z) from windward end and*
** these values apply to remaining length
h is the height to eaves at lower side
L is the greater horizontal dimension of a building and
W is the lesser horizontal dimension of the building.

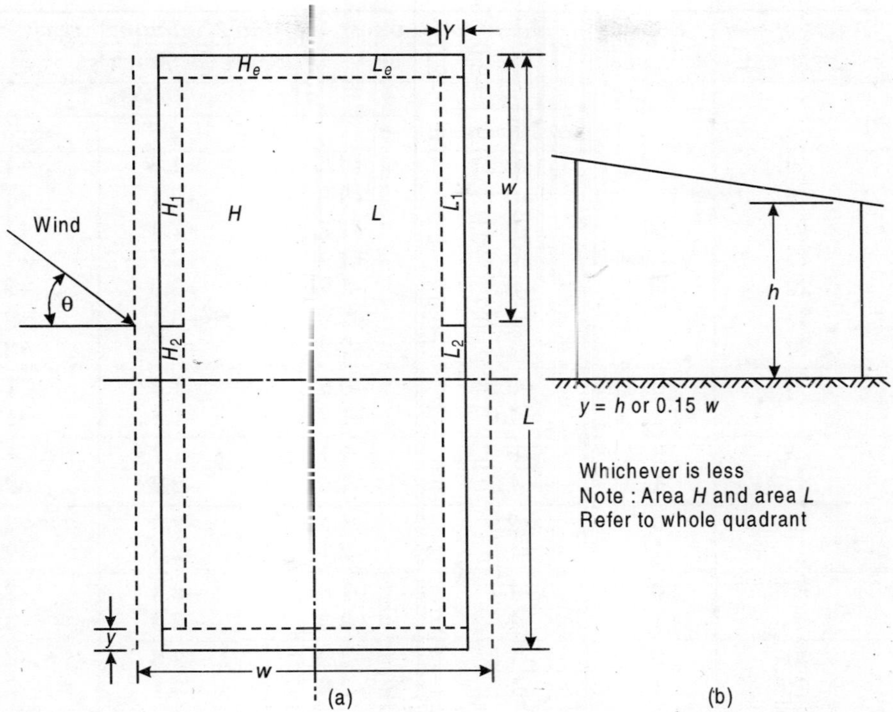

Fig. 9.9 *Mono slope roofs for rectangular clad buildings with (h/w) < 2*

9.11.4 Canopy Roofs with [(*L*/4) < (*h*/*w*) < 1 and 1(< *L*/*W*) < 3]

The pressure coefficients are given in Tables 9.9 and 9.10 separately for mono-pitch and double-pitch canopy roofs (as shpwn in Figs 9.10 and 9.11, respectively) such as open-air parking garages, shelter areas, outer areas, railway platforms, stadiums and theatres. These coefficients take into consideration the combined effect of the wind exerted on and under the roof for all wind direction; the resultant is to be taken normal to the canopy.

In case the local coefficients overlap, the greater of the two given values should be considered. However, the effect of partial closure of one side and/or both sides, such as those due to trains, buses and stored materials shall be foreseen and considered.

The **solidity ratio ϕ** is equal to the area obstruction under the canopy divided by the gross area under the canopy, both areas normal to wind direction ($\phi = 0$ represents a canopy with no obstruction under ncath) and $(\phi) = 1$ represents the canopy fully blocked with contents to the downward caves).

Table 9.9 *Pressure coefficients for mono-slopes free roofs (as per IS : 875 (Part 3) – 1937)*

Angle of roof (degrees)	*Solidity ratio*	*Maximum (largest +ve) and Minimum (argest –ve) Pressure coefficients*			
		Overall coefficients	*Local coefficients*		
			1	2	3
0	All values of φ	+0.2	+0.5	+1.8	+1.1
5		+0.4	+0.8	+2.1	+1.3
10		+0.5	+1.2	+2.4	+1.6
15		+0.7	+1.4	+2.7	+1.8
20		+0.8	+1.7	+2.9	+2.1
25		+1.0	+2.0	+3.1	+2.3
30		+1.2	+2.2	+3.2	+2.4
0	0	–0.5	–0.6	–1.3	–1.4
	1	–1.0	–1.2	–1.8	–1.9
5	0	–0.7	–1.1	–1.7	–1.8
	1	–1.1	–1.6	–2.2	–2.3
10	0	–0.9	–1.5	–2.0	–2.1
	1	–1.3	–2.1	–2.6	–2.7
15	0	–1.1	–1.8	–2.4	–2.5
	1	–1.4	–2.3	–2.9	–3.0
20	0	–1.3	–2.2	–2.8	–2.9
	1	–1.5	–2.6	–3.1	–3.2
25	0	–1.6	–2.6	–3.2	–3.2
	1	–1.7	–2.8	–3.5	–3.5
30	0	–1.8	–3.0	–3.8	–3.6
	0	–1.8	–30	–3.8	–3.6

Note. *For mono-pitch canopies, the centre of pressure should be taken to act at 0.3W from the windward edge.*

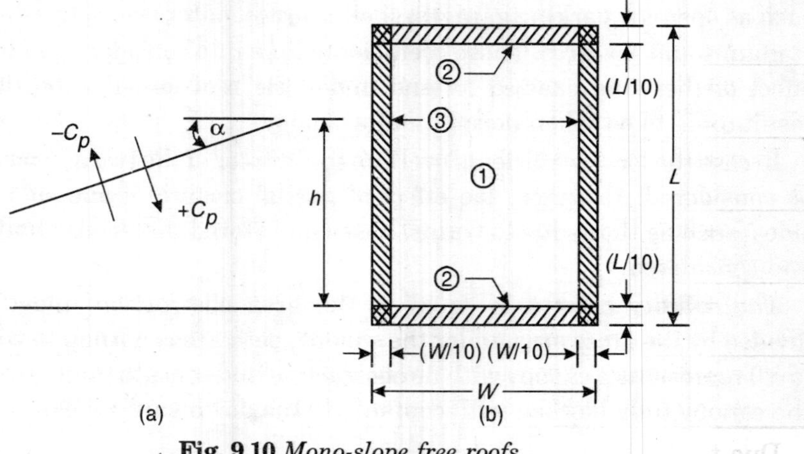

Fig. 9.10 *Mono-slope free roofs*

The values of C_p for intermediate solidities may be linearly interpolated between these two extremes and apply upwind of the position of maximum lockage only. Downwind of the position of maximum blockage, the coefficients for $\phi = 0$ may be used.

Table 9.10 *Pressure coefficients for free standing double sloped roofs (as per IS : 875 (Part 3) – 1987)*

Angle of roof (degrees)	*Solidity ratio*	*Maximum (largest] +ve) and Minimum (argest –ve) Pressure coefficients*				
		Overall coefficients	*Local cofficients*			
			1	*2*	*3*	*4*
–20	All values of ϕ	+0.7	+0.8	+1.6	+0.6	+1.7
–15		+0.5	+0.6	+1.5	+0.7	+1.4
–10		+0.4	+0.6	+1.4	+0.8	+1.1
–5		+0.3	+0.5	+1.5	+0.8	+0.8
+5		+0.3	+0.6	+1.8	+1.3	+0.4
+10		+0.4	+0.7	+1.8	+1.3	+0.4
+15		+0.4	+0.9	+1.9	+1.4	+0.4
+20		+0.6	+1.1	+1.9	+1.5	+0.4
+25		+0.7	+1.2	+1.9	+1.6	+0.5
+30		+0.9	+1.3	+1.9	+1.6	+0.7
–20	0	–0.7	–0.9	–1.3	–1.6	–0.6
	1	–0.9	–1.2	–1.7	–1.9	–1.2
–15	0	–0.6	–0.8	–1.3	–1.6	–0.6
	1	–0.8	–1.1	–1.7	–1.9	–1.2
–10	0	–0.6	–0.8	–1.3	–1.5	–0.6
	1	–0.8	–1.1	–1.7	–1.9	–1.3
–5	0	–0.5	–0.7	–1.3	–1.6	–0.6
	1	–0.8	–1.5	–1.7	–1.9	–1.4
+ 5	0	–0.6	–0.6	–1.4	–1.4	–1.1
	1	–0.9	–1.3	–1.8	–1.8	–2.1
+ 10	0	–0.7	–0.7	–1.5	–1.4	–1.4
	1	–1.1	–1.4	–2.0	–1.8	–2.4
+ 15	0	–0.8	–0.9	–1.7	–1.4	–1.8
	1	–1.2	–1.5	–2.2	–1.9	–2.8
+ 20	0	–0.9	–12	–1.8	–1.4	–2.0
	1	–1.3	–1-7	–2.3	–1.9	–3.0
+ 25	0	–1.0	–1.4	–1.9	–1.4	–2.0
	1	–1.4	–1.9	–2.4	–2.1	–3.0
+ 30	0	–1.0	–1.4	–1.9	–1.4	–2.0
	1	–1.4	–2.1	–2.6	–2.2	–3.0

Due to wind pressure on any fascia and to friction over the surface of the canopy, the horizontal loads also act in addition to the pressure forces normal to the canopy. For any wind direction, only the greater of these two forces need be considered. Fascia loads should be found on the area of the surface facing the wind, using a force coefficient of 1.3.

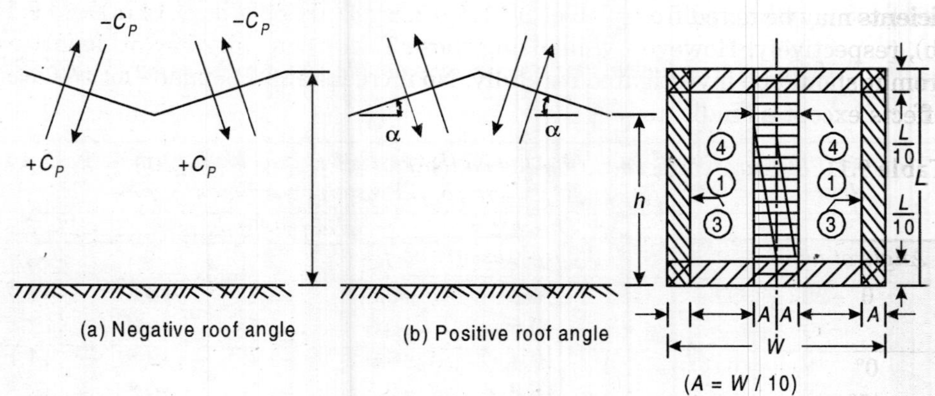

Fig. 9.11 *Free standing double-sloped roofs*

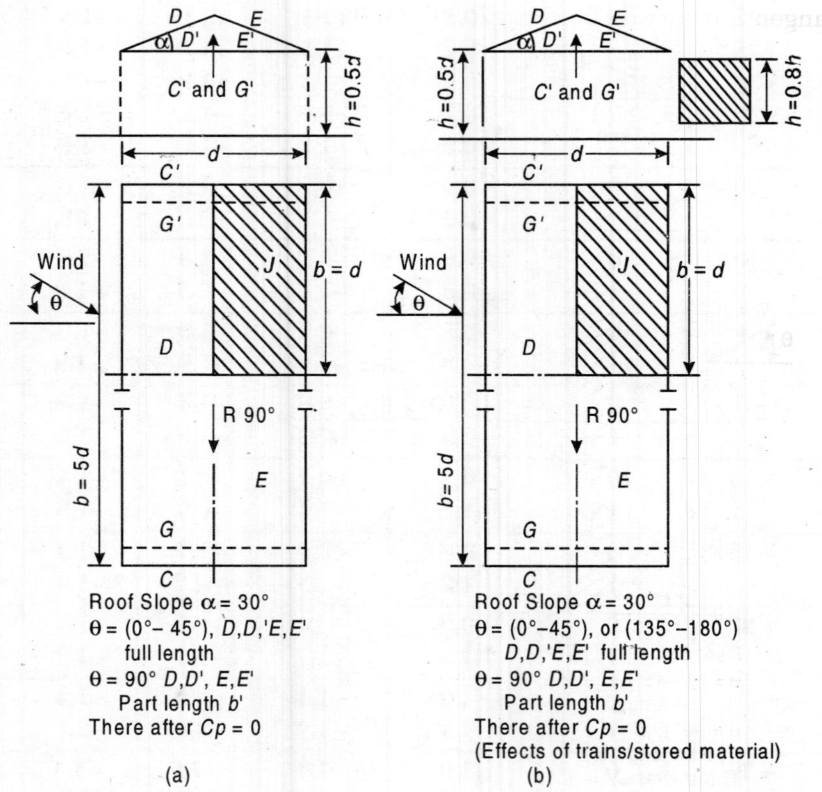

Fig. 9.12 *Pitched free roofs* $\alpha = 30°$

For pitched and toughed free roofs for some specific cases for which aspect ratios and roof slopes have been specific as shown in Fig. 9.12 (a), and (b), Fig. 9.13 (a) and (b) and Fig. 9.14 (a) and (b) the **internal and external pressure coefficients** may be noted from Tables 9.11, 9.12, 9.13 (a), 9.13 (b), 9.14 (a) and 9.14 (b), respectively. However while using these Tables, any significant departure from it should be investigated carefully. No increase shall be made for the local effects except as indicated.

Table 9.11 *Pressure coefficients (top and bottom) for pitched roofs, a = 30°*
(As per IS : 875 (Part 3) – 1987)

Angle of wind θ	Pressure coefficients, C_p							
	D	D′	E	E′	End Surfaces			
					C	C′	G	G′
0°	0.6	–1.0	– 0.5	– 0.9	—	—	—	—
45°	0.1	–0.3	–0.6	–0.3	—	—	—	—
90°	–0.3	–0.4	–0.3	–0.4	–0.3	0.8	0.3	0.4

For j : C_p (top) = –1.0; C_p (bottom) = – 0.2
Tangentially acting friction : R 90° = 0.05 p_d . bd

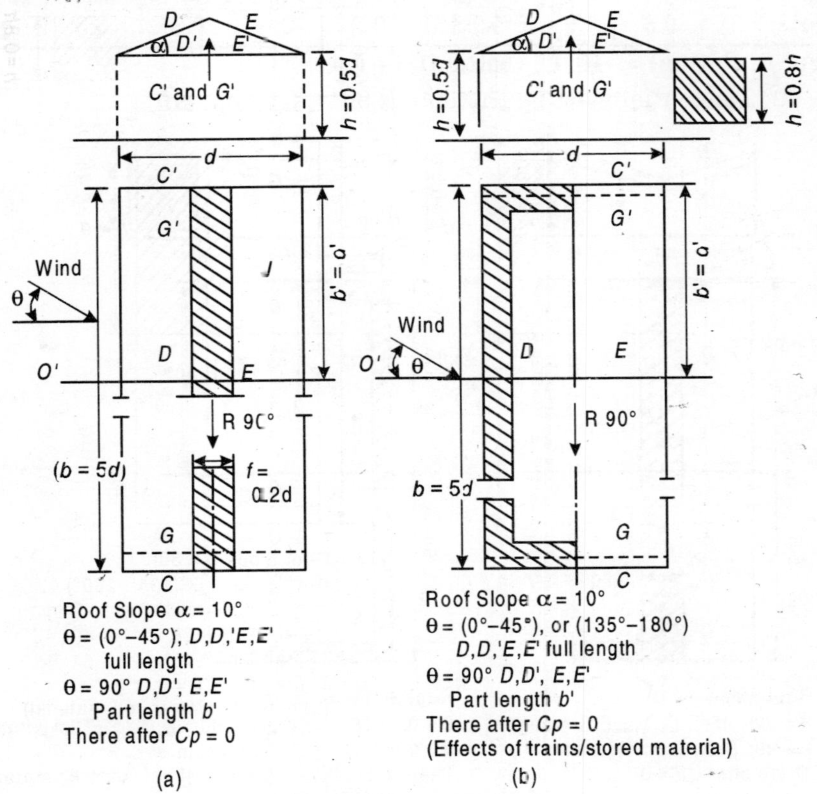

Fig 9.13 *Pitched free roofs, α = 10°*

Table 9.12 *Pressure coefficients (top and bottom) for pitched roofs, a = 30°*
With effect of train or stored material (As per IS : 875 (Part 3) – 1987)

Angle of wind	Pressure coefficients, C_p							
θ	D	D′	E	E′	End Surfaces			
					C	C′	G	G′
0°	0.1	+0.8	+ 0.7	+ 0.9	—	—	—	—
45°	0.1	+0.5	–0.8	+0.5	—	—	—	—
90°	–0.4	–0.5	–04	–0.5	–0.3	0.8	0.3	0.4
180°	–0.3	–0.6	–04	–0.6	—	—	—	—

45° For f C_p (top) = –1.5; C_p (bottom) = – 0.5
Tangentially acting friction : R 90° = 0.05 $p_d \cdot bd$

Table 9.13 (a) *Pressure coefficients (top and bottom) for pitched roofs, a = 10°*
(As per IS : 875 (Part 3) – 1987)

Angle of wind	Pressure coefficients, C_p							
	D	D′	E	E′	End Surfaces			
					C	C′	G	G′
0°	–1.6	+0.3	– 0.5	– 0.2	–	–	–	–
45°	–0.3	+0.1	–0.3	–0.2	–	–	–	–
90°	–0.3	0	–0.3	0	–0.4	0.8	0.3	0.6

0° For f : C_p (top) = –10; C_p (bottom) = + 0.4
0° – 90° Tangentially acting friction : R 90° = 0.01 $p_d \cdot bd$

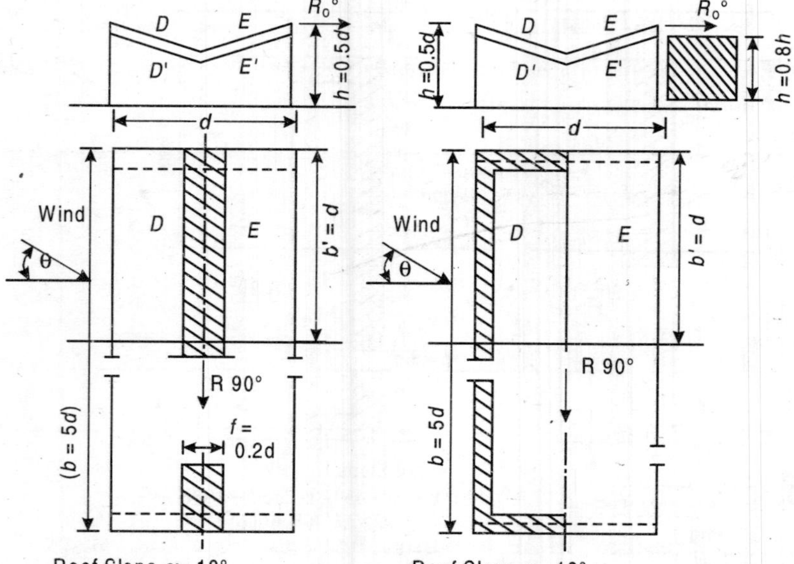

Fig. 9.14 *Troughed free roofs α = – 10°*

Table 9.13 (b) *Pressure coefficients (top and bottom) for pitched roofs, a = 10° With effect of train or stored material (As per IS : 875 (Part 3) – 1987)*

Angle of wind, θ	*Pressure coefficients,* C_p							
	D	*D′*	*E*	*E′*	*End surfaces*			
					C	*C′*	*G*	*G′*
0°	–0.3	+0.8	– 0.6	+0.7	—	—	—	—
45°	–0.5	+0.4	–0.3	+0.3	—	—	—	—
90°	–0.3	0	–0.3	0	0.4	+0.8	+0.3	–0.6
180°	–0.4	–0.3	0.6	0.3	—	—	—	—

0° For *f* : C_p (top) = –1.6; C_p (bottom) = + 0.9

0° – 180° Tangentially acting friction : R 90° = 0.01 p_d . bd

Table 9.14 (a) *External pressure coefficients for troughed free roofs, α = 10° (As per IS : 875 (Part 3) – 1987)*

Angle of wind, θ	*Pressure coefficients,* C_p			
	D	*D′*	*E*	*E′*
0	0 3	–0.7	0.2	–0.9
45	0	–0.2	0.1	–0.3
90	–0.1	+0.1	– 0.1	0.1

0° For *f* : C_p (top) = 04, C_p (bottom) = –1.5

0°–90° Tangentially acting friction : R 90° = 0.1 p_{dd}.bd

Table 9.14 (b) *Pressure coefficients (top and bottom) for troughed free roofs, a = 10° with effects of train or stored materials (As per IS : 875 (Part 3) – 1987)*

Angle of wind, θ	*Pressure coefficients,* C_p			
	D	*D′*	*E*	*E′*
0	–0.7	0.8	–0.9	0.6
45°	–04	0.3	–0.2	0.2
90°	–0.1	0.1	–0.1	0.1
180°	–0.4	–0.2	–0.6	–0.3

0°

0°–180°

9.12 INTERNAL PRESSURE COEFFICIENTS

The internal air pressure in a building depends upon the **degree of permeability** of the structure (cladding) to the flow of air. The internal air pressure may be *positive* or *negative* (that is, a *positive pressure or suction*) depending upon the direction of flow of air in relation to the openings in the buildings. Four types of building having different degrees of permeability and internal pressures shall be considered.

9.12.1 Building Having a Small Degree of Permeability

In the case of buildings where the cladding permits the flow of air with openings not more than about 5 percent of the wall area but where there are no large openings, it is necessary to consider the possibility of the internal pressure being positive or negative. Following two design conditions shall be examined.

1. A building with an internal pressure coefficient of + 0.2, and
2. A building with an internal pressure coefficient of – 0.2.

9.12.1.1 Design Wind Pressure on Roofs

The internal pressure coefficient is algebraically added to the external pressure coefficient and the analysis which indicates more distress of the member shall be adopted. In most situations, a simple inspection of the sign of external pressure will at once indicate the proper sign of the internal pressure coefficient to be taken for design.

The term **normal permeability** relates to the flow of air commonly afforded by the structures (claddings). It includes the flow of air not only through open windows and doors, but also through the slits round the closed windows, doors, ventilators, chimneys and through the joints between the roof coverings, the total open area being less than 5 percent of area of the walls having the openings.

9.12.2 Buildings Having Medium Degree of Permeability

Buildings with medium openings may also show either positive or negative internal pressure depending upon the direction of wind. The buildings with medium openings between about 5 to 20 percent of wall area shall be designed for an internal pressure coefficient of + 0.5 and again with an internal pressure coefficient of – 0.5. The analysis shall be done for maximum distress of the members.

9.12.3 Buildings Having Large Degree of Permeability

Buildings with large openings, that is, openings larger than 20 percent of the wall area shall be designed once with an internal pressure coefficient of + 0.7 and again with an internal pressure coefficient of – 0.7. The analysis shall be carried out for maximum distress on the members.

9.12.4 Buildings Having High Degree of Permeability

Buildings with one open side or having opening exceeding 20 percent of wall area may be assumed to be subjected to internal positive pressure or suction (internal negative pressure) similar to those for buildings with large openings. The buildings with one sided openings are shown in Fig. 9.15. The values of internal pressure coefficients with respect to the direction of wind are also indicated.

The values of intenal pressure coefficients for all the four above categories of buildings have been noted briefly in Table 9.15.

Table 9.15 *Internal pressure coefficients, C_{pi} (as per IS : 875 (Part 3) –1987)*

S. No.	*Categories of buildings*	C_{pi}
1.	Buildings with small degree of permeability (openings ≯ percent)	+0.2 – 0.2
2.	Buildings with medium degree of permeability (openings 5 to 20 percent)	+0.5 – 0.5
3.	Buildings with large degree of permeabilities (opening > 20 percent)	+0.7 – 0.7
4.	Buildings with one open side or (opening > 20 percent)	Fig. 9.15

Note. *For (B/L) equal to unity, average values may be used.*

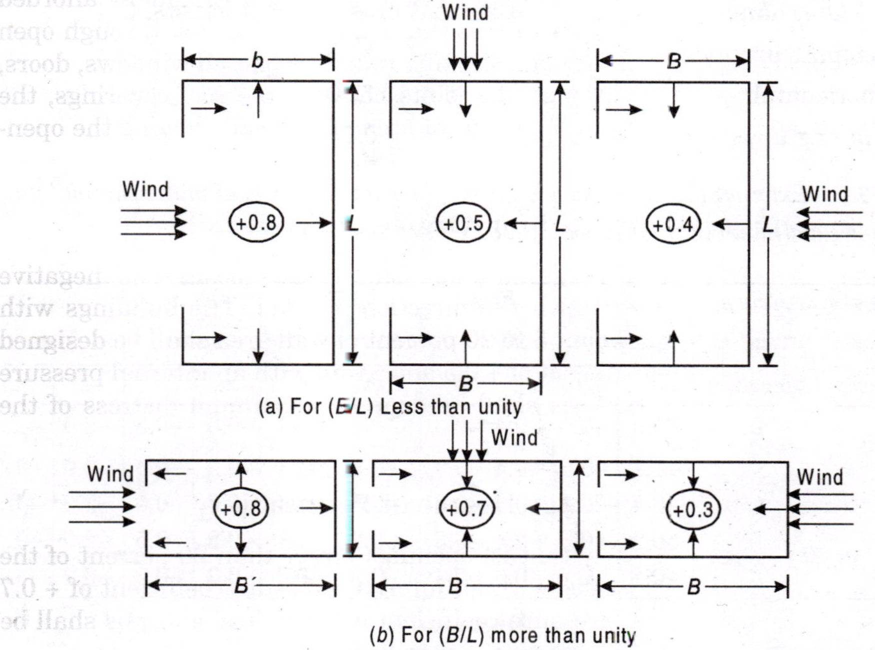

(a) For (B/L) Less than unity

(b) For (B/L) more than unity

(*Note* : for (B/L) = 1 average values are used)

Fig. 9.15

9.13 EXTERNAL PRESSURE COEFFICIENTS FOR SOME OTHER ROOFS

The external pressure coefficients for the following more types of roofs are adopted as specified in IS : 875 (Part 3)–1987.

9.13.1 Pitched and Saw-tooth Roofs of Multi-span Building

For the pitched and saw-tooth roofs of multispan buildings, as shown in Figs. 9.16 and 9.17 (see page 602–603) the external average pressure coefficients and pressure concentration coefficients may be noted from Tables 9.16 and 9.17, respectively provided all the spans are equal and the height to eaves shall not exceed the span.

9.13.2 Pressure Coefficients on Overhangs from Roofs

The pressure coefficients on the top overhanging portion of the roof are taken to be the same, as that of the nearest top portion of the non-overhanging portions of the roofs. The pressure coefficients for the underside surface of the portions are taken as follows and shall be taken as *positive* in case the overhanging portion is on the windward side.

Overhangs	External pressure coefficients, C_{pc}
(a) sloping downward	1.25
(b) horizontal	1.00
(c) loping upward	0.75

Table 9.16 *External pressure coefficients, (C_{pe}) for pitched roofs of multispan buildings (all spans equal) with (h > w)(As per IS : 875 (Part 3) –1987)*

Angle of roof, a (degrees)	*Angle of wind θ (degress)*	*First span*		*First interspan*		*Other intermediate*		*End span*		*Local coefficients*	
		a	*b*	*c*	*d*	*m*	*n*	*x*	*z*	*1*	*2*
5	0	–0.9	–0.6	–0.4	–0.3	–0.3	–0.3	–0.3	–0.3	–2.0	–1.5
10		–1.1	–0.6	–0.4	–0.3	–0.3	–0.3	–0.3	–0.4	–2.0	–1.5
20		–0.7	–0.6	–0.4	–0.3	–0.3	–0.3	–0.3	–0.5	–2.0	–1.5
30		–0.2	–0.6	–0.4	–0.3	–0.2	–0.3	–0.2	–0.5	–2.0	–1.5
45		+0.3	–0.6	–0.6	–0.4	–0.2	–0.4	–0.2	–0.5	–0.2	–1.5

		Distance		
Angle of roof, α	Angle of wind θ	h_1	h_2	h_3
upto 45°	90°	–0.8	–0.6	–0.2

Note: *It is to note that when the angle of wind q is 0°, the horizontal forces due to frictional drag are allowed for in the above values. And when the angle of wind is 90°, the above values are allowed for frictional drag as specified in IS : 875 (Part 3) 1987.*

Table 9.17 *External pressure coefficients, (C_{pe}) for saw-tooth roofs of multi-span buildings (all span equal) with ($h \not> w$) (as per IS : 875 (Part 3) – 1987)*

Angle of wind, θ (degrees)	*First span*		*First intermediate span*		*Other intermediate span*		*End spans*		*Local coefficients*	
	a	b	c	*d*	*m*	*n*	*x*	*z*	*1*	*2*
0	+0.6	–0.7	–0.7	–0.4	–0.3	–0.2	–0.1	–0.3	–2.0	–1.5
180	–0.5	–03	–0.3	–0.3	–0.4	–0.6	–0.6	–0.1	–2.0	–1.5

	Distance		
Angle of wind, θ	h_1	h_2	h_3
90°	–0.8	–0.6	–0.2
270°	similarly, but handed		

Note: *It is to note that when the angle of wind, θ is zero, the horizontal forces due to frictional drag are allowed for in the above values. And when the angle of wind, 8 is 90°, the frictional drag is allowed as recommended in IS : 875 (Part 3) – 1987.*

It is to note that for the overhanging portions on the sides othe than the windward side, the average pressure coefficients on adjoining walls may be used.

9.13.3 Combined Roofs and Roofs with a Sky-light

The average external pressure coefficients for combined roofs with a sky light as shown in Figs. 9.18 and 9.19 may be noted from Tables 9.18 and 9.19, respectively.

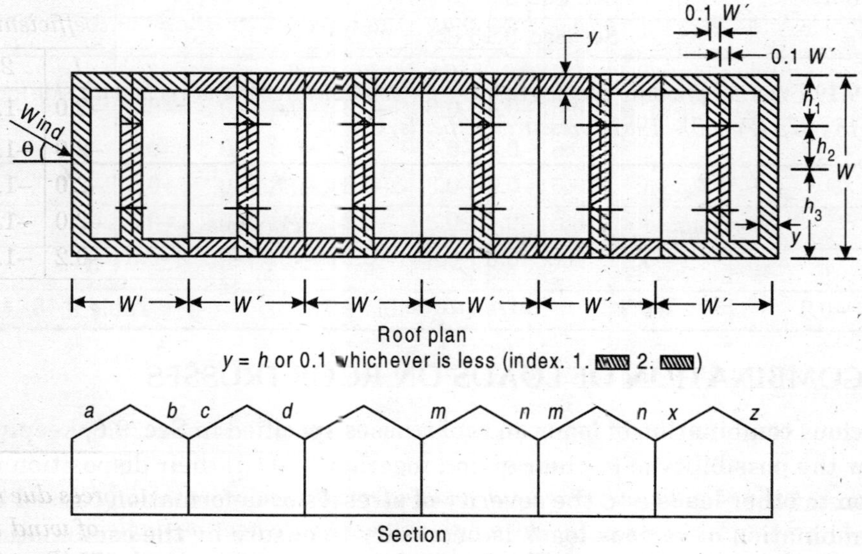

Fig. 9.16 *Pitched roofs of multispan buildings (all spans equal with h ($\not>$ w′)*

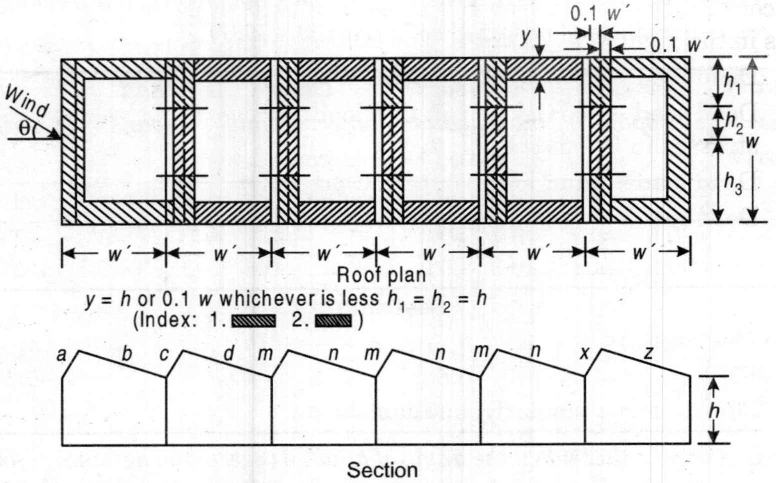

Fig. 9.17 *Saw-tooth roofs of multispan buildings (all spans equal with (h > w′)*

Table 9.18 *External pressure coefficients, C_{pe} for combined roofs and roofs with sky-light per IS : 875 (Part 3) – 1987) values of C_{pe}*

Potion	*Direction 1*	*Direction 2*
a	From Fig. 9.18	
b	$C_{pe} = -0.5\ (h_1/h_2) \leq 1.5$ $C_{pe} = -0.7\ (h_1/h_2) > 1.5$	– 0.4
c and *d*	See Table 9	
e	See Sec. 9.13 (2)	

Table 9.19 *External pressure coefficients, C_{pe} for combined roofs and roofs with sky-light (as per IS : 875 (Part 3)– 1987) pressure coeffients, C_{pe}*

$b_1 > b_2$		$b_1 \leq b_2$	
Portions		*Portions*	
a	*b*	*a*	*b*
–0.6	+0.7	Table 9.18	Table 9.18

9.14 COMBINATION OF LOADS ON ROOF TRUSSES

A judicious combination of loads on roof trusses specified in Sec. 9.6, keeping in view the possibility of (i) their action together and (ii) their disposition in relation to other loads and the severity of stresses or deformation caused by the combination of various loads is necessary to ensure by the combination of various loads is necessary to ensure the required safety and economy in the design of roof trusses.

The combination of loads on roof trusses is used to determine the maximum stresses in the members of a truss. The following combination of loads are used:

(i) Dead load + snow load, if any.

(ii) Dead load + partial or full live load whichever causes the maximum stresses in the members.

(iii) Dead load + wind load + internal pressure.

(iv) Dead load + wind load + internal suction.

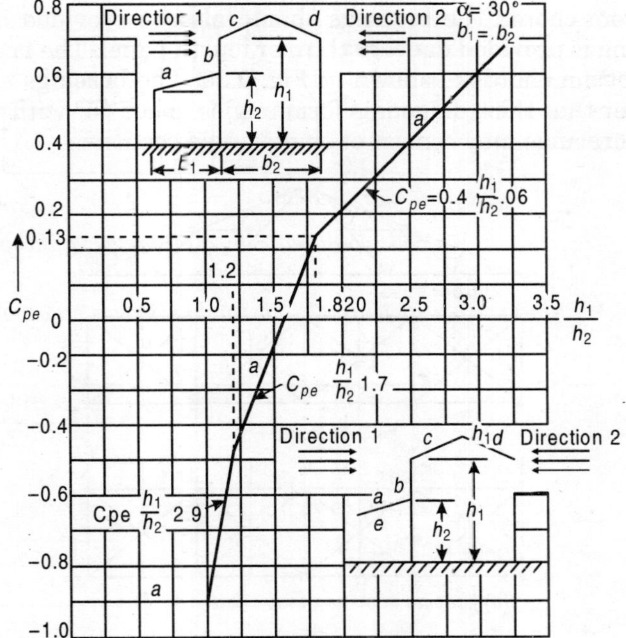

Fig. 9.18 *Combined roofs and roofs with a sky light*

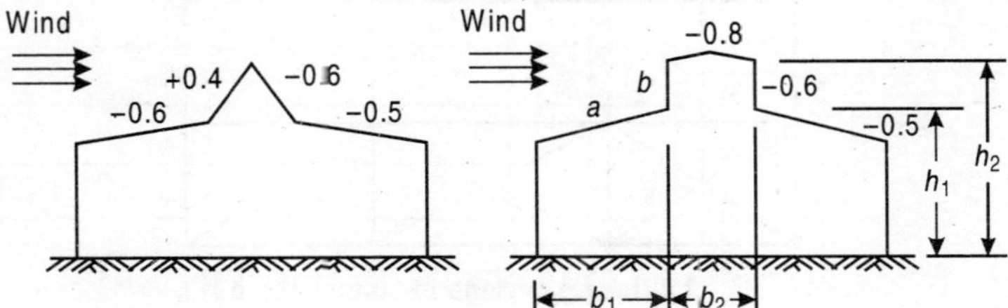

Fig. 9.19 *Combined roofs and roofs with a sky light*

9.15 BRACING OF ROOF TRUSSES

The bracing of roof trusses is done to provide a stiff rigid structures so that structure is not subjected to vibrations due to rolling loads of machinery. In building structures, the roof trusses are supported on masonry walls. The purlins at ends also rest on end masonry walls. For such structures, lateral bracing is not necessary. The purlins provide sufficient lateral support. The roof trusses are also securely fixed with anchor bolts in the masonry walls.

When the roof trusses are supported over steel columns, then lateral bracing should be provided for stability against lateral forces. In absence of lateral bracing, the structure will collapse in a high wind storm. The structure may collapse due to stress and vibrations due to rolling loads. The lateral bracing provided should be such that it transmits lateral forces directly as possible to the walls or foundations of the buildings.

The roof trusses should be braced in the plane of top chord and also in the plane of bottom chord. The bracings should also be provided in the plane of column bracing is provided in every third or fourth panel. The bracing of a truss in top and bottom chords is shown in Fig. 9.20. The bracings are provided in such a manner that their diagonals form angles about 45° with the loads to be carried. The bracings may consist of single angle sections.

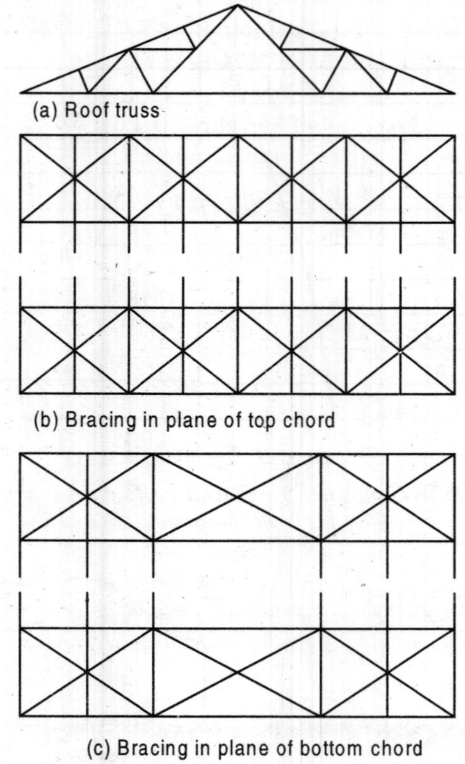

Fig. 9.20 *Bracing of roof trusses*

9.16 ASSUMPTIONS IN ANALYSIS OF ROOF TRUSSES

The analysis of roof trusses consist of two parts. Part one consists of determination of loads and reactions. The other part consists of determination of the internal forces in the members of roof trusses. The following assumptions are made for determining the forces (stresses) in the members :

1. The roof truss is not restrained by the reactions.
2. The axes of members meeting at a joint intersect at a common point.
3. The riveted joints in roof trusses act as frictionless hinges.

9.17 PURLINS

The rolled steel sections and rectangular wooden beams are used as purlins in the roof trusses. The purlins are placed with their webs or sides perpendicular to the top chords of roof as shown in Fig. 9.21. The principal axes of rolled steel I-section, channel section, rectangular wooden section shown in Fig. 9.21 (a), (b) and (c) respectively are parallel to the sides of sections. Whereas Fig. 9.21 (d) shows rolled steel angle section. One leg of angle section is normal to the top chord of roof truss. It supports the normal component of loading. The principal axes of angle section are not parallel to its sides. In general, when a structural member is subjected to transverse loading, the plane of bending (or plane of loading coincides with one of the principal axes). The neutral axis of the section coincides with other principal axes. Whereas in case of purlin, the line of action of resultant of vertical loads and normal component of wind load do not coincide with any of the principal axes. Therefore, the bending of purlins occurs in a plane other than the principal plane of section. Such a bending is known as *unsymmetrical* bending. The purlins are subjected to unsymmetrical bending.

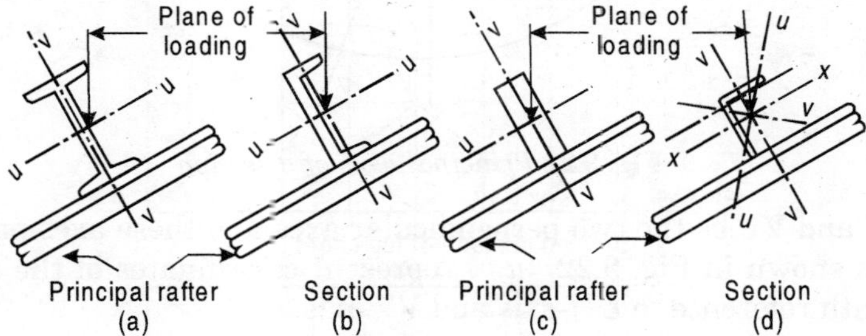

Fig. 9.21 *uu, vv principal axes*

I-scction shown in Fig. 9 21 (a), rectangular section shown in Fig. 9.21 (c) have two axes of symmetry. The channel section shown in Fig. 9.21 (b) has only one axis of symmetry. Whereas angle section shown in Fig. 9.21 (d) has no axes of symmetry.

9.18 PRINCIPAL AXES OF SECTION

The moment of inertia of a section about any axis is defined as the second moment of an area about that axis. Figure 9.22 shows any plane section. The point O represents centre of gravity of the section. xx-axis and yy-axis represent two co-ordinates axes. The co-ordinates axes pass through the point O, and these are perpendicular to each other. (x, y) represent co-ordinates of any elementary area δA.

Then, $I_{xx} = \Sigma y^2 \, \delta A$, $I_{yy} = \Sigma x^2 \, \delta A$, $I_{xy} = \Sigma xy \, \Sigma A$

where, I_{xx} = Moment of inertia of the section about xx-axis

I_{yy} = Moment of inertia of the section about yy-axis

I_{xy} = Product of inertia about xx-axis and yy-axis of section.

The product of inertia of a section is determined by taking moment of the area about one axis, and then taking the moment of this moment about other axis.

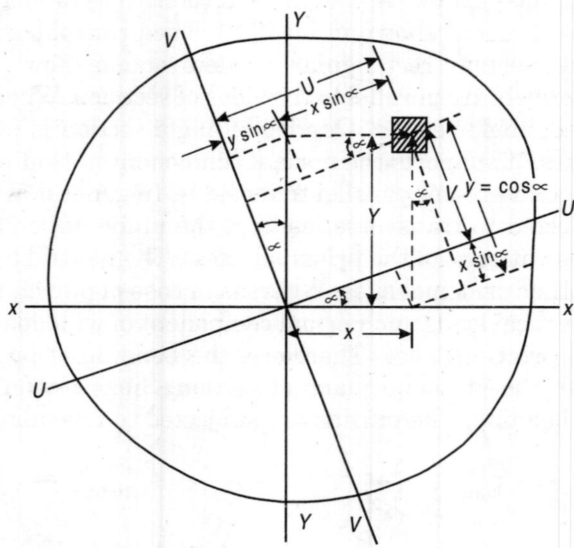

Fig. 9.22 *Principal axes of a section*

Let UU and VV be the two perpendicular axes and these axes pass through point O as shown in Fig. 9.22. (u,v) represent co-ordinates of the elementary area ΣA with reference to UU-axis and VV-axis.

Then; $I_{uu} = \Sigma v^2 \delta A$, $I_{vv} = \Sigma u^2 \delta A$, $I_{uv} = \Sigma uv\delta 4$

where, I_{uu} = Moment of inertia about UU-axis

I_{vv} = Moment of inertia about VV-axis

I_{uv} = Product of inertia about UU-axis and VV-axis.

The *principal axes* of a plane section are defined as two *perpendicular axes* in its plane passing through the centre of gravity of the section, such that the product of inertia is zero.

If the product of inertia, I_{uv} is zero, then, UU-axis and VV-axis are known as *principal axes* of the plane section. The angle of inclination of principal axes with XX-axis and YY-axis is as shown in Fig. 9.22. If any plane section has an axis of symmetry, then the product of inertia about that axfs is zero. The axis of symmetry represents one of the principal axis. The other principal axis is at right angle to this axis of symmetry. If any plane section has two axes of symmetry, then both the axes of symmetry represent principal axis of the plane section. I-section and rectangular section are shown in Fig. 9.21 (a) and Fig. 9.21 (c). UU-axis and VV-axis are two axes of symmetry. These axes are principal axes of the sections. The channel section shown in Fig. 9.21 (b) has one axis of symmetry. The axes of symmetry *UU is* principal axis. The other principal axis *VV* is at right angle to this axis and passes through C.G. of section. Angle section shown in Fig. 9.21 (d) has no axis of symmetry. Such I sections are called unsymmetrical section. The principal axes for such sections are located either by analytical method or by graphical method.

9.18.1 Analytical Method

The principal axes of a plane section shown in Fig. 9.22 can be located by determining angle of inclination, a with *UU*-axis and *XX*-axis as below. The co-ordinates (u, v) of the elementary δA with reference to *UU*-axis and *W*-axis may be expressed in terms of co-ordinates (x, y) and α

$$u = (x \,.\, \cos\alpha + y \,.\, \sin\alpha) \quad \text{...(i)}$$

$$v = (y \,.\, \cos\alpha - x \,.\, \sin\alpha) \quad \text{...(ii)}$$

The product of inertia about *UU*-axis and *VV*-axis,

$$I_{UV} = \Sigma uv\delta A \quad \text{...(iii)}$$

$$= \Sigma (x \,.\, \cos\alpha + y \,.\, \sin\alpha)(y \,.\, \cos\alpha - x \,.\, \sin\alpha)\,\delta A$$

$$= (\Sigma xy \,.\, \cos^2\alpha \,.\, \delta A) - (\Sigma x^2 \,.\, \sin\alpha\cos\alpha \,.\, \delta A)$$

$$- (\Sigma xy \,.\, \sin^2\alpha\,\delta A) + (\Sigma x^2 \,.\, \sin\alpha\cos\alpha \,.\, \delta A)$$

$$I_{UV} = (\cos^2\alpha - \sin^2\alpha)\,\Sigma xy \,.\, \delta A + \sin\alpha \,.\, \cos\alpha\,(\Sigma y^2 \,.\, \delta 4 - \Sigma x^2 \,.\, \delta A)$$

$$= (\cos 2\alpha \,.\, \Sigma xy \,.\, \delta A) + \frac{1}{2}\sin 2\alpha \,.\, (\Sigma y^2 \,.\, \delta A - \Sigma x^2 \,.\, \delta A)$$

$$= \cos 2\alpha° \,.\, I_{xy} + \left(\frac{I_{xx} - I_{yy}}{2}\right)\cdot\sin 2\alpha$$

For the principal axes, the product of inertia of the plane section about those axes is zero.

Therefore,
$$I_{UV} = 0 \quad \text{... (iv)}$$

$$\cos 2\alpha \,.\, I_{xy} + \left(\frac{I_{xx} - I_{yy}}{2}\right)\cdot\sin 2\alpha = 0 \quad \text{...(v)}$$

$$\tan 2\alpha = \left(\frac{-2I_{xx}}{I_{xx} - I_{yy}}\right) \quad \text{...(9.14)}$$

From this, two values, (α and 90° + α) are obtained. *UU*-axis is inclined to *XX*-axis by an angle α, and *VV*-axis is inclined to the same axis by an angle (90°+ α). Thus, two principal axes *UU* and *VV* of the plane section are located with reference to *XX*-axis.

The values of principal moment of inertia l_{UU} and I_{VV} can be found in terms of I_{xx}, I_{yy} and I_{xy} as below :

$$I_{UU} = \Sigma v^2 - \delta A$$

$$= \Sigma (y \,.\, \cos\alpha - x \,.\, \sin\alpha)^2 \,.\, \delta A$$

$$= (\Sigma \cos^2\alpha \,.\, y^2\,\delta A) + (\Sigma \sin^2\alpha\, x^2\,\delta A) - (\Sigma 2\sin\alpha\cos\alpha \,.\, xy\,\delta A)$$

$$= I_{xx} \cdot \cos^2 \alpha + I_{yy} \cdot \sin^2 \alpha - I_{yy} \cdot \sin 2\alpha \quad ...(9.15)$$

$$= I_{xx} \cdot \left(\frac{1+\cos 2\alpha}{2}\right) + I_{yy} \cdot \left(\frac{1-\cos 2\alpha}{2}\right) - I_{ay} \sin 2\alpha$$

$$= \left(\frac{I_{xx}+I_{yy}}{2}\right)\left(\frac{I_{xx}-I_{yy}}{2}\right)\cos 2\alpha - I_{xy} \sin 2\alpha$$

From value of tan 2α, Eq. 9.6, we have

$$\sin 2\alpha = \frac{-I_{xy}}{\left[\left(\frac{I_{xx}-I_{yy}}{2}\right)^2 + I_{xy}^2\right]^{1/2}} \quad ...(i)$$

$$\cos 2\alpha = \frac{\frac{1}{2}(I_{xx}-I_{yy})}{\left[\left(\frac{I_{xx}-I_{yy}}{2}\right)^2 + I_{xy}{}^2\right]^{1/2}} \quad ...(ii)$$

Substituting the values of sin 2α and cos 2α, we have

$$I_{UU} = \left(\frac{I_{xx}+I_{yy}}{2}\right)\left[\left(\frac{I_{xx}-I_{yy}}{2}\right)^2 + I_{xy}^2\right]^{1/2} \quad ...(iii)$$

Similarly,

$$I_{VV} = \Sigma u^2 \, \delta A$$

$$= \Sigma(x \cdot \cos \alpha + y \cdot \sin \alpha)^2 \, \delta A$$

$$= (\Sigma \cos^2 \alpha \cdot x^2 \, \delta A) + (\Sigma \sin^2 \alpha \, y^2 \, \delta A) - (\Sigma\, 2 \sin \alpha \cos \alpha \cdot x \cdot y \cdot \delta A)$$

$$= I_{yy} \cdot \cos^2 \alpha + I_{xx} \cdot \sin^2 \alpha - I_{xy} \cdot \sin 2\alpha \quad ...(9.16)$$

$$= I_{yy} \cdot \left(\frac{1+\cos 2\alpha}{2}\right) + I_{xx} \cdot \left(\frac{1-\cos 2\alpha}{2}\right) - I_{xy} \sin 2\alpha$$

$$= \left(\frac{I_{xx}+I_{yy}}{2}\right) - \left(\frac{I_{xx}-I_{yy}}{2}\right)\cos 2\alpha + I_{xy} \sin 2\alpha$$

$$= \left(\frac{I_{xx}-I_{yy}}{2}\right) - \left[\left(\frac{I_{xx}-I_{yy}}{2}\right)^2 + I_{xy}^2\right]^{1/2} \quad ...(iv)$$

Thus, the principal moment of inertias are

$$I_{UU} = \left(\frac{I_{xx}+I_{yy}}{2}\right) + \left[\left(\frac{I_{xx}-I_{yy}}{2}\right)^2 + I_{xy}^2\right]^{1/2} \quad ...(9.17)$$

$$I_{VV} = \left(\frac{I_{xx}+I_{yy}}{2}\right) - \left[\left(\frac{I_{xx}-I_{yy}}{2}\right)^2 + I_{xy}^2\right]^{1/2} \quad ...(9.18)$$

If, the values of I_{UU} and I_{VV} are added, then, we have

$$(I_{UU} + I_{VV}) = (I_{xx} + I_{yy}) \qquad ...(919)$$

Thus, it is seen that the sum of moment of inertias about any two perpendicular axes of a plane section, passing through the centre of gravity of section remains unchanged.

9.18.2 Graphical Method

The principal axes *UU* and *VV* of a plane section shown in Fig. 9.23 can also be located by graphical method. The values of principal moment of inertias I_{UU} and I_{VV} can also be found graphically. Following are the graphical methods for locating the principal axes and determining the values of I_{UU} and I_{VV}.

(i) Mohr's circle method, and (ii) Circle of inertia method.

1. Mohr's circle method :

Let *OX* and *OY* be the two co-ordinate axes, passing through point *O*, as origin, as shown in Fig. 9.23.

Plot $OP = I_{xx}$ and $OQ = I_{yy}$ on *x*-axis.

Draw $PR = I_{xx}$ perpendicular to *OX*-axis at point *P*. *PR* is plotted upwards to *OX*-axis, if the value of I_{xy} is positive and it is plotted downwards to *OX*-axis, if its value is negative. The point *C* is obtained by bisecting the distance *PQ*. Joint

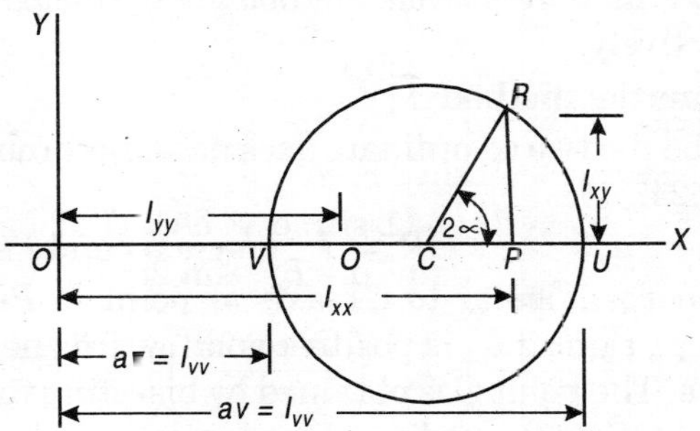

Fig. 9.23 *Mohr circle method*

C and *R*. With *C* as the centre, and *CR* as radius draw a circle as shown in Fig, 9.23. The circle is known as *Mohr's circle.* The circle intersects *OX*-axis at two points *U* and *V*.

Distance $\quad PQ = (I_{xx} - I_{yy})$

In the triangle *PCR*,

$$\therefore \qquad PC = \left(\frac{I_{xx} - I_{yy}}{2}\right), \; PR = I_{xy}$$

$$CR = \left[\left(\frac{I_{xx} - I_{yy}}{2}\right)^2 + I_{xy}^2\right]^{1/2}$$

$$\therefore \qquad \tan(\angle PCR) = \frac{PR}{PC} = \frac{I_{xx}}{\frac{1}{2}(I_{xx} - I_{yy})} = \left(\frac{2I_{xy}}{I_{xx} - I_{yy}}\right)$$

$$= \tan(-2\alpha)$$

$$\therefore \qquad \angle PCR = 2\alpha$$

where, α represents the angle of inclination of principal axis *UU* of the plane section. The negative sign shows that angle is to be plotted in the negative direction, i.e., clockwise. In Mohr's circle, if *CR* appears anti-clockwise to *OX*-axis, then angle is plotted clockwise. If *CR* appears clockwise to *OX*-axis, then angle is plotted anti-clockwise.

Distance $\qquad OU = (OC + CU) = (OC + CR) \qquad (\because CU = CR)$

$$\left(\frac{I_{xx} + I_{yy}}{2}\right) + \left[\left(\frac{I_{xx} + I_{yy}}{2}\right)^2 + I_{xy}^2\right]^{1/2} = I_{UU} \qquad (\because CV = CR)$$

Distance $\qquad OV = (OC - CV) = (OC - CR) \qquad (\because CV = CR)$

$$= \left(\frac{I_{xx} + I_{yy}}{2}\right) - \left[\left(\frac{I_{xx} - I_{yy}}{2}\right)^2 + I_{xy}^2\right]^{1/2} = I_{VV}$$

Thus, *OU* and *OV* in Mohr's circle method give principal moment of inertias I_{UU} and I_{VV}, respectively.

(ii) Circle of inertia method :

Let *OX* and *OY* be the two co-ordinate axes passing through point *O*, as origin as shown in Fig. 9.24.

Plot $\qquad OP = I_{xx}$ on *y*-axis and $PQ = I_{yy}$ on *x*-axis.

Draw $PR = I_{xy}$ perpendicular to *CY*-axis at point *P*. *PR* is plotted towards positive direction i.e., right if I_{xy} is positive and towards negative direction, i.e., left if I_{xy} is negative. The point *C* is obtained by bisecting the distance *OQ*. With *C* as the centre and *OC* as the radius, draw a circle as shown in Fig. 9.24. The circle is known as *circle of inertia*. Join *C* and *R* and produce it on both sides. It intersects the circle at points *U* and *V*.

Distance $\qquad OQ = (I_{xx} + I_{yy})$

$$OQ = CQ = \left(\frac{I_{xx} + I_{yy}}{2}\right) = \text{Radius of circle}$$

Then, distance $\qquad PC = (CQ + PQ)$

$$\left(\frac{I_{xx} + I_{yy}}{2} - I_{yy}\right) = \left(\frac{I_{xx} - I_{yy}}{2}\right)$$

In ΔPCR,

$$CP = \left(\frac{I_{xx} - I_{yy}}{2}\right) \text{ and } PR = I_{yy}$$

$$CR = \left[\left(\frac{I_{xx} - I_{yy}}{2}\right)^2 + I_{xy}^2\right]^{1/2}$$

$$\tan \angle PCE = \frac{PR}{PC} = \frac{I_{xx}}{\left(\dfrac{I_{xx} - I_{yy}}{2}\right)}$$

$$\tan (\angle -PCR) = \frac{I_{xx}}{(I_{xx} - I_{yy})}$$

$$\therefore \quad \angle PCR = 2\,\alpha$$

where, α represents the angle of inclination of principle axis UU of the plane section. The negative sign shows the angle is to be plotted in the negative direction, i.e., clockwise. OU and OV give the directions of the principal axes UU and VV.

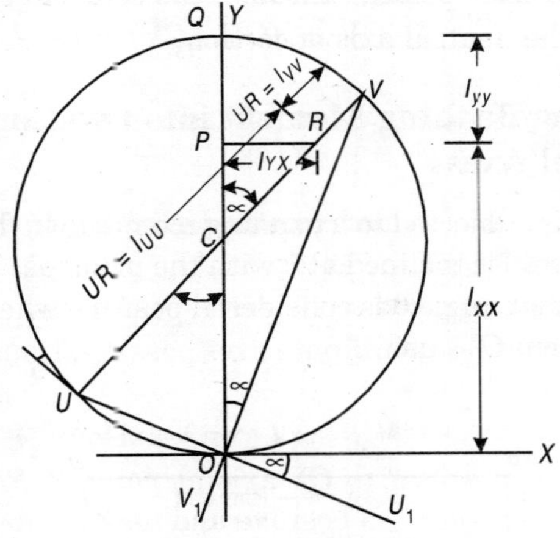

Fig. 9.24 *Circle of inertia method*

Distance $$UR = (UC + CR) = (OC + CR) \qquad (\because UC = OC)$$

$$= \left(\frac{I_{xx} + I_{yy}}{2}\right) + \left[\left(\frac{I_{xx} - I_{yy}}{2}\right)^2 + I_{xy}^2\right]^{1/2} = I_{UU}$$

Distance $VR = (VC - CR) = (OC - CR)$ $\qquad (\because VC = OC)$

$$= \left(\frac{I_{xx} - I_{yy}}{2}\right) - \left[\left(\frac{I_{xx} - I_{yy}}{2}\right)^2 + I_{xy}^2\right]^{1/2} = I_{VV}$$

Thus, UR and VR in circle of inertia method give principal moment of inertias I_{UU} and I_{VV} respectively.

9.19 MAXIMUM BENDING STRESS DUE TO UNSYMMETRICAL BENDING

When a structural member is subjected to transverse loading such that the bending of beam occurs in the principal plane of section, then the neutral axis of the section of beam is perpendicular to the plane of bending. The neutral axis passes through the centroid of section. If the transverse loading is such that the bending of beam occurs in a plane other than the principal plane of section, the beam section is subjected to *unsymmetrical bending.* The direction of neutral axis is not perpendicular to plane of bending. The maximum bending stress in a beam-section subjected to unsymmetrical bending may be determined by the following methods:

1. By resolving bending moment into two components along the principle axes.
2. By resolving bending moment in two components along any two perpendicular axes passing through the centroid of the section.
3. By locating the neutral axis of section.

9.19.1 By Resolving Bending Moment into two Components Along the Principal Axes

Let the beam section is subjected to a bending moment, *M*. The plane of bending (i.e., the plane of loading) is inclined at θ with the principal *VV*-axis as shown in Fig. 9.25. It is noted that angle θ is considered positive when measured in anti-clockwise direction from *OV*-axis.

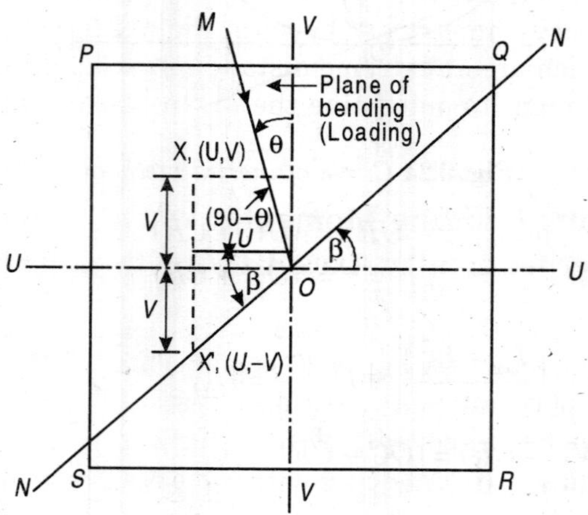

Fig. 9.25 *Unsymmetrical bending*

The bending moment, *M* may be resolved into two components along the principal axes *UU* and *VV*. The component of bending moment *M* along *UU*-axis is *M* sin θ, and that along*VV*-axis *M* cos θ. After resolving the bending moment, into its two components, the theory of simple bending can be applied. For the component, M sin θ, the bending occurs in the principal plane passing through UU-axis, and VV-axis becomes a neutral axis. For the component M cos θ, the bending occurs in the principal plane passing throughVV-axis and UU-axis becomes a neutral axis.

The bending stress at any point X, may be found by algebraic sum of the stress due to M . sin θ and M . cos θ. The bending stress at point X is given by

$$\sigma_b = \left(\frac{M \cdot \cos\theta}{I_{UU}} v + \frac{M \cdot \sin\theta}{I_{VV}} u\right) \quad ...(9.20)$$

where, I_{UU} = Principal moment of inertia of the section about UU-axis
I_{VV} = Principal moment of inertia pf the section about VV-axis
u, v = Co-ordinates of point, X in the beam section with reference to principal axes UU and VV.

The co-ordinates (u, v) are assumed positive as shown in Fig. 9.25. Due to M.cos θ the portion of beam section above UU-axis is subjected to compressive bending stress and that below UU-axis is subjected to tensile bending stress. Due to M . sin θ, the portion of beam section to the left of VV-section is subjected to compressive bending stress and that to the right of VV-axis is subjected to tensile bending stress. Thus, it is seen that point P is subjected to maximum tensile stress. The co-ordinates of point P and R are known. The maximum compressive and maximum tensile bending stresses in a beam section can be found by Eq. 9.20.

The method is suitable for beam section which has at least one axis of symmetry. The axis of symmetry is also a principal axis. The points of extreme distances from principal axes may be seen by visual inspection. When a beam section has no axes of symmetry as shown in Fig. 9.21 (d), it is not possible to locate the points which are at extreme distances by visual inspection, and this method is not convenient to determine the maximum bending stress for such sections.

9.19.2 By Resolving Bending Moment into two Components along any two Perpendicular Axes, Passing through the Centroid of Section

Let the beam section shown in Fig. 9.26 is subjected to a bending moment M. XX and YY are two perpendicular axes and pass through the centroid of the beam section, O. The bending stress at any point can be determined as under:

The plane of loading is inclined at θ with YY-axis as shown in Fig. 9.26. The bending moment M may be resolved into two components

$M \cdot \cos\theta\, M$. sin θ along YY and XX axes respectively.

The bending of beam due to M . cos θ takes place about XX-axis, and that due to M . sin θ along YY and XX axes, respectively.

Then, $$M_{xx} = M . \cos\theta \quad ...(i)$$

and $$M_{yy} = M \cdot \sin\theta \qquad ...(ii)$$

where, M_{xx} is the component of bending moment, M, about XX-axis, and M_{yy} is the component of bending moment, about YY-axis.

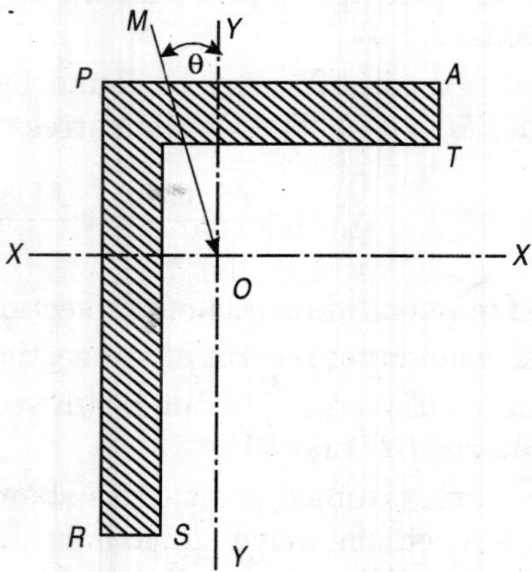

Fig. 9.26 *Angle-section (maximum bending stress) on unsymmetrical bending*

The variation of bending stresses at any point whose co-ordinate (x, y), is linear

then $$\sigma_b = (a_1 \cdot x + a_2 \cdot y) \qquad ...(iii)$$

where, σ_b = Bending stress at point, (x, y)

$a_1\ a_2$ = Any constants

then $$M_{xx} = \int \sigma_b \cdot dA \cdot y \qquad ...(iv)$$

where, dA = Any elementary area of beam section

$$\therefore \quad M_{xx} = \int (a_1 x + a_2 y)\, y dA = a_1 \int xy dA + a_2 \int y^2 dA$$

$$= (a_1 . I_{xy} + a_2 . I_{xx}) \qquad ...(v)$$

where, I_{xx} is the moment of inertia of beam section about XX-axis and I_{xx} is the product of inertia of beam section about XX-axis and YY-axis.

And $$M_{yx} = \int \sigma_b \cdot dA \cdot = \int (a_1 x + a_2 y)\, x dA \qquad ...(vi)$$

$$= a_1 \int x^2 dA + a_2 \int xy dA$$

$$= (a_1 \cdot I_{xy} + a_2 \cdot I_{xx}) \qquad ...(vii)$$

where, I_{yy} is the moment of inertia of beam section about YY-axis.

From (v) and (vii), we have

$$a_1 = \left(\frac{M_{yy} I_{xx} - M_{xx} I_{xy}}{I_{xx} I_{yy} - I_{xy}^2} \right) \qquad ...(viii)$$

$$a_2 = \left(\frac{M_{xx} I_{yy} - M_{yy} I_{xy}}{I_{xx} I_{yy} - I_{xy}^2} \right) \qquad ...(ix)$$

Therefore, substituting from (viii) and (ix) and values of a_1 and a_2 in (iii),

$$\sigma_b = \left(\frac{M_{xx}I_{yy} - M_{xx}I_{xy}}{I_{xx}I_{yy} - I_{xy}^2}\right)\cdot x + \left(\frac{M_{xx}I_{yy} - M_{yy}I_{xy}}{I_{xx}I_{yy} - I_{xy}^2}\right)\cdot y$$

$$\sigma_b = \left(\frac{M_{xx}I_{yy} - M_{yy}I_{xy}}{I_{xx}I_{yy} - I_{xy}^2}\right)\cdot y + \left(\frac{M_{yy}I_{xx} - M_{xx}I_{xy}}{I_{yy}I_{xx} - I_{x}^2}\right)\cdot x \qquad ...(9.21)$$

Equation (9.21) is specially useful for beam sections, in which the web and flanges or legs of angle sections are parallel to *XX* and *YY* axes. The values of bending stress at different corner points are found. The maximum bending stress in the beam section, then be obtained.

9.19.3 By Locating the Neutral Axes of Beam Section

When a beam section is subjected to unsymmetrical bending moment *M*, inclined at θ, with the principal axis *VV* as shown in Fig. 9.25, then the neutral axis is not perpendicular to the plane of bending. The neutral axis of beam section may be located by

(i) *Analytical method* and (ii) *Graphical method.*

(i) Analytical method. The bending stress at any point *X*, is given by Eq. 9.12.The bending stress at any point X_1 on the neutral-axis is zero. Therefore, from Eq. (9.12), we have

$$\left[\frac{M\cdot\cos\theta}{I_{UU}}\cdot V + \frac{M\cdot\sin\theta}{I_{VV}}\cdot u\right] = 0$$

$$v = -u\,.\left(\frac{I_{UU}}{I_{VV}}\right).\tan\theta \qquad ...(9.22)$$

Equation (9.22) is an equation of straight line, and represents the equation of neutral *NN*-axis of the beam section. From this equation it is seen that it passes through the centroid of the section *O*.

Let β be the angle of inclination of neutral *NN*-axis with the principal *xx*-axis. Then, from Fig. 9.25

$$\tan\beta = \left(\frac{v}{u}\right) \qquad ...(i)$$

From Eq. 9.21, we have

$$-\left(\frac{v}{u}\right) = \left(\frac{I_{UU}}{I_{VV}}\right).\tan\theta \qquad ...(i)$$

Substituting the value from this, we have

$$\tan\beta = \left(\frac{I_{UU}}{I_{VV}}\right).\tan\theta \qquad ....(9.23)$$

From Eq. 9.23, it is seen that angle β in general, is not same as θ. If the angle θ is zero, the bending moment *M* will be acting along the principal plane, and the neutral axis will become perpendicular to the plane of bending. The angle β will also be equal to θ in case

$\therefore \quad (I_{UU} = I_{VV}) \therefore (\tan\beta = \tan\theta), \angle\beta = \angle\theta$

In such cases, the ellipse of inertia of the beam section becomes a circle. Any pair of two perpendicular axes passing through the centroid can be considered as principal axes. Therefore, the neutral axis remains perpendicular to the plane of bending.

UU and *VV* axes are the principal. Therefore, for product of inertia I_{UV} about *UU*-axis and *VV*-axis is zero. The moment of inertia of beam section about any axis inclined β with the axis-*UU* is obtained from Eq. 9.15.

Therefore,
$$I_{NN} = (I_{UU} \cdot \cos^2\beta + I_{VV} \cdot \sin^2\beta) \quad \text{...(9.24)}$$

The angle of inclination of plane of loading with the neutral axis-*NN*, is equal to (90 – θ + β) or (90 + β – θ).

The angle of inclination of plane of loading with the line perpendicular to neutral axis-*NN* will therefore be equal to (β – θ).

The component of bending moment, *M*, with the axis

$$= M \cdot \cos(\beta - \theta)$$

If, any point is at a maximum distance, Y_N, from the neutral, axis-*NN*, then, the maximum bending stress

$$\sigma_b = \left[\frac{M \cdot \cos(\beta - \theta)}{I_{NN}}\right] \cdot Y_N \quad \text{...(9.25)}$$

The bending stress will be compressive or tensile depending upon the position of point relative to the neutral axis and direction of loading.

(ii) Graphical method. The radii of gyration about principal axes are known as principal radii of gyration. Let r_{UU} and r_{VV} be the principal radii of gyration about *UU* and *VV* principal axes of a beam-section. The moment of inertia about neutral axis I_{NN} may be found by determining the radius of gyration r_{NN} about it. The radius of gyration r_{NN} about neutral axis-*NN* is found graphically by constructing *momental ellipse.* The momental ellipses is also known as ellipse of *inertia.*

Draw *UU* and *VV* principal axes of a beam section passing through point *O* as shown in Fig. 9.27. The point *O* represents centroid of beam-section. Plot *OA* = r_{UU} on *UU*-axis and *OB* = r_{VV} on *VV*-axis on some suitable scale. Draw two concentric circles with point *O* as centre and, *OA* and *OB* as radii, respectively. Draw *MM* at angle θ, with *VV*-axis. *MM* represents the plane of loading. Draw momental ellipse with *OA* as length of major axis and *OB* is length of minor axis as shown in Fig. 9.27. Draw a line *M′ M′* tangential to the momental ellipse such that it is parallel to *MM.* This line is tangent to the moment ellipse at point *S.* Join points *O* and *S,* and produce it on both sides.

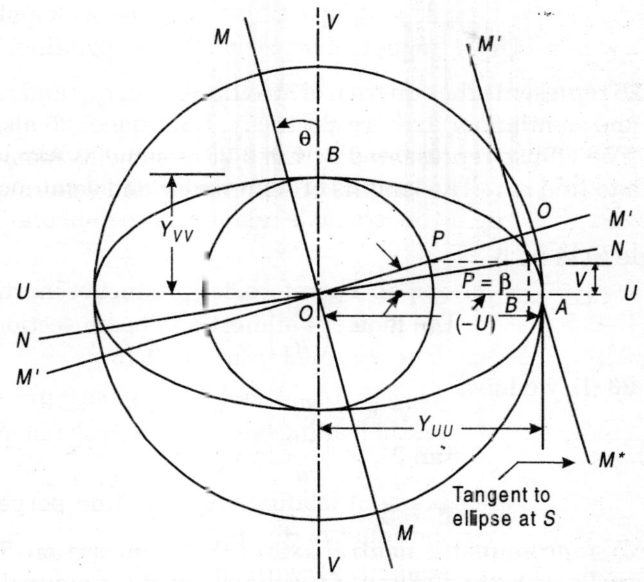

Fig. 9.27 *Momental elllipse*

This locates neutral axis-*NN* of beam section. The radius of gration r_{NN} about the neutral axis, *AW*, is given by *OS*.

Then, $I_{NN} = A.r^2_{NN}$

where, *A* is the area of beam section. This can be proved as under :

Draw *M′M′* perpendicular to *MM* and passing through point *O* as shown in Fig. 9.27. *M′ M′* intersects the concentric circles at points *P* and *Q*, respectively. From point *Q*, draw a perpendicular *QR* to *OA*. From the point *P* draw a perpendicular to *PS* to *QR*.

Let $(-u, v)$ be the co-ordinates of point *S* with reference to *UU* and *VV* principal axes. The co-ordinates of points *S* are as $(-u, v)$. Since co-ordinates of point *X* in Fig. 9.25 are assumed as (u, v) in the positive direction of axes, i.e., *u* and *v* are positive in second quadrant. Then

$$-u = OR = OQ \,.\, \cos\theta = OA \,.\, \cos\theta$$

$$= r_{UU} \,.\, \cos\theta \; (OQ = OA)$$

$$u^2 = r_{UU}^2 \,.\, \cos^2\theta$$

$$y = SR = PT = OP \,.\, \sin\theta = OB \,.\, \sin\theta$$

$$(\because OP = OB)$$

$$= r_{VV} \,.\, \sin\theta$$

$$v^2 = (r_{VV}^2 \sin^2\theta)$$

From (i) and (ii), we have

$$\left(\frac{u^2}{r_{UU}^2} + \frac{v^2}{r_{VV}^2}\right) = (\cos^2\theta + \sin^2\theta)$$

$$\therefore \qquad \left(\frac{u^2}{r_{UU}^2}+\frac{v^2}{r_{VV}^2}\right) = 1 \qquad ...(9.26)$$

Equation 9.26 represents the equation of an ellipse, with r_{UU} and r_{VV} as lengths of semi-major and semi-minor axes, respectively. Equation 9.26 also represents the locus of *S*. The ellipse represented by Eq. 9 26 is same as momental ellipse which is drawn to find r_{NN}. The lengths of semi-major and semi-major axes are the same.

Let the angle *SOA* be β′

$$\tan \beta' = \left(\frac{-v}{u}\right)$$

From Eq. 9.23 (i), we have

$$\tan \beta' = \left(\frac{-v}{u}\right)$$

Therefore, $\beta' = \beta$

Therefore, *NN* represents the neutral axis of the beam section. The line *MM* may be considered as direction of one of the diameter of momentral ellipse. The line *NN* passes through middle point of *MM* and point, *S*, the point on tangent drawn parallel to *MM*. Line *NN* shows the direction of conjugate diameter, therefore, *MM* and *NN* form conjugate diameter of the ellipse.

The slope of *MM* with *UU*-axis is tan (90° + θ) or cot θ and that of *NN* with *UU*-axis is tan β. From the properties of conjugate diameters of an ellipse, we have

$$-\cot\theta \,.\, \tan\beta = \left(\frac{r_{UU}^{\ 2}}{r_{VV}}\right) \qquad ...(i)$$

$$\tan\beta = \left(+\frac{A\cdot r_{UU}^{\ 2}}{A\cdot r_{VV}}\right)\cdot\tan\theta = \left(+\frac{I_{UU}^{\ 2}}{I_{VV}}\tan\theta\right) \qquad ...(ii)$$

$$\therefore \qquad \tan\beta = \left(\frac{-v}{u}\right)$$

Therefore, $$v = -u\left(\frac{I_{UU}^{\ 2}}{I_{VV}}\right)\cdot\tan\theta \qquad ...(iii)$$

This relation obtained is same as Eq. 9.22 and it is equation of the neutral axis. Further from Fig. 9.27,

$(OS)^2 = (OR)^2 + (RS)^2 = (OQ.\ \cos\beta)^2 + (OP\ .\ \sin\beta)^2$ ... (i)

$(\because OQ = OA, OP = OB)$

$(OS)^2 = (OA\ .\ \cos\beta)^2 + (OB\ .\ \sin\beta)^2$

$= (r_{UU}^{\ 2}.\ \cos^2\beta + r_{VV}^{\ 2}\ .\ \sin^2\beta)$ ...(ii)

Multiplying both sides by A, the area of cross-section of beam section, we have

$$A.(OS)^2 = (A.r_{uu}^2 \cos^2 \beta + A.r_{vv}^2 \sin^2 \beta)$$

$$= (I_{UU} \cdot \cos^2 \beta + I_{VV} \cdot \sin^2 \beta) \quad ..(iii)$$

$$I_{VV} = A.r_{UU}^2, I_{VV} = A.r_{VV}^2 \text{ and } I_{VV} = 0$$

From Eq. (9.16), therefore

$$A \cdot (OS)^2 = I_{NN}$$

$$(OS)^2 = \left(\frac{I_{NN}}{A}\right) = r^2_{NN} \quad ...(iv)$$

$$OS = r_{NN} \quad ...(v)$$

After determining the value of radius of gyration graphically the moment of inertia I_{NN} about neutral axis is determined. The maximum bending stress at the extreme distance from the neutral axis may be found from Eq. 9.25.

9.20 THE Z-POLYGON

The strength of a beam section in bending depends upon the modulus of section Z. Along with the geometrical properties of a beam section, the value of modulus of section Z also depends upon plane of bending (or plane of loading). The value of modulus of section Z, varies as the position of the plane of bending changes. The maximum and minimum values of modulus of section, Z, may be found by determining the corresponding positions of plane of bending.

Consider any point A, whose co-ordinates are (u_A, v_A) with reference to principal axes UU and VV. The beam section is subjected to bending moment M. The plane of bending is inclined at θ with VV-axis, as shown in Fig. 9.28. From Eq. 9.20 the bending stress at point is given by

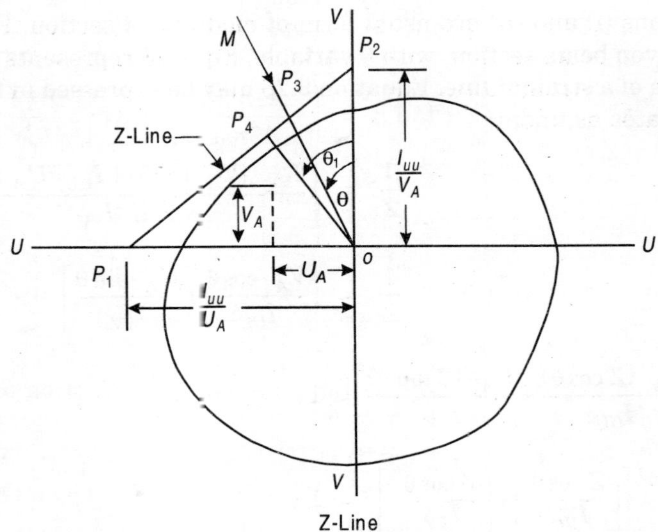

Fig 9.28

$$\sigma_b = \left(\frac{M\cos\theta}{I_{UU}}\right)\cdot v_A + \left(\frac{M\sin\theta}{I_{VV}}\right)\cdot u_A = M\left[\frac{v_A\cdot\cos\theta}{I_{UU}} + \frac{u_A\cdot\sin\theta}{I_{VV}}\right]$$

$$\sigma_b = \left[\frac{I_{VV}\cdot(v_A\cdot\cos\theta) + I_{UU}\cdot(u_A\cdot\sin\theta)}{I_{UU}\cdot I_{VV}}\right]$$

$$\sigma_b = \left[\frac{M}{\dfrac{I_{UU}\cdot I_{VV}}{I_{VV}\cdot(v_A\cdot\cos\theta) + I_{UU}\cdot u_A\cdot\sin\theta}}\right]$$

$$\therefore \qquad \sigma_b = \left(\frac{M}{Z}\right) \qquad \text{...(9.27)}$$

where, Z is the modulus of section of the beam section, and

$$Z = \left[\frac{I_{UU}\cdot I_{VV}}{I_{VV}\cdot(V_A\cdot\cos\theta) + I_{UU}\cdot U_A\cdot\sin\theta}\right] \qquad \text{...(9.28)}$$

In Eq. 9.28, I_{UU} and I_{VV} are principal moment of inertias of the beam section about *UU* and *VV* axes, respectively. For a given beam section, the value of I_{UU} and I_{VV} are constant. For a given point, *A* on the beam section u_A and v_A are also constant. Thus, it is seen that the value of Z depends upon θ only.

When $\qquad (\theta = 0°),\ Z = \left(\frac{I_{UU}}{I_{VV}}\right) \qquad$ ...(i)

and when $\qquad (\theta = 90°),\ Z = \left(\frac{I_{UU}}{U_A}\right) \qquad$ ... (ii)

Expressions (i) and (ii) are usual form of modulus of section. For any given point of a given beam section, with θ variable, Eq. 9.28 represents polar form of the equation of a straight line. Equation 9.28 may be expressed in the rectangular co-ordinates as under:

$$\frac{1}{Z} = \left[\frac{I_{VV}\cdot(V_A\cdot\cos\theta) + I_{UU}\cdot U_A\cdot\sin\theta}{I_{UU}\cdot I_{VV}}\right] \qquad \text{...(iii)}$$

or

$$\frac{1}{Z} = \left[\frac{V_A\cdot\cos\theta}{I_{UU}} + \frac{U_A\cdot\sin\theta}{I_{VV}}\right] \qquad \text{...(iv)}$$

or

$$\left[\frac{V_A\cdot(Z\cos\theta)}{I_{UU}} + \frac{U_A\cdot(Z\sin\theta)}{I_{VV}}\right] = 1 \qquad \text{...(v)}$$

or

$$\left[\left(\frac{Z\cos\theta}{\dfrac{I_{UU}}{V_A}}\right) + \left(\frac{Z\cos\theta}{\dfrac{I_{VV}}{U_A}}\right)\right] = 1 \qquad \text{...(vi)}$$

Put $Z \cdot \cos\theta = v$, and $Z \cdot \sin\theta = u$

Then $$\left[\left(\frac{u}{\frac{I_{VV}}{u_A}}\right)+\left(\frac{u}{\frac{I_{UU}}{V_A}}\right)\right] = 1 \qquad ...(9.29)$$

This represents equation of a straight line, which is known as Z-line. The length of interception on *UU*-axis is $\left(\frac{I_{VV}}{u_A}\right)$ and the length of interception on *VV*-axis as $\left(\frac{I_{UU}}{u_A}\right)$.

Equation 9.29 can also be expressed as below :

$$\left(\frac{v}{\frac{I_{UU}}{v_A}}\right) = -\left(\frac{u}{\frac{I_{VV}}{u_A}}\right) + 1 \qquad ...(i)$$

$$v = -u\left(\frac{u_A}{v_A}\right)\cdot\left(\frac{I_{UU}}{I_{VV}}\right)+\left(\frac{I_{UU}}{v_A}\right) \qquad ...(ii)$$

When the plane of bending passes through the point *A*,

$$\left(\frac{u_A}{v_A}\right) = \tan\theta$$

Therefore, $$v = -u\left(\frac{I_{UU}}{I_{VV}}\right).\tan\theta + \left(\frac{I_{UU}}{v_A}\right) \qquad ...(9.30)$$

By comparing Eq. 9.22 and Eq. 9.30 for the plane of bending passing through given point, it is seen that Z-line is parallel to the neutral axis.

The *Z*-line represented by Eq. 9.29 is plotted by taking $OP = \left(\frac{I_{UU}}{u_A}\right)$ and $OP_2 = \left(\frac{I_{UU}}{u_A}\right)$ along *UU*-axis and *VV*-axis respectively. The points P_1 and P_2 are joined. The line P_1P_2 represents *Z*-line. The *Z*-line intersects the plane of bending at P_3. OP_3 represents the modulus of section, *Z*. the bending stress at point *A* is given by

$$\sigma_b = \left(\frac{M}{OP_3}\right)$$

Let OP_4 represent perpendicular from point *O*, the centroid of beam section to the Z-line. The minimum distance from point *O* to Z-line is given by perpendicular distance OP_4 Then OP_4 represents the minimum value of *Z*. Let

θ_1 be the angle of OP_4 measured from OV-axis. The bending stress at point A is maximum when the plane of loading is inclined at 1 to principal axis W.

The maximum bending stress at the point A

$$\sigma_{b(\max)} = \left(\frac{M}{OP_4}\right) \text{ where } \theta = \theta_1.$$

The maximum distance of any point on Z-line from the centroid O, of the beam section, depends, upon the lengths of intercepts on UU-axis and VV-axis. If OP_2 is greater than OP_1, then OP_2 is the maximum distance. The maximum value of modulus of section Z is given by OP_2. The bending stress at point O is minimum. The plane of bending coincides with the principal axis VV.

Minimum bending stress at point A

$$\sigma_{b(\min)} = \left(\frac{M}{OP_2}\right)$$

where $\theta = \theta$

$$OP_2 = \left(\frac{I_{VV}}{v_A}\right)$$

It is necessary to draw Z-lines for critical points, (the points which are at extreme distance from C.G. of beam section) in order to determine maximum bending stress. The Z-lines drawn critical points form a polygon. This polygon is called Z-polygon. The values of Z for these critical points for a given plane of bending may be compared. The maximum bending stress occurs at a critical point having least value of Z. The Z-polygon for a rectangular beam section is drawn as below:

The principal axes UU and VV are drawn passing through point O, the centroid of rectangular beam. The rectangular beam section $ABCD$ is drawn as shown in Fig. 9.29. Let the width of section be b and depth of section be h. The rectangular beam section $ABCD$ is drawn as shown in Fig. 9.29. The rectangular section is symmetrical section. Therefore, co-ordinate axes XX and YY coincide with the principal axes of the section.

Z-line for point A is drawn first. The co-ordinates of point A, lying in first quadrant are $u_A = \frac{1}{2} b . v_A = \frac{1}{2} h$.

$$I_{UU} = I_{xx} = \left(\frac{1}{12}\right) bh^3 \text{ and } I_{VV} = I_{yy} = \left(\frac{1}{12}\right) hb^3$$

The length of intercepts of Z-line along UU-axis is

$$OP_1 = \left(\frac{I_{VV}}{u_A}\right) \text{ and } OP_1 = \left[\frac{\left(\frac{1}{12}\right) hb^3}{\frac{1}{2} b}\right] = \left(\frac{1}{6}\right) hb^2$$

$$= Z_{yy} = Z_{VV} \qquad (\because Z_{yy} = Z_{VV})$$

The length of intercepts of Z-line along VV axis is

$$OP_2 = \left[\frac{\left(\frac{1}{12}\right)bh^3}{\frac{1}{2}h}\right] = \left(\frac{1}{6}\right)bh^2 = Z_{xx} = Z_{UU}$$

$$(\because Z_{xx} = Z_{UU})$$

Equation 9.29 becomes

$$\left[\frac{u}{Z_{VV}} - \frac{v}{Z_{UU}}\right] = 1$$

On UU-axis, plot $OP_1 = \left(\frac{1}{6}\right) hb^2$ and on VV-axis plot $OP_2 = \left(\frac{1}{6}\right)bh^2$ to a suitable scale. Points P_1 and P_2 are joined. P_1P_2 represents Z-line for the point A. Similarly, Z-lines P_2P_3, P_3P_4 and P_4P_1 are drawn for points B, C and D respectively. $P_1P_2P_3P_4$ represents Z-polygon. The Z-polygon for a rectangular beam section is a rhombus.

The position of plane of bending for maximum and minimum strength of a section can be found by inspection of Z-polygon. For example, for minimum strength of the section, the plane of bending is along VV-axis.

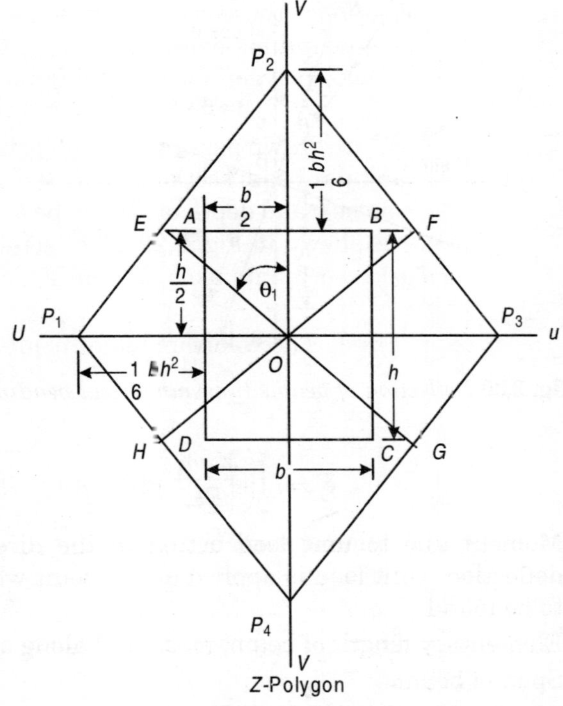

Fig. 9.29

The value of Z is maximum along this line. Similarly, for minimum strength of the section, perpendiculars OE, OF, OG and OH are drawn from point, O, to Z-lines as shown in Fig. 9.29. the strength of beam section is minimum along EG and FH planes. The values of Z for plane bending along these planes is minimum. The inclination of plane EG is θ_1. The value of θ_1 is obtained by

$$\tan\theta_1 = \left(\frac{OP_2}{OP_1}\right) = \left(\frac{Z_{UU}}{Z_{VV}}\right) = \left(\frac{\frac{1}{6}\cdot bh^2}{\frac{1}{6}\cdot hb^2}\right) = \frac{h}{b}, \qquad \therefore \theta_1 = \tan^{-1}\left(\frac{h}{b}\right)$$

9.21 DEFLECTION OF BEAMS UNDER UNSYMMETRICAL BENDING

The plane of bending or the plane of loading is inclined at an angle θ, with principal axis-VV as shown in Fig. 9.30. The bending moment M_1, acting along the plane of bending may be resolved into two components along W-axis and UU-axis, respectively. The deflection of beam subjected to unsymmetrical bending may be found as under :

From unit load method, the deflection of beam subjected to moment M_1, is given by

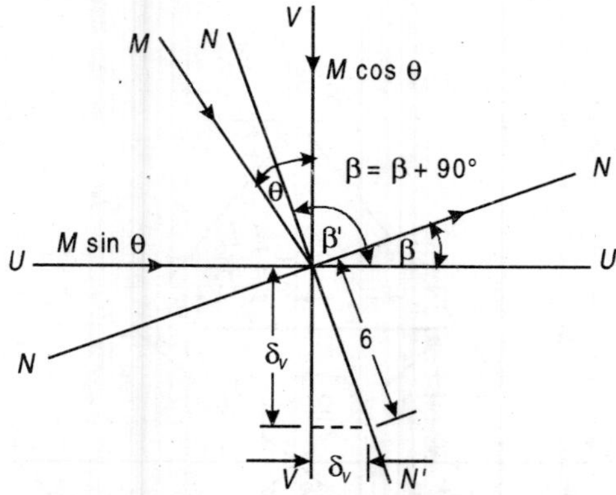

Fig. 9.30 *Deflection of beams (unsymmetrical bending)*

$$\delta = \int_o^L \left(\frac{M_1 m_1}{EI}\right) dx$$

where, m_1 = Moment due to unit load acting in the direction of desired deflection. Unit load is applied at the point where deflection is to be found

dx = Elementary length of beam, measured along span of beam

L = Span of beam.

The deflection in the direction of VV-axis is given by

$$\delta_V = \int_o^L \left(\frac{M\cos\theta}{EI_{UU}}\right)\cdot m_v \cdot dx \qquad ...(i)$$

The deflection in the direction of UU-axis is given by

$$\delta_V = \int_o^L \left(\frac{M\sin\theta}{EI_{VV}}\right)\cdot m_u \cdot dx \qquad ...(ii)$$

The resultant deflection is found by adding these deflections vectorially. Therefor, resultant deflection

$$\delta = \left[(\delta_U)^2 + (\delta_V)^2\right]^{1/2} \qquad ...(iii)$$

$$\delta = \int_o^L \left(\frac{M_m}{E}\right)\left[\left(\frac{\cos\theta}{I_{UU}}\right)^2 + \left(\frac{\sin\theta}{I_{VV}}\right)^2\right]^{1/2} \qquad ...(iv)$$

Since $\qquad m_V = m_U = m$

or

$$\delta = \int_o^L \left(\frac{M_m}{E}\right)\left[\frac{I_{UU}{}^2\cos\theta + I_{UU}{}^2\sin\theta}{I_{UU}{}^2 I_{VV}{}^2}\right]^{1/2} dx \qquad ...(9.31)$$

Let β' be the angle which the resultant deflection makes with the UU-axis.

$$\tan\beta' = \left(\frac{-\delta_U}{\delta_V}\right) = \frac{\int_o^L \left(\frac{M_m\sin\theta}{EI_{VV}}\right)dx}{\int_o^L \left(\frac{M_m\cos\theta}{EI_{UU}}\right)dx}$$

or

$$\tan\beta = -\left(\frac{I_{UU}}{I_{VV}}\right)\tan\theta \qquad ...(9.32)$$

From Eq. 9.23, we have

$$\tan\beta = +\left(\frac{I_{UU}}{I_{VV}}\right)\tan\theta$$

By comparing Eq. 9.23, and Eq. 9.23, it is seen that

$$\tan\beta' = \tan(90° + \beta) \text{ and } \beta' = 90° + \beta$$

Therefore, it is seen that the resultant deflection occurs in the direction $N'N'$, which is perpendicular to neutral axis-NN for any given plane of loading.

If the beam is subjected to uniformly distributed load w, then

$$\delta_{UU} = \frac{5}{384}\left(\frac{w\cdot\sin\theta\cdot L^4}{EI_{VV}}\right) \qquad ...(i)$$

$$\delta_{VV} = \frac{5}{384}\left(\frac{w\cdot\cos\theta\cdot L^4}{E\cdot I_{UU}}\right) \qquad ...(ii)$$

The resultant defletion of the beam

$$\delta = \left[(\delta_{UU})^2 + (\delta_{VV})^2\right]^{1/2} \qquad \text{...(iii)}$$

$$\delta = \frac{5}{384}\frac{wL^4}{E}\left[\frac{\sin^2\theta}{I^2_{VV}} + \frac{\cos^2\theta}{I^2_{UU}}\right]^{1/2}$$

$$= \frac{5}{384}\left(\frac{wL^4\cos\theta}{E \cdot I_{UU}}\right)\left[1 + \left(\frac{I_{UU}}{I_{VV}}\tan\theta\right)^2\right]^{1/2}$$

...(iv)

From Eq. 9.23, $\tan\beta = \left(+\frac{I_{UU}}{I_{VV}}\tan\theta\right)$

Therefore, $\delta = \frac{5}{384}\left(\frac{wL^4\cos\theta}{E \cdot I_{UU}}\right)\left[1 + \tan^2\beta\right]^{1/2}$ ...(v)

or $= \frac{5}{384}\cdot\left(\frac{wL^4}{EI_{UU}}\right)\cdot\cos\theta\cdot\sec\beta = \frac{5}{384}\cdot\left(\frac{wL^4}{EI_{uu}}\right)\cdot\frac{\cos\theta}{\cos\beta}$

or $= \frac{5}{384}\cdot\left(\frac{wL^4}{EI_{uu}}\right)\cdot\left(\frac{\cos\theta}{\cos\beta}\right)\cdot\left(\frac{\cos(\beta-\theta)}{\cos(\beta-\theta)}\right)$

or $= \frac{5}{384}\cdot\left(\frac{wL^4}{EI_{uu}}\right)\cdot\left(\frac{\cos\theta.\cos(\beta-\theta)}{\cos\beta[\cos\beta\cos\theta + \sin\beta\sin\theta]}\right)$

$$= \frac{5}{384}\cdot\left(\frac{wL^4}{EI_{UU}}\right)\cdot\left(\frac{\cos(\beta-\theta)\cos\theta}{\cos\beta\cdot\cos\beta\cos\theta[1 + \tan\beta\tan\theta]}\right) \qquad \text{...(vi)}$$

From Eq. 9.23, $\tan\beta = \left(\frac{I_{UU}}{I_{VV}}\right)\cdot\tan\theta$

Value of $\tan\theta$ is substituted in (vi)

$$\therefore \quad \delta = \frac{5}{384}\cdot\left(\frac{wL^4}{EI_{UU}}\right)\left(\frac{\cos(\beta-\theta)}{\cos^2\beta\cdot\left[1 + \frac{I_{UU}}{I_{VV}}\tan^2\beta\right]}\right)$$

$$\text{or } \delta = \frac{5}{384}\cdot\left(\frac{wL^4}{EI_{UU}}\right)\left(\frac{\cos(\beta-\theta)}{\cos^2\beta\cdot\left[\frac{I_{UU}\cos^2\beta + I_{VV}\sin^2\beta}{I_{UU}\cdot\cos^2\beta}\right]}\right)$$

$$\delta = \left(\frac{wL^4\cos(\beta-\theta)}{E\left[I_{UU}\cdot\cos^2\beta + I_{VV}\sin^2\beta\right]}\right) \qquad \text{...(viii)}$$

From Eq. 9.24, $$I_{NN} = \left(I_{UU} \cdot \cos^2\beta + I_{VV} \sin^2\beta\right)$$

$$\therefore \qquad \delta = \frac{5}{384}\left(\frac{w \cdot \cos(\beta - \theta)L^4}{E \cdot I_{NN}}\right) \qquad \text{...(9.33)}$$

In Eq. 9.33, $w \cos(\beta - \theta)$ is the component of resultant uniformly distributed load along the direction $N'N'$ perpendicular to the neutral axis NN, for plane bending (or plane loading) inclined θ with VV-axis.

9.22 DESIGN OF PURLINS SUBJECTED TO UNSYMMETRICAL BENDING

The rectangular section, I-section, channel section and angle section are used for purlins of roof trusses. The rectangular section and I-section have two axes of symmetry. The channel section has at least one axis of symmetry. The axes of symmetry locate the principal axes in case of rectangular section, I-section and channel section.

All the purlins are designed in accordance with the requirements for uncased beams. However, the limitations of bending stress based on lateral instability of the compression flange and limitation of deflection are waived for the design of purlins. The calculated deflections should not exceed those permitted for the type of roof cladding used. In calculating the bending moment, the advantage may be taken of the continuity of the purlin over the supports. The bending stresses about the two axes should be determined separately and checked by the combined interaction equations.

The design of purlins, in cases where principal axes are located directly from axes of symmetry of a section, is done as follows :

The span of the purlin is centre to centre distance between adjacent trusses, (i.e., it is equal to the spacing of trusses).

The spacing of purlins should should not exceed 1.4 m so that ACC sheets do not fail.

Step 1. Roof sheets are supported by the purlins. These root sheets transfer the loads to the purlins. The various loads acting over the purlins are weight of roof-sheets and fixtures, self-weight of the purlins, live load from the roof-sheets and the wind load. Except the wind loads, other loads are the gravity loads. These gravity loads act vertically. The vertical loads and wind load acting over the purlins are shown in Fig. 9.31. The vertical load acts downward and the wind load acts normal to the sheets either inward or outward depending upon the wind pressure or suction. The purlins are connected to the rafters of the roof trusses. The major principal axis of the purlin-UU remains parallel to the principal rafter. Therefore, the vertical load remains inclined with the major principal axes. As a result of which unsymmetrical bending (that is, biaxial bending) of the purlins occur.

The appropriate combination of loads on purlins are resolved into two components perpendicular to and parallel to the roof slope.

Step 2. The principal axes *UU* and *VV* of the section coincide with the co-ordinate axes *XX* and *YY* respectively. The maximum bending moments are found about these axes. Let the maximum bending moment about *UU*-axis (i.e., *XX*-axis) be M_{UV} and that about *VV*-axis (i.e.,*YY*-axis) *be* M_{VV}.

Step 3. The maximum bending stress at any point on a purlin section is given by

$$\sigma_b = \left(\frac{M_{UU}}{I_{VV}}\right)\cdot u + \left(\frac{M_{VV}}{I_{VV}}\right)\cdot u \quad \text{...(i)}$$

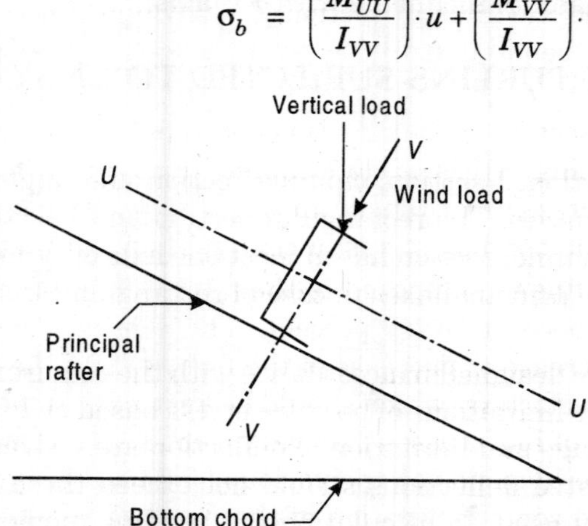

Fig. 9.31 *Channel section purlin*

If *M* is the maximum bending moment caused due to resultant loading, and if a plane of bending (i.e., plane of loading) is inclined at θ then,

$$M_{UV} = M\cos\theta \text{ and } M_{VV} = M\sin\theta$$

Expression (i) is in the same form as Eq. 9.20. In above expression

$$I_{UU} = I_{XX} \text{ and } I_{VV} = I_{YY}$$

Expression (i) can be written as

$$\sigma_b = \left[\frac{M_{UU}}{\frac{I_{UU}}{v}} + \frac{M_{VV}}{\frac{I_{VV}}{u}}\right] = \left(\frac{M_{UU}}{Z_{UU}}\right) + \left(\frac{M_{VV}}{Z_{VV}}\right)$$

or
$$\sigma_b = \frac{M_{UU}}{Z_{UU}}\cdot\left(1 + \frac{M_{VV}}{M_{UU}}\frac{I_{UU}}{Z_{VV}}\right) \quad \text{...(9.34)}$$

or
$$Z_{UU} = \frac{M_{UU}}{\sigma_b}\left(1 + \frac{M_{VV}}{M_{UU}}\frac{Z_{UU}}{Z_{VV}}\right) \quad \text{...(9.35)}$$

In Eq. 9.35 it is to note that Z_{UU} and Z_{VV}, the moduli of section about *UU*-axis and *VV*-axis, respectively, are the same as Z_{xx} and Z_{yy}, the moduli of section about *XX*-axis and *YY*-axis, respectively. In Eq. 9.35 suitable value of ratio $\left(\frac{Z_{uu}}{Z_{vv}}\right)$ is assumed in order to determine, modulus of section Z_{UU}. For rectangular section, the value of $\left(\frac{Z_{UU}}{Z_{VV}}\right)$ may be assumed as 5. For I-sections, this ratio may be assumed 5 to 7 for lightly loaded purlins, and 7 to 10 for heavily loaded purlins. For channel section, the value of $\left(\frac{Z_{UU}}{Z_{VV}}\right)$ may be assumed as 8. From Eq. 9.35 the required modulus of section Z_{UU} is determined by assuming σ_b, equal to permissible bending stress. A trail section may be selected from ISI Handbook No. 1.

Step 4. This trial section for purlins is checked for maximum bending stress. The maximum bending stress is determined from Eq. 9.34. The maximum bending stress compressive or tensile in beam section should not exceed the maximum permissible bending stress in compression or in tension.

The weight of purlins as compared to the total weight of the steel structure may vary from 10 percent to 25 percent. The economic impact (i.e., cost), durability of roof, increase or decrease of the fixtures and the vibrations of roof sheets are the various factors which should be kept in view while designing the purlins. The spacing of roof trusses may vary from 3 m to 12 m. The economical spacing of roof trusses depends on the various factors (e.g., type of soil, bearing capacity of soil, wind zone, inclination of the roof and the functional aspects). The various sections (e.g., channels, angles, tubes, Z-shape or cold formed, Warren or N-type truss, triangular or open section) may be used for the purlins. The channels or angles should be so placed that the accumulation of dust, moisture and water is minimum or avoided. The purlin sections should be connected with the principal rafters of the roof trusses in their correct orientations. For example, the flanges in case of channels may be kept sloping downward with the web normal to the rafter and upward. So also, in case of the angle sections, one leg is kept and connected normal to the rafter while the other leg is kept sloping downward. The durability, that is, the life of purlin reduces due to incorrect position (i.e., wrong orientation) of the purlin sections.

9.22.1 Design of Angle Iron Purlins

The angle-section are most commonly used for purlins. The angle sections do not have any symmetric axis. Therefore it becomes essential to locate the principal axes of the angle-section. The principal axes may be located, if the geometrical properties of angle section are known. Depending upon the resultant load, the trial section is selected for purlins. Then, general procedure for design of purlins mentioned above may be followed.

If the slope of roof truss is not greater than 30° and the steel is conforming to Grades Fe 410-O or Fe 410-W, then, the angle purlins may be designed as an alternate to the general design procedure, as recommend by IS: 800–1984. However, the following requirements which are based on a maximum imposed load of 0.75 kN/m^2 are fulfilled.

Step 1. The depth of angle purlin in the plane approximate to the incidence of the maximum load or maximum component of the load is not less than $\left(\frac{1}{45}\right)$th of the length of purlin.

Step 2. The width of angle purlin is not less than $\left(\frac{1}{60}\right)$th of the length of purlin.

Knowing the depth of purlin, and width of purlin, a trial angle selection is selected from ISI Hand-book No. 1.

Step 3. The maximum bending moment in a purlin is taken as

$$M_{max} = \left(\frac{WL}{10}\right)$$

where, W is total uniformly distributed load on the purlin including wind load and L is the distance centre to centre of the rafters or other supports of the purlins.

The loads are assumed as acting normal to the roof surface. The bending of the purlin about the minor axis is neglected. In calculating the bending moment, the advantage of continuity of the purlins over roof trusses has been taken.

Step 4. The maximum bending stress calculated for M_{max} should not exceed the maximum permissible bending stress in compression or in tension. The design or purlins takes into account the wind load. Therefore, the permissible stress in bending in compression or tensile are 1.33 times the usual permissible values, (viz, dead load plus live load, dead load plus half the live load plus the wind load or the dead load and wind load) shall be considered. It is to note that the dead load and live load cause biaxial bending, whereas the wind load produces bending about major principal axis. The bending moments due to dead loads, due to live load and due to wind load are calculated.

These purlins may be treated as continuous beams supported over principal rafters of the roof trusses. The moments at the supports and mid-spans of the end and intermediate spans may be calculated as recommended in the code.

For end span (one end simply supported and other end continuous)

(a) Bending moent due to dead loads

$$\text{mid-span } M = +\left(w_d \cdot \frac{L^2}{16}\right), \text{ and} \qquad \text{...(i)}$$

$$\text{support } M = \left(w_d \cdot \frac{L^2}{9}\right) \qquad \text{...(ii)}$$

(b) Bending moment due to live loads

$$\text{mid-span } M = +\left(w_L \cdot \frac{L^2}{9}\right) \text{ and} \qquad \text{....(iii)}$$

$$\text{support } M = \left(w_L \cdot \frac{L^2}{9}\right) \qquad \text{...(iv)}$$

It is to note that for the intermediate spans, the bending moments are less than these values. It is recommended to use the value $\left(w_L \dfrac{L^2}{10}\right)$.

Following guidelines may be kept in view while selecting a particular section for the purlins. These guide lines will be advantageous in selecting the purlin sections.

1. The angle or channel sections may be used suitably and economically upto 5 m span without sag bars except in heavy snow zones.
2. The tubular sections may be used upto 6 m span. A minimum of 3 mm thickness of the tube is essential for the structural welding. In case the thickness of tube is small, it will be rusted soon. As far as possible, tubes with thickness more than 3.5 mm may be used.
3. In case of snow zones, sag bars are used for purlins for 4 m and more spans for angle sections and channel sections. The channel sections even sag bars for purlins of spans beyond 8 m are not suitable.
4. For purlins of spans 3 m and more, the triangulation or truss purlins may be used economically. However, their fabrication is slightly difficult.

The durability, maintenance aspect fabrications and the erection convenience should also be considered.

Example 9.1 *A roof truss-shed is to be built in Jodhpur city area for an industrial use. Determine the basic wind pressure. The size of shed 18 m × 30 m.*

Solution

Step 1: Basic wind speed

From IS : 875 (Part 3) – 1987 (from wind zone map of the country or Table 9.2, the basic wind speed in Jodhpur,

$$V_d = 47 \text{ m/sec} \qquad \text{...(i)}$$

Step 2: Risk coefficient (factor – k_1)

The design life for roof shed for industrial use may be assumed 50 years. From IS : 875 (Part 3) – 1987 or Table 9.3, the risk coefficient (factor – k_1), for V_b = 47 m/sec,

$$k_1 = 100 \qquad \text{...(ii)}$$

Step 3 : Terrain, height and structure size factor (factor-k_2)

The terrain is Jodhpur city industrial area. Therefore, the area is of type *category* 3. The size of shed shall be 18 m × 30 m. Therefore, the maximum horizontal dimension is between 20 m and 50 m. Therefore, the roof shed structure is of type class *B*. In category 3, the terrain with numerous closely spaced obstructions having the size of building structures upto 10 m in height with or without a few isolated tall structures, therefore, from IS : 875 (Part 3) 1987 or Table 9.4,

$$k_2 = 0.98 \quad \text{...(iii)}$$

Step 4 : Topography factor, k_3

Near the roof-shed, the ground is assumed to be plain. Therefore, the topography factor from Eq. 9.6,

$$k_3 = (1 + C \, . \, S)$$

where $C = \left(\frac{Z}{L}\right) = 0$

$$k_3 = 1.00 \quad \text{...(iv)}$$

Step 5 : Design wind sped (Eq. 9 3)

$$V_z = k_1 \, . \, k_2 \, . \, k_3 \, . V_b$$

$$V_z = (1 \times 0.98 \times 1 \times 47)$$

$$= \mathbf{46.06 \ m/sec.}$$

Step 6 : Basic wind pressure (from Eq. 9.12)

$$p_z = 0.0006 \times V_z^2 \text{ kN/m}^2$$

$$= 0.0006 \times (46.06)^2$$

$$= \mathbf{1.2729 \ kN/m^2}$$

Example 9.2 *An industrial roof shed of size 20 m × 30 m is proposed to be constructed at Manglore near a hillock of 160 m and the slope is 1 in 2.8. The roof shed is to be built at a height of 120 m from the base of the hill. Determine the design wind pressure on the slope. The height of roof shed shall be 12 m.*

Solution

Step 1: Basic wind speed

From IS : 875 (Part 3) – 1987 or Table 9.2, tne basic wind speed V_b is 39 m/sec.

Step 2: Risk coefficient (factor-k_1)

For 50 years of design life of the structure and all general buildings for V_b = 39 m/sec, *risk coefficient,*

$$k = 100 \quad \text{...(i)}$$

Step 3: Terrain factor, k_2

Near the hillock side, the terrain shall be open with well scattered obstructions having heights generally batween 15 m to 10 m. Therefore, the terrain is of type *category* 2. The size of roof shed is 20 m × 30 m. Therefore, it is of type *class B*. (since maximum horizontal dimension is between 20 m and 50 m). The height of roof-shed shall be 12 m. Therefore, the *terrain, height and structure size* factor, from IS : 875 (Part 3) – 1987 or Table 9.2

$$k_2 = \left[0 \cdot 98 + \frac{(1 \cdot 02 - 0 \cdot 98) \times 2}{5}\right] = 0.996$$

Step 4 : Topography factor k_3

The effective height of the hill, z is 160 m.

The slope is 1 in 2.8. Therefore,

$$\theta = \tan^{-1}\left(\frac{1}{2\cdot 8}\right) = 19.654°$$

Length of upward slope

$$L = (160 \times 2.8) = 448 \text{ m}$$

For $\theta > 17°$, effective horizontal length of the hill

$$Le = \left(\frac{Z}{0\cdot 3}\right) = \left(\frac{160}{0\cdot 3}\right) = 533.33 \text{ m}$$

For ($\theta = 19.654°$) > 17°, from IS : 875 (Part 3) – 1987, Appendix C.

$$C = 0.36$$

$$k_3 = (1 + C\,.\,S)$$

Let x be the horizontal distance of the building from the crest. It is measured *positive* towards leeward side and negative towards the windward side.

$$x = -(L - 120)$$

$$= -(448 - 120) = -328 \text{ m}$$

$$\frac{x}{Le} = \left(\frac{-328}{533\cdot 33}\right) = -0.615$$

$$\frac{H}{Le} = \left(\frac{12}{533\cdot 33}\right) = 0.0225$$

From Fig. 9.6 (b), for $\left(\frac{x}{Le}\right) = -0.615$ and $\left(\frac{H}{Le}\right) = 0.0225$

$$S = 0.2$$

From Eq. 9.6, for (b) = 19 654°) > 17; $c = 0.36$

$$k_3 = (1 + C.S)$$

$$k_3 = (1 + 0.36 \times 0.2) = 1.072$$

Step 5: Design wind speed (from Eq. 9.3)

$$V_z = k_1.k_2.k_3.V_b$$

$$V_z = (1 \times 0.996 \times 1.072) \times 39$$

$$= \mathbf{41.641 \ m/sec}$$

Step 6: Design wind pressure (from Eq. 9.12)

$$p_z = 0.0006\, V_2^{\,2} \text{ kN/m}^2$$

$$p_z = (0.0006 \times 41.641^2) = \mathbf{1.04 \ m/sec}$$

Example 9.3 *A hospital building of size 50 m × 100 m and a height 10 m is proposed to be built at Poona on a hill top. Determine the design wind pressure on the building. The height of hill is 320 m with a slope of 1 in 4. The hospital is proposed at a distance 120 m from the crest on the downward slope.*

Solution

Step 1: Basic wind speed: From IS : 875 (Part 3) – 1987 or Table 9.2, the basic wind speed at Poona is 39 m/sec.

Step 2: Risk coefficient (factor –k_1)

For important buildings and structures, such as hospitals for which the design life be assumed 100 years, the risk coefficient from IS : 875 (Part 3) –1987 or Table 9.3, for V_b = 39 m/sec

$$k_1 = 1.06 \quad \text{...(i)}$$

Step 3: Terraine, height and structure size factor, k_2

The height of hospital shall be 10 m. The open terraine with well scattered obstruction having heights generally between 1.5 m to 10 m belongs to *category* 2. The size of hospital dimension is more than 50 m. As such, it belongs to *class* (c). From IS : 875 (Part 3) – 1987 for Table 9.4, for 10 m height

$$k_2 = 0.93 \quad \text{...(ii)}$$

Step 4: Topography factor k_3

The effective height of the hill, Z is 320 m. The slope is θ = 1 in 4

$$\theta = \tan^{-1}\left(\frac{1}{4}\right) = 14.036°$$

Length of upward slope

$$L = (320 \times 4) = 1280 \text{ m}$$

Effective horizontal length of the hill, for (θ = 14.036°) < 17°,

$$Le = L = 1280 \text{ m}$$

Value of C for θ 17°

$$C = 1.2 \times \left(\frac{Z}{L}\right) = \left(\frac{1\cdot 2 \times 320}{1280}\right) = 0.3 \quad \text{...(iv)}$$

Height of the hospital building shall be 10 m. Let x be the distance from the crest, (then, x = + 120 m). Non-dimensional factors

$$\left(\frac{x}{Le}\right) = \left(\frac{120}{1280}\right) = +0.09375$$

$$\left(\frac{H}{Le}\right) = \left(\frac{10}{1280}\right) = +0.0078$$

From Fig. 9.6 (b), the value of S for ridge and hill

$$S = 1.00 \quad \text{...(v)}$$

From Eq. 9.6,

$$k_3 = (1 + C.S) = (1 + 0.3 \times 1.00) = 1.3 \quad \text{...(vi)}$$

Step 5: Design wind speed (from Eq. 9.3)

$$V_z = k_1 . k_2 . k_3 . V_b$$

$$V_z = [(1.06) . (0.93) . (1.3) . (39)]$$

$$= \mathbf{49.98 \ m/sec}$$

Step 6 : Design wind pressure (from Eq. 9.12)

$$p_z = 0.0006 \,.\, V_z^2$$

$$p_z = 0.0006 \times (49.98)^2 = \mathbf{1.4988\ kN/m^2}$$

Example 9.4 *A communication tower of 80 m height is proposed to be built over hill top of height 520 m with a gradient of 1 in 5. The horizontal approach distance is 2.8 km from the level ground. The tower is proposed at Abu mount. Determine the design wind pressure.*

Solution

Step 1 : Basic wind speed: From IS : 875 (Part 3) – 1987, or Table 9.2. the basic wind speed at Abu mount (wind zone 5),

$$V_b = 39 \text{ m/sec} \quad ..(i)$$

Step 2 : Risk coefficient (factor – k_1)

For important buildings and structures such as the communication towers, the design life may be assumed 100 years. From IS : 875 (Part 3) – 1987, or Table 9.3, for V_b = 39 m/sec.

$$k_1 = 1.06 \quad ...(ii)$$

Step 3 : Terraine, height and structure size factor, k_2

The height of communication tower is 80 m. The structure shall be having maximum vertical dimension more than 50 m. The structure belongs to *class* $\left(\frac{C}{1}\right)$. The exposed open terraine with few or no obstruction and in which the average height of any object surrounding the structure is less than 15 m. The terraine belongs to *category-1.* The terraine factor

$$k_2 = \left[1 \cdot 14 + \frac{(1 \cdot 20 - 1 \cdot 14) \times 30}{50}\right] = 1.176$$

Step 4 : Topography factor k_3

The effective height of the hill, Z is 520 m. The slope is 1 in 5

$$\theta = \tan^{-1}\left(\frac{1}{5}\right) = 11.31°$$

The length of upward slope, *L* is 2.800 km, that is, 2.800 m. The effective length of hill,

for $\theta = 11.31° < 17°$, $Le = L = 2800$ m

Value of *C* for θ 17°

$$C = 1.2 \times \left(\frac{Z}{L}\right) = \left(\frac{1 \cdot 2 \times 520}{2800}\right) = 0.222857$$

For $H = 80$ m, and $x = 0$,

$$\left(\frac{x}{Le}\right) = 0 \text{ and}$$

$$\left(\frac{H}{Le}\right) = \left(\frac{80}{2800}\right) = 0.02857$$

From Fig. 9.6 (b), or IS : 875 (Part 3) – 1987,

$$S = 1.00$$
$$k_3 = (1+C.S)$$
$$k_3 = (1 + 0.22857 \times 1) = 1.222857$$

Step 5: Design wind speed (from Eq. 9.3)

$$V_z = k_1 . k_2 . k_3 . V_b$$
$$V_z = (1.06) . (1.176) . (1.222857) \times (39))$$
$$= \mathbf{59.45\ m/sec}$$

Step 6: Design wind pressure (from Eq. 9.12)

$$p_z = 0.0006 . V_z^2 \text{ kN/m}^2$$
$$p_z = (0.0006 \times 59.45^2) = \mathbf{2.121\ kN/m^2}$$

9.23 DESIGN OF ROOF TRUSSES

The following are usual steps in the design of roof trusses :

Step 1. The type of truss which suits the requirements of span and roof covering to be used is selected.

Step 2. The general proportion for the type of truss selected is fixed as discussed in Art. 9.3. The line diagram of roof truss is drawn.

Step 3. The dead and live loads acting on the roof truss are determined as discussed in Art. 9.6. The wind load on roof truss depending upon slope of the truss and other conditions is found as discussed in Sec. 9.6. Thus, various loads acting on the roof truss are estimated.

Step 4. The purlins are ordinarily provided at the panel points. The purlin section is designed as discussed in Sec. 9.22.

Step 5. After the design of purlins, the loads acting at the panel points of the roof truss are determined.

Step 6. The forces in various members of the roof truss are determined for various combinations of loads as discussed in Sec. 9.14. The forces in roof truss may be found analytically or graphically by drawing force (stress) diagram. A design table for forces in various members of the roof truss is prepared. In this design table, the forces in various members for different combinations of loads are tabulated. The design forces for various members of the roof truss are noted.

Step 7. The compression members of roof truss are designed as discussed in Chapter 3. The principal rafters of roof truss are designed as continuous struts as discussed in Sec. 3.11. Other comspression members, have their lengths in between two joints and these are not continuous. Such members are designed as discontinuous struts as discussed in Sec. 3.11. The double angle sections will be designed for principal rafter and single angle sections are designed as far as possible for other members. The unequal angle sections are selected for double angle sections. The long legs of angles sections are connected together. The equal angle sections are selected for single angle section. The angle sections smaller in size than 50 mm × 50 mm × 6 mm will not be used. The members are so designed that their slenderness ratios are not greater than 180. The effective length of compression members in roof truss is adopted as 0.7 to 1.0 times the distance between centre to centre of intersections of longitudinal axes of members at the joint.

Step 8. The tension members are designed as discussed in Chapter 5. The double angle sections are used for all tension members except those which are lightly stressed. The single angle sections are objectionable. The single angle sections have a tendency to twist the truss and produce eccentric forces at the joint. The long legs are kept outstanding. The angle sections smaller in size than 50 mm × 50 mm × 6 mm will not be used. If reversal of force takes place in tension members, their slenderness ratio should be less than 350.

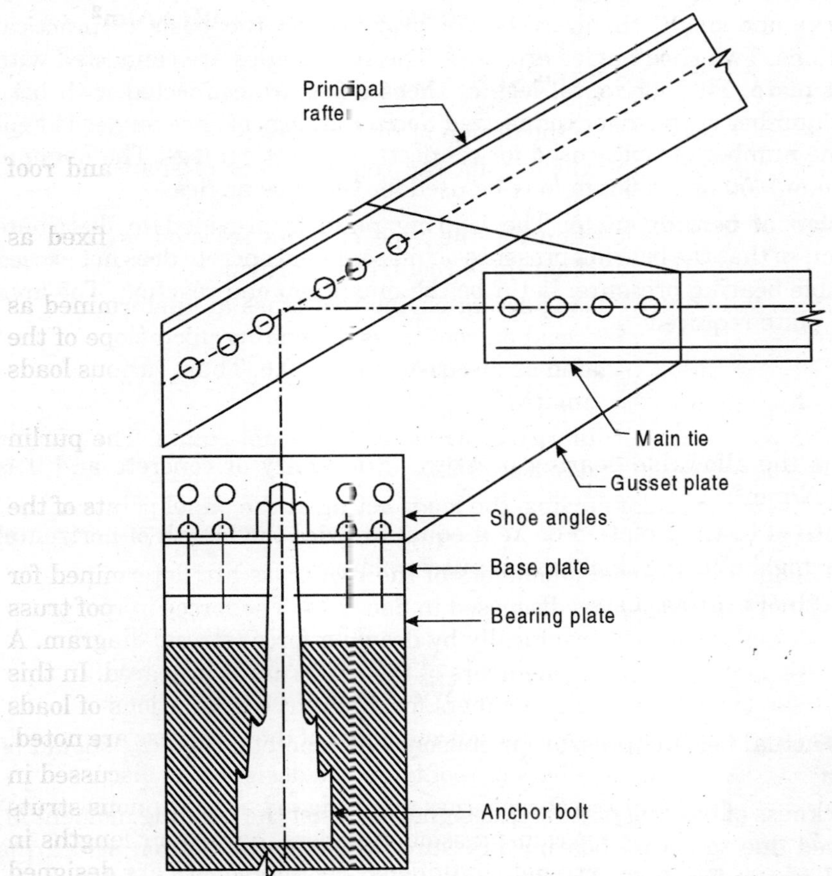

Fig. 9.32 *Elements at end support of a roof truss*

Step 9. The members meeting at a joint should have their centroidal axes intersecting at one point so as to avoid eccentricity effect. The joints are designed for the forces in the members. When the members are continuous over a joint, the forces on both sides of the joints act in opposite directions. The rivets are provided for the difference of their values. The rivets are placed symmetrical to the joints. The rivet of nominal diameter less than 16 mm will not be used. The thickness of gusset plate at any joint will not be less than 6 mm.

Step10. The lateral bracings for roof trusses are provided if necessary.

Step11. Design of end support. The various elements used at the end supports of truss are shown in Fig. 9.32. The design ni end supports consists of (i) design of shoe angle, (ii) design of bearing plate, and (iii) design of anchor bolt.

(i) Design of shoe angles. The shoes angles are designed to accommodate number of rivets necessary to transmit maximum end reactions. The number of rivets are found by dividing maximum reaction by rivet value. If the number of rivets works out small, then, rivets are provided on the basis of practical considerations. Two shoe angles are used. The shoe angles are connected with the gusset plate. The horizontal legs of shoes agles are connected with base plate. The number of rivets for connecting horizontal legs of shoe angles is kept equal to the number of rivets used for connection of vertical legs. The nominal size of 75 mm × 50 mm × 6 mm may be used for the shoe angles.

(ii) Design of bearing plate. The bearing plate is provided to distribute end reaction so that the bearing pressure on masonry or concrete does not exceed the allowable bearing pressure. Let R be the maximum end reaction. The area of bearing plate required

$$A = \left(\frac{R}{\sigma_p}\right)$$

where σ_p is the allowable bearing pressure in masonry or concrete and it is taken as 4 N/mm^2.

The width of bearing plate b be kept equal to twice the length of horizontal leg of shoe angle plus the thickness of gusset plate.

The length of bearing plate.

$$l = \left(\frac{\text{Area of bearing plate}}{\text{Width of bearing plate}}\right)$$

If σ_{P1} is actual bearing pressure in masonry or concrete, then σ_{P1} should be less than σ_p.

The thickness of bearing plate t, is designed to resist the bending moment M, at xx caused due to actual bearing pressure acting at the bottom of bearing plate, as shown in Fig. 9.33. The moment of resistance of bearing plate.

$$M_1 = (\sigma_{BS} \times z) = \left(\sigma_{bs} \times \frac{1}{6} \times x \times t_1^2\right) \quad (\because M = M_1)$$

Therefore,

$$t_1 = \left(\frac{6M}{x \times \sigma_{bs}}\right)^{\frac{1}{2}} \quad \text{...(9.36)}$$

where, Z = Modulus of section of bearing plate

σ_{bs} = Allowable bearing pressure in steel bearing plate

= 185 N/mm^2 (MPa)

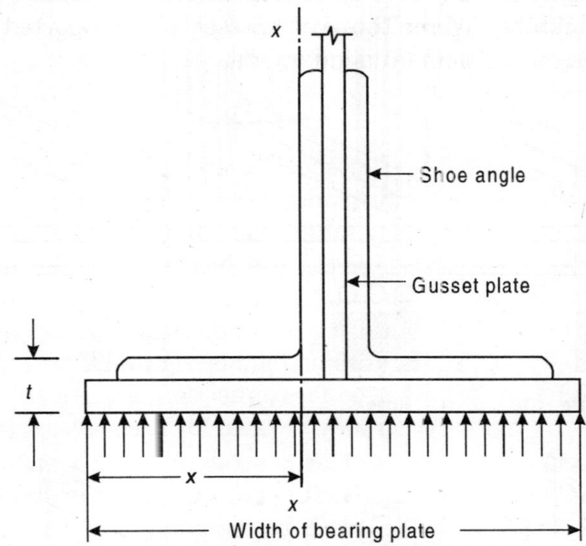

Fig. 9.33 *Shoe angles and bearing plate*

∴ t = (t thickness of angle)

A base plate is provided under the bearing plate. The thickness of base plate is kept equal to the thickness of bearing plate.

(iii) Design of anchor bolt. The anchor bolts are designed to resist net uplift pressure acting on the roof truss.

One end of the truss is kept fixed, while slots of elliptical shape are provided at the other end. The elliptical slots are provided in the horizontal legs of those angles and in the bearing plate. This provides allowance for the expansion or contraction of the truss. The expansion or contraction of truss may take place due to rise or fall of temperature. The bearing plate can slide over the base plate.

Step 12. After the design of all the elements of roof truss, a complete drawing is prepared. The complete drawing is necessary for fabrication, The complete drawing is also useful for estimating the weight of truss.

9.24 TYPICAL DETAILS OF ROOF TRUSSES

In Fig. 9.34 details at the shoe joint of a roof truss have been shown. The roof trusses have horizontal movement at their support due to expansion and contraction of its members under change of temperature. In general provision for horizontal movement at support is made at one support either by providing a sliding plate as shown in Fig. 9.34 (a) or by providing a roller as shown in Fig. 9.34 (c). The joints in roof trusses are so designed that lines of action of forces meet at joint to avoid effect of eccentricity. There is eccentricity in the connections shown in Fig. 9.34 (a), (b) and (c). If these connections are made as shown in Figure 9.34 (g), (h) and (j) ; this eccentricity may be avoided. Figure 9.34 (b), (d) and (f) show trusses supported on columns. When the roof trusses are supported over columns, provision for horizontal moment is not made.

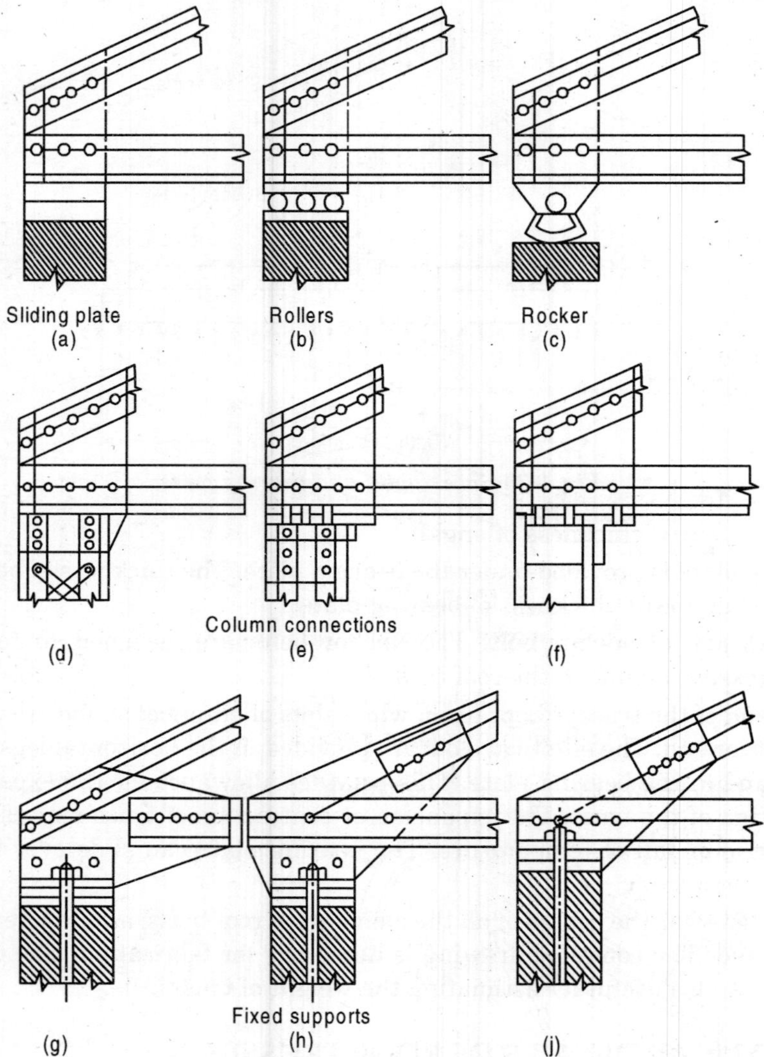

Fig. 9.34 *Details at shoe of a roof truss*

The typical details at the ridge and eves of roof trusses are shown in Fig. 9.35. A.C.C. sheets are shown as roofing material over the trusses. A.C.C. sheets are connected to the angle purlins by hook bolts. The bitumen washers are provided under G.I. washers to make the joint tight.

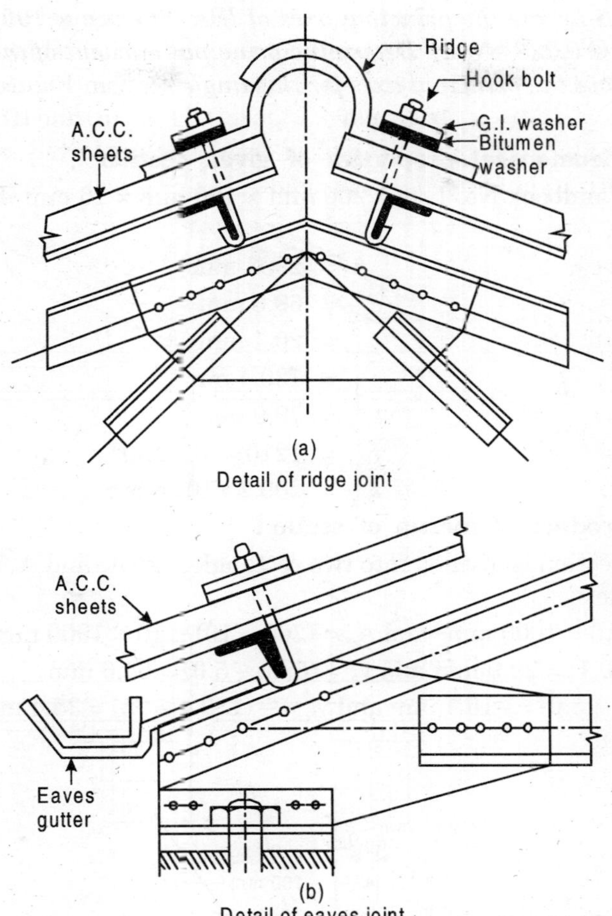

Fig. 9.35 *Typical details of A.C.C. sheeting*

Figure 9.36. shows the details of northlight glazing.

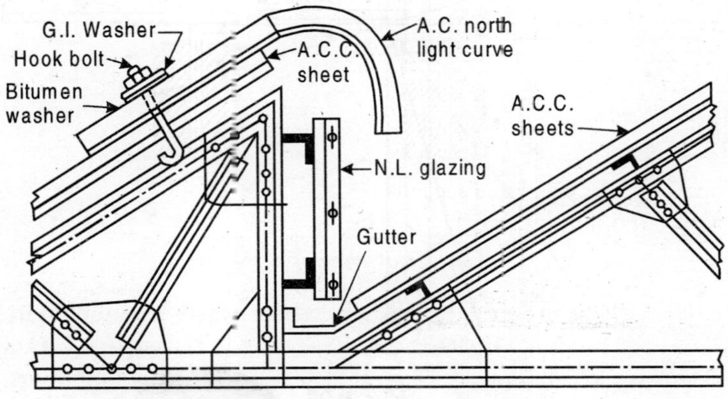

Fig. 9.36 *Detail at north light glazing*

Example 9.5 *Locate the principal axes of ISA 200 mm × 100 mm × 10 mm (ISA 200 100, @ 0.228 kN/m). Determine principal moment of inertia and radii of gyration about the principal axes for this angle section.*

Solution

Step 1 : Geometrical properties of given section

From ISI Handbook No, 1, ISA 200 mm × 100 mm × 10 mm (ISA 200 100, @ 0.228 kN/m)

Sectional area, $A = 2903 \text{ mm}^2$

$c_{xx} = 69.6$ mm

$c_{yy} = 20.1$ mm

$e_{xx} = 130.4$ mm

$e_{yy} = 79.9$ mm

$I_{xx} = 1210 \times 10^4 \text{ mm}^4$

$I_{yy} = 209.2 \times 10^4 \text{ mm}^4$

Step 2 : Product of inertia of section

The angle section is divided into two rectangle. Let A_1 and A_2 be the areas of these rectangles

$A_1 = 100 \times 10 = 1000 \text{ mm}^2$ and $A_2 = (200 - 10) \times 10 = 1900 \text{ mm}^2$

$x_1 = (50 - 20.1) = 29.9$ mm and $y_1 = (69.6 - 5.0) = 64.6$ mm

$x_2 = -(20.1 - 5.0) = -15.1$ mm and $y_2 = -(130.4 - 95) = 35.4$ mm

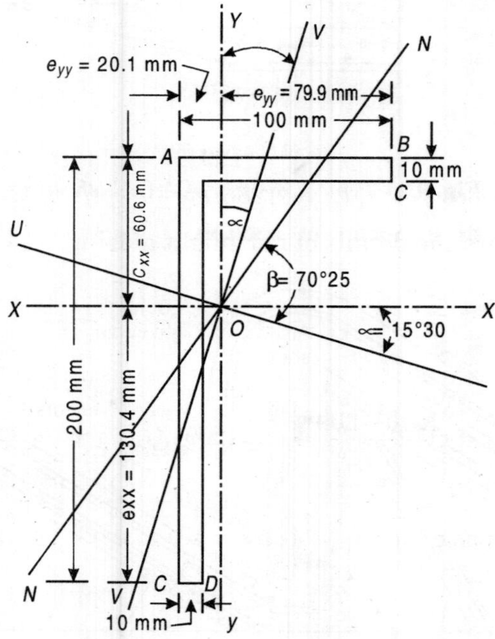

Fig. 9.37

∴ Product of inertia I_{xx} about xx-axis and yy-axis

$$I_{xy} = (A_1 \cdot x_1y_1 + A_2 \cdot x_2y_2)$$

$$I_{xy} = (10 \times 2.99 \times 6.46) + \{19 \times (-1.51) \times (3.54)\} \times 10^4 \text{ mm}^4$$

$$= 295.3 \times 10^4 \text{ mm}$$

From Eq. 9.6.

$$\tan 2\alpha = \left(\frac{-2I_{xy}}{I_{xx} - I_{yy}}\right)$$

$$= \left(\frac{(-2 \times 295 \cdot 3) \times 10^4}{(1210 - 209 \cdot 2) \times 10^4}\right) = -0.59$$

$$\therefore \quad 2\alpha = -31°, \alpha = -15° \, 30'$$

$$\tan \alpha = -0.275$$

(From ISI Handbook No. 1, $\tan \alpha = -0.27$)

The principal axes are as shown in Fig. 9.37.

The negative sign indicates that angle α is to be measured clockwise from xx-axis.

Step 3 : Principal moment of inertia of section (From Eq. 9.9),

$$I_{UU} = \left(\frac{I_{xx} + I_{yy}}{2}\right) + \left[\left(\frac{I_{xx} - I_{yy}}{2}\right)^2 + I_{yy}^{\ 2}\right]^{1/2}$$

$$= \left(\frac{1210 + 209 \cdot 2}{2}\right) + \left[\left(\frac{1210 - 209 \cdot 2}{2}\right)^2 + (295 \cdot 3)^2\right]^{1/2} \times 10^4 \text{ mm}^4$$

$$I_{UU} = 1291.6 \times 10^4 \text{ mm}^4$$

(From ISI Handbook No. 1, $I_{UU} = 1286.7 \times 10^4 \text{ mm}^4$)

Step 4: Radius of gyration of section

$$I_{VV} = \left(\frac{I_{xx} - I_{yy}}{2}\right) - \left[\left(\frac{I_{xx} - I_{yy}}{2}\right)^2 + I_{yy}^2\right]^{1/2}$$

$$= \left(\frac{1210 + 209 \cdot 2}{2}\right) - \left[\left(\frac{1210 - 209 \cdot 2}{2}\right)^2 + (209 \cdot 3)^2\right]^{1/2} \times 10^4 \text{ mm}^4$$

$$= 127.6 \times 10^4 \text{ mm}^4$$

(From ISI Handbook No.1 $I_{VV} = 132.5 \times 10^4 \text{ mm}^4$)

$$r_{UU} = \left(\frac{I_{UU}}{A}\right)^{1/2} = \left(\frac{1291 \cdot 6 \times 10^4}{2903}\right)^{1/2} = 66.6 \text{ mm}$$

(From ISI Handbook No. 1, $r_{UU} = 66.6$ mm)

$$r_{VV} = \left(\frac{I_{VV}}{A}\right)^{1/2} = \left[\frac{127 \cdot 6 \times 10^4}{2902}\right]^{1/2} = \mathbf{21.0\ mm}$$

(From ISI Handbook No. 1, r_{VV} = 21.4 mm)

The actual angle section is rounded at the toe and at the root. In above calculations, angle section is not rounded. As a result of this, there are small differences in the values calculated and the values noted from Handbook No. 1.

Example 9.6 *In Example 9.5, the angle section is subjected to bending moment 6 kN-m acting in the vertical plane through the centroid of section. Determine the maximum bending stress induced in the angle section.*

Solution :

Step 1: Principal moment of inertias of section

From Example 9.5,

$$I_{UU} = 1291.6 \times 10^4 \text{ mm}^4$$
$$I_{VV} = 127.6 \times 10^4 \text{ mm}^4$$

The plane of loading is vertical and passes through the centroid of the angle section. From Eq. 9.15 for neutral axis

$$\tan \beta = \left(\frac{I_{uu}}{I_{vv}}\right) \tan \theta \left(\frac{1291 \cdot 6 \times 10^4}{127 \cdot 6 \times 10^4}\right) \times \tan 15°30' = +\ 2.81$$

$$\therefore \qquad \beta = +\ 70°\ 25'$$

The value of β is positive. It is plotted anti-clockwise from principal axis *UU* is shown in Fig. 9.37. From Fig. 9.37, it is seen that the point *D* is at extreme distance from the neutral axis.

Step 2 : Co-ordinates of farthest point *D*

$$x_D = -\ (20.1 - 10) = -10.1 \text{ mm}$$
$$y_D = -\ 130.4 \text{ mm}, \alpha = -\ 15°\ 30'$$
$$u_D = (-x_D \cos \alpha + y_D \sin \alpha)$$
$$= [-10.1 \times \cos(-15°\ 30') + (-\ 1\ 30.4) \times \sin(15°\ 30')]$$
$$= +\ 25.1 \text{ mm}$$
$$v_D = (y_D \cos \alpha - x_D \sin \alpha)$$
$$= [-130.4 \cos(-15°\ 30') - (10.1) \times \sin(-\ 15°30')]$$
$$= -\ 128.5 \text{ mm}$$
$$M_{VV} = M \,.\, \cos \theta \text{ and } M_{VU} - M \,.\, \sin \theta$$

Step 3 : Maximum bending stress at point *D* (From Eq. 9.12),

$$\sigma_b = \left(\frac{M \cos \theta}{I_{UU}}\right) \cdot v_D + \left(\frac{M \sin \theta}{I_{vv}}\right) \cdot u_D$$

Both the components of bending stresses are additive

$$\sigma_b = \left(\frac{6 \times 10^6 \cos 15°30'}{1291 \cdot 6 \times 10^4}\right) \times 128 \cdot 5 + \left(\frac{6 \times 10^6 \sin 15°30'}{127 \cdot 6 \times 10^4}\right) = 88.8 \text{ N/mm}^2$$

The maximum bending stress at point *D* is tensile.

(i) Alternative. The moment of inertia of the section about neutral axis fromEq. 9.16

$$I_{NN} = (I_{UU}.\cos^2\beta + I_{VV}.\sin^2\beta)$$
$$= [1291.6 \times \cos^2 70° \; 25' + 127.6 \sin^2 70° \; 25'] \times 10^4 \text{ mm}^4$$
$$= 258.3 \times 10^4 \text{ mm}^4$$
$$(\beta - \theta) = (70° - 25' - 15° \; 30')$$
$$= 54°55'$$
$$Y_N = (-v_D . \cos\beta - u_D . \sin\beta)$$
$$\therefore \quad Y_N = (-128.5 \times 0.335 - 25.1 \times 0.942)$$
$$= -66.6 \text{ mm}$$

From Eq. 9.17, maximum bending stress at point D

$$\sigma_b = \left[\frac{M \cdot \cos(\beta - \theta)}{I_{NN}}\right] . Y_N$$

$$= \left(\frac{6 \times 10^6 \cos 54°55' \times 66.6}{258 \cdot 3 \times 10^4}\right) = 88.7 \text{ N/mm}^2$$

The maximum bending stress at the point D is tensile.

(ii) Alternative. The maximum bending stress at point D may be found by Eq. 9.13. Eq. 9.13 is useful specially for beam sections in which less of angle section are parallel to xx and yy axes. From Eq. 9.13, the bending stress at point D,

$$\sigma_b = \left(\frac{M_{xx}I_{yy} - M_{yy}\,I_{xy}}{I_{xx}I_{yy} - I_{xy}^2}\right) \cdot y_D \left(\frac{M_{yy}I_{xx} - M_{xx}\,I_{xy}}{I_{yy}I_{xx} - I_{xy}^2}\right) x_D, \text{ and } \phi = 0°$$

$M_{xx} = M \cos\phi = 6 \times 10^6$ kN-m and $M_{yy} = M \sin\phi =$ Zero.

$$a_2 = \left(\frac{M_{xx}I_{yy} - M_{yy}I_{xy}}{I_{xx}I_{yy} - I_{xy}^2}\right) = \left(\frac{6 \times 10^6 \times 209 \cdot 2 \times 10^4}{\left[1210 \times 209 \cdot 2 - (29 \cdot 3)^2\right] \times 10^8}\right) = 0.754$$

$$a_2 = \left(\frac{M_{yy}I_{xx} - M_{xx}I_{xy}}{I_{yy}I_{xx} - I_{xy}^2}\right) = \left(\frac{-6 \times 10^6 \times 295 \cdot 3 \times 10^4}{\left[1210 \times 209 \cdot 2 - (295 \cdot 3)^2\right] \times 10^8}\right) = -1.06$$

$$\sigma_b = (0.754\, y_D - 1.06\, x_D)$$
$$= [0.754\,(-130.4) - (1.06)(-10.1)] = -87.8 \text{ N/mm}^2.$$

The maximum bending stress at point D is tensile and it has been worked out from Eqs. 920, 921 and 925. The values differ slightly.

Example 9.7 *LB 200, @ 0.198 kN/m is subjected to bending moment 12 kN-m. The plate of loading passes through centroid of beam section, and it is inclined 8° with the YY–axis in the anti-clockwise direction. Locate the neutral axis. Determine maximum bending stress induced in the beam section.*

Solution

Step 1. Geometrical properties of section

From IS I Handbook No. 1, for LB 200, @ 0.198 kN/m

Depth of section, $h = 200$ mm

Width of section, $b = 100$ mm

$$I_{UU} = I_{xx} = 1696.6 \times 10^4 \text{ mm}^4$$

$$I_{VV} = I_{yy} = 115.4 \times 10^4 \text{ mm}^4$$

It is to note that I-section shown in Fig. 9.33 is a symmetrical section. The axes of symmetry XX and YY are principal axes UU and VV

From Eq. 9.23, for neutral axis

$$\tan \beta = +\left(\frac{I_{UU}}{I_{VV}}\right) \tan \theta$$

or

$$\tan \beta = +\left(\frac{1696 \cdot 6}{115 \cdot 5}\right) \tan 8°$$

$\beta = 64° \, 7'$. The value of β is positive. It is plotted anti-clockwise from principal axis-UU as shown in Fig. 9.38.

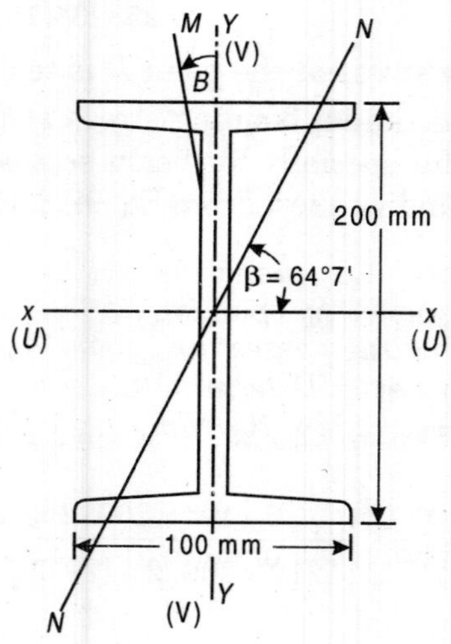

Fig. 9.38 *I-section purlin*

Step 2. Moment of inertia of beam section about neutral axis (from Eq. 9.24)

$$I_{NN} = (I_{UU} \cdot \cos^2 \beta + I_{VV} \cdot \sin^2 \beta)$$

or

$$I_{NN} = [1696.6 \cos^2 64° \, 7' + 115.4 \times \sin^2 64° \, 7'] \times 10^4 \text{ mm}^4$$

$$= 417.5 \times 10^4 \text{ mm}^4$$

$$(\beta - \theta) = [(64°7') - 8°] = 56°7'$$

Step 3. Co-ordinates of points *A* and *D* are at extreme distance from the neutral axis.

$$\therefore \quad y_N = [100 \cos \beta - (-50) \sin \beta]$$
$$= [(100 \cos 64° 7' + 50 \sin 64° 7')]$$
$$= 88.6 \text{ mm}$$

Step 4. Maximum bending stress at point A (from Eq. 9.1)

$$\sigma_b = \left[\frac{M \cdot \cos(\beta - \theta)}{I_{NN}}\right] \cdot y_N$$
$$= \left(\frac{12 \times 10^6 \cos 56°7'}{417 \cdot 5 \times 10^4}\right) \times 88.6$$
$$= 141.8 \text{ M/mm}^2$$

The maximum bending stress at point A is compressive and that at point D is tensile.

Alternative. The maximum bending stress at point A, can be found by resolving bending moment along VV-axis and UU-axis. From Eq. 9.20, bending stress at point A,

$$\sigma_b = \left(\frac{M \cos 8°}{1_{UU}}\right) \cdot v_A + \left(\frac{M \sin 8°}{I_{VV}}\right) u_A$$

$$\text{or } \sigma_b = \left(\frac{121 \times 10^6 \times 0 \cdot 99 \times 100}{1696 \cdot 6 \times 10^4}\right) + \left(\frac{12 \times 10^6 \times 1 \cdot 39 \times 50}{115 \cdot 54 \times 10^4}\right) \text{N/mm}^2$$
$$= \mathbf{141.8 \ N/mm^2}.$$

The maximum bending stress at point A is compressive and that at the point D is tensile.

Example 9.8 *In Example 9.7, the beam section is carrying uniformly distributed load 6 kN/m. The effective span is 4 m. The beam is simply supported at both ends. The plane of loading passes through centroid of beam section and it is inclined 8 ° with the yy-axis. Determine the magnitude and direction of maximum deflection. Take E = 2.04 × 10⁵ N/mm².*

Solution

Step 1. Properties of section

From Example 9.7, the inclination of neutral axis from UU-axis

$$\beta = +64.7'$$

Moment of inertia about neutral axis

$$I_{NN} = 417.5 \times 10^4 \text{ mm}^4$$
$$(\beta - \theta) = [(64° 7') - 8°] = 56°7'$$

Step 2. Maximum deflection for the beam section (from Eq. 9.33),

$$\delta = \frac{5}{384} \cdot \left(\frac{w \cdot l^4 \cos(\beta - \theta)}{E \cdot I_{NN}}\right) \text{mm}$$

$$\delta = \frac{5}{384} \times \left(\frac{6 \times 10^3 \times 4 \times (4000)^3 \times \cos 56° 7'}{2.04 \times 10^5 \times 417.5 \times 10^4}\right)$$
$$= \mathbf{13.2 \ mm}$$

The maximum deflection occurs in the direction perpendicular to **neutral** axis *NN*. It occurs in the direction β′measured from *UU* principal axis,

$$\beta' = (\beta + 90°)$$
$$= (64° 7' + 90°) = \mathbf{154°7'}$$

Example 9.9 *Draw Z polygon for ISA 200 mm × 100 mm × 10 mm. The angle section is subjected to bending moment 6 kN-m acting in the vertical plane through the centroid of section. Determine the maximum bending stress induced in the angle section. Also determine the absolute maximum bending stress for the given moment and corresponding inclination of plane of loading with YY-axis.*

Solution

Step 1 : Geometrical propertion of given section

From ISI Handbook No. 1 for ISA 200 mm × 100 mm × 10 mm. (ISA 200 100, @ 0.228 kN/m)

Sectional area, $A = 2903 \text{ mm}^2$

$c_{xx} = 69.6 \text{ mm}^2$, $c_{yy} = 20.1$ mm

$e_{xx} = 130.4$ mm, $c_{yy} = 79.9$ mm

$I_{xx} = 1210 \times 10^4 \text{ mm}^4$, $I_{yy} = 209.2 \times 10^4 \text{ mm}^4$

$I_{uu} = 132.5 \times 10^4 \text{ mm}^4$ $I_{vv} = 1286.7 \times 10^4 \text{ mm}^4$,

$\tan \alpha = -0.270$, $\therefore \alpha = -15.7'$

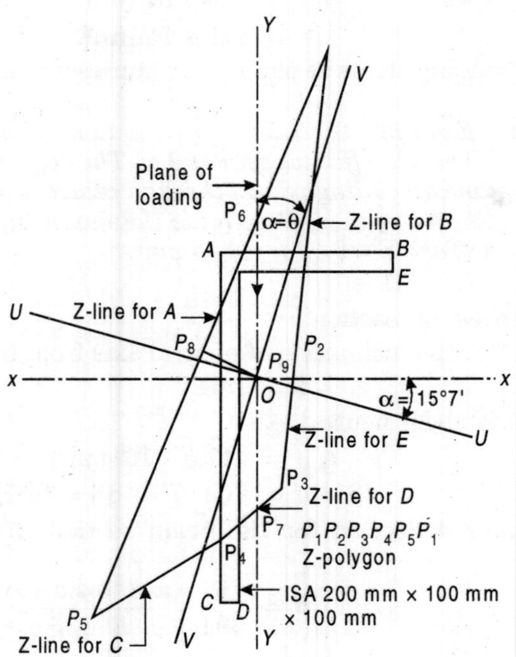

Fig. 9.39

It is to note that angle α is negative. It is plotted in clock-wise direction from XX-axis as shown in Fig. 9.39. *UU* and *VV* are principal axes.

Step 2 : Co-ordinates of corner points with reference to *UU* and *VV*-axes.

(i) **Corner point *A*,**

$x_A = -20.1$ mm, $y_A = +69.6$ mm

$u_A = (x_A \cdot \cos\alpha + y_A \cdot \sin\alpha)$

$= [(-20.1)\cos(-15.7') + 69.6\sin(-15.7')]$

$= -37.6$ mm

$v_A = (y_A \cdot \cos\alpha - x_A \cdot \sin\alpha)$

$= (69.6 \times 0.956 - (-20.1)(-0.261))$

$= 51.9$ mm

(ii) Corner point *B*,

$x_B = +79.6$ mm, $y_B = +69.6$ mm

$u_B = (x_B \cdot \cos\alpha - y_B \cdot \sin\alpha)$

$= (79.9 \times 0.965 - 69.6 \times 0.261) = +58.8$ mm

$v_B = (y_B \cdot \cos\alpha - x_B \cdot \sin\alpha)$

$= [(69.6 \times 0.965) - (79.9)(-0.261)] = 88.0$ mm

(iii) Corner point *C*,

$x_D = -20.1$ mm, $y_c = -130.4$ mm

$u_C = (x_C \cdot \cos\alpha - x_C \cdot \sin\alpha)$

$= [(-20.1 \times 0.965) + (-130.4)(-0.261)]$

$= +14.7$ mm

$v_C = (y_C \cdot \cos\alpha - x_C \cdot \sin\alpha)$

$= [-130.4 \times 0.965 - (-20.1)(-0.261)]$

$= -131.3$ mm

(iv) Corner point *D*,

$x_D = 10.1$ mm, $y_D = -130.4$ mm

$u_D = (x_D \cdot \cos\alpha - y_D \cdot \sin\alpha)$

$= [(-10.1)(0.965) + (-130.4)(-0.261)]$

$= +24.4$ mm

$v_D = (y_D \cdot \cos\alpha - x_D \cdot \sin\alpha)$

$= [(-130.4)(0.965) - (-10.1)(-0.261)]$

$= 128.6$ mm

(v) Corner point *E*,

$x_E = (+79.9$ mm, $y_E = (69.6 - 10))$

$= +59.6$ mm

$u_E = (x_E \cdot \cos\alpha - y_E \cdot \sin\alpha)$

$= 79.9 \times 0.965 + (59.6)(-0.261)$

$= +61.4$ mm

$v_E = (y_E \cdot \cos\alpha - x_E \cdot \sin\alpha)$

$= [59.6 \times 0.965 - (79.9)(-0.261)]$

$= 78.4$ mm.

Step 3. Equation of Z-lines for corner points. *It is to note that, now co-ordinates (u, v) are taken positive in second quadrant.*

(i) **Corner point** *A*, from Eq. 9.29

$$\frac{u}{\left(\frac{I_{VV}}{u_A}\right)}+\frac{v}{\left(\frac{I_{UU}}{V_A}\right)} = 1, \frac{u}{\left(\frac{132.5}{37.6}\right)}+\frac{v}{\left(\frac{1286.7}{51.9}\right)}=1\times10^4$$

or $$v = (-7.06u + 24.8 \times 10^4) \quad ...(i)$$

(ii) Corner point *B*

$$\frac{u}{\left(\frac{I_{VV}}{u_B}\right)}+\frac{v}{\left(\frac{I_{UU}}{V_B}\right)} = 1, \frac{u}{\left(\frac{132.5}{-58.8}\right)}+\frac{v}{\left(\frac{1286.7}{88.0}\right)}=1\times10^4$$

or $$v = (6.5u + 14.6 \times 10^4) \quad ...(ii)$$

(iii) Corner point *C*

$$\frac{u}{\left(\frac{I_{VV}}{u_C}\right)}+\frac{v}{\left(\frac{I_{UU}}{V_C}\right)} = 1, \frac{u}{\left(\frac{132.5}{-14.7}\right)}+\frac{v}{\left(\frac{1286.7}{-131.3}\right)}=1\times10^4$$

or $$v = (1.08u - 9.78 \times 10^4) \quad ...(iii)$$

(iv) Corner point *D*

$$\frac{u}{\left(\frac{I_{VV}}{u_D}\right)}+\frac{v}{\left(\frac{I_{UU}}{V_D}\right)} = 1, \frac{u}{\left(\frac{132.5}{-24.4}\right)}+\frac{v}{\left(\frac{1286.7}{-128.6}\right)}=1\times10^4$$

or $$v = (1.84u - 10.9 \times 10^4) \quad ...(iv)$$

(v) Corner point *E*

$$\frac{u}{\left(\frac{132.5}{-61.4}\right)}+\frac{v}{\left(\frac{1286.7}{78.4}\right)} = 1 \text{ or } v = (7.46\, u - 16.4 \times 10^4) \quad ...(v)$$

Step 4. Z-polygon

(i) The intersection of *Z*-lines for points *A* and *B* locales point P_1 of *Z*-polygon.

Z – line for point *A*, $v = (-7.06u + 24.8 \times 10^4)$ [from Eq. (i)]

Z – line for point B_v, $v = (6.5u + 14.6 \times 10^4)$ [from Eq. (ii)]

Solving equations (i) and (ii),

Co-ordinates of P_1 are (75.4, 1948).

(ii) The intersection of *Z*-lines for points *B* and *E* locates point P_2 of *Z* polygon.

Z – line for point *B*, $v = (6.5u + 14.6 \times 10^4)$ [from Eq. (ii)

Z-line for point *E*, $u = (7.4u + 16.4 \times 10^4)$ [from Eq. (v)

Solving Eqs., (ii) and (v), co-ordinates of point P_2 are (–188, 240).

(iii) The intersection of Z-lines for points E and D, locates P_3 of Z-polygon.

Z-line for point E, $v = (7.46u + 16.4 \times 10^4)$ [from Eq. (v)]

Z-line for point D, $v = (-1.84u - 10.0 \times 10^4)$ [from Eq. (iv)]

Solving Eqs. (v) and (iv), co-ordinates of point P_4 are (–283, –580).

(iv) The intersection of Z-lines for points D and C, locates point P_A of Z-polygon.

Z-line fur point D, $v = (-1.84\ u - 10.0 \times 10^4)$ [from Eq. (iv)]

Z-line for point C, $v = (-1.08\ u - 9.78 \times 10^4)$ [from Eq. (iii)]

Solving Eqs. (iv) and (iii), co-ordinates of point P_5 are (–289, 946.8).

(v) The intersection of Z-lines for points C and A, locates point P_5 of Z polygon.

Z-line for point C, $v = (-1.08\ u - 9.78 \times 10^4)$ [from Eq. (iii)]

Z-line for point A, $v = (-7.06\ u + 2.48 \times 10^4)$ [from Eq. (i)]

The points P_1, P_2, P_3, P_4 and P_5 are plotted with reference to UU and VV principal axes. Join P_1 and P_2, P_2 and P_3, P_3 and P_4, P_4 and P_5 and P_5 and P_1. $P_1P_2P_3P_4P_5P_1$ represents Z-polygon.

The bending moment M is acting in the vertical plane passing through the centroid of section. The vertical plane through the centroid intersects Z-line of point A at P_6 and Z-line of point D at P_7.

From Fig. 9.39, $OP_6 = 88 \times 10^3$ mm^3, $OP_7 = 68 \times 10^3$ mm^3

Therefore, point D is subjected to maximum bending stress

Step 5. Maximum bending stress

$$\sigma_b = \left(\frac{M}{OP_7}\right) = \left(\frac{6 \times 10^6}{68 \times 10^3}\right)$$

$$= \mathbf{88.235\ N/mm^2}$$

The bending stress at point D is tensile.

For absolute maximum bending stress draw perpendiculars OP_8 and OP_9 from centroid O. OP_9 gives minimum modulus of section. $OP_9 = 24 \times 10^3$ mm^3. Absolute maximum bending stress for the given moment

$$\sigma_b = \left(\frac{6 \times 10^6}{24 \times 10^3}\right) = \mathbf{250\ N/mm^2}$$

For absolute maximum bending stress, the inclination of plane of loading is 96° clockwise from vertical.

Example 9.10 *A roof truss-shed is to be built in Jodhpur city area for an industrial use. The size of shed shall be 18 m × 30 m. Determine the basic wind pressure (Example 9.1).*

Roof trusses of 18 m span and 4.5 rise shall be provided at 4 m spacing centre to centre. The height of cladding or eave level is 6 m. Determine the wind load parallel to the ridge causing a total outward normal wind pressure on the interior purlins, for (i) small permeability (ii) medium permeability and (iii) high permeability.

Solution

Step 1 : From Exmaple 9.1,

The basic wind speed, V_b is 47 m/sec. The design wind speed (as worked out), V_z is 41.641 m/sec. The design wind pressure. p_z is 1.04 kN/m^2.

Step 2 : Geometry of truss

Span of the roof-trusses to be used is 18 m its rise shall be 4.5 m. Slope of the roof

$$\theta = \tan^{-1}\left(\frac{4\cdot 5}{9}\right) = 26.565° \quad \text{...(i)}$$

Length of the principal rafter

$$= [(4.5)^2 + (9)^2]^{1/2} = 10.06 \text{ m} \quad \text{...(ii)}$$

Panel length should not be more than 1.4 m, so that the failure of roof covering sheets does not occur. A compound French roof truss may be used with 8 (eight) panels @ 1.25 m, so that there shall be 9 (nine) panel joints. The remaining 6.23 mm shall be left near the ridge and near the eaves.

Total length covered on the principal rafter

$$= (2 \times 31.15 + 8 \times 1250)$$
$$= 10062.3 \text{ mm}$$

The spacing of purlins shall be 1.25 m.

Step 3 : Wind load

Let the height of plinth be 600 mm and the projection of eave girder behond centre line be 400 mm. The eave height of the roof shed above ground level.

$$h = (6 + 0.600) = 6.6 \text{ m}$$

Gross-width of the roof-shed

$$w = (18 + 2 \times 0.400) = 18.8 \text{ m}$$

Ratio of height to width of roof shed

$$\left(\frac{h}{w}\right) = \left(\frac{6\cdot 6}{18\cdot 8}\right) = 0.351 < 0.5$$

(A) External air pressure coefficients. The wind pressure coefficients are to be adopted from a set $\left(\frac{h}{w}\right) < 0.5$ and for slope $\theta = 26.565°$. The wind pressure coefficients are noted from IS : 875 (Part 3) – 1987, or from Table 9.7 (External pressure coefficients, C_{pe} for pitched roofs of rectangular clad buildings.

Side of pitched roof	*Wind angle, θ*			
	θ = 0°	*	θ = 90°	*
Windward	– 0.1374	–.8687	– 0.7	–
Leeward	– 0.4	–1.1344	– 0.7	–

(*Local effects)

(B) Internal pressure coefficients. Internal air pressure in a building depends upon the degree of permeability of cladding to the flow or air.

(i) For small permeability $C_{pi} = \pm 0.2$

(ii) For medium permeability $C_{pi} = \pm 0.5$

(iii) For high permeability $C_{pi} = \pm 0.7$

Note. *The internal pressure coefficient is added algebraically to the external pressure coefficient and the worst combination shall be considered.*

Step 4. Total wind load parallel to ridge

(i) For small permeability :

External + internal pressure coefficients

$$(C_{pe} - C_{pi}) = [-0.7 - (+0.2)] = 0.9$$

Total wind load

$$W_w = (C_{pe} - C_{pe}) . p_d \cdot A$$

$$W_w = \mathbf{-0.9\ p_d . A}$$

(ii) For medium permeability :

External + internal pressure coefficients

$$(C_{Pe} - C_{pi}) = [-0.7 - (+0.5)] = -1.2$$

Total wind load

$$W_w = \mathbf{-1.2\ p_d . A}$$

(iii) For high permeability :

External + internal pressure coefficients

$$(C_{pe} - C_{pi}) = [-0.7 - (+0.7)1 = -1.4$$

Total wind load

$$W_w = \mathbf{-1.4\ p_d . A}$$

Note. p_d *(design wind pressure) is 104 kN/m². A (effective exposed area, that is, spacing of purlins multiplied by the length of purlin is (1.25 × 4) = 5 m². Therefore, total wind load parallel to ridge for*

(i) small permeability:

$$W_w = [(-0.9).(1.04).(5)] = \mathbf{-4.68\ kN}$$

(ii) medium permeability:

$$W_w = [(-1.2).(1.04).(5)] = \mathbf{-624\ kN}$$

(iii) high permeability:

$$W_w = [(-1.4).(1.04).(5)] = \mathbf{-7.28\ kN}$$

Figures 9.40 (a), (b) and (c) show external air pressure coefficients, C_{pe} for pitched roofs, and internal air pressure coefficients *(positive* and *negative)* for small permeability buildings, medium permeability buildings and high permeability buildings, respectively.

Example 9.11 *Design an I-section purlin to support galvanised corrugated iron sheet roof. The purlins are 1.25 m apart over roof trusses spaced 5 m centre to centre. The roof surface has an inclination of 30° to the horizontal. The weight of corrugated iron sheet is 0.1331 kN/m². The weight of fixtures etc. 0.053 kN/m².*

The design wind pressure for medium permeability is 1.20 kN/m² (outward) parallel to ridge.

Solution

Design :

Step 1. Dead load on purlin per metre length

Weight of galvanized corrugated iron sheets

$$(0.1331 \times 1.25 \times 1) = 0.1664 \text{ kN/m}$$

Extra load for overlap and fixtures

$$(0.053 \times 1.25 \times 1) = 0.0663 \text{ kN/m}$$

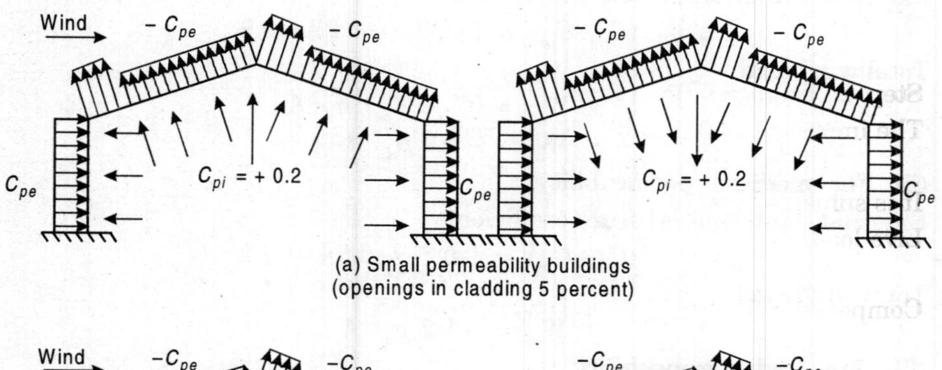

(a) Small permeability buildings
(openings in cladding 5 percent)

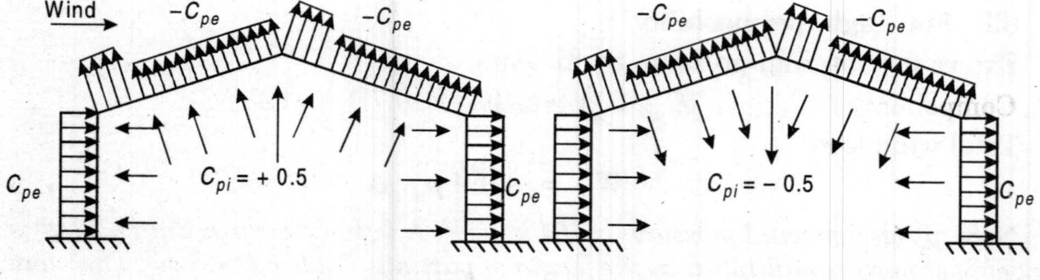

(b) Medium permeability buildings
(Openings in cladding 5 to 20 percent)

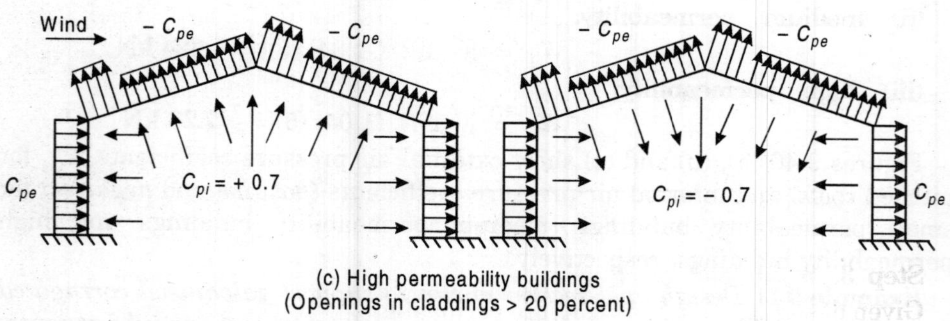

(c) High permeability buildings
(Openings in claddings > 20 percent)

Fig. 9.40 *External and internal wind pressure coefficients*

Self-weight of purlin (assumed)

$$= 0.120 \text{ kN/m}$$

The dead load = 0.3527 kN/m (say 0.360 kN/m)

Dead load acts vertically downward

Component of dead load normal to the roof

$$(0.360 \times \cos 30°) = \left(\frac{0 \cdot 360 \times \sqrt{3}}{2}\right)$$

$$= 0.312 \text{ kN/m}$$

Component of dead load parallel to the roof

$$(0.360 \times \sin 30°) = \left(\frac{0 \cdot 360 \times 1}{2}\right) = 0.180 \text{ kN/m}$$

Step 2. Imposed (live) load

The imposed (live) load for sloping roof with slope grater then 10°

$$[0.75 - (30 - 10)(0.020)] = 0.35 \text{ kN/m}^2$$

It is subjected to a minimum of 04 kN/m^2

Live load

$$(0.4 \times 1.25 \times 1) = 0.5 \text{ kN/m}$$

Component of live load normal to the roof

$$0.5 \times \cos 30° = \left(\frac{0 \cdot 5 \times \sqrt{3}}{2}\right) = 0.433 \text{ kN/m}$$

Component of live load parallel to the roof

$$(0.5 \times \sin 30°) = \left(\frac{0 \cdot 5 \times 1}{2}\right) = 0.25 \text{ kN/m}$$

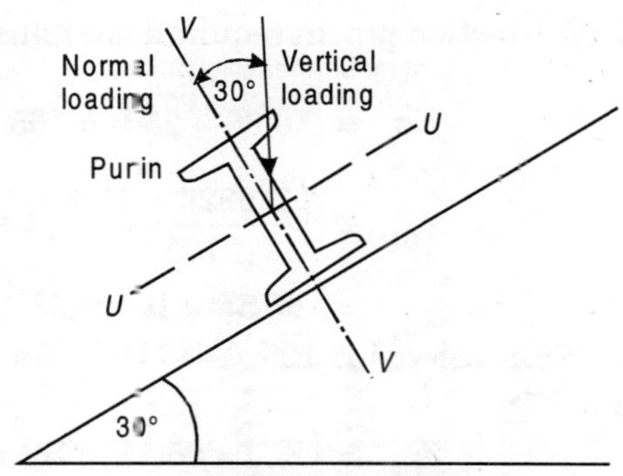

Fig. 9.41 *I-Section purlin*

Step 3. Wind load on Purlin

Given design pressure for medium permeability (parallel to ridge) is 1.20 kN/m^2 (outward or negative). That is, it acts upward. Wind load on purlin parallel to the ridge.

Step 4. Combination of loads

The purlins are designed for worst combination of the load and wind from any direction. The worst combinations of loads for design of purlin are

(i) dead load *plus* live load, and

(ii) dead load *plus* live load plus wind load (paralle) to the ridge). In case, the design wind pressure acts outward (negative), the imposed live load shall not be considered.

Step 5. Design of purlin for dead load plus live load

Bi-axial bending of the purlin occurs due to dead load plus live load. The purlins remain continuous and the end span has the maximum support (negative) moment.

1. Bending moment due to dead load *plus* live load parallel to the major principal axis (*UU*-axis, Fig. 9.40)

$$M_{UU} = \left[\left(\frac{0\cdot312\times5}{10}\right)\times5+\left(\frac{0\cdot25\times5}{9}\right)\times5\right]$$

$$= (0.78 + 1.2028) = 1.9828 \text{ kN-m}$$

2. Bending moment due to dead load *plus* live load parallel to the minor principal axis (*VV*-axis, Fig. 9.40)

$$M_{VV} = \left[\left(\frac{0\cdot180\times5}{10}\right)\times5+\left(\frac{0\cdot25\times5}{9}\right)\times5\right]$$

$$= (0.45 + 0.694) = 1.144 \text{ kN-m}$$

Section modulus required for purlin Eq. 9.35

$$Z_{UU} = \frac{M_{UU}}{\sigma_b}\left(1+\frac{M_{VV}}{M_{UU}}\cdot\frac{Z_{UU}}{Z_{VV}}\right) \text{ mm}$$

Assuming $\left(\frac{Z_{UU}}{Z_{VV}}\right)$ 7 for I-section purlin required modulus of section for purlin

$$(\sigma_b = (0.66 \times 250) = 165 \text{ N/mm}^2)$$

$$Z_{UU} = \left(\frac{1\cdot9828\times10^6}{165}\right)\cdot\left(1+\frac{1\cdot144}{1\cdot9828}\times5.61\right)$$

$$= 60.55 \times 10^3 \text{ mm}^3$$

From ISI Handbook No.1, select LB 125, @ 0.119 kN/m

Modulus of section

$$Z_{UU} = Z_{XX} = 65.11 \times 10^3 \text{ mm}^3 \text{ and}$$

$$Z_{VV} = Z_{YY} = 11.6 \times 10^3 \text{ mm}^3$$

It is to note that *XX* and *YY* axes are also the axes of symmetry for I-section. Therefore, these are also principal axes of the section.

$$\left(\frac{Z_{UU}}{Z_{VV}}\right) = \left(\frac{65\cdot1\times10^3}{11\cdot6\times10^3}\right) = 5.61$$

Step 6. Check for bending stress

Maximum bending stress in purlin from Eq. 9.34

$$\sigma_b = \frac{M_{UU}}{Z_{UU}}\left(1+\frac{M_{VV}}{M_{UU}}\cdot\frac{Z_{UU}}{Z_{VV}}\right)$$

$$\sigma_b = \left(\frac{1\cdot 9828\times 10^6}{65.11\times 10^3}\right)\cdot\left(1+\frac{1\cdot 144}{1\cdot 9828}\times 5\cdot 61\right)$$

$$= 129.02 \text{ N/mm}^2$$

$$< 165 \text{ N/mm}^2\text{. Hence satisfactory}$$

Step 7. Design of purlin for dead load plus wind load

The wind load (given) acts outward. Therefore, the live load is not considered as worst combination of load. From step 5, moment due to dead load parallel to the major axes, *UU*

$$M_{U1U1} = +\,0.780 \text{ kN-m}$$

Moment due to dead load parallel to the minor axis, *VV*

$$M_{V1V1} = +\,0.45 \text{ kN-m}$$

Bending moment due to wind load acting outward about major principal -axis, *UU*

$$\left(\frac{1\cdot 50\times 5}{10}\right)\times 5 = -3.750 \text{ kN-m}$$

Total design moments

$$M_{UU} = (+\,0.780 - 3.750) = -\,2.97 \text{ kN-m}$$

Section modulus required for purlin from Eq. 9.35

$$Z_{UU} = \frac{M_{UU}}{\sigma_b}\left(1+\frac{M_{VV}}{M_{UU}}\cdot\frac{Z_{UU}}{Z_{VV}}\right) \text{ mm}^3$$

Ratio $\left(\frac{Z_{UU}}{Z_{VV}}\right)$ is assumed as 7 for I-section

$$Z_{UU} = \left(\frac{2\cdot 97\times 10^6}{1\cdot 33\times 165}\right)\cdot\left(1+\frac{0\cdot 450}{2\cdot 97}\times 7\right)$$

$$= 27.888 \times 10^3 \text{ mm}^3$$

Step 8. Check for bending stress

Maximum bending stress in the purlin from Eq. 9.34

$$\sigma_b = \frac{M_{UU}}{\sigma_b}\left(1+\frac{M_{VV}}{M_{UU}}\cdot\frac{Z_{UU}}{Z_{VV}}\right)$$

$$\sigma_b = \left(\frac{2.97\times 10^6}{65\cdot 11\times 10^3}\right)\cdot\left(1+\frac{1\cdot 144}{1\cdot 9828}\times 5\cdot 61\right)$$

$$= 84.388 \text{ N/mm}^2$$

$$< 165 \text{ N/mm}^2\text{. Hence satisfactory}$$

Step 9. Local effects

It is to note that the local effects are for the design of fixtures and not for the design of trusses or even purlins. The net wind pressure at the ridge due to local effects as the wind is parallel to the ridge (From Example 9.10)

$$W_w = -(1.1344 + 0.5)\, p_d \,.\, A$$
$$= -1.6344 \,.\, p_d \,.\, A$$

This is $\left(\dfrac{1\cdot 6344}{1\cdot 20}\right)$ = 1362 times that on the normal purlin.

Example 9.12. *In Example design a channel-section purlin.*

Solution

Design : In Example 9.11, the dead load and its components normal to and parallel to the roof have been calculated in step 1. The imposed (live) load and its components normal to and parallel to the roof have also been determined in step 2. The wind load parallel to the roof acting normal to the roof have been found in Step 3. The worst combinations of loads for the design of purlin have been described in step 4. The section modulus of section for the purlin for dead load and live load combinations from Step 5, Example 9.11

$$Z_{UU} = 60.55 \times 10^3 \text{ mm}^3$$

From IS : 808, select MC 125, @ 0.131 kN/m section modulus

$$Z_{UU} = Z_{XX} = 68.1 \times 10^3 \text{ mm}^3$$
$$Z_{VV} = Z_{YY} = 13.4 \times 10^3 \text{ mm}^3$$

It is to note that *XX* and *YY* axes are also the axes of symmetry for channel-section. Therefore, there are also principal axes of the action.

$$\left(\frac{Z_{UU}}{Z_{VV}}\right) = \left(\frac{68\cdot 1\times 10^3}{13\cdot 4\times 10^3}\right)$$
$$= 5.082$$

Step 1: Check for the bending stress

Maximum bending stress in the purlin from Eq. 9.34

$$\sigma_b = \frac{M_{UU}}{\sigma_{VV}}\cdot\left(1+\frac{M_{VV}}{M_{UU}}\cdot\frac{Z_{UU}}{Z_{VV}}\right)$$

$$\sigma_b = \left(\frac{1\cdot 9828\times 10^6}{68\cdot 1\times 10^3}\right)\cdot\left(1+\frac{1\cdot 144}{1\cdot 9828}\times 5\cdot 082\right)$$
$$= 144.489 \text{ N/mm}^2$$

< 165 N/mm^2 . Hence satisfactory

Step 2. Design of purlin for dead load plus wind load

From Example 9.11 total design moment

$$M_{UU} = -2.97 \text{ kN-m}$$
$$M_{VV} = 0.45 \text{ kN-m}$$

Step 3. Check for bending stresses

Maximum bending stress in the purlin from Eq. 9.34

$$\sigma_b = \frac{M_{UU}}{\sigma_b}\left(1+\frac{M_{VV}}{M_{UU}}\cdot\frac{Z_{UU}}{Z_{VV}}\right)$$

$$\sigma_b = \left(\frac{2\cdot 97\times 10^6}{68\cdot 1\times 165}\right)\cdot\left(1+\frac{0\cdot 450}{2\cdot 97}\times 5\cdot 082\right)$$

$$= 77.194 \text{ N/mm}^2$$

$$< 165 \text{ N/mm}^2. \text{ Hence satisfactory}$$

Provide MC 125, @ 0.131 kN/m.

Step 4. Local effects

It is to note that the local effects are for the design of fixtures and not for the design of trusses and even for the purlins. The net wind pressure at the ridge due to local effects as the wind as parallel to the ridge, from Example 9.11

$$W_w = -(1.1344 + 0.5)\, p_d \cdot A$$

$$= -1.6344 \,.\, p_d \cdot A$$

This is $\left(\frac{1\cdot 6344}{1\cdot 20}\right)$ = 1.362 times that on the normal purlin.

Example 9.13 *Design an angle iron purlin for a trussed roof from the following data :*

Span of roof truss = *12 m*
Spacing of roof trusses = *5 m*
Spacing of purlins along the slope of roof truss = 1.2 m
Slope of roof trass = 1 vertical to 2 horizontal
Wind load on roof surface normal to roof = 1.04 kN/m²
Vertical loadfrom roof sheeting etc. = 0.200 kN/m².

Solution

Design :

Step 1: Slope of roof truss

Slope of roof truss = 1 vertical to 2 horizontal

$\therefore \tan\theta = \frac{1}{2}, \theta = 26.565° = 26° 34' < 30°$

The angle iron purlin may be designed according to IS: 800–1984.

Step 2. Vertical load on purlin per metre length

Vertical load from roof sheeting etc. = (0.200 × 1.2 × 1) = 0.240 kN/m

Assume self-weight of angle purlin = 0.120 kN/m

Total vertical load = 0.360 kN/m

Step 3. Wind load on roof surface normal to roof

(1.04 × 1.2 × 1) = 1.248 kN/m

Step 4: Total load (assumed to act normal to roof)

Therefore, load on purlin normal to roof

$$(1.248 + 0.360) = 1\ 608 \text{ kN/m}$$

The maximum bending moment in purlin,

$$M = \left(\frac{WL}{10}\right) = \left(\frac{1 \cdot 608 \times 5 \times 5000}{10}\right) = 4.02 \text{ kN-m}$$

The required modulus of section

$$Z = \left(\frac{4 \cdot 02 \times 10^6}{1 \cdot 33 \times 165}\right) = 18.318 \times 10^3 \text{ mm}^3$$

The depth of angle purlin

$$\frac{L}{45} = \left(\frac{5000}{45}\right) = 111.11 \text{ mm}$$

The width of angle purlin

$$\frac{L}{60} = \left(\frac{5000}{60}\right) = 83.33 \text{ mm}$$

From ISI Handbook No. 1, select ISA 125 mm × 95 mm × 6 mm. (ISA 125 95, @ 0.129 kN/m)

Modulus of section provided

$= 23.4 \times 10^3$ mm^3. Hence, the section is satisfactory.

Provide ISA 125 mm × 95 mm × 6 mm (ISA 125 95, @ 0·129 kN/m) for angle purlin for trussed roof.

Example 9.14 *A 10 m × 40 m godown is to be constructed. The steel roof trusses will be used for roofing. The trusses will be supported over masonry walls 300 mm thick. Galvanized corrugated iron sheets will be used for covering. Propose a suitable type of roof truss.*

The basic wind pressure is 1.04 kN/m^2, and there is no snowfall. Determine the load at each point.

Solution

Design :

Step 1. Preliminary geometry of truss

Clear spacing between the masonry walls = 10 m

Thickness of masonry = 0.30 m

Centre to centre spacing between the masonry walls = 10.30 m

Therefore, span of roof truss = 10.30 m

For galvanized corrugated iron sheet,

Provide pitch of roof truss = $\frac{1}{4}$

Rise of roof truss $= \frac{1}{4} \times \text{span} = \left(\frac{1}{4} \times 10.30\right) = 2.575$ m

Let θ be slope of roof truss for pith of $\frac{1}{4}$

$$\tan\theta = \frac{1}{2}, \qquad \therefore \theta = 26.565° = 26° 34'$$

Length along the sloping roof

$$= \left[\left(\frac{10.30}{2}\right)^2 + (2.575)^2\right]^{1/2} = 5.76 \text{ m}$$

The length along the sloping roof may be divided into six equal panels.

Length of each panel $= \left(\frac{5.76}{6}\right) = 0.960$ m (> 1.40 m)

Therefore, for above proportions, a French roof truss as shown in Fig. 9.41 (a) may be selected. These trusses may be provided at 4 m spacing. 11 trusses will be provided over the entire length of godown. The purlins will be provided at each panel point.

Step 2 : Load at each panel point is determined as under :

1. Dead load

Weight of galvanized corrugated iron sheets = 0.133 kN/m^2

Weight of purlins (assumed) = 0.150 kN/m^2

Weight of bracing (assumed) = 0.015 kN/m^2

Self weight of roof truss

$$w = \frac{1}{100}\left(\frac{l}{3}+5\right) = \frac{1}{100}\cdot\left(\frac{10.30}{3}+5\right)$$

$$= 0.084 \text{ kN/m}^2$$

Total dead load = (0.133 + 0.150 + 0.015 + 0.084)

= 0.382 kN/m^2

Spacing of trasses = 4 m

Panel length in plan = 0.960 cos (26° 34′) = 0.858 m

Load at each intermediate panel due to dead load, w_1

(0.382 × 4 × 0.858) = 1.311 kN

Load at end panel due to dead load $\left(\frac{w_1}{2}\right)$ = 0.656 kN.

2. Live load

Access is not provided for roofing except for maintenance. Slope of roof truss

= 26° 34′ = 26.565°

∴ Live load = (0.750 – 0.020 × 16.565) = 0.4187 kN/m^2

Load each intermediate panel due to the live load, w_2

0.4187 ×4 × 0.960 = 1.61 kN

Load at each end panel due to live load $\left(\frac{w_2}{2}\right)$ = 0.804 kN

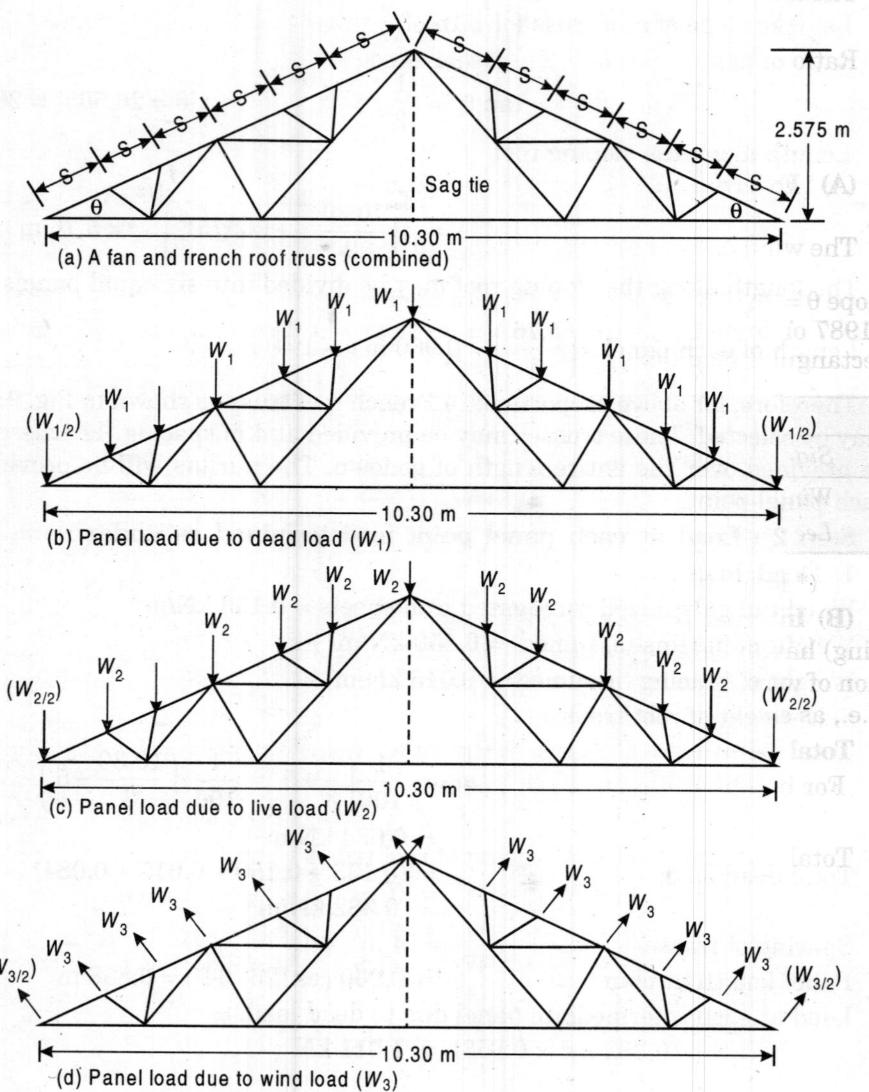

Fig. 9.42 *Dead live and wind loads at panel points of roof truss*

3. Wind load

Slope of roof trass = 26.565°

Intensity of design wind pressure = 1.4 kN/m²

For pitched roof, the height of eave, and the width of cladding, w are also necesary data. Let the height of plinth be 600 mm and the projection of eave girder beyond the centre line be 400 mm. The gross width of the roof-shed.

$$w = (10 + 2 \times 0.400)$$

$$= 10.80 \text{ m assumed}$$

The height of eave of the roof-shed above ground level

$$h = (3.18 + 0.60) = 3.78 \text{ m assumed}$$

Ratio of height to the width of roof-shed

$$\left(\frac{h}{w}\right) = \left(\frac{3.78}{10.8}\right) = 0.35 < 0.5$$

(A) External air pressure coefficients

The wind pressure coefficients are to be adopted from a set $\left(\frac{h}{w}\right) < 0.5$ and for slope $\theta = 26.565°$. The wind pressure coefficients are noted from IS : 875 (Part 3) –1987 or from Table 9.7 (External presure coefficients, C_{pe} for pitched roofs of rectangular clad buildings).

	Wind angle, θ			
Side of pitched roof	*θ = 0°*	*	*θ = 9°*	*
Wind ward	–0.1374	–0.8687	–0.7	—
Lee ward	– 0.4	–1.1344	–0.7	—

(* local effects)

(B) Internal pressure coefficients. The godown shall be a building (cladding) having large opening. Let the large opening be in 40 m length and direction of wind be in the direction perpendicular to this length having the opening (i.e., as shown in Fig. 9.43).

Total wind load C_{pi} = + 0.8 parallel to the ridge

For buildings having large, opening

$$(c_{pe} - c_{pi}) = [-0.7 - (+0.8)] = -1.5$$

Total wind load (let the spacing of roof trusses be 4 m)

$$W_3 = (C_{pe} - C_{pi}).\, p_d .\, A$$

$$W_3 = -1.5 .\, p_d .\, A$$

$$W_3 = -1.5 \times 1.04 \times (0.960 \times 4) = 5.99 \text{ kN}$$

$$\left(\frac{W_3}{2}\right) = 2.995 \text{ kN}$$

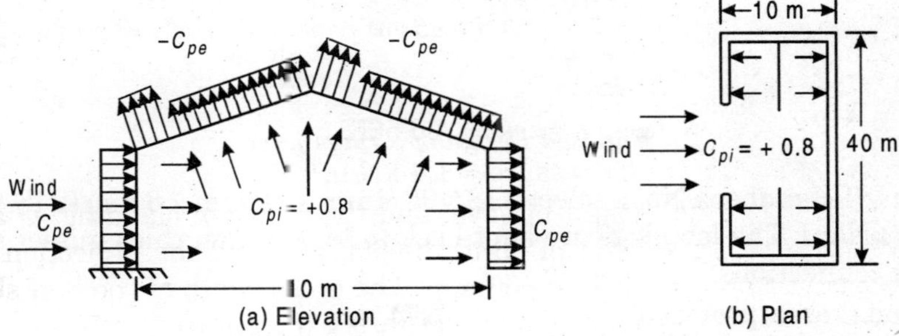

Fig. 9.43 *Buildings with large areas of openings on one side*

The dead load, live load and wind load at each panel point of truss are shown in Fig. 9.42 (b), (c) and (d) respectively.

Note. *For the design of various members, the stresses in members can be found for each loading either graphically or analytically. The maximum force in each member may be found for the combination of loads whichever is severe. The members may be designed for maximum force in them.*

Example 9.15 *At the apex joint A of a Fink roof truss, four members AB, AC, AD and AE meet so that ∠BAC = ∠EAD = 30° and ∠CAD = 60°. The lengths of members between the panel points are AB = AE = 3 metres, AC = AD = 3.4 metres and the stresses carried are AB = AE : 170 kN compression, and AB = AD : 100 kN tension. Design the members and detail the joint A.*

Solution

Design :

Step 1 : Design of compression members :

Force in *AB* = *AE* = 170 kN compression.

AB and *AE* are rafter members in Fink roof truss. The rafter members in a truss are continuous struts.

Length of members $AB = AE = 3$ m

Effective length $AB = AE = 0.85 \times 3$ m, ($l = 2.55$ m)

Let the value of yield stress for the structural steel be 260 N/mm², and the slenderness ratio of the compression member be 100. Therefore, the permissible stress in axial compression from Table 5.1, IS: 800–1984

$$\sigma_{bc} = 82 \text{ N/mm}^2$$

Area required for the compression member

$$\left(\frac{170 \times 1000}{82}\right) = 2073.17 \text{ mm}^2$$

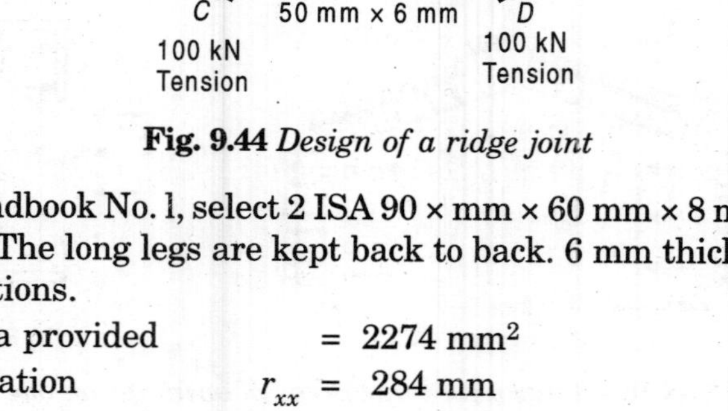

Fig. 9.44 *Design of a ridge joint*

From ISI Handbook No. 1, select 2 ISA 90 × mm × 60 mm × 8 mm (2 ISA 90 60, @ 0.089 kN/m). The long legs are kept back to back. 6 mm thick gusset plate is used for connections.

Sectional area provided = 2274 mm²

Radius of gyration r_{xx} = 284 mm

Radius of gyration r_{yy} = 24.5 mm

Minimum radius of gyration r_{min} = 24.5 mm

Maximum slenderness ratio

$$\left(\frac{l}{r_{min}}\right) = \left(\frac{2.55\times1000}{24.5}\right) = 104$$

From IS : 860–1984, allowable stress in axial compression

$$= 78.4 \text{ N/mm}^2$$

Safe force carrying capacity of each compression member

$$\left(\frac{78.4\times2274}{1000}\right) = 178.28 \text{ kN} > 170 \text{ kN}$$

(Compression in members). Hence, safe.

Provide 2 ISA 90 mm × 60 mm × 8 mm (2 ISA 90 60, @ 0089 kN/m) for *AB* and *AE*.

Step 2: Design of tension members

Force in members

$$AC = AD = 100 \text{ kN}$$

The allowable stress in axial tension is (0.6 × 260) N/mm²

$$= 156 \text{ N/mm}^2$$

Net area required for each member

$$\left(\frac{100\times1000}{156}\right) = 641.025 \text{ mm}^2$$

From IS1 Handbook No. 1, select 2 ISA 50 mm × 50 mm × 6 mm. (2 ISA 50 50, @ 0045 kN/m). The angles are provided on both sides of a gusset plate, 16 mm diameter rivets are used in one row.

Gross area of angle sections = 1136 mm²

Net area provided = (1136 – 2 × 23.5) = 776 mm²

Allowable axial tension in each member

$$\left(\frac{156\times776}{10}\right) = 121.056 \text{ kN} > 100 \text{ kN}$$

(Tension in member). Hence, safe.

Provide 2 ISA 50 mm × 50 mm × 6 mm (2 ISA 50 50, @ 0.045 kN/m) for *AC* and *AD*.

Step 3 : Design of Joint A

16 mm diameter power driven rivets are used for connections. Gross diameter of rivet

$$(16 + 1.5) = 17.5 \text{ mm}$$

Strength of rivet in double shear

$$\left[2\times\frac{\pi}{4}\times(17.5)^2\times\frac{100}{1000}\right] = 48.1 \text{ kN}$$

Strength of rivet in bearing

$$\left(17.5 \times 6 \times \frac{300}{1000}\right) = 31.5 \text{ kN}$$

Rivet value $R = 31.5$ kN

Number of rivets required to connect AB and AE individually

$$\left(\frac{170}{31.5}\right) = 5.397 \text{ (say 6)}$$

Number of rivets required to connect AC and AD individually 3.17 (say 4)

The complete design of joint A is shown in Fig. 9.44.

Example 9.16 *The sliding end of a roof truss rests on 450 mm brick wall through a concrete bearing pad. The maximum normal reaction on the bearing is 125 kN. The principal rafter is inclined at 30 ° to the main tie which is horizontal. If the panel length of the principal rafter is 1.38 metre, design the rafter, the tie and the sliding joint. Also design the anchor bolt for a pull of 7.50 kN.*

Solution

Design :

Normal reaction on bearing = 125 kN

Force in principal rafter = 125 sec 60° = 250 kN

Force in member = 250 cos 30° = 216.50 kN

Step 1 : Design of compression member

Force in principal rafter = 250 kN

Let the value of yield stress for the structural steel be 260 N/mm^2 and the slenderness ratio for the compression member be 100. Therefore, the permissible stress in axial compression from Table 5.1, IS: 800–1984.

$$\sigma_{bc} = 82 \text{ N/mm}^2$$

Area required for the compression member

$$\left(\frac{250 \times 1000}{82}\right) = 3048.78 \text{ mm}^2$$

From ISI Handbook No. 1, select 2 ISA 90 mm × 60 mm × 12 mm. (2 ISA 90 60, @ 0.130 kN/m) 10 mm thick gusset plate is used.

Sectional area provided = 3314 mm^2

Radius of gyration r_{xx} = 27.9 mm

Minimum radius of gyration r_{min} = 27.0 mm

Actual length of member = 2.60 m

The principal rafter is a continuous strut.

Effective length of member = 0.85 × 1.38

= 1.173 m

$$\text{Maximum slenderness ratio} = \left(\frac{1.173 \times 1000}{27.0}\right) = 43.44$$

The allowable stress in axial compression in 141.9 N/mm^2

The safe force carrying capacity of member

$$\left(\frac{3314 \times 141.9}{1000}\right) = 470.27 \text{ kN} > 250 \text{ kN}$$

(Actual force in member). Hence, sale.

Provide 2 ISA 90 mm × 60 mm × 12 mm (2 ISA 9060, @ 0.130 kN/m) for principal rafter.

Step 2 : Design of tension member

Force in the member = 216.50 kN

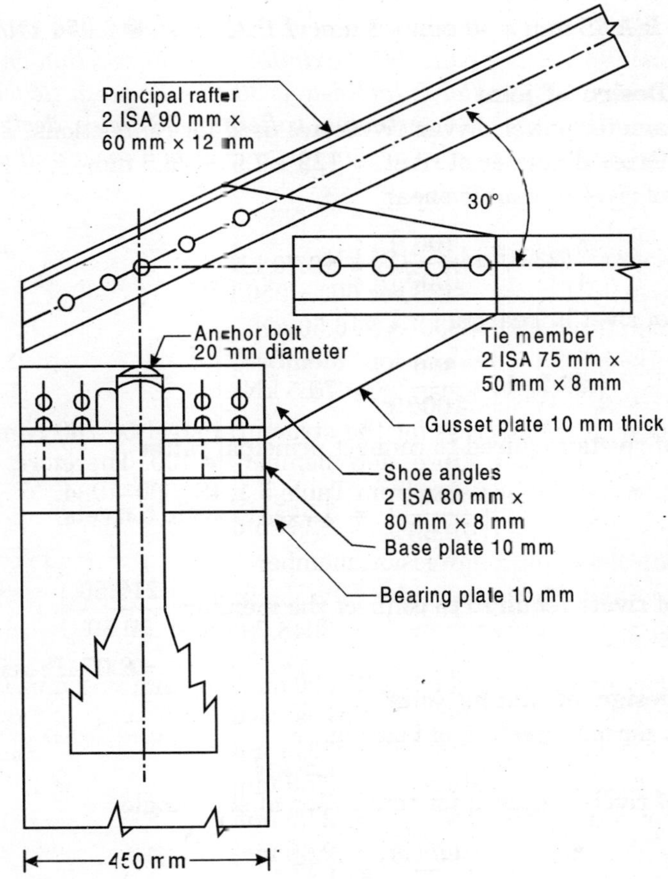

Fig. 9.45 *Design of sliding end of a roof truss*

Allowable stress in axial tension = (0.6 × 260) = 156 N/mm^2

Net area required for each member

$$\left(\frac{216.50 \times 1000}{156}\right) = 1387.82 \text{ mm}^2$$

From ISA Handbook No. 1, select 2 ISA 75 mm × 50 mm × 8 mm. (2 ISA 75 50, @ 0.074 kN/m).

The angles are provided on both sides of gusset plate. 22 mm diameter rivets are used in one row for connections.

$$\text{Gross area provided} = 1876 \text{ mm}^2$$

$$\text{Net area provided} = (1876 - 2 \times 8 \times 23.5)$$

$$= 1516 \text{ mm}^2$$

Allowable axial tension in member

$$\left(\frac{156 \times 1516}{1000}\right) = 236.496 \text{ kN} > 21650 \text{ kN. Hence, safe.}$$

Provide 2 ISA 75 mm × 50 mm × 8 mm (2 ISA 75 80, @ 0.074 kN/m) for main tie.

Step 3 : Design of joint

22 mm diameter power driven rivets are used for connections.

$$\text{Gross diameter of rivet} = (22 + 1.5) = 23.5 \text{ mm}$$

Strength of rivet in double shear

$$\left[2 \times \frac{\pi}{4}(23.5)^2 \times \frac{100}{1000}\right] = 86.70 \text{ kN}$$

Strength of rivet in bearing

$$\left(23.5 \times 10 \times \frac{300}{1000}\right) = 70.5 \text{ kN}$$

Number of rivets required to connect principal rafter

$$\left(\frac{250}{70.50}\right) = 3.545 \text{ (Provide 5 rivets)}$$

Number of rivets required to connect the member $\left(\frac{216.50}{70.50}\right)$

$$= 3.07 \text{ (Provide 4 rivets)}$$

Step 4: Design of sliding joint

Maximum normal reaction of bearing

$$= 125 \text{ kN}$$

Number of rivets required for connection of shoe angles

$$\left(\frac{125}{70.50}\right) = 1.77$$

4 rivets are provided to connect shoe angles with gusset plate. 4 rivets are also provided to connect shoe angles with base plate. 2 ISA 80 mm × 80 mm × 8 mm, 450 mm long are used for shoe angles.

1. Bearing plate

Normal reaction = 125 kN

Length of base plate = 45 mm

Width of bearing base plate

$$(80 + 80 + 10) = 170 \text{ mm}$$

Bearing pressure on concrete bearing pad

$$\left(\frac{125 \times 1000}{450 \times 170}\right) = 1.634 \text{ N/mm}^2$$

< Allowable bearing pressure in concrete pad

Consider 1 mm strip of base plate. Bending moment about xx as shown in Fig. 9.38

$$M = \left(\frac{1.634 \times (80-8)^2}{2}\right)$$

$$= 4235.33 \text{ N-mm}$$

Moment of resistance of the base plate

$$\left(185 \times 1 \times \frac{t^2}{6}\right) = 4235.33,$$

$$\therefore \quad t = 11.72 \text{ mm}$$

Therefore, the thickness of base plate required

$$t_1 = (11.72 - 8) = 3.72 \text{ mm}$$

Provide 6 mm thick base plate 450 mm × 170 mm × 6 mm bearing plate below the base plate. An elliptical hole is kept on each side of shoe angles and base plate. The base plate can slide over bearing plate.

2. Anchor bolt

Pull in anchor bolt = 7.50 kN

Allowable axial tension in anchor bolt is 0.6 × 260 N/mm²

$$= 156 \text{ N/mm}^2$$

Area required at the roof of thread

$$\left(\frac{7.50 \times 1000}{156}\right) = 48.07 \text{ mm}^2$$

Two 20 mm nominal diameter anchor bolts are provided on each side of shoe angles. A complete sliding joint is shown in Fig. 9.43.

PROBLEMS

9.1. Locate the principal axes of ISA 90 mm × 60 mm × 10 mm. [ISA 9060, @ 0.110 kN/m]. Determine the principal moment of inertias and radii of gyration about the principal axes for the angle section. Compare these values with those given in ISI Handbook No. 1. Other geometrical properties may be noted from handbook.

9.2. In Problem 9.1, the angle section is subjected to bending moment 2 kN-m acting in the vertical plane through the centroid of section. Determine the maximum bending stress induced in the section.

9.3. An *MC*, 350. @ 0.421 kN/m is subjected to bending moment 8 kN-m. The plane of loading is inclined 10° with the *YY*-axis in anti-clockwise direction, and it passes through the centroid of channel section. Determine the maximum bending stress induced in the section. Locate the neutral axis.

9.4. In Problem 9.3, the channel section is carrying uniformly distributed load 5 kN/m. The effective span is 5 m. The section is simply supported at both ends. The plane of loading is inclined 10° with *YY*-axis and it passes through the centroid of beam section. Determine the magnitude and direction of maximum deflection. Take $E = 2.04 \times 10^5$ N/mm^2.

9.5. Draw Z-polygon for *MC*, 350 @ 0421 kN/m. Determine the maximum bending stress induced in the section for data given in Problem 9.3. Also determine the absolute maximum bending stress for the given moments and corresponding inclition of plane of loading with *YY*-axis.

9.6. It is proposed to construct a shed in Jodhpur. The average height of roof above ground level is 12 m. Determine the wind load for the following :

(i) When degree of permeability is small. (ii) When shed is having normal permeability. (iii) When shed is having large opening.

9.7. A gable with corrugated asbestos roofing is supported on steel trusses of span 9 m, rise 2.25 m, spaced at 3 m c/c. The dead load is 1.220 kN/m^2 normal to roof. Taking purlins one at each end of the slope and the others equally spaced, design an unequal angle purlin with long leg perpendicular to the slope of roof.

9.8. Design an angle iron purlin for a roof with the following data :

Span of truss = 8 m

Spacing of trusses = 4 m c/c

Pitch of truss = 1/4

Spacing of purlins along the slope of truss = 120 m c/c

Roof covering = Asbestos sheets

Basic wind pressure = 1.04 kN/m^2.

9.9. A 10 m × 50 in godown is to be coverted by roof trusses. Propose a suitable type of roof truss for the purpose and design the various members of the truss.

9.10. The forces in the members of the roof truss at joint *A* , *AB* compression 500 kN and *AD* tension 600 kN. Design the members *AB* and *AD* and the joint *A* as shown in Fig. P.9.10. Sketch the joint *A* including shoe detail.

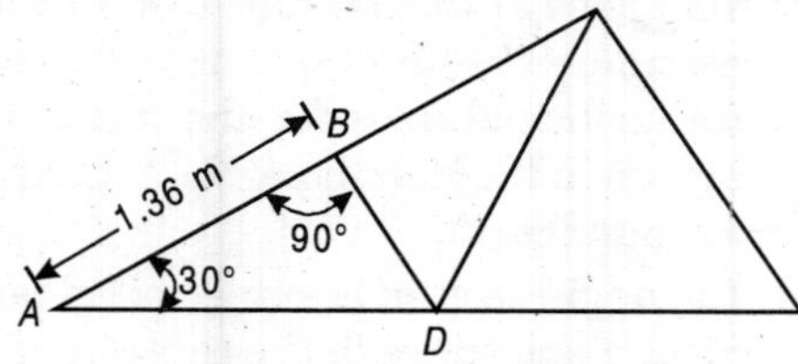

Fig. P 9.10

9.11. In steel roof truss, the vertical reaction at the left hand support is 500 kN. The principal rafter at this support is inclined at an angle of 30° to the horizontal. Design the main tie, the principal rafter, and the riveted joint. The panel length of the principal rafter may be assumed to be 1.28 m. Draw a fully dimensioned neat sketch.

9.12. For a steel roof with an asbestos cement sheet covering design an I-section purlin from the following data :

Spacing of trusses = 4 m

Spacing of purlins of principal rafter = 1.32 m

Inclination of principal rafter to the main tie = 30°

Wind pressure of roof = 1.000 kNm^2.

Chapter 10

Design of Welded Joints

10.1 INTRODUCTION

The *welding* is one of the methods of connecting the structural members. In the welding, a metallic link is made between the structural members. The *weld* is defined as a union between two pieces of metal at faces rendered plastic or liquid by heat or by pressure or both. A typical weld is shown in Fig. 10.1. This shows the various zones of the weld. In the *fusion zone,* the portion of the structural members has been fused (i.e., made plastic or liquid by heat). In the weld metal zone, the portion of structural members to be connected and *filler metal* are melted or made plastic in making a weld. The portion of the structural members to be connected and *filler metal* melted or made plastic is termed as *weld metal.* The metal added in the welding is termed *as filler metal.* In the *heat-affected zone,* the parent metal is metallurgically affected by the heat of welding.

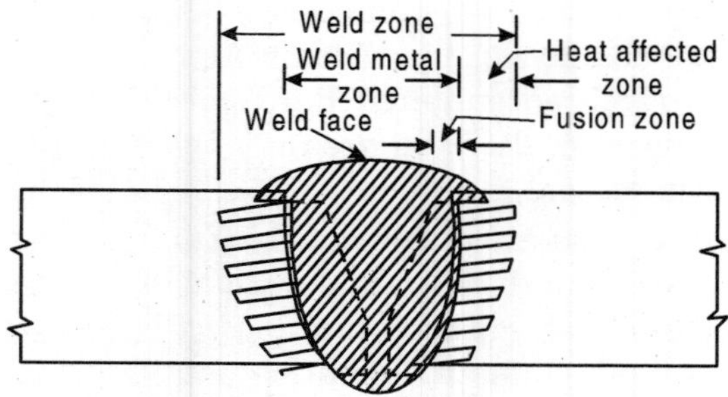

Fig. 10.1 *Various zones of a typical weld*

The welding process in which the weld is made between metals in a state of fusion without hammering or pressure is known as *fusion welding.* The welding process in which the weld is made by sustained pressure while the surfaces to be united are plastic is known as *pressure welding*. The welding is done by several processes. These processes are listed below :

10.1.1 Fusion Welding

(i) *Arc Welding*

(a) Metal arc welding (Shielded/or unshielded)

(b) Carbon arc welding (Shielded/or unshielded)

(ii) *Gas welding*

(a) Oxy-Acetylene welding

(b) Oxy-Hydrocarbon welding

10.1.2 Pressure Welding

(i) Resistance welding

(ii) Pressure welding

10.1.3 Thermit Welding

10.1.4 Union Melt Welding

In general, the process of "metal arc welding" is used for the structure steel connections. The procedure for metal arc welding is covered by IS : 823 –1964.

In the metal arc welding, the electrodes are used. When considering electrods, it is necessary always to bear in mind the fact that heated metal combines chemically with oxygen and nitrogen from the atmosphere forming oxides and nitrides, which result in brittle poor quality weld metal. For this reason, it is essential to provide some means of preventing the atmosphere from reacting the hot weld area. This can be done either by shrouding the area with an inert gas or by the use of a suitable *flux. The flux is fusible material used in welding or oxygen-cutting to dissolve and facilitate removal of oxides and other undesirable substances.*

The metal arc welding was originally done with bare electrodes consisting of a piece of wire or rod of the same composition as to the metal to be welded. For the reasons mentioned above, the welds were of poor quality, and thus bare electrodes are now rarely encountered. Bare wire is of course used for automatic, welding, but in this case, the weld area is protected from the atmosphere by powdered flux or inert gas.

Most modern electrodes are of the coated or covered type consisting of a metal core wire surrounded by a thick coating applied by extrusion, winding or other process. The success of welding operation depends on the composition of coating, which is varied to suit different conditions and metals. The coating on the electrodes play the following function :

1. To stabilise the arc and enable the use of alternating current (if necessary)
2. To flux away any impurities present on the surface being welded.

3. To form a slag over the weld to
   (a) Protect the molten metal from the atmospheric contamination,
   (b) Slow the rate of cooling of the weld and to reduce chance of brittleness, and
   (c) Provide a smooth surface by reducing the ripples caused by welding operations.
4. To add certain constituents to the weld metal to compensate for the loss of any volatile alloying elements or any constituent lost by oxidation.
5. To speed up the welding operation by increasing the rate of melting.

The various types of electrodes are :

(i) the normal penetration electrodes,
(ii) the deep penetration electrodes,
(iii) the automatic electrodes,
(iv) shielded arc electrode, and
(v) low hydrogen electrodes.

The electrodes are used to suit the different conditions. The electrodes have the following types of flux covering. The numbers indicating the digit refer to type of covering.

Digit (type of covering)

1. Having a high cellulose content.
2. Having a high content of titania and producing a fairly viscous slag.
3. Containing an appreciable amount of titania and producing a fluid slag.
4. Having a high content of oxides/or silicates of iron and manganese and producing an inflated slag.
5. Having an high content of iron oxide and/or silicates and producing a heavy solid slag.
6. Having a high contents of calcium carbonate and fluoride.
7. Having a high content of iron powder.
8. A covering of any type not classified above.

The code number of covered electrodes shall consist of seven essential and one or two additional symbol, each indicating a specific characteristic of the electrode. The essential symbol shall consist of the letter *M* indicating the suitability of the electrode for metal arc welding and six digits in Hindu – Arabic numerical each denoting the following :

(a) Type of covering material (mentioned above).
(b) Welding position in which the electrode may be used.
(c) Welding current conditions.
(d) Ultimate tensile strength of deposited metal.
(e) Percentage elongation in tensile test of disposited metal, and
(f) Impact test values of deposited metal.

The essential symbols are followed by letter *P* or *A* or both, denoting the electrode for deep penetration, or automatic welding or for both purposes. The size of electrode is designated by the diameter of the core wire in mm.

It became possible to provide the perfect rigid joints in the rigid frame structures for the designers with the advent of welding. There is no loss of cross-section of a member in a welded joint. In riveted joint, the cross-section area is reduced due to rivet holes. The welded joints are capable to transmit axial forces, shears and moments.

10.2 ADVANTAGES OF WELDING

The use of welding is increasing rapidly because of its several advantage over riveting. The advantages of welding are as under:

1. There is *silence* in the process of welding.

   When the riveting is done, the rivet head is prepared by hammering. The hammering causes noise in the process of riveting.

2. There is *safety* of welding operator in the welding.

   The riveting is done cold as well as hot. While hammering, rivets may fly away and injure the persons. When the hot riveting is done, the rivets are made red hot and handed over to place them in position. During transactions, the red hot rivets may fall down and strike the workers.

3. The welding may be done *quickly* in comparison to the riveting.
4. The welded joints have better *appearance* than riveted joints.
5. The welded joints are more *rigid* than the riveted joints.
6. The welding is more *adaptable* than riveting.

   In the complicated structures, riveting may be extremely difficult or not possible but the welding may be done easily. For circular steel pipes and tube connections, riveting is practically impossible, but welding may be adopted conveniently. The welding is also convenient for additions in the existing structures.

7. The welded joints have more *efficiency* than riveted joints. The efficiency of a joint is the ratio of strength of the joint to the strength of the solid plate. In the welded joints, holes are not made as it is done in riveted joints. The strength of joint even in a tension member may be obtained equal to the strength of solid member. As such, in welded joint efficiency of joint may be achieved equal to 100 per cent. In the riveted joint, the efficiency of a joint in general is about 80 percent only.

8. The welding is *economical* in comparison to the riveting.

   In the welding connecting angles, gusset plates, splice plates and rivet head are not required. Thus, it reduces the weight of structure also. In the riveted connections, the minimum size is governed by the diameter of the rivet besides the requirement of force to be transmitted. Even if a smaller size of a member is sufficient to transmit the force, but the minimum size of member is to be used, which may provide minimum edge distance and minimum pitch for connecting rivets. In the welding, in absence of such restrictions, small size members may be used.

In the welding, the cost of labour is less than the riveting. The cost of painting and other maintenance is also less.

Overlapping of mild steel bars at joints in reinforced cement concrete structures is not necessary, if the bars are welded.

10.3 DISADVANTAGES OF WELDING

Besides the several advantages, there are disadvantages of welding. The disadvantages of welding are as under:

1. The members are likely to *distort* in the process of welding.
   Uneven heating and cooling are caused due to welding. As a result of this, the distorsion of members may take place.
2. A welded joint fails earlier than riveted joint, if the structure is under *fatigue stresses.*
3. There is a greater possibility of *brittle fracture* in welding than the riveting.
4. The *inspection* of welded joint is more difficult and more expensive than the riveted joint.
   The inspection of riveted joint may be done simply by tapping with a hammer. For inspection of a welded joint, a stethoscope and a light hammer is required. Some of the defects in a weld may be detected by this. The inspection of welded joint may also be done by the magnetic power method, radiography by means of X-rays, Gamma rays or by trepanning.
5. A *more skilled person* is required in the welding than in the riveting.
6. The welded joints are over rigid.

10.4 TYPES OF WELDED JOINTS

The various types of welded joints are as follows :

1. Butt weld
2. Fillet weld
3. Slot weld and Plug weld
4. Spot weld
5. Seam weld
6. Pipe weld

10.5 BUTT WELD

A *butt weld* is also termed as groove weld. The butt weld is used to joint structural members carrying direct compression or tension. It is used to make tee-joint and butt-joint. Various types of butt joint are as under. These are named depending upon shape of the grove made for welding.

10.5.1 Square Butt Weld

A *square butt weld* is a weld in the preparation which the fusion faces lie approximately at right angles to the surfaces of the components to be joined and are substantially parallel to one another. The square butt weld is shown in Fig. 10.2.

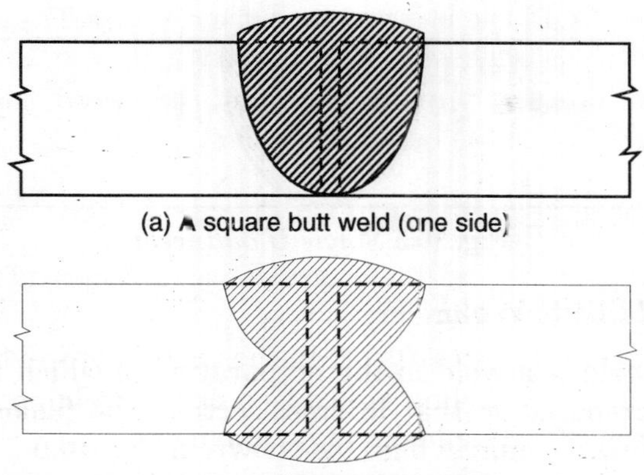

(a) A square butt weld (one side)

(b) A square butt weld (both sides)

Fig. 10.2

10.5.2 Single-V Butt Weld

A *single-V butt weld is a weld* in the preparation of which the edges of both components are prepared so that in the cross-section, the fusion faces form a 'V' as shown in Fig. 10.3.

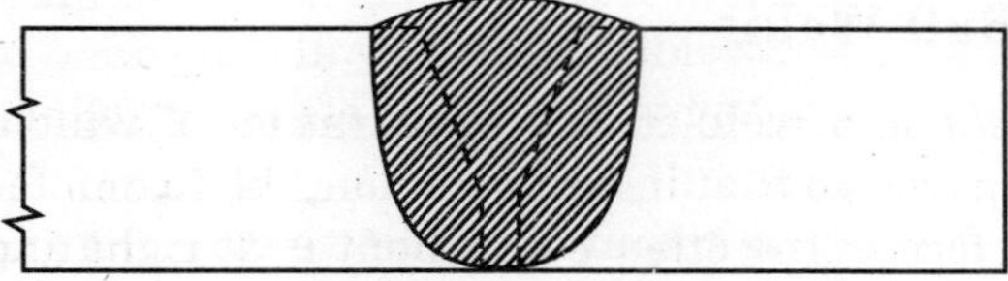

Fig.10.3 *Single-V butt weld*

10.5.3 Double-V Butt Weld

A *double-V butt weld* is a weld in the preparation of which the edges of both components are double bevelled so that (in cross-section, the fusion faces form two opposing 'V's as shown in Fig. 10.4.

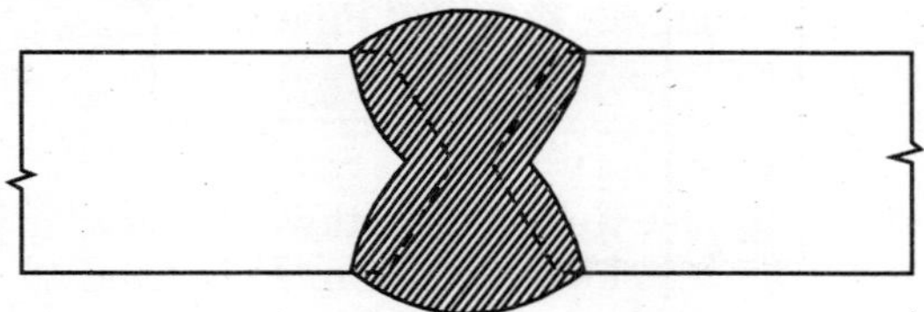

Fig. 10.4 *Double-V butt weld*

10.5.4 Single-U Butt Weld

A *single-U butt weld* is a weld in the preparation of which the edges of both components are prepared so that in cross-section, the fusion faces form a 'U' as shown in Fig. 10.5.

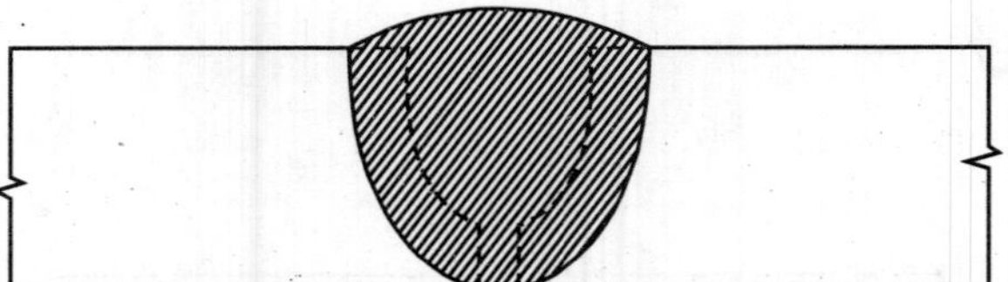

Fig. 10.5 *Single-U butt weld*

10.5.5 Double-U Butt Weld

A double-U butt weld is a weld in the preparation of which the edges of both components are prepared so that in cross-section, the fusion faces form two opposing 'U's' having a common base, as shown in Fig. 10.6.

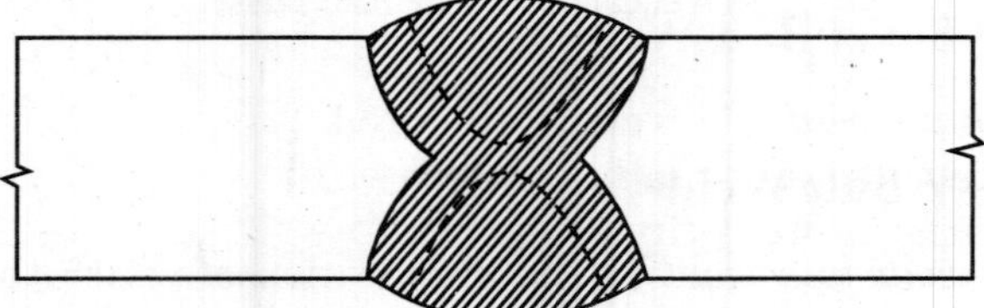

Fig. 10.6 *Double -U butt weld*

10.5.6 Single-J Butt Weld

A *single-J butt weld* is a weld in the preparation of which the edges of one component are prepared so that in cross-section, the fusion face is in the form of a *T* and the fusion face of the other component is at right angles to the surface of the first component as shown in Fig. 10.7.

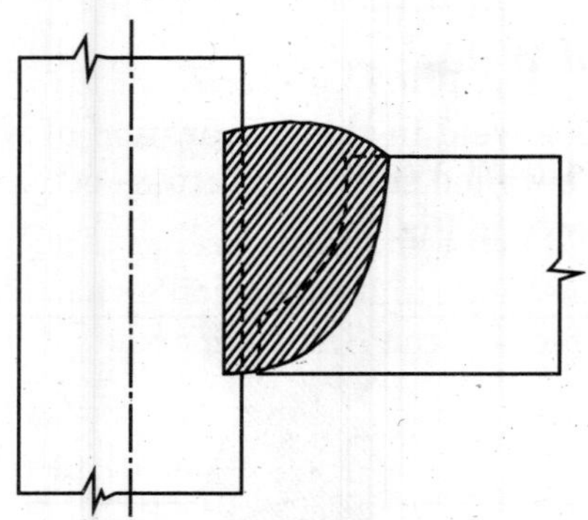

Fig. 10.7 *Single-J butt weld*

10.5.7 Double-J Butt Weld

A *double-J butt weld* is a weld in the preparation of which the edges of one component are prepared so that in cross-section, the fusion face is in the form of two opposing '*J*'s and the fusion face of the other component is at right angles to the surfaces of the first component as shown in Fig. 10.8.

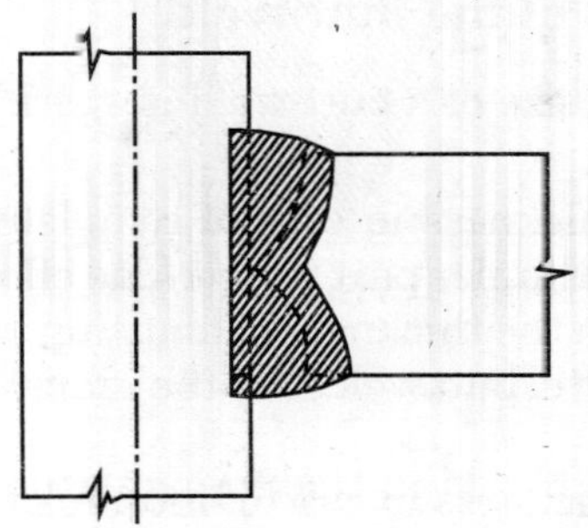

Fig. 10.8 *Double-J butt weld*

10.5.8 Single-Bevel Butt Weld

A *single-bevel butt weld* is a weld in the preparation of which the edge of one component is bevelled and the fusion face of the other component is at right angles to the surfaces of the first component as shown in Fig. 10.9.

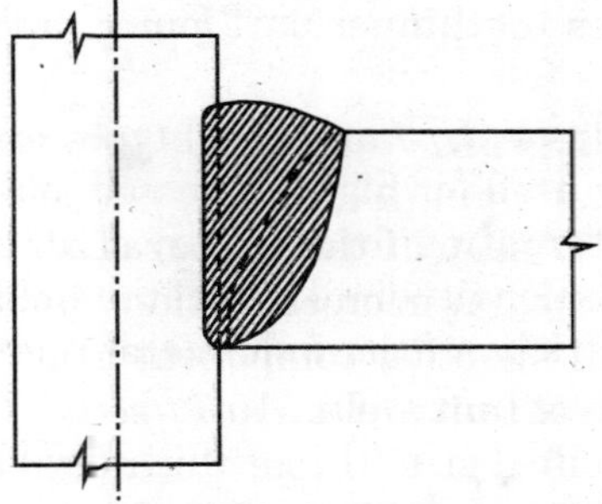

Fig. 10.9 *Single-bevel butt weld*

10.5.9 Double-Bevel Butt Weld

A *double-bevel butt weld* is a weld in the preparation of which the edges of one component are double bevelled and the fusion face of the other component is at right angles to the surfaces of the first components as shown in Fig. 10.10.

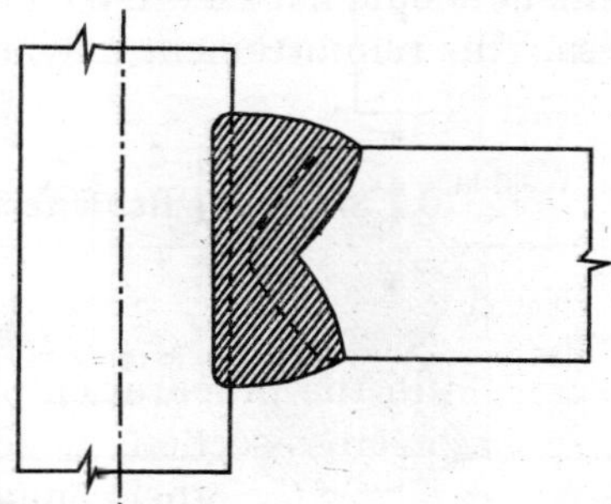

Fig. 10.10 *Double-bevel butt weld*

10.5.10 Specifications of the Butt Weld

(i) Size of butt weld. The size of a butt weld is specified by *the effective throat thickness.*

The effective throat thickness in case of complete penetration butt weld is taken as the thickness of thinner part joined. The effective throat thickness of *T* or *L* butt-joints is equal to the thickness of abutting part. The double-*V*, double-*U*, double-*J* and double-bevel butt welds are the examples of complete penetration butt weld.

The effective throat thickness in case of incomplete penetration butt weld is taken as 7/8th of the thickness of the thinner part joined. But for the purpose of stress calculation, a required effective throat thickness not exceeding 5/8th of the thickness of thinner part joined should be used. An incomplete penetration butt-weld is also termed as *unsealed single butt weld.* Single-*V*, single-*U*, single-*J* and single- bevel butt joints are the examples of incomplete penetration butt weld. In incomplete penetration butt weld, the weld metal is not deposited intentionally through the full thickness of the joint. The unwelded portion in incomplete penetration butt weld welded from both sides shall not be greater than 1/4th of the thickness of thinner part joined, and should be central in the depth of the weld.

The unsealed butt welds *V, U, J* and bevel types and incomplete penetration butt welds should not be used for highly stressed joints and joints subjected to dynamic, repeated or alternating forces. They shall also not be subjected to a bending moment about the longitudinal axis of the weld other than that normally resulting from the eccentricity of the weld metal relative to the parts joined.

(ii) Effective length of butt weld. *The effective length of butt weld* is the length for which the specified size (throat thickness) of the weld exists.

(iii) Effective area of butt weld. *The effective area of a butt weld* is taken as the product of the effective throat thickness and the effective length of butt weld.

(iv) Reinforcement. The extra metal deposited proud of the surface of the parent metal as shown in Fig. 10.11, is called reinforcement. This reinforcement is provided to provide sufficient surfaces convexity and to ensure full effectiveness at the joint. This requires a minimum practical surface convexity of 1.0 mm. This reinforcement should not exceed 3.0 mm. This is not considered as part of throat thickness. This reinforcement may also be removed if a flush surface is desired.

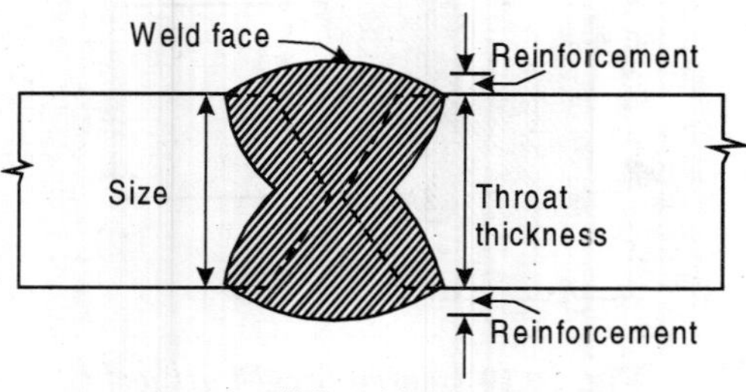

Fig. 10.11

Intermittent butt welds are used to resist shear only. The effective length of intermittent weld is taken not less than four times the thickness of thinner part joined. The longitudinal space between the effective lengths of welds is taken not more than 16 times the thickness of thinner part joined. The intermittent weld is not used in position subjected to dynamic, repetitive and/or alternate stresses.

When the structural members of unequal thickness are butt welded and difference in thickness of members exceeds 25 per cent of the thinner part or 3.0 mm in metal are welding and 6.0 mm or more in oxy acetylene welding, the thicker part is bevelled so that the slope of the surface from one part to the other is not steeper than one in five as shown in Fig. 10.12 (a). Where this arrangement is not practicable, the weld metal should be built-up at the junction with the thicker part to dimension at least 25 per cent greater than that of the thinner part (in metal are welding) as shown in Fig. 10.12 (b), alternatively, the weld metal should be built-up to the dimensions of thicker members as shown in Fig. 10.12 (c).

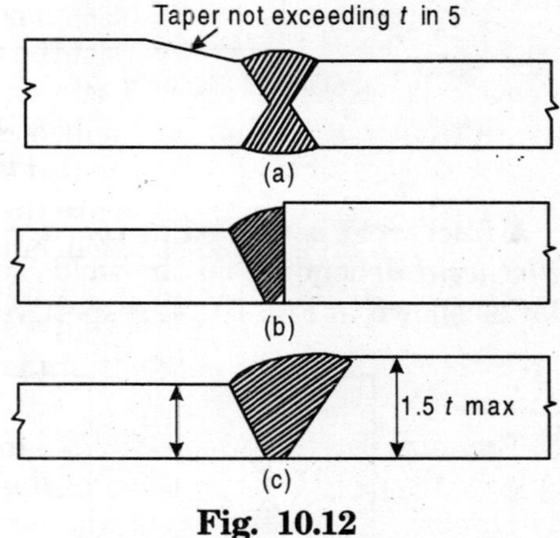

Fig. 10.12

In case of complete penetration butt weld, generally, design calculations are not necessary, as these will usually provide the strength at the joint equal to the strength of the member connected.

10.6 FILLET WELD

A fillet weld is a weld of approximately triangular cross-section joining two surfaces approximately as right angles to each other in lap joint, or tee-joint. A fillet weld is shown in Fig. 10.13 (a).

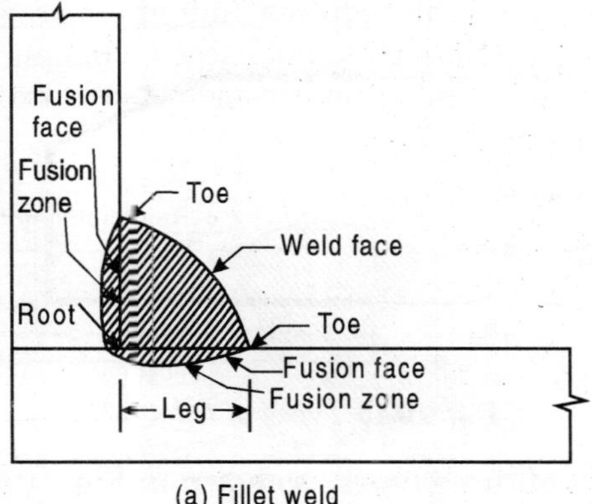

(a) Fillet weld

When the cross-section of fillet weld is 45°, isosceles triangle as shown in Fig. 10.13 (b) (i), it is known as a *standard fillet weld.* The standard 45° fillet welds are generally used. When the cross-section of the fillet weld is 30° and 60° triangle as shown in Fig. 10.13 (b) (ii), it is known as a *special fillet weld.*

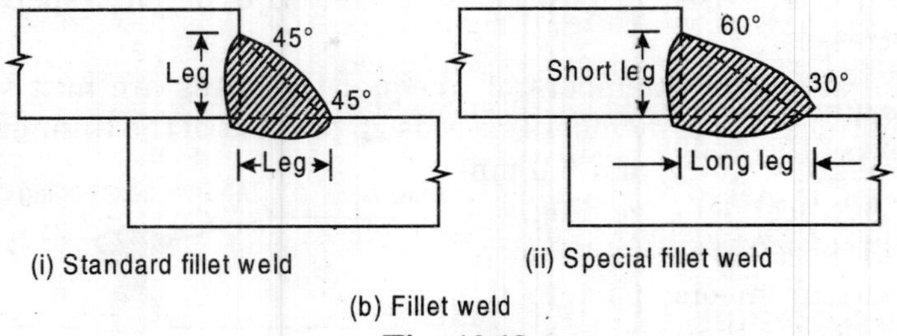

(b) Fillet weld

Fig. 10.13

A fillet weld is termed as *concave fillet weld* or *convex fillet weld* or *mitre fillet weld* depending on the weld face is concave or convex or approximately flat as shown in Fig. 10.14, respectively.

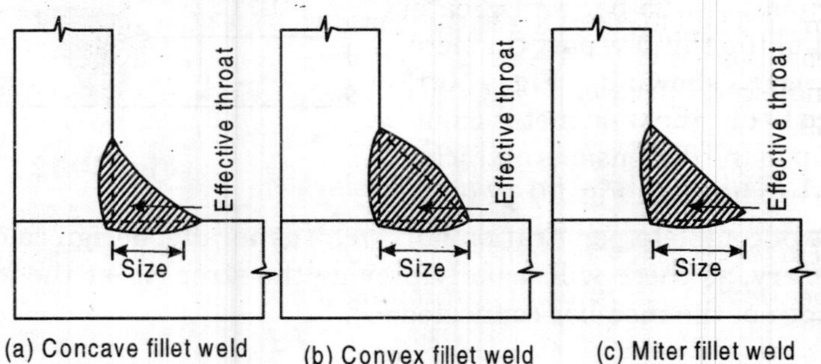

Fig. 10.14 *Fillet welds*

A fillet weld is termed as *normal fillet weld* or *deep penetration fillet weld* depending upon the depth of penetration beyond the root is less than 2.4 mm or (2.4 mm or more), respectively.

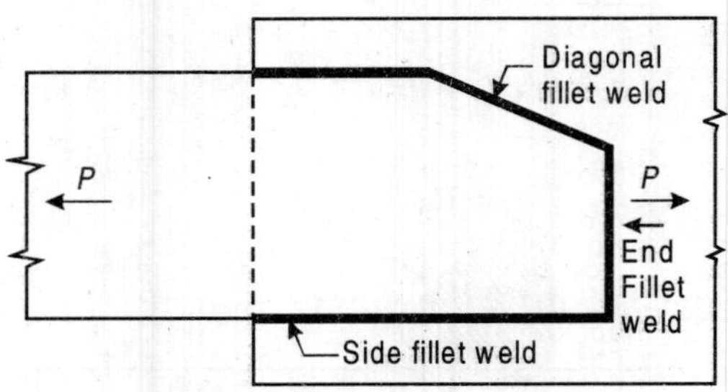

Fig. 10.15 *Types of fillet weld*

The fillet welds are of three types as shown in Fig. 10.15.

1. Side fillet weld

2. End fillet weld
3. Diagonal fillet weld.

1. Side fillet weld is a fillet weld stressed in longitudinal shear, i.e., a fillet weld, the axis of which is parallel to the direction of these applied load. It is also termed as *longitudinal fillet weld.*

2. End fillet weld is a fillet weld stressed in transverse shear, i.e., a fillet weld, the axis of which is at right angles to the direction of the applied load. It is also termed as *transverse fillet weld.*

3. Diagonal fillet weld is fillet weld the axis of which is inclined to the direction of the applied load.

A fillet weld is placed on the sides or end of the base metal and it is subjected to shear along with tension or compression and usually bending.

10.6.1 Specifications of Fillet Weld

10.6.1.1 Size of Fillet Weld

The size of normal fillet weld is specified as minimum leg length of a convex or mitre fillet weld or 1.414 times the effective throat thickness of a concave fillet weld. The size of deep penetration fillet weld is specified as minimum leg length plus 2.4 mm. The length of leg is the distance from the root to the toe of a fillet weld, measured along the fusion face as shown in Fig. 10.13 (a). The minimum size of single run fillet weld is adopted as per Table 10.1 as recommended in IS : 816 –1969.

Table 10.1 *Minimum size of first run or of a single run fillet weld*

Thickness of thicker part		*Minimum size*
Over (mm)	*Upto and including (mm)*	*(mm)*
—	10	3
10	20	5
20	32	6
32	50	8 First run 10 mm size of fillet

Notes 1. *When the minimum size of the fillet weld is greater than the thickness of the thinner part, the minimum size of the weld should be equal to the thickness of thinner part. The thicker part shall be adequately preheated to prevent cracking of the weld.*

2. *Where the thicker part is more than 50 mm thick, special precaution like preheating will have to be taken.*

10.6.1.2 Effective Throat Thickness

The effective throat thickness of a fillet weld is the perpendicular distance from the root to the hypotenuse of the largest isosceles right-angled triangle that can be inscribed within the weld cross-section as shown in Fig. 10.14. The effective throat thickness of a fillet weld shall not be less than 3 mm and shall generally not exceed $0.7t$ and $1.0t$ under special circumstances where t is the thickness of thinner part.

$$\text{Effective throat thickness} = \frac{1}{\sqrt{2}} \times \text{size of weld} = 0.7 \times \text{Size of weld}$$

In general, for the purpose of stress calculations, the effective throat thickness

$$= K \,.\, \text{Size of weld}$$

where, K is a constant.

The value of constant K for different angles between fusion faces is adopted as per Table 10.2 as recommended in IS : 816 –1969.

Table 10.2 *Value of K for different angles between fusion faces*

Angle between fusion	*Value of constant K*
60°–90°	0.70
91°–100°	0.65
101°–106°	0.60
107°–113°	0.55
114°–120°	0.50

A fillet weld is not used for joining parts, if the angle between fusion faces is greater than 120° or less than 60°.

10.6.1.3 Effective Length

The effective length of the weld is the length of the weld for which the specified size and throat thickness i.e., correctly proportioned cross-section of the weld, exists. *It is taken as the actual length minus twice the size of weld,* since the specified size and throat thickness do not exist at the ends. The effective length of the weld is shown on the drawings. In practice the actual length of weld is made equal to the effective length shown on the drawing plus twice the weld size.

The effective length of fillet weld should not be less than four times the size of the weld.

When the ends are returned, as shown in Fig. 10.16 then the ends should be carried continuous around the corners for distance not less than twice the size of weld. This should be applied particularly to side and top fillet welds in tension.

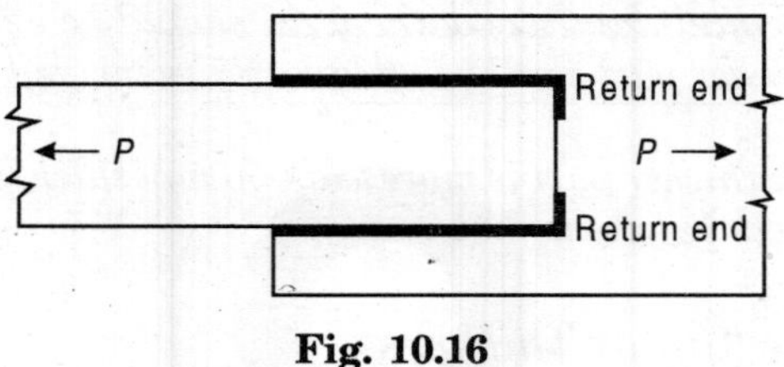

Fig. 10.16

10.6.1.4 Effective Area

The effective area of a fillet weld is effective is taken as the product of effective length and effective throat thickness.

10.6.2 Intermittent Fillet Welds

The intermittent fillet welds may be used to transfer calculated stress across a joint when the strength required is less than that developed by continuous fillet weld of the smallest allowable size for the thickness of the parts joined. Any section of intermittent fillet welding shall have an effective length of not less than four times the weld size with a minimum of 40 mm except for plate girder stiffeners.

The clear spacing between the effective lengths of intermittent fillet welds carrying calculated *stress shall* not exceed the following number of times the thickness of the thinner plate joined and shall in no case be more than 200 mm.

(i) 12 times for compression, and

(ii) 16 times for tension

The longitudinal fillet welds at the ends of built-up members shall have an effective length of not less than the width of the component part joined unless end transverse welds are used, in which case, the sum of the end longitudinal and end transverse welds shall be not less than twice the width of the component part.

The chain intermittent welding is to be preferred to staggered intermittent weld. Where the staggered intermittent welding is used, the ends of the component parts shall be welded on both the sides.

In a line of intermittent fillet welds, the welding should extend to the ends of parts connected for welds staggered about two edges, this applies generally to both the edges.

The intermittent fillet welds are not used in the case of main members of structures directly exposed to weather.

10.7 SLOT WELD AND PLUG WELD

The slot weld is the weld used to join two touching contiguous components by a fillet weld round the periphery of a slot in one component so as to join it to the surface of the other component exposed through slot as shown in Fig. 10.17.

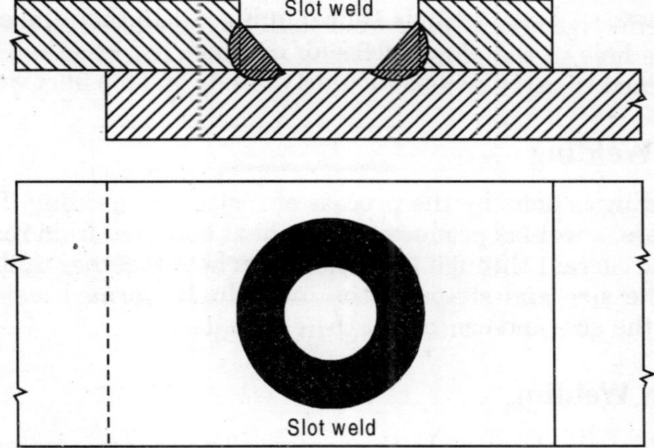

Fig. 10.17 *Slot weld*

The plug weld is the weld made by filling in a hole in one part with filler metal so as to join it to the surface of contiguous part exposed through a hole as shown in Fig. 10.18.

In case where there is insufficient space to accommodate the required length of fillet weld, plug weld or slot weld is used, to provide extra strength. These are also used in addition to the fillet weld when plates are welded. These welds balance the stresses in the plates and avoid buckling.

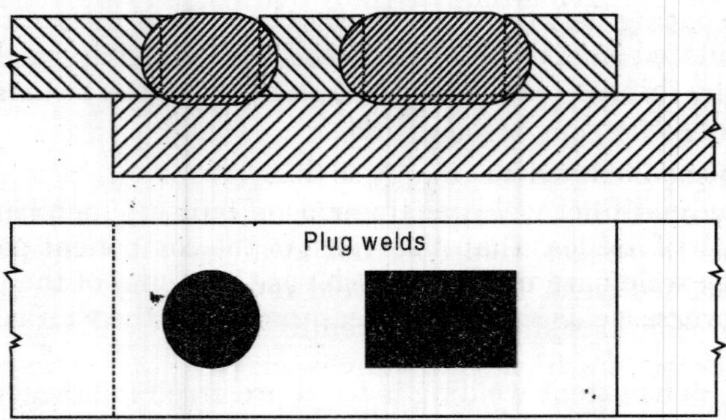

Fig. 10.18 *Plug welds*

Specifications

The dimensions of slot or hole should comply with the following limits in terms of the thickness of the part in which the slot or hole is made :

(i) The width or diameter should not be less than three times the thickness or 25 mm whichever is greater.

(ii) The corners at the enclosed ends of slots should be rounded with a radius not less than $1\frac{1}{2}$ times the thickness or 12 mm whichever is greater.

(iii) The distance between the edge of the part and the edge of the slot or hole or between the adjacent slot or holes should not be less than twice the thickness and not less than 25 mm for holes.

(iv) The effective area of plug weld shall be considered as the nominal area of the hole in the plane of facing surface. The plug welds shall not be designed to carry stress.

10.7.1 Slot Welding

The spot welding is done by the process of resistance welding. In a resistance welding process, a weld is produced by the heat obtained from resistance to the flow of electric current through the work parts held together under pressure by electrodes. The size and shape of the individually formed welds are limited primarily by the size and contour of the electrodes.

10.7.2 Seam Welding

The seam welding is also done by the process of resistance welding. The welding in which the pressure is applied continuously, and current impulsively to produce a linear weld, the work piece being between two electrodes wheels or between an electrode wheel and electrode bar. The electrode wheels apply the pressure and may be rotated continuously or stopped during the passage of the current.

10.7.3 Pipe Welding

The pipe welding is used in pipe manufacture, in pipe fabrication and in pipe joint. The steel pipe is manufactured by five different processes (viz., butt weld, lap weld, electric weld, hammer weld and seamless weld).

10.8 IMPERFECTIONS IN WELD

The welding should be done carefully so that there are no *defects* or *imperfections* in it. The strength of an imperfect or a defective weld is less than that of a perfect weld. Imperfect weld does not attain the strength equal to that of a properly designed weld. Imperfect weld remains weak and unsafe. Therefore, necessary precautions should be taken to avoid imperfections in welding. The various imperfections in welding may be classified in two categories:

1. Surface imperfections in the weld or adjacent metal.
2. Internal imperfections in the weld or adjacent metal.

10.8.1 Surface Imperfections in Weld

The surface imperfections are generally removed in final dressing of the surface of weld. Sometimes, surface imperfections cannot be eliminated completely. The following are main surface imperfections in the weld or adjacent metal:

1. Edge of the plate melted off
2. Exposed inclusion
3. Exposed porosity
4. Incompletely filled groove
5. Overlap
6. Undercut

10.8.1.1 Edge of the Plate Melted Off

This type of imperfection occurs in a fillet weld, the free edge of plate may be melted off as shown in Fig. 10.19. This may happen because the free edge may not be sufficiently built-up.

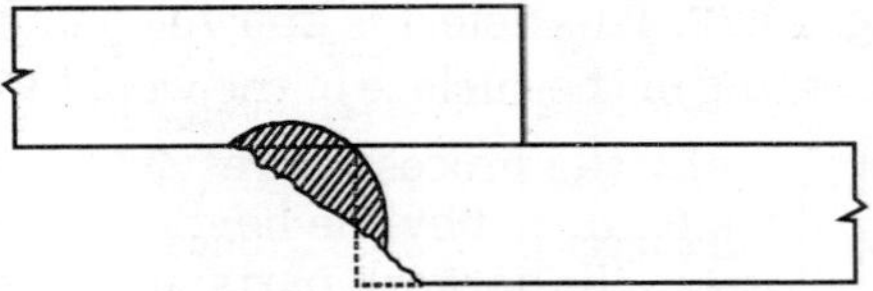

Fig. 10.19 *Edge of plate melted off*

10.8.1.2 Exposed Inclusion

A slag or any other foreign matter may be entrapped at the surface. This type of defect is called exposed inclusion.

10.8.1.3 *Exposed Porosity*

A group of gas pores may remain at the surface of weld. These gas pores may not be completely removed while final dressing of the surface of weld.

10.8.1.4 *Incompletely Filled Groove*

This type of defect occurs in butt weld. A continuous or an intermittent channel may remain in the face of a butt-weld as shown in Fig. 10.20. As a result of this, the thickness of throat of the weld remains less than that of the parental metal.

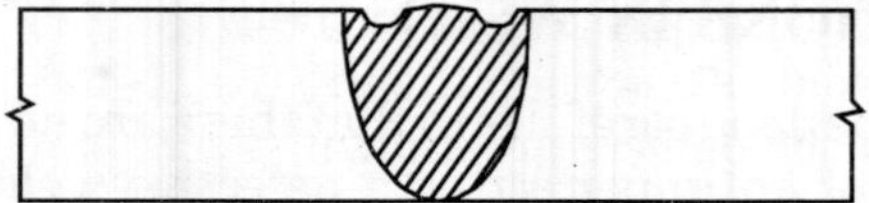

Fig. 10.20 *Incompletely filled groove*

10.8.1.5 *Overlap*

When the overflow of weld metal takes place over the surface of weld and fusion of the parent metal does occur, then, this defect is caused at toe or root of the weld as shown in Fig. 10.21.

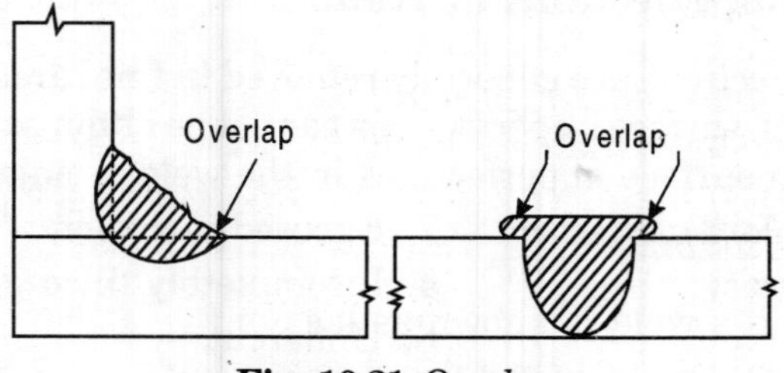

Fig. 10.21 *Overlap*

10.7.1.6 *Undercut*

A groove may form in the base metal along the toe of a weld because of excess heat of arc in the metal arc welding. The groove may remain unfilled by weld metal as shown in Fig. 10.22. This defect is known as undercut. This defect may be removed in final dressing of the surface of the weld by depositing weld metal.

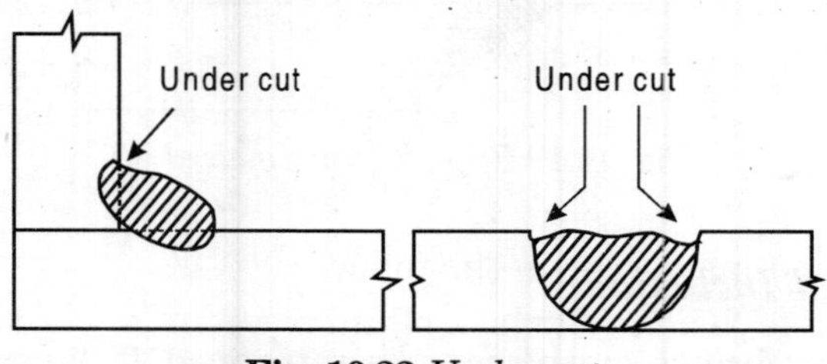

Fig. 10.22 *Under cut*

10.8.2 Internal Imperfections in Weld

The internal imperfections in the weld or adjacent metal cannot be removed. The following are main internal imperfections in the weld :

1. Gas pore
2. Inclusion
3. Lack of fusion
4. Incomplete penetration
5. Porosity

10.8.2.1 Gas Pore

While welding, gas may be entrapped in the weld. As a result of this, a small cavity occurs in the weld as shown in the Fig. 10.23. The cavities upto 1.6 mm in diameter included in the defects are known as gas pore.

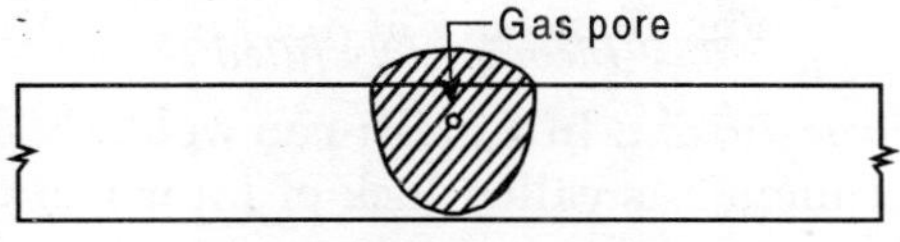

Fig. 10.23 *Gas pore*

10.8.2.2 Inclusion

While welding, gas or other foreign matter may be entrapped in the weld. This effect is known as inclusion. The inclusions may appear generally in the form as *large isolated inclusions, clusters of small inclusion* or *line inclusion.* This defect is more irregular in shape than gas cavity.

10.8.2.3 Lack of Fusion

The lack of union in a weld is known as lack of fusion. The lack of union in a weld may remain between weld metal and parent metal, or between parent metal and parent metal or between weld metal and weld metal as shown in Fig. 10.24.

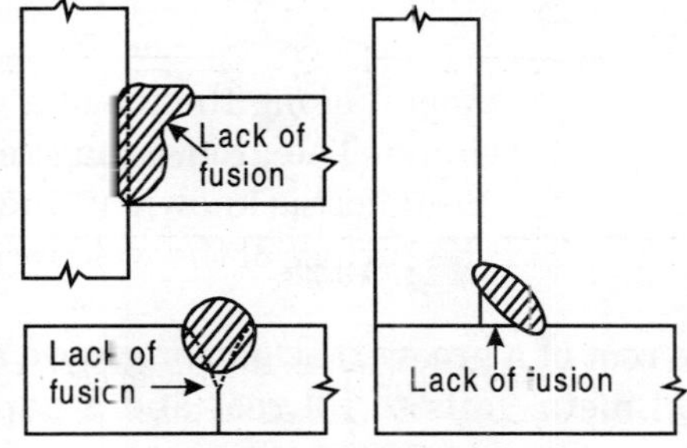

Fig. 10.24 *Lack of fusion*

The lack of fusion may occur in the following forms.

(i) *Lack of side fusion* The lack of union between weld metal and parent metal at a side of a weld outside the root is called lack of side fusion. It is shown in Fig. 10.25.

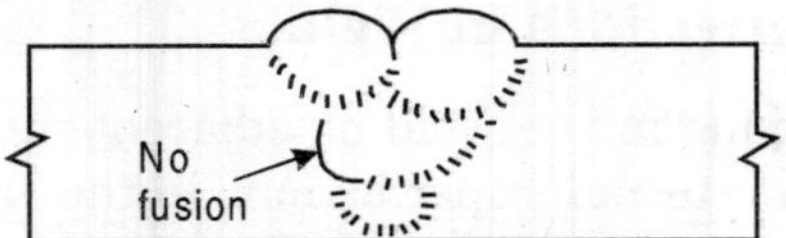

Fig. 10.25 *Lack of fusion*

(ii) *Lack of root fusion.* The lack of union between weld metal and parent or between adjacent faces of the parent metal at the root is called lack of root fusion. It is shown in Fig. 10.26.

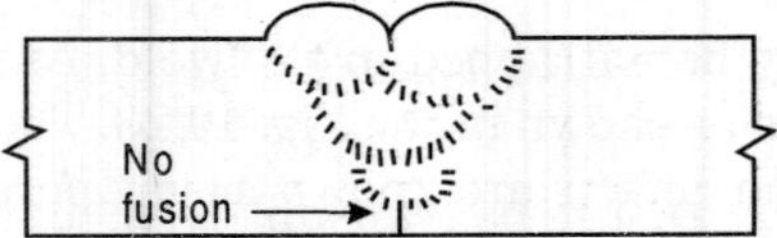

Fig. 10.26 *Lack of root fusion*

(iii) *Lack of inter-run fusion.* In a multi-run weld, the lack of union between adjacent runs of weld metals is called lack of inter-run fusion. It is shown in Fig. 10.27.

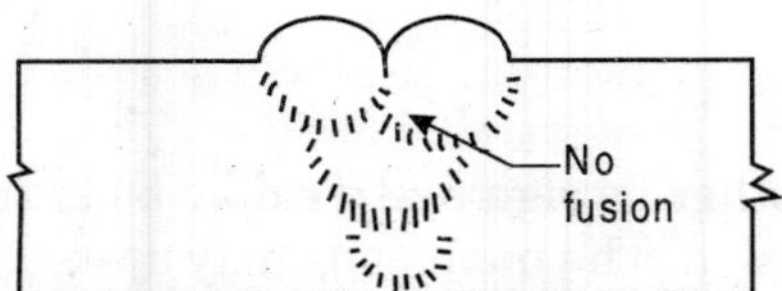

Fig. 10.27 *Lack of inter-run fusion*

10.8.2.4 Incomplete Penetration

In case of multi-run weld, small fissures may form by previous runs or run. These small fissures may usually occur at the toes of the underlying runs or run.

These small fissures may remain unfilled. At such positions, a gap may occur. This defect is known as *complete inter-run penetration.* It is shown in Fig. 10.28.

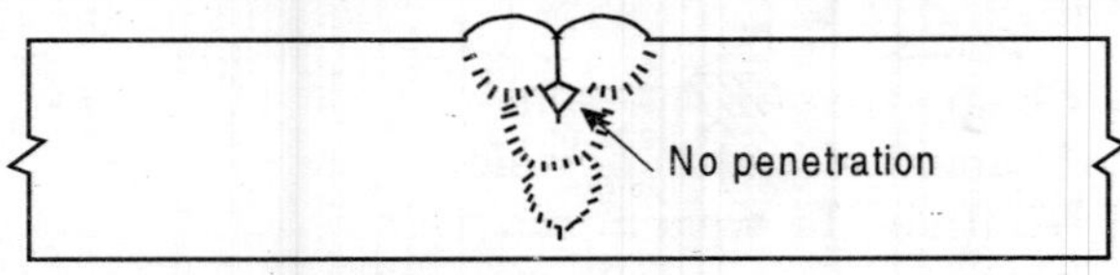

Fig. 10.28

Sometimes, at the root of a groove, the gap or groove angle is insufficient. As a result of this, weld metal fails to fill root and a gap is left. Such defect is known as *incomplete roof penetration.*

10.8.2.5 Porosity

A group of gas pores may occur in a weld. Such defect is known as porosity in weld. Depending upon size of pores, it is called fine or medium or coarse. The porosity may occur as clusters or chains or it may be scattered.

10.9 STRESS IN WELDS

10.9.1 Butt Weld

A complete penetration butt weld provides strength equal to that of the member itself. In general, there is no necessity to determine stress in the weld. There is also no necessity to determine the size of weld so long as the weld metal possesses the physical properties of the plate. In case of incomplete penetration, the stress in direct compression or tension is given by

$$p = \left(\frac{P}{A}\right) \qquad \text{...(10.1)}$$

where, P = Pull or thrust to be transmitted by the weld
l = Effective length of butt weld
t = Effective throat thickness

Stress in binding

$$\sigma = \left(\frac{M}{I}\right) \times y \qquad \text{...(10.2)}$$

where, M = Bending moment at the weld section
I = Moment of inertia of weld group
y = Distance of the fibre of weld under consideration

Stress in torsion

$$\tau = \left(\frac{T}{J}\right) \times r \qquad \text{...(10.3)}$$

where, τ = Shear stress due to torsion
T = Torsion at the weld section
b = Distance to the fibre of weld under consideration
J = Torsional moment of inertia

Stress in beam shear

$$\tau_s = \left(\frac{FA\bar{y}}{I \cdot b}\right) \qquad \text{....(10.4)}$$

where, τ_s = Shear stress in the weld
I = Moment of inertia of weld group
b = Width of the weld at the weld section under consideration
$A\bar{y}$ = Statistical moment of area of weld above the fibre under consideration.

The above expressions for stresses in butt weld in direct compression or tension, or bending or torsion or in beam-shear are based upon simple analysis, and are satisfactory for purpose of design.

10.9.2 Fillet Weld

The stress in a fillet weld is considered as shear on the throat regardless of the direction of the applied load. In design, it is convenient to use longitudinal weld. The distribution of stress in weld is shown in Fig. 10.29. The stresses in weld near the ends are high.

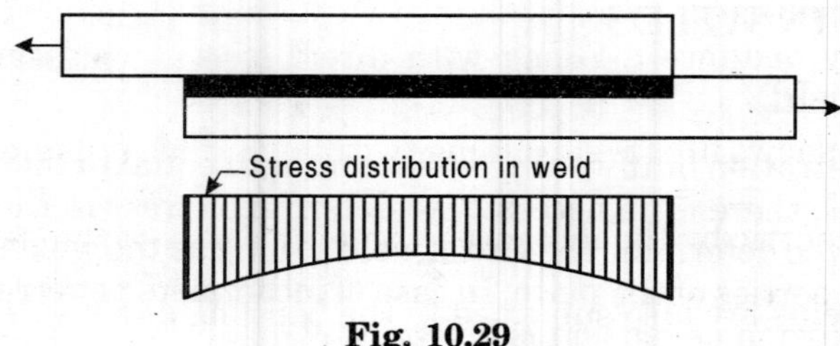

Fig. 10.29

For the structural welds, the heavily coated electrodes are generally used. The heavily coated electrodes produce more *ductile weld.* Because of plastic deformation near the ends, the shear stress in the weld becomes more or less uniform before failure. The following formulae are considered sufficient and satisfactory for purpose of design :

1. Direct stress formula, $\left(\sigma = \dfrac{P}{A}\right)$

2. Flexure formula, $\left(\sigma_b = \dfrac{M}{I} \times y\right)$

3. Torsion formula, $\left(\tau = \left(\dfrac{T}{J}\right) \times r\right)$

4. Beam-shear formula $\tau_s = \left(\dfrac{FA\bar{y}}{I \cdot b}\right)$

10.9.3 Direct Stress Formula

It is assumed that when the weld is subjected to direct compression or tension, the stress in weld is obtained by dividing the load by effective throat area. The effective throat area is product of effective length and effective throat thickness of weld. The heavily coated electrodes are used. These electrodes produce ductile weld. The *ductility of weld* allows equalization of stress in such weld before failure. Therefore, the use of direct stress formula is satisfactory for purpose of design.

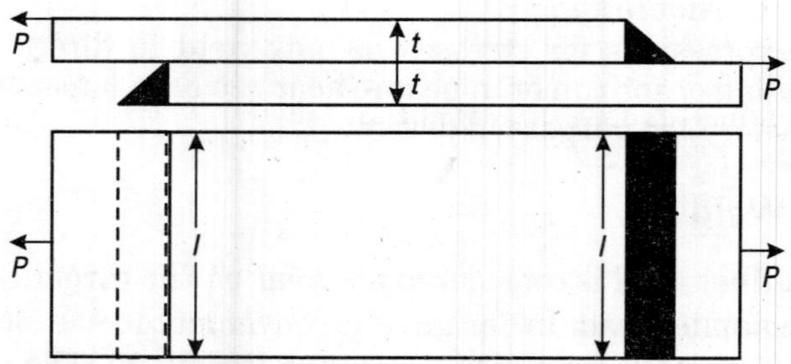

Fig. 10.30 *Transverse fillet weld*

Figure 10.30 shows fillet welds joining two plates in lap joint. The loads acting are eccentric by distance *t*. Along with direct stress, welds are subjected to bending and shear.

Figure 10.31 shows fillet weld joining two plates at right angles. The resisting loads $\frac{P}{2}$ are eccentric by the distance $\frac{t}{2}$ on either side of the plate. Along with direct stress, welds are also subjected to the bending and shear.

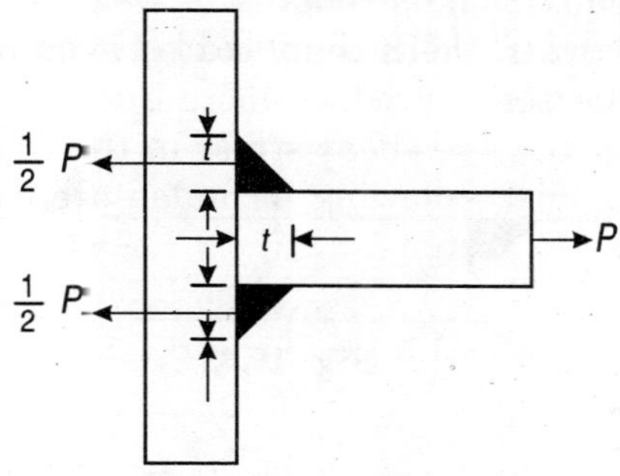

Fig. 10.31

Figure 10.32 shows fillet welds joining two plates in lap joint. The longitudinal fillet weld have been used.

Fig. 10.32 *Longitudinal fillet weld*

It is common practices to treat welds shown in Figs. 10.30, 10.31 and 10.32, stressed uniformly.

In Fig. 10.30, direct stress

$$\sigma = \left(\frac{P}{A}\right) = \left(\frac{P}{\left(\frac{1}{\sqrt{2}}\right) \cdot 2l \cdot s}\right)$$

where, *l* is the effective length of weld and *s* is the size of weld In Fig. 10.31, direct stress

$$\sigma = \left(\frac{P}{A}\right) = \left(\frac{P}{\left(\frac{1}{\sqrt{2}}\right) \cdot 2l \cdot s}\right)$$

In Fig. 10.32, direct stress

$$\sigma = \left(\frac{P}{A}\right) = \left(\frac{P}{\left(\frac{1}{\sqrt{2}}\right)\cdot 2l \cdot s}\right)$$

The direct stresses in all these three cases are equal in intensities. Since the allowable stress in shear is less, it controls the design.

In Fig. 10.33, the thickness of two plates connected by transverse fillet weld in lap joint are not equal. The strengths of weld per unit length are equal. Therefore, the load between welds is spread between plates is in proportion to their respective thicknesses.

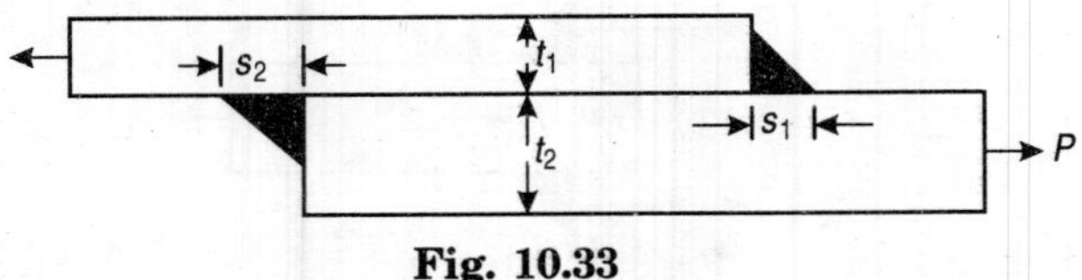

Fig. 10.33

Load in plate $A, P_1 = \left(\dfrac{P \cdot t_1}{(t_1 + t_2)}\right)$

where, t_1 is the thickness of plate A and t_2 is the thickness of plate B

Stress in weld, $\sigma = \left(\dfrac{P}{\left(\frac{1}{\sqrt{2}}\right)\cdot l \cdot s_1}\right)$

$$\sigma = \left(\frac{P \cdot t_1}{(t_1 + t_2)} \cdot \frac{\sqrt{2}}{l \cdot s_1}\right) \quad ...(10.5)$$

where, l is the effective length of weld and s_1 is the size of weld in plate A

Load in plate $B, P_2 = \left(\dfrac{P_2 \cdot t_2}{(t_1 + t_2)}\right)$

Stress in weld, $\sigma = \left(\dfrac{P_2}{\left(\frac{1}{\sqrt{2}}\right)\cdot l \cdot s_2}\right) = \left(\dfrac{P \cdot t_2}{(t_1 + t_2)}\right) \cdot \dfrac{\sqrt{2}}{l \cdot s_2}$ ...(10.6)

where, s_2 is the size of weld in plate B.

10.9. 4 Flexure Formula

Figure 10.34 shows fillet weld subjected to longitudinal flexure. The throat section is subjected to critical stress. The failure will occur at this section. The moment M, is resisted by width of weld $2.\dfrac{s}{\sqrt{2}}$ and depth of weld as l. The bending stress

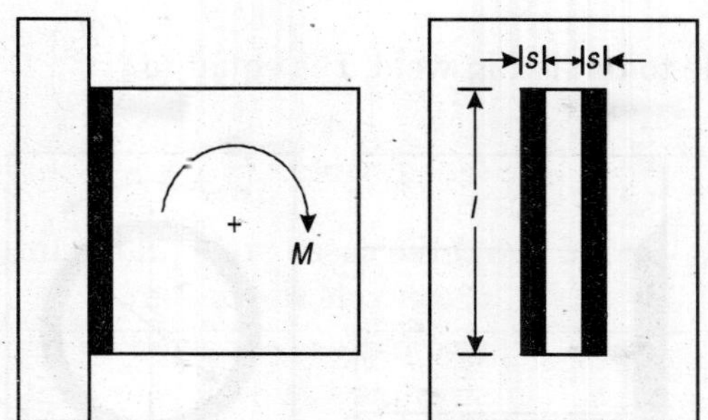

Fig. 10.34 *Longitudinal flexure*

$$\sigma = \frac{M}{I} \cdot y = \left[\frac{M}{\frac{1}{12}\left(2 \cdot \frac{s}{\sqrt{2}} \cdot (l^3)\right)} \right] \cdot \frac{l}{2} = \left[\frac{6M}{\frac{1}{\sqrt{2}} \cdot l^2 2s} \right] \qquad ...(10.7)$$

Figure 10.35 shows fillet welds subjected to cross flexure. The foot of weld is subjected to maximum shear stress and maximum tensile stress.

The fillet weld subjected to cross flexure may be analysed by steel theory. The moment M, is resisted by a couple $C\,(h + s)$ or $T\,(h + s)$. The top weld is subjected to tension T, while bottom weld is subjected to thrust C, and both forces are equal.

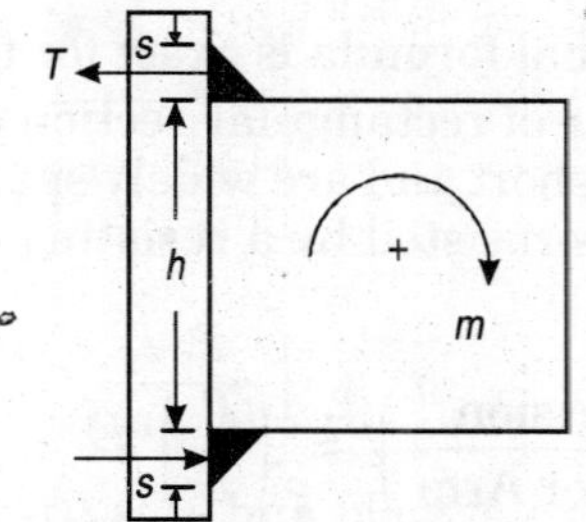

Fig. 10.35 *Cross flexure*

Force in weld, $C = \left(\frac{M}{(h+s)}\right)$ and $T = \left(\frac{M}{(h+s)}\right)$

where, $(h + s)$ is the lever arm.

$$\text{Sttess in weld } f = \left(\frac{\text{Force}}{\text{Throat Area}}\right) = \left(\frac{M}{(h+s) \cdot \left(\frac{s}{\sqrt{2}}\right) \cdot l}\right) \qquad ...(10.8)$$

where, l is the effective length of weld and s is the size of weld.

10.9.5 Torsion Formula

Figure 10.36 shows a circular pipe connected to a plate by a fillet weld. The weld is subjected to torsion, T The throat of weld is subjected to critical stress.

The shear stress due to torsion in weld, x is equal to $\left(\frac{T}{J}\right) \cdot r$

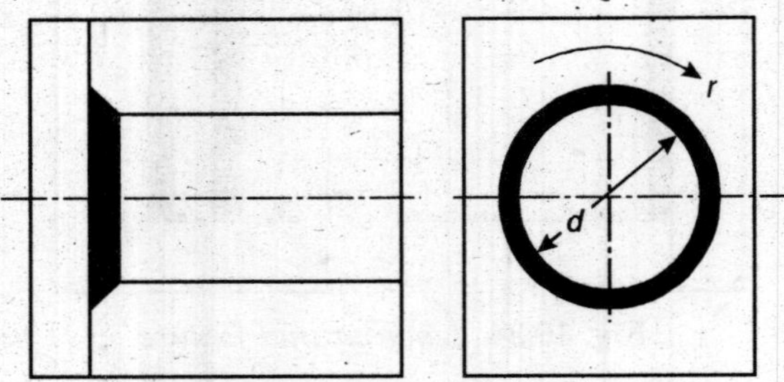

Fig. 10.36 *Torsion of circular fillet weld*

$$J = \pi d\left(\frac{s}{\sqrt{2}}\right)\left(\frac{d}{2}\right)^2 = \left(\frac{\pi \cdot s \cdot d^3}{4\sqrt{2}}\right) \quad \left(\because r = \frac{d}{2}\right)$$

$$\therefore \qquad \tau = \left(\frac{T \cdot \left(\frac{d}{2}\right)}{\left(\frac{\pi \cdot s \cdot d^3}{4\sqrt{2}}\right)}\right) = \left(\frac{2\sqrt{2}T}{\pi \cdot s \cdot d^2}\right) \qquad \text{...(10.9)}$$

It is seen that the torsional formula is exact for fillet weld of circular section.

Figure 10.37 shows a bar of rectangular section connected to a plate by fillet welds. The fillet welds are short and are widely spaced. The welds are subjected to torsion, T. The torsion is resisted by a resisting couple.

Force in each weld

$$= \left(\frac{\text{Torsion}}{\text{Lever Arm}}\right) = \left(\frac{T}{L}\right)$$

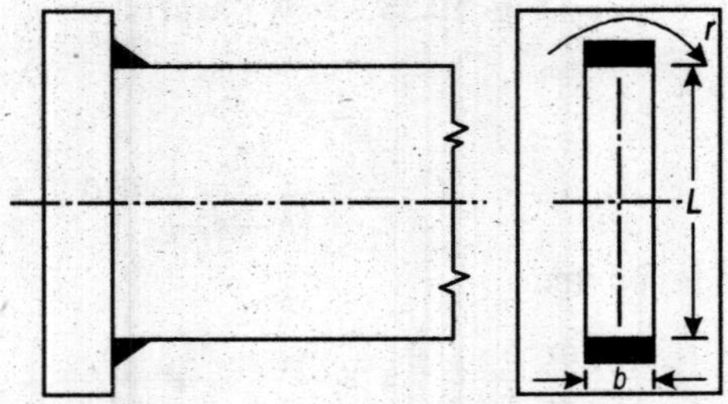

Fig. 10.37 *Torsion of widely spaced short fillet weld*

Shear stress in weld

$$\tau_s = \frac{T}{L}\left(\frac{1}{\left(\frac{s}{\sqrt{2}}\right) \cdot b}\right) = \left(\frac{T\sqrt{2}}{L \cdot s \cdot b}\right) \qquad \text{...(10.10)}$$

Figure 10.38 shows a plate of rectangular cross-section. It is connected to another plate by two adjacent long and narrow fillet welds. The welds are subjected to torsion T. The polar moment of inertia J is approximately equal to moment of inertia I_{xx}. The distance from centre of rotation (e.g., of weld group) to the extreme fibre of weld is approximately equal to half the depth of weld.

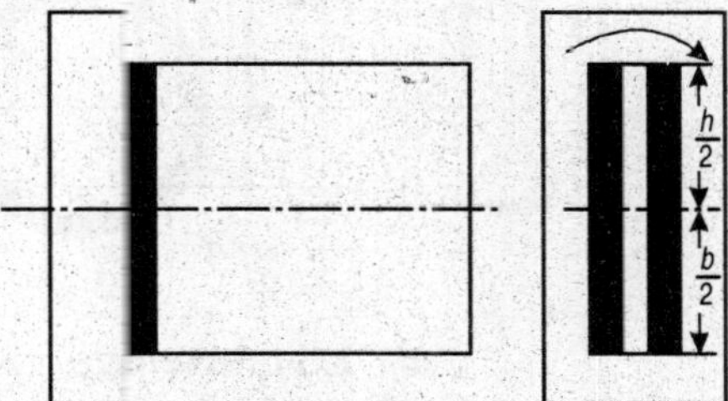

Fig. 10.38 *Torsion of adjacent fillet weld*

Therefore, shear stress in such weld

$$f_s = \frac{T}{I_{xx}} \cdot \left(\frac{h}{2}\right) = \left[\frac{T}{\frac{1}{12} \cdot 2\left(\frac{s}{\sqrt{2}}\right)h^3}\right] \cdot \frac{h}{2}\left(\frac{3\sqrt{2}T}{s \cdot h^2}\right) \qquad \text{...(10.11)}$$

It is seen that the torsional formula for long and narrow fillet weld becomes equivalent to the flexural formula.

10.9.6 Beam Shear Formula

Figure 10.39 shows a rectangular bar connected to a plate by fillet weld. The bar carries vertical load at its free end. The weld is subjected to shear stress and bending stress. The distribution of shear stress is parabolic. The shear stress is maximum at the neutral axis and it is 1.5 times the average shear stress.

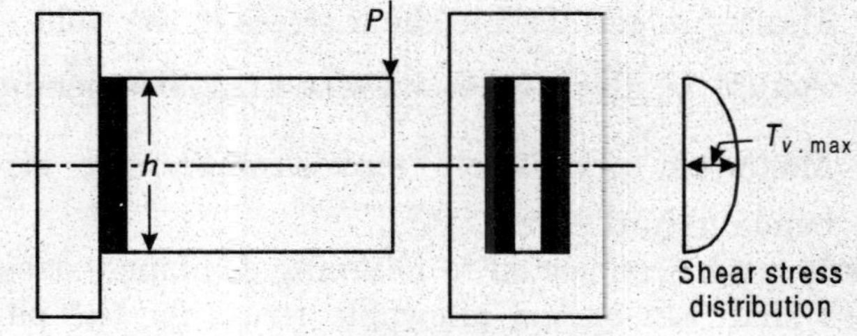

Fig. 10.39

The shear stress, $\tau_s = \left(\frac{FA\bar{y}}{I \cdot b}\right)$

Statical moment of area of the weld above neutral axis about xx-axis

$$A \cdot \bar{y} = 2 \times \left(\frac{s}{\sqrt{2}}\right) \times \left(\frac{h}{2}\right) \times \left(\frac{h}{4}\right) = \frac{s \cdot h^2}{4\sqrt{2}}$$

$$I_{xx} = \frac{1}{12}\cdot 2\left(\frac{s}{\sqrt{2}}\right)(h^3)\frac{1}{6\sqrt{2}} s\cdot h^3$$

Width of weld at $xx = \left(\frac{2s}{\sqrt{2}}\right)$

$$\tau_{s(\max)} = \left[\frac{F\cdot\dfrac{sh^2}{4\sqrt{2}}}{\dfrac{1}{6\sqrt{2}}sh^3\left(\dfrac{2s}{\sqrt{2}}\right)}\right] = \frac{3}{2}\times\frac{F}{\left(\dfrac{2s}{\sqrt{2}}\cdot h\right)}$$

$$= 1.5\, f_s \text{ (average)}$$

10.10 COMBINED STRESSES IN WELDS

10.10.1 Butt Weld

A structural material may be subjected to any possible combined stress. When two structural members are connected by butt weld, it takes the place of equivalent amount of base metal. Therefore, the butt weld may also be subjected to any possible combined stress. In butt weld, the allowable stresses for compression, tension and shear are different. Therefore, maximum values of each stress are found to determine the critical stress. When butt weld is subjected to tensile or compressive stress (axial and/or bending) along with direct shear stress, the welds are proportioned so that quantity.

$$\left[\frac{p_s}{p_s}\right]^2 + \left[\frac{p_{t/c}}{P_{t/c}}\right]^2 \text{ does not exceed unity.}$$

where, p_s = Actual shear stress in the weld

P_s = Maximum permissible shear stress in the weld

$p_{t/c}$ = Actual tensile/compressive stress (Axial or bending) in the weld

$P_{t/c}$ = Maximum permissible tensile/compressive stress (axial or bending) in the weld

When the butt weld is subjected to following combined stresses, then, the combined stress shall not exceed allowable stress for the parent metal or equivalent stress, $\sigma_{e\ cal}$ which shall not exceed 110 N/mm².

(A) For shear combined with bending compressive or tensile.

$$\sigma_{e.cal} = \left[\sigma^2_{bt\cdot cal} + 3\tau^2_{vm.cal}\right]^{\frac{1}{2}} \qquad ...(10.12\ a)$$

or

$$\sigma_{e.cal} = \left[\sigma^2_{bc\cdot cal} + 3\tau^2_{vm.cal}\right]^{\frac{1}{2}} \qquad ...(10.12\ b)$$

(B) For the combined bearing, bending and shear stresses

$$\sigma_{e.\text{cal}} = \left[\sigma^2_{bt\cdot\text{cal}} + \sigma^2_{p\cdot\text{cal}} + \sigma_{bt\cdot\text{cal}}\sigma_{p\cdot\text{cal}} + 3\tau^2_{vm\text{cal}}\right]^{\frac{1}{2}} \qquad \text{...(10.12 c)}$$

$$\text{or } \sigma_{e.\text{cal}} = \left[\sigma^2_{bc\cdot\text{cal}} + \sigma^2_{p\cdot\text{cal}} + \sigma_{bc\cdot\text{cal}}\sigma_{p\cdot\text{cal}} + 3\tau^2_{vm\text{cal}}\right]^{\frac{1}{2}} \qquad \text{...(10.12 d)}$$

where, $\sigma_{bc.\text{cal}}$; $\tau_{vm\text{cal}}$ and $\tau_{vm\text{cal}}$, are the numerical values of the co-existing bending (tensile or compressive), shear and bearing stress, respectively. $\sigma_{e.cal}$ should not exceed $\sigma_{e.\max}$ permissible equivalent stress.

10.10.2 Fillet Weld

When a fillet weld is subjected to two load systems, which produce collinear stress, then these are added directly to determine combined stress. When a fillet weld is subjected to two load systems which produce stresses in two perpendicular direction, the maximum combined stress is computed as the vector sum and that should not exceed the maximum permissible stress in shear.

Figure 10.40 shows a rectangular bar connected to plate by fillet welds. The fillet weld is subjected to direct stress due to load P, and bending stress due to moment M. The two load systems produce collinear stresses. The combined stress in weld is found by adding these stresses directly.

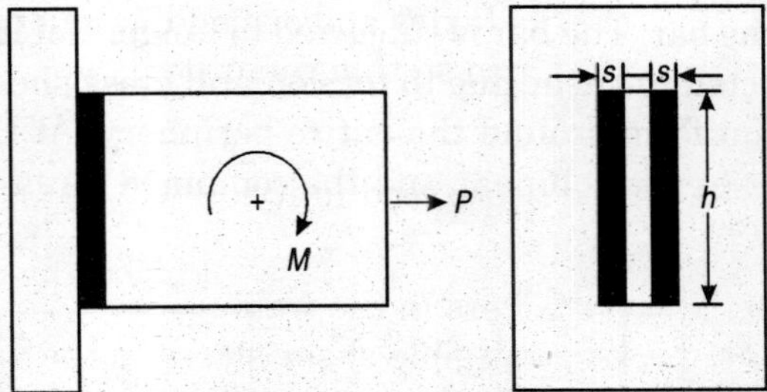

Fig. 10.40 *Direct stress and flexure*

$$\sigma_{b(\max)} = \left(\frac{P}{A} + \frac{My}{I_{xx}}\right) = \left[\frac{p}{2\dfrac{s}{\sqrt{2}}h} + \frac{M\cdot\dfrac{h}{2}}{2\cdot\dfrac{1}{2}\cdot\dfrac{s}{\sqrt{2}}h^3}\right]$$

or

$$\sigma_{s(\max)} = \left(\frac{P\sqrt{2}}{2s\cdot h} + \frac{3\sqrt{2}M}{s\cdot h^2}\right) \qquad \text{... (10.13)}$$

Figure10.41 shows rectangular bar connected to a plate by fillet welds. The weld is subjected to an eccentric load. The load system produces shear stress and bending stress. The stresses produced are collinear. The combined stress is obtained by two stresses

$$f_{s(\max)} = \left(\frac{P}{A}+\frac{My}{I_{xx}}\right) = \left(\frac{P}{2\frac{s}{\sqrt{2}}h}+\frac{P\left(L+\frac{h}{2}\right)\cdot\frac{h}{2}}{2\times\frac{1}{12}\times\frac{s}{\sqrt{2}}h^3}\right)$$

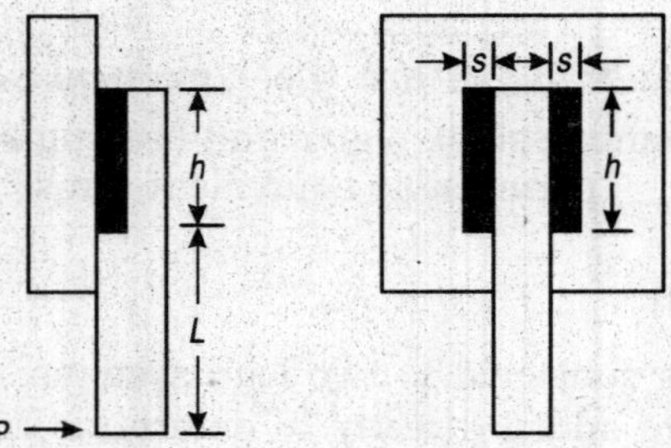

Fig. 10.41 *Cross shear and flexure*

$$f_{s(\max)} = \left(\frac{P\sqrt{2}}{2s\cdot h}+\frac{3\sqrt{2}P\cdot\left(\frac{L+h}{2}\right)}{s\cdot h^2}\right) \quad \text{...(10.14)}$$

Figure 10.42 shows a circular bar connected to a plate by fillet weld around the periphery of the bar. The bar is subjected to tangential load P at its free end. The weld is subjected to shear due to torsion and cross shear. The shear stress due to torsion is uniform around the entire periphery. At the upper end of the bar, the two stresses are collinear and the combined stress is found by adding them directly.

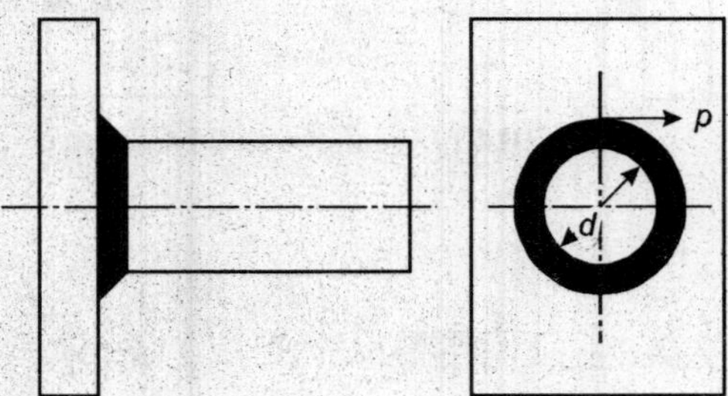

Fig. 10.42 *Cross shear and torsion*

Figure 10.43 shows a rectangular bar connected to a plate by fillet welds. The bar is carrying a vertical load, P, at its free end. The weld is subjected to shear stress and bending stress. The stresses are acting in two perpendicular directions. The combined stress is obtained by vectorial sum of two stresses.

Shear stress, $$\tau_s = \frac{P}{A}$$

$$= \left(\frac{P}{2\left(\frac{s}{\sqrt{2}}\right)h} \right) = \left(\frac{P\sqrt{2}}{2h} \right)$$

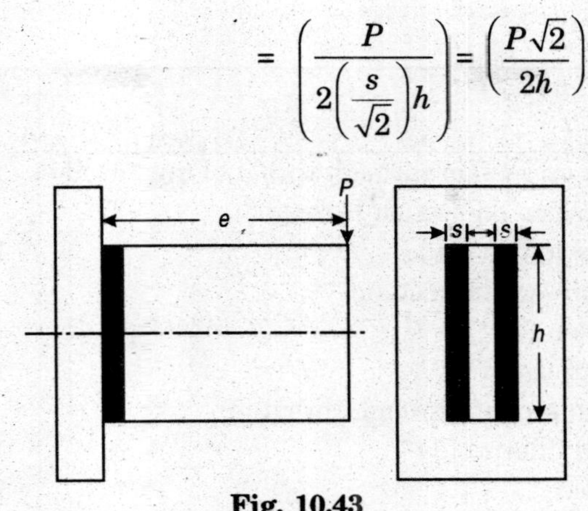

Fig. 10.43

The maximum bending stress

$$\sigma_b = \left(\frac{My}{A} \right) = \left(\frac{P \cdot e \cdot \frac{h}{2}}{2 \cdot \frac{1}{12} \cdot \frac{s}{\sqrt{2}} h^3} \right)$$

$$= \left(\frac{P \cdot e (3\sqrt{2})}{s \cdot h^2} \right)$$

The combined stress, $\sigma = [\tau_s^2 + \sigma_b^2]^{1/2}$

When the fillet weld is subjected to following combined stress, then the equivalent stress, $\sigma_{e.cal}$ is calculated, which should not exceed 110 N/mm².

(A) Combined bending and shear in fillet weld

$$\sigma_{e.cal} = \left[\sigma_{e \cdot cal}^2 + 3\tau_{vm \cdot cal}^2 \right]^{1/2} \quad \text{...(10.14 a)}$$

$$\sigma_{e.cal} = \left[\sigma_{bc \cdot cal}^2 + 3\tau_{vm \cdot cal}^2 \right]^{1/2} \quad \text{...(10.14 b)}$$

Equations (10.14 a) and (10.14 b) are used when the stress $\frac{\sigma_{bt \cdot cal}}{\sigma_{bc \cdot cal}}$ and $\tau_{vm.cal}$ are mutually perpendicular to each other.

(B) Combined axial and shear in fillet weld

$$\sigma_e = [\sigma^2 + 1.8\, \tau^2]^{1/2} \quad \text{...(10.14 c)}$$

where, σ is the axial stress in the weld, σ_{bt}, σ_{bc} and τ have been defined in (10.12 a), (10.12 b) and (10.12 c), respectively.

10.11 WORKING STRESSES IN WELDS

Working stresses in welds, when welded joints are constructed with mild steel confoming to IS : 226–1962 as parent metal and with electrodes conforming to IS : 814–1974 are adopted as per Table 10.3 are recommended in IS : 816–1969.

Table 10.3 *Working stresses in welds*

S.No.	*Kind of stresses*	*Max. permissible values*
1.	Tension on section through throat of butt weld	142 N/mm^2
2.	Compression on section through throat of butt weld	142 Nm^2
3.	Fibre stresses in bending	
	(a) tension	157.5 N/mm^2
	(b) compression	157.5 N/mm^2
4	Shear on section through throat of butt and fillet welds	110 N/mm^2
5.	Plug welds	110 N/mm^2

Note. *IS 816–1969 is yet to be revised in S.I. units. as such the values of the maximum permissible stresses have been converted from those given in M.K.S. units.*

The maximum permissible value of stresses of shear and tension are reduced to 80 per cent of those given in Table 10.3, in case, the welding in done at site, (field).

When the effects of wind or earthquake forces are considered, then, maximum permissible values of stresses are increased by 25 per cent.

It is to note that maximum permissible stresses given in Table 10.3 are same as for the parent metal (mild steel IS : 226 –1962).

Example 10.1 *Two plates 16 mm thick are joined by (i) a double-U butt weld, (ii) a-single-U butt weld. Determine the strength of the welded joint in tension in each case. Effective length of weld is 150 mm. Allowable stress in butt weld in tension is 142 N/mm^2.*

Solution

(i) In case of double-U butt weld, complete penetration of weld takes place. Effective throat thickness of weld = 16 mm

Effective length of weld= 150 mm

Strength of single -*U* butt weld

$$\left(\frac{142 \times 150 \times 16}{1000}\right) = 340.8 \text{ kN}$$

(ii) In case of single-U butt weld, incomplete penetration of butt weld takes place.

$$\text{Effective throat thickness} = \frac{5}{8} \times 16 = 10 \text{ mm}$$

Effective length of weld =150 mm

Strength of single-U butt weld

$$\left(\frac{142 \times 150 \times 10}{1000}\right) = \mathbf{213.0 \text{ kN}}$$

Example 10.2 *In a truss girder of a bridge, a tie as shown in Fig. 10.44 is connected to the gusset plate by fillet weld. Determine the strength of the weld. The size of the weld in the fillet weld is 6 mm.*

Solution

Size of weld = 6 mm

Effective throat thickness

$$= 0.7 \times 6 = 4.2 \text{ mm}$$

Effective length of fillet weld

$$= (200 + 200 + 200) = 600 \text{ mm}$$

Strength of fillet weld

$$= \left(\frac{110 \times 600 \times 4 \cdot 2}{1000}\right) = \mathbf{277.2\ kN}$$

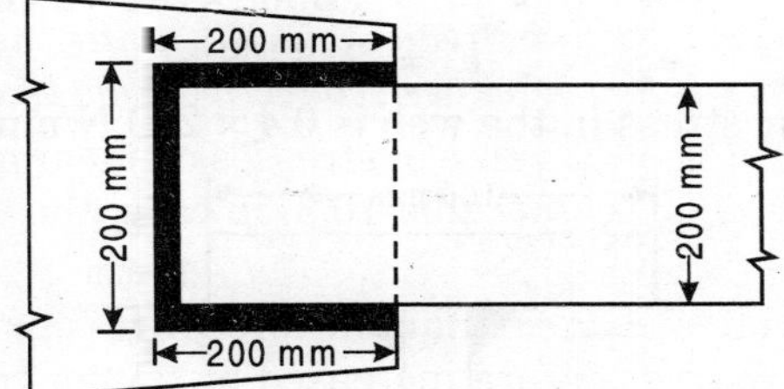

Fig. 10.44

Example 10.3 *In Example 10.2, the pull to be transmitted by the tie is 300 kN. Determine the necessary overlap of the tie.*

Solution

Size of weld = 6 mm

Effective throat thickness is 4.2 mm

The pull transmitted by the end fillet weld

$$\left(\frac{110 \times 200 \times 4 \cdot 2}{1000}\right) = 9.4 \text{ kN}$$

Let l be the necessary overlap required. The pull transmitted by the side fillets

$$\left(\frac{2 \times 110 \times l \times 4 \cdot 2}{1000}\right) = 0.924\, l \text{ kN}$$

$$(92.4 + 0.924 \,.\, l) = 300, \qquad \therefore l = 224.7 \text{ mm}$$

Necessary overlap of the tie is 224.7 mm

Example 10.4 *The web plate of a built-up welded I-section is 200 mm × 12 mm, and the flange plates are 100 mm × 12 mm. The size of fillet weld is 6 mm. Compute the maximum shear force that may be allowed at any section, if the average allowable shear in the web is 0.4 f_y and maximum allowable shear in the weld is 110 N/mm².*

Solution

Step 1: Moment of inertia of the built-up section (about *xx*-axis)

$$I_{xx} = \frac{1}{12}[10 \times 22.4^3 - 8.8 \times 20^3] \times 10^4 = 3508 \times 10^4 \text{ mm}^4$$

Step 2: Intensity of shear stress (at the weld section)

$$\tau_s = \left(\frac{F \cdot A\bar{y}}{I_{xx} \cdot (2t)}\right)$$

where, $A\bar{y}$ = Moment of the area above the section about xx-axis

F = Shear force at the section

t = Effective throat thickness of one weld

$$\tau_s = \left(\frac{F \times 100 \times 12 \times (100+6)}{3508 \times 104 \times 2 \times 0 \cdot 7 \times 6}\right)$$

$$\therefore \quad F = 238 \text{ kN}$$

The average shear stress in the web is 0.4 × 250 N/mm³

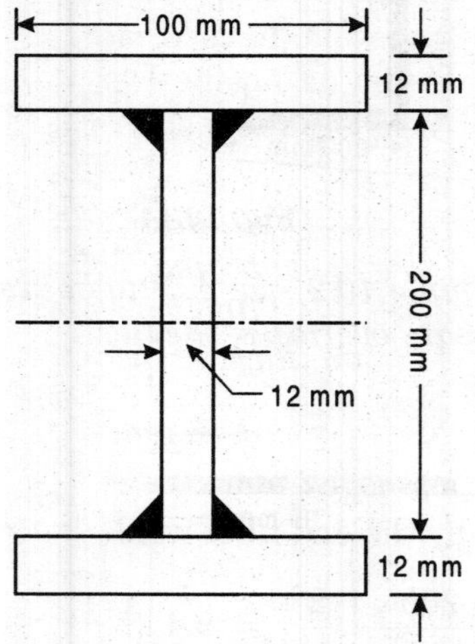

Fig. 10.45

Step 3: Allowable shear force in the web

$$F_1 = \left(\frac{0 \cdot 4 \times 240 \times 200 \times 12}{1000}\right) = \mathbf{240 \text{ kN}}$$

10.12 DESIGN OF WELDED JOINTS SUBJECTED TO AXIAL LOAD

The complete penetration butt weld as mentioned earlier, does not require design calculations. In case of incomplete penetration butt weld, effective throat thickness of the weld is computed, and welding is done upto the required length.

In case of fillet weld, size of the weld is fixed keeping in view the minimum size of the weld as per Table 10.1. IS : 816 –1969 recommends that when fillet weld is applied to the square edge of member, the maximum size of weld should be less than the edge thickness by at least 15 mm, as shown in Fig. 10.46. This avoids the washing down of edges of weld. When fillet weld is applied to the round toe of rolled steel sections, the maximum size of the weld should not exceed $\frac{3}{4}$ of the thickness of the section at the toe. When fillet weld is used for lap joint, then overlap of the members connected as shown in Fig. 10.46, should not be less then five times thickness of thinner part.

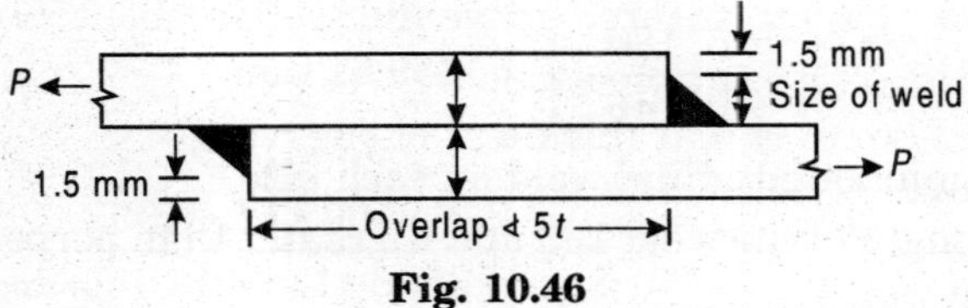

Fig. 10.46

The strength of the fillet weld is determined per mm length for the size of the weld adopted. The effective length of the weld is then computed for the full or thrust to be transmitted by the weld. In case, only side fillet welds are applied, the length of the each weld should not be less than perpendicular distance between them, and spacing between them shall not be more than 16 times the thinner part.

Example 10.5 *Design a suitable longitudinal fillet weld to connect the plates as shown in Fig. 10.47, and to transmit a pull equal to the full strength of thin plate. Allowable stress in the weld is 110 N/mm*^{2} *and tensile stress in the plate* $0.6f_y$ *Nlmm*^{2}*. The plates are 10 mm thick.*

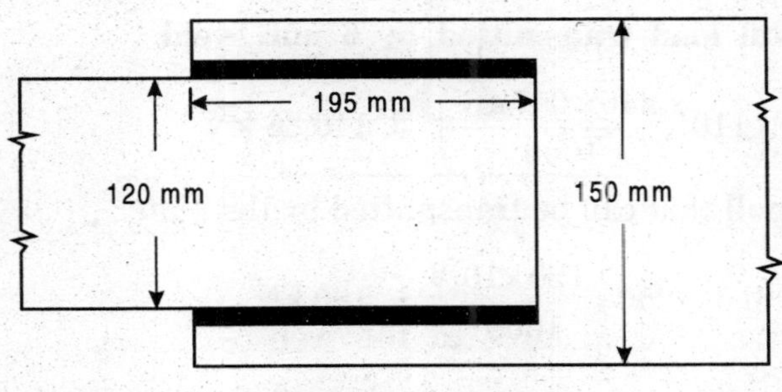

Fig. 10.47

Solution

Design :

Step 1 : Size of weld

The maximum size of weld required for thickness upto 19 mm is 5 mm. The maximum size of fillet weld is limited by the thickness of the plate i.e.,(10 –1.5) = 8.5 mm

Provide 6 mm fillet weld

Step 2 : Pull transmitted by 1 mm weld

$$\left(\frac{110\times 1\times 0\cdot 7\times 6)}{1000}\right) = 0.462 \text{ kN}$$

Step 3 : Tensile strength of thin plate

$$\left(0\cdot 6\times 250\times\frac{120\times 10}{1000}\right) = 180 \text{ kN}$$

Step 4 : Necessary length of the weld

$$\left(\frac{180}{0\cdot 462}\right) = 389.61 \text{ mm}$$

Provide 195 mm longitudinal weld on each side.

Check: (*a*) Length of the weld 195 mm is greater than perpendicular distance 12 mm between welds.

(*b*) Spacing between weld = 120 mm < 16 × 10 mm.

Example 10.6. *Two plates 120 mm × 10 mm are overlapped and connected together by transverse fillet weld to transmit pull equal to full strength of the plate. Design the suitable welding. Allowable stress in the weld is 110 N/mm². Allowable stress in tension in the plate is 0.4* f_y *N/mm².*

Solution

Design :

Step 1: Size of weld

Minimum size of weld = 5 mm

Maximum size of weld = (10 –1.5) = 8.5 mm

Total length of two welds = 240 mm

Step 2: Total load transmitted by 6 mm weld

$$\left(110\times\frac{240\times 0\cdot 7\times 6}{1000}\right) = 110.88 \text{ kN}$$

Maximum pull that can be transmitted by the plate

$$\left(0\cdot 6\times 250\times\frac{120\times 10}{1000}\right) = 180 \text{ kN}$$

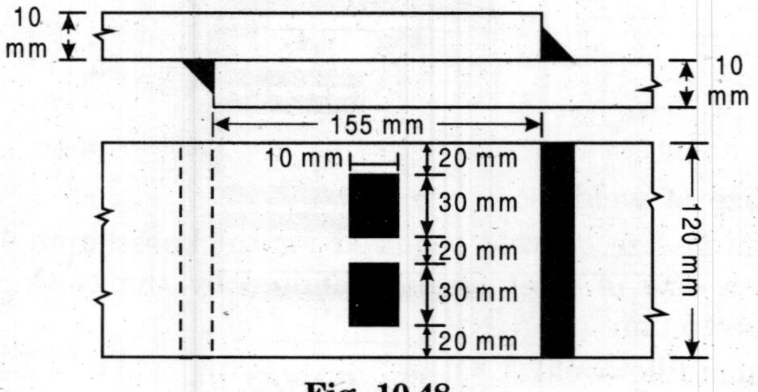

Fig. 10.48

To transmit the pull equal to the full strength of plate, provide additional weld by plug weld. Provide two rectangular plug welds 30 mm × 15 mm as shown in Fig. 10.48 which satisfies the specification. Strength of two plug welds.

$$\left(2\times110\times\frac{30\times15}{1000}\right) = 99 \text{ kN}$$

Total pull now transmitted

$$(110.88 + 99) = 209.88 \text{ kN} > 180 \text{ kN}$$

Hence, satisfactory.

Example 10.7 *A tie member consists of two MC 225, @ 0.259 kN/m. The channels are connected to either side of a gusset plate 12 mm thick. Design the welded joint to develop the full strength of the tie. The overlap is limited to 400 mm.*

Solution

Design :

Step 1 : Properties of section

From **ISI** Handbook No. 1, for MC 225, @ 0.259 kN/m

Thickness of web = 6.4 mm

Thickness of flange = 12.4 mm

Sectional area = 3301 mm^2

Step 2 : Tensile strength of each channel section

$$\left(\frac{0\cdot6\times250\times3301}{1000}\right) = 495.15 \text{ kN}$$

Provide 4 mm weld.

Step 3 : Strength of weld per mm length

$$\left(\frac{110\times0\cdot7\times4}{1000}\right) = 0.308 \text{ kN}$$

Step 4 : Total length of fillet weld

Total length of fillet weld necessary to connect one channel section

$$\left(\frac{495\cdot15}{0\cdot308}\right) = 1607.6 \text{ mm}$$

The overlap of channels is limited to 400 mm.

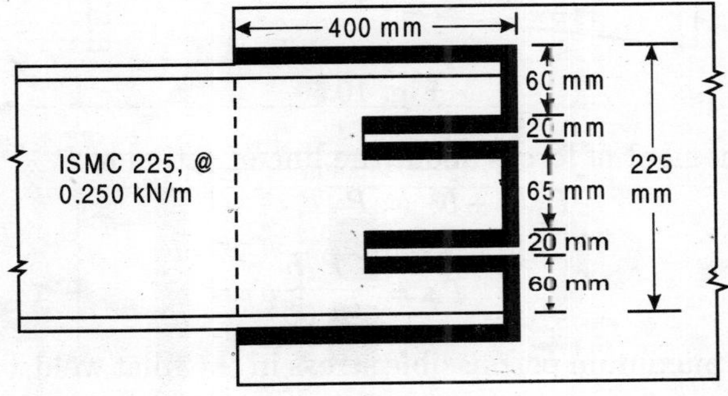

Fig. 10.49

Step 5: Arrangement of slots

The width of slots should not be less than 3 times thickness (3 × 6.4 mm). Provide two slots 20 mm wide. The distance between edge of the slot and edge of channel or between adjacent slots also should not be less than twice the thickness (i.e. 2 × 6.4 mm).

Provide these distances as shown in Fig. 10.49. Let x be the length of the slot.

Total length of the weld.

$(800 + 225 + 4x - 2 \times 20) = 1607.6$ mm

$\therefore \quad x = 155.65$ mm

160 mm long fillet welding is done as shown in Fig. 10.49.

10.13 DESIGN OF WELDED JOINTS FROM UNSYMMETRICAL SECTIONS SUBJECTED TO AXIAL LOAD

In order to avoid the effect of eccentricity, the load is applied along the neutral axis of unsymmetrical sections (e.g., angle-section, tees). When the unsymmetrical sections are connected by welding, then, fillet welding is so applied that centre of gravity of the welds lies on the line of action of action of the load i.e., it coincides with the neutral axis. An angle section connected to a gusset plate is shown in Fig. 10.50.

The angle section carries and axial pull p. Let l_1 and l_2 be the lengths of welds applied at the sides. These welds resist P_1 and P_2 forces as shown in 10 .50. The lengths l_1 and l_2 may be determined as under.

Equating the moment of forces about the line of action of P_2

$$P_1 \cdot (a + b) = P.b$$

$$\therefore \quad P_1 = \frac{P \cdot b}{(a+b)} \quad \text{...(10.15)}$$

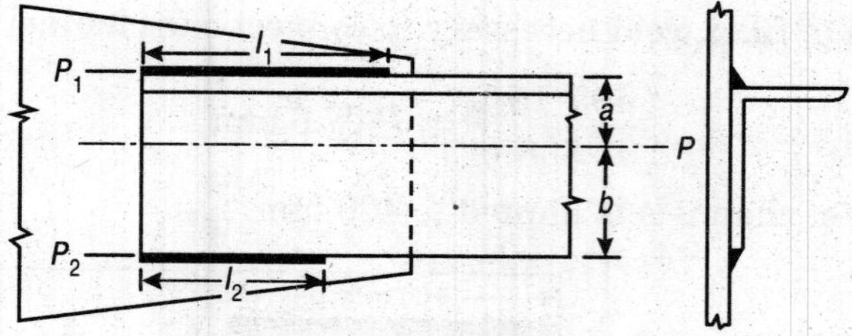

Fig. 10.50

Equate the moment of forces about the line of action of P_1

$$P_2 .(a + b) = P . a$$

$$\therefore \quad P_2 = \frac{P \cdot b}{(a+b)} \quad \text{...(10.16)}$$

Let τ_{vm} be the maximum permissible stress in the fillet weld, and s be the size of the weld.

The strength of weld per mm length

$$= (\tau_{vm} \times 0.7\,s)$$

$$l_1 = \left(\frac{p_1}{\tau_{vm} \times 0 \cdot 7s}\right) = \left(\frac{P \cdot b}{\tau_{vm} \times 0 \cdot 7s(a+b)}\right) \quad ...(10.17)$$

$$l_2 = \left(\frac{p_2}{\tau_{vm} \times 0 \cdot 7s}\right) = \left(\frac{P \cdot a}{\tau_{vm} \times 0 \cdot 7s(a+b)}\right) \quad ...(10.18)$$

The lengths of the welds should further satisfy the specifications of the filler weld. The total length of the weld required to apply is computed by dividing the load transmitted by the strength of the weld per mm length. In case the lengths available at the sides become insufficient or to make the joint more effective, end fillet weld is also applied as shown in Fig. 10.51. In such cases, the lengths of welds l_1 and l_2 to be applied at the sides are computed for the load to be transmitted less the strength of end fillet weld.

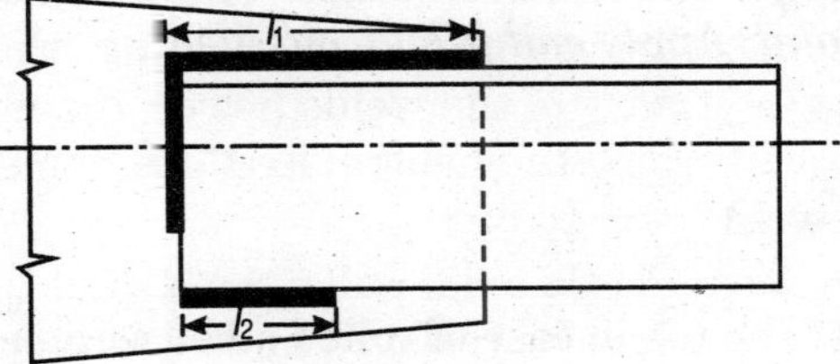

Fig. 10.51

Example 10.8 *A tie member of a roof truss consists of 2 ISA 90 mm × 60 mm × 10 mm. The tie member is subjected to put of 250 UN. The angles are connected either side of a gusset plate 10 mm thick. Design the welded connection.*

Solution

Design:

Step 1: Size of weld

The size of weld should not exceed $\frac{3}{4}$th thickness of the rolled steel section at the toe i.e., $\frac{3}{4} \times 10 = (7.5 \text{ mm})$

Step 2: Lengths of welds

The fillet weld of 6 mm size is provided on both the sides. Each angle section carries a pull of 125 kN. From Eq. 10.17,

$$l_1 = \left(\frac{P \cdot b \times 1000}{(a+b) \times 0 \cdot 7 \times 6 \times 110}\right) = \left(\frac{125 \times 61 \cdot 3 \times 1000}{90 \times 4 \cdot 2 \times 110}\right) = 184.28 \text{ mm}$$

From Eq. 10.18,

$$l_2 = \left(\frac{125 \times 28 \cdot 7 \times 1000}{90 \times 0 \cdot 7 \times 6 \times 10}\right) = 86.279 \text{ mm}$$

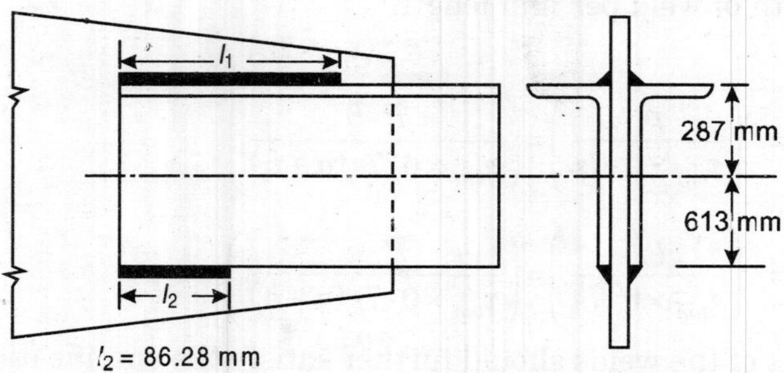

Fig. 10.52

These are the effective lengths of welds. Add twice the size of weld to have actual lengths of the welds.

The effective lengths of welds are shown in Fig. 10.52.

Example 10.9 *In a roof truss, a tie members ISA 110 mm × 110 mm × 8 mm carries a pull of 210 kN. The tie is connected with the gusset plate 8 mm thick. Design the welded joint. Apply end fillet weld also.*

Solution

Design :

Step 1 : Size of weld

The size of weld at the toe in case of rolled steel section should not exceed $\frac{3}{4}$ to thickness ($\ngtr \frac{3}{4} \times 8$ mm).

The maximum size of weld connecting vertical leg of the angle with gusset (8 – 1.5) = 6.5 mm

Provide a uniform size of 6 mm for the weld.

Distance of neutral axis is 30 mm from top.

Provide an end fillet weld of 60 mm effective length, symmetrical to the neutral axis.

Step 2 : Strength of end fillet weld

$$\left(\frac{110 \times 0 \cdot 7 \times 6 \times 60}{1000}\right) = 27.72 \text{ kN}$$

Step 3 : Pull resisted by side welds

$$(210 - 27.72) = 182.28 \text{ kN}$$

Step 4 : Length of welds

FromEq. 10.17,

$$l_1 = \left(\frac{182 \cdot 28 \times 80 \times 1000}{100 \times 0 \cdot 7 \times 6 \times 110}\right) = 286.94 \text{ mm}$$

From Eq. 10.18

$$l_2 = \left(\frac{182 \cdot 28 \times 30 \times 1000}{100 \times 0 \cdot 7 \times 6 \times 110}\right) = 107.5 \text{ mm}$$

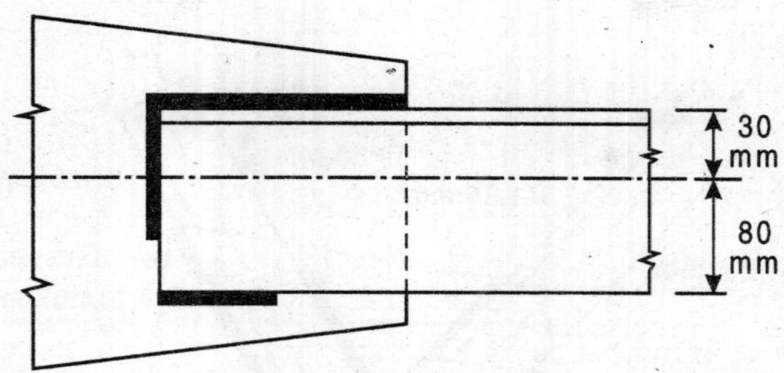

Fig. 10.53

The effective lengths of side welds are shown in Fig. 10.53.

Example 10.10 A *joint in a weld truss is as shown in Fig. 10.54. Design the welded connections.*

Solution

Design :

Step 1 : Size of weld

$$\text{Force in member} = 50 \text{ kN}$$

Size of weld should not exceed $\frac{3}{4}$ th thickness of rolled steel-section at the toe.

Thickness of angle section 8 mm

$$\therefore \quad \text{Size of weld} \not> \frac{3}{4} \times 8 \text{ mm} \not> 6 \text{ mm}$$

Fillet weld of 6 mm size is provided as shown in Fig. 10.54. From ISI Handbook No. 1, for ISA 75 mm × 50 mm × 8 mm (ISA 7550, @ 0.074 kN/m)

$$a = 25.2 \text{ mm, and } b = 49.8 \text{ mm}$$

$$(a + b) = 75.0 \text{ mm}$$

Step 2 : Effective length of weld, from Eq. 10.17

$$l_2 = \left(\frac{P \cdot b}{\tau_{vm} \times 0 \cdot 7s(a+b)}\right) = \left(\frac{50 \times 1000 \times 49 \cdot 8}{110 \times 0 \cdot 7 \times 6 \times 75}\right) = 71.86 \text{ mm}$$

Effective length of weld from Fig. 10.18.

$$l_2 = \left(\frac{P \cdot b}{\tau_{vm} \times 0 \cdot 7s(a+b)}\right) = \left(\frac{50 \times 1000 \times 25 \cdot 2}{110 \times 0 \cdot 7 \times 6 \times 75}\right) = 71.86 \text{ mm}$$

Step 3 : Actual length of welds

$$l_1' = (71.86 + 2 \times 6) = 83.86 \text{ mm}$$

$$l_2' = (36.36 + 2 \times 6) = 48.36 \text{ mm}$$

Effective lengths of weld are shown in Fig. 10.54.

$$\text{Force in member} = 80 \text{ kN}$$

The thickness of angle section is 8 mm

$$\text{Size of weld} \not> \frac{3}{4} 8 \not> 6 \text{ mm.}$$

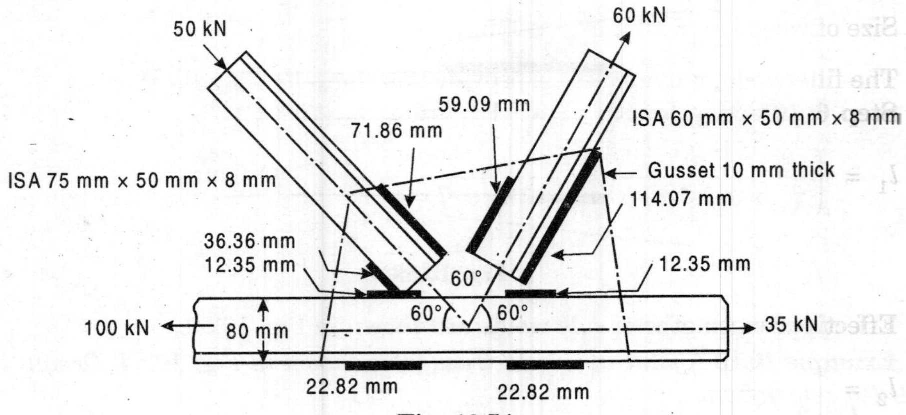

Fig. 10.54

The fillet weld of 6 mm size is provided as shown in Fig. 10.54.

From ISI Handbook No. 1, for ISA 80 mm × 50 mm × 8 mm (ISA 8050, @ 0.077 kN/m)

$$a = 27.2 \text{ mm and } b = 52.7 \text{ mm}$$

$$(a + b) = 80.0 \text{ mm}$$

Step 4 : Effective length of weld, from Eq. 10.17,

$$l_1 = \left(\frac{P \cdot b}{\tau_{vm} \times 0 \cdot 7s(a+b)}\right) = \left(\frac{80 \times 1000 \times 52 \cdot 7}{110 \times 0 \cdot 7 \times 6 \times 80}\right) = 114.07 \text{ mm}$$

Effective length of weld, from Eq. 10.18,

$$l_2 = \left(\frac{P \cdot b}{\tau_{vm} \times 0 \cdot 7s(a+b)}\right) = \left(\frac{80 \times 1000 \times 27 \cdot 3}{110 \times 0 \cdot 7 \times 6 \times 80}\right) = 59.09 \text{ mm}$$

Step 5 : Actual length of welds

$$l_1' = (114.86 + 2 \times 6) = 126.07 \text{ mm}$$

$$l_2' = (59.09 + 2 \times 6) = 71.09 \text{ mm}$$

The effective lengths of welds are shown in Fig. 10.54.

The horizontal member is a continuous tie member. It carries pull of 100 kN on one side and 35 kN on the other side. The weld is designed for difference of forces i.e., for (100 – 35) = 65 kN

There are two angles sections. The welding is done on both angle sections. Force in weld on one face

$$\frac{1}{2} \times 65 = 32.5 \text{ kN}$$

From ISI Handbook No 1, for ISA 80 mm × 50 mm × 10 mm (ISA 8050, @ 0.094 kN/m)

$$a = 28.1 \text{ mm and } b = 51.9 \text{ mm}$$

$$(a + b) = 80 \text{ mm}$$

Size of weld $\not> \frac{3}{4} \times 10$ mm $\not>$ 7.3 mm.

The fillet weld of 6 mm size is provided as shown in Fig. 10.54

Step 6: Effective length of weld l_1 from Eq. 10.17,

$$l_1 = \left(\frac{P \cdot b}{\tau_{vm} \times 0 \cdot 7s(a+b)}\right) = \left(\frac{32 \cdot 5 \times 1000 \times 51 \cdot 9}{110 \times 0 \cdot 7 \times 6 \times 80}\right) = 45.627 \text{ mm}$$

$$\frac{1}{2} l_2 = 22.82 \text{ mm}$$

Effective length of weld l_2 from Eq. 10.18,

$$l_2 = \left(\frac{P \cdot b}{\tau_{vm} \times 0 \cdot 7s(a+b)}\right) = \left(\frac{32 \cdot 5 \times 1000 \times 28 \cdot 1}{110 \times 0 \cdot 7 \times 6 \times 80}\right) = 24.71 \text{ mm}$$

$$\frac{1}{2} l_2 = 12.35 \text{ mm}$$

Step 7: Actual length of welds

$$l_1' = (45637 + 2 \times 6) = 57.637 \text{ mm}$$

$$l_2' = (24.71 + 2 \times 6) = 36.71 \text{ mm}$$

Half-length of welds are placed symmetrically as shown in Fig. 10.54.

10.14 DESIGN OF WELDED JOINTS SUBJECTED TO ECCENTRIC LOAD

When the C.G. of group of the weld does not lie on the line of action of applied load, then the welded joint is subjected to eccentric load. Similar to the riveted bracket connections, the welded joints subjected to eccentric load are also of two types as under:

1. The C.G. of group of the weld lies in the plane of line of action of the applied load.
2. The C.G. of group of the weld does not lie in the plane of line of action of the applied load.

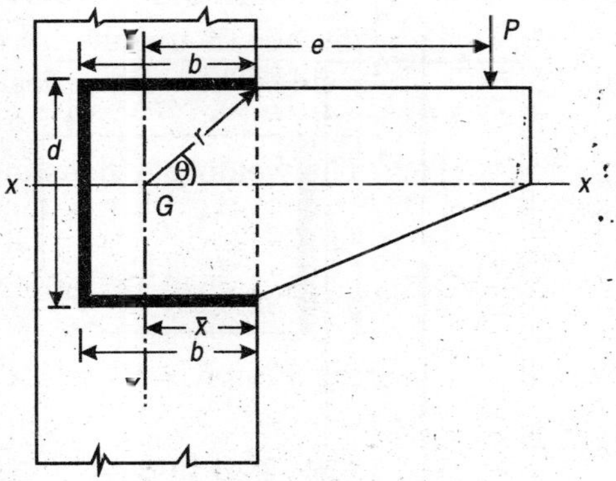

Fig. 10.55

Type 1. Figure 10.55 shows a fillet welded bracket connection. A vertical load, P is acting at a distance e from the C.G. of the group of the weld.

The eccentric load P is considered equivalent to direct load P passing through C.G. of the group of the weld and twisting moment '$P \times e$'.

Let a weld of uniform size be applied throughout, and t be the effective throat thickness.

The direct shear stress in the weld

$$p_s = \left(\frac{P}{(2b+d)\times t}\right)$$

The stress in the weld due to twisting moment is maximum in the weld at the extreme distance from the C.G. of group for the weld and acts in the direction perpendicular to the radius vector. The maximum stress due to twisting

$$p_b = \left(P \cdot e \cdot \frac{r}{I_{zz}}\right)$$

where, r = Distance to the extreme weld from the C.G., of weld group

I_{zz} = Polar moment of inertia

$I_{zz} = (I_{xx} + I_{yy})$

The vector sum of two stresses is given by

$$p = [(p_s)^2 + (p_s)^2 + 2\,p_s p_s \cos\theta]^{1/2} \quad ...(10.19)$$

The resulting stress p, is not to exceed the maximum permissible stress in the weld. In this type of eccentric welded joint, the stresses p_s and p_b are not normal to each other. As such, the combined stress is determined by using Eq. (10.19).

The thickness of gusset plate (bracket plate) should not be less than the size of the fillet weld. It should also be thicker than the thickness of flange of the column section.

Type 2. In this type C.G. of group of weld does not lie in the plane of action of the applied load. In such cases, butt weld can be applied or filled weld can also be used.

Figure 10.56 shows a bracket connection in which complete penetration *butt weld* has been applied.

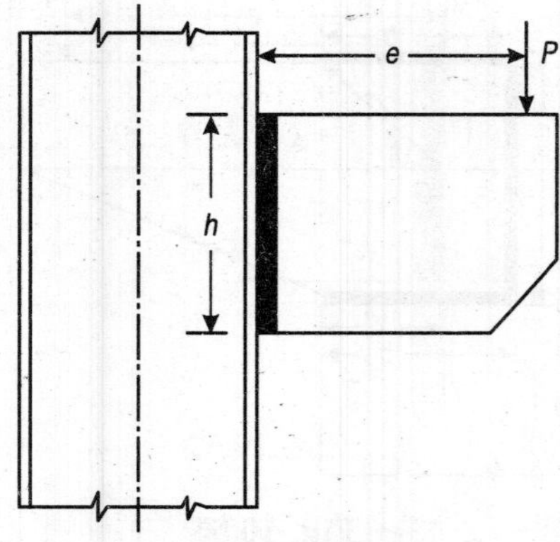

Fig. 10.56

The load P is acting at an eccentric distance e. The weld is subjected to direct shear stress as well as bending stress. Let t be the effective throat thickness of the weld.

Direct shear stress, $p_s = \dfrac{P}{(h \cdot t)}$

where, h = Effective depth of the weld.

The actual bending stress at the extreme distance of the weld

$$p_{t/e} = \left(\frac{M}{Z} = \frac{M}{\frac{1}{6}t \cdot h^2} = \frac{6M}{t\,h^2} = \frac{6P \cdot e}{t \cdot h^2}\right)$$

where, $p_{t/e}$ = Actual tensile/compressive bending stress in the weld,

For the combined stress in the butt weld, the expression $\left[\dfrac{p_s}{P_s}\right]^2 + \left[\dfrac{p_{t/e}}{P_{t/e}}\right]^2$ should not exceed unity.

For the purpose of design, the effective depth weld is computed from the following approximate depth formula

$$d = \left[\frac{6M}{t \times P_{t/e}}\right]^{1/2} \qquad \text{...(10.20)}$$

where, $P_{t/c}$ is the maximum permissible tensile or compressive bending stress in the weld.

The effective throat thickness of the weld is know, knowing the thickness of the member transmitting the load. The butt weld so designed is checked for combined stresses as above.

Figure 10.57 shows a bracket connection in which fillet weld has been applied on both the sides of the plate.

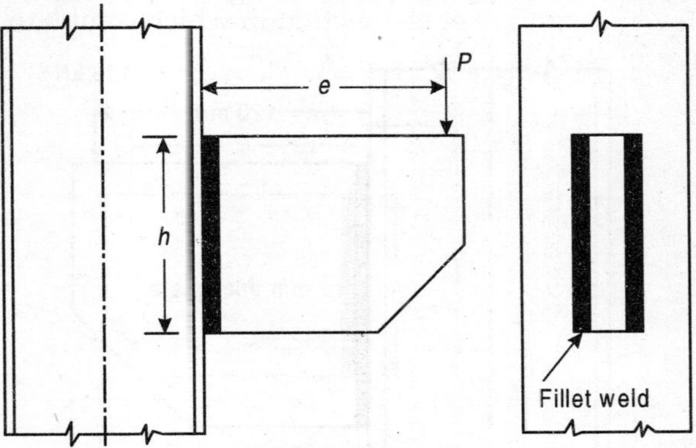

Fig. 10.57

The fillet welds are subjected to direct shear stress and bending stress. Let t be the effective throat thickness of the weld.

The direct shear stress in the weld

$$\tau_v = \left(\frac{P}{2t \cdot h}\right)$$

The bending stress in the weld

$$p_b = \left(\frac{M}{Z} = \frac{6P \cdot e}{2 \cdot t \cdot h^2}\right)$$

The combined stress in the fillet weld is obtained by the vector sum of these stresses

$$P = [(\tau_v)^2 + (p_b)^2]^{1/2} \quad \text{...(10.21)}$$

The resulting stress p should not exceed the maximum permissible shear stress in the weld.

For the purpose of the design, size of the weld is fixed depending upon thickness of the member transmitting the load. The effective throat thickness of the weld is thereby known.

The depth of fillet weld applied *on both the sides* is obtained by

$$h = \left[\frac{6M}{2 \cdot t \cdot P_s}\right]^{1/2} \quad \text{...(10.22)}$$

where, P_s is the maximum permissible shear stress in the weld.

In case, the depth of weld is limited, the combined stress in the weld for shear and bending is computed assuming the effective throat thickness of the fillet weld. The combined stress in the fillet weld is not to exceed maximum permissible shear in the weld. Equating the combined stress in the weld with the maximum permissible shear stress in the weld, effective throat thickness of the weld is found. The size of the weld is computed knowing the effective throat thickness of the weld.

Example 10.11 *Design : (i) a suitable butt weld, (ii) a fillet weld for the bracket connection as shown in Fig. 10.58.*

Allowable stress on the butt weld in bending is 157.5 N/mm².

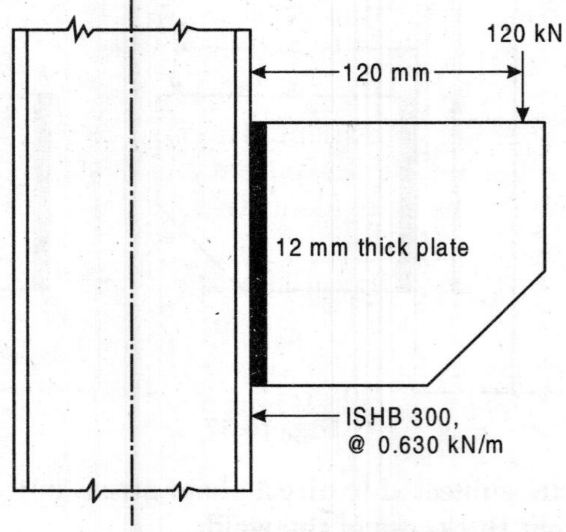

Fig. 10.58

Solution

Design :

(i) **Butt weld**

Step 1: Depth of weld

Provide a double *V* butt weld. The effective throat thickness of weld is 12 mm.

Let *h* be the approximate depth of the weld. From Eq. 10.20,

where, M = Bending moment

t = Effective throat thickness of weld

$P_{t/c}$ = Allowable tensile or compressive stress in the weld in bending

$$\therefore \quad h = \left[\frac{6\times120\times1000\times120}{12\times157.5}\right]^{1/2} = 213.81 \text{ mm}$$

Try 230 mm depth of the weld

Step 2: Stresses in welds

Direct shear stress in the weld

$$\tau_v = \left(\frac{120\times1000}{230\times12}\right) = 43.478 \text{ N/mm}^2$$

Actual bending stress

$$P_{t/c} = \left(\frac{6\times120\times1000\times120}{12\times230^2}\right) = 136.11 \text{ N/mm}^2$$

Check

Step 3: For combined stress in the butt weld,

$$\left[\frac{p_s}{p_s}\right]^2 + \left[\frac{p_{t/e}}{p_{t/e}}\right]^2 = \left[\frac{43.478}{110}\right]^2 + \left[\frac{136.1}{157.5}\right]^2 = 0.903 < 1.$$

Hence, safe.

Step 4: Check for equivalent stress

For butt weld, the equivalent stress $\sigma_{e\ cal}$ due to coexisting bending stress (tension or compression) and shear stress is given by

$$\sigma_{e\cdot cal} = [\sigma_{bc\cdot cal}{}^2 + 3\tau_{vm\cdot cal}{}^2]^{1/2} = [(136.1)^2 + 3 \times (43.478)^2]^{1/2}$$
$$= 155.54 \text{ N/mm}^2$$

As per IS 816–1969, the equivalent stress $\sigma_{e.cal}$ shall not exceed the values allowed for the parent metal. For the parent metal (structural steel having yield stress, f_y as 250 N/mm²), the equivalent stress, σ_e (as per IS : 800–1984) is given by

$$\sigma_e = (0.9 \times 250)$$
$$= 225 \text{ N/mm}^2; (\sigma_{e.cal} \not> \sigma_e).$$

Hence, the design is safe

Provide 230 mm depth of double *V* butt weld.

(ii) **Fillet weld**

Step 1: Depth of weld

Apply 8 mm fillet weld on each side of plate.

Effective throat thickness of fillet weld

$$(0.7 \times 8) = 5.6 \text{ mm}$$

Strength of 1 mm fillet weld

$$\left(\frac{110 \times 5 \cdot 6 \times 1}{1000}\right) = 0.616 \text{ kN}$$

Approximate depth of fillet weld applied on two faces

$$h = \left[\frac{6 \times 120 \times 120}{2 \times 0 \cdot 616 \times 1}\right]^{\frac{1}{2}} = 264.82 \text{ mm}$$

Try 300 mm depth for fillet welds.

Step 2: Stresses in welds

Direct shear stress in fillet weld

$$\tau_v = \left(\frac{120 \times 1000}{2 \times 300 \times 5 \cdot 6}\right) = 35.71 \text{ N/mm}^2$$

Bending stress in fillet weld

$$p_b = \left(\frac{6 \times 120 \times 1000}{2 \times 5 \cdot 6 \times 3002}\right) = 85.71 \text{ N/mm}^2$$

Step 3: Check for combined stress in fillet weld

$$\begin{aligned} p &= [35.71^2 + 85.71^2]^{1/2} \\ &= 92.85 < 110 \text{ N/mm}^2. \end{aligned}$$

Hence, safe.

Step 4: Check for equivalent stress

For fillet weld, the equivalent stress $\sigma_{e \cdot \text{cal}}$ due to co-existing bending stress and shear stress is given by

$$\begin{aligned} \sigma_{eca \cdot \text{cal}} &= \left[\sigma_{bc \cdot \text{cal}}{}^2 + 1 \cdot 8 t_{vm \cdot \text{cal}}{}^2\right]^{\frac{1}{2}} \\ &= [85.71^2 + 1.8 \times 35.71^2]^{1/2} \\ &= 98.19 \text{ N/mm}^2 \ngtr (\sigma_e = 110 \text{ N/mm}^2) \end{aligned}$$

Hence, the design is safe.

Provide 300 mm depth of 8 mm fillet weld.

Example 10.12 *Plates have been connected with the flanges of an I-section by applying 8 mm fillet weld as shown in Fig. 10.59. Compute the maximum load which may be placed at a distance of 100 mm from the flanges.*

Solution :

Step 1: Size of weld

Consider one face,

Size of weld $= 8$ mm

Effective throat thickness $= (0.7 \times 8)$

$= 5.6$ mm

Step 2: Properties of welds

Let $\bar{x}$ be the distance of centroid 'G' of weld group from left hand edge of the plate

$$\bar{x} = \frac{(2\times200\times5\cdot6\times100)+(200\times5\cdot6\times0)}{(2\times200\times5\cdot6)+(200\times\cdot6)}$$

$$= 66.7 \text{ mm}$$

Moment of inertia of weld group about xx-axis

$$I_{xx} = \left[2\times20\times0\cdot56\times100+\frac{1}{2}\times0\cdot56\times20^3\right]\times10^4$$

$$= 4480 \times 10^4 \text{ mm}^4$$

Moment of inertia of weld group about yy-axis

$$I_{yy} = \left[\frac{1}{2}\times2\times0\cdot56\times20^3+2\times20\times0\cdot56(20-6\cdot67)^2+20\times0\cdot56\times6^2\right]10^4 \text{ mm}^4$$

$$= 1493.2 \times 10^4 \text{ mm}^4$$

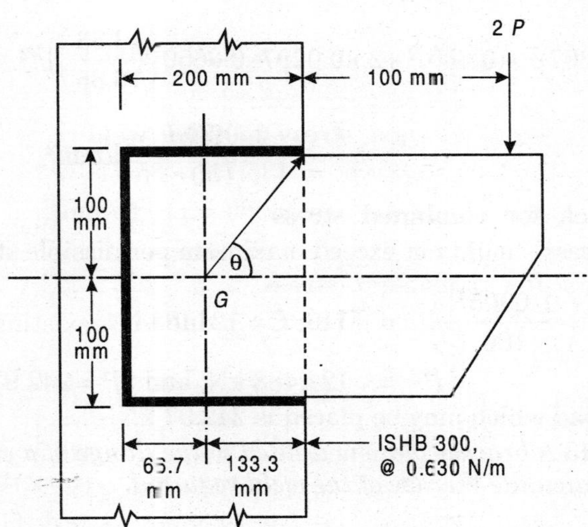

Fig. 10.59

Polar moment of inertia of weld group

$$I_{xx} = (I_{xx}+I_{yy}) = (4480 + 1493.2)$$

$$= 5973.2 \times 10^4 \text{ mm}^4$$

Distance to the extreme weld from the centroid of weld group

$$r = [100^2 + 133\cdot3^2]^{1/2}$$

$$= 166.64 \text{ mm}$$

$$\cos\theta = \left(\frac{13\cdot33}{16\cdot65}\right)$$

Let $2P$ be the maximum load which can be placed on the bracket. Load transmitted by each face = P

Step 3 : Stresses in welds

Direct shear stress

$$\tau_v = \left(\frac{P}{(2\times 200+200)\times 5\cdot 6}\right) = \left(\frac{(0\cdot 0297P)}{100}\right)\text{Nmm}^2$$

Twisting moment resisted by the weld group

$$T = P \times (133.3 + 100)$$
$$= 233.3\, P \text{ N-mm}$$

Maximum shear stress due to twisting

$$p_b = \left(\frac{233\cdot 3P\times 166\cdot 64}{5973\cdot 2\times 104}\right) = \left(\frac{0.0650P}{100}\right)$$

Combined stress in the weld group, from Eq. 10.19,

$$p = \left(\frac{P}{100}\right)$$

$$\left(\frac{P}{100}\right)\times[(0.0297)^2 + (0.650)^2 + 2\times 0.0297\times 0.0650\left(\frac{13.33}{16.65}\right)]^{1/2}$$

$$= \left(\frac{0\cdot 0905P}{100}\right)\text{N/mm}^2$$

Step 4: Check for combined stress

Combined stress should not exceed maximum permissible stress 110 N/mm^2

$$\therefore \quad \left(\frac{0\cdot 0905P}{100}\right) = 110,\ P = 121488 \text{ N}$$

$$\therefore \quad P = 121.488 \text{ kN, and } 2P = 242.97 \text{ kN}$$

Maximum load which may be placed is 242.97 kN.

Example 10.13 *A bracket plate is welded to the flange of a column as shown in Fig. 10.60. Calculate the size of the weld required.*

Solution :

Step 1: Properties of welds

Let t be the effective throat thickness of the weld group, and $\bar{x}$ be the distance of its centroid from left hand edge of the plate ,

$$\bar{x} = \left(\frac{2\times 180\times t\times 90+250\times t\times 0}{2\times 180\times t+250\times t}\right)$$
$$= 53.1 \text{ mm}$$

Moment of inertia of weld group above xx-axis

$$I_{xx} = \left[2\times 18\times t\times 12\cdot 5^2 + \frac{1}{2}\times t\times 25^3\right]\times 10^4 \text{ mm}^4$$
$$= 6942t \times 10^4 \text{ mm}^4$$

Moment of inertia of weld group about y-axis

$$I_{yy} = \left[\frac{1}{2}\times 2\times t\times 18^3 + 2\times 13\times t(9-5\cdot 31)^2 + 25t\times 5\cdot 31^2\right]\times 10^4$$

$$= 1884.5\ t \times 10^4\ \text{mm}^4$$

Polar moment of inertia of weld group

$$I_{xx} = (6942 + 1884.5)t \times 10^3\ \text{mm}^4$$
$$= 8826.5t \times 10^3\ \text{mm}^4$$

Distance to the extreme weld from the centroid of weld group

$$r = [(125)^2 + (126.9)^2]^{1/2}$$
$$= 178.8\ \text{mm}$$

$$\cos\theta = \left(\frac{126\cdot 9}{178\cdot 8}\right) = 0{:}71$$

Step 2: Stres in welds

Direct shear stress

$$\tau_v = \left(\frac{100\times 1000}{(2\times 180\times 250)\cdot t}\right) = \left(\frac{163\cdot 93}{t}\right)\text{N/mm}^2$$

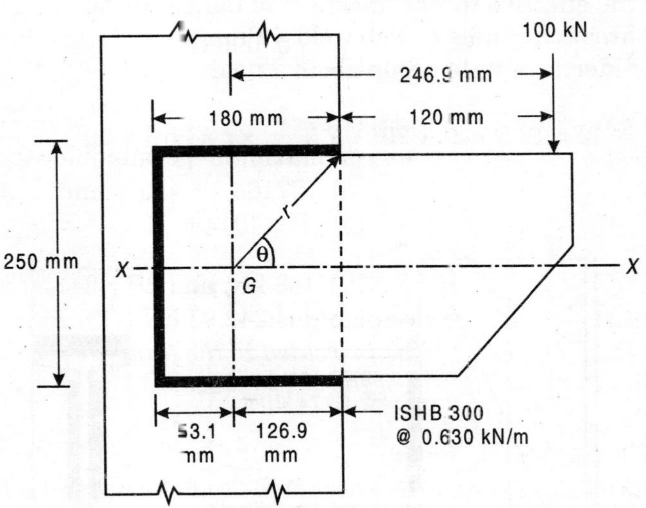

Fig. 10.60

Maximum shear stress due to twisting

$$p_b = \left(\frac{100\times 1000\times 246\cdot 9\times 178\cdot 8}{8826\cdot 5t\times 10^3}\right)\text{N/mm}^2$$

$$= \left(\frac{500\cdot 15}{t}\right)\text{N/mm}^2$$

Step 3: Combined stress in the weld group, from Eq. 10.19,

$$P = \left[\left(\frac{163\cdot 93}{t}\right)^2 + \left(\frac{500\cdot 15}{t}\right)^2 + 2\times\left(\frac{163\cdot 93}{t}\right)\left(\frac{500\cdot 15}{t}\right)\left(\frac{500\cdot 15}{t}\right)\times 0\cdot 73\right]^{\frac{1}{2}}$$

$$= \frac{627\cdot 25}{t}\ \text{N/mm}^2$$

Combined stress in the fillet weld should not exceed 110 N/mm².

$$\therefore \qquad t = \left(\frac{627\cdot 25}{110}\right) = 5.7\ \text{mm}$$

$$\text{Size of weld required} = \left(\frac{5\cdot 7}{0\cdot 7}\right) = 8.146\ \text{mm}$$

Provide 9 mm fillet weld.

Example 10.14 *Calculate the size of the weld required for the joint cutting used as a bracket loaded as shown in Fig. 10.61. The flange welds are of double the size of welds.*

Solution :

Step 1 : Properties of welds

Let 2 t be the effective throat thickness of flange welds.

Effective throat thickness of web weld = t mm

Moment of inertia of weld group about xx-axis

$$I_{xx} = [2 \times 12.5 \times 2t \times 20^2 + 2 \times \frac{1}{12} \times t \times 35^3] \times 10^3$$

$$= (27166 \times t \times 10^3)\ \text{mm}^4$$

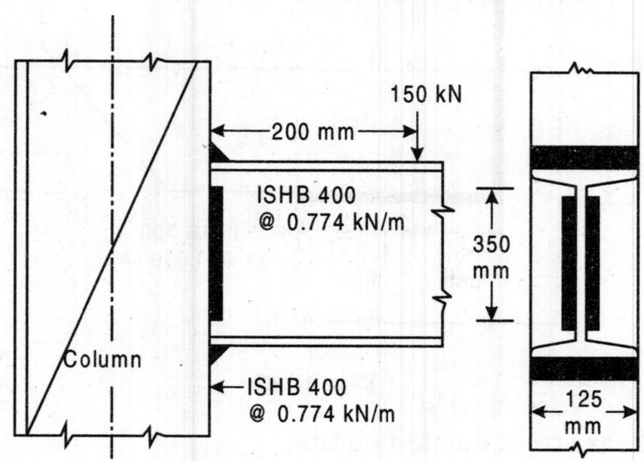

Fig. 10.61

Step 2 : Stresses in welds

1. Direct shear stress in weld

Direct load shared by welds in the proportion to the areas of welds. Total area of welds = $(2 \times 125 \times 2t + 2 \times 350 \times t) = 1200\ t$ mm²

Load shared by flange welds $= \left(\frac{500t}{1200t} \times 150\right) = 62.50$ kN

Load shared by web weld $= 87.50$ kN

Shear stress in the flange weld

$$p_s = \left(\frac{62 \cdot 50 \times 1000}{2 \times 350 \times t}\right) = \left(\frac{125}{t}\right) \text{N/mm}^2$$

Shear stress in the web weld

$$p_{s1} = \left(\frac{87 \cdot 50 \times 1000}{2 \times 350 \times t}\right) = \left(\frac{125}{t}\right) \text{N/mm}^2$$

2. Bending stress in welds

The bending stress in the welds is in proportion to the distance from the neutral axis of weld group.

Bending stress in the weld

$$Pb = \left(\frac{M}{I_{xx}}\right) \cdot y$$

$$Pb = \left(\frac{150 \times 1000 \times 200 \times 200}{27166 \times t \times 10^3}\right)$$

$$= \left(\frac{220 \cdot 86}{t}\right) \text{N/mm}^2$$

Bending stress in the extreme fibre of web welds

$$Pb_1 = \left(\frac{150 \times 1000 \times 200 \times 175}{27166 \times t \times 10^3}\right)$$

$$= \left(\frac{193 \cdot 26}{t}\right) \text{N/mm}^2$$

Combined stress in flange welds. From Eq. 10.21.

$$p = \left[\left(\frac{125}{t}\right)^2 + \left(\frac{220 \cdot 86}{t}\right)^2\right]^{\frac{1}{2}}$$

$$= \left(\frac{253 \cdot 78}{t}\right) \text{N/mm}^2$$

Combined stress in web weld

$$p_1 = \left[\left(\frac{125}{t}\right)^2 + \left(\frac{193 \cdot 26}{t}\right)^2\right]^{\frac{1}{2}}$$

$$= \left(\frac{230 \cdot 16}{t}\right) \text{N/mm}^2$$

Combined stress in the weld should not exceed 110 N/mm^2

$$\therefore \quad t = \left(\frac{253\cdot78}{110}\right) = 2.307 \text{ mm}$$

Size of web welds $= \left(\frac{2\cdot307}{0\cdot7}\right)$ 3 296 mm. Adopt 5 mm weld.

Size of flange welds

$$2s = 6.59 \text{ mm. Adopt } 2 \times 5 = 10 \text{ mm weld.}$$

Example 10.15 *A bracket plate is welded to the flange of a column as shown in Fig. 10.62. The bracket plate carries a load of 250 kN at a distance of 500 mm from the flange of column. Provide intermittent fillet weld on both sides of the bracket plate. Design the welded connection.*

Solution

Design :

Step 1 : Number of intermittent fillet welds

The number of weld lengths is found in the same manner as the number of rivets necessary to resist the moment caused. From Sec. 2.17, eccentric riveted connections, (design of bracket connections type 1) relation

$$n = \left[\frac{6M(n-1)}{p \cdot R \cdot n}\right]^{\frac{1}{2}}$$

may be used. Factor $\frac{(n-1)}{n}$ is assumed to be nearly unity, therefore

$$n = \left[\frac{6M}{p \cdot R}\right]^{\frac{1}{2}}$$

The intermittent fillet welds are provided on both the sides of bracket plate. Therefore,

$$n = \left[\frac{6M}{2p \cdot R}\right]^{\frac{1}{2}}$$

where, n = Number of weld lengths in one line

p = Strength of one weld length.

Let the size of weld for 12 mm thick bracket plate be 8 mm

The effective length of intermittent weld should be four times the size of weld or 38 mm whichever is more.

Therefore, effective length of weld = 38 mm

Step 2: Spacing between welds

Assume pitch of welds, i.e., distance between centre to centre of adjacent intermittent welds

$$p = 80 \text{ mm}$$

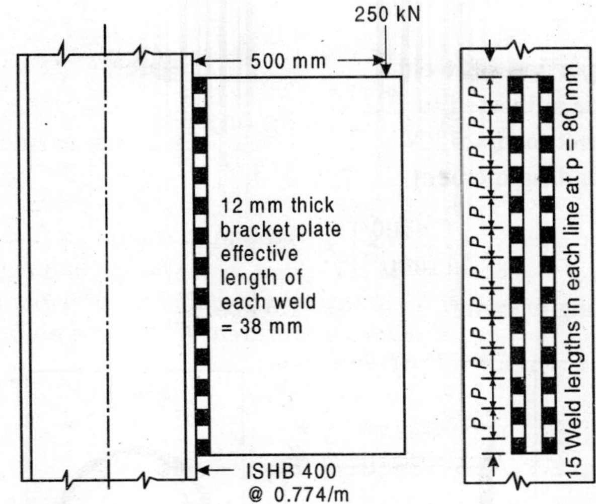

Fig. 10.62

Clear spacing between the effective length of welds

$$= \left(80 - 2\times\frac{38}{2}\right) = 42 \text{ mm.}$$

The clear spacing does not exceed 16 t for compression and 24 t for tension.

The moment at weld section

$$= \left(250\times\frac{500}{100}\right) = 125 \text{ kN-m}$$

Strength of one 38 mm weld

$$R = \left(0\cdot 7\times 8\times\frac{38\times 110}{1000}\right) = 23.41 \text{ kN}$$

Number of weld lengths in one line

$$n = \left[\frac{6M}{2p\cdot R}\right] = \left[\frac{6\times 125\times 1000}{2\times 80\times 23\cdot 41}\right]^{\frac{1}{2}} = 14.15$$

Provide 15 intermittent weld lengths on either face of bracket plate as shown in Fig. 10.62. the effective length of each weld

$$= 38 \text{ mm}$$

The pitch of weld length is 80 mm.

Example 10.16 *A 150 mm diameter mild steel pipe 0.50 metre long is welded to a vertical plate 12 mm thick, with the axis of pipe at right angles to the face of the plate. The pipe is subjected to a twisting moment of 1.5 kN-m and carries a vertical load of 5 kN acting at the end of pipe. Design the welded connection. Assume thickness of pipe 6 mm. Allowable stress in the weld is 110 N/mm².*

Solution

Design :

Step 1: Properties of weld

The welding is subjected to

(i) a direct load = 5 kN

(ii) a bending moment

$$\left(\frac{5\times500}{1000}\right) = 2.5 \text{ kN-m}$$

and (iii) a twisting moment $= 1.5$ kN-m

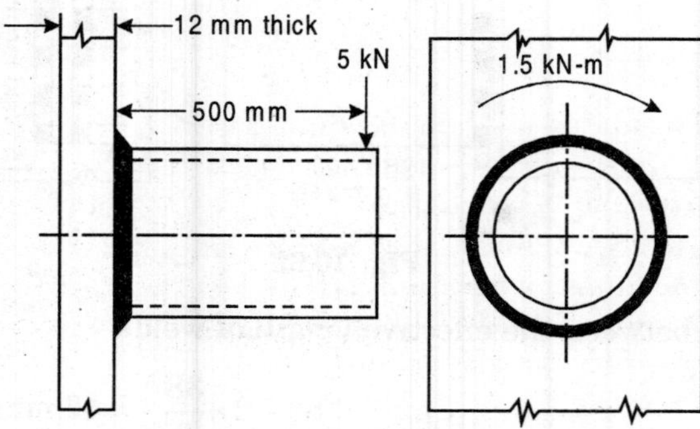

Fig. 10.63

Let t mm be the effective throat thickness of weld.

The approximate polar moment of inertia of weld group

$$I_{xx} = \Sigma\delta a \times r^2$$

where, δa is the elementary area of weld and r is the distance from the centre

$$I_{zz} = (\pi rt \times r^2 = 2\,\pi\, r^3 t)$$
$$= (2\pi \times 75^3 \times t \times 10^3)$$
$$= 2640\, t \times 10^3 \text{ mm}^4$$

Moment of inertia of weld about xx-axis

$$I_{xx} = \frac{1}{2} I_{zz} = 1320t \times 30^3 \text{ mm}^4$$

Step 2 : Stresses in welds

Shear stress in the weld due to twisting

$$p_{tw} = \left(\frac{1\cdot5\times10^6\times75}{2640t\times10^3}\right) = \left(\frac{42\cdot6}{t}\right)\text{N/mm}^2$$

Stress in the weld due to bending

$$p_b = \left(\frac{2\cdot5\times10^6\times75}{1320\times t\times10^3}\right) = \left(\frac{142}{t}\right)\text{N/mm}^2$$

The stress in the weld due to shear is maximum at the centre. But average shear stress is considered. Average shear stress remains uniform.

$$p_s = \left(\frac{5\times1000}{2\pi\times75t}\right) = \left(\frac{10\cdot62}{t}\right)\text{N/mm}^2$$

Combined stress in the weld group at the top

$$\left[\left(\frac{10\cdot62}{t}\right)^2 + \left(\frac{142}{t}\right)^2 + \left(\frac{42\cdot6}{t}\right)^2\right]^{\frac{1}{2}} = \frac{148\cdot63}{t}\text{ N/mm}^2$$

The combined stress in the fillet weld is not to exceed maximum permissible stress in shear

$$\therefore \qquad t = \left(\frac{148\cdot63}{110}\right) = 1.35 \text{ mm}$$

Step 3 : Size of weld $\left(\frac{1\cdot38}{0\cdot7}\right) = 193$ mm

Provide 3 mm size of fillet weld.

Example 10.17 *An ISA 125 mm × 95 mm × 12 mm (ISA 12595, @ 0.196 kN/m) is welded with the flange of a column HB 300, @ 0.630 kN/m as shown in Fig. 10.64. The bracket carries a load of 100 kN at a distance of 60 mm from the face of column. Design the bracket connection.*

Solution

Design :

Step 1: Throat thickness of weld

Width of flange of column HB 300, @ 0.630 kN/m

$$b = 250 \text{ mm}$$

The horizontal fillet welds of 250 mm length each, are provided as shown in Fig. 10.64.

Let t mm be the throat thickness of weld. The weld is subjected to direct shear and bending.

Stpe 2: Stresses in welds

Shear force in weld

$$\tau_{vm.\text{cal}} = \left(\frac{100\times100}{2\times t\times250}\right)\text{N/mm}^2$$

$$= \left(\frac{200}{t}\right)\text{ N/mm}^2$$

Moment about the faces of column

$$= \left(100\times100\times\frac{60}{10^6}\right)$$

$$= 6 \text{ kN-m}$$

The bending moment is resisted by a couple. The weld at the top is subjected to pull and that at the bottom is subjected to thrust. This is a case of discontinuous weld i.e., cross flexure

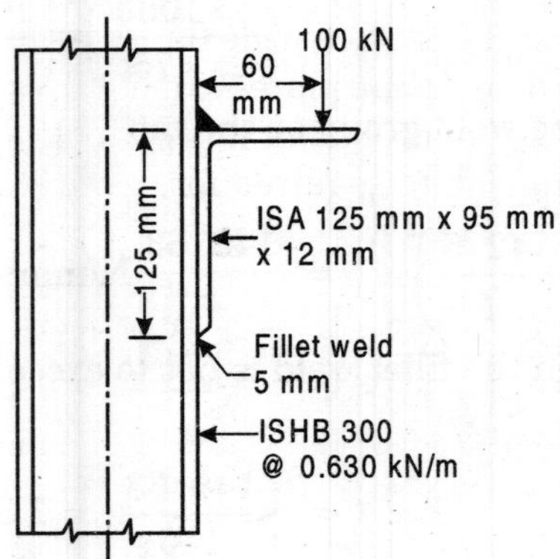

Fig. 10.64

From Eq. 10.8,

Stress in weld,
$$f_s = \left(\frac{M}{(h+s)\cdot\left(\frac{s}{\sqrt{2}}\right)\cdot l}\right)$$

$$f_t = \left(\frac{6\times10^6}{125\times t\times 250}\right)$$

$(h+s)$ is taken as h only

$$f_t = \left(\frac{192}{t}\right)\text{N/mm}^2$$

Step 3: Resultant stress in the weld

$$f = [fs^2 + f_t^2]^{1/2}$$

$$110 = \frac{1}{t}[(200)^2 + (192)^2]^{1/2}$$

$$t = \frac{1}{100}[(200)^2 + (192)^2]^{1/2} = 2.52 \text{ mm}$$

Size of weld
$$s = \frac{2\cdot52}{0\cdot7} = 3.6 \text{ mm}$$

The size of weld is very small. Provide 5 mm weld as shown in Fig. 10.64.

10.15 WELDED PLATE GIRDER

In the welded plate girders, whole of the section is effective in resisting the loads. Therefore, it is efficient than the riveted plate girder. Figure 10.65 (a) and (b) show the sections of welded plate girder commonly used in section. Figure 10.65 (c) shows the elevation of a plate girder with bearing stiffeners at the ends and intermediate stiffeners. In the welded plate girder, flange angles are not used. The flange plates are directly welded to the web of the girder as shown in Fig. 10.65 (a). The cut sections from wide flange beams are also used as shown in Fig. 10.65 (b). The stacked flange plates are not used for welded plate girders because of multiple welding involved, but it is possible to vary the cross-sectional area by flanking the flange plate required for maximum bending moment with

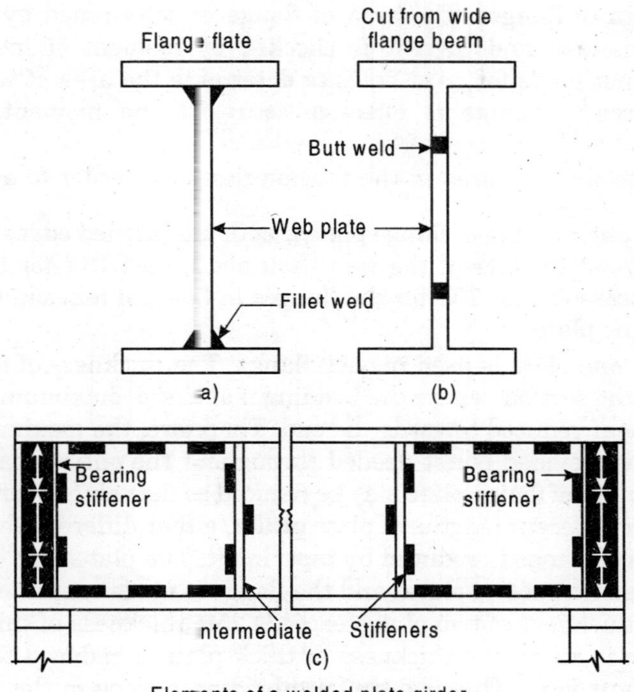

Fig. 10.65 *Welded plate girder*

one or more successively thinner plates, butt-welded end to end. In welded plate girder, it is not necessary to make an allowance for rivet holes to determine the net section for tension flange. There is saving in the section of tension flange. Generally, the welded plate girders weigh less than riveted plate girders. The saving may vary from 5 to 15 per cent depending on the governing specifications. The welded plate girder is economical in material and cost. The section of welded plate girder is proportioned by the same principles as in the case of riveted plate girder. Instead of rivets, welds are used for connection.

(i) **Design of web.** The depth of plate girder is adopted as $\frac{1}{8}$ to $\frac{1}{12}$ of the span of girder. The web plate is designed for maximum shear. The web plate designed should also satisfy the requirements of local buckling. The economical depth of web is given by

$$d_e = 5\left[\frac{M}{f_b}\right]^{\frac{1}{3}} \quad ...(10.23)$$

where, M is the maximum bending moment and f_b is the maximum permissible bending stress (150 N/mm^2).

(ii) **Design of flanges.** The area of flange is determined by approximate flange area method and then it is checked by moment of inertia method. Allowance is not made for rivet holes to determine the area of tension flange. The gross area of flange is effective to resist the moment. The flange includes $\left(\frac{1}{6}\right)$th the web area in the tension flange. In order to avoid the local buckling, the outstanding of flange plates, with unstiffened edges, that is, their projection beyond the face of the web shall not exceed 16 t for the flanges in flexural compression and 20 t for the flanges in flexural tension, where t is the thickness of the plate.

In general, one plate is used in each flange. The thickness of flange plate is maximum at the section, where the bending moment is maximum. The value of bending moment reduced towards the end. Therefore, the maximum thickness of flange plate provided is not needed throughout the entire span. Either the thickness or width of flange plate may be reduced as the value of bending moment decreases. For different lengths of plate girder, either different thicknesses are used or width of flange is reduced by tapering it. The plates are joined by butt weld at the junction of two plates and the flange is made continuous. When the difference in thickness of two plates exceeds 25% thickness of thinner plate or 3.2 whichever is more, the thickness of thick plate is reduced to that of thin plate at the butt joint. This is done by providing a slope in the thicker plate. This slope should not be greater than one in five. This is shown in Fig. 10.12 (a). In case, the thick plate cannot be reduced, the weld metal is built up 25% greater than the thickness of thinner plate as shown in Fig. 10.12 (c). The thickness of thicker plate in such case should not be greater than 50% the thickness of thiner plate.

(iii) **Design of connection of flange plate and web.** The welds connecting flange plates with the web plate may be applied continuously or intermittently on both the sides of web. When continuous weld is used, the size of fillet weld is changed in different portions of girder. The fillet weld is designed for horizontal shear per unit length. The horizontal shear per unit length for the loaded flange is given by

$$f_{sh} = \left[\frac{F \cdot A\bar{y}}{I_{xx} \cdot (2t)} \times (2t \times 1)\right]$$

$$\therefore \quad f_{sh} = \left[\frac{F \cdot A\bar{y}}{I_{xx}}\right] \quad ...(10.24)$$

where, F = Shear force at the section

$A\bar{y}$ = Moment of the area above the horizontal section considered about xx-axis (neutral axis)

t = Effective throat thickness of the weld

I_{xx} = Moment of inertia of the welded plate girder about xx-axis (neutral axis)

Let w be vertical load per unit length over the loaded flange. The combined stress per unit length is given by

$$f = \left[(f_{sh})^2 + w^2\right]^{\frac{1}{2}} \quad ...(10.25)$$

For the unloaded flange, w is zero.

The size of fillet weld to be applied should be such that the strength of weld per unit length is equal to the combined stress per unit length. If s is the size of fillet weld in mm, strength of weld per mm length

$$0.707 \times (2s) \times 110 = 155.54s \text{ N}$$

Therefore, $155.54\, s = f_{sh}$

$$s = \left(\frac{f_{sh}}{155 \cdot 4}\right)$$

The size of welds should not be less than minimum size of weld required as per Table 10.1.

In case intermittent fillet welds are provided, the pitch of weld is found similar to that or rivets connecting flanges with the web. The minimum size of weld depends upon thickness of thicker plate to be connected. The thickness of plate is maximum near the centre where the bending moment is maximum. Near the supports, the thickness of the plate is small, but shear force is maximum. A suitable uniform size of weld is selected for the entire span.

The effective length of intermittent fillet weld should not be less than four times the size of weld or 38 mm, whichever is more. The clear spacing between the weld should not be greater than 12 t or 200 mm for compression flange and 16 t or 200 mm for tension flange, where t is thickness of thinner plate.

If R is the strength of one weld, the pitch of weld for loaded flange

$$p = \frac{R}{\left[(f_{sh})^2 + w^2\right]^{\frac{1}{2}}} \quad ...(10.26)$$

For the unloaded flange, $w = 0$.

The weld is provided on both faces, therefore, the weld is provided at a distance equal to twice the pitch calculated. The size of intermittent weld adopted used to be large. It is economical to use continuous fillet weld for connections.

The gap between the web plate and the flange plates shall be kept to a minimum, and for fillet welds shall not exceed 1 mm at any point before welding. The dispersion of loads through the flange is assumed at 30° to the horizontal.

(iv) **Design of bearing stiffener.** The stiffeners consist of flat plates instead of angles as in case of riveted plate girder as shown in Fig. 10.65 (c). The ends of plates are machined to provide direct bearing so that the load is transmitted by bearing. The welds connecting the stiffeners to the web are designed to transmit full load or support reaction. Besides this, the bearing stiffeners are designed similar to those in riveted plate girder.

(v) **Intermediate stiffeners.** The stiffeners consist of flat plates instead of angles as incase of riveted plate girder as shown in Fig. 10.65 (c). The outstand of stiffeners of flat sections shall be not more than 12 *t,* where *t* is the thickness of flat section used. The intermediate stiffeners are designed similar to those in riveted plate girder. The weld connecting stiffeners and web are designed to resist horizontal shear $\left(\frac{1\cdot 5t_w^2}{h}\right)$ kN/m length, where t_w is the thickness of web in mm and h is the outstand of the stiffener in mm. The minimum size of weld depends upon the thickness of stiffener or thickness of web.

In case intermittent welds are used, the instance between the effective lengths of any two welds, even if staggered on opposite side should not exceed 16 times the thickness of stiffener nor 300 mm. In case intermittent welds are provided on one side of stiffener only or both sides but taggered of where single plate stiffeners are butt welded to the web, the effective length of each weld should not be less than 10 times the thickness of stiffener. In case, intermittent welds are provided in pairs on both sides, the effective length of each weld should not be less than *four times* the thickness of the stiffener.

Example 10.18 *A welded plate girder simply supported at two ends has an effective span of 28 m. It carries a uniformly distributed load of 25 kN per metre. Design the maximum section of welded plate girder. Show the reduction of flange plate.*

Solution

Design :

Step 1: Maximum bending moment and shear force

The effective span of welded plate girder is 28 m and the uniformly distributed load, w is 25 kN/m.

Self weight of girder,

$$w_1 = \left(\frac{wL}{300}\right) = \left(\frac{25\times 28}{300}\right) = 2.33 \text{ kN-m}$$

Total uniformly distributed load

$$(25 + 23.3) = 27.33 \text{ kN/m}$$

Maximum bending moment,

$$M = \left(\frac{27\cdot 33\times 28\times 28}{8}\right) = 2678.34 \text{ kN-m}$$

Maximum shear force,

$$F = \left(\frac{27\cdot 33\times 28\times 28}{2}\right) = 382.62 \text{ kN}$$

Stpe 2 : Economical depth of web plate from Eq. 10.23.

$$d_s = 5\left(\frac{M}{f_b}\right)^{1/3} = 5\left(\frac{2678 \cdot 34 \times 10^6}{0 \cdot 66 \times 250}\right)^{\frac{1}{3}}$$

$$= 1265.99 \text{ mm}$$

Adopt depth of web = 1200 mm

Let t_w be the thickness of web

Thickness of web, $= \dfrac{F}{f_s d} = \left(\dfrac{382 \cdot 62 \times 1000}{(0 \cdot 4 \times 250) \times 1200}\right)$

$= 3.1885$ mm

The thickness obtained is very small. Adopt 10 mm thickness of web.

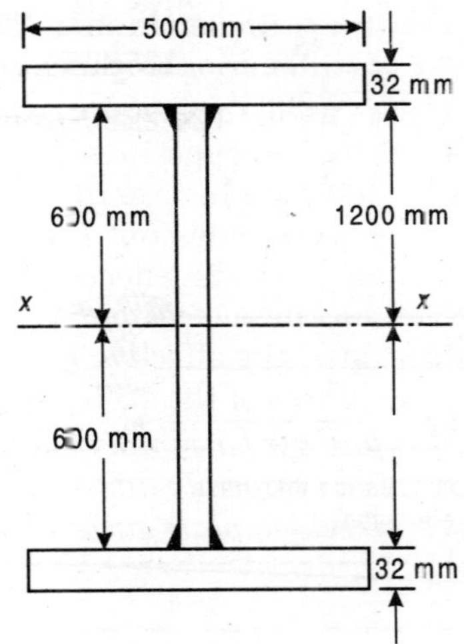

Fig. 10.66 *Welded plate girder*

Stpe 3 : Design of flange plate

Assume thickness of flange 30 mm. Distance between centre of gravity of flanges

$= (1200 + 15 + 15 = 1230)$ mm

Area of flange inclusive of web equivalent

$$A'_f = \left(\frac{2678 \times 10^6}{(0 \cdot 66 \times 250) \times 1230}\right) = 13197 \text{ mm}^2$$

Web equivalent

$$\frac{1}{6}A_w = \frac{1}{6} \times (10 \times 1200) = 2000 \text{ mm}^2$$

Area of flange $= (13197 - 2000) = 11197 \text{ mm}^2$

Thickness of flange plates

$= 32$ mm

Therefore, width of flange plate required = $\frac{11197}{32}$ = 349.906 mm

Provide 500 mm × 32 mm flange plates one at the top and one at the bottom of web plate as shown in Fig. 10.66.

Area of flange plate provided

$$= 500 \times 32 = 16000 \text{ mm}^2$$

Step 4: Check for bending stress

Moment of inertia $I_{xx} = \frac{1}{12} \times 1 \times (120)^3 + 2 \times 3.2 \times 50 \times (61.60)^2] \times 10^4$

$$= 1358259.2 \times 10^4 \text{ mm}^4$$

Maximum bending stress

$$f_b = \left(\frac{2678 \cdot 34 \times 10^6 \times 630}{1358259 \cdot 2 \times 10^4}\right)$$

$$= 124.228 \text{ N/mm}^2$$

Hence, satisfactory.

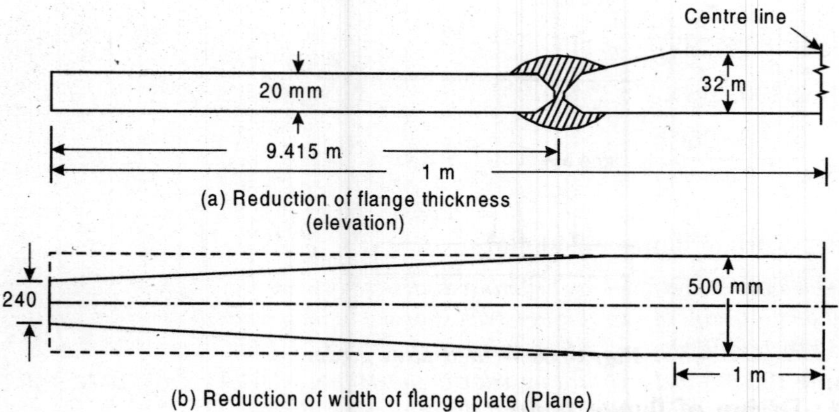

Fig. 10.67

Step 5: Reduction of flange plate

The thickness of flange plate is reduced to 20 mm.

Moment of inertia of welded plate girder with 500 mm × 20 mm flange plates

$$I_{xx} = [\frac{1}{2} \times 1 \times (120)^3 + 2 \times 2 \times 50 \times (61)^2] \times 10^4$$

$$= 888200 \times 10^4 \text{ mm}^4$$

Moment of resistance

$$M_1 = \left(\frac{(0 \cdot 66 \times 250) \times 888200 \times 10^4}{610 \times 10^6}\right)$$

$$= 2402.51 \text{ kN-m}$$

The equation of parabolic bending moment diagram with left hand support as origin is $y = k \cdot x \cdot (L - x)$

The maximum bending moment occurs at centre

$\therefore \quad y_{max} = 2678.34$ kN-m

At $x = 14$ m, $\therefore k = 13.665$

$\therefore \quad y = 13.665 \cdot x \cdot (L - x)$

At $y = 239.15$, $x = 9.415$ m

Alternative. Instead of reducing the thickness of flange plate, it is convenient to reduce the width of flange plate. The tapering width is provided as shown in Fig. 10.67 (b). The equation of parabolic bending moment diagram curve is

$$y = 13.665\, x\, (28 - x)$$

At

$$x = 6.5 \text{ m},$$

$$y = (13.665 \times 6.5 \times 21.5)$$

$$= 1909.68 \text{ kN-m}$$

The bending moment at x equal to 6.5 m

$$M_2 = 1909.68 \text{ kN-m}$$

Let b be the width of flange at x equal to 6.5 m

Moment of resistance

$$M_x = f \cdot \frac{I}{y} = \frac{(0 \cdot 66 \times 250)}{615} \times \left[\frac{1}{12} \times 1 \times (120)^3 + 2 \times 3 \times b (61 \cdot 6)^2\right] \times 10^4$$

Equating 1909.68×10^6

$$= \frac{(0 \cdot 66 \times 250)}{615} \times \left[\frac{1}{12} \times 1 \times (120)^3 + 2 \times 3 \times b (61 \cdot 6)^2\right] \times 10^4$$

$$\therefore \quad b = (23.65 \times 10) = 236.5 \text{ mm}$$

Provide, width of flange = 240 mm

Note. *The width of flange may be found at different sections. The flange plate may be tapered on both edges from one end such that the taper accommodates the required width at those sections.*

Example 10.19 *A welded plate girder is made of a web 2000 mm deep and 20 mm thick and flange 500 m wide and 40 mm thick. The span of the girder is 28 m and total load per metre inclusive its own weight is 27 kN per metre. Design a suitable welded connection between the web and the flange.*

Solution

Design :

Step 1: Maximum shear force

Total uniformly distributed load is 27 kN/m

Maximum shear force,

$$F = \left(\frac{27 \times 28}{2}\right) = 378 \text{ kN}$$

Load on compression flange (loaded flange) per mm length

$$W = \left(\frac{27}{1000}\right) = 0.0270 \text{ kN/mm}$$

Step 2: Properties of plate girder

Area of flange plate

$$A = (500 \times 40)$$
$$= 20000 \text{ mm}^2$$

Moment of this area about xx-axis (neutral axis)

$$A\bar{y} = (20000 \times 1020)$$
$$= 20.4 \times 10^6 \text{ mm}^3$$

Moment of inertia of welded plate girder

$$I_{xx} = \left(\frac{1}{12} \times 1 \times 200^3 + 2 \times 50 \times 4 \times 102^2\right) \times 10^4 \text{ mnm}^4$$
$$= 5494900 \times 10^4 \text{ mm}^4$$

Step 3: Horizontal shear per mm length

$$f_{sh} = \left(\frac{F \cdot A\bar{y}}{I_{xx}}\right) = \left(\frac{328 \times 10^3 \times 20 \cdot 4 \times 10^6}{549400 \times 10^4}\right)$$
$$= 140.33 \text{ N/mm}$$

Step 4: Combined stress for per 1 mm

$$f = [(0.14033)^2 + (0.0270)^2]^{1/2}$$
$$= 0.1429 \text{ kN/mm}$$

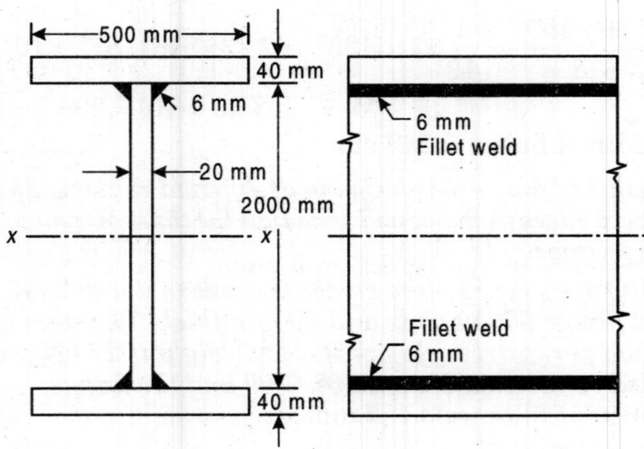

Fig. 10.68 *Welded connection (flange and web)*

Step 5 : Continuous weld

Size of weld, $$s = \left(\frac{0 \cdot 1429 \times 1000}{2 \times 0 \cdot 7 \times 110}\right)$$
$$= 0.928 \text{ mm}$$

Minimum size of filet weld $= 6$ mm

Provide 6 mm fillet weld for connection of the flange plate and the web plate as shown in Fig. 10.68.

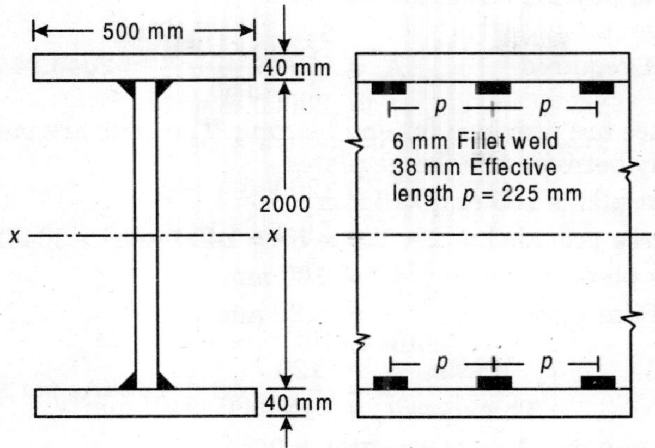

Fig. 10.69 *Intermittent welded connection flange and web*

Step 6 : Intermittent weld

Let 6 mm size of fillet weld be used

Effective length = 4 × Size of weld or 38 mm

= 24 mm or 38 mm

(whichever is more)

Hence the effective length of fillet weld is adopted as 38 mm

Strength of weld on both the faces

$$R = \left(\frac{2 \times 38 \times 0 \cdot 7 \times 6 \times 110}{1000}\right) = 35.11 \text{ kN}$$

Pitch of weld,

$$p = \frac{R}{f_{sh}} = \left(\frac{35\ 11}{0 \cdot 1429}\right) = 245.71 \text{ mm}$$

Provide intermittent weld at 225 mm pitch

Allowable clear spacing

$= 12 \times t = 12 \times 20 = 240$ mm

Hence the clear spacing provided is satisfactory.

The intermittent fillet weld of 6 mm size is provided at a pitch of 225 mm as shown in Fig. 10.69.

Example 10.20 *Design the end and intermediate stiffeners for the girder in Example 10.10. Assumed shear in weld is not to exceed 110 N/mm^2.*

Solution

Design of end stiffeners

Step 1: Bearing area required

The uniformly distributed load inclusive of self weight of the welded plate girder is 27 kN/m and the span of girder is 28 m.

Support reaction $= \left(\frac{27 \times 28}{2}\right) = 378$ kN

The allowable bearing stress is 185 N/mm^2

Bearing area required $= \left(\dfrac{378 \times 1000}{185}\right) = 2043.24$ mm^2

The flat plates are provided for end bearing. The ends are machined so that plates fit tightly between the flange plates

Provide 2 flat plates 160 mm ×10 mm

% Bearing area provided = (2 × 120 × 10) = 2400 mm^2 > 204324 mm^2

Width of flat plate = 120 mm

Thickness of flat plate = 120 mm

$$\left(\frac{\text{Width}}{\text{Thickness}}\right) = \frac{120}{10} = 12 \not> 12\,t \text{ (As per IS : 800–1984)}$$

Step 2 : Check for bearing stiffener

The bearing stiffener is checked as a compression member. This column section consists of the pair of stiffeners together with a length of wed on each side of the centre line of the stiffeners and equal to 20 times the thickness of web, if available.

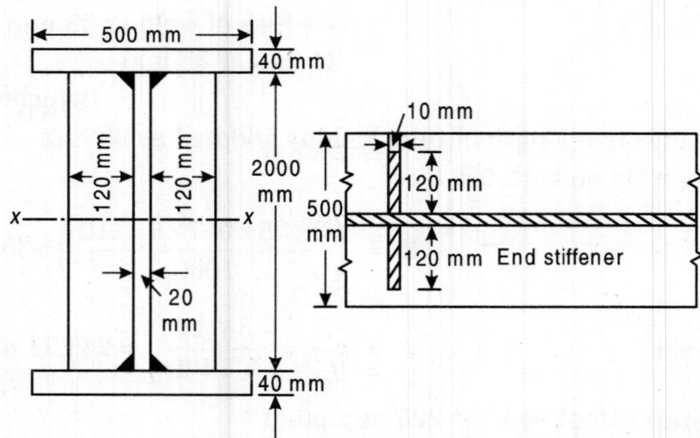

Fig. 10.70 *Bearing stiffener (Welded plate girder)*

Area of cross-section

= (2 × 20 × 20 + 2 × 120 × 10)

= 3200 mm^2

Moment of inertia of the stiffeners about centre line

$$2 \times \frac{1}{2} \times 10 \times 170^3 \times 10^4 = 818.83 \times 10^4 \text{ mm}^4$$

Radius of gyration $r = \left[\dfrac{818 \cdot 83}{3200} \times 10^4\right]^{\frac{1}{2}} = 50.58$ mm

Effective length = (0.7 × 2000) = 1400 mm

Slenderness ratio $= \left(\dfrac{1400}{50 \cdot 58}\right) = 20.7$

From IS : 800–1984, allowable stress in axial compression for the structural steel having yield stress as 250 N/mm²

Safe load carrying capacity

$$\left(\frac{147 \cdot 79 \times 3200}{1000}\right) = 472.93 \text{ kN} > 378 \text{ kN}$$

Hence the design as satisfactory.

Provide 120 mm × 10 mm two flat plates for end stiffeners. The flat plates are connected by 66 mm minimum size of fillet weld as shown in plates, Fig. 10.70.

Step 3. Design of intermediate stiffener

The clear distance between flanges of welded plate girder is 2000 mm and the thickness of web, t_w is 20 mm.

In case, the web plate of the welded plate girder remains unstiffened, the thickness of web plate shall be not less than $t_{w.\min}$ found as under :

Average shear stress at the support

$$\tau_{va.\text{cal}} = \left(\frac{378 \times 1000}{20 \times 2000}\right)$$

$$= 9.45 \text{ N/mm}^2$$

(i) $$\tau_{w.\min} = \left(\frac{d_1 \cdot t_{va \cdot \text{cal}}}{816}^{1/2}\right) = \left(\frac{2000 \times 9 \cdot 45^{1/2}}{816}\right)$$

$$= 7.5 \text{ mm}$$

(ii) $$\tau_{w.\min} = \left(\frac{d_1 \cdot f_y^{1/2}}{1344}\right)\left(\frac{200 \times 250^{1/2}}{1344}\right)$$

$$= 23.528 \text{ mm}$$

(iii) $$\tau_{w.\min} = \left(\frac{d_1}{85} = \frac{2000}{85}\right) = 23.53 \text{ mm.}$$

Actual thickness of the web plate is 20 mm and it is less than 23.53 mm. Therefore, it is necessary to provide vertical stiffeners to stiffener the web plate.

The shear stress is very small. The intermediate stiffeners may be provided at maximum spacing i.e., 1.5 d

$$1.5 \times 2000 = 3000 \text{ mm}$$

Moment of inertia required for intermediate stiffener

$$\left(\frac{1 \cdot 5 \times (2000^3 \times 20^3)}{30002}\right) = 1066.66 \times 10^4 \text{ mm}^4$$

Provide 2 flat plates 10 mm ×110 mm, as shown in Fig. 10.71

Moment of inertia provided

$$\frac{1}{12} \times 10 \times (110 + 110 + 20)^3 = 1152 \times 10^4 \text{ mm}^4$$

$$> 1066.66 \times 10^4 \text{ mm}^4.$$ Hence, satisfactory.

Shear force on the weld connecting intermediate stiffeners

$$\left(\frac{125\, t_w^2}{h}\right) = \left(\frac{125 \times 20 \times 20}{110}\right)$$

$$= 454.545 \text{ kN}$$

where, h is outstand of stiffeners in mm

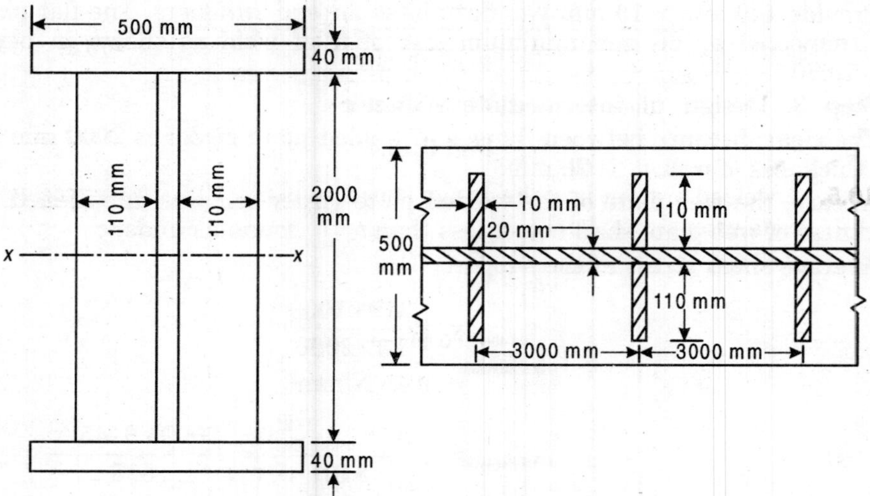

Fig. 10.71 *Welded plate girder*

Size of weld required

$$\left(\frac{454.54 \times 1000}{2 \times 0.7 \times 100}\right) = 2.95 \text{ mm}$$

The size of weld obtained is very small. The intermittent fillet welds of 3 mm size are provided for the connection.

PROBLEMS

10.1. Two plates 14 mm thick are joined by (i) a double V butt weld, (ii) a single-V butt weld. Determine the strength of the welded joint in tension in each case. The effective length of weld is 200 mm.

10.2. Design a suitable side fillet weld to connect two plates 100 mm × 10 mm and 120 mm × 12 mm, and to transmit pull equal to the full strength of thin plate.

10.3. Two plates 100 mm × 10 mm are connected in a lap joint by means of end fillet weld. Design suitable weld to transmit pull equal to full strength of the plate.

10.4. A truss joint is as sketched in Fig. P.10.4. Design and detail the welds connecting the members to the gusset.

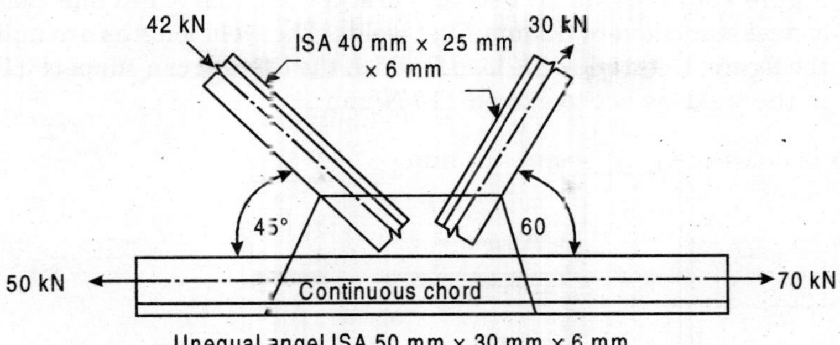

Fig. P.10.4

10.5. A welded bracket is of the design shown in Fig. P.10.5. Design the welded connections completely.

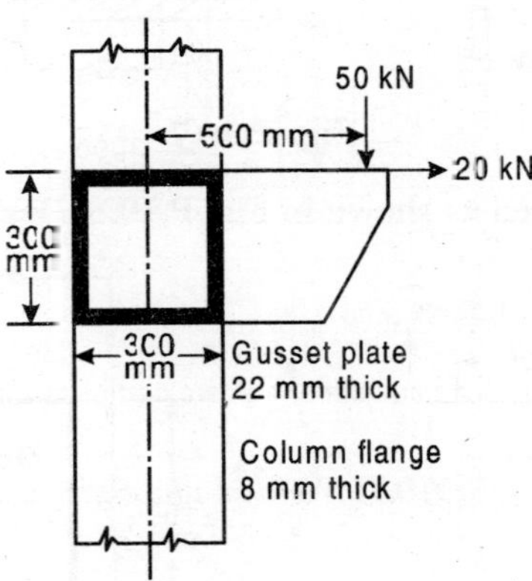

Fig. P.10.5

10.6. An ISA 100 mm × 75 mm × 10 mm (ISA 100 75, @ 0130 kN/m) is welded with the flange of a column HB 350, @ 674 N/m as shown in Fig. P.10.6. The bracket carries a load of 80 kN at a distance of 40 mm from the face of column. Design the bracket connection.

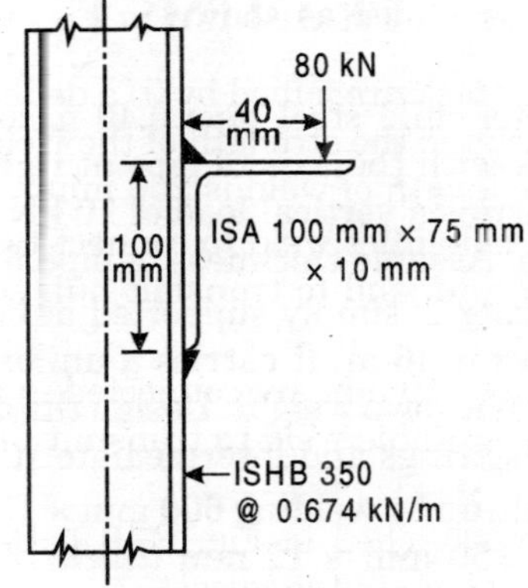

Fig. P.10.6

10.7. Figure P.10.7 shows an I-section bracket 350 mm × 140 mm connected to a steel stanchion by 12 mm fillet welds. The weld lengths are indicated on the figure. Determine the load P which the bracket can support if the stress in the welds is not to exceed 110 N/mm².

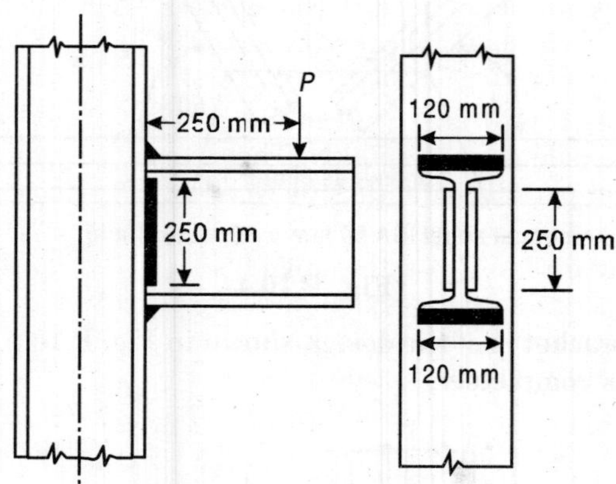

Fig. P.10.7

10.8. A bracket is loaded as shown in Fig. P.10.8. Design the welded connection.

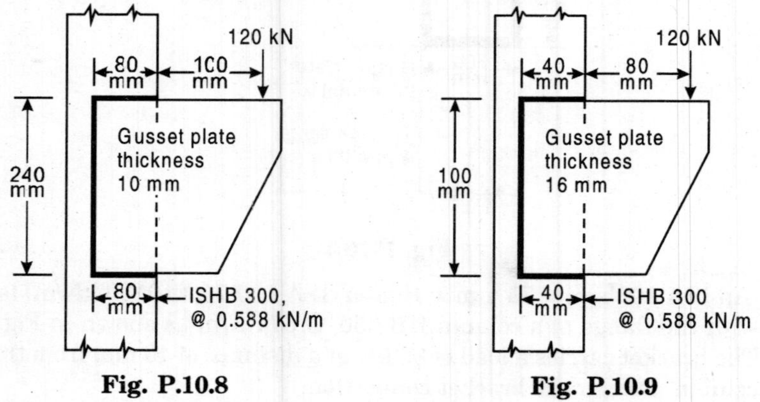

Fig. P.10.8 **Fig. P.10.9**

10.9. A welded bracket is loaded as shown in Fig. P.10.9. Design the suitable weld.

10.10. A 100 mm diameter mild steel pipe 0.40 m long is welded to a vertical plate 10 mm thick with the axis of pipe at right angles to the face of the plate. The pipe carries a vertical load of 10 kN at its free end. Design the welded connection. Assume thickness of pipe 6 mm.

10.11. A welded plate girder is simply supported at both the ends. The effective span of plale girder is 16 m. It carries a uniformly distributed load of 30 kN/m of exclusive of its own weight. Design the maximum section of welded plate girder, end bearings and intermediate stiffeners.

10.12. An I-section is built-up by welding 600 mm × 15 mm (thick) web plate and two flange plates 150 mm × 12 mm (thick). Design the welded joint to develop full strength of the section. Adopt safe stresses as per I.S. code.

10.13. A circular shaft of diameter 100 mm is welded to a rigid plate by a fillet weld 6 mm size. If a torque of 6 kN-m is applied to the shaft, find the maximum stress in the weld.

10.14. An angle section ISA 40 mm × 25 mm × 6 mm (ISA 4025, @ 0028 kN/m) is to be joined to a gusset plate with the longer leg attached. The joint is to be designed for maximum load carrying capacity of the angle iron in tension. The thickness of the gusset plate is 8 mm. Design the welded joint.

10.15. Two angles of ISA 90 mm × 90 mm × 8 mm (ISA 9090, @ 0108 kN/m) transmit at tensile force of 250 kN. The angles are connected one on each side of a gusset plate 10 mm thick by welding. Design the joint using 6 mm weld. The C.G. line for each angle $C_{xx} = C_{yy} = 28.7$ mm.

10.16. A welded plate girder of effective span 25 m is made of web 2000 mm deep and 20 mm thick and flanges 500 mm wide and 30 mm thick. It carries a uniformly distributed load of 30 kN/m including its own weight. Design the intermittent weld connection between the web and the flange.

10.17. A welded plate girder has a uniform section as shown in Fig. P.10.17. The plate girder carries a uniformly distributed load of 40 kN/m over the entire span of 25 m. The girder is simply supported at both the ends. Design the welded connection between flanges and web plate.

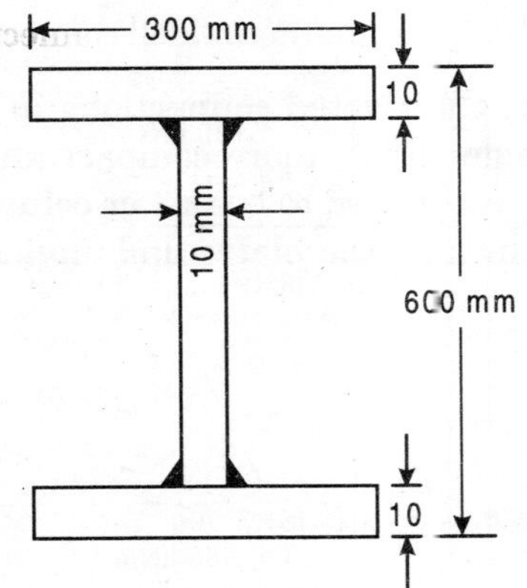

Fig. P.10.17

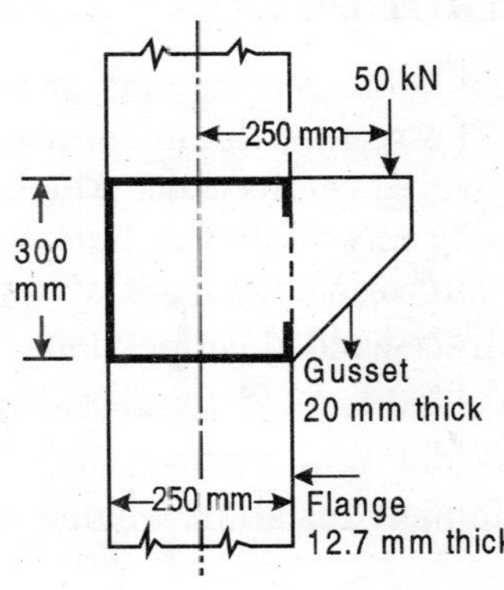

Fig. P.10.18

10.18. Design the bracket connection (Fig. P.10.18) assuming that the loads shown are for one gusset plate only.

10.19. Design a single angle tension member of a roof truss carrying an axial load of 100 kN. The member is connected to a gusset by welds on end and sides. Design the welded connection.

10.20. A bracket web plate 12 mm thick transmits a load of 200 kN at an eccentricity of 350 mm to flange of a column. Design the fillet welds on either side of web.

10.21. A 12 m span welded plate girder is subjected to a uniformly distributed load of 30 kN/m along with a concentrated load of 150 kN at 4 m from one of the supports. Design

(i) the cross-section of the plate girder assuming the web to be 6 mm thick,

(ii) the welded joint for connecting the flange plates with the web

(iii) the web stiffeners including bearing stiffeners at supports.

Chapter 11

Welded Beam Connections

11.1 INTRODUCTION

The welded beam connections are similar to the riveted connections in many respects. The welded beam connections are generally more compact and free than the riveted connections. The beams are connected to beams or columns at their ends by direct fillet or butt welds, or through the plates and angles. The welded beam connections are of four types :

1. Direct welded connections
2. Welded framed connections
3. Welded seat connections
4. Moment-resistant welded connections.

11.2 DIRECT WELDED CONNECTIONS

The beams are connected to beams or columns at their ends by direct fillet welds or butt welds. The fillet welds are applied on both the sides of the web of beam as shown in Fig. 11.1. The web is connected directly to the face of supporting member. The fillet welds are designed for vertical reaction at the end of beam. The restraining moment existing at the weld section is not considered. As far as possible, the fillet welds are applied near the top. The minimum size of fillet weld is kept equal to the thickness of web or 7 mm whichever is more. The welds, web of beam and element of supporting member are checked for shear. Let V be the vertical reaction at the end of beam. Then, for shop welds, the shear stress

$$\left(\frac{V}{2\times 0\cdot 7s\cdot h}\right) < 110\ \text{N/mm}^2 \qquad \text{...(11.1)}$$

Average shear stress in web of beam

$$\left(\frac{V}{t_w h}\right) \leq 0.4 f_y \text{ N/mm}^2 \qquad \text{...(11.2)}$$

The shear at supporting element is resisted at two sections. Therefore, the shear stress in the element of supporting member

$$\left(\frac{V}{2t \cdot h}\right) \leq 0.4 f_y \text{ N/mm}^2 \qquad \text{...(11.3)}$$

From Eq. (11.1) and Eq. (11.2)

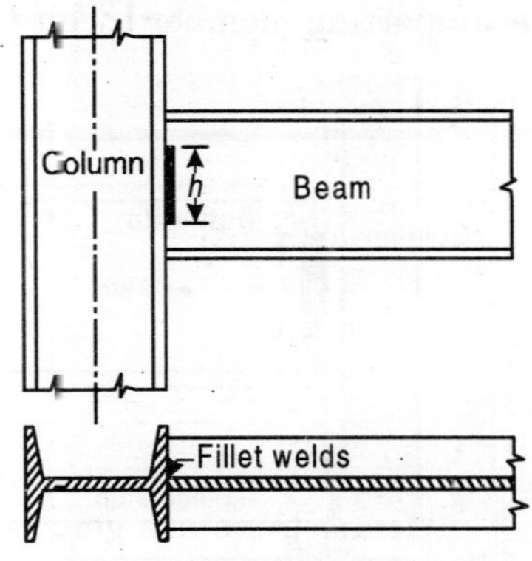

Fig. 11.1

$$\left[\frac{\dfrac{V}{2 \times 0 \cdot 7s \cdot h}}{\dfrac{V}{t_w h}}\right] < \left(\frac{110}{0 \cdot 4 f_y}\right)$$

$$s \leq \left(\frac{0 \cdot 4 f_y}{11 \times 1 \cdot 4} \cdot t_w\right)$$

$$s \leq \left(\frac{0 \cdot 4 f_y \cdot t_w}{154}\right) \qquad \text{...(11.4)}$$

From Eq.(11.1) and Eq.(11.3)

$$\left[\frac{\left(\dfrac{V}{2 \times 0 \cdot 7s \cdot h}\right)}{\left(\dfrac{V}{2s \cdot h}\right)}\right] \leq \left(\frac{110}{0 \cdot 4 f_y}\right)$$

$$\left(s < \frac{0 \cdot 4 f \times 2}{110 \times 1 \cdot 4} \cdot t ; s < \frac{0.8 f_y}{154} \cdot t\right) \qquad \text{...(11.5)}$$

The rotation of end of beam takes place. This rotation may take place by yielding of web or yielding of weld. The size of weld is made larger than the minimum size necessary, so that yeilding will occur in the web plate. In order to provide adequate flexibility, the depth of weld should not be greater than one-half to two-third the depth of beam, and size of the weld is kept at least four-fifth the thickness of web. However, size of weld equal to two-third the thickness of web is considered effective to determine depth of weld.

When the beams are connected to beams or columns at their ends directly or columns at their ends directly by *butt weld*, a backing strip is used as shown in Fig. 11.2. The backing strip is welded with the web in the shop. Then the web of beam is connected to the supporting member by butt weld.

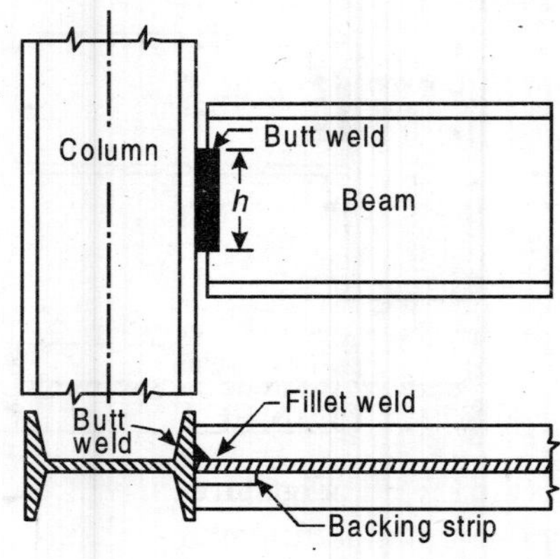

Fig. 11.2

The extra weld metal is deposited on the butt weld. This provides extra strength to butt weld. The depth of butt weld is kept small as far as possible in order to allow end rotation. The butt weld is designed to resist the vertical reaction at the end of beam. If V be the vertical reaction, then, the shear stress in butt weld.

$$\left(\frac{V}{d \times t_w}\right) \leq f_{sh}(110 \text{ N/mm}^2)$$

The direct fillet weld or butt weld needs preparation of accurate length of beam. The rotation of end of beam takes place by yielding of web. Therefore, as far as possible, these types of connections are not used.

Example 11.1 *A MB 500, @ 0.869 kN/m transmits an end reaction of 130 kN to the flange of a column MB 250, @ 0.510 kN/m. The fillet welds are applied directly on both the sides of web of the beam. Design the connections.*

Solution

Design:

Step 1: Size of weld

From ISI Handbook No. 1, for MB 500, @ 0.869 kN/m beam

The thickness of web, t_w is 10.2 mm

For MB 250, @ 0.510 kN/m column,

The thickness of flange, t_f is 9.7 mm

Size of fillet weld = $\times \frac{4}{5}$ 10.2 = 8.16 mm

Provide 9 mm fillet yield.

Effective size of fillet weld from Eq. 11.4

$$s = \left(\frac{0 \cdot 4f_y \cdot t_w}{154}\right) = \left(\frac{0 \cdot 4 \times 250 \times 10 \cdot 2}{154}\right)$$

$$= 6.62 \text{ mm}$$

Step 2: Depth of weld $= \left(\frac{130 \times 1000}{0 \cdot 7 \times 6 \cdot 62 \times 10}\right) = 272.31$ mm

Provide 140 mm deep fillet welds on each side of web.

Effective size of weld

$$< \left(\frac{0 \cdot 8f_y \cdot t}{154}\right) = \left(\frac{0 \cdot 8 \times 250 \times 9 \cdot 7}{154}\right) = 12.6 \text{ mm}$$

Therefore, the flange of column is not overstressed.

11.3 WELDED FRAMED CONNECTIONS

The direct fillet weld or butt weld needs preparation of accurate length of beam. The gap between end of beam and supporting member must be short. For fillet weld, the end of beam must remain completely flush with the element of supporting member. Butt weld may be provided even if the gap is upto the extent of 3 mm. But, when beams are manufactured for ordered length, the rolling mill allows a tolerance of 12 mm either way. The gap may be greater at one flange of beam, than at the other. The beam end may be prepared by milling, cold sawing or flame cutting. All these operations are fairly expensive. To avoid this difficulty, welded framed connections are used. In the welded framed connections, two plates or two angles are used. The plate connection as shown in Fig. 11.3 may be considered as a special case. The plates make the connection somewhat flexible, than the direct welded connection.

One plate is used on one side of web of the beam and other plate is attached on the other side. For connection, one plate is welded in the shop with the beam and other plate is welded in the shop with the element of supporting member. When both the members are kept in position, erection bolts are inserted and welding is done. To allow rotation at the end of beam, the lengthof plates is limited to two-third the depth of beam. The size of weld connecting plates to the web of beam is kept equal to four-fifth of thickness of web.

The welds connecting plate to the supporting member are designed to resist the vertical reaction V, at the end of beam, Whereas, the weld connecting plate to the web of beam are designed to resist vertical reaction, V, moment, $V.l$ where l is distance between two welds. The length of plates is kept about

5 mm. The plates may be thicker than the welds connecting plates and web of beam by 16 mm. In that case, the plates will also be safe, if the welds connecting plates and web ofbeam are safe.

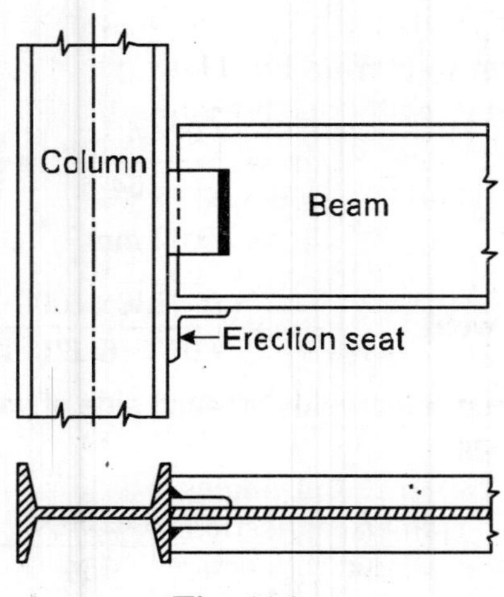

Fig. 11.3

Example 11.2 *In Example 11.1, two plates are used for connection on both the sides of web of a beam. Design the welded connections*

Solution

Design

Step 1: Moment to be resisted

The vertical reaction at the end of beam, V is 130 kN

Take two plates 50 mm long

Moment
$$M = \left(\frac{130 \times 50}{1000}\right) = 6.5 \text{ kN-m}$$

Step 2: Strength of 1 mm weld : weld value

1. Weld connecting plates and web in the shop

For web thickness of 10.2 mm, 9 mm fillet welds are adopted.

$$R = \left(\frac{0 \cdot 7 \times 9 \times 110}{1000}\right) = 0.693 \text{ kN/m}$$

The fillet welds are provided on both the sides of web

The depth of weld

$$d = \left[\frac{6M}{2R}\right]^{\frac{1}{2}} = \left[\frac{6 \times 6 \cdot 5 \times 1000}{2 \times 0 \cdot 693}\right]^{\frac{1}{2}} = 167.75 \text{ mm}$$

Try 250 mm deep welds

Vertical shear per mm length,

$$f_{sv} = \left(\frac{130 \times 1000}{2 \times 250}\right) = 260 \text{ N/mm}$$

Horizontal shear per mm length,

$$f_{sv} = \left(\frac{6 \times 6 \cdot 5 \times 106}{2 \times 250 \times 250}\right) = 312 \text{ N/mm}$$

Resultant shear, $f = ((260)^2 + (312)^2)^{1/2} = 406$ N/mm < 693 N/mm. Hence, safe.

2. Weld connecting plate to the supporting member in the field

Let s be the size of weld

Then $0.7 \times s \times (0.1 \times 110) > 2 \times 250 = 130 \times 1000$

$$s = \left(\frac{130 \times 1000}{0 \cdot 7 \times 0 \cdot 8 \times 110 \times 500}\right) = 4.22 \text{ mm}$$

Provide 6 mm size for the welds

Thickness of plates = (9 + 1.6) =10.6 mm

Provide 12 mm × 50 mm plates.

The connections of two framing angles are similar to the riveted frame connections. The welded framed connections with two framing angles are more flexible than other type of welded connections. The bending of outstanding legs of the angles provides flexibility to the connection. During the erection, the erection bolts are provided through the connection angles.

A welded framed connection is shown in Fig. 11.4. In order to provide flexibility to the connections, thickness of angle is kept less than the size of weld.

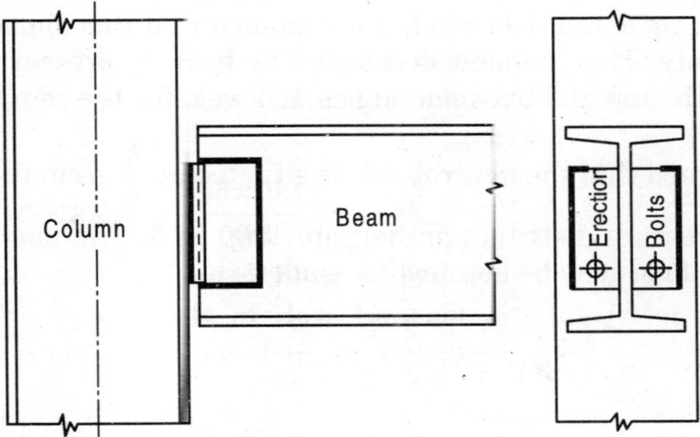

Fig. 11.4

For economical consideration, the length of leg of angle connecting the web of beam should be short. The welds are usually returned to the end of beam on this leg, as shown in Fig. 11.4. This welding is done in the shop. This will now be referred as shop weld. The outstanding legs are connected to the supporting member in the field and the weld will be referred as field weld. The field welds are returned at the top for short distance.

Figure 11.5 shows a beam with two frame angles attached to it. The beam is assumed simply supported at the face of outstanding legs of connecting angles. Therefore, the field welds are not subjected to any moment. Both the angles are connected individually to the web of beam. These angles are not connected together. Each field weld is subjected to the vertical reaction $\frac{R}{2}$ as shown in Fig. 11.5. Therefore, each shop weld is subjected to vertical reaction $\left(\frac{R}{2}\right)$, and a twisting moment, $\left(\frac{R}{2}\right) . l_2$.

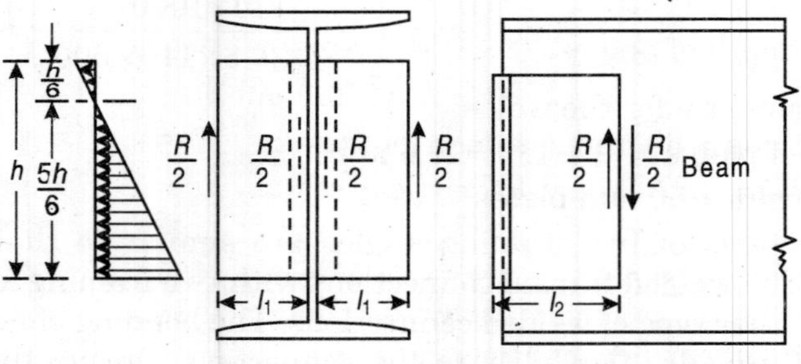

Fig. 11.5

Each angle is also subjected to a moment $\frac{R}{2} . l_1$. Though this moment is resisted by both shop weld and field weld, it is assumed that this moment is resisted by field weld only. This moment is resisted by bearing pressure between angles and web at the top and between angles and weld for the remaining length.

It is assumed that the neutral axis is situated at $\frac{h}{6}$ from the top of angles as shown in pressure distribution diagram, Fig. 11.5. The maximum horizontal shear in the field may be obtained as under:

$$\left(\frac{1}{2} f_{sh} \cdot \frac{5}{6} h\right) \cdot \frac{2}{3} h = \frac{R}{2} \cdot l_1$$

$$f_{sh} = \frac{5}{9} \cdot \left(\frac{R \cdot l_1}{h_2}\right) \quad \text{...(11.6)}$$

The resultant of horizontal shear and shear due to vertical reaction $\frac{R}{2}$ is found by vector sum. The resultant shear stress must be less than the allowable shear. The depth of angle is kept about one-half to two-third the depth of the beam, 2 ISA 90 mm × 60 mm are used for framing angles. The thickness of angles is determined after designing the welds.

Example 11.3 *In Example 11.1, two framing angles are used for connections on both the sides of web of beam. Design the welded connections.*

Solution

Design

Step 1 : Properties of section

$$\text{Depth of beam} = 500 \text{ mm}$$

$$\text{Half depth of beam} = 250 \text{ mm}$$

$$\frac{2}{3} \text{ depth of beam} = 333.3 \text{ mm}$$

The depth of angles is adopted between one-half to two-thirds the depth of beam.

Assume depth of angles,

$$h = 300 \text{ mm}$$

2 ISA 90 mm × 60 mm are used for framing angles.

1. Design of field weld vertical reaction on each weld

$$\frac{R}{2} = \frac{130}{2} = 65 \text{ kN}$$

Horizontal shear per mm length

$$f_{sh} = \left(\frac{9}{5} \times \frac{R \cdot l_1}{h^2}\right)$$

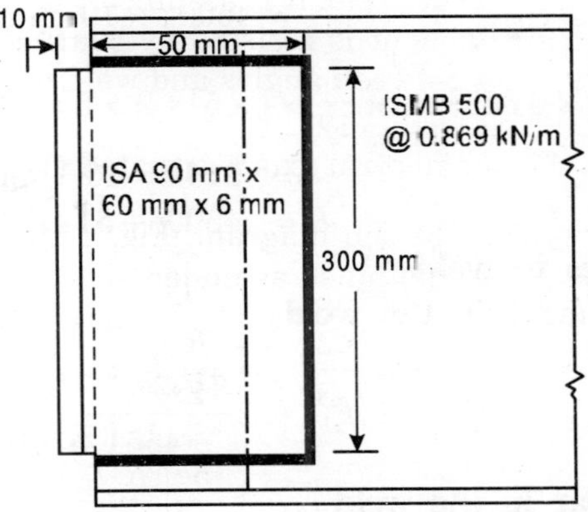

Fig. 11.6

$$f_{sh} = \frac{5}{9} \times \left(\frac{130 \times 1000 \times 90}{300 \times 300}\right) = 234 \text{ N/mm}$$

Vertical shear per mm length

$$f_{sv} = \left(\frac{65 \times 1000}{300}\right) = 216.6 \text{ N/mm}$$

The resultant shear stress

$$t_s = [f_{sv})^2 + (f_{sh})^2]^{1/2} = [(216.6)^2 + (234)^2]^{1/2} = 320 \text{ N/mm}$$

Let s be the size of field weld. The strength of 1 mm weld is $0.7s \times (0.8 \times 0.8) \times 1 = 320$

$$s = \left(\frac{330}{0.7 \times 0.8 \times 110}\right) 5.195 \text{ mm}$$

Provide 6 mm size for fillet weld.

2. Design of shop weld

$$\bar{x} = \left(\frac{2 \times 50 \times \frac{50}{2} \times t}{2 \times 50 \times t + 300 \times t}\right) = 6.25 \text{ mm}$$

Therefore, the twisting moment

$$M = \frac{130}{2} \times (60 - 6.25) = 3493.75 \text{ kN-m}$$

Step 2: Properties of welds

Let t mm be the throat thickness

$$I_{xx} = \frac{1}{12} [t \times 30^3 + 2 \times 5 \times t \times 15^2] \times 10^3 \text{ mm}^4 = (2250 + 2250) \times t \times 10^3$$

$$= 4500\, t \times 10^3 \text{ mm}^4$$

$$I_{yy} = [30 \times t \times 0.625^2 + 2 \times \frac{1}{12} \times t \times 5^3 + 2 \times 5 \times t\, (2.5 - 0.625)^2] \times 10^3 \text{ mm}^4$$

$$= (514\, t \times 10^3) \text{ mm}^4$$

$$J = (4500t + 51 \cdot 4t) = (4551.4\, t \times 10^3) \text{ mm}^4$$

Step 3: Stresses in welds

1. Horizontal shear in the weld

$$f_{sh} = \left(\frac{3493.75 \times 10^3 \times 150}{4551.4t \times 10^3}\right) = \frac{115}{t} \text{ kN/mm}^2$$

2. Vertical shear in the weld

$$f_{sv} = \left(\frac{65000}{2 \times 50 \times t + 300t} + \frac{3493.75 \times 10^3 \times 43.75}{4551.3t \times 10^3}\right) = \frac{196.1}{t}$$

3. Resultant shear stress

$$f = \frac{1}{t}\left[115^2 + (196 \cdot 1)\right]^{\frac{1}{2}} = \frac{227 \cdot 33}{t} \text{ N/mm}^2$$

$$\therefore \quad \frac{227 \cdot 33}{t} = 110, \; t = 2.066 \text{ mm} \qquad \therefore s = 2.98 \text{ mm}$$

Provide 6 mm size for shop welds. The thickness of angle is kept equal to the size of weld i.e., 6 mm.

11.4 WELDED SEAT CONNECTIONS

When a beam is connected to a beam or a column by means of an angle at the bottom of the beam, which is shop welded to the supporting member and an angle ISA 100 mm × 100 mm × 6 mm at top which is field welded, the connections are known as seated connections. The top angle provides stability to the beam. The welded seat connections provide some flexibility as the welded framed connections. The welded seat connections provide satisfactory and efficient support to the beams. Similar to the riveted connections, the welded seat connections are also unstiffened and stiffened.

11.4.1 Unstiffened Welded Seat Connections

The unstiffened welded seat connections are shown in Fig. 11.7. In Fig. 11.7, one beam is supported at the flange of a column and other beam is supported at the web. The unstiffened seat connections are used to support light loads. The

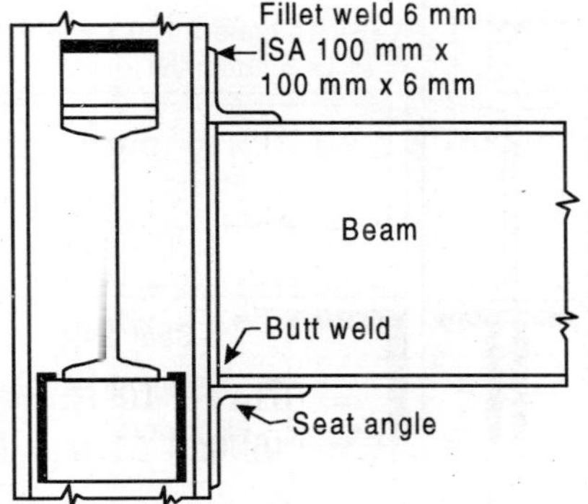

Fig. 11.7 *Unstiffened welded connections*

vertical welds are used to connect seat angles. These welds are returned at the top of seat angles. The return ends of weld are not considered to carry-loads. The top angles are welded along the toss by 6 mm welds. Therefore, it remains free to bend and provides flexibility to the connection.

The bearing length b of seat angle is determined in the manner seat similar to that of the riveted connection.

The bearing length

$$b = \left(\frac{F}{f_b \cdot t_w} - \sqrt{3} \cdot h_2\right) \nless \frac{1}{2} \frac{1}{f_b \cdot t_w}$$

The end reaction F is considered as uniformly distributed on the outstanding leg of seat angle over the length b. A clearance of 10 mm is provided between the end of beam and column. The critical section is assumed at the edge of filler.

The distance of end reaction from critical section of the outstanding leg of seat angle is determined, and bending moment M is found at the critical section. The length of seating angle is assumed equal to width of the flange of beam. The thickness of seating angle is found by equating moment of resistance to the moment at the critical section.

The neutral axis for vertical welds is assumed to be situated at mid-height, The horizontal shear per mm and vertical shear per mm of weld are determined and combined resultant shear per mm length should be less than or equal to allowable shear per mm of welds.

11.4.2 Stiffened Welded Seat Connections

The design of stiffened seat connections is somewhat simpler than that of riveted connections, In the stiffened welded seat connections, two plates welded together forming a tee-section are used. A split-beam section is also used to form a seat in stiffened welded seat connections. The stiffened welded seat connections are shown in Fig. 11.8. The thickness of stiffening plate is kept equal to thickness of

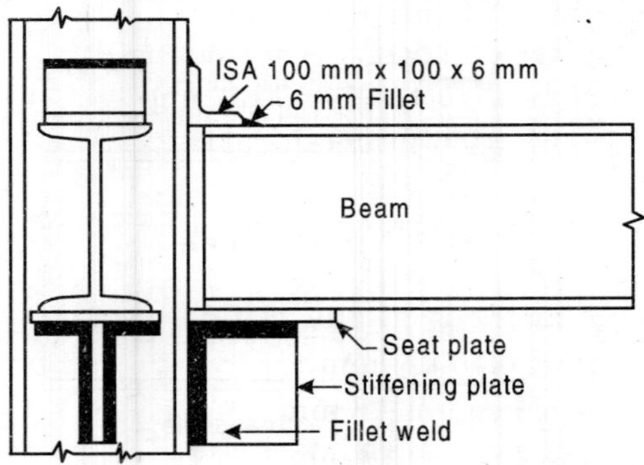

Fig. 11.8 *Stiffened welded seat connection*

web of the beam. The thickness of seat plate is kept equal to flange thickness. The stiffening plate is welded along both the vertical sides. The horizontal welds are provided along the under-side of the seat plate. The length of each horizontal weld is kept one-fifth to one-half the length of vertical leg. The horizontal welds and vertical welds form a double angle section as shown in Fig. 11.8.

The bearing length b is determined as for the unstiffened welded seat connections. For the stiffened seat connections the eccentricity of beam reaction is more than the unstiffened one. The position of beam reaction is assumed to lie at the mid-point of bearing length. The bearing length is measured from outer end of seat plate.

Example 11.4 *A MB 500, @ 0.869 kN/m transmits an end reaction of 130 kN to the flange of column HB 250, @ 0.510 kN/m. Design an unstiffened welded seat connection.*

Solution

Design

Step 1: Bearing length of seat connection

The end reaction of beam, F is 130 kN

From ISI Handbook No. 1, for MB 500, @ 0.869 kN/m

Width of flange, b_t = 180 mm and thickness of web,

t_w = 10.2 mm

Thickness of flange, t_f = 17.2 mm, h_2 = 37.95 mm

The bearing length

$$b = \left(\frac{F}{f_b . t_w} - \sqrt{3} \cdot h_2\right) = \left(\frac{130 \times 1000}{185 \times 10 \cdot 2} - \sqrt{3} \times 37 \cdot 95\right) \text{mm} = 3.16 \text{ mm}$$

$$\text{But } b \nless \left(\frac{1}{2}\frac{F}{f_b . t_w}\right) \text{or} \nless \frac{1}{2} \times \left(\frac{130 \times 1000}{185 \times 10 \cdot 2}\right) \nless 34.45 \text{ mm}$$

Let the bearing length of seating angle b be 34.45 mm

Try seating angle ISA 150 mm × 115 mm × 12 mm. (ISA 150 115, @ 0.238 kN/m) The long leg is kept vertical.

Radius at root, r_1 = 11.0 mm

The distance of end reaction from critical section

$$\left[\left(1 \cdot 0 + \frac{1}{2} \times 3 \cdot 37\right) - (1 \cdot 2 + 1 \cdot 1)\right] \times 10 = 3.85 \text{ mm}$$

Step 2: Bending moment at the critical section

(3.85 × 130 × 1000) = 500.5 kN-mm

The width of flange of beam is 18 mm

The length of seating angle is 18 mm

The moment of resistance of seating angle

$$M_R = \left(\frac{0 \cdot 66 \times 250 \times 180 \times 12^2}{6}\right) = 712.8 \text{ kN-mm}$$

Hence, the seating angle is safe.

Step 3: Design of welds

The length of vertical welds is 150 mm. The vertical shear per mm welds

$$\left(\frac{130 \times 1000}{2 \times 150}\right) = 433.33 \text{ N/mm}$$

The distance of end reaction from the weld

$$\left(\frac{34 \cdot 45}{2} + 10\right) = 27.225 \text{ mm}$$

Step 4: Bending moment at the weld,

M = 130 × 27.225 = 3539.25 kN-mm

Step 5: Horizontal shear per mm,

$$f_{sh} = \left(\frac{3539\cdot 25\times 1000}{2\times\frac{1}{6}\times 450\times 150}\right) = 471.9 \text{ N/mm}$$

Step 6: Resultant shear stress per mm,

$$f_s = \left[(433\cdot 33)^2 + (471\cdot 9)^2\right]^{\frac{1}{2}}$$
$$= 640.675 \text{ N/mm}$$

Let s be the size of the weld

Then, $(0.7 \times s \times 1 \times 110) = 640.675$

$\therefore$ $s = 8.235$ mm

Provide 9 mm fillet weld and ISA 150 mm × 115 mm × 12 mm (ISA 150 115, @ 0.238 kN/m) seating angle. ISA 100 mm × 100 mm × 6 mm (ISA 100 100, @ 0.092 kN/m) is used at the top and 6 mm weld is applied to the toes.

Example 11.5 *A MB 500, @ 0.869 kN/m transmits an end reaction of 240 kN to the flange of a column HB 250, @ 0.510 kN/m. Design stiffened seat connections.*

Solution

Design

Step 1: Bearing length of seat connection

End reaction of beam

$$F = 240 \text{ kN}$$

From ISI Hand book No. 1.

For MB 500, @ 0.869 kN/m

Width of flange, $b_f = 180$ mm

Thickness of web, $t_w = 10.2$ mm

Thickness of flange, $t_f = 17.2$ mm

$h_2 = 37.95$ mm

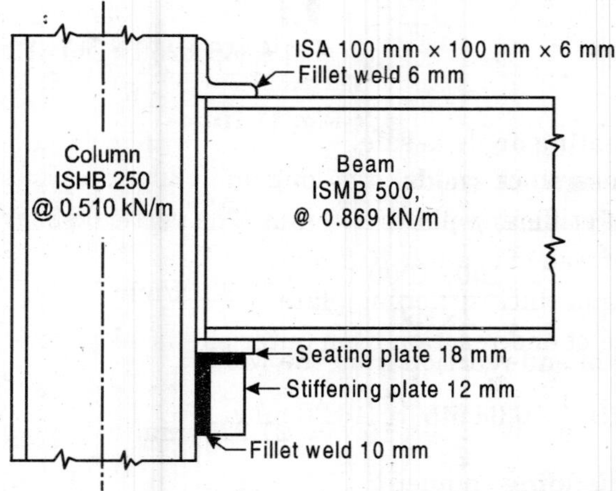

Fig. 11.9 *Stiffened welded seat connections*

The bearing length required

$$b = \left(\frac{F}{f_b.t_w} - \sqrt{3}\cdot h_2\right)$$

$$= \left(\frac{240\times10\cdot2}{185\times10\cdot2} - \sqrt{3}\times37\cdot95\right)$$

$$= 61.455 \text{ mm}$$

$$b \nless \left(\frac{1}{2}\frac{1}{f_b.t_w}\right) \text{or} \nless \frac{1}{2}\times\left(\frac{240\times1000}{185\times10\cdot2}\right) \nless 63\cdot59 \text{ mm}$$

∴ The bearing length = 63.59 mm

10 mm clearance is provided between the end of the beam and flange of the column

Width of seat plate = (63.59 + 10)
= 73.59 mm

Adopt width of a seat plate = 80 mm

Thickness of a seat plate = Thickness of flange of beam
= 17.2 mm

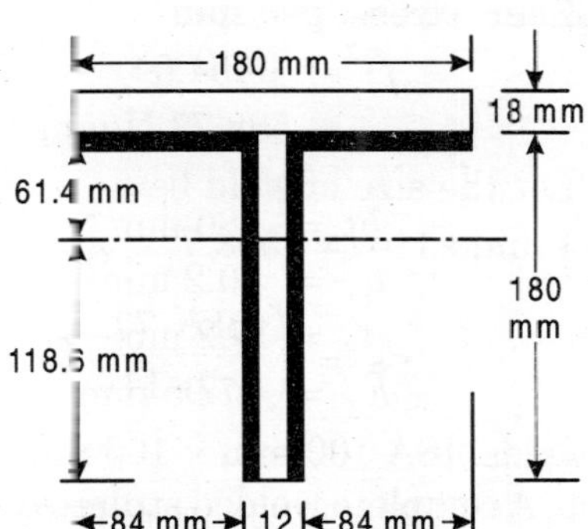

Fig. 11.10

Provide 18 mm thick and 80 mm long and 180 mm wide seat plate.
Thickness of stiffening plate
Thickness of web of beam = 10.2 mm
Provide 12 mm thick stiffening plate,
The distance of end reaction from outer end of seat plate

$$\left(80\times\frac{1}{2}\times63\cdot59\right) = 48.20 \text{ mm}$$

Step 2: Bending moment,

M = (240 × 48.20) =11568 kN-mm

Step 3: Properties of welds

$$\bar{y}_1 = \left(\frac{2\times180\times90}{2\times180+168}\right) = 61.4 \text{ mm},$$

and $\bar{y}_2 = 118.6$ mm

$$I_{xx} = \left[16\cdot8\times6\cdot14^2+\frac{1}{12}\times18^3+2\times18\times(11\cdot86-9)^2\right]\times10^4 = 1898\times10^4 \text{ mm}^4$$

Step 4: Stresses in welds

The vertical shear per mm of weld

$$f_{sh} = \left(\frac{240\times1000}{2\times180\times168}\right) = 454.54 \text{ N/mm}$$

The horizontal shear per mm of weld

$$f_{sh} = \left(\frac{11568\times1000\times614}{1898\times10^4}\right) = 374.22 \text{ N/mm}$$

Since the connection will not allow failure of weld in thrust.

$$\bar{y} = 61.4 \text{ mm}$$

Step 5: Resultant shear stress per mm

$$f_s = [(454.54)^2 + (374.22)^2]^{1/2}$$
$$= 588.77 \text{ N/mm}$$

Step 6: Size of weld. Let the size of weld be

$$0.7 \times s \times 1 \text{ mm} \times 110 = 588.77$$

$$s = \left(\frac{588\cdot77}{0\cdot7\times110}\right) = 7.646 \text{ mm}$$

Provide 10 mm fillet welds. ISA 100 mm × 100 mm × 6 mm (ISA 100 100; @ 0.092 kN/m) is used at top. A complete welded stiffened seat connection designed is shown in Fig. 11.9.

11.5 MOMENT RESISTANT WELDED CONNECTIONS

When the ends of beam are subjected to moments and shears, moments resistant welded connections as shown in Fig. 11.11 are used. The end moments of beam are resisted by top plate and bottom flange. When the moment of beam, say, on the left side of connection is clockwise, the top plate carries tensile force and the bottom flange of beam carries compressive force. The forces are equal in magnitude, and opposite in direction. These forces form a resisting couple. When the direction of moment is anticlockwise, these elements of connections carry equal forces of opposite sign. The top plate is butt welded to the flange of column. The butt weld needs a 5 mm minimum gap between plate and the flange of column. It requires a backing strip. The backing strip is tack welded to the flange. The length of top plate is kept large. The entire length of top plate is not welded. A length equal to one-half to one times the width of the plate is kept

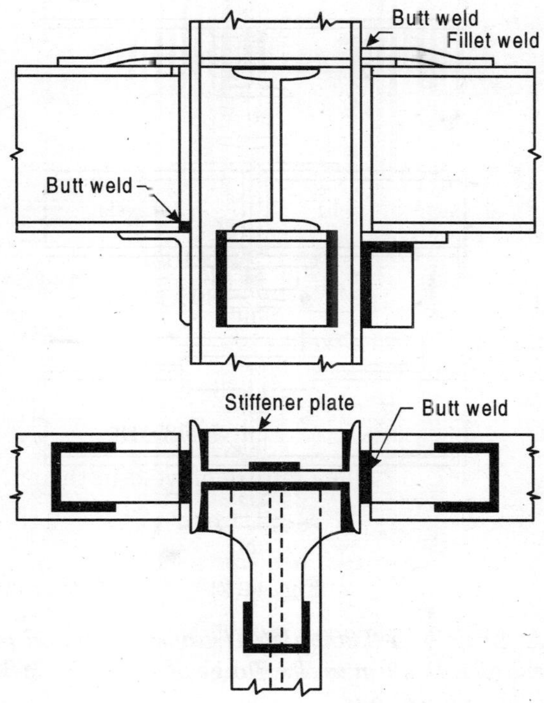

Fig. 11.11

unwelded. This portion of plate is kept unwelded so that if yielding occurs, it occurs in this, rather than in weld. This provides ductility to the connections. The size of fillet weld should be such that the length of side welds remains about one-half to one times the width of plate. In case, the moment resistance connections are made with the web of column, the top plate is flared. This allows side welds to the column flange.

The welded seat connections are designed to resist the vertical end reactions of beam. This welded seat connections are kept unstiffened or stiffened depending upon the magnitude of end reactions.

Figure 11.12 shows moment resistant welded connections for beams to beams. In case, the top flanges of two beams to be connected are at the same level of supporting beam, then a plate is welded as shown in Fig. 11.12 (a). In case, the top flange of two beams to be connected are not at the same level as that of the flange of supporting beam, then, the top plates are butt welded to the web as shown in Fig. 11.12 (b). As an alternative to this a slot may be cut in a channel to clear the flange of supporting beam and it is connected to the flanges as shown in Fig. 11.12 (c).

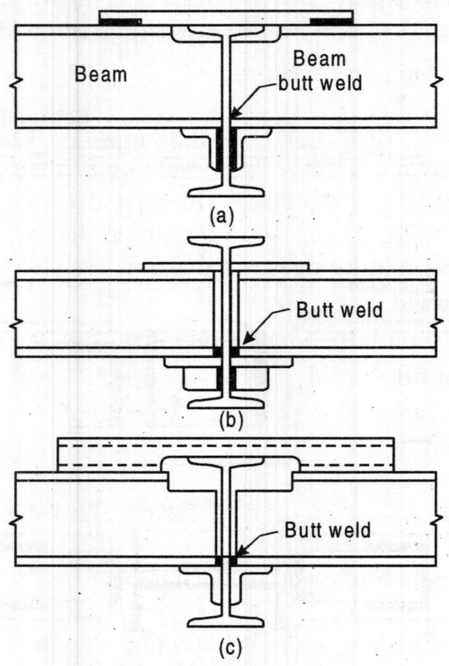

Fig. 11.12

Example 11.6 *A MB 500, @ 0.869 kN/m transmits an end reaction of 130 kN and an end moment of 120 kN-m to the flange of a column HB 250, @ 0.510 kN/m. Design the welded connections.*

Solution

Design

Top plate

Step 1: Propertion of section

From ISI Handbook No. 1, for MB 500, @ 0.869 kN/m

Depth of beam, h = 500 mm

Width of flange, b_f = 180 mm

Thickness of web, t_w = 10.2 mm

Thickness of flange, t_f = 17.2 mm

Step 2: Thickness of plate

Force in the top plate $= \left(\dfrac{120\times10^3}{500}\right) = 240$ kN

Area required for the top plate

$$\left(\frac{240\times1000}{0\cdot66\times250}\right) = 1454.54 \text{ mm}^2$$

Width of flange of beam

$$b_f = 180 \text{ mm}$$

Adopt width of the top plate

$$= 160 \text{ mm}$$

Thickness of the top plate required

$$\left(\frac{1451 \cdot 54}{130}\right) = 8.08 \text{ mm}$$

Provide 160 mm × 10 mm top plate. The top plate is connected to the flange of column HB 250, @ 0.510 kN/m. 10 mm thick stiffener plates are provided, so that the flange of the column does not deform.

Step 3: Fillet weld

For 10 mm thick top plate adopt 8 mm size of weld

Strength of 1 mm fillet weld

$$(0.7 \times 8 \times 1 \times 110) = 616 \text{ N/mm}$$

Length of the fillet weld

$$\left(\frac{240 \times 1000}{616}\right) = 389.61 \text{ mm}$$

Provide 120 mm side fillet weld on each side of the top plate and 160 mm end fillet weld.

The total length of weld provided is 400 mm. The length of unwelded portion is kept equal to the width of plate (viz., 160 mm).

Therefore, the total length of top plate is 290 mm.

The bottom flange of beam may be directly welded to the flange of column. The welded unstiffened seat connections for end reaction 130 kN designed in Example 11.4 are adopted. The connections designed are shown in Fig. 11.13.

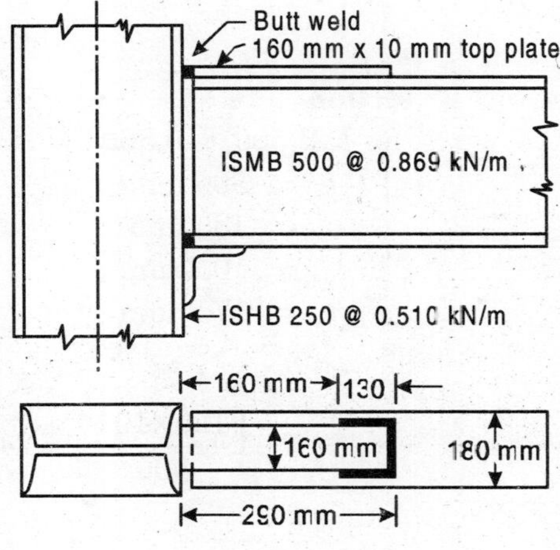

Fig. 11.13

PROBLEMS

11.1 A LB 300, @ 0.377 kN/m transmits an end reaction of 80 kN to the flange of a column HB 200, @ 0400 kN/m. The fillet welds are applied directly on both the sides of web of the beam. Design the connections.

11.2 In Problem 11.1 if two plates are used for connections on both the sides of web of beam, then, design the welded connections.

11.3 In Problem 11.1, if two framing angles are used for connections on both the sides of web of a beam, then, design the connections.

11.4 A LB 400, @ 0.569 kN/m transmits and end reaction of 110 kN to the flange of a column HB 350, @ 0.674 kN/m. Design unstiffened welded seat connections.

11.5 A HB 450, @ 0.653 kN/m transmits an end reaction of 220 kN to the flange of a column HB 300, @ 0.630 kN/m. Design stiffened welded seat connections.

11.6 A MB 300, @ 0.442 kN/m transmits an end reaction of 11 kN and an end moment of 80 kN-m to the fladge of a column HB 300, @ 0.630 kN/m. Design the welded connections.

Chapter 12

Design of Round Tubular Structures

12.1 INTRODUCTION

The round tubular sections are used as structural components since long. A large number of tubular structures have been constructed in the past. The tubular sections have been used for columns, compression members, tension members, roof trusses, mill buildings, air plane hangers, cross bracings and beams. The tubular sections are also used for single storey industrial buildings, warehouses, and shopping centres where long span column free areas are required. The tubular sections are used to advantages in structures designed for material handling equipments (e.g., a bridge, derrick and tower cranes) where weight savings may be very substantial economic consideration. The tubular sections are effectively used in large space frame, lattice structures for arenas, stadium exhibition halls where appearance as well as weight become an important design consideration. The masts and transmission towers are other examples where tubular sections are utilised effectively. In the past, the use of tube was hampered because of connection details. A major contribution to the solution of this problem has undoubtedly been the development of fully automatic oxyacetylene tube cutting machines, which not only cut tubes to fit flat surfaces but also cut them to fit cylindrical surfaces such as tube as well. The machine cuts the tubes to the correct profile with a bevelled edge to simplify the welding process at the joint.

The round tubular sections have as much as 30 to 40 percent less surface area than that of an equivalent rolled steel shape. Therefore, the cost of maintenance, cost of painting, fire proofing and/or other protective coatings reduce considerably. The moisture and dirt do not collect on the smooth external surface of the tubes. Therefore, the possibility of corrosion also reduces. The ends of tubes are sealed. As a result of this, the interior surface is not subjected to corrosion. The interior surfaces do not need any protective treatment.

The use of round tubular members is becoming more generally adopted for structures. The tubes are of special interest to the architects from an effective view point and to the engineers from a structural effectiveness view point.

The tubular sections have more torsional resistance than the other sections of the equal weights. The tube sections have a higher frequency vibrations under dynamic loading than the other sections including the solid round one. The round tubular sections offer less resistance to wind. Some of the codes allow a *one-third* reduction in wind load compared to equivalent projected area. IS : 875 –1984 recommends a shape factor of 0.7 for round tubular sections.

The tubular sections are most efficient sections which are adoptable to many different situation. The tubular sections cannot be surprised in its efficiency by other sections.

12.2 ROUND TUBULAR SECTIONS

The round tubular sections are manufactured from steel which when analysed in accordance with the methods in IS : 228, Methods of Chemical Analysis of Pig Iron, Cast Iron and Plain Carbon and Low Alloy Steels shall show not more than 0.06 percent sulphur, not more than 0.06 percent phosphorus. The plain carbon steel tubes as specified in IS : 1161–1963 are used for the structural purpose. The sizes and properties of steel tubes for structural purpose are given in Tables 12.1 (a) and (b) as per IS : 1161–1963. These tubes are of following types :

1. HFW : Hot finishes welded
2. HFS : Hot finished seamless
3. ERW : Electric resistance or induction butt welded.

The hot rolled carbon steel structural tubes are made in round shapes from a continuous welding or seamless process. In the continuous weld process coils of steel called skelp are welded and to end to form a continuous band of steel and passed through furnaces. As the skelp exists from the furnace, it is formed into a round pipe and welded by pressure. The pipe is then placed into sketch reducing mill, where, it is brought to the desired diameter and wall thickness.

In seamless tubes, a solid round bar of redetermined size is treated and then pierced by a mandrel while rotating at high speed. The tube is then placed to other rolling operations which bring it to the proper diameter and wall thickness.

The structural members consisting of round steel tubes are designed as per specification given in IS : 806 –1957 (Code of Practice for Use of Steel Tubes in General Building Construction).

Table 12.1 (a) *Sizes and properties of steel tubes for structural purposes*

Nominal bore	*Outside diameter*	*Class*	*Thickness*	*Weight*	*Area of X-section*	*Internal volume*	*Surface*		*M.I.*	*Z*	*r*	r^2
							Ext.	*Int.*				
mm	*mm*		*mm*	*N/m*	mm^2	mm^3/m	mm^2/m	mm^2/m	m^4	cm^3	*cm*	cm^2
1	*2*	*3*	*4*	*5*	*6*	*7*	*8*	*9*	*10*	*11*	*12*	*13*
15	21.3	Light	2.0	96.2	121	23500	66900	54300	0.57	0.54	0.69	0.47
		Medium	2.6	12.1	153	20400		50600	0.68	0.64	0.67	0.45
		Heavy	3.2	14.4	182	17400		46800	0.77	0.72	0.65	0.42
20	26.9	Light	2.3	14.1	178	39100	84500	70100	1.36	1.01	0.87	0.76
		Medium	2.6	15.7	198	37000		68200	1.48	1.10	0.86	0.75
		Heavy	3.2	18.9	238	32000		64400	1.70	1.27	0.85	0.71
25	33.7	Light	2.6	20.1	254	63800	105900	89500	3.00	1.84	1.10	1.22
		Medium	3.7	24.2	307	58500		85800	3.60	2.14	1.08	1.18
		Heavy	4.0	29.5	373	51900		80700	4.19	2.49	1.06	1.12
32	42.4	Light	2.6	25.7	325	108700	133200	116900	6.46	3.05	1.41	1.99
		Medium	3.2	31.1	394	101800		113000	8.99	4.24	1.36	1.93
		Heavy	4.0	38.1	483	91900		108100	8.99	4.24	1.36	1.86
40	40.3	Light	2.9	32.7	414	141900	151700	133500	10.70	4.43	1.61	2.59
		Medium	3.2	35.9	453	137900		131600	11.59	4.80	1.60	2.56
		Heavy	4.0	44.1	550	127600		126600	13.77	5.70	1.57	2.47

Contd.

Table 12.1 (a) Contd.

1	*2*	*3*	*4*	*5*	*6*	*7*	*8*	*9*	*10*	*11*	*12*	*13*
50	60.3	Light (1)	2.9	41.4	523	233300	189400	171200	21.59	7.16	2.03	4.13
		Light (2)	3.2	45.4	574	228200		169300	23.47	7.18	2.02	4.09
		Medium	3.6	50.7	641	221500		166800	25.87	8.58	2.01	4.03
		Heavy	4.5	61.7	788	206700		161200	30.90	10.20	1.98	3.92
65	76.1	Light	3.2	58.0	733	381600	239100	219000	48.78	12.8	2.58	6.66
		Medium	3.6	64.9	820	372800		216500	54.01	14.2	2.57	6.59
		Heavy	4.5	79.7	1010	353600		210800	65.12	17.1	2.54	9.48
80	88.9	Light	3.2	68.1	862	534600	279300	259200	79.21	17.8	3.03	9.19
		Medium	4.0	84.3	1070	514000		254200	96.34	21.7	3.00	8.03
		Heavy	5.0	103.0	1320	488900		247900	116.04	26.2	2.97	8.83

(where M.I. is moment of inertia, Z is modulus of section, r is the radius of gyration and r^2 square of radius of gyration)

Table 12.1 (b) *Sizes and properties of steel tubes for structural purposes*

Nominal Bore	*Outside diameter*	*Class*	*Thickness*	*Weight*	*Area of X-section*	*Internal volume*	*Surface*		*M.I.*	*Z*	*r*	r^2
							Ext.	*Int.*				
mm	*mm*		*mm*	*N/m*	mm^2	mm^3/m	mm^2/m	mm^2/m	m^4	cm^3	*cm*	cm^2
1	*2*	*3*	*4*	*5*	*6*	*7*	*8*	*9*	*10*	*11*	*12*	*13*
90	101.6	Light	3.6	87.6	1110	699900	319200	296600	133.2	26.2	3.47	12.02
		Medium	4.2	97.0	1230	688100		294100	146.3	28.8	3.45	11.98
		Heavy	5.0	119.0	1520	659000		287800	177.5	34.9	3.42	11.70
100	114.3	Light	3.6	99	1250	900900	359100	336500	192.0	33.6	3.92	15.33
		Medium	4.5	121	1550	870900		330800	234.3	41.0	3.89	15.10
		Heavy	5.4	145	1850	841300		335200	274.5	48.0	3.85	14.86
110	127.0	Light	4.5	122	1730	1093600	399000	370700	325.3	51.2	4.33	18.78
		Medium	5.0	150	1920	1075100		367600	357.1	56.2	4.32	18.64
		Heavy	5.4	162	2060	1060500		865100	382.0	60.2	4.30	18.52
125	139.7	Light	4.5	149	1910	1341700	438900	410600	437.2	62.6	4.78	22.87
		Medium	5.0	166	2120	1321200		407500	480.5	68.8	4.77	22.71
		Heavy	5.4	179	2280	1305000		405000	514.5	73.7	4.75	22.58
135	152.4	Light	4.5	164	2090	1615100	478800	450500	572.2	75.1	5.23	27.37
		Medium	5.0	182	2320	1592500		447400	629.5	82.6	5.21	27 .19
		Heavy	5.4	195	2490	1574800		447900	674.5	88.5	5.20	27.05
150	165.3	Light	4.5	178	2270	1913800	518700	490400	732.6	88.7	5.78	32.27
		Medium	5.0	197	2510	1889400		487300	806.6	97.7	5.66	32.07
		Heavy	5.4	212	2710	1869900		484700	864.7	105.0	5.65	31.92

Contd.

Table 12.1 (b) Contd.

1	2	3	4	5	6	7	8	9	*10*	*11*	*12*	*13*
150	168.3	Light	4.5	181	2320	1993100	528700	500500	777.2	92.4	5.79	33.56
		Medium	5.0	201	2570	1968100		497300	855.8	102.0	5.78	33.36
		Heavy (1)	5.4	217	2760	1948300		494800	917.7	109.0	5.76	33.21
		Heavy (2)	6.3	253	3210	1904000		489100	1053.4	125.0	5.73	32.85
175	193.7	Light	5.0	233	2960	2650400	608500	577100	132.0	136	6.67	44.54
		Medium	5.4	250	3190	2627400		574600	141.7	146	6.66	44.36
		Heavy	5.9	273	3480	2598700		571600	153.6	159	6.64	44.13
200	219.1	Light	5.0	264	3360	3434000	688300	650000	191.8	176	7.57	57.33
		Medium	5.6	294	3760	3894700		653100	214.2	195	7.55	57.02
		Heavy	5.9	310	3950	3375100		651300	224.7	205	7.54	56.86
225	244.5	Heavy	5.9	342	4420	4252900	768100	731100	314.9	258	8.44	71.71

12.3 PERMISSIBLE STRESSES

The *direct stress in axial tension* on the net cross-sectional area of tubes shall not exceed the values of σ_t given in Table 12.2 as per IS : 806 –1957.

Table 12.2 *Permissible axial stress in tension*

IS : 1161 Grade	σ_t *in N/mm*^{2}
St. 35 (Yst. 22)	125.0
St. 40 (Yst. 25)	150.0
St. 55 (Yst. 32)	190.0

Note. *The values of permissible stress in axial tension have been converted in S.I. units*

The *direct stress in compression* on the cross sectional area of axially loaded steel tubes shall not exceed the values of σ_c given in Table 12.3 as per IS : 806 – 1957. (**Note** : These values have been converted in S.I. units).

Table 12.3 *Permissible axial stress in compression*

Slenderness ratio $\left(\frac{l}{r}\right)$	σ_c *in N/mm*^{2}		
	IS : 1161 Grade St. 35 (Yst. 22)	*IS : 1161 Grade St. 40 (Yst. 25)*	*IS : 1161 Grade St. 55 (Yst. 32)*
0	125.0	150.0	190.0
10	121.7	144.8	182.1
20	117.5	140.0	175.0
30	113.1	135.2	167.9
40	108.8	130.3	161.0
50	104.6	125.5	153.9
60	100.2	120.7	146.8
70	97.0	115.5	137.5
80	92.9	108.8	126.3
90	87.6	103.3	112.8
100	81.4	91.0	98.9
110	74.5	81.3	86.9
120	67.4	72.1	75.8
130	60.3	63.8	65.5
140	54.0	56.5	58.4
150	49.0	50.3	51.7
160	43.2	44.3	45.0
170	38.1	39.2	39.8
180	33.9	34.8	35.3

Contd.

Table 12.3 Contd.

190	30.4	31.1	31.5
200	27.1	27.8	28.0
210	24.3	24.9	25.0
220	21.9	22.5	22.7
230	19.8	20.4	20.5
240	18.0	18.5	18.7
250	16.2	16.7	16.7
300	10.6	10.6	10.6
350	7.1	7.1	7.2

Notes. 1. *Intermediate values may be obtained by linear interpolation.*

2. *The values of allowable stress in axial compression has been found from secant formula.*

The *tensile and compression bending stress in the extreme fibres* of the tubes shall not exceed the values a_b given in Table 12.4 as per IS : 806–1953.

(**Note** : These values have been converted in S.I. units).

Table 12.4 *Permissible axial stress in extreme fibres in tension and compression*

IS : 1161 Grade	σ_t *in N/mm*^{2}
St. 35 (Yst. 22)	140.0
St. 40 (Yst. 25)	155.0
St. 55 (Yst. 32)	205.0

The *maximum shear stress* in a tube calculated by dividing the total shear by an *area equal to half the net cross-sectional area of tube* shall not exceed the value of σ_b given in Table 12.5 as per IS : 806 –1957. The net cross-sectional area shall be derived by deducting areas of all holes from the gross cross-sectional area (**Note** : These values have been converted in S.I. units).

Table 12.5 *Permissible axial stress in tension*

IS : 1161 Grade	σ_t *in N/mm*^{2}
St. 35 (Yst. 22)	190.0
St. 40 (Yst. 25)	110.0
St. 55 (Yst. 32)	135.0

The *average bearing stress* on the net projected area of contact shall not exceed the values σ_b given in Table 12.6 as per IS : 800 –1987.

(**Note** : These values have been converted in S.I. units).

Table 12.6 *Permissible axial stress in tension*

IS : 1161 Grade	σ_t *in N/mm*^{2}
St. 35 (Yst. 22)	170.0
St. 40 (Yst. 25)	190.0
St. 55 (Yst. 32)	250.0

The members subjected to both bending and axial stresses shall be so proportioned that the quantity $\left[\left(\frac{f_c'}{f_a}\right)+\left(\frac{f_b'}{f_b}\right)\right]$ does not exceed unity.

where, f_a' = Calculated axial stress
= Axial load divided by appropriate area of member
f_a = Permissible stress in member for the axial load
f_b' = Calculated bending stress in the extreme fibre
f_b = Permissible bending stress in the extreme fibre

When the bending occurs about both the axes of the members, then f_b shall be taken as

$$f_b = [(f_{bx})^2 + (f_{by})^2]^{1/2} \quad \text{...(12.1)}$$

where f_{bx} and f_{by} are the two calculated unit fibre stresses. The *equivalent stress,* σ_e due to co-existing bending shear stresses shall not exceed the value given in Table 12.7 as per IS : 806 –1957.

(**Note** : These values have been converted in S.I. units).

Table 12.7 *Permissible axial stress in tension*

IS : 1161 Grade	σ_t *in N/mm²*
St. 35 (Yst. 22)	190.0
St. 40 (Yst. 25)	228.0
St. 55 (Yst. 32)	290.0

The equivalent stress σ_b is obtained from the following formula

$$\sigma_e = [\sigma_b{}^2 + 3\tau_s{}^2]^{1/2} \quad \text{...(12.2)}$$

where, σ_b is the calculated bending stresses at the point under consideration and τ_s is the calculated actual shear stress at the point under consideration.

12.4 TUBE COLUMNS AND COMRESSION MEMBERS

The round tubular sections provide the most efficient cross-sectional shape for the columns and compression members having lateral restraint in all directions

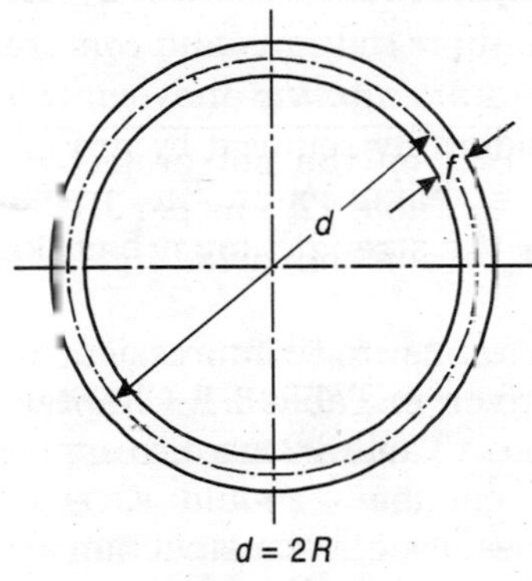

Fig. 12.1

normal to the axis of member. The diameter of such a member should be as large as possible with the additional requirement that d/t (see Fig. 12.1) should be small enough to assure that pressure failure by local buckling will not occur.

The local buckling strength of very short perfect tubes depends primarily on L/d ratio. In case of the medium tubes, the local buckling strength is primarily a function of $\left(\frac{d}{t}\right)$ ratio. For long tubes, the local buckling failure should be avoided so that the general column buckling is the primary problem. The medium length tubes are most frequently used in practice and the theoretical local buckling stress based on small deflection theory for solution of Donell's equation

$$\sigma_b = \frac{C \cdot E}{\left(\frac{R}{t}\right)} \quad \text{...(12.3)}$$

where, $C = [3(1 - \upsilon^2)]^{-1/2}$ (≈ 0.6 when $\upsilon = 0.3$) and

R and t are defined in Fig. 12.1. ...(12.4)

The medium length tube solution defined by Eq. 12.3 is applied for value of Z greater than 2.85, whereas Z is given by Batdorf

$$Z = \frac{L^2}{R \cdot t}\left[(1 - \upsilon^2)\right]^{\frac{1}{2}} \quad \text{...(12.5)}$$

where, Z is the limiting constant for medium length tubular, columns.

The experiments show that as $\left(\frac{d}{t}\right)$ increases, failure usually occurs at a stress increasing lower than that obtained from Eq. 12.3. This is because residual stresses and/or small irregularities in shape have an extremely detrimental effect on the compressive strength of round tubular section. The value of constant C depends to some degree on the tube manufacturing process and should be based on the recommendations made by the particular manufacturer.

In the design of steel tubular columns and compression members the attention must be paid to two points : *crinkling* and *heat treatment*. The yield strength of mild steel may be considerably reduced by any heat treatment which it receives such as welding. On this account, the precautions should be taken to prevent heat treatment or the strength must be taken as that of the annealed material.

For the round tubular columns, and compression members, the effective length of member is adopted as given in Table 3.1, Chapter 3. The limits specified for maximum slenderness ratio in Table 3.2 are also applicable for the tube members. In addition to this, the members should satisfy the *minimum thickness* requirement. The steel tubes used for construction exposed to the weather shall be not less than 4 mm thick and for construction not so exposed shall be not less than 3.2 mm thick when the structures are not readily accessible for maintenance, the minimum thickness shall be 5 mm. Notwithstanding the above, the thickness of tube subjected to compressive stress shall not be less than that given by the formula

$$t = 0.865\,[D]^{1/3} \qquad \text{...(12.6)}$$

where, D is the outside diameter of tube in mm. The ends of the members are sealed.

12.5 CRINKLING

When a steel tubular member is subjected to compression, then, the tube may *crinkle* (i.e., the walls of the tubes may cave in and form folds after the manner of concertina). These folds may be circular, oval or polygonal and they may occur after or before the longitudinal stress reaches the yield point.

Professor R.V. Southwell investigated the matter and deduced the following formula analytically

$$p = E \cdot \frac{t}{R}\left[\frac{m^2}{3(m^2-1)}\right]^{\frac{1}{2}} \qquad \text{...(12.7)}$$

where, p = Stress causing the collapse
t = Thickness of the tube
R = Mean radius of tube
$1/m$ = Poisson's ratio.

The stress which will produce the lobbed failures is $\left[\left(\dfrac{K^2-1}{K^2+1}\right)\right]$ times that given by Eq.12.7, where K is the number of lobes formed. This formula applies only if p is less than the yield stress. Thus, for a mild steel tube having a yield point of 315.0 N/mm^2 the formula will only apply when $\left(\dfrac{t}{R}\right)$ ratio is less than about $\left(\dfrac{1}{400}\right)$.

12.6 TUBE TENSION MEMBERS

The effectiveness of a member subjected to axial tension depends on its cross-sectional area, grade of steel and its method of connection. Therefore, a round tube or any rolled steel shape with the same net cross-sectional area and the same material would have the same equivalent resistance to the force. The tubular section do not have any advantage as tension member. The tubular sections have higher cost of production than other rolled steel sections.

12.7 TUBULAR ROOF TRUSSES

The design of roof trusses has been discussed in Chapter 9. The roof trusses are subjected to external loads. The external loads. The external loads, in general, are applied at the joints. Therefore, the different members of the roof trusses are subjected to axial compressive and tensile forces only. The tubular compression members have been discussed in Sec. 12.4. The tubular tension members have been discussed in Sec. 12.6. The members of tubular trusses are generally joined by welding. A tubular roof truss is shown in Fig. 12.2.

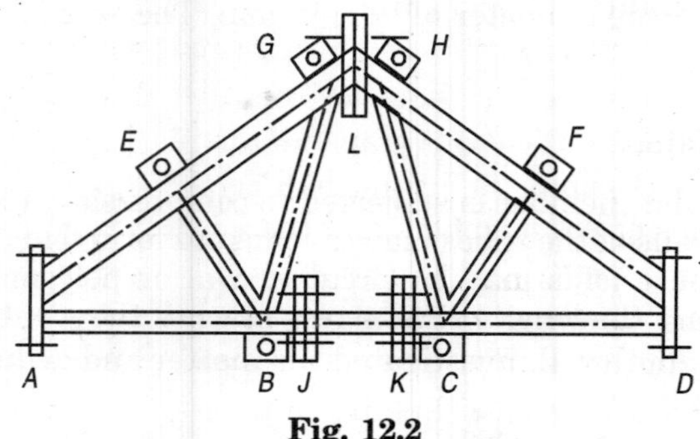

Fig. 12.2

In order to facilitate the erection of the trusses, the complete truss is splitted in different parts (viz., *AEGB, DFHC* and *JK*). These parts are then assembled after the erection. These parts are then connected by bolts and nuts. The arrangements for bolt connections made are as shown in Fig. 12.2 at *J, K* and *L*. The rectangular pieces of plates are welded to the projected end of the parts of the truss. The bolts are placed in the holes in both the plates and then, the nuts are tightened. The different parts of the truss are then assembled and a complete truss is built. The rectangular pieces with holes are also welded with the principal rafters at the joints and these pieces are kept projected as shown in Fig. 12.2 at *E, K, G* and *H*. These projected pieces are used to provide tubular purlins. The tubular purlins are provided with holes at the ends with internal threads. The tubular purlins are placed from one side. The studs are used for the connections of the purlins. The studs are placed through the holes of projected plates and tightened in the internally threaded ends of the purlins. The purlins are placed and fixed with the other ends of the studs. Similar arrangements are also made for tubular bracings with the bottom tie of the truss as shown in Fig. 12.2 at *B* and *C*.

In addition to the type of truss shown in Fig. 12.2, the tubular trusses of various other types are also made with suitable arrangements for the erection.

12.8 JOINTS IN TUBULAR TRUSSES

The welding is generally adopted for connections in tubular steel construction. The welded connections provide rigid joints. The rigid joints results in saving in materials and greater overall economy. The actual condition of rigidity should be taken into consideration while designing these types of joints.

The welded connections in the tubular structures may, however, be designed on a simple design basis comparable with IS : 800. In such cases, the secondary stresses may be neglected in the design of roof trusses except where the axes of the members do not meet at a point where there is such eccentricity, the effects of the eccentricity should also be considered.

In joints in compression members, the ends of the members are faced for complete bearing over their whole area. The welding and joining materials are kept sufficient to retain the members accurately in place and to resist all forces other than direct, compression including those arising during transit, unloading and erection.

Where such members are not faced for complete bearing, the welding and joining material shall be sufficient to transmit all the forces to which the joint is subjected. Wherever possible, the joints shall be proportioned and arranged so that the gravity axes of the members and the joints are in line, so as to avoid eccentricity.

Initially, the tube members in tubular trusses were welded to the gusset plates as shown in Fig. 12.3.

The welded joint with the gusset plates in the tubular trusses were found to be unsatisfactory for severe load conditions. The difficulty with this type of connection arises from the concentrated stress flow between the tube walls and the gusset plate even though the gusset plate is capable of transferring the forces effectively.

As far as possible, tube connections between the members are made directly tube to tube without gusset plates and other attachments. A weld connecting two tubes end to end shall be full penetration butt weld. The effective throat thickness of the weld shall be taken as the thickness of the thinner part joined.

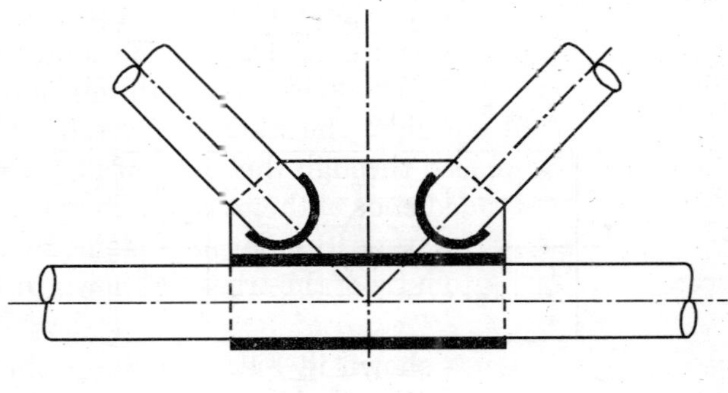

Fig. 12.3

A weld connecting the end of one tube (branch tube) to the surface of another tube (main tube) with their axes at an angle of not less than 30° shall be one of the following :

I. A butt weld throughout.

II. A fillet weld throughout.

III. A fillet-butt weld, the weld being a fillet weld in one part and a butt weld in another with a continuous change from the one form to the other in the intervening portions.

Type I may be used whatever the ratio of the diameter of the tubes joined provided complete penetration is secured either by the use of the backing material, or by deposing a sealing run of metal on the back of joint or by scme special method of welding. Type I weld is shown in Fig. 12.4. When Type I is not employed then, Type II should be used whether the diameter of branch tube is less than $\frac{1}{3}$rd of the diameter of the main tube. Type II weld is shown in Fig. 12.4. When the diameter of branch tube is equal or greater than $\frac{1}{3}$rd of the diameter of the main tube, then, the Type III weld as shown in Fig. 12.4 is used.

For the purpose of stress calculation, the throat thickness of the butt weld portion shall be taken as the thickness of the thinner part joined. The throat thickness of the fillet weld and fillet-butt weld is taken as the minimum effective throat thickness of the fillet or fillet butt weld.

A weld connecting the end of one tube to the surface of another with the axes of the tubes intersecting at an angle less than 30° shall be permitted only if adequate efficiency of the junction has been shown.

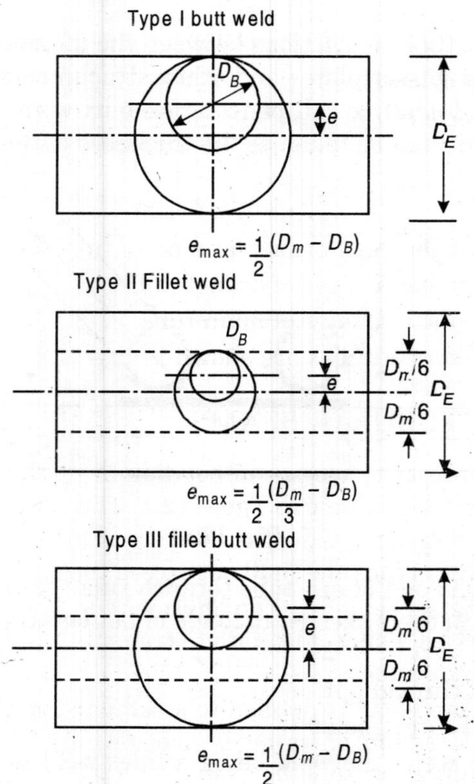

Fig. 12.4

The length of curve of intersection of the tube with another may be determined from the following formula :

$$L = \left(a + b + 3\sqrt{a^2 + b^2}\right) \qquad ...(12.8)$$

where

$$a = \left(\frac{d}{2}\right) . \operatorname{cosec} \theta$$

For intersection with a tube :

$$b = \left[\frac{d}{3} \cdot \frac{3 - \left(\frac{d}{D}\right)^2}{2 - \left(\frac{d}{D}\right)^2}\right] \qquad ...(12.9)$$

Alternatively,

$$b = \left(\frac{d}{4}\right) . \phi \qquad ...(12.10)$$

where, $\sin\left(\frac{\phi}{2}\right) = \left(\frac{d}{D}\right)$ ...(12.11)

where ϕ is measured in radians.

For intersection with flat plates :

$$b = \left(\frac{d}{2}\right) \qquad ...(12.12)$$

where θ = Angle between branch and main tubes

d = Outside diameter of branch tube

D = Outside diameter of main tube.

The above parameters of tubes are shown in Fig. 12.5.

12.9 DESIGN OF TUBULAR BEAMS

The tensile and compressive stresses in bending in the extreme fibres of tubes shall not exceed the values given in Table 12.4. The maximum shear stress in tubes in flexure calculated by diving the total shear by an area equal to half the net cross-sectional area of the tube shall not exceed the values given in Table 12.5.

When the tubular steel beam rests on abutment or other supporting member, it shall be provided with a shoe adequate to transmit the load to the abutment and to stiffen the end of tube.

Where a concentrated load is applied to a tubular member transverse to its length or the effect of load concentration is given by the intersection of triangular truss member, consideration shall be given to the local stresses set up and the method of application of the load and stiffening shall be provided as necessary to prevent the local stresses from being excessive. The increase in the intensity of local bending stresses caused by concentrated loads particularly marked if either the diameter of connected member or the connected length of a gusset or the like is small in relation to the diameter of the tubular member to which it is connected.

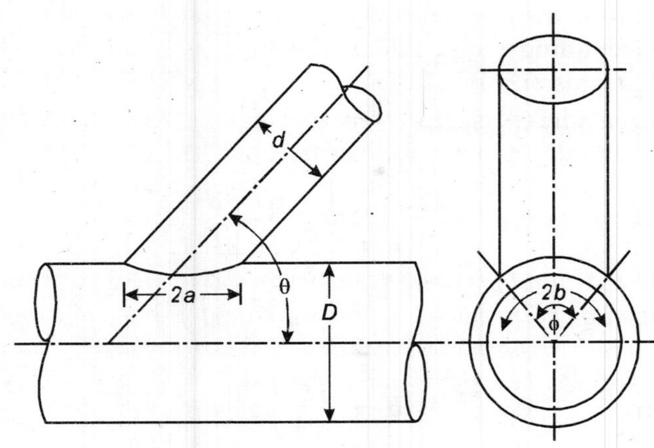

Fig. 12.5

The deflection of a member shall not be such as to impair the strength, efficiency or appearance of the structure or lead to damage to fittings and finishings. Generally, the maximum deflection should not exceed $\frac{1}{325}$ of the span for simply supported beams. This requirement may be deemed to be satisfied if the bending stress in compression or in tension does not exceed $3153\left(\frac{D}{l}\right)$ N/mm^2, where D is the outer diameter of the tube in mm and l is the effective length of the beam in mm.

12.10 DESIGN OF TUBULAR PURLINS

The requirement mentioned in Sec. 12.9 regarding limiting deflection may be waved in the design of simple tubular purlins provided the following requirements are satisfied:

	Simply supported	Effectively continuous
1. Name of end fixing	Simply supported	Effectively continuous
2. Maximum value of section modulus (mm^3)		
(a) IS : 1161 grade St. 35 (Yst. 22)	$\left(\frac{WL}{11200}\right) \times 10^3$	$\left(\frac{WL}{16800}\right) \times 10^3$
(b) IS : 1161 grade St 40 (Yst. 25)	$\left(\frac{WL}{13230}\right) \times 10^3$	$\left(\frac{WL}{19840}\right) \times 10^3$

3. Minimum outside diameter
IS : 1161 grade St. 35 or

St. 40 (mm) (Yst 32) $\left(\dfrac{L}{40}\right) \times (10)$ $\left(\dfrac{L}{70}\right) \times (10)$

where, W is the total distributed load in N on the purlins arising from dead load and snow load excluding wind, and L is the distance in mm between the centres of the steel principals or other supports.

A purlins is considered as effectively continuous at any intermediate point of support, if it is actually continuous over that point or it has there a joint able to provide a fixing moment of not less than $\left(\dfrac{WL}{2}\right)$.

Example. 12.1 *A tubular column consists of IS : 1161 grade St. 35 steel. The column is hinged at both the ends. The outside diameter of tube is 219.1 mm. The weight of 1 metre length of tube is 310 N. The length of column is 4.5 metre. Determine the safe load carrying capacity of the column.*

Solution

Step 1: Slenderness ratio

The column is 4.5 m long and it is hinged at both ends. The effective length of column

$$l = (4.5 \times 1000) = 4500 \text{ mm}$$

The outside diameter of tube is 219.1 mm, the weight of 1 metre length of tube is 310 N. The column belongs to heavy class. The radius of gyration of the tube from IS : 1161–1963.

$$r = 75.4 \text{ mm}$$

The slenderness ratio of the column

$$\frac{l}{r} = \left(\frac{4500}{75.4}\right) = 59.68 \not> 180$$

Step 2: Allowable stress in axial compression

The steel tube is IS : 1161 grade St. 35. The maximum allowable stress for $\left(\dfrac{l}{r}\right) = 59.58$ from IS : 806 –1957

$$\sigma_c = 101.282 \text{ N/mm}^2$$

The cross-sectional area of the tube column from IS : 1161–1963

$$A = 3950 \text{ mm}^2$$

Step 3: Safe load carrying capacity of the column

$$P = \left(\frac{101 \cdot 282 \times 3950}{1000}\right) = \mathbf{400.064 \text{ kN}}$$

Example 12.2 *The principal rafter in a round tubular truss carries a maximum force of 84 kN. A tension member meeting at right angles to the* *principal rafter carries a force of 20 kN. Design the member using IS : 1161 grade St. 35 steel for the tube. The panel length along the principal rafter is 1.80 m.*

Solution

Step 1: Slenderness ratio

The maximum force in the principal rafter of a tubular truss

$$P_c = 84 \text{ kN}$$

Assuming the maximum allowable stress for the tubular compression member

$$\sigma_c = 800 \text{ N/mm}^2$$

The cross-sectional area required for the column

$$A = \left(\frac{80 \times 1000}{80}\right) = 1050 \text{ mm}^2.$$

From IS : 1161–1963, provide a tube of outside diameter 88.9 mm, and 84.3 N/m weight. The properties of this tube member are as follows:

Cross-sectional area, $A = 1070 \text{ mm}^2$.

Radius of gyration, $r = 30$ mm

Thickness of tube, $t = 4$ mm

Internal diameter of the tube $= (88.9 - 2 \times 4) = 80.9$ mm

Pannel length of principal rafter = 1.80 m

The effective length of the member

$$l = (0.85 \times 1.8 \times 1000) = 1530 \text{ mm}$$

The slenderness ratio of tube member

$$\frac{l}{r} = \left(\frac{1530}{30}\right) = 51 \not> 180$$

Step 2: Allowable stress in axial compression

The tube consists of IS : 1161 grade St. 35 steel. The allowable stress for tube column from IS : 806 –1967

$$\sigma_c = 10416 \text{ N/mm}^2$$

Step 3: Safe load carrying capacity of member

$$P_{c1} = \left(\frac{1070 \times 104 \cdot 16}{1000}\right) = 111.45 \text{ kN} > 14 \text{ kN}$$

Hence, the design is satisfactory.

The thickness of tube 4 mm is equal to minimum thickness required for exposed situation. Hence satisfactory. Provide tube of outside diameter 88.9 mm and 86 5 N/m weight.

Design of tension member

The allowable stress in steel IS : 1161 grade St. 35 in tension

$$\sigma_a = 125 \text{ N/mm}^2$$

Since, the welding is used for the joint, the net area of tube is equal to gross area. Therefore, the area required for the tube

$$A_t = \left(\frac{20\times1000}{125}\right) = 160 \text{ mm}^2$$

Provide tube of 26.9 mm outside diameter and 14.1 N/m weight. The cross-sectional area provided is 178 mm^2 > 160 mm^2.

Example 12.3 *Design the welded joint for the tubular truss members in Example 12.2.*

Solution

Step 1: Length of curve of intersection

The principal rafter is the main tube member. The outside diameter of main tube from Example 12.2, D is 88.9 mm.

The outside diameter of branch tube (i.e., tension member)

$$d = 26.9 \text{ mm}$$

The tension member is meeting the principal rafter at right angles. Therefore, the angle between main tube and branch tube

$$\theta = 90°$$

$$\therefore \quad a = \left(\frac{d}{2}\right).\ \text{cosec}\ \theta = \left(\frac{26\cdot9}{2}\times1\right) = 13.45 \text{ mm}$$

From Eq. 12.9

$$b = \left[\frac{d}{3}\times\frac{3-\left(\frac{d}{D}\right)^2}{2-\left(\frac{d}{D}\right)^2}\right] \text{mm}$$

or

$$b = \frac{269\cdot3}{3}\times\left[\frac{3-\left(\frac{26\cdot9}{88\cdot9}\right)^2}{2-\left(\frac{26\cdot9}{88\cdot9}\right)^2}\right] = 13.68 \text{ mm}$$

The length of curve of intersection from Eq. 12.8,

$$L = a + b + 3(a^2 + b^2)^{1/2}$$

or

$$L = (13.45 + 13.68 + 3\ [(13.45^2 + 13.68^2)]^{1/2}$$

$$= 84.67 \text{ mm}$$

Let t be the throat thickness of fillet weld. The allowable stress in shear in weld from IS: 816 –1969

$$f_s = 110 \text{ N/mm}^2$$

Force to be transmitted is 20 kN. Therefore, throat thickness required

$$t = \left(\frac{20 \times 1000}{110 \times 84 \cdot 67}\right) = 2.147 \text{ mm}$$

The size of weld required

$$s = \left(\frac{2 \cdot 147}{0 \cdot 7} \times 10\right) = 3.06 \text{ mm}$$

Provide 4 mm fillet weld for the joint.

PROBLEMS

12.1 Two members meeting at a joint at 30° inclination in a tubular truss carry 120 kN compression and 36 kN tension. The length of compression member from centre to centre of joint is 2 m. The members are made of IS : 1161 grade St. 35 steel. Design the members.

12.2 Design the welded joint for the tubular truss in Problem 12.1.

PART 2

Timber Structure

Chapter 13

Design of Timber Structures

13.1 INTRODUCTION

The wood is one of many building materials used in building construction and various types of structures. The wood is an organic material. It is generally used in its natural state. When the wood is used as a structural material, it is called timber. The timber is used for temporary structures mainly because of its low cost. The temporary structures are those structures which are dismantled immediately after their use. The concrete form work, scaffolding, shuttering, and structures of exhibitions are the examples of temporary structures. The timber is also used for the permanent structures. The timber is used for various structural components such as beams, stringers, purlins, columns etc. The various types of timber structures are roof trusses, electric transmission posts, trestles and other types of frames. The various types of timbers which are used for structural purposes are teak, sal, deodar, rosewood etc. In addition to these, there are many other types of timbers which are also used for structural purposes. There are eighty-seven varieties of timbers which are used for structural purposes. The life of timber structure is long if it is maintained either continuously dry or wet. The life of a timber structure is short if it remains alternately wet and dry.

All the structural members, assembles or framework in a building in combination with the floors, walls and other structural parts of the building shall be *strong* (viz., capable of sustaining the whole dead and superimposed loads including other types of loads referred to IS : 875 –1984, with due *stability* and *stiffness* without exceeding the limits of stress specified.

13.2 DEFFECTS IN TIMBER

Any irregularity which occurs on or in a specimen of a timber and affects its strength or durability or appearance is known as defect. The timber is a natural product. It is seldom entirely free from defects. The defects in timber may be classified in two following groups :

1. Natural defects 2. Other defects.

13.2.1 Natural Defects

There are some factors which influence the growing tissues of a living tree. As a result of this, the defects are developed. Such defects are called natural defects. The natural defects are described below.

(i) **Knots.** When the portion of a branch or a limb embedded in the body of a tree is cut, then knot is formed. Sometimes lower branches of a tree die and fall down because of insufficient supply of sunlight. As a result of this, knot is formed in the body of a tree as shown in Fig. 13.1. Ordinarily, the knots are composed

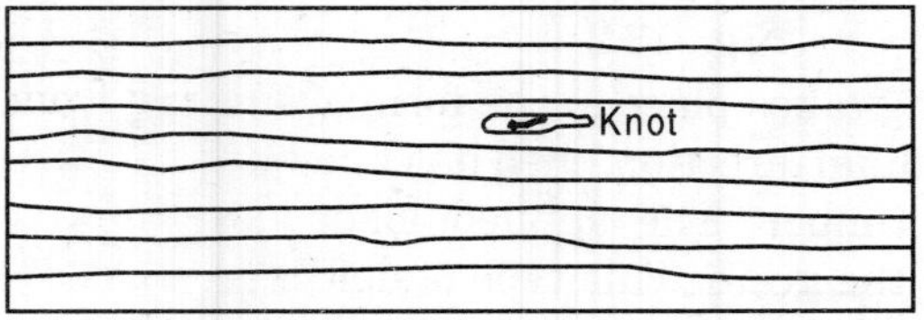

Fig. 13.1

of inferior wood than the normal wood. In the vicinity of knots, the grains are irregular and distorted. The knots reduce the strength of timber. The reduction in the strength of timber depends upon size, position and type of knots.

The strength of beams is reduced considerably, when the knots are situated in the bottom of a beam (the portion of beam in tension). When the knots are situated near the top of a beam (the portion of beam in compression) then it has less weakening effects. Knots have still less weakening effects when these occur near the centre of depth. The strength of timber is considerably decreased in tension because of knots. The strength of timber reduces to a small extent in compression because of knots. The knots do not affect the stiffness of timber appreciably.

(ii) **Wane.** A wane is bark or lack of wood on the edge or corner of a piece of timber. It is shown in Fig. 13.2. Wane affects the strength of a timber member only because the cross-sectional area is reduced. It can be avoided by properly squaring the edge.

13.2.2 Other Defects

Some defects develop because of the activity of external agents and subsequent treatment of sawn timber. As a result of faulty seasoning techniques, permanent distortion of timber may take place. The permanent distortions result in various types of warping and rupture of tissues. These types of defects are described below.

(i) **Warping.** The *warping* is caused as a result of permanent distortion of timber. The warping includes *cupping, twisting* and *bowing*. When the warping takes place across the width of a timber board, then it is known as cupping. When the warping takes place in spiral form in a longitudinal direction, then it is known as twisting. The twisting in extreme cases renders the timber useless.

Fig. 13.2

Fig. 13.3

When the warping takes place in the form of sagging from one end to the other along the longitudinal direction, then it is known as bowing. The warping reduces the strength of timber. The applied loads do not remain parallel with or perpendicular to the grain. This type of defect is most serious in long columns, loaded parallel to the grains.

(ii) **Rupture of tissues.** The rupture of tissues takes place as a result of permanent distortion of timber. The rupture of tissues includes *checks, shakes* and *splits. A check* is a crack or lengthwise separation of the timber. It usually occurs across the annual rings. It is shown in Fig. 13.3. During drying from outside towards centre, the outer fibres shrink more than the inner fibres. As a result of this cracks develop across the annual rings.

A *shake* is separation along the grains of the timber. It usually occurs between and parallel to the annual rings, as shown in Fig. 13.4. The shakes do not develop because of drying. The shakes originate because of careless felling. When the living trees are felled, the internal stresses are released which result in separation along the grains.

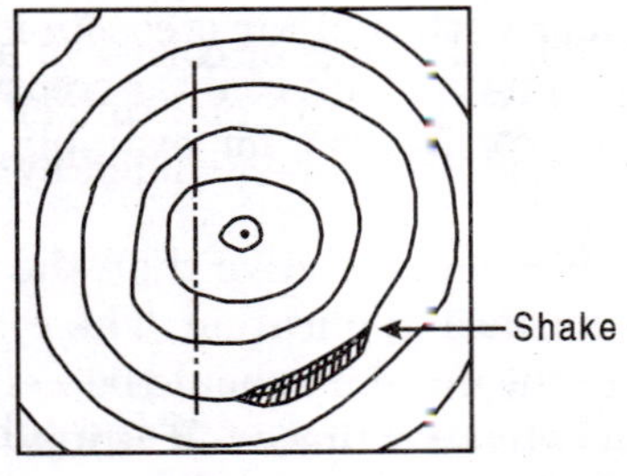

Fig. 13.4

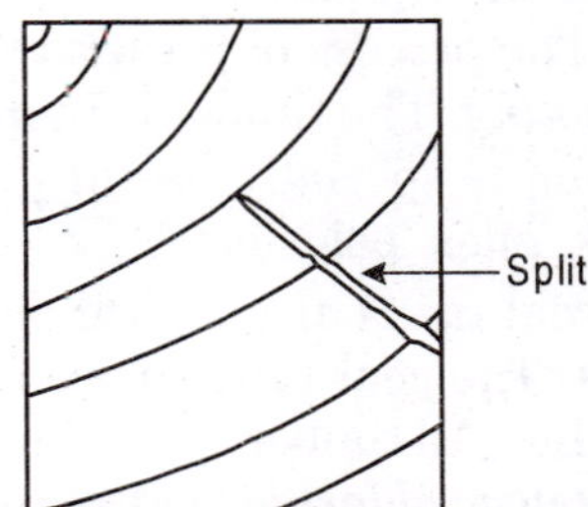

Fig. 13.5

A *split* is separation of the timber which extends from one face to the other. When it occurs at the end, it is known as end split. It is shown in Fig. 13.5.

The rupture of tissues (checks, shakes and splits) reduce the resistance of timber against shear. The strength of a beam is reduced considerably, if these occur in a horizontal plane and specially at mid-height. At mid-height of a beam, the shear stress is more than those at other heights. These have little effect, when these occur in a vertical plane.

13.3 DECAY OF TIMBER

The disintegration of timber substance due to the action of timber destroying agents is known as *decay*. The various timber destroying agents are fungi, marine borers, and insects. The fungi is most destructive. The fungi develops in damp places. The fungi is a kind of plant life. It grows in the form of hair like threads. The fungi travels through the timber. It renders the timber ultimately to powder. When the timber is maintained either completely dry or completely saturated, then, it lasts long. If the timber remains alternately wet and dry, and used in dark, damp and unventilated conditions, then it deteriorates very soon. If timber is used in ground it also deteriorates soon. The timber structures which are used for marine works, are situated in sea-water. The marine borers attack these structures. They bore into the timber structures and reduce or entirely destroy the strength of timber. The decay of timber also takes place by rot. There are two types of rots, dry rot and wet rot. When the timber is used in unventilated places where air is confined, then, the decay of timber takes place by dry rot. Wet rot attacks living trees. The timber also deteriorates by the attack of white ants or termites. The white ants get into structures and destroy the timber.

13.4 PRESERVATION OF TIMBER

The preservation of timber is essential to avoid the decay of timber. The pores of timber are filled up by preservatives. The preservatives exclude moisture from penetrating below the surface. The various antiseptic chemical compounds are injected to fill up the pores of timber. The coal tar creosote is most important preservative. The coal tar creosote is also known as *creosote oil or dead oil of coal tar or creosote coal tar solutions.* The composition of coal tar varies widely. The coal tar creosote used for preservation of timber should conform to IS : 218 –1952. The process of preservating the timber with coal tar creosote is known as *creosoting.* The timber is thoroughly impregnated with coal tar creosote. The creosoting is specially useful for timber for exterior use, for example railway sleepers, piles, poles etc.

The coal tar is also a good preservative. It is less effective than the coal tar creosote. The coal tar is more suitable for surface application. The coal tar is applied hot. All timbers placed in contact with the masonry should be well tarred. The *mercuric chloride* is also used as preservative for timber. It is injected into the pores of timber. The mercuric chloride is used in the form of corrosive sublimate, one part of sublimate to 150 parts of water. It prevents the destruction of timber by ants and sea-worms. The mercuric chloride is also effective against dry rot. The process of preserving the timber by mercuric chloride is known as *kyanizing. Sodium flouride, zinc chloride* and *copper sulphate* are also used as timber preservatives. When the timber is thoroughly impregnated with copper sulphate, then it is prevented from dry rot. Copper sulphate is not suitable for timber exposed to the action of water. It does not prevent timber from white ants.

A new preservative, called *Ascu* has been developed at the Forest Research Institute and College, Dehradun. It is made up of three chemicals proportioned as under:

One part by weight of $As_2O_5.2H_2O$

Three parts by weight of $CuSO_4.5H_2O$

Four parts by weight of $K_2Cr_2O_7$.

Ascu is available in the powder form. A solution if prepared by mixing six parts by weight of the powder in a hundred parts by weight of water. The solution is odourless. The wood treated with it can be painted, varnished, polished or waxed. Ascu solution can be applied or sprayed on in two coats or the wood pieces can be soaked in the solution tank or the wood pieces can be impregnated with the solution under pressure. The method to be adopted depends on the degree of immunity required, and the nature of wood. The last method is most effective.

When the untreated timber is used contact with the masonry foundations, then, termite shields are provided. These are provided at the top of masonry foundation and below the timber work as shown in Fig. 13.6 as per IS : 883–1961. The termite shield is provided by galvanized iron sheet or copper sheet

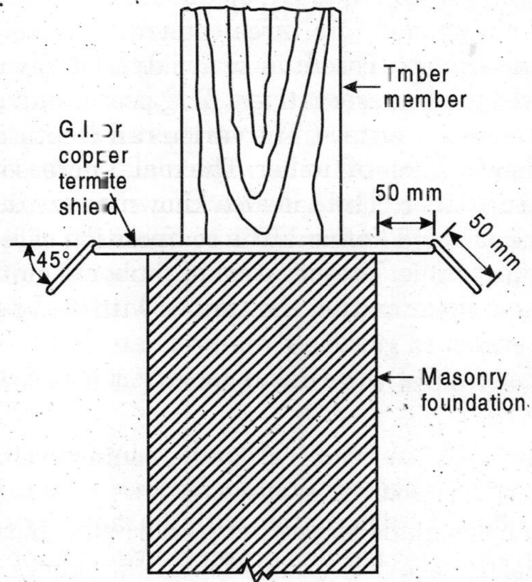

Fig. 13.6 *Termite shield*

not less than 0.45 mm thick. The galvanized iron sheets are laid on a damp proof course. When the copper sheets are used, damp proof course is not necessary. It prevents termite and white ants to enter the timber structures.

13.5 SEASONING OF TIMBER

The seasoning is the process of either expelling or drying up the sap remaining in the green timber. Seasoning of timber is done by exposing it in air, or placing it in water or in hot air. Seasoning of timber by exposing it in air, is known as natural seasoning. In natural seasoning, bulks of timber are stacked in layers under cover. Free circulation of air is allowed to pass around each bark. Natural seasoning takes considerable time. Seasoning of timber by placing it in water is known as water seasoning. In water seasoning, timber is placed in water for 15 days. It is preferable to place the timber in streams. Timber is then stacked under cover. It is allowed to dry thoroughly. The sap of timber is diluted and carried away by water. It reduces the time necessary to season, when stacked. Seasoning of timber by keeping it in hot air is known as *dessicating seasoning.* Timber is placed in chambers. Hot air is circulated around the timber. This method also reduces the time necessary for seasoning when stacked. The strength of timber increases because of seasoning. Timber becomes more durable and stiff than green timber.

13.6 GRADING OF TIMBER

The method of designating the quality of piece of timber is known as *grading.* While grading the timber, the defects in timber, their size, their number, and their locations are taken into consideration. The effect of density of timber and effect of slope of grains of timber are also taken into account while grading the timber. The timber is graded into three categories, *selected timber, standard timber* and *common timber.* The cut sizes of structural timber is graded after seasoning into three grades, *selected grade,* grade I, and grage II in IS : 883–1994. The selected timber is either free from the defect or it has minimum defects. The standard timber has defects within specified limits. The common timber is inferior to standard timber. The safe working stresses in compression, shear and bending for the selected limber are 1.16 times the safe working stresses for timber of the standard grade. The safe working stresses for timber of the common grade in compression, shear and bending are 0.84 times the safe working stresses for timbers of the standard grade.

All the grades of timber with the following defects are prohibited for the structural use:

(a) The timber with loose grains, splits, compression wood in coniferous structural timber, heartwood rot, saprot and warp.

(b) The worm holes made by powder post beetles and pitch pockets.

The following defects are permissible for all the grades of timber :

(a) The wanes are permitted provided they are not combined with knots and the reduction in strength on account of the wanes is not more than the reduction with the maximum allowable knots. The wanes may also be permitted provided there is no objection to its use as bearing area, nailing edge and affects general appearance.

(b) The worm holes other than those due to powder post beetles located and grouped to reduce the strength of timber shall be evaluated in the same way as the knots.

(c) All other defects which do not affect any of the mechanical properties of timber.

In addition to the permissible defects mentioned above, the defects listed in Table 13.1 (as per IS : 883–1994) are permitted for the selected grade, grade I and grade II timber for structural purposes.

Table 13.1 *Defects permissible for different grades of timber for structural purposes (As per IS : 883–1994)*

Selected I Grade		*Grade I*				*Grade II*
	B	*C*		*D*	θ	
		C_1	C_2			
1	2	3	4	5	6	7
	mm	mm	mm	mm		
Knots, checks and shakes shall be of	75	19	19	25	1 in	Knots, check and shakes
half the size for grade I	100	25	25	35	15	shall be 1.5 the size for
	150	38	38	50	for	grade
Slope of grain shall be not more than 1 in 20.	200	44	50	65	all	Slope of grain shall be not
	250	50	57	81		more than 1 in 12.
	300	54	75	100		
	350	57	81	115		
	400	63	87	131		
	450	66	93	150		
	500	69	100	165		
	550	72	103	181		
	600	75	106	200		

where, B = Width of face

C = Location and sizes of live knots

C_1 = Narrow face and $\left(\frac{1}{4}\right)$ of the width of wide face close to top and bottom edges

C_2 = Central half of the width of wide face

D = Depth of checks and shakes

θ = Slope of grain.

13.7 SAP WOOD AND HEART WOOD

The size of a tree increases by the growth of new wood on the outer surface below the bark. A layer of new wood is formed every year on the periphery of the trees. The layer is called annual ring. The trunk of a young tree consists entirely of light colour wood. Light colour wood is known as *sap wood.* The sap wood carries sap from roots of the tree to leaves. The sap wood contains the living cells. With the increase of age of tree and addition of new layers every year, the sap wood gradually changes to *heart wood* and new sap wood is formed. The older wood ceases to take part in the growth activity. The cells in older wood die. The various substances are formed in the older wood. As a result of these substances, the heart wood has distinct dark colour, and heart wood is more resistant to decay. The heart wood is more durable than sap wood. The heart wood is less porous than the sap wood. The sap wood is stronger and heavier than the heart wood, provided the moisture content is same. The sap wood can be impregnated with preservatives more readily because of its absorptiveness. Properly treated with preservative, the sap wood also becomes more durable than even the most durable wood like teak, sal and deodar, heart wood.

There are two choices, as regards suitability in respect of durability and treatability for the structural timber. The timber spices in *the first choice* shall be of any one of the following :

(a) Untreated heart wood of high durability.

(b) Treated heart wood of moderate and low durability and class A and class B treatability.

(c) Heart wood of moderate durability and class C treatability after pressure impregnation.

(d) Sap wood of all classes of durability after thorough treatment with preservatives.

The durability and treatability for the various species of timber have been given in a table in IS : 883 –1994.

The timber species in the *second choice* shall be of heart wood of moderate durability and Class D treatability.

13.8 GRAINS OF TIMBER

The *grain* is a term used to denote the direction of fibre of timber in relation to the main axis of a tree or member. It also denotes the relative width of annual rings. When the direction of fibres of a tree is more or less parallel to the axis of tree (or longitudinal axis), then the grains are called *straight grains.* The straight grains are also known as *parallel grains.* The straight or parallel grains of a limber member are shown in Fig. 13.7. The strength of timber along the straight grains is more than other directions. The directions along the grains and across the grains are also shown in Fig. 13.7. When the direction of fibres is at varying and irregular inclinations in relation to the main axis, it is known as irregular grain. The irregular grains occur locally. These reduce the strength of timber. When the fibres take a more or less winding or spiral course, the grains are called *spiral grains.* The spiral grains give the whole tree a twisted appearance. The spiral grains are considered as serious structural defect. The spiral grained

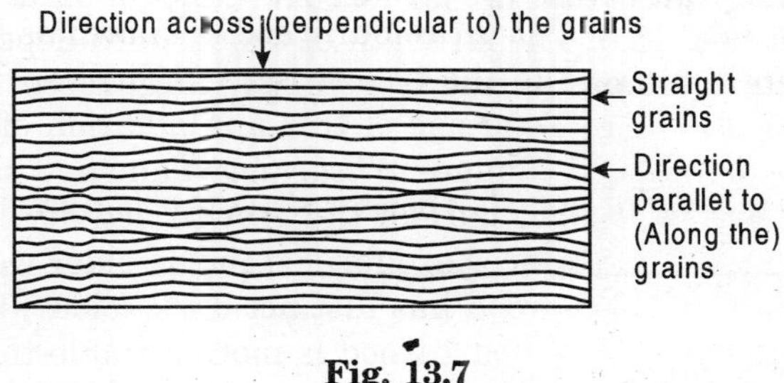

Fig. 13.7

timber is weak. Some trees have spiral grains in one direction for number of years. The slope of grains reverses in direction around the trees in the succeeding annual layers of growth. Such fibres are called *interlocked grains.* The interlocked grained timber is not so weak as spiral grained timber.

When the fibres take the form of waves and undulations, then these are called *wavy grains.* When the fibre are distorted to that they have a curled direction, then these are termed as *curly grains.* When the grains are straight but inclined to the main axis, these are termed at *diagonal grains.* In the diagonal grains, fibres extend at an inclination (i.e., diagonally across a piece). It is a sawing defect. This defect is produced when a straight grained timber is not cut parallel to the grains. All the deviations from straight or parallel grains (including spiral and diagonal grains) are termed as *cross grains.*

The slope of grains is defined as the ratio of unit deviation to the distance within which this deviation occurs. In case of spiral and diagonal grains, the effective slope is found by taking square root of sum of the squares of two types of cross-grains.

13.9 STRENGTH OF TIMBER

The strength of timber is influenced by moisture content, density, presence of defects, type of grains and size. There is free moisture in the green wood within the cell spaces. There is also absorbed moisture in the cell walls. When green wood dries, free moisture evaporates. On the evaporation of all the free moisture, *the fibre saturation point* is reached, and absorbed moisture from cell wall starts evaporating. The fibre saturation point differs for different timbers. The fibre saturation point is the point at which the weight of moisture in the wood is between 21 percent and 28 percent of oven dry weight of wood. When the moisture content in the wood is more than this point, the strength of wood is practically constant. When the moisture content in the wood is less than this point and decreases, the strength of wood increases and wood starts to shrink. The strength of wood continues to increase until the wood is bone dry.

The density of wood measures amount of wood substance in a given volume. The strength of wood depends on the amount of wood substance. The strength of wood is more for dense wood than for light wood. The strength of timber reduces due to the presence of defects. The influence of particular type of defect, and position of defects have been described in Sec. 13.2. The strength of timber also depends upon type of grain of timber. The strength of timber along the grains is more than that across the grains. The effect of each type of grain has been described in Sec. 13.8. The strength of timber also depends upon size of timber. The defects developed during seasoning in large pieces are more than in smaller pieces.

Table 13.2 *Factors of safety to be applied to basic stress to obtain safe working stresses*

S.No.	*Stress N/mm²*	*Standard Grade*		
		Inside location	*Outside location*	*Wet location*
1.	Fibre stress in beams for broad leaved species, max.	5	6	7.5
2.	Fibre stresses in beams in conifers, max.	6	7	8.5
3.	Shear along grain	7	7	7
4.	Horizontal shear in beams	10	10	10
5.	Compressive stress parallel to grain	4	4.5	5.5
6.	Compressive stress perpendicular to grain, max.	1.75	2.25	2.75
7.	Modulus of elasticity for calculation of allowable deflection in beams	1.25	1.75	2

13.10 FACTOR OF SAFETY IN TIMBER

The factor of safety depends upon strength of timber, effect of moisture, the presence of defects and location of structure. The timber structure may be situated in inside location, or in outside location or in wet location. When the timber structures are situated in inside location they remain continuously dry and protected from moisture. When the timber structures are situated in outside location, they remain occasionally wet but dry quickly, for example, timber bridges, trestles, framework of open sheds etc. Sometimes the timber structures are situated in such a location, where they remain usually wet, for example, timber structures exposed to tide water or waves, or in contact with water of moist ground. The factor of safety for different locations to be applied to basic structures to obtain safe working stresses for standard grade of timbers is given in Table 13.2 as per IS : 883–1961.

13.11 WORKING STRESSES IN TIMBER

The working stresses for the timber depend on many factors. The working stresses are based upon the basic stresses and appropriate factor of safety. The basic stresses are obtained by tests on small clear specimen of standard grades

Table 13.3 *Safe working stresses for indian timbers*

S.No.	Trade Name of Timber	Ave. unit wt at 12% moisture	All grades, all location Modulus of Elas (kN/mm²)	Working stress in N/mm² for Standard Grade (Grade I)											
				Bending and Tension along grade (Extreme fibre stress)			Shear		Compression						
							Horizontal[1]	Along grain[2]	Parallel to grain			Perpendicular to grain			
				Inside location	Outside location	Wet location	All location	All location	Inside location	Out side location	Wet location	Inside location	Out side location	Wet location
1	2	3	4	5	6	7	8	9	10	11	12	13	14	15
	Group A													
1.	Bullet wood	11.05	17.40	22.8	19.0	15.2	1.5	2.1	14.4	12.6	10.2	11.2	8.8	7.1
2.	Ballagi	11.35	16.30	22.6	18.6	15.2	1.5	2.2	14.8	13.0	10.6	8.7	6.7	5.5
3.	Dhaman	7.85	14.80	18.2	15.2	12.4	1.3	1.9	12.0	10.6	8.8	6.0	4.6	3.8
4.	Kala Siris	7.35	13.60	18.6	15.4	12.6	1.5	2.7	13.4	12.0	9.4	7.2	5.6	4.6
5.	Sal	8.65	12.70	16.8	14.0	11.2	0.9	1.3	10.6	9.4	7.8	4.5	3.5	2.9

Notes. 1. The timber of group A (viz. super group) have E greater than or equal to 12600 N/mm² and those of group B (standard group) have E greater than 9800 N/mm². Whereas the timbers of group C (viz. ordinary group) have E greater than 5600 N/mm².

2. Above values of the working stresses have been converted in S.I. units.

Table 13.3 Contd.

1	2	3	4	5	6	7	8	9	10	11	12	13	14	15
	Group B													
1.	Indian Oak	8.65	12.5	14.8	12.4	9.8	1.2	1.7	9.2	8.0	6.6	4.5	3.5	2.9
2.	Bullet wood	8.80	12.4	17.2	14.4	11.6	1.3	1.8	10.8	9.8	8.8	5.5	4.3	3.5
3.	Dhaman (U.P. M.P.)	7.70	12.0	15.5	13.0	10.2	1.4	2.0	9.1	8.1	6.7	4.1	3.2	2.6
4.	Eucalyptus	8.50	11.5	16.6	13.8	10.8	1.2	1.7	11.2	10.2	8.0	7.6	5.9	4.8
5.	Jaman	8.50	11.2	15.1	12.6	10.2	1.2	1.7	9.1	8.4	6.7	5.8	4.5	3.6
6.	Babul	7.85	10.8	18.2	15.4	12.4	1.5	2.2	11.2	10.2	8.0	6.5	5.0	4.1
7.	Teak	6.40	9.6	14.0	11.6	9.4	1.0	1.4	8.8	7.8	6.4	4.0	3.1	2.5
	Group C													
1.	Indian chestnut	6.40	9.8	10.6	8.8	7.0	0.8	1.2	6.4	5.6	4.6	2.6	2.0	1.7
2.	Deodar	5.45	9.5	10.7	8.8	7.0	0.7	1.0	7.8	7.0	5.6	2.6	2.1	1.7
3.	Rose wood	7.55	9.3	16.8	14.0	11.2	1.2	1.7	10.6	9.2	7.8	6.4	5.0	4.1
4.	Sissoo	7.85	8.9	15.2	12.6	10.7	1.7	1.7	9.4	8.4	6.6	4.6	3.6	2.9
5.	Safed Siris	6.40	9.0	13.4	11.2	8.8	1.0	1.4	8.4	7.8	6.4	4.3	3.3	2.7
6.	Chir	5.75	9.8	8.4	7.0	6.0	0.6	0.9	6.4	5.6	4.6	2.7	1.7	1.4

Notes. 1. The values of horizontal shears are to be used for beams.

2. In all other cases, the shear along grains are to be used.

3. Above values of the working stresses have been converted in S.I. units.

in the laboratories for standard conditions of service. The safe working stresses for 72 species of timbers of standard grade are given in IS : 883 –1994. The actual values of safe working stresses for the few species of timber are given (for standard or grade I) as per IS : 883 – 1994 in Table 13.3 for reference purpose, for inside, outside and wet locations.

The permissible stresses for groups A, B and C for different locations applicable to grade I structural timber are given in Table 13.4 as per IS : 883 –1994 provided that the following conditions are fulfilled :

(a) The timber should be of high or moderate durability and be given the suitable treatment where necessary. They may be used in any location. If the location is inside and not in contact with the ground, low durability timber may be used after proper seasoning and preservative treatment are given, and

(b) The loads should be continuous and permanent.

The permissible stresses for timbers of selected grade, are 1.16 times the stresses for grade I timbers given in Table 13.4. The permissible stresses for timber of grade II are 0.84 times the stresses for grade I timbers given in Table 13.4. When the low durability timbers are used on outside locations, the permissible stresses for all grades of timber found are multiplied by 0.80.

In *inside location,* the timbers inside the buildings remain continuously dry or protected from the weather. In *outside locations,* the timbers either in open sheds or in outside exposed structure are occasionally subjected to wetting and drying. Whereas in *wet locations,* the timbers are almost continuously damp or wet or in contact with earth or water such as piles and timber foundation.

Table 13.4 *Permissible stresses for grade I timber (As per IS : 883–1994)*

S. No.	*Types of stresses*	*Permissible stress in N/mm²¶*		
		Group A	*Group B*	*Group C*
1	2	3	4	5
1.	**Bending and tension along grain**			
	(i) Inside location	18.2	12.3	8.4
	(ii) Outside location	15.2	10.2	7.0
	(iii) Wet location	12.0	8.1	6.0
2.	**Shear**			
	(i) Horizontal* all locations	1.2	0.9	0.6
	(ii) Along grain** all locations	1.7	1.3	0.9
3.	**Compression parallel to grain**			
	(i) Inside location	12.0	7.0	6.4
	(ii) Outside location	10.6	6.3	4.6
	(iii) Wet location	8.8	5.8	4.6

Contd.

* The values of horizontal shears are to be used for beam.

** In all other cases, the values of shear along the grains are to be used.

¶ Above values have been converted in S.I. units.

Contd. Table 13.4

1	2	*Group A* 3	*Group B* 4	*Group C* 5
4.	**Compression perpendicular to grain**			
	(i) Inside location	6.0	2.2	2.2
	(ii) Outside location	4.6	1.8	1.7
	(iii) Wet location	3.8	1.5	1.4

Modifications factors for permissible stresses

The permissible stresses given in Table 13.4 are modified by multiplying with the modification factors for the following:

(i) **For change in slope of grain.** When the timber has major defects like slope of grains, knots, and checks or shakes (but not beyond permissible value), the permissible stresses given in Table 13.4 are multiplied by the modification factors K_1 for different slopes of grain as given in Table 13.5 as per IS : 883 –1994.

Table 13.5 *Modification factor K_1 to allow for change in slope of grain IS : (883–1994)*

Slope	K_1	
	Strength of beams ; joints and Ties	*Strength of Posts or Column*
1	2	3
1 in 10	0.80	0.74
1 in 12	0.90	0.82
1 in 14	0.98	0.87
1 in 15	1.00	1.00

Table 13.6 *Modification factor K_2 to allow for change in slope of grain (883–1994)*

S.No.	*During of loading*	K_2
1	2	3
1	Continuous	1.0
2	Two months	1.15
3	Seven days	1.25
4	Wind and earthquake	1.33
5	Instantaneous or impact	2.00

(i) **For change in the duration of load.** For the duration of design load other than continuous, the permissible stresses given of Table 13.4 are multiplied by the modification factors K_2 given in Table 13.6 as per IS : 883 –1994.

13.12 BEARING STRESS IN TIMBER

The bearing stress or compressive stress on a surface depends upon the inclination of surface with the direction of grain, bearing length, and distance from the end of a structural member. At any bearing of the side grain of timber, the permissible stress in compression perpendicular to the grain, f_{cn} is dependent on the length and position of bearing. The permissible stresses given in Table 13.4 for compression perpendicular to the grain are also the permissible stresses for bearings of any length at the ends of members and for the bearings 150 mm or more in length at any other position. For the bearings less than 150 mm in length and located 75 mm or more from the end of a member, the permissible compressive stress perpendicular to the grain may be multiplied by the modification factor K_7 given in Table 13.7 as per IS : 883 –1994. The allowance is not made for difference in intensity of the bearing stress due to bending of a beam. For the bearing stress under washer or a small plate, the same co-efficient recommended in Table 13.7 may be taken for bearing with a length equal to the diameter of the washer or the width of the small plate.

Table 13.7 *Modification factor K_7, for bearing stresses*

Length of bearing in mm	*Co-efficient*
15	1.67
25	1.40
40	1.25
50	1.20
75	1.13
100	1.10
150	1.00

When the direction of stress is at an angle to the direction of grain in any structural member as in roof trusses, as shown in Fig. 13.8, the permissible

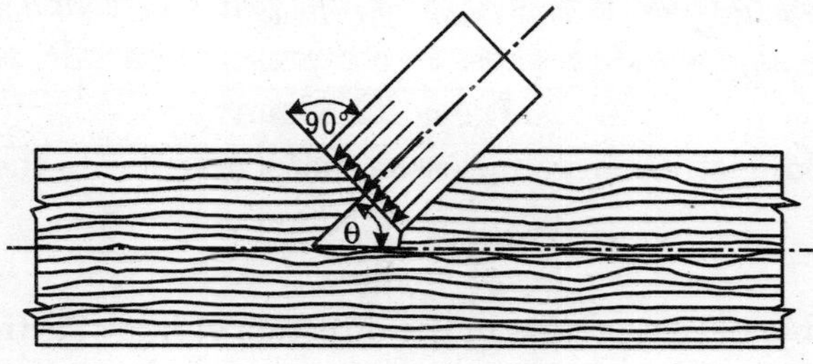

Fig. 13.8

bearing stress in that member is calculated from Hankinson's formula given below

$$f_{ce} = \left(\frac{f_{cp} \cdot f_{cn}}{f_{cp} \sin^2 \theta + f_{cp} \cos^2 \theta} \right) \qquad ...(13.1)$$

where, f_{ce} = permissible bearing stress in the direction of the line of action of load

f_{cp} = permissible compressive stress parallel to the grain
f_{cn} = permissible compressive stress perpendicular to the grain
θ = Angle of load with the direction of grain

Example 13.1 *The safe working stress in compression of a standard timber (dhaman) for inside location, parallel to the grain is 12 N/mm² and that perpendicular to the grain is 6 N/mm². Determine the safe working stress for this timber, if the timber is of selected grade and if the timber is of common grade (i.e., grade II).*

Solution

For standard grade. The safe working stress in compression parallel to the grain is 12 N/mm². The safe working stress in compression perpendicular to the grain is 6 N/mm².

For selected grade. The safe working stress in compression parallel to the grain

$$(1.16 \times 12) = \mathbf{13.92\ N/mm^2}$$

The safe working stress in compression perpendicular to the grain

$$(1.16 \times 12) = \mathbf{6.96\ N/mm^2}$$

For common grade. The safe working stress in compression parallel to the grain

$$(0.84 \times 12) = \mathbf{10.08\ N/mm^2}$$

The safe working stress in compression perpendicular to the grain

$$(0.84 \times 6) = \mathbf{5.04\ N/mm^2.}$$

Example 13.2 *The safe working stress in compression of a timber (dhaman) for inside location parallel to the grain is 12 N/mm² and that perpendicular to the grain is 6 N/mm². Determine the permissible bearing stress in the direction of line of action of the load acting at 30° to the grain direction.*

Solution The safe working stress in compression parallel to the grain

$$f_{cp} = 12\ \text{N/mm}^2$$

The safe working stress in compression perpendicular to the grain

$$f_{cn} = 6\ \text{N/mm}^2,$$
$$\theta = 30°.$$

The permissible bearing stress in the direction of line of action of the load as per Eq. 13.1

$$f_{ce} = \left(\frac{f_{cp} \cdot f_{cn}}{f_{cp} \cdot \sin^2\theta + f_{cp} \cdot \cos^2\theta}\right)$$

$$= \left(\frac{12 \times 6}{12 \times \frac{1}{4} + 6 \times \frac{3}{4}}\right) = \mathbf{9.6\ N/mm^2}$$

13.13 SOLID WOOD COLUMNS

The columns are defined as the structural members which support load primarily by inducing compressive stress along the grain.

The solid wood columns consist of a single piece of wood. Generally, these are of rectangular cross-section. The columns of round or circular cross-section are also known as solid wood columns. The round wood columns are not commonly used. The load carrying capacities of columns of round and square wood columns of equal cross-sectional area are same. The *slenderness ratio* $\frac{s}{d}$ in case of solid wood columns is defined as the ratio of unsupported length (effective length), s, of the column to the dimension of least side, d. For a rectangular section, the dimension of least side, d is the dimension of narrowest of two faces. For a circular section, the dimension of least side, d is the dimension of side of a square column of an equivalent cross-sectional area. For tapered column, the least dimension, d is taken as sum of the corresponding least dimensions at the small end of column and one-third of the difference between the least dimensions at the large end and small end. If D is the least dimension at the large end and D_2 is the least dimension at the small end, then the least dimension of tapered column, d is $[D_2 + \frac{1}{3}(D_1 - D_2)]$. The least dimension of tapered column should not be greater than one and a half times the least dimension at the small end (i.e. $d \not> 1.5\, D_2$). The induced stress at the small end tapered column should not exceed the safe compressive stress in the direction parallel to the grain. The solid wood columns are classified into the following three categories:

1. Short columns 2. Intermediate columns 3. Long columns.

The solid columns are formed of any section having a solid core throughout.

1. Short columns. When the slenderness ratio, $\left(\frac{s}{d}\right)$ is not greater than 11, then the columns are known as *short columns.* For short columns, the permissible stress is calculated as follows :

$$f_c = \left(\frac{P}{A}\right) = f_{c.p} \qquad ...(13.2)$$

where, P = Load of the column in N

A = Area of cross-section of column in mm^2

$f_{c \cdot p}$ = Stress in compression parallel to the grains.

2. Intermediate columns. When the slenderness ratio, $\left(\frac{s}{d}\right)$ is greater than 11, and less than or equal to K_8 then the columns are known as *intermediate columns.* For intermediate columns, the permissible stress is found as follows.

$$f_c = \left(\frac{P}{A}\right) = f_{c.p}\left[1 - \frac{1}{3}\left(\frac{s}{k_8 d}\right)^4\right] \qquad ...(13.3)$$

where, K_8 = Constant. It depends upon E and f_{cp}

$$K_8 = \left[\frac{E}{f_{c \cdot p}}\right]^{\frac{1}{2}}$$

E = Modulus of elasticity in N/mm².

3. Long columns. When the slenderness ratio, $\left(\frac{s}{d}\right)$ is greater than K_8, then the columns are called *long columns.* For long columns, the permissible stress is determined as follows :

$$f_c = \left(\frac{P}{A}\right) = \left(\frac{0.329E}{\left(\frac{s}{d}\right)^2}\right) \quad \text{...(13.6)}$$

In solid wood columns, the slenderness ratio $\frac{s}{d}$ should not exceed 50.

13.14 BUILT-UP AND BOX COLUMNS

Built up wooden columns consist of wooden pieces joined together with spikes, bolts, nails, screws or glue or with other mechanical fasteners. Built-up wooden columns are shown in Fig. 13.9. The strength of a built-up column isnot equal to that of a solid wood column. The built-up column should be fastened together at intervals not exceeding six times the thickness of each individual compression member by bolts and a limber connector or which are braced in direction parallel to the least dimension of the individual member.

Fig.13.9 *Built-up and box columns*

A rectangular box column is formed of members and a *hollow* core. The members arc joined with one another forming box provided with solid blocks at ends and at intermediate points.

The box columns are also classified into short, intermediate and long columns as follows :

(a) The built-up columns are called as the short columns when

$$\left[\frac{s}{\left(d_1^2 + d_2^2\right)^{1/2}}\right] \text{ is less than 8.}$$

where, d_1 = Least overall width of box column in mm

d_2 = Least overall dimension of core in box column in mm

(b) The built-up columns are called as the intermediate columns when

$$\left[\frac{s}{\left(d_1^2+d_2^2\right)^{1/2}}\right] \text{ is between 8 and } K_9 \text{ and}$$

(c) The built-up columns are called as the long columns when

$$\left[\frac{s}{\left(d_1^2+d_2^2\right)^{1/2}}\right] \text{ is greater than } K_9$$

where, K_9 is a constant

or
$$K_9 = \frac{\pi}{2}\left[\frac{U \cdot E}{5 \cdot q \cdot f_{cp}}\right]^{-1/2} \qquad \text{...(13.6)}$$

For the short columns, the permissible compressive stress is calculated as follows :

$$f_s = q \cdot f_{c.p} \qquad \text{...(13.7)}$$

For the intermediate columns the permissible compressive stress is calculated as follows :

$$f_s = q \cdot f_{cp}\left[1-\frac{1}{3}\left(\frac{s}{K_9 \cdot \left[d_1^2+d_2^2\right]^{\frac{1}{2}}}\right)^4\right] \qquad \text{...(13.8)}$$

For the long columns, the permissible compressive stress is calculated by using the following formula :

$$f_c = \frac{0 \cdot 329\, U \cdot E}{\left[\dfrac{s}{\left[d_1^2+d_2^2\right]^{\frac{1}{2}}}\right]^2} \qquad \text{...(13.9)}$$

where q and U are constants for the particular thickness of the plank. The following values of constant q and U are used depending upon plank thickness, t

t	U	q
25 mm	0.80	1.00
50 mm	0.60	1.00

13.15 SPACED COLUMNS

The spaced columns consist of two or more wooden members with their longitudinal axes parallel joined at their ends and at intermediate points by block pieces. The members are separated at ends by end blocks and at middle by spacer block. These are connected at the ends by timber connectors and bolts. Spaced columns are used as compression members in trusses in which timber connectors are used. Spaced column is shown in Fig. 13.10.

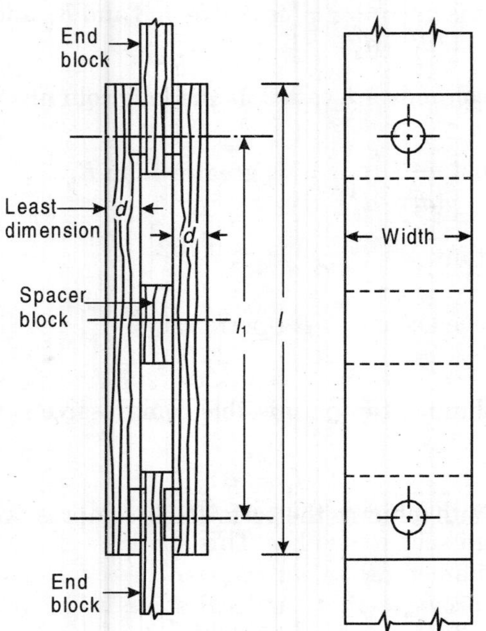

Fig. 13.10 *Spaced column*

The spacer block used for separation of members is placed at middle 10 per cent of length. The timber connectors are not required for a single spacer block. In case more than one block is used, the distance between two adjacent spacer blocks should not be greater than half the length between the end connectors. The timber connectors are necessary in case more than one spacer block are used. The end connectors may be provided either within 5 per cent of actual length. The dimensions of end blocks arc kept to accommodate the timber connectors.

In case the end blocks are provided within 5 per cent of actual length of individual members, then the effective length of columns is $\frac{l_1}{\sqrt{2 \cdot 5}}$. In case the end blocks are provided at a distance 5 to 10 per cent of actual length of individual members, then the effective length of column is $\frac{l_1}{\sqrt{3}}$. The slenderness ratio of a spaced column is the ratio of its effective length to the least dimension of its individual members. The safe load carrying capacity of spaced column is the sum of safe load carrying capacity of individual members.

The formulae for solid column specified in Sec. 13.13 are applicable to spaced column with a restraint factor of 2.5 or 3, depending upon the distance of end connectors in the column. For intermediate spaced column, the permissible compressive stress is given by

$$f_c = f_{c.p}\left[1-\frac{1}{3}\left(\frac{s}{k_{10}d}\right)^4\right] \quad ...(13.10)$$

where, K_{10} is a constant

or

$$K_{10} = 0.702\left[\frac{2\cdot 5E}{f_{c\cdot p}}\right]^{-1/2} \quad ...(13.11)$$

For the long spaced columns the permissible compressive stress shall be

$$f_c = \left(\frac{0\cdot 329E\times 2\cdot 5}{\left(\frac{s}{d}\right)^2}\right) \quad ...(13.12)$$

The compression members shall be notched, where it is necessary to pass service line through such a member. This shall be affected by means of a bored hole not larger in diameter than *one-quarter* the width of the face through which the hole is bored provided that the local stress is calculated and found to be within the permissible stress specified. The distance from the edge of the hole to the edge of member shall not be less than *one-quarter* of the hole of the face. The above recommendations and formulae are as per IS : 883 – 1970.

13.16 COLUMNS SUBJECTED TO COMBINED STRESS

Generally, the purlins are placed at panel points of the roof trusses. Sometimes, the purlins are placed at the points other than panel points. In such cases, principal rafters of the roof truss are subjected both to axial compression and bending. When a structural member is subjected both to axial compression and bending, then the compression,

$\left[\left\{\frac{f_{ac}}{f_{c\cdot p}}+\frac{f_{ab}}{f_b}\right\}\right]$ should not be greater than one.

i.e.,

$$\left[\left\{\frac{f_{ac}}{f_{c\cdot p}}+\frac{f_{ab}}{f_b}\right\}\right] \not> 1.00 \quad ...(13.13)$$

where, P = Total axial load in N

A = Area of cross-section in mm^2

$f_{ac} = \left(\frac{P}{A}\right)$ = Actual average stress in compression in the member

f_{cp} = Permissible stress in compression parallel to the grain

M = Bending moment

Z = Section modulus of the member in mm^2

$f_{ab} = \frac{M}{Z}$ = Actual maximum bending stress in the member

f_b = Maximum permissible stress in member.

Equation 13.13 is applicable to spaced column only when the bending moment is acting in the direction parallel to the mutually contacting surfaces of the blocks and individual members.

Example 13.3 *A column 150 mm × 150 mm is made of babul wood. The unsupported length of column is 1.50 m. Determine safe axial load on the column.*

Solution

Step 1: Slenderness ratio

Unsupported length of column, s = 1.50 m

Least dimension of column, d = 150 mm

Maximum slenderness ratio

$$\frac{s}{d} = \left(\frac{10 \times 1000}{150}\right) = 10 < 11$$

Step 2: Safe working stress

The column is a short column. Assuming that the column is used for inside location and wood is of standard grade. Safe working stress in compression parallel to the grain for babul wood

$$f_{cp} = 11.2 \text{ N/mm}^2$$

Step 3: Safe axial load in column

$$P = \left(\frac{11 \cdot 2 \times 150 \times 150}{1000}\right) = \mathbf{252.5\ kN}$$

Example 13.4 *In Example 13.3. If the unsupported length of column is 3 m, then, determine safe axial load on the column.*

Solution

Step 1: Slenderness ratio

Unsupported length of column, s = 3 m

Least dimension of column, d = 150 mm

Maximum slenderness ratio

$$\frac{s}{d} = \left(\frac{3 \times 1000}{150}\right) = 20$$

Step 2: Safe working stress

For babul wood, safe working stress in axial compression parallel to the grain,

$$f_{cp} = 11.2 \text{ N/mm}^2$$

and $$E = 10800 \text{ N/mm}^2$$

$$K_8 = 0.702\left[\frac{E}{f_{cp}}\right]^{\frac{1}{2}}$$

$$= 0.702\left[\frac{10800}{11.2}\right]^{\frac{1}{2}} = 21.8$$

The slenderness ratio of the column is greater than 11 and it is less than K_8. The column is considered as an intermediate column,

$$f_c = 11 \cdot 2\left[1 - \frac{1}{3}\left(\frac{3000}{21 \cdot 8 \times 150}\right)\right] = \mathbf{8.55\ N/mm^2}$$

Step 3: Safe axial load on the column

$$P = \left(\frac{8 \cdot 55 \times 150 \times 150}{1000}\right) = 192.5 \text{ kN}$$

Example 13.5 *In Example 13.4. If the effective length column is 3.60 m, then determine the safe axial load on the column.*

Solution

Step 1: Slenderness ratio

Effective length of column, $s = 3.6$ m

Least dimension of column, $d = 150$ mm

Maximum slenderness ratio,

$$\frac{s}{d} = \left(\frac{3 \cdot 60 \times 1000}{150}\right) = 24$$

For Example 13.4 $K_s = 21.8$

The slenderness ratio of the column is greater than K_8. The column is treated as a long column.

Step 2: Permissible stress on the column (from Eq. 13.5)

$$f_c = \frac{0 \cdot 329E}{\left(\frac{s}{d}\right)^2} = \left(\frac{0 \cdot 329 \times 10800}{(24)^2}\right)$$

$$= 6.18 \text{ N/mm}^2$$

Step 3: Safe axial load on the column

$$P = \left(\frac{6 \cdot 18 \times 150 \times 150}{1000}\right) = \mathbf{139\ kN}$$

Example 13.6 A *column 150 mm in diameter is made of deodar wood. The effective length of column is 1.20 mm. Determine the safe axial load of the round column. The column in situated in outside location and subjected to alternate wetting and drying.*

Solution

Step 1:. Slenderness ratio

Sectional area of round column

$$\frac{\pi}{4} \times (150)^2 = 16900 \text{ mm}^2$$

Let d be the dimension of a square column of equivalent cross-sectional area to that of round column

Sectional are of square column $= d^2$ mm^2

$$d^2 = 16900 \qquad \therefore d = 130 \text{ mm}$$

The effective length of column, s is 1.20 mm

The least dimension of the column, d is 130 mm

Maximum slenderness ratio

$$\frac{s}{d} = \left(\frac{120\times1000}{130}\right) = 9.24 < 11$$

Step 2: Safe working stress

The column is a short column. For deodar wood, safe working stress in axial compression parallel to the grains for outside location

$$f_{cp} = 7 \text{ N/mm}^2$$

Step 3: Safe axial load for the round column

$$P = \left(\frac{7\times130\times130}{1000}\right) = \mathbf{118.3\ kN}$$

Example 13.7 *A column carries an axial load of 500 kN inclusive of self weight. The effective length of column is 3.50 m. Design the wood column.*

Solution

Design :

The effective length of column, s is 3.50 m

Let d be the least dimension of the column. For a short column, $\left(\frac{s}{d}\right) \leq 11$

$$d \geq \left(\frac{3\cdot50\times1000}{11}\right) \geq 318.1 \text{ mm}$$

Adopt the least dimension of the column as 320 mm

Let fir wood be used for the solid column

Step 1: Safe working stress in compression parallel to the grain for firewood

$$f_{cp} = 5.2 \text{ N/mm}^2$$

Step 2: Sectional area required for column

$$\left(\frac{500\times1000}{5\cdot2}\right) = 96153.846 \text{ mm}^2$$

The least dimension of the column, d is 320 mm

Other dimension of the square column

$$\left(\frac{96153 \cdot 846}{320}\right) = 300.48 \text{ mm}$$

Adopt 320 mm × 320 mm fir wood column

Step 3: The load carrying capacity of the column

$$P = \left(\frac{52 \times 320 \times 320}{1000}\right) = 533 \text{ kN} > 500 \text{ kN}$$

Hence, safe.

Example 13.8 *A built-up sal wood column consists of a solid core 200 mm × 200 mm and four planks 50 mm × 50 mm. All the pieces are spiked together. The effective length of column is 3 m. Determine the safe axial load on the column.*

Solution :

Step 1: Slenderness ratio

The built-up column is shown in Fig. 13.11. All the elements of built-up column are joined together. The least overall width of box column, d_1 is 300 mm and the least overall dimension of core, d_2 in the box column is 200 mm. The slenderness ratio of the column

$$\frac{s}{(d_1^2 + d_2^2)^{\frac{1}{2}}} = \left(\frac{300}{(300^2 + 200^2)^{\frac{1}{2}}}\right) = 8.32$$

$$K_9 = \frac{\pi}{2}\left[\frac{U \cdot E}{5qf_{cp}}\right]^{\frac{1}{2}}$$

$$= \frac{\pi}{2}\left[\frac{0 \cdot 60 \times 127 \times 1000}{5 \times 1 \times 10 \cdot 6}\right]^{1/2}$$

$$= 18.825 \; (8 > 8.32 < K_9)$$

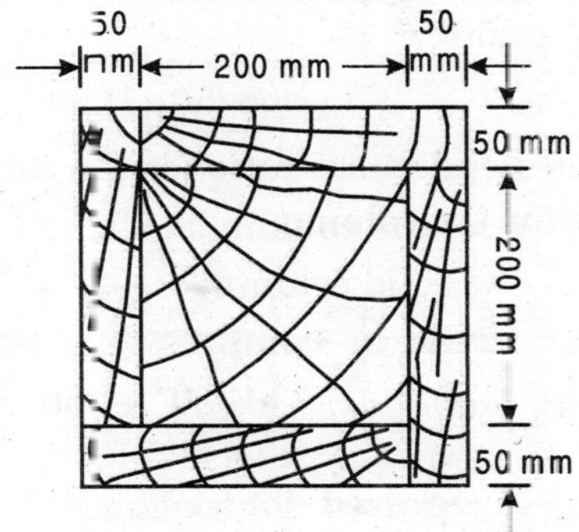

Fig.13.11 *Built-up column*

Step 2: Safe working stress

Hence, the built-up column is an intermediate column. The permissible compressive stress for the built-up intermediate column

$$f_c = q \cdot f_{cp}\left[1-\frac{1}{3}\left(\frac{s}{k_9\left(d_1^2+d_2^2\right)^{\frac{1}{2}}}\right)^4\right]$$

$$f_c = 1 \cdot 00 \times 10 \cdot 6\left[1-\frac{1}{3}\left(\frac{8 \cdot 32}{18 \cdot 825}\right)^4\right]$$

$$= 10.465 \text{ N/mm}^2$$

Step 3: Safe axial load, which may be carried by the built-up column

$$P = \left(\frac{10 \cdot 465 \times 300 \times 300}{1000}\right) = \textbf{941 kN}$$

Example 13.9 *A column carries an axial load of 1000 kN inclusive of self weight. The effective length of column is 3.60 m. Design a built-up sal wood column.*

Solution

Design:

Step 1: Slenderness ratio

The effective length of column, s is 3.60 m. Let the least overall width of box column be d_1 and the least overall dimension of core d_2 be 250 mm in the box column. In case, the column is designed as a short column, then

$$\left(\frac{s}{\left(d_1^2+d_2^2\right)^{\frac{1}{2}}}\right) \le 8$$

$$\left(\frac{3600}{\left(d_1^2+250^2\right)^{\frac{1}{2}}}\right) \le 8$$

$$\therefore \quad d_1 = 374 \text{ mm}$$

Let d_1 be 350 mm. The slenderness ratio for the column

The slenderness ratio for the column

$$\left(\frac{s}{\left(d_1^2+d_2^2\right)^{\frac{1}{2}}}\right) = \left(\frac{300}{\left(350^2+250^2\right)^{\frac{1}{2}}}\right) = 8.37$$

$$K_9 = \frac{\pi}{2}\left[\frac{U \cdot E}{5q \cdot f_{cp}}\right]^{\frac{1}{2}}$$

Let the sal wood be used for the column,

$$K_9 = \frac{\pi}{2}\left[\frac{0\cdot 60\times 127\times 1000}{5\times 1\times 10\cdot 6}\right]^{\frac{1}{2}} = 18.825$$

$$(8 > 8.37 > 18.825\ )$$

Step 2: Safe working stress

Hence, the built-up column shall be intermediate column. The permissible compressive stress for the built-up intermediate column

$$f_c = q\cdot f_{cp}\left[1-\frac{1}{3}\left(\frac{s}{k_9\left(d_1^2+d_2^2\right)^{\frac{1}{2}}}\right)^4\right]$$

$$f_c = 1\times 10\cdot 6\left[1-\frac{1}{3}\left(\frac{8\cdot 37}{18\cdot 825}\right)^4\right]$$

$$= 10.4619\ \text{N/mm}^2$$

Step 3: Area required for the column

$$A = \left(\frac{1000\times 1000}{10\cdot 4619}\right) = 95584.8\ \text{mm}^2$$

Provide 2 planks 350 mm × 50 mm and 7 planks 250 mm × 50 mm as shown in Fig. 13.12. The area provided is 350 mm × 350 mm (i.e., 122500 mm²)

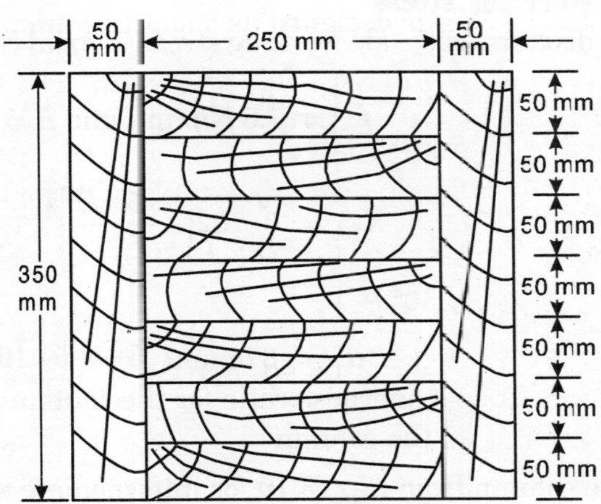

Fig. 13.12 *Built-up column*

Step 4: Safe axial load

$$P = \left(\frac{122500\times 10\cdot 4619}{1000}\right)$$

$$= 1281.59\ \text{kN} > 1000\ \text{kN}$$

Example 13.10 *A spaced column is 3.6 m long. It carries an axial compression of 45 kN. Design the column. Use deodar wood for the spaced column.*

Solution

Design

Step 1: Slenderness ratio

Actual length of spaced column,

$$l = 3.60 \text{ m}$$

Provide end connectors at 5 per cent distance

Distance of end connectors

$$\left(\frac{5}{100}\times 3\cdot 60\times 1000\right) = 180 \text{ mm}$$

Distance between centre to centre of connectors

$$l_1 = (3600 - 2 \times 180) = 3240 \text{ mm}$$

Effective length of column

$$s = \left(\frac{l_1}{\sqrt{2\cdot 5}}\right) = \left(\frac{3240}{\sqrt{2\cdot 5}}\right) = 2050 \text{ mm}$$

Let the least dimension of the spaced column be 60 mm.

Slenderness ratio of column

$$\left(\frac{2050}{60}\right) = 34.16 > 11$$

Step 2: Safe working stress

From code, for deodar wood, safe working stress in axial compression parallel to the grain

$$f_{cp} = 7.8 \text{ N/mm}^2, \text{ and } E = 9.5 \text{ kN/mm}^2$$

$$K_{10} = 0\cdot 702\left[\frac{2\cdot 5\times E}{f_{cp}}\right]^{\frac{1}{2}}$$

$$= 0\cdot 702\left[\frac{2\cdot 5\times 9\cdot 5\times 10^3}{7\cdot 8}\right]^{\frac{1}{2}} = 38.74$$

$\left(\frac{s}{d}\right) < K_{10}$. The column from Eq. 13.10 for intermediate spaced column

$$f_c = \left[1-\frac{1}{3}\left(\frac{34\cdot 16}{38\cdot 76}\right)^4\right] = 6.228 \text{ N/mm}^2$$

Step 3: Area required for column

$$\left(\frac{45\times 1000}{6\cdot 228}\right) = 7225.24 \text{ mm}^2$$

Use 2 planks. Area required for one plank

$$\left(\frac{7225\cdot 24}{2}\right) = 3612.62 \text{ mm}^2$$

Least dimension assumed = 60 mm

Step 4: Width of plank required

$$\left(\frac{3612\cdot 62}{60}\right) = 60.21 \text{ mm}$$

Provide 2 planks 80 mm × 60 mm as shown in Fig. 13.13. The spacer block is placed at the centre. The spacer block does not require connectors. The spacer

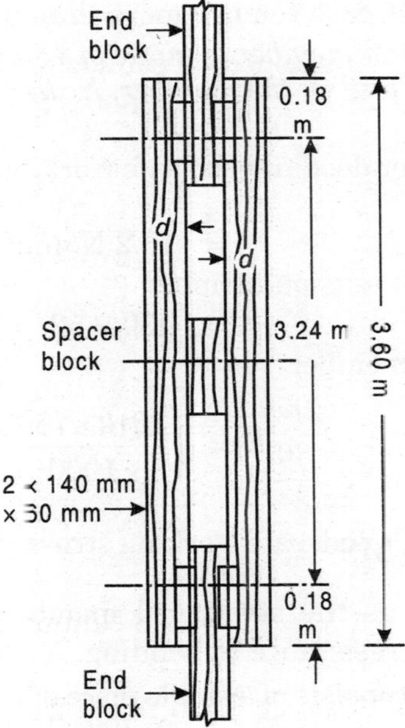

Fig. 13.13 *Spaced column*

block is kept in position by a bolt. The end blocks are provided at the ends with suitable connectors.

13.17 TENSION MEMBERS

Ordinarily, the wood is not used for tension members because steel tie rods are more economical. With the use of modern timber connectors it is possible to have wooden tension members. The safe working stresses in axial tension along the grain for different timbers are given in IS : 883–1970.

When a structural member is subjected both to axial tension and bending, then, the expression

$$\left[\frac{f_{at}}{f_t} + \frac{f_{ab}}{f_b}\right] \text{ is not greater than one,}$$

where, f_{at} is actual average stress in tension in the member and f_t is permissible stress in axial tension

$$\left[\frac{f_{at}}{f_t} + \frac{f_{ab}}{f_b}\right] \not> 1.0 \qquad \text{...(13.14)}$$

The other notations are same as mentioned in Sec. 13.6.

The net section is used for the tension member. The net section is obtained by deducting from the gross-section the projected area of all material removed by boring, grooving or other means. The net area used in calculating load carrying capacity of a member shall be the least net section determined as above passing a plane for a section or a series of connected planes transversely through the members.

The notches shall in no case remove more than one quarter of the section.

Example 13.11 *A tension member is made of 150 mm × 50 mm deodar wood. Determine the safe axial pull in the member. Nails are used for connections.*

Solution

From IS : 883 –1970, for deodar wood safe working stress in tension along the grain for inside location

$$= 10.2 \text{ N/mm}^2$$

Cross-sectional area of tension member

$$= (150 \times 50) = 7500 \text{ mm}^2$$

Safe axial pull in the member

$$P = \left(\frac{10 \cdot 2 \times 7500}{1000}\right) = \mathbf{76.5\ kN}$$

13.18 WOODEN BEAMS

The beams are defined as the structural members which support the load primarily by its internal resistance to bending.

Wooden beam usually consists of a single piece of rectangular section as shown in Fig. 13.14.

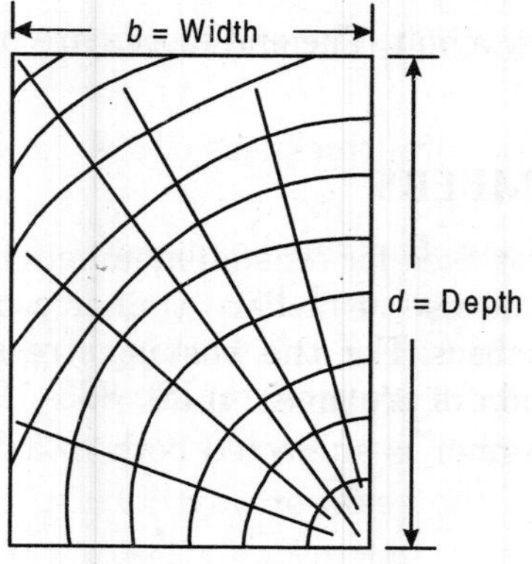

Fig. 13.14 *Wooden beam*

The **effective span** of beams and other flexural members shall be taken as the distance from face of the supports plus one half of the required length of bearing at each end except that for continuous beams and joists, the span may be measured from centre of bearing at those supports over which the beam is continuous.

When the span of beam and load combination is such that the beams of large dimensions are required. It is not possible to have wooden beams of large dimensions. The *built-up* wooden beams are formed by connecting a number of smaller beams together by bolts, screws or spikes as shown in Fig. 13.15. Bolts are staggered and spaced longitudinally at a distance less than four times the depth of beams. The large dimension of individual beam is kept vertical.

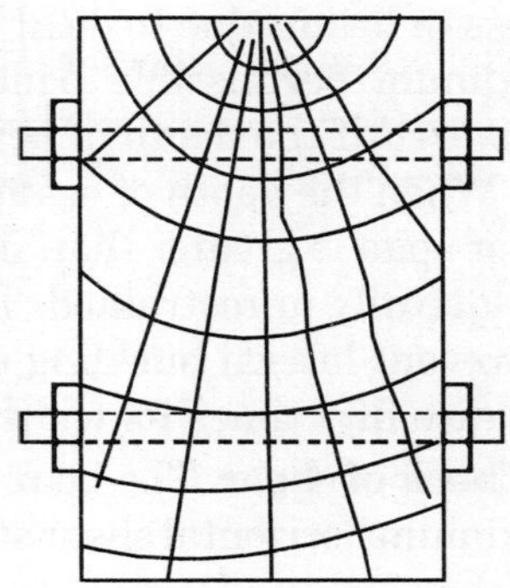

Fig. 13.15 *Built-up beam*

The following *form factors* are applied to the bending stress for the following cross-sections of the beams as per IS : 883 –1970.

(a) **Rectangular section.** For rectangular sections for different depth of beams, the form factor K_3 shall be taken as

$$K_3 = 0.81\left(\frac{D^2+89400}{D^2+55000}\right) \quad ...(13.15)$$

It is to note that form factor K_3 shall not be applied for beams having depth less than or equal to 300 mm.

(b) **Box beams and I-beams.** For the box beams and I-beams the form factor K_4 shall be obtained by using the formula

$$K_4 = 0\cdot8+0\cdot8\gamma\left(\frac{D^2+89400}{D^2+5500}-1\right) \quad ...(13.16)$$

where $y = p_1^2(6-8p_1+3p_1^2)(1-q_1)+q_1$ ...(13.17)

p_1 = Ratio of thickness of the compression flange to the depth of the beam.

q_1 = Ratio of the total thickness of web or webs to the overall width of the beam.

(c) **Solid cross sections.** For the beams of solid circular cross-sections, the form factor K_5 shall be taken as 1.18.

(d) **Square cross sections.** For the beams of square cross-section where the load is in the direction of diagonal, the form factor K_6 shall be taken as 1.414.

The minimum width of the beam or any flexural member shall not be less than 50 mm or $\frac{1}{50}$ of the span whichever is greater. All the flexural member having a depth exceeding three times its width and/or a span exceeding fifty times its width shall be laterally restrained from twisting or buckling and the distance between such restraints shall not exceed 50 times its width.

The wooden beams are designed for maximum bending stress. The bending stress in beams due to dead load and super-imposed load should not exceed the maximum permissible bending stress. The beams are considered *laterally supported* if the depth of beam, d is less than three times its width b, (i.e., $d < 3b$). When the depth of beam d, is greater than three times its width b, (i.e., $d > 3d$) or span is greater than fifty times its width b, (i.e., $l > 50b$), then, the beams are laterally unrestrained. The lateral buckling occurs in such beams. In order to prevent lateral buckling of such beams, lateral restraints are provided at a distance fifty times its widths.

Check of shear. The beams designed for bending are checked for shear. The maximum horizontal shear stress occurs at the neutral axis and it can be obtained by

$$f_{sh} = \left(\frac{V \cdot Q}{I \cdot b}\right) \qquad \text{...(13.18)}$$

where, f_{sh} = Horizontal shear stress in beams in N/mm^2
V = Vertical shear at the section in N
b = Width of beam section in mm
D = Depth of beam section in mm
I = Moment of inertia of section in mm^4
Q = Statical moment of area above the level under consideration about neutral axis in mm^3

Equation 13.18 gives the general formula for calculating horizontal shear. For the following types of beams, the maximum shear stress may be found by using the following formulae :

(a) **Rectangular beams**

$$f_{sh} = \frac{3}{2} \cdot \left(\frac{V}{b \cdot D}\right) \qquad \text{...(13.19)}$$

(b) **Notched beams, notched at tension face at the support**

$$f_{sh} = \frac{3}{2} \cdot \left(\frac{V \cdot D}{b \cdot D_1^2}\right) \qquad \text{...(13.20)}$$

where D_1 = Depth of beam at notch in mm
D_2 = Depth of notch.

(c) **Notched at the compression face (where, $e > D$)**

$$f_{sh} = \frac{3}{2} \cdot \left(\frac{V}{b \cdot D_1}\right) \qquad \text{...(13.21)}$$

(d) **Notched at compression face (where, $e > D$)**

$$f_{sh} = \left(\frac{3V}{2b\left[D - \left(\frac{D_2}{D}\right)e\right]} \right) \quad ...(13.22)$$

where, e = Length of the notch along the beam span from the inner edge of the support to the farthest edge of the notch in mm.

The maximum horizontal shear when the load on a beam moves from the support towards the centre of the span, and the load is at a distance of *three* to *four* times the depth of beam from the support shall be calculated from the formula

$$f_{sh} = \left(\frac{V \cdot Q}{I \cdot b}\right)$$

For rectangular beam

$$Q = \left(\frac{1}{2} \times b \times D \times \frac{D}{4}\right) = \frac{1}{8} b \cdot D^2 \text{ and}$$

$$I = \frac{1}{12} b \cdot D^3$$

$$\therefore \quad f_{sh} = \left(\frac{V \cdot Q}{I \cdot b}\right) = \frac{3}{2}\left(\frac{V}{b \cdot D}\right)$$

The value of V shall be calculated from the following formula.

For concentrated loads

$$V = \left(\frac{10C(l - x)\left(\frac{x}{D}\right)^2}{9l\left[2 + \left(\frac{x}{D}\right)^2\right]} \right) \quad ...(13.23)$$

where, C = Concentrated load in N

l = Span of beam in mm

x = Distance from reaction to load in mm

For uniformly distributed loads

$$V = \frac{W}{2}\left(l - \frac{2D}{l}\right) \quad ...(13.24)$$

The maximum horizontal shear stress should not exceed the maximum allowable shear stress in the wood.

The maximum shear in a simply supported beam carrying uniformly distributed load occurs at the edge of the support. When the beams are carrying concentrated loads, then these concentrated loads which are acting in the vicinity of the supports, are reduced by the percentage mentioned in Table 13.8.

Table 13.8 *Percentage of reduction for concentrated loads in the vicinity of supports*

Distance of load from the nearest support	*1.5 D or less*	*2D*	*2.5D*	*3D or more*
Percentage reduction	60	40	20	No reduction

D = Depth of beam

All the uniformly distributed loads within a distance equal to the depth of the beam from edge of the nearest support may be neglected.

13.18.1 End Bearing of Beams

The ends of beams are supported in recess and these are not enclosed. The recess provides proper ventilation. The bearing stress perpendicular to the grain at supports and under concentrated loads should not be greater than safe working stress in compression across the grain. When the beams are supported directly on masonry or concrete', then a minimum of 75 mm bearing length should be provided. The timber joists or floor planks supported on top flange of steel beams should rest on and be secured to a timber plate having a minimum cross-section of 75 mm × 40 mm. The timber joists should not be supported on the top flange of steel beams unless the bearing stress calculated on the net bearing as shaped to fit the beam, is not greater than the permissible compressive stress perpendicular to the grain.

13.18.2 Check for Deflection

The deflection in case of beam joists, purlins, battens, and other flexural members supporting brittle material like gypsum ceiling slates, tiles and asbestos sheets should not be greater than $\left(\frac{1}{360}\right)$ of the span. The deflection in case of other flexural members should not be greater than $\frac{1}{240}$ of span. For the cantilever beams, the deflection should not be greater than $\left(\frac{1}{180}\right)$ or span. For checking up the deflection in case of beams and joists, the loads taken shall be twice the dead load plus 0.75 times live load.

Notched beams. When a groove is cut either at the ends or at the middle of span or anywhere in between support in the timber beams, then beams are known as *notched beams.* The beams are cut or notched at the ends to reduce the depth of floors. The ends of beams are also notched to bring top surfaces level with adjacent beams. Sometimes, the beams are notched to increase the room clearance. The beams are notched at the middle or anywhere between the supports to provide space for pipes or support to other beams and frames. The notched beams are shown in Fig. 13.16. The cross sectional area of beams at notches are reduced.

When the beams are carrying uniformly distributed loads only, then it is not necessary to calculate the effect of notches if their depth and situations are within the limits specified by IS : 883 –1970. Unless the local stress is calculated and found to be within the permissible stress, flexural member shall not be cut, notched or bored except as follows :

The notches cut at the top or bottom of the beams should not be deeper than one-fifth of depth of the beam. The notches should not be cut at distances more than one-sixth of the span from the edges of support. If notches occur at a distance greater than three times the depth of beam from the edge of nearest support

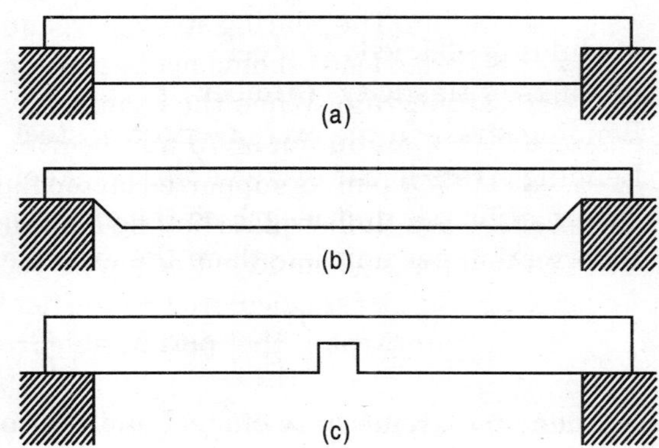

Fig. 13.16 *Notched beams*

the net depth should be used in determining the bending strength, since the modulus of section at the notch reduces. The holes at larger in diameter than *one-quarter* of the depth may be bored in the middle third of the depth and length.

The shear strength of notched beams decreases. The decrease in strength of notched beams depend upon the shape of notch and on the relation of the depth of notch to the depth of beam. Concentration of stress occurs at the notch. The shear stress in case of square notch at the ends as shown in Fig. 13.16 (a) is given by Eqs. 13.20, 13.21 and 13.22.

Flitched beams. The flitched beams consist of wooden beams and steel beams joined together by means of bolts or screws. The flitched beams are shown in Fig. 13.17. When the timber sections are joined together with steel plates, then the deformations in the fibres of timber and steel in flitched beams are equal.

Consider a flitched beam shown in Fig. 13.17 (b). The modulus of elasticity of a material is defined as stress divided by unit deformation.

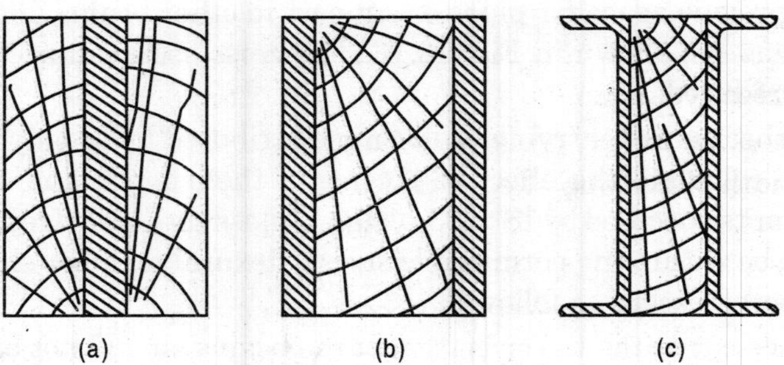

Fig. 13.17 *Flitched beams*

Then
$$E_s = \frac{f_s}{\Delta_s} \text{ and } E_w = \frac{f_w}{\Delta_w}$$

where, E_s = Modulus of elasticity of steel
E_w = Modulus of elasticity of timber
f_s = Bending stress in the extreme fibre of steel
f_w = Bending stress in the extreme fibre of steel
Δ_s = Deformation per unit length on the extreme fibre of steel
Δ_w = Deformation per unit length on the extreme fibre of timber

or
$$\Delta_s = \left(\frac{f_s}{E_s}\right) \text{ and } \Delta_w = \left(\frac{f_w}{E_w}\right)$$

The unit deformation on extreme fibre of steel and that on extreme fibre of timber are equal

$$\Delta_s = \Delta_{w,} \left(\frac{f_s}{E_s} = \frac{f_w}{E_w}\right) \quad \text{...(13.25)}$$

$$\therefore \quad f_w = f_s \times \left(\frac{E_w}{E_s}\right)$$

The ratio of $\left(\frac{E_w}{E_s}\right)$ is known as the modular ratio.

From Eq. 13.25, it is seen that bending stress on the extreme fibre of timber is equal to the bending stress on the extreme fibre of steel multiplied by $\left(\frac{E_w}{E_s}\right)$. When the bending stress in extreme fibre of steel is maximum, then the bending stress in timber on the extreme fibre should be less than or equal to the maximum allowable bending stress in the timber.

Example 13.12 *A deodar timber beam carries a uniformly distributed load 16 kN/m inclusive of self-weight of the beam. The beam is simply supported at both ends. The clear span of beam is 5 m. Design the timber beam.*

Solution

Design

Step 1: Effective span

Clear span of beam $= 5$ m

Assume width of bearing at each end

$= 300$ mm

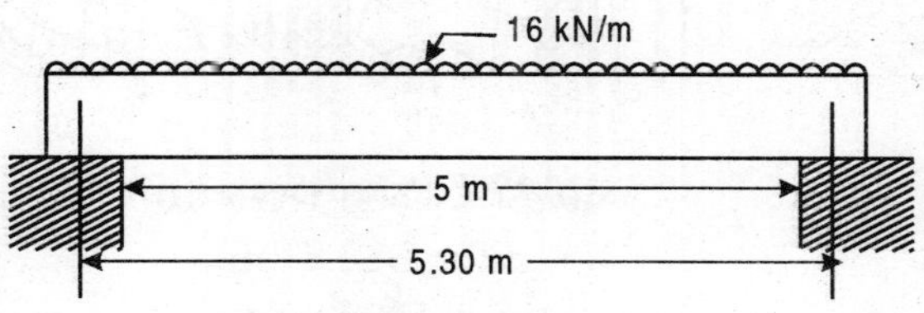

Fig. 13.18

Effective span of beam

$$= \left(5+\frac{1}{2}\times 0\cdot 30+\frac{1}{2}\times 0\cdot 30\right) = 5.30 \text{ mm}$$

Step 2: Maximum bending moment

$$M = \left(16\times\frac{5\cdot 30\times 5\cdot 30}{8}\right) = 56.18 \text{ kN-m}$$

The maximum allowable bending stress along the grain for inside location for deodar wood of standard grade

$= 10.2$ N/mm^2

From factor for rectangular section : K_3

Assuming the depth of beam as 400 mm, from Eq. 13.15

$$K_3 = 0.81\left(\frac{D^2+89400}{D^2+55000}\right)$$

$$K_3 = 0.81\left(\frac{160000+89400}{160000+55000}\right) = 0.9396$$

Maximum allowable bending stress, (since $D > 300$ mm)

$= (0.9396 \times 10.2) = 9.584$ N/mm^2

Step 3: Section modulus required

$$Z = \left(\frac{5\cdot 618\times 1000\times 1000}{9\cdot 584}\right)$$

$= 5861.85 \times 10^3$ mm^3

The beam is considered laterally supported if

$$b > \frac{1}{50}\times \text{span, and } b > \frac{d}{3}$$

Width of beam required

$$b = \left(\frac{1}{50}\times 5\cdot 30\times 1000\right) = 106 \text{ mm}$$

Adopt $b = 240$ mm

The modulus of section of rectangular beam

$$\frac{1}{6}b\cdot d^2 = \left(\frac{1}{6}\times 240\times d^2\right) = 40\,d^2$$

Therefore,

$$40\,d^2 = 586.185 \times 10^3$$

$$d = 382.81 \text{ mm}$$

Adopt $d = 400$ mm

$$\frac{d}{3} = \left(\frac{400}{3}\right) = 133.3 \text{ mm} < b$$

Step 4: Check for shear

Maximum shear force at the edge of the support

$$V = \left(\frac{16\times 5\cdot 30}{2} - (0\cdot 15 + 0\cdot 4)\times 16\right)$$

$$= 33.6 \text{ kN}$$

Maximum shear stress in the beam, from Eq. 13.19

$$f_{sh} = \frac{3}{2}\times\left(\frac{V}{b\cdot d}\right)$$

$$= \frac{3}{2}\times\left(\frac{33\cdot 6\times 1000}{240\times 400}\right) = 0.5166 \text{ N/mm}^2$$

Step 5: Check for deflection

Maximum deflection

$$y_{max} = \frac{5}{384}\times\left[\frac{(16\times 5\cdot 30)(5\cdot 30)^3\times(1000)^3}{9\cdot 5\times 1000\times\frac{1}{12}\times 240\times(400)^3}\right] = 13.518 \text{ mm}$$

Allowable deflection

$$\left(\frac{1}{240}\times 5\cdot 30\times 1000\right) = 22.1 \text{ mm} > y_{max}. \text{ Hence, satisfactory.}$$

Step 6: Check for bearing

Reaction at the support

$$\left(\frac{16\times 5\cdot 30}{2}\right) = 42.40 \text{ kN}$$

Bearing stress at the support

$$\left(\frac{42\cdot 40\times 1000}{300\times 240}\right) = 0.578 \text{ N/mm}^2$$

Safe working stress in compression perpendicular to the grain

$= 2.6$ N/mm²

$>$ Bearing stress. Hence, satisfactory.

Provide a rectangular beam 240 mm × 400 mm.

Example 13.13 *A simply supported timber beam carries a total uniformly distributed load of 50 kN inclusive of self weight. The effective span of beam is 8 m. The timber beam is made of sal wood. 300 mm × 50 mm planks are only available. Design a built-up beam.*

Solution

Design :

Step 1: Effective span

The strength of a built-up timber beam is equal to sum of the strength of each element of the built-up beam.

Total uniformly distributed load inclusive of self weight

$= 50$ kN

Effective span $= 8$ m

Step 2: Maximum bending moment

$$M = \left(\frac{WL}{8}\right) = \left(\frac{50\times 8000}{8\times 1000}\right) = 50 \text{ kN-m}$$

From IS : 883 –1970, safe working stress in bending for inside location and standard grade sal wood = 16.8 N/mm².

Step 3: Sectional modulus required

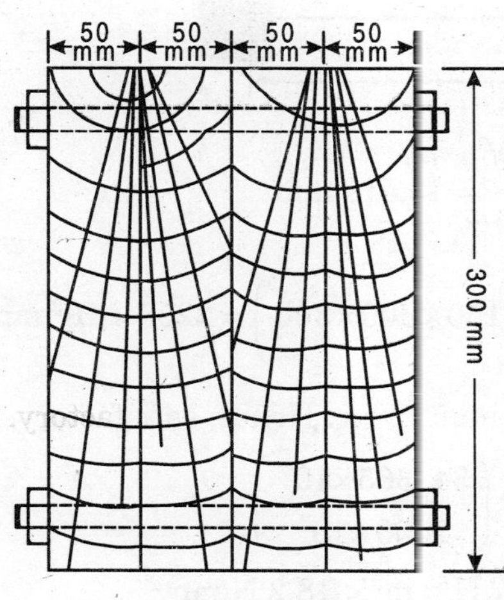

Fig. 13.19

$$Z = \left(\frac{50\times 10^6}{16\cdot 8}\right) = 2976.19 \times 10^3 \text{ mm}^3$$

Section modulus of 300 mm × 50 mm plank

$$Z_1 = \left(\frac{1}{6}\times 50\times 300\times 300\right)$$

$$= 750 \times 10^3 \text{ mm}^3$$

Number of planks required

$$\frac{Z}{Z_1} = \frac{2976\cdot 19\times 10^3}{750\times 10^3} = 3.968$$

Provide 4 planks of 300 mm × 50 mm as shown in Fig. 13.19. The planks are bolted together. The bolts are staggered. The bolts are spaced 1.10 m apart (less than four times the depth of the beam).

Example 13.14 *A 150 mm × 300 mm timber beam has an effective span of 5.30 m. It carries a concentrated load of 25 kN at the centre. The beam has 150 mm × 50 mm notches at the ends as shown in Fig. 13.20. Verify that whether the beam is safe in bending and shear or not. The beam consists of dhaman wood of Group A.*

Solution

Step 1: Effective span and load

Effective span of beam = 5.30 m

Concentrated load at the centre = 25 kN

Self weight of beam $= \left(\dfrac{150 \times 300 \times 5 \cdot 30 \times 7 \cdot 85}{1000 \times 1000}\right) = 1.872$ kN

Step 2: Bending moment

Maximum bending moment due to concentrated load

$$M_1 = \left(\frac{25 \times 5 \cdot 30}{4}\right) = 38.125 \text{ kN-m}$$

Maximum bending moment due to self-weight

$$M_2 = \left(\frac{1 \cdot 872 \times 5 \cdot 30}{8}\right) = 1.24 \text{ kN-m}$$

Total bending moment at the centre

$$M = M_1 + M_2 = (33.125 + 1.24) = 34.365 \text{ kN-m}$$

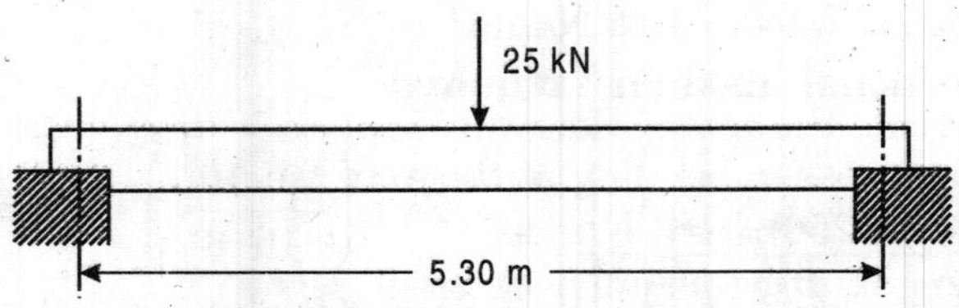

Fig. 13.20 *Notched beam*

Step 3: Section modulus of beam

$$Z = \left(\frac{1}{6} \times 150 \times 300 \times 300\right) = 2250 \times 10^3 \text{ mm}^3$$

Actual stress in bending

$$f_b = \frac{M}{Z} = \left(\frac{34 \cdot 365 \times 10^6}{2250 \times 10^3}\right)$$

$$= 15.27 \text{ N/mm}^2 < 18.2 \text{ N/mm}^2$$

(Safe working stress in bending in dhaman wood)

Hence, the beam is safe in bending.

Step 4: Maximum shear force

$$V = \left(\frac{25}{2} + \frac{1 \cdot 872}{2}\right) = 13.436 \text{ kN}$$

Total depth of beam d is 300 mm.

Size of notch = 150 mm × 50 mm

Depth of beam above notch

$$d_s = 250 \text{ mm}$$

Step 5: Shear stress in beam, from Eq. 13.20

$$f_s = \left(\frac{3}{2}\frac{V}{b} \cdot \frac{D}{D_1^2}\right)$$

$$f_s = \frac{3}{2} \times \left(\frac{13 \cdot 436 \times 1000 \times 300}{150 \times 250 \times 250}\right)$$

$$= 0.645 \text{ N/mm}^2 < 1.3 \text{ N/mm}^2$$

(Safe working stress in shear in shear in dhaman wood)

Hence, the beam is safe in shear. The beam is found safe in bending and shear both.

Example 13.15 A *beam is simply supported at its both the ends. The effective span of beam is 6 mm. It consists of 200 mm × 300 mm teak wood with 300 mm × 12 mm steel plates bolted to its sides as shown in Fig. 13.21. Determine the safe uniformly distributed load, which the beam will support.*

Solution :

Step 1: Modulus of section of two steel plates

$$Z = \left(2 \times \frac{1}{6} \times 12 \times 300^2\right) = 360 \times 10^3 \text{ mm}^3$$

It is assumed that the beam is laterally supported. For the laterally supported beam, allowable stress in bending at the extreme fibre, in steel

$$f_b = (0.66 \times 250) = 165 \text{ N/mm}^2$$

Step 2: Moment of resistance of steel plates

$$M = f_b . Z = 165 \times \left(\frac{360 \times 10^3}{1000 \times 1000}\right)$$

$$= 59.4 \text{ kN-m}$$

Step 3: Load supported by beam

Let W be the uniformly distributed load, supported by the steel plates.

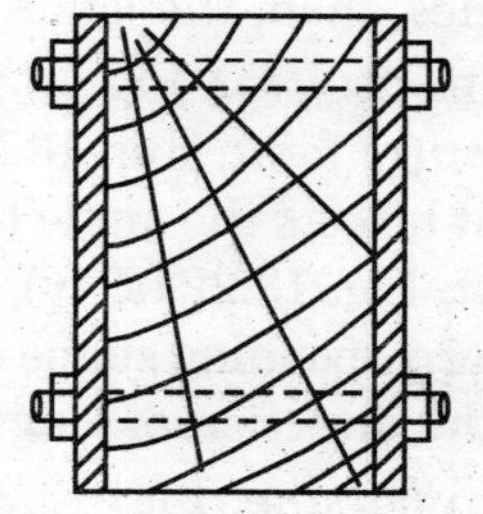

Fig. 13.21 *Flitched beam*

$$\therefore \quad \frac{WL}{8} = 59.4 \times 10^6$$

$$\therefore \quad W = \left(\frac{59 \cdot 4 \times 10^6 \times 8}{6000 \times 1000}\right) = 79.2 \text{ kN}$$

From IS : 883 –1970, for teak wood, modulus of elasticity = 9600 N/mm². Modulus of elasticity for steel = 2.06×10^5 N/mm².

Step 4: Stress in the extreme fibre of wood, (from Eq. 13.25)

$$f_w = f \times \frac{E_w}{E_s} = 165 \times \left(\frac{9600}{2 \cdot 06 \times 10^5}\right)$$

$$= 7.689 \text{ N/mm}^2 < 14.0 \text{ N/mm}^2$$

(Allowable stress in bending for teak-wood)
Modulus of section of wooden beam

$$Z_1 = \left(\frac{1}{6} \times 200 \times 300^2\right) = 3000 \times 10^3 \text{ mm}^3$$

Moment of resistance of wooden beam

$$M_1 = f_w \times Z_1 = \left(\frac{7 \cdot 689 \times 3000 \times 10^3}{1000 \times 1000}\right)$$

$$= 23.067 \text{ kN-m}$$

Let W_1 be the uniformly distributed load, supported by wooden beam

$$\frac{W_1 L}{8} = 23.067 \times 10^6$$

or

$$W_1 = \left(\frac{23 \cdot 067 \times 10^6 \times 8}{6000 \times 1000}\right) = 30.756 \text{ kN}$$

The uniformly distributed load inclusive of the self-weight of flitched beam

$$W + W_1 = (79.2 + 30.756) = \mathbf{109.956 \text{ kN}}$$

13.19 FRAMED JOINTS

Two pieces of timber are connected by *framed joints* as shown in Fig. 13.22. In framed joints, when two pieces are joined at their ends, then the joints are known as *corner joints* or *angle joints* as shown in Fig. 13.22 (a). When two pieces are joined in between the length of one or both the members at right angles, then, the joints are known at *T joints* as shown in Fig.13.22 (b).

In *butt joint,* Fig. 13.22 (a) (i), one piece comes squarely against another piece. In *mitre joint,* Fig. 13.22 (ii), two pieces are bevelled and joined. The plane of joint bisects the angle between them as at the cornei of photo frame. In *dovetail joint,* Fig. 13.22 (a) (v), one piece has the shape of wedge. It fits into the groove of corresponding shape in another piece. Wedge shape piece cannot be withdrawn in the direction of length.

In *halving,* Fig. 13.22 (b) (i), both pieces are cut to one-half of their depth. When both pieces are joined, the lower and upper surfaces remain flush. In *bevelled halving,* Fig, 13.22 (b) (ii), both pieces are cut and their surfaces of contact are kept inclined. In *tenon* and mortise joint, Fig. 13.22 (b) (iii), one piece has a tenon or rectangular projection on its end. It is inserted into a socket or mortise in another piece. In *double notching,* Fig. 13.22 (b) (iv) notches are

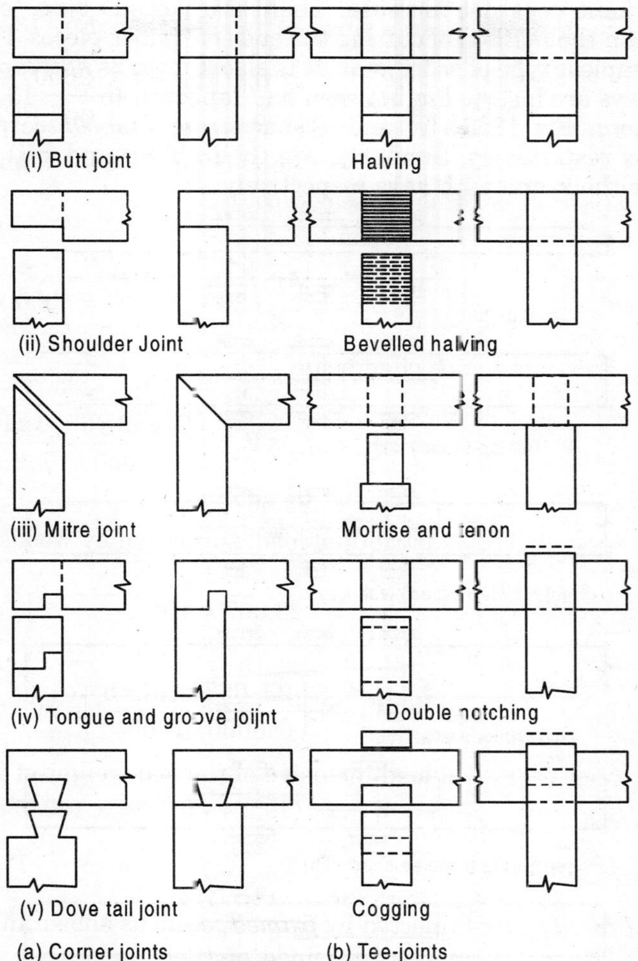

Fig. 13.22 *Framed joints*

prepared by cutting grooves in the pieces. On piece is inserted into another piece. In *cogging,* Fig. 13.22 (b) (v), groove does not extend entirely across the width of one piece. These joints are used in the frames of doors, windows and ventilators.

13.20 LAP, SCARF AND FISH PLATE JOINTS

Two pieces of timber are joined along the longitudinal direction by *lap joint* or *scarf joint* or *fish plate joint.* When joint is provided along the length of a member, then it is known as *splice* or *lengthening joint.* In lap joint, one piece of timber is placed over another piece and the pieces are joined together by nails or bolts as shown in Fig. 13.23 (a). Lap joint is used to transmit tension or compression of small values only. Lap joint is not commonly used. In *scarf joint,* two timber pieces are cut away at their ends on opposite sides so that these may be lapped and joined together by nails or bolts as shown in Fig. 13.23 (b). In scarf joint, the thickness at joint does not increase. There are various types of scarf joints depending upon the method of cutting the ends of timber pieces. Figure 13.23 b (i) shows a simplest type of scarf joint. It is also known as *half lap joint.* Figure 13.23 b (ii), keys are inserted in between half-lap joint. In Fig. 13.23 (iii) shows *oblique scarf joint.* Fig. 1323 b (iv) and (v) show *straight tabled scarf joint* without and with keys respectively. Similarly, Fig. 13.23 b (vi) and (vii) show *oblique tables scarf* without and with keys respectively.

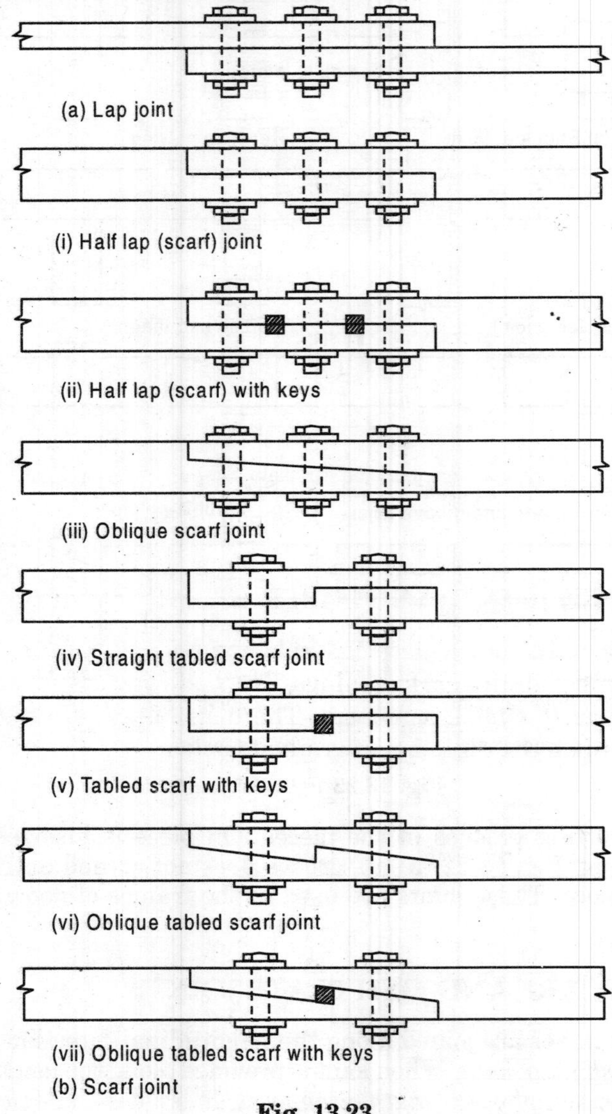

Fig. 13.23

The *fish plate joint* is also known as butt-joint. In the fish plate joints, two pieces of timber are placed end to end and cover plates or fish-plates are placed on them, and these are joined together as shown in Fig. 13.24. The fish-plates are made of timber. The steel fish-plate are also used for connection. Figure

13.24 (i) shows *plane fish plate joint.* Figure 13.24 (ii) shows *tabled fish-joint.* Figure 13.24 (iii) shows *indented fish-plate joint.* The fish-plate joint is commonly used or splice in timber member.

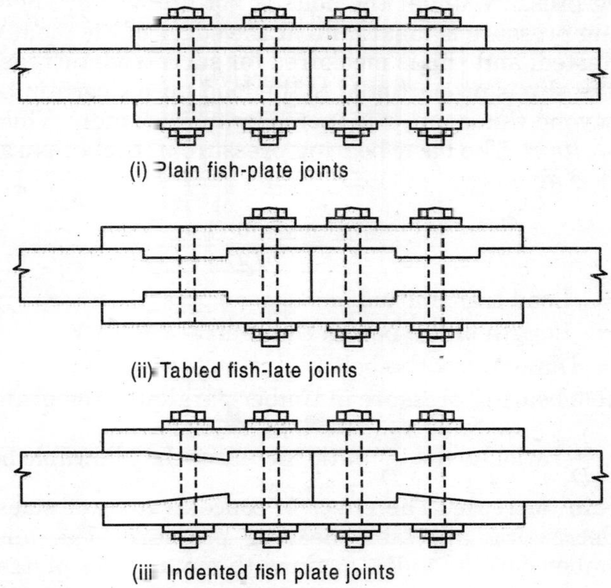

(i) Plain fish-plate joints

(ii) Tabled fish-late joints

(iii Indented fish plate joints

Fig. 13.24 *Fish-plate joints*

13.21 BOLTED JOINTS IN TIMBER

The bolts are used for connecting timber members. The bolts are used for connecting timber-to-timber fastenings. The bolts are also used for connecting plate fastenings to timber members. The distribution of pressure under the bolts depend upon the direction of grain of timber.

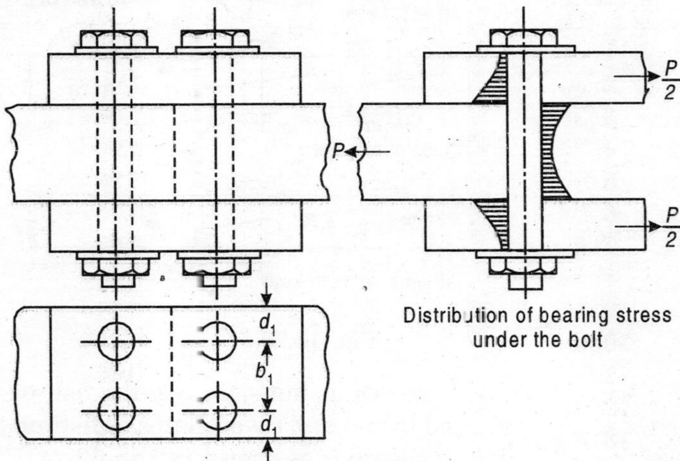

Fig 13.25 *Bolted joints in timber*

Figure 13.25 (a) shows a fish-plate joint. The distribution of pressure under the bolts is parallel to the grains of timber. The surfaces of member under the bolts are subjected to bearing pressure.

The bearing pressure under the bolts is not uniform as shown in Fig. 13.25 (b). The bearing pressure is maximum at the edges of the members. If this fish-plate joint is tested, and slip is measured for successive increase of load, it will be seen that the slip is proportional to the load upto a certain point. If the load is increased beyond this point, slip increases more rapidly. This point is known *as proportional limit.* The basic bearing pressure in timber parallel to the grain under the bolt is given by

$$p = \left(\frac{P}{LD}\right) \quad ...(13.26)$$

where, P = Load transmitted

L = Length of the bolt in the central member

D = Diameter of the bolt

The allowable bearing pressure in timber parallel to the grain under the bolt depends upon $\frac{L}{D}$ ratio. As the $\frac{L}{D}$ ratio increases the allowable bearing pressure parallel to grain decrease. The effect of concentration of stress is taken into account by decreasing allowable bearing pressure. The allowable bearing pressure in timber parallel to the grain with metal cover plates is given by

$$p_1 = K_1 . p \quad ...(13.27)$$

where, K_1 = Constant. The values of K_1 depend upon $\frac{L}{D}$ ratio.

The values of K_1 are given in Table 13.9.

Table 13.9 *Values of* K_1

$\frac{L}{D}$ *Ratio*	*Values of* K_1
1	1.00
2	1.00
3	1.00
4	0.90
5	0.86
6	0.75
7	0.64
8	0.58
9	0.51
10	0.46
11	0.42
12	0.38
13	0.34
14	0.30

It is to note that when wooden cover plates are used instead of metal cover plates, each being half the thickness of the main member, then the allowable bearing stress in the direction parallel to the grain is taken as 80 per cent of calculated stress.

For the loading parallel to the grain, the net sectional area should be greater than 80 per cent of bearing area of bolt in case of soft wood, and 100 per cent of bearing of bolt in case of hard wood.

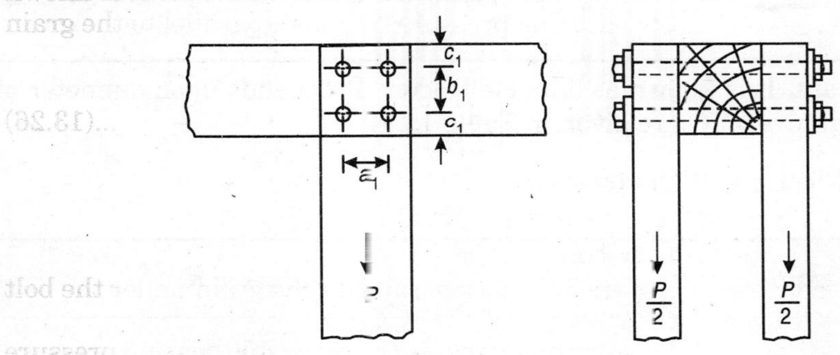

Fig. 13.26

The distribution of pressure under the bolt in the vertical member is shown in Fig. 13.26 is parallel to the grain and that in the horizontal member is perpendicular to the grain. The allowable bearing pressure under the bolt in timber perpendicular to the grain with metal cover plates is obtained by

$$q_1 = K_2.K.q \qquad \text{...(13.28)}$$

where, q = Basic stress in timber across the grain

K_2 = Constant. It depends upon $\frac{L}{D}$ ratio. The values of K_2 are given in Table 13.10.

Table 13.10 *Values of K_2*

$\frac{L}{D}$ *Ratio*	*Values of K_1*
1	1.00
2	1.00
3	1.00
5	1.00
6	1.00
7	1.00
8	0.93
9	0.88
10	0.80

Contd.

Contd. Table 13.10

$\frac{L}{D}$ *Ratio*	*Values of K_1*
11	0.68
12	0.61
13	0.55
14	0.51

K_2 = Constant. It is known as diameter factor. It depends upon diameter of the bolt. The values of K_3 are given in Table 13.11.

Table 13.11 *Values of K_3 (Diameter factor)*

$\frac{L}{D}$ *Ratio*	*Values of K_1*
6	2.50
10	1.95
13	1.68
16	1.52
19	1.41
22	1.33
25	1.27
32	1.19
38	1.14
42	1.10
50	1.07
64	1.03
75 and above	1.00

It is to note that when wooden cover plates are used instead of metal cover plates, each being half the thickness of the main member, then the allowable bearing stress in the direction perpendicular the grain is not reduced.

The distribution of pressure under the bolt in the inclined members as shown in Fig. 13.27 is parallel to the grain, and that in the horizontal member is neither perpendicular nor parallel to the grain. The allowable bearing pressure under the bolt in the horizontal member is obtained by Hankinson's formula

$$N_1 = \left(\frac{p_1 \cdot q_1}{p_1 \cdot \sin^2\theta + q_1 \cdot \cos^2\theta}\right) \quad ...(13.29)$$

where, p_1 = Allowable bearing pressure in the timber parallel to the grain in bolted joints

q_1 = Allowable bearing pressure in the timber perpendicular to the grain in bolted joints.

Pitch of bolts. When the direction of loading is parallel to the grain, then the minimum distance a_1, (the centre to centre distance between the bolts in the direction parallel to the loading, i.e., pitch) is 4 times the diameter of bolt. When the direction of loading is perpendicular lo the grain, then, the minimum distance as shown in Fig. 13.26 is also 4 times the diameter of bolt.

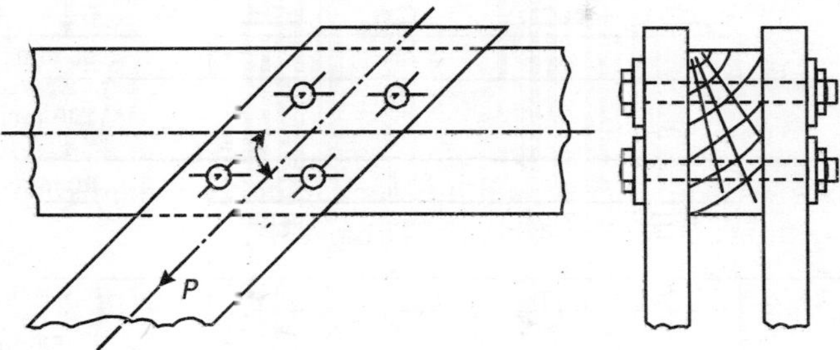

Fig. 13.27

13.21.1 Gauge of Bolts

When the direction of loading is parallel to the grain, the maximum distance b_1, (the centre to centre distance between the bolts in the direction perpendicular to the loading, i.e., gauge) as shown in Fig.13.25, depends upon allowable bearing pressure and net sectional area. The distance, b_1 should be such that net sectional area should be less than 80 per cent of gross sectional area. When the direction of loading is perpendicular to the grain, then the minimum distance, b_1 as shown in Fig. 13.26 is 2.5 times the diameter of bolt when L/D ratio is upto 2 and 5 times the diameter of bolt when L/D ratio is three or more.

13.21.2 End Distance

The distance from end of timber to the centre of bolt is knows as end distance. When the load is tensile and parallel to the grain, then the end distance 7 times the diameter of bolt for soft wood and 5 times the diameter of bolt for hard wood. When the load is compressive, then, the end distance is 4 times the diameter of bolt. When the direction of loading is perpendicular to the grain, then the end distance is 4 times the diameter of bolt.

13.21.3 Edge Distance

The distance from edge of timber to the centre of bolt is known as edge distance. When the load is parallel to the grain whether it is compressive or tensile, then the edge distance d_1, as shown in Fig 13.25 is 1.5 times the diameter of bolt for $\frac{L}{D}$ ratio less than or equal to 6. When $\frac{L}{D}$ ratio is greater than 6, edge distance is taken as half gauge distance. When the direction of loading is perpendicular to the grain, then, the edge distance is 4 times the diameter of bolt.

Example 13.16 *Two 50 mm × 150 mm wooden fish plates are used to splice a 100 mm × 150 mm timber tension member. Deodar wood is used for all elements. Four 22 mm diameter bolts are used on each side of the splice. Determine the maximum load that the splice will carry.*

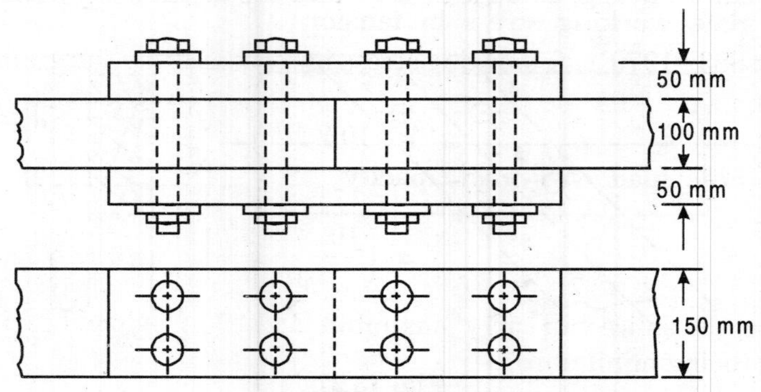

Fig. 13.28

Solution

Design :

Length of bolt, L = Thickness of main timber member
= 100 mm

Diameter of bolt, D = 22 mm

$$\frac{L}{D} \text{ ratio} = \left(\frac{100}{20}\right) = 4.55$$

From Table 13.9,
$$K_1 = \left[0\cdot90 - \frac{0\cdot55}{1\cdot00}\times(0\cdot90 - 0\cdot86)\right] = 0.88$$

Step 1: Safe working stress in compression

From IS : 883 –1970, safe working stress in compression (bearing) parallel to the grain for deodar wood

= 7.8 N/mm²

Allowable bearing stress parallel to the grain with metal cover plates

p_1 = (0.88 × 7.8) = 6.864 N/mm²

Since wooden cover plates are used,

p_1 = (0.80 × 6.684) = 5.49 N/mm²

Load carrying capacity of one bolt

$$\left(5\cdot49\times\frac{22\times100}{1000}\right) = 12.078 \text{ kN}$$

Step 2: Load carrying capacity of 4 bolts

(4 × 12.078) = 48.312 kN

The bearing area of 4 bolts

(4 × 22 × 100) = 8800 mm²

80 per cent of bearing of bolts

$= 0.80 \times 88 = 7040 \text{ mm}^2$

Diameter of bolt hole $= (22 + 3) = 25$ mm

Net sectional area of member

$100 \times (150 - 2 \times 25) = 10000 \text{ mm}^2 > 7040 \text{ mm}^2$

Step 3: Safe working stress in tension

From IS : 883 –1970, safe working stress in tension along the grain for deodar wood

$= 10.2 \text{ N/mm}^2$

Step 4: Safe load carrying capacity

$$= \left(\frac{10 \cdot 2 \times 10000}{1000}\right) = 102 \text{ kN}$$

Therefore, the splice will carry maximum load equal to load bearing capacity of the bolts (being minimum)

$= \mathbf{48.312\ kN}$

Example 13.17 *A 150 mm tension member carrier a load of 150 kN. Design suitable splice for the member. Use 22 mm diameter bolts. The tension member is made of babul wood.*

Solution

Design :

Length of bolt, L = Thickness of main timber member

= 150 mm

Diameter of bolt, D = 22 mm

$$\frac{L}{D} \text{ ratio} = \left(\frac{150}{22}\right) = 6.83$$

From Table 13.9, $$K_1 = \left[0 \cdot 75 - \frac{0 \cdot 83}{1 \cdot 00} \times (0 \cdot 75 - 0 \cdot 64)\right] = 0.66$$

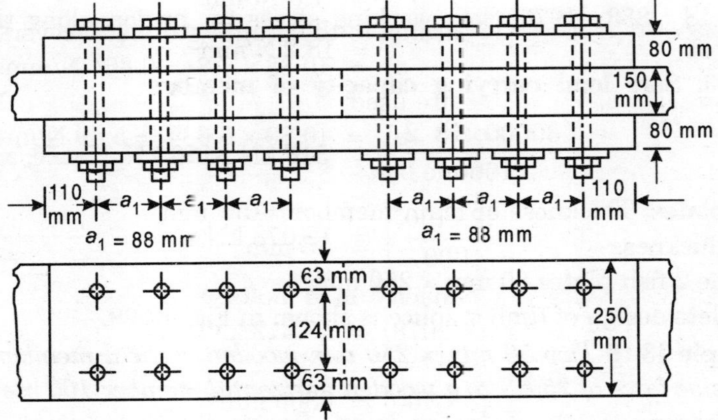

Fig. 13.29

Step 1: Safe working stress in compression

From IS : 883 –1970, safe working stress in compression (bearing) parallel to the grain for babul wood = 11.2 N/mm^2

Allowable bearing stress parallel to the grain with metal cover plates

$$(0.66 \times 11.2) = 7.392 \text{ N/mm}^2$$

Since, wooden cover plates are used

$$(0.80 \times 7.392) = 5.914 \text{ N/mm}^2$$

Step 2: Load carrying capacity of one bolt

$$\left(\frac{5 \cdot 914 \times 150}{1000}\right) = 19.514 \text{ kN}$$

Number of bolts required

$$\left(\frac{150}{19 \cdot 514}\right) = 7.686$$

Provide 8 bolts in 2 rows on either side of splice as shown in Fig. 13.29.

Bearing area of 8 bolts = 8 × 22 × 150 = 26400 mm^2

Diameter of bolt hole = (22 + 3) = 25 mm

Net sectional area of member

$$(250 - 2 \times 25) \times 150 = 30000 \text{ mm}^2$$

$$> 80\% \text{ Gross-sectional area}$$

Net sectional area is greater than 100% of total bearing area of the bolt. It suits bolt requirement for hard wood (Babul).

Pitch of bolts, $a_1 = 4 \times D = 4 \times 22$ mm = 88 mm

Provide $a_1 = 90$ mm

Gauge distance $d_1 = 124$ mm

Edge distance $d_1 = \frac{1}{2} \times 124 = 62$ mm

Provide $d_1 = 63$ mm

End distance = $5D$ for hard wood = 5 × 22 = 110 mm

From IS : 883 –1970, Safe working stress in tension along the grain for babul wood = 18.2 N/mm^2

Step 3: Safe load carrying capacity of member

$$\left(\frac{30000 \times 18 \cdot 2}{1000}\right) = 516 \text{ kN} > 150 \text{ kN. Hence, safe.}$$

Fish-plates. Thickness of main member =150 mm

Half-thickness = 75 mm

Provide 2 fish-plates 80 mm × 250 mm

Complete design of timber splice is shown in Fig. 13.29.

Example 13.18 *Two 50 mm × 250 mm wooden vertical member transmit a compressive force of 25 kN to a wooden horizontal member 100 mm × 250 mm. The members consists of deodar wood. Design the bolted joint. Use 19 mm diameter bolt.*

Solution

Design

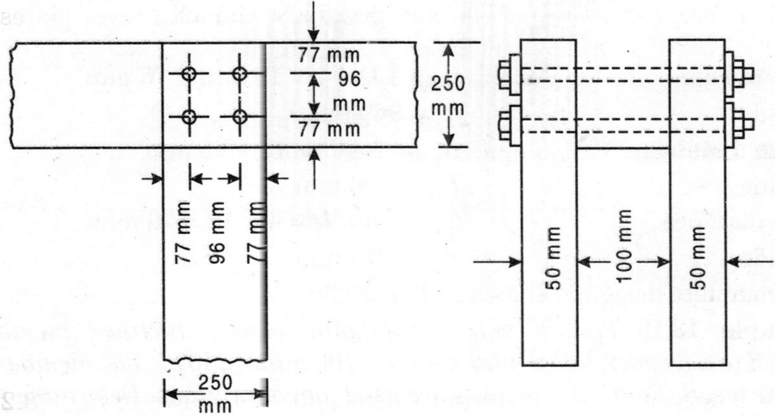

Fig. 13.30

Length of bolt, L = Thickness of central member = 10 mm

Diameter of bolt, D = 19 mm

$$\frac{L}{D} \text{ ratio} = \left(\frac{100}{19}\right) = 5.26$$

From Table 13.10, For $\frac{L}{D}$ ratio = 5.26

Value of K_2 = 1.00

From Table 13.11 for 19 mm diameter of bolt, diameter factor

K_2 = 1.41

Load transmitted is perpendicular to the horizontal member.

Step 1: Safe working stress in compression

From IS : 883 –1970, the safe working stress in compression (bearing across the grain in deodar wood)

q = 2.65 N/mm²

Allowable bearing pressure under the bolt

$$q_1 = K_2K_3q = (1.00 \times 1.41 \times 2.65)$$
$$= 3.737 \text{ N/mm}^2$$

Step 2: Load carrying capacity of one bolt

$$\left(\frac{3\cdot737 \times 19 \times 100}{1000}\right) = 7.099 \text{ kN}$$

Number of bolts required

$$\left(\frac{25}{7\cdot099}\right) = 3.052$$

Provide 4 bolts as shown in Fig. 13.30

Step 3: Load carrying capacity of the member

$$\left(\frac{2\cdot 65\times 100\times 250}{1000}\right) = 66.25 \text{ kN} > 25 \text{ kN}$$

Pitch of bolts $a_1 = 4\,D = 4 \times 19 \text{ mm} = 76 \text{ mm}$

Provide $a_1 = 96 \text{ mm}$

Gauge distance $b_1 = 5 \times D \times 19 = 95 \text{ mm}$

Provide $b_1 = 96 \text{ mm}$

Edge distance $C_1 = 4 \times D = 4 \times 19 = 76 \text{ mm}$

Provide $c_1 = 77 \text{ mm}$

The complete design is shown in Fig. 13.30.

Example. 13.19 *Two 50 mm × 250 mm wooden inclined members are connected to a wooden horizontal member 100 mm × 250 m. The member consist of deodar wood. Four 19 mm bolts are used for connections. Determine the total load transmitted by sloping member if the inclination of members with the horizontal is 45 °*

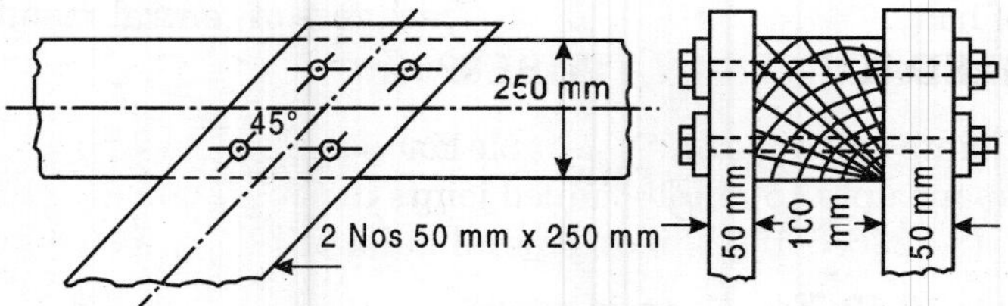

Fig. 13.31

Solution

Length of bolt, $L = 100 \text{ mm}$

Diameter of bolt, $D = 19 \text{ mm}$

$$\frac{L}{D} \text{ ratio} = \left(\frac{100}{18}\right) = 5.26$$

$$\text{From Table 13.9, value of } K_1 = \left[0\cdot 86 - \frac{0\cdot 26}{1\cdot 00}\times(0\cdot 86 - 0\cdot 75)\right] = 0.83$$

From Table 13.10, value of K_2 is 100 From Table 13.11, diameter factor K_3 is 1.41

Step 1: Safe working stress in compression

From IS : 883 – 1970.

Safe working stress in compression along the grain for deodar wood

$$p = 7.8 \text{ N/mm}^2$$

Safe working stress in compression across the grain for deodar wood

$$p = 2.65 \text{ N/mm}^2$$

Allowable bearing pressure along the grain

$$p_1 = K_1 \cdot p = 0.83 \times 7.8 = 6.474 \text{ N/mm}^2$$

It is reduced to 80 per cent since wooden cover plates are used

$$(6.474 \times 0.8) = 5.18 \text{ N/mm}^2$$

Allowable bearing pressure across the grain

$$q_1 = K_2 K_3 \cdot q = (1.00 \times 1.41 \times 2.65) = 3.74 \text{ N/mm}^2$$

Allowable bearing pressure along the sloping member

$$\frac{5 \cdot 18 \times 3 \cdot 74}{\left(5 \cdot 18 \times \frac{1}{4} + 3 \cdot 74 \times \frac{1}{4}\right)} = 8.69 \text{ N/mm}^2$$

Step 2: Load transmitted by one bolt

$$\left(\frac{8 \cdot 69 \times 19 \times 100}{1000}\right) = 16.5 \text{ kN}$$

Step 3: Load transmitted by 4 bolts

$$4 \times 16.5 = \mathbf{66 \text{ kN}}$$

13.22 NAILED JOINTS IN TIMBER

The nailed joints in timber are suitable for light timber frames such as roof trusses for spans upto 15 m. The nailed joints in timber are designed as per IS : 2366 –1963, IS : 2363 –1963 gives general guidance and furnishes the strength

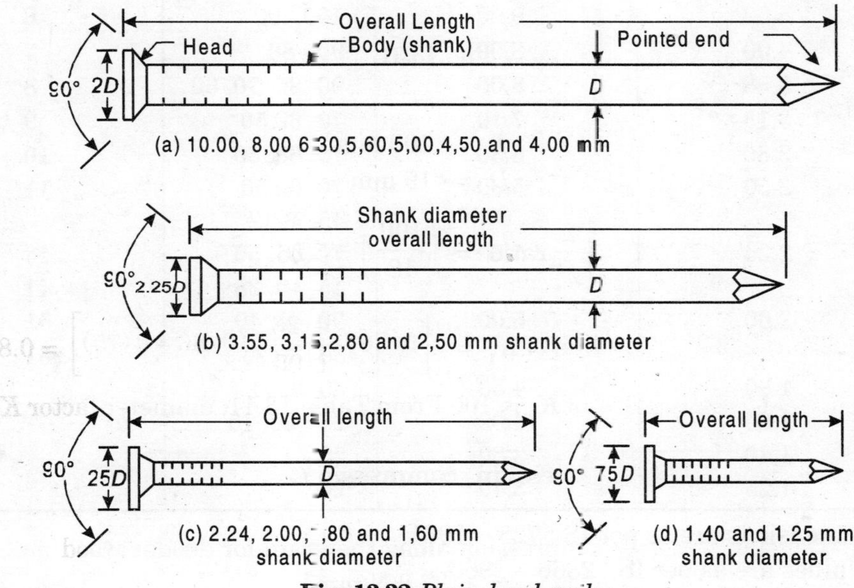

Fig. 13.32 *Plain head nails*

data. The strength data were made available by Forest Research Institute and Colleges, Dehradun. The nailed joints in soft timber are safe and economical to transmit load upto 25 kN and in hard wood upto 50 kN. The nails are used in nailed joints to connect two pieces of timber. A nail consists of a head at one end a slender body either tapering or pointed at other end in Fig. 13.22 (a). The head of nail is used to drive the nail through and into the wood. The nails as classified as *cut nails, wrought nails* and *wire nails. Cut nails* are of rectangular cross-section and are cut from a metal strip. *Wrought nails are* forged. *Wire nails* are made directly from wire by machine which forms both head and point. Wire nails are made of mild steel. Wire nails used for nailed joints in timber are plain-head nails specified in IS : 723 –1961 as shown in Fig. 13.32. Wire nails may be diamond pointed, or blunt and tapered to blunt-pointed or long sharp pointed. Diamond pointed nails are used in nailed joints. A diamond point nails has a point which commonly resembles as octahedron and less commonly a tetrahedron.

The dimensions of plains head round mild steel wire nails are given in Table 13.12.

Table 13.12 *Dimensions of round mild steel wire nails plain head nails*

*Diameter of shank mm**	*Diameter of head mm**	*Length mm***	*Corresponding SWG***
10.0	20.00	250	–
8.00	16.00	250, 200	–
6.30	12.00	175, 150	–
5.60	11.60	150, 125	–
5.00	10.20	125, 100	–
4.50	9.00	100, 90	6
4.00	8.00	90, 80	7
3.55	8.00	90, 80, 70, 60	8
3.15	7.10	70, 60,50	9
2.80	6.30	70, 60, 50	10
2.50	5.60	70, 60,50	11
		45, 4012	
2.24	5.60	70, 60, 50	13
		45, 40, 35	41
2.00	5.00	50, 45, 40	51
		35, 30,25	61
1.80	4.50	30, 2571	
1.60	4.00	25, 20, 15	
1.40	3.80	20	
1.25	3.40	20	

* Dimensions are as per IS : 723 – 1961.

**Values are as per IS : 2366 – 1963.

Note. *Rationalized metric values of diameter of shank of nails approximately equivalent to SWO numbers are shown in Table 13.12.*

When a nail is double shear, the depth of penetration of the nail in the main central member shall be not less than two-thirds of the total penetration in the side members. For nails in single shear, the depth of penetration in the member containing the nail point shall be not less than two-thirds length of the nail driven into the joint. For penetration of nails in soft woods. pre-boring is not necessary. For penetration of nails woods, pre-boring is necessary. In hard woods, holes are drilled and pre-boring is done and then, the nails are driven. The diameter of pre-bore should not be greater than 4/5th diameter of the nail.The nail points after nailing are finished either by clenching the nails across the grain as shown in Fig. 13.33 (a) or clenching the nails along the grains as shown in Fig. 13.33 (b) or by protruding from the surface of the joint and cut so as to be flush with the joint face. Turning over the projecting point of a nail so as to be flush with the surface of member is known as *clenching*. Clenching increases the holding power of nails and makes withdrawal more difficult. Clenching perpendicular to (across the) grain gives greater power to nails than parallel to the grain.

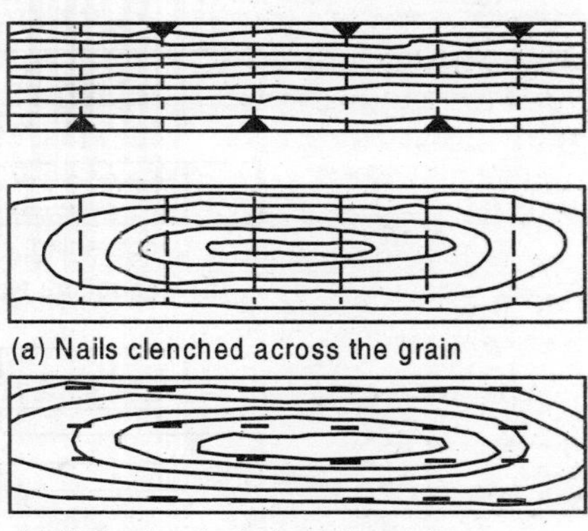

Fig. 13.33

The nailed joints in timber used for structural joints are of various types. Figure 13.34 shows monochord type and split-chord type lap joints, used for lengthening

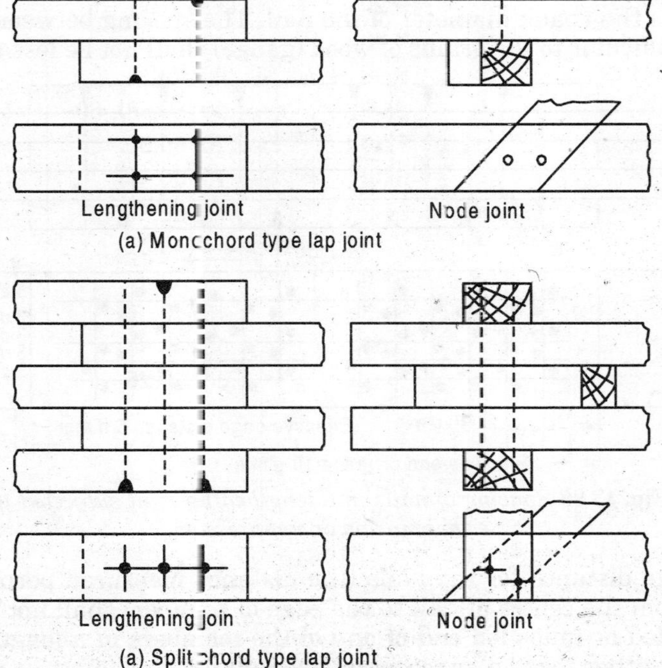

Fig. 13.34

joints and node joints. Figure 13.35 shows monochord type and split-chord type butt joints used for lengthening joints and node joints. In monochord type joint, the main member transmitting load consists of a single member. In split chcrd type joint, the main member consists of more than one member.

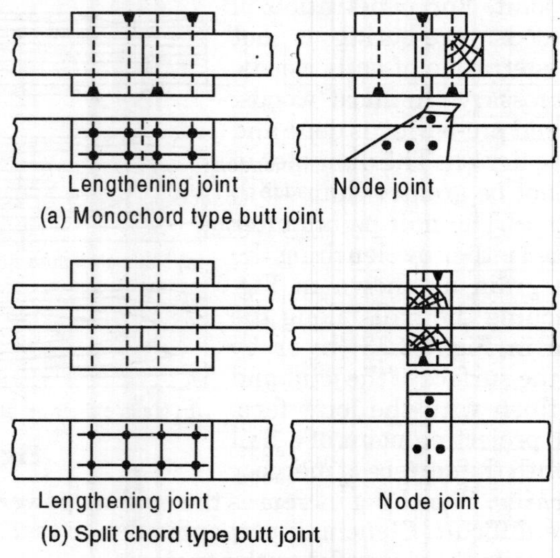

Fig. 13.35

The nails are arranged in a *lengthening joint* as shown in Fig. 13.36. The spacing of nails in the direction of grains of wood (pitch) shall not be less than 10 *d* where, *d* is the shank diameter of the nail. The spacing between the rows of nails perpendicular to the grains of wood (gauge) shall not be less than 3*d*. The

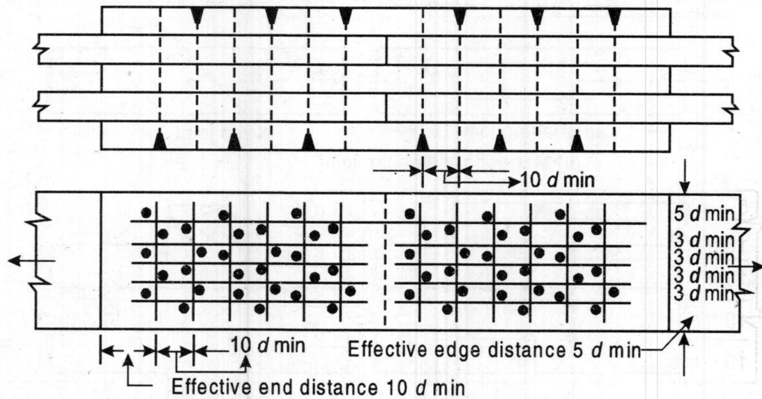

Fig. 13.36 *Spacing of nails in a lengthening joint subjected to either tension or compression*

effective edge distance for the nails (the distance measured perpendicular to the grain from the centre of nail to the edge of number) shall not be less than 5*d*. The loaded or unloaded end of any of the members in a lengthening joint subjected to either tension or compression shall have an effective end distarce (the distance measured parallel to the grain from the centre of nail to the square end of member) of not less than 10*d*. These recommendations are as per IS : 2366 – 1963.

The nails are arranged in a node joint members are at right angles to one another as shown in Fig. 13.37 as per IS : 2366 –1963.

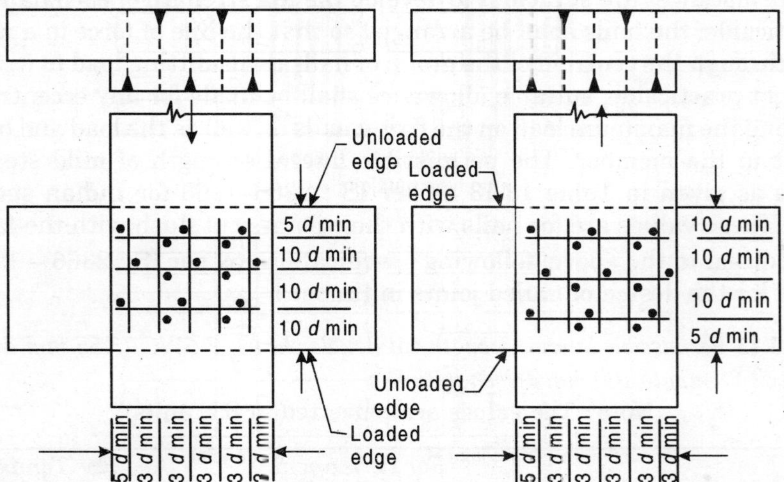

Fig. 13.37. *Spacing of nails where members at right angles to one another*

The nails are arranged in a node joint where members are inclined to one another at angles other than 90° as shown in Fig. 13.38 as per IS : 2366 –1963.

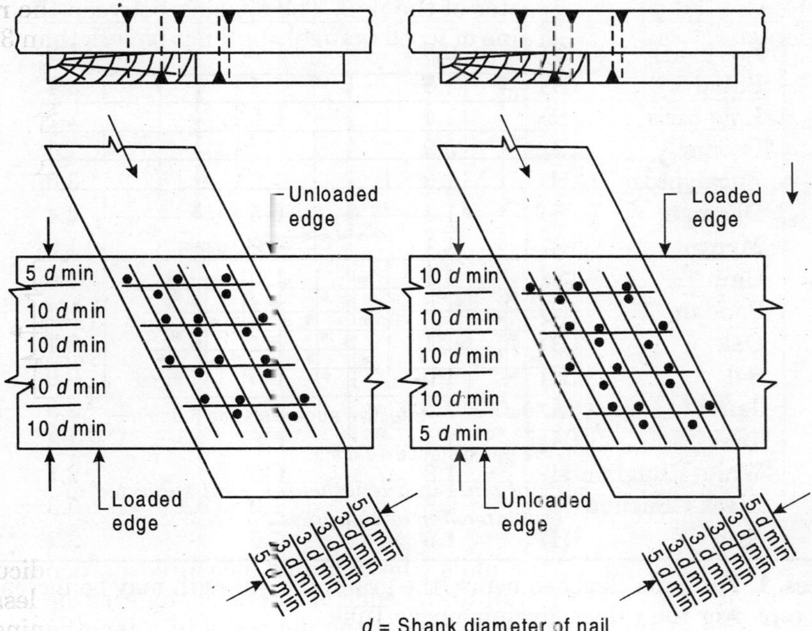

Fig. 13.38 *Spacing of nails where members are inclined to one another at angles other than 90°*

Strength of nailed joints. Where number of nails are used in a joint, the allowable load in withdrawal or lateral resistance is sum of the allowable loads for the individual nails, provided that the centroid of the group of fixing nails lies on the axis of the member and the spacing between nails, end distances, and edge distances are sufficient to develop the full strength of each nail. As far as practicable, the nails shall be arranged so that the line of force in a member passes through the centroid of the group of nails transmitting load to it. Where, this is not practicable, suitable allowance shall be made for any eccentricity in computing the maximum load on the fixing nails as well as the load and bending moment in the member. The permissible lateral strength of mild steel-nails shall be as given in Table 13.13 as per IS : 2366–1963 for Indian species of timber. These values are for nails with their points cut flush with the faces.

In addition to the above, following specifications as per IS : 2366 – 1963 are adopted for the design of nailed joints in timber :

Table 13.13 *Permissible lateral strength (in double shear), 9 SWG (3.55 mm diameter) nails of size 80 mm to 100 mm driven in timber*

(**Note :** The values are converted in S.I. units)

S. No.	*Species of wood*		*For Permanent Construction*		*For Temporary Construction strength per nail. (For both lengthening and node joint (kN)*
			Lengthening joint (kN)	*Node joint (kN)*	
1*	Fir	(S)	0.8	0.2	1.2
2	Babul	(H)	1.5	1.1	3.4
3	Kala Siris	(H)	1.4	0.5	2.2
4*	Deodar	(S)	0.9	0.4	1.5
5	Eucalyptus	(H)	1.0	1.0	3.0
6*	Dhaman	(H)	1.3	0.5	2.4
7*	Mango	(S)	1.1	0.9	1.6
8*	Chir	(S)	1.1	1.0	1.6
9	Sandan	(H)	1.7	1.1	1.8
10	Oak	(H)	1.1	1.1	2.7
11	Sal	(H)	1.0	0.5	0.9
12	Jamun	(H)	1.5	1.2	2.5
13	Teak	(H)	1.4	0.8	1.3
14	White Chuglam	(H)	1.8	0.9	2.1
15*	Black Chuglam	(H)	2.3	1.0	3.3
16	Sain	(H)	1.6	1.6	2.2

Notes. 1. For nails clenched across the grain, the strength may be increased by 20 per cent over the values given in above Table.

* 2. Species required no pre-boring for nail penetration.

3. H = Hard wood S = Soft wood

Dimensions of member

1. The minimum thickness of any individual piece of timber (that, is any single member) shall be 15 mm.

2. The maximum thickness of any individual piece of timber (other than spacer block) shall be 100 mm.

3. Normally the width of any individual piece of timber shall not exceed eight times the thickness of member.

Fish plates. Wooden plates which are used in joint to hold the members in alignment and also to stiffen the joints, are known as *fish plates.* Fish plates ransmit load from one member to another. Fish-plates are also known as *timber gussets* or *timber splice plates.*

1. The total combined thickness of the gusset or splice plates (fish-plates) on either side of a joint in a monochord type construction shall not be less than 5 times the thickness of the main members.

2. The total combined thickness of all spacer block and/or plates including outer splice plates, at any joint in a split-chord type construction shall not be less than 15 times the total thickness of all the main members at that joint.

Nails

1. The diameter of nails shall be between $\frac{1}{6}$ and $\frac{1}{4}$ of the thickness of the main member.

2. In a lengthening joint, in order to avoid eccentricity, if any, due to the hinge action, a minimum member number of 8 nails (i.e., 4 nails on either side of the joint) shall be provided.

3. In a node joint, a minimum number of 2 nails shall be provided to keep the member in position.

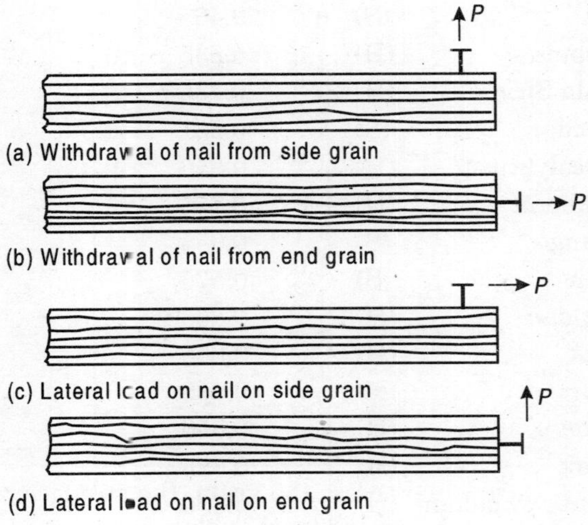

Fig. 13.39

The holding power (strength) of nail depends upon many factors including kind of nails, its diameter, depth of penetration, the density of timber, the moisture content of timber, its state of seasoning, thickness of the actual rings. the proportion of hard wood, the mode of driving nails, the mode of withdrawing the nails relative to the direction of grain, use of pre-bore holes. and manner of clenching the nails. Figure 13.39 (a) and (b) show the withdrawal of nails from side grain and grain, respectively. Figure 13.39 (c) and (d) show the lateral load on nail on side grain and on end grain respectively.

Withdrawal resistance of nails from side grain and lateral resistance of nails on side grain are important for structural considerations. The roof covering materials are fixed to the purlins by nails then, the suction tends to withdraw the nails. The timber members in roof trusses carry axial forces. The nails in nailed joints in timber members are subjected to lateral loads. The permissible withdrawal resistance per nail is given by

$$R = K \,.\, G^{25} \,.\, D \qquad ...(13.30)$$

where, G = Specific gravity of oven dry timber

D = Diameter of nails in mm

K = Constant $= \left(\dfrac{972 \cdot 5}{1000}\right) = 0.9725$

R = Permissible withdrawal resistance per nail in kN per mm of penetration. Permissible withdrawal resistance of nail for various timber species are given in Table 13.14.

Table 13.14 *Permissible withdrawal resistance for nail per 10 mm penetration*

S. No.	*Species of Timber*		*Specific gravity (G)*	*Withdrawal Resistance 0.9725 G^{25} (kN)*
1	Fir	(S)	0.465	0.143
2	Babul	(H)	0.835	0.619
3	Kala Siris	(H)	0.735	0.449
4	Deodar	(S)	0.561	0.228
5	Eucalyptus	(H)	0.850	0.646
6	Dhaman	(H)	0.755	0.482
7	Mango	(S)	0.655	0.339
8	Chir	(S)	0.575	0.244
9	Sandan	(H)	0.865	0.674
10	Oak	(H)	0.865	0.674
11	Sal	(H)	0.800	0.556
12	Jamun	(H)	0.850	0.647
13	Teak	(H)	0.625	0.300
14	White Chuglam	(H)	0.690	0.382
15	Black Chuglam	(H)	0.835	0.619
16	Sain	(H)	0.880	0.705

The permissible lateral strength (in double shear) for nail that has its point cut flush with the surface is given by the following :

1. For lengthening joint for permanent construction

$$R_1 = \left(\frac{k_1 \cdot D}{1000}\right) \quad ...(13.31)$$

2. For node joint permanent construction

$$R_2 = \left(\frac{k_2 \cdot D}{1000}\right) \quad ...(13.32)$$

3. For lengthening joint and node joint for temporary construction

$$R_3 = \left(\frac{k_3 \cdot D}{1000}\right) \quad ...(13.33)$$

where, $R_1 R_2 R_3$ = Permissible lateral strengths per nail (in double-shear)
$K_1\ K_2,\ K_3$ = Constants depending upon the type of timber
D = Diameter of nails in mm

The nails are cut flush with the surface of timber.

In case the nails are driven in unseasoned timber, which will dry and season under the load, the permissible withdrawal resistance per nail as obtained by Eq. 13.30 is reduced to 75 percent and the permissible resistance per nail as obtained by Eqs. 13.31,13.32 and 13.33 are reduced to 25 percent.

The values given in Table 13.13 for permissible lateral strength (in double shear) are applicable for 9 SWG (3.55 mm) diameter nails. In general the permissible lateral strength (in double shear) for nails may be found from Eqs. 13.31, 13.32 and 13.33. The values of constants $K_1\ K_2$ and K_3 for some species of wood have been given in Table 13.15.

Example 13.20 *A 35 mm × 150 mm tension member carries an axial load of 50 kN. Design the suitable splice for the member. The tension member is made of babul wood. Use standard wire nails.*

Solution

Design :

Thickness of tension member = 35 mm

Diameter of nail $= \frac{1}{6}$ and $\frac{1}{4}$ th thickness of member

$$= \left(\frac{35}{6} \text{to} \frac{35}{4}\right) = (5.83 \text{ to } 8.75) \text{ mm}$$

Use 8 mm diameter wire nails, with 200 mm length. The excess length of nail is clenched across the grain. Assume that lengthening joint is used in permanent construction. From Eq. 13.31, the permissible lateral strength (in double shear)

$$R_1 = \left(\frac{K_1 \cdot D}{1000}\right) \text{kN.}$$

Note. *For nails clenched across the grain, the values of constants may be increased by 20 per cent over the values given in Table 13.15.*

Table 13.15 *Values for constant K_1, K_2 and K_3 or lateral strength of nails in double shear*

S. No.	Species of wood		For Permanent Construction		For Temporary Construction strength per nail. (For both lengthening and node joint (kN)
			Lengthening joint (kN)	Node joint (kN)	
1	Fir	(S)	2.26	5.65	3.38
2	Babul	(H)	4.22	3.10	9.59
3	Kala Siris	(H)	3.94	1.41	6.20
4	Deodar	(S)	2.54	1.125	4.22
5	Eucalyptus	(H)	4.79	2.82	8.43
6	Dhaman	(H)	3.66	1.41	6.76
7	Mango	(S)	3.10	2.54	4.52
8	Chir	(S)	3.10	2.82	4.52
9	Sandan	(H)	4.79	3.10	5.08
10	Oak	(H)	3.10	3.10	7.62
11	Sal	(H)	2.82	1.41	5.36
12	Jamun	(H)	4.22	3.38	7.04
13	Teak	(H)	3.95	2.26	3.66
14	White Chuglam	(H)	5.08	2.54	7.92
15	Black Chuglam	(H)	6.49	2.82	9.30
16	Sain	(H)	4.54	4.50	8.18

H = Hard wood S = Soft wood

From Table 13.15, for babul wood,

$$K_1 = 4.220$$

$$R_1 = (4.220 \times 0.8) = 3.376 \text{ kN}$$

The nails are clenched across the grain. Therefore, the permissible lateral strength is increased by 20 percent.

Step 1: Lateral strength per nail in double shear

$$(1.20 \times 3.376) = 4.05 \text{ kN}$$

Force in member $= 50$ kN

Number of nails $= \left(\dfrac{50}{4\cdot50}\right) = 12.34$

Provide 15 nails on either side in three rows as shown in Fig. 13.40.

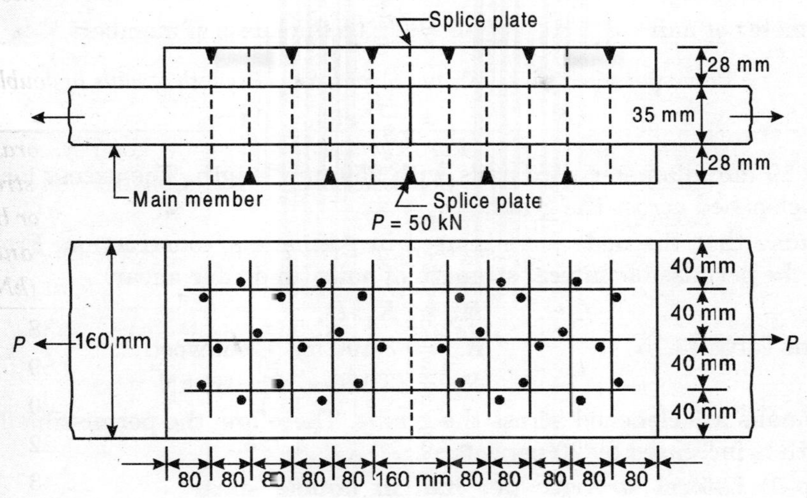

Fig. 13.40

Step 2: Arrangement of nails

Effective end distance = 10 × Diameter of nail (Min.)

= (10 × 8) = 80 mm

Effective edge distance = 5 × Diameter of nail (Min.)

Pitch of nails = 10 × Diameter of nail (Min.)

= (10 × 8) = 80 mm

Gauge distance for nails

3 × Diameter of nail (Min.) = 3 × 8 = 24 mm

Provide 40 mm gauge distance of nails. Thickness of splice plates

= 28 mm

Step 3: Safe working stress in tension

Babul wood does not require preboring. From IS : 883 –1970, safe working strength in tension along the grain for babul wood

= 18.2 N/mm²

Step 4: Strength of tension member

$$\left(\frac{(18\cdot 2\times 35\times 160)}{1000}\right) = 101.92 \text{ kN} > 50 \text{ kN}$$

The complete design of splice (lengthening joint) is shown in Fig. 13.40.

Example 13.21 *Two 45 mm × 250 mm wooden vertical members transmit a compression force 25.4 kN to a wooden horizontal member 60 mm × 250 mm. The members consist of babul wood. Design the nailed joint.*

Solution

Design:

Thickness of member carrying compression perpendicular to the grain

= 60 mm

Diameter of nail $= \frac{1}{6}$ th thickness of member

$= \frac{1}{6} \times 60 = 10$ mm

Use 10 mm diameter wire nails with 250 mm length. The excess length of nail is clenched across the grain.

Assume that the node joint is used in permanent construction. From Eq. 13.32, the permissible lateral strength of nails (in double shear)

$$R_2 = K_2 . D$$

From Table 13.15, $K_2 = 3.100$ for babul wood

$$R_2 = (3.100 \times 10) = 31 \text{ kN}$$

The nails are clenched across the grains. Therefore, the permissible lateral strength is increased by 20 percent.

Step 1: Lateral strength per nail in double shear

$= (1.20 \times 310) = 3.72$ kN

Force in member $= 25.40$ kN

Number of nails $= \left(\frac{25 \cdot 40}{37 \cdot 2}\right) = 0.683$

Provide 8 nails as shown in Fig. 13.41.

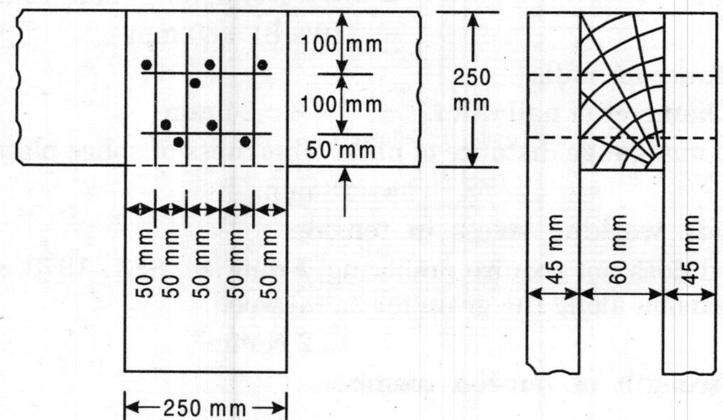

Fig. 13.41

Step 2: Arrangement of nails

Effective edge distance

5 × Diameter of nails (Min.) $= 5 \times 10 = 50$ mm

Pitch of nails $= 10 \times$ diameter of nail (Min.)

$= 10 \times 10 = 100$ mm

Gauge distance of nails

3 × Diameter of nails (Min.) $= 3 \times 10 = 30$ mm

Provide 50 mm gauge distance for nails. The horizontal member is subjected to compression perpendicular to the grain. From IS : 883 –1970, safe vorking stress in compression perpendicular to the grain for babul wood = 6.55 N/mm²

Step 3: Strength of member

$$\left(6 \cdot 55 \times \frac{60 \times 250}{1000}\right) = 98.25 \text{ kN} > 25.40 \text{ kN}$$

The complete design of nailed joint is shown in Fig. 13.41.

13.23 DISC DOWELLED JOINTS IN TIMBER

In disc-dowelled joints in timber, circular grooves are made in the members transmitting the load and a circular disc is placed in the grooves as shown in Fig. 13.42.

Figure 13.42 (a) shows a lap joint using a circular disc, and Fig. 13.42 (b) shows a butt-joint using circular disc in disc-dowelled joints in timber. Hard timber like *babul* is used for preparing the discs. The directions of grains of discs are kept parallel to the direction of loading. The force is transmitted in disc-dowelled joint partly by shear along the middle surface and partly by bearing along the end. Bolt is used to keep the members together. Bolt carries tensile load. Disc-dowelled joints are used to transmit large forces.

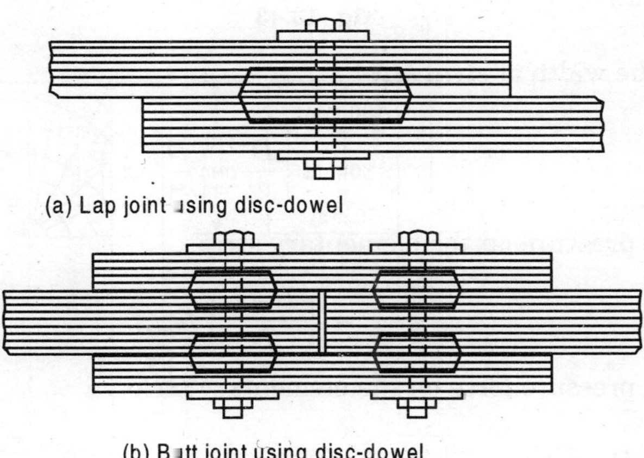

(a) Lap joint using disc-dowel

(b) Butt joint using disc-dowel

Fig. 13.42 *Disc-dowelled joints*

Consider a lap joint using a disc-dowel as shown in Fig. 13.43, to transmit force from one member to the other. The forces, *P*, *P* in members tend to rotate the dowel in anti-clockwise direction. As a result of this, dowel tends to separate the members. Bolt prevents the separation of members and keep the member together. Surface *AB* and *CD* of the dowel are subjected to bearing pressure. The bearing pressure is zero at the toe *A* of surface *AB* and toe *D* of surface *CD*. The points *B* and *C* of surfaces *AB* and *CD* are subjected to maximum bearing pressure $f_{b\,max}$. The bolt is subjected to tensile load. The maximum bearing pressure in dowel and tensile load in bolt can be found as under.

Consider an elementary strip dx along EF at a distance X from F as shown in Fig. 13.43 (b). The distance of elementary strip from centre, O is $\left(\dfrac{D}{2} - x\right)$

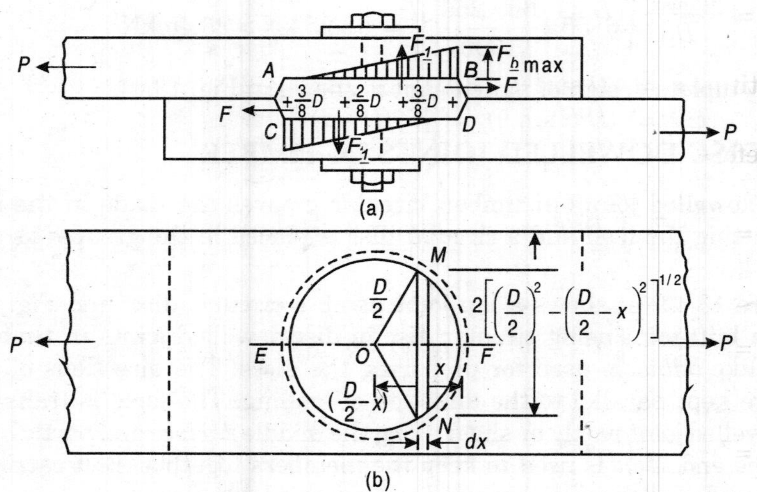

Fig. 13.43

Therefore, the width of strip MN

$$= 2\left[\frac{D^2}{2} - \left(\frac{D}{2} - x\right)^2\right]^{1/2} \qquad \text{...(i)}$$

The bearing pressure on the elementary strip

$$= \left(\frac{D - x}{D}\right) \cdot f_{b\,\max} \qquad \text{...(ii)}$$

The bearing pressure force on the elementary strip

$$\delta F_1 = \left(\frac{D-x}{D}\right) \cdot f_{b\,\max}\, 2\left[\left(\frac{D}{2}\right)^2 - \left(\frac{D}{2} - x\right)^2\right]^{1/2} dx \qquad \text{...(iii)}$$

Total bearing force acting on the surface AB

$$F_1 = \frac{2 f_{b\,\max}}{D} \int_0^D (D - x)\left[\left(\frac{D}{2}\right)^2 - \left(\frac{D}{2} - x\right)^2\right]^{1/2} dx \qquad \text{...(iv)}$$

$$= \frac{2 f_{b\,\max}}{D} \int_0^D (D - x)\left[(Dx - x^2)\right]^{1/2} dx$$

$$= \frac{2f_{b\max}}{D}\int_0^D (D-x)^{1/2}\ (D-x)x^{1/2}\cdot dx$$

$$= \frac{2f_{b\max}}{D}\int_0^D (D-x)^{3/2}\cdot x^{1/2}\cdot dx \qquad \text{...(v)}$$

Substitute $x = D\sin^2\theta$, $dx = (2D\sin\theta\,.\,\cos\theta).d\theta$

$x^{1/2} = (D^{1/2}\sin\theta)$, $(D-x) = D.\cos^2\theta$

Therefore,

$$F_1 = \frac{2f_{b\max}}{D}\int_0^{\pi/2} D^{3/2}(\cos^2\theta\cdot D^{1/2}\sin\theta\cdot 2D\sin\theta\cdot\cos\theta)\cdot d\theta$$

$$= \frac{4f_{b\max}}{D}\int_0^{\pi/2} D^{3/2}\sin\theta\cdot\cos^4\theta)\cdot d\theta$$

$$= 4f_{b\max}\cdot\left(D^2\frac{\pi}{32}\right) = \left(\frac{\pi\cdot f_{b\max}\cdot D^2}{8}\right) \qquad \text{...(vi)}$$

Similarly a force F_1 is also acting in opposite direction on the surface CD.

Two forces act at a distance $\left(\frac{3}{8}\right)Z$ from the edges. These forces act at a distance $\frac{2D}{8}$ apart. Two equal and opposite forces acting at a distance $\frac{2D}{8}$ apart from a couple M.

$$M = F_1\cdot\frac{2D}{8} = \left(\frac{\pi\cdot f_{b\max}\cdot D^2}{8}\right)\cdot\frac{2D}{8} \qquad \text{...(vii)}$$

$$\therefore \quad M = \left(\frac{\pi\cdot f_{b\max}\cdot D^3}{32}\right) \qquad \text{...(viii)}$$

This couple is balanced by bearing forces, F, F acting on the ends of dowel at a distance $\frac{t}{2}$ apart.

$$\therefore \quad F\cdot\frac{1}{2} = \left(\frac{\pi\cdot f_{b\max}\cdot D^3}{32}\right) \qquad \text{... (ix)}$$

$$\therefore \quad f_{b\max} = \left(\frac{16}{\pi}\cdot\frac{F\cdot t}{D^3}\right) \qquad \text{... (13.34)}$$

where t = Thickness of dowel.

The tensile load in the bolt is equal to F_1

From expression (ix)

$$\left(\frac{\pi \cdot f_{b\max} \cdot D^3}{32}\right) = \left(\frac{F \cdot t}{2}\right)\left(\frac{\pi \cdot f_{b\max} \cdot D^2}{8}\right) = \frac{2F \cdot t}{D} \quad \text{...(x)}$$

For expression (vi)

$$F_1 = \left(\frac{\pi \cdot f_{b\max} \cdot D^3}{32}\right), \therefore F_1 = \left(\frac{2F \cdot t}{D}\right) \quad \text{...(13.35)}$$

In lap joint, $F = P$

In a butt joint, dowels are placed on both the sides of the members. The force transmitted by each dowel $F = \frac{F}{2}$. The maximum bearing pressure on dowel and tension on the belt are given by Eqs. 13.34 and 13 .35, respectively.

The maximum bearing stress on the dowel should not exceed the safe bearing stress of the timber. To utilize the full strength of dowel, the strength of dowel in shear and in bearing are kept equal. For this purpose, the diameter of dowel is kept 3.0 to 3.5 times the thickness of dowel. The minimum thickness of dowel kept is 25 mm. The following specifications are followed for disc-dowelled joint in timber in addition to above :

(1) The effective edge distance (the distance measured from the edge of member to centre of dowel) is $\left(\frac{D}{2}+12\right)$ mm to $\left(\frac{D}{2}+20\right)$ mm depending upon the size of dowel.

(2) The effective end distance (the distance measured from the end of member to dowel) is 1.5D in case of tension member and 1.25D in case of compression members. For timber, which are weak in shearing strength, the effective end distance is determined from safe working stress in shear for those timbers.

(3) The minimum thickness of a member in case of lap joint is $\left(\frac{1}{2}t+12\right)$ mm and in case of butt joint, it is $(t + 12)$ mm, where t is the thickness of the dowel.

Example 13.22 *A teak wood member 80 mm × 240 mm carries an axial pull of 40 kN. Design a suitable disc-dowelled joint for the member.*

Solution

Design :

A butt joint as shown in Fig. 13.44 with dowels, is used for the splice in the member.

Thickness of member = 80 mm

Maximum thickness of member in case of butt joint

$$= (t + 12) \text{ mm}$$

where, t = Thickness of dowel

Let the thickness of dowel

$$t = 40 \text{ mm}$$

Minimum thickness of member

$$(40 + 12) = 52 \text{ mm} < 80 \text{ mm}$$

The diameter of dowel

$$D = 3 \times \text{thickness} = (3 \times 40) = 120 \text{ mm}$$

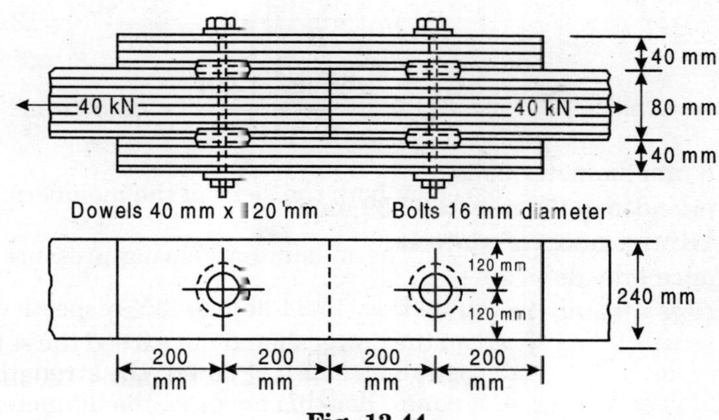

Fig. 13.44

Step 1: Safe working stress

Let the dowel be made of babul wood. From IS : 883–1970 safe working stress in shear along the grain in babul

$$= 2.22 \text{ N/mm}^2$$

Strength of dowel in shear along the grain

$$\left(\frac{\pi}{4} \times \frac{(120)^2 \times 2.22}{1000}\right) = 25.095 \text{ kN}$$

From IS : 883–1970,

Safe working stress in bearing in teak wood

$$= 8.8 \text{ N/mm}^2$$

Strength of dowel in bearing

$$\left(120 \times \frac{40}{2} \times \frac{8.8}{1000}\right) = 21.12 \text{ kN}$$

Dowel value $= 21.12$ kN

Step 2: Number of dowels required

$$\left(\frac{40}{21.12}\right) = 1.894$$

Provide two dowels on each side as shown in Fig. 13.44.

Force in each dowel

$$F = \frac{40}{2} = 20 \text{ kN}$$

Maximum bearing pressure in the dowel from Eq. 13.34,

$$f_{b\,\max} = \left(\frac{16}{\pi} \times \frac{F \cdot t}{D^3}\right) = \left(\frac{16 \times 20 \times 1000 \times 40}{\pi \times 120^3}\right)$$

$$= 2.36 \text{ N/mm}^2$$

From IS : 883–1970, safe bearing pressure perpendicular to the grain in teak wood

$$= 4 \text{ N/mm}^2 > 2.36 \text{ N/mm}^2. \text{ Hence, safe.}$$

Tensile force in the bolt from Eq. 13.35,

$$F_1 = \left(\frac{2F \cdot t}{D}\right) = \left(\frac{2 \times 20 \times 40}{120}\right) = 13.33 \text{ kN}$$

$$\text{Net area required} = \left(\frac{13.33 \times 1000}{(0.6 \times 250)}\right) = 88.89 \text{ mm}^2$$

Provide 16 mm diameter bolts

Net area provided = 156 mm^2 > 88.89 mm^2

Step 3: Arrangement of dowels

Edge distance provide =120

Effective edge distance required

$$\left(\frac{D}{2} + 20\right) = \left(\frac{120}{2}\right) + 20 = 80 \text{ mm} < 120 \text{ mm}$$

Effective end distance required for tension member

$$1.5D = 15 \times 120 = 180 \text{ mm}$$

Provide effective end distance = 200 mm

Step 4: Tensile strength of member

$$\left(\frac{14(240 \times 80 - 40 \times 120)}{1000}\right) = 201.6 \text{ kN} > 40 \text{ kN}$$

Provide 2 fish-plates 40 mm × 240 mm on both the sides as shown in Fig. 13.44.

13.24 METAL CONNECTORS IN TIMBER

The *metal connectors* in timber are also known as *modern timber connectors.* The metal connectors are basically metal rings or fabricated plates. The metal rings or fabricated plates are embedded in the faces of contact of two adjacent members. The members are held together by small diameter bolts. These connectors are used to transmit the load from one member to the other. The various metal connectors in timber are described below.

13.24.1 Split Rings

The split rings used as metal connectors in timber as shown in Fig. 13.45. The split rings are most commonly used. The split rings are made of low-carbon steel. The split rings are placed between two pieces of timber. The circular grooves for split rings are pre-cut in the members by a special tool. The diameter of grooves are slightly more than diameters of rings.

The rings are spread and then inserted in the grooves. The rings remain tight in the grooves. The half width of ring remains in one member and other half width remains in the other member. The inner surface of ring has bearing against the core left in the groove. The outer surface of ring has bearing against

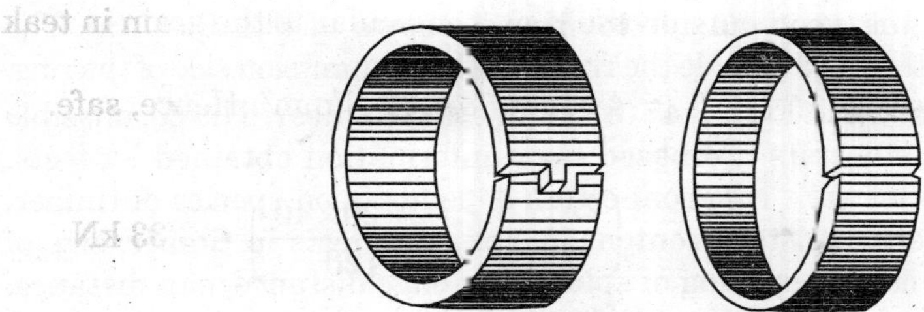

Fig. 13.45 *Split rings*

the outer wall of groove. After embedding the rings into the grooves, the timber members are held together with small diameter bolts and washers. The bolts are kept axial with the rings. The single split ring and multiple split rings used to transmit the load are shown in Fig. 13.46 (a), (b), respectively.

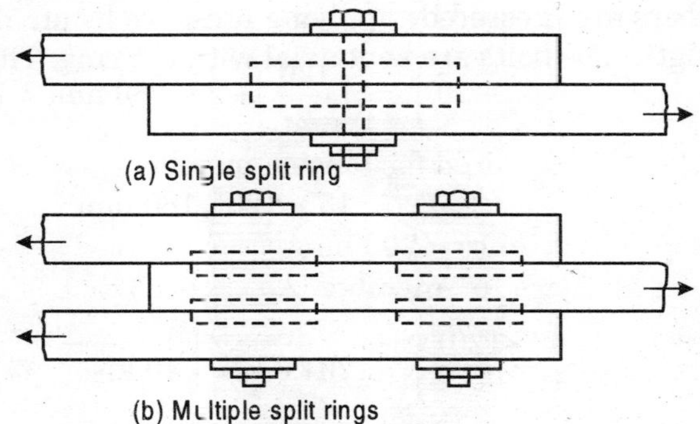

Fig. 13.46 *Multiple split rings*

Figure 13.47 shows a section through split ring connectors joining two pieces of timber. The distribution of bearing pressure of the ring against the core inside

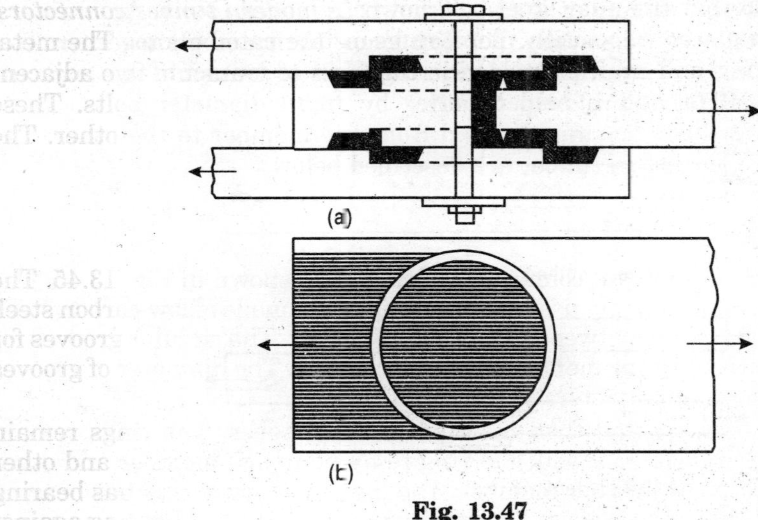

Fig. 13.47

the ring and against the wood outside the ring is shown in Fig. 13.47 (a). The bearing area against the wood outside the ring is at the compression side of the ring.

The shaded area shown in Fig. 13.47 (b) is subjected to shear. The permissible loads for split ring connectors are based upon information obtained by tests. The permissible load for split ring connectors depends upon species of timber, size of timber members, moisture content in timber, defects in timber, size of split rings, diameter of bolt, spacing of split rings, edge distance, end distance, and direction of load relative to the grain of timber.

13.24. 2 Toothed Ring

The toothed ring used as metal connector in timber is shown in Fig. 13 .48. The toothed ring transmits the load in the manner similar to the split-ring. The groove is placed in between surfaces of contact of two adjacent timber members. The timber members are pressed by applying pressure by an alloy steel bolt of high tensile strength. The bolts are kept axial with the ring. The sharp teeth of

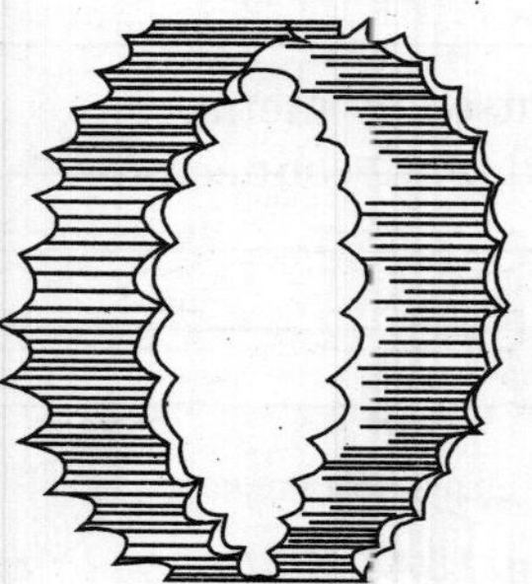

Fig. 13.48

toothed ring penetrated into the timber members. The hole of bolt is made 2 mm over size. So that the alloy steel bolt can be removed easily. An ordinary steel bolt is then placed in position. The bolt keeps the members together. The single toothed ring and multiple toothed ring used to transmit the load are shown in Fig. 13.49 (a) and (b) respectively.

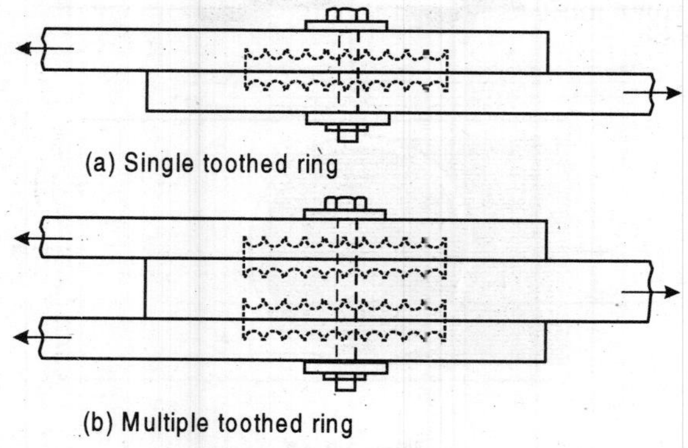

Fig. 13.49

13.24.3 Bulldog Plates

The *bulldog plates* used as metal connectors are shown in Fig. 13 50. The teeth in bulldog plates are distributed over its entire surface. The bulldog plates are

Fig. 13.50

embedded in timber member by bolt pressure in the manner similar to that of toothed ring. The bolt is kept in the centre of the connector.

13.24.4 Shear Plate Connectors

The *shear plate connectors* used as metal connectors are shown in Fig. 13.51. These are also called flanged plate connectors. The shear plate-connectors are

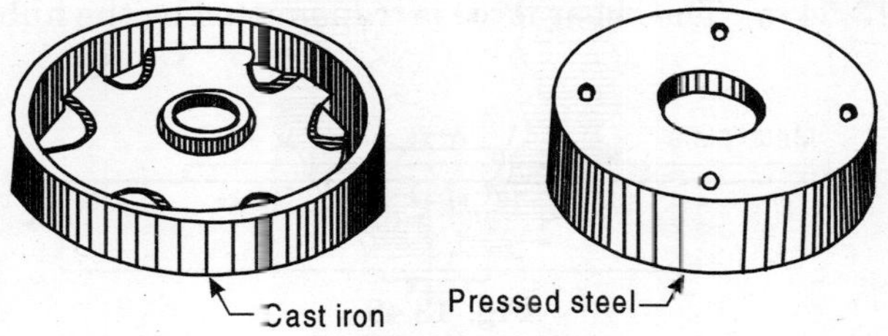

Fig. 13.51

made of pressed steel or cast iron. A hub is provided in the shear plate connector for bearing against the bolt. The grooves are precut in timber members to embed the shear connectors.

The shear plate connectors are used singly in a timber to metal joint or back for a timber to timber joint as shown in Fig. 13.52 (a) and (b), respectively.

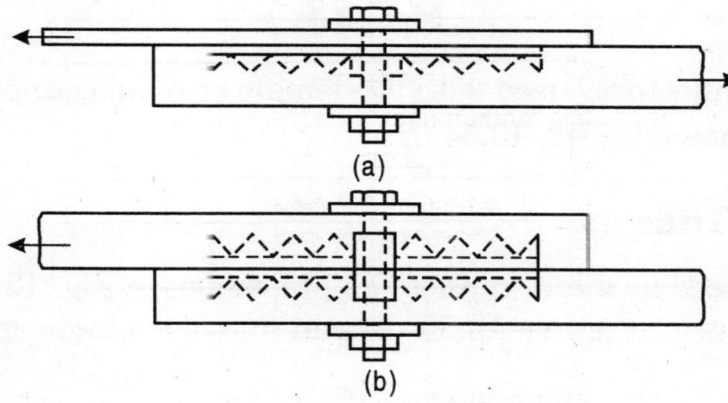

Fig. 13.52

13.24.5 Claw-Plate Connectors

The *claw-plate connectors* used as metal connectors are as shown in Fig. 13.53. These are made of cast iron. These connectors transmit the load in the same manner as shear plate connectors. The grooves are precut in the surfaces of timber members. The teeth of claw plates are pressed by bolt pressure. The connectors are made as male type connector and female type connector.

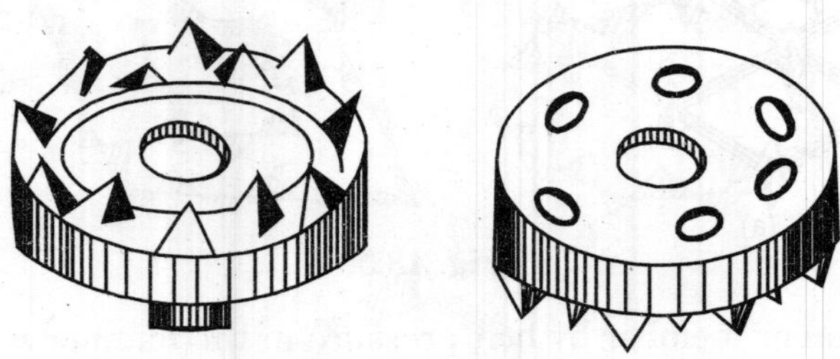

Fig. 13.53 *Claw-plate connectors*

The male type connector has a hub. It is used for *timber to metal joint* as shown in Fig. 13.54 (a). The shear force is transmitted by the hub and not by the bolt. The male connector is used with the female or plain connector for *timber to timber joint* as shown in Fig. 13.54 (b).

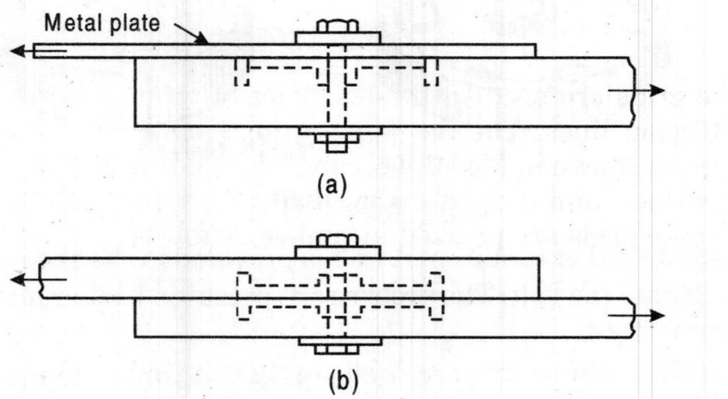

Fig. 13.54

13.24.6 Spike Grids

The *spike grids* used as metal connectors are shown in Fig. 13.55. Fig 13.55 (a) shows flat spike grid. Figures 13.55 (b) and (c) show single curved spike grid, respectively.

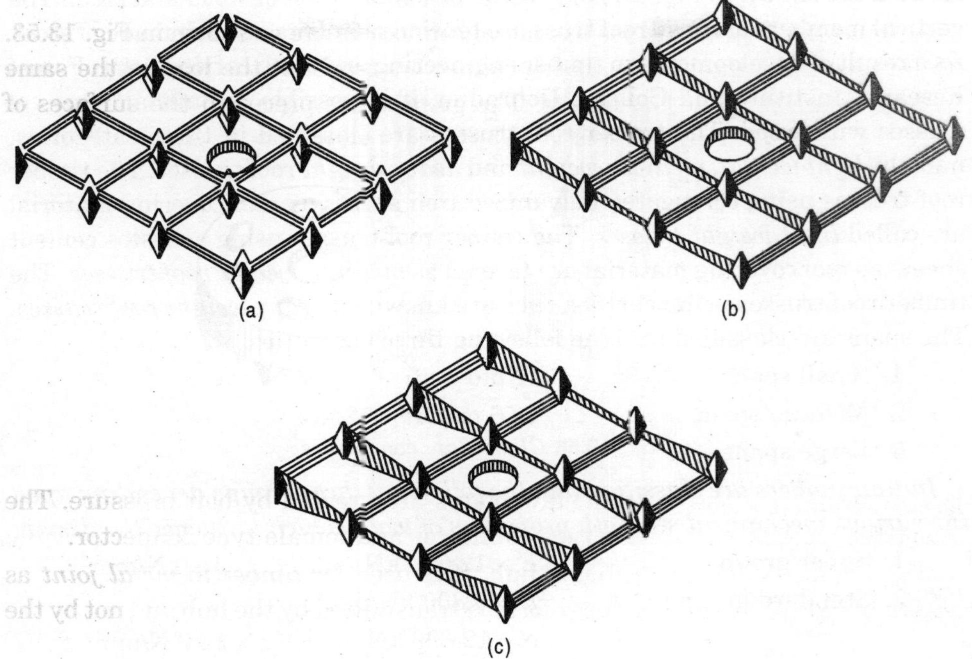

Fig. 13.55

The spike grids are used with piles, poles of trestles, piers whavers, jetties and transmission lines. The flat spike grid is used to connect sawn faces of timber pieces as shown in Fig. 13.56. Single curved spike grid is used to join flat and a curved face, and a double curved spike grid is used to joint two curved faces. The spike grids are pressed in timber members.

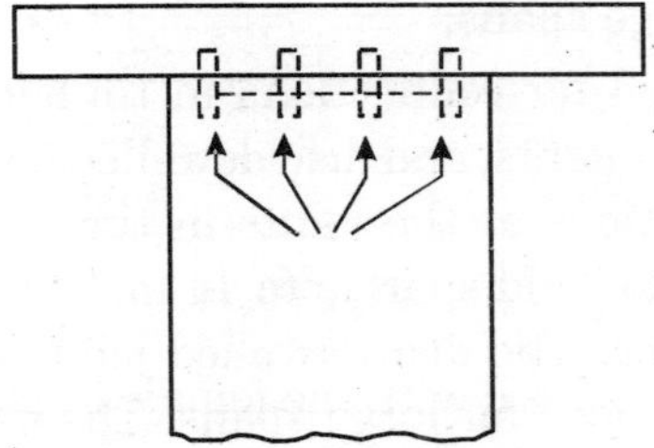

Fig. 13.56 *Spike grid*

13.25 TIMBER ROOF TRUSSES

The timber roof trusses are built principally of wood. In the past, timber roof trusses were used to employ steel rods for its tension web members. Now-a-days timber trusses can be built-in which all the members are of timber. King post truss and Queen post truss shown in Fig. 9.1 (a) and (b) are wooden trusses. Howe truss shown in Fig. 9.1 (b) is made of combination of wood and steel. The vertical members of Howe roof truss are tension members and are made of steel. As a result of developments in timber engineering and work carried out at Forest Research Institute and College, Dehradun, it is possible to built all types of trusses with wood. The timber roof trusses are classified in three categories, namely, *light weight, medium weight* and *heavy weight* roof trusses. The timber roof trusses using corrugated galvanized iron sheets as roof covering material are called *light weight trusses.* The timber roof trusses using asbestos cement sheets as roof covering material are termed as *medium weight roof trusses.* The timber roof trusses using earthen tiles are known as *heavy weight roof trusses.* The spans are classified into the following three categories:

1. Small spans (Upto 6 m)
2. Medium spans (> 6 m, but < 12 m)
3. Large spans (>12 m)

Indian timbers are classified into the following three groups depending upon the various mechanical strength properties of wood which influence the design.

1. Super group $E > 12.600$ kN/mm², f_t, ≥ 18.0 N/mm²
2. Standard group $E > 9.800$ kN/mm², $\alpha \leq 12.600$ kN/mm², f_t, ≥ 12.0 N/mm²
3. Ordinary group $E > 5.600$ kN/mm², $\alpha \leq 9.800$ kN/mm², f_t, ≥ 8.5 N/mm²

where, E = modulus of elasticity of timber species

α = modulus of rupture of timber species

f_t = extreme fibre stress in bending and tension along the grain.

The timbers of ordinary group are recommended for small spans. The timbers of standard group are recommended for medium spans and those of super group are recommended for large spans.

The various joints used for connections in timber roof trusses are framed joints, bolted joints, nailed joints, and disc-dowelled joints. The metal connectors are also used for connections at the joints in timber roof trusses. The nailed joints are suitable for loads and span upto 14 m. The bolted joints are suitable for medium to large spans. The disc-dowelled joints or metal connection with small diameter bolts are used for large spans. The web compression members (discontinuous struts) are made of single members and are inserted in between the other members at the joints in trusses. The web tension members are made of double members. The members in principal rafter (continuous struts) and lower chord members (main ties) are also made of double members. The members in principal rafters and in main tie members are spliced if necessary in between the panel points. The various types of joints at the support used for timber trusses are shown in Fig. 13.57.

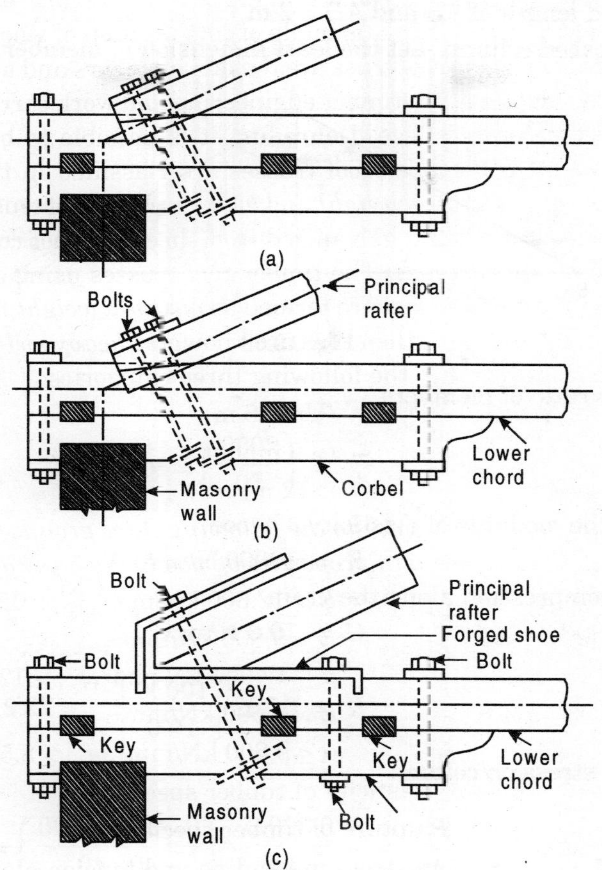

Fig. 13.57

The principal of design of trusses and various loads which will be acting on the roof trusses were discussed in Chapter 9. The order of design of various elements of roof trusses was also discussed in Sec. 9.16.

Example 13.23. *At the apex A of a wooden roof truss, three member AB, AC and AD meet as shown. Forces in AB and AD arc 42.26 kN (compressive) and in AC 11.31 kN tensile. The unsupported length of the members AB and AD are 2 metres. Design and sketch the wooden members the joint.*

Take the following allowable stresses in timber :

Compression along the grain	= *10.0 N/mm²*
Compression across the grain	= *4.0 N/mm²*
Tension along the grain	= *16.0 N/mm²*
Shear stress	= *1.6 N/mm²*

Solution Design

Step 1: Principal Rafters

Members *AB* and *AD* are principal rafters. Forces in *AB* and *AD* = 42.36 kN (compression).

Unsupported length of AB and AD = 2 m

Provide a spaced column. Let the least dimension of member d be 50 mm.

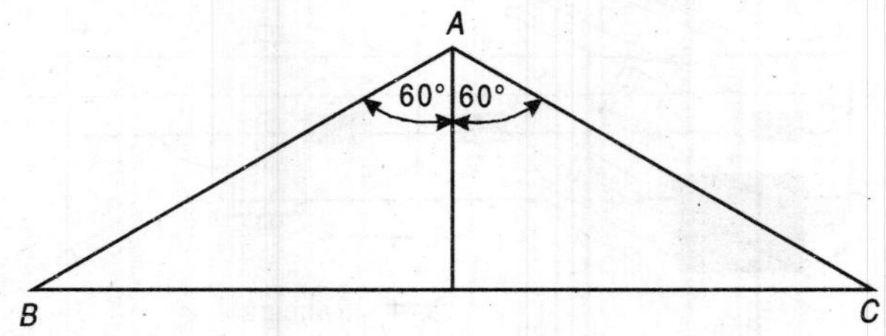

Fig. 13.58

Slenderness ratio of members,

$$\frac{s}{d} = \left(\frac{2000}{50}\right) = 40 > 11$$

Assuming, the modulus of elasticity of wood,

$$E = 10000 \text{ N/mm}^2$$

Allowable compression along the grain

$$C = 10.0 \text{ N/mm}^2$$

$$K = 0.702\left[\frac{10000}{10.0}\right]^{1/2} = 22.2, \ \frac{s}{d} > K$$

Permissible stress on column

$$\frac{P}{A} = \frac{0.329E}{\left(\frac{s}{d}\right)^2} = \left(\frac{0.329 \times 10000}{40 \times 40}\right) = 2.056 \text{ N/mm}^2$$

Area required for column

$$\left(\frac{42.36 \times 1000}{2.056}\right) = 20603.11 \text{ mm}^2$$

Area required for plank is 10301.555 mm^2

Least dimension required is 50 mm

Width of plank required

$$\left(\frac{13301.555}{50}\right) = 206.03 \text{ mm}$$

Provide 2 planks 50 mm × 210 mm

Step 2: Tension members

Force in member AC = 11.31 kN (Tensile). The nailed joint is used for connection. Assuming that preboring is not necessary for driving the nails. Therefore, the gross area and net area of members are same.

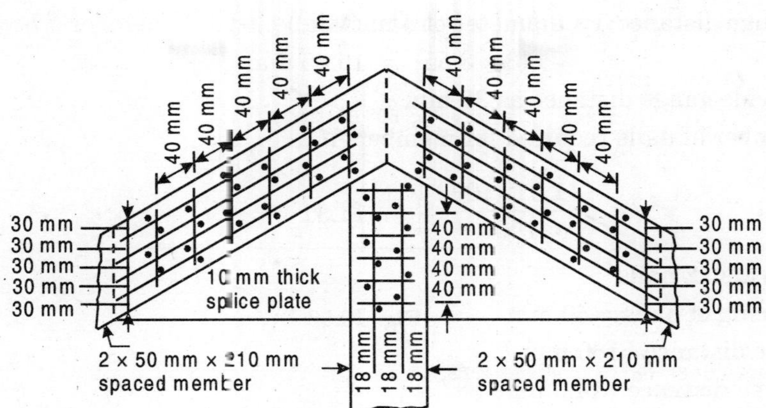

Fig. 13.59

Allowable tension along the grain is 16.0 N/mm^2

Area required for member

$$\frac{11.31 \times 1000}{16.0} = 706.88 \text{ mm}^2$$

Provide 50 mm × 20 mm member

Area provided = 50 × 20 = 1000 mm^2 > 706.88 mm^2

Step 3: Design of joint A. Thickness of member *AB*, and

$$AD = 20 \text{ mm}$$

$$\text{Diameter of nails} = \frac{1}{4} \text{ to } \frac{1}{6} \text{ the thickness}$$

$$\left(\frac{20}{4} \text{to} \frac{20}{6}\right) \text{mm} = (5 \text{ to } 3.33) \text{ mm}$$

Provide 3.55 mm diameter (9 SWG) wire nails. Assuming the lateral strength of nails of size 9 SWG in double shear for permanent construction = 1000 N

Number of nails required for members

$$AB \text{ (and } AD) = \left(\frac{42.36 \times 1000}{1000}\right) = 42.36$$

Provide 48 nails

Step 4: Arrangement of nails

Spacing of nails (pitch)

10 × diameter of nail (Min.) = 10 × 3.55 = 3.55 mm

Provide nails at 40 mm pitch

Edge distance 5 × diameter of nail (Min.)

$$(5 \times 3.55) = 17.75 \text{ mm}$$

Provide nails at 30 mm from edge

Gauge distance 3 × diameter of nail (Min.)

$$(3 \times 3.55) = 10.65 \text{ mm}$$

Provide gauge distance as 30 mm

Number of nails required for member *AC*

$$\left(\frac{11.31 \times 1000}{1000}\right) = 11.31$$

Provide 12 nails

Spacing of nails = 40 mm

Edge distance =180 mm

Gauge distance =14.0 mm

10 mm thick splice plates are provided. A sketch of the joint is shown in Fig. 13.59.

PROBLEMS

13.1 The safe working stress in compression for deodar wood for inside location, parallel to the grain is 7.8 N/mm^2 and that perpendicular to the grain is 2.65 N/mm^2. Determine the permissible bearing stress in the direction of line of action of the load. The load is acting at 40° to the grain direction. The wood is of common grade.

13.2 Determine the safe axial load on a deodar wood column, size 120 mm × 120 mm if its unsupported length is (i) 2.80 m, (ii) 4.20 m, (iii) 6.20 m and it has an inside location.

13.3 A column 100 mm in diameter is made of sal wood. The column is situated in outside location and subjected to alternate wetting and drying. The effective length of column is 1.65 m. Determine the safe load carrying capacity of the round column.

13.4 A sal wood column is to be designed for an outside location with a selected grader timber. If the axial load on the column is 200 kN and the height of the column is 3 m, determine the size of the section to be provided.

13.5 A column carries an axial load of 750 kN inclusive of self-weight. The effective length of column is 3 m. Design a built-up dhaman wood column.

13.6 A spaced column carries an axial compression of 36 kN. The effective length of column is 2.50 m. Design the column using babul wood.

13.7 Design a simply supported deodar wood beam for the following data :

Grade of timber	: Common
Location	: Inside
Span	: 4 m
Load u.d.l.	: 15 kN/m.

13.8 A simply supported timber beam carries a total uniformly distributed load of 42 kN inclusive its own weight. The effective span of beam is 7'20 m. The beam is made of deodar wood. Design a built-up beam. 60 mm × 240 mm planks are available.

13.9 A 120 mm × 300 mm timber beam has an effective span 4.80 m. It carries a connected load of 20 kN at the centre. The beam has a 120 mm × 40 mm notch at the ends. Verify that whether the beam is safe in bending and shear or not. The beam consists of deodar wood.

13.10 A simply supported beam is 560 m long. It consists of two 100 mm × 300 mm wooden planks and one 10 mm × 300 mm steel plate in the centre. Determine the safe load carrying capacity of the beam.

13.11 A 120 mm × 240 mm tension member carries an axial pull of 80 kN. The tension member is made of dhaman wood. Design the lengthening joint. Use 19 mm diameter bolt.

13.12 Two 40 mm × 240 mm wooden vertical members transmit a compression of 20 kN to a wooden horizontal 80 mm × 240 mm. The members consists of dhaman wood. Design the bolted joint.

13.13 Two 40 mm × 240 mm wooden inclined members are connected to a wooden horizontal member 80 mm × 240 mm. The members consists of dhaman wood. Determine the load transmitted by the sloping member if four 22 mm diameter bolts are used for connections. The inclination of the members with the horizontal is 30°.

13.14 In Problem 13.11, design the lengthening joint using standard wire nails.

13.15 In Problem 13.12, if standard wire nails are used, then, design the nailed joint.

13.16 A dhaman wood member 60 mm × 180 mm carries an axial pull of 36 kN. Design a suitable disc-dowelled joint for the member.

13.17 Design a timber joint for the ridge of a king-post truss where the rafters, having a slope of 30° to the horizontal, transmit a load of 300 kN axially and the vertical member 100 kN axially.

13.18 A timber king-post truss has a span of 6 metres and a rise of 2 metres. It is supported on 450 mm walls. The maximum forces in the principal rafter and the horizontal tie are 47.50 kN and 50.00 kN, respectively. Design the principal rafter, the tie and their joint. Allowable stresses in timber are :

Compressive stress along the grains = 10.0 N/mm^2

Compressive stress across the grain = 4.0 N/mm^2

Tensile stress along the grain = 16.0 N/mm^2

Shear stress = 1.0 N/mm^2

13.19 A cycle stand consists of timber frame as shown in Fig. P. 13.60. The frames are spaced at 2.44 m centres and the roof covering consists of A.C. sheets over wooden purlins.

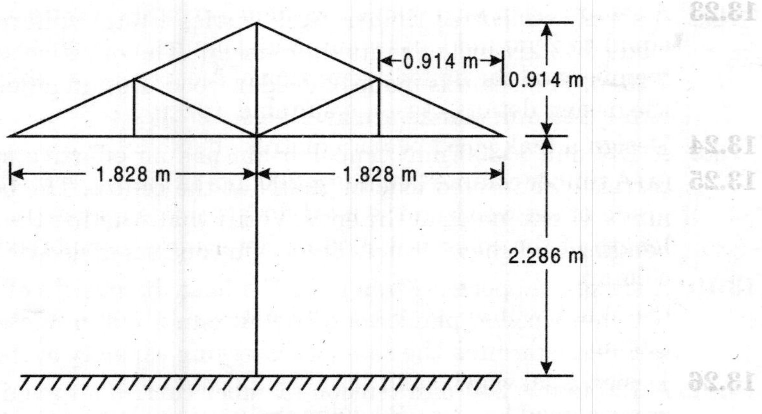

Fig. P 13.60

(a) Design the members of the frame including vertical support if the wind pressure is equal to 0.735 kN/m² on the projected surface vertically or horizontally.

Permissible stress the nail diameter in mm. Show the details of the joint by means of neat sketch.

13.21 A deodar sal wood tension member 50 mm × 150 mm in section carries a pull of 35 kN. Design a double cover bull joint to splice the member, using the standard wire nails. The coefficient for lateral resistance in double shear with clenching is 300 N/mm of nail diameter. What is the efficiency of the joint designed.

Gauge of nail	*Nail data*	
	Diameter (mm)	*Available length (mm)*
2	7.0	228
3	6.5	178
4	5.9	152

13.22 The members *AB* and *AE* of a timber truss as shown in Fig. P. 13.61 are 1.16 m long and carry axial compression of 10.30 kN. The members *AC* and *AD* carry axial pull of 4.90 kN. Design the members and the joint *A*.

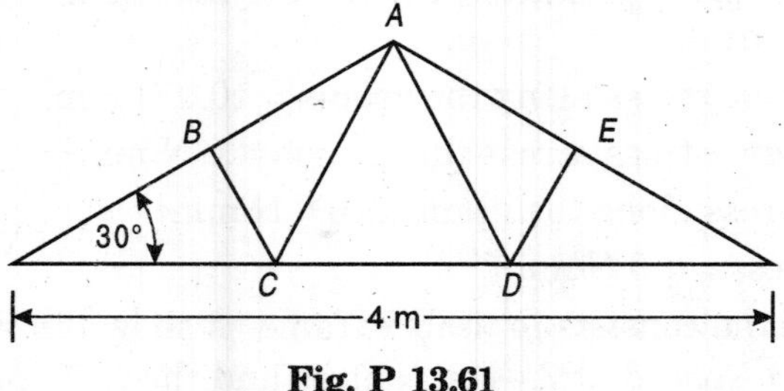

Fig. P 13.61

13.23 A compression member of a wooden roof truss has unsupported length equal to 2 m and carries an axial compression of 30 kN. Design the member and the nailed connection of the member to a fish plate. Shown the design details. Assume suitable working stress.

13.24 Design a teak wood beam 5 metres effective span.

13.25 (a) A timber column 200 mm × 200 mm in section is made from selected grade of sal wood and is used for an outside location. If the effective length of the column is 3.00 m, determine the safe axial load for the column.

(b) What will be the safe load if the effective length is increased to 5.00 m.

13.26 Design a sal wood beam simply supported over a clear span of 2.4 m to carry a dead load of 0.80 kN/m including self-weight, and a live load of 0.40 kN/m. The deflection should not exceed 1/240 of span. Permissible stresses are :

Bending stress = 14.0 N/mm^2
Horizontal shear stress = 0.7 N/mm^2
Bearing stress = 0.40 N/mm^2
E = 1.30 × 105 N/mm^2
Unit weight of sal = 8.70 kN/m^3

13.27 A beam of 8 m span has to support a load of 30 kN/m. Using 200 mm × 300 mm sal wood pieces design a depend section of not more than 600 mm depth for the beam. Also design suitable connectors for it so that the pieces act as one unit. Assume for sal wood.

Permissible shear stress along the grain = 1.5 N/mm^2
Permissible compressive stress along the grain = 10.0 N/mm^2
Permissible bending stress along the grain = 15.0 N/mm^2
Permissible tension in steel bolts = 0.6 f_y

where, f_y, is the yield stress for the structural steel to be used in N/mm^2.

Masonry Structure

PART 3

Chapter 14 Design of Masonry Structures

14.1 INTRODUCTION

The masonry walls, masonry retaining walls, masonry dams, and masonry chimneys are some of the examples of masonry structures. The *masonry retaining walls* are the masonry structures which are used for maintaining the ground surfaces at different elevations on either side of it. The material supported by the retaining wall is known as the *backfill*. The backfill of a retaining wall may have its surface horizontal or inclined. The position of backfill lying above a horizontal plane at the level of top of a retaining wall is termed as *surcharge*. The inclination of the backfill with the horizontal plane is known as the *angle of surcharge*.

The *masonry dams* are the masonry structures which are used to retain water on one side. The masonry chimneys are the masonry structures which are used to escape fuel gases in the atmosphere at high levels. The masonry chimneys have been discussed in Chapter 15. The various forces acting on these masonry structures are the vertical forces and lateral forces. The vertical forces are mainly due to their self-weights. The wind forces or earthquake forces and the earth or the water pressure act as lateral forces on structures. The wind forces have been discussed in Chapter 1 and Chapter 9. The magnitudes and directions of wind forces are adopted from IS : 875 –1984. The earthquake forces have been discussed in Chapter 1. The earthquake forces are adopted from IS : 1893–1962. The earth pressure and water pressure acting on the different masonry structures have been discussed in the subsequent articles.

The masonry structures are generally designed so that there is only compressive strength between the blocks of which the structure is composed. Although, the mortar has some tensile strength, it is usual to assume that the mortar bear no tensile stress (i.e., the tensile strength of the mortar is neglected) and the adhesion between the masonry and the mortar is negligible. Therefore, the shear of tangential force on the masonry at any level must not be greater than the natural friction between masonry and masonry. The stability of the masonry structures depends on their self-weight only. The dimensions of the masonry structures are designed, so that the masonry structures remain stable when subjected to lateral forces.

14.2 GENERAL CONDITION OF STABILITY OF MASONRY STRUCTURES

Following general conditions should be satisfied in the masonry structures for their stability :

1. There must be no tension across the cross-section.
2. The maximum compressive stress must be within the safe stress for the material.
3. The shear force must not be greater than the natural friction between the masonry.
4. The restoring moment must be greater than overturning moment.

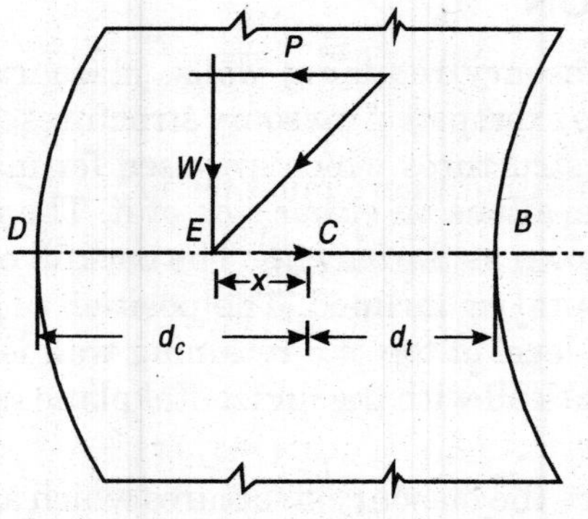

Fig. 14.1

Condition 1. Figure 14.1 shows the cross-section of a masonry structure The point C is the centroid of the cross-section. Let the lateral pressure acting be P ; and gravity force be W, and line of pressure cuts the cross-section at the load point E. The resultant force acting on the cross-section is R. Let the area of cross-section of the masonry structure be A.

The direct compressive stress at the section under consideration

$$f_c = \left(\frac{W}{A}\right) \quad \text{...(i)}$$

The cross section is also subjected to bending moment M, due to eccentricity of the gravity force W

$$M = W \,.\, EC = W \,.\, x \quad \text{...(ii)}$$

The compressive stress due to bending moment

$$f_{bc} = \frac{M}{I} \cdot y_c = \left(\frac{W \cdot x \cdot d_c}{Ak^2}\right) \quad \text{...(iii)}$$

The tensile stress due to bending moment

$$f_{bt} = \frac{M}{I} \cdot y_c = \left(\frac{W \cdot x \cdot d_c}{Ak^2}\right) \quad \text{...(iv)}$$

where, k = Radius of gyration of the cross-section
The combined compressive stress

$$f_1 = f_c + f_{bc} = \left(\frac{W}{A} + \frac{W \cdot x \cdot d_c}{Ak^2}\right)$$

or $$f_1 = \frac{W}{A}\left(1 + \frac{x \cdot d_c}{k^2}\right) \quad \text{...(14.1)}$$

The combined tensile stress

$$f_2 = \left(\frac{W \cdot x \cdot d_c}{A} - \frac{W}{A}\right) = \frac{W}{A}\left(\frac{x \cdot d_c}{k^2} - 1\right) \quad \text{...(14.2)}$$

According to first condition, there must be no tensile stress. Therefore,

$$\left(\frac{x \cdot d_t}{k^2} - 1\right) \text{must be negative}$$

or $$\frac{x \cdot d_t}{k^2} < 1, \text{ or } \left(x > \frac{k^2}{d_t}\right)$$

The maximum possible value of x is given by

$$x = \left(\frac{k^2}{d_t}\right) \quad \text{...(14.3)}$$

Now consider the following cross-sections.

(a) **A solid rectangular cross-section.** It is the most general case in the masonry structure. Let the width of structure $DB = B$, then

$$k^2 = \frac{B^2}{12'} \text{ and } d_t = \frac{B}{2}$$

$$\therefore \quad x = \left(\frac{B^2}{12'} \times \frac{2}{B}\right) = \frac{B}{6} \quad \text{...(14.4)}$$

The point E may lie anywhere between *middle third of cross-section.*

(b) **A solid circular cross-section.** Let the diameter of the circular cross-section be D, then

$$k^2 = \frac{D^2}{12} \text{ and } d_c = d_t = \frac{D}{2}$$

$$x = \left(\frac{D^2}{16} \times \frac{2}{D}\right) = \frac{D}{8} \quad ...(14.5)$$

The point E may lie anywhere between *middle quarter* of the cross-section.

The value of x may be found similarly for hollow rectangular, hollow circular and any other cross-section.

Condition 2. When the line of pressure is in the limiting position, and the section is symmetrical, so that, $d_c = d_t$, then the combined bending stress is given by

$$f_1 = \frac{W}{A}\left(1 + \frac{x \cdot d_c}{k^2}\right) \quad \left[\because \frac{x \cdot d_c}{k^2} = 1\right]$$

$$\therefore \quad f_1 = \frac{W}{A}(1+1) = \frac{2W}{A} \quad ...(14.6)$$

The compressive stress f_1 should be less than the maximum allowable compressive stress for the masonry.

Condition 3. Let μ be the coefficient of friction for the material. Then, the frictional force

$$F = \mu W, \; F > P \text{ or } \mu W > P$$

or

$$\mu > \left(\frac{P}{W}\right) \text{ or } \mu > \tan\theta \quad ...(14.7)$$

But if $\tan \phi = \mu$ then, the condition may be expressed as follows :

The angle of friction for the masonry on masonry must be greater than θ.

Condition 4. The lateral force P has a tendency to overturn the masonry structure about the edge. The moment of the forces tending to over-turn the structure is called as **overturning moment.** The moment of the forces tending to restore the position or restraining the overturning is known as **stability moment.** The stability factor or factor of safety against overturning is defined as the ratio of the stability moment to the overturning moment. When the line of pressure is just at the middle third, then, this factor is 3.

14.3 EARTH PRESSURE

The retained mass of backfill or earth exerts lateral pressure on the retaining walls. The lateral pressure exerted by the earth used as backfill material is known as *lateral earth pressure*. The problem of determining the lateral earth pressure against the retaining wall is one of the oldest in the civil engineering field. A lot of theoretical and experimental work has been conducted in this field. Many theories and hypothesis have been developed to determine the magnitude of lateral earth pressure. Rankine's theory of earth pressure is most commonly used to find the lateral earth pressure. The earth pressure does not act equally in all the directions.

The lateral earth pressure is determined as per *Rankine's theory* of earth pressure in its original form. Rankine's theory of earth pressure is applicable to uniform cohesionless soil only. The following assumptions were made in Rankine's theory:

1. The soil mass is cohesionless, homogeneous and semi-infinite.
2. The ground surface in the backfill is a plane. It may be horizontal or inclined.
3. The back of retaining wall is vertical and smooth. There are no other shear stresses between the wall and the soil. The stress relationship for any element adjacent to the wall is the same as for any other element far away from the wall.
4. The wall yields at the base.

The retaining walls are made in masonry or cement concrete. The back of retaining wall does not remain smooth. As a result of this, the frictional forces develop between the soil and the wall surface. According to Rankine's assumptions of non-existence of frictional forces at the wall surface, the resultant pressure must be parallel to the surface of backfill. The existence of friction makes the resultant pressure inclined to the normal to the wall at the angle that approaches the friction angle between the soil and the wall.

The various cases of cohesionless backfill have been considered for the determination of lateral earth pressure by Rankine's theory in the subsequent articles.

If a bank of earth be left to itself, it will crumble down under the action of weather, until it has taken up a certain slope. The angle of inclination at which such crumbling cases is known as *angle of repose*. It depends on the nature of earth and its wetness.

14.4 LATERAL EARTH PRESSURE ON RETAINING WALLS WITH DRY AND MOIST BACKFILL WITH NO SURCHARGE

Consider an element of earth in the form of a cube at a depth z, below the ground surface as shown in Fig. 14.2 (a). Let p be the normal pressure on the

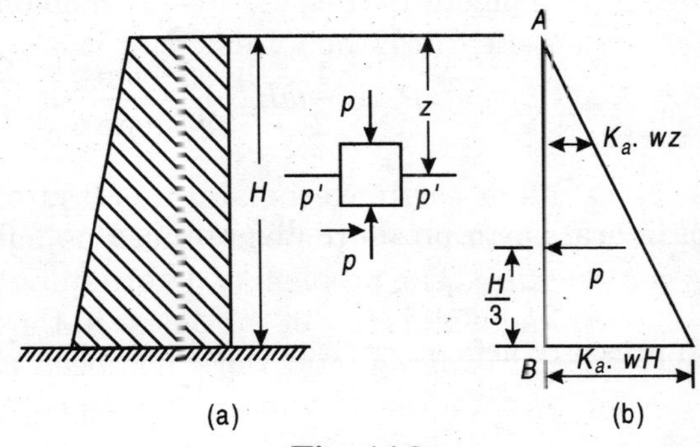

Fig. 14.2

two horizontal faces. The horizontal intensity of pressure developed on the vertical face of cube is p' is less than p. Rankine gave the relationship between p and p' as below :

$$p' = p \cdot \left(\frac{1-\sin\phi}{1+\sin\phi}\right) \quad \text{...(14.8)}$$

where, ϕ = Angle of repose of the backfill material

Equation 14.8 may also be written as below :

$$p' = K_p \cdot p \quad \text{...(14.9)}$$

where, K_a = Coefficient of active earth pressure.

The total earth pressure on the retaining wall may be determined by considering 1 metre length of the retaining wall retaining earth at its vertical face. The normal intensity of pressure varies with its depth from the surface

$$p = w \cdot z \quad \text{...(14.10)}$$

where, w = Unit weight of the material

Therefore, Eq. 14.8 may be written as

$$p' = w \cdot z \cdot \left(\frac{1-\sin\phi}{1+\sin\phi}\right) \quad \text{...(i)}$$

or

$$p' = K_a \cdot w \cdot z \quad \text{...(ii)}$$

The expression (ii) shows that the horizontal intensity of pressure varies linearly with the depth z of the element. At the top, at $z = 0$,

$$p' = 0, \text{ and}$$

At

$$z = H, \quad p' = K_a \cdot w \cdot H \quad \text{...(iii)}$$

The distribution of horizontal intensity of earth pressure is shown in Fig.14.2 (b). The distribution diagram is a triangle ABC. The total earth, pressure P on 1 metre length of the retaining wall is given by the area of triangle ABC.

Therefore,

$$P = \frac{1}{2} \times H \times (K_a \cdot w \cdot H)$$

or

$$P = \frac{1}{2} K_a \cdot w \cdot H^2 \quad \text{...(14.11)}$$

or

$$P = \frac{1}{2} w H^2 \cdot \left(\frac{1-\sin\phi}{1+\sin\phi}\right) \quad \text{...(14.12)}$$

The centroid of lateral earth pressure diagram is at $\frac{H}{3}$ from the base. The total lateral earth pressure acts at $\frac{H}{3}$ from the base.

14.5 LATERAL EARTH PRESSURE ON RETAINING WALLS WITH SUBMERGED BACKFILL

When the backfill of retaining wall is saturated with water then, it is known as submerged backfill. In this case, the lateral pressure is made of two components:

1. Lateral pressure due to submerged weight of the soil, w.
2. Lateral pressure due to water.

The lateral pressure at the base of retaining wall with submerged backfill as shown in Fig. 14.3 (a) is given by

$$p' = (E_a \cdot w' \cdot H + w_w \cdot H) \quad \text{...(14.13)}$$

where, w' = Unit weight of submerged soil

w_w = Unit weight of water

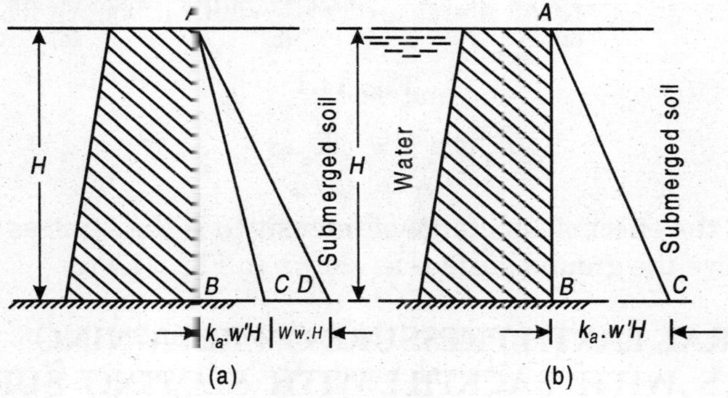

Fig. 14.3

The lateral pressure distribution diagram is shown in Fig. 14.3 (a) by triangles *ABC* and *ACD*. When the water stands on both the sides of wall as shown in Fig. 14.3 (b), then, water pressure is not considered. Then,

$$p' = K_a w'. H \quad \text{...(14.14)}$$

14.6 LATERAL EARTH PRESSURE ON RETAINING WALLS WITH UNIFORM SURCHARGE

When the backfill is horizontal and it carries a surcharge of uniform intensity q per unit area as shown in Fig. 14.4 (a), then, the vertical intensity of pressure at any depth z increases by q. The increase in the lateral pressure is equal to $K_a .q$.

The lateral pressure at any depth z is given by

$$p' = (K_a .w . z + K_A.q) \quad \text{...(14.15)}$$

The lateral pressure at the base of retaining wall is given by

$$p' = (K_a .w.H + K_a. q) \quad \text{...(14.16)}$$

The lateral pressure distribution diagram is (*ABC* + *BCDE*) as shown in Fig. 14.4 (a). Figure 14.4 (b) shows the alternative method of plotting the lateral pressure diagram. The lateral pressure increment K_a, q due to surcharge is the same at every point of the backfill of the wall and does not vary with the depth z. The surcharge intensity q may be expressed by equivalent height of backfill z, such that

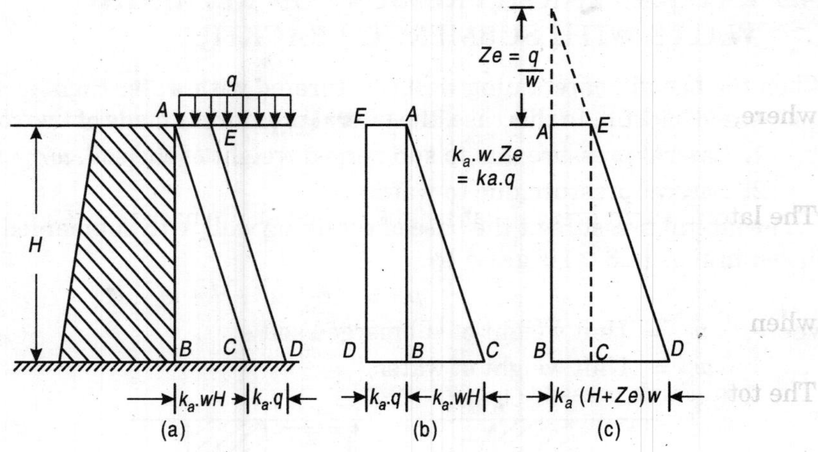

Fig. 14.4

$$(k_a.w.z_e) = (K_a.q)$$

$$\therefore \quad z_e = (q/w) \quad ...(14.17)$$

Therefore, the effect of surcharge of intensity q is the same as that of a fill of height z_e, above the ground surface as shown in Fig. 14.4 (c).

14.7 LATERAL EARTH PRESSURE ON RETAINING WALLS WITH BACKFILL WITH SLOPING SURFACE

The backfill of retaining wall is inclined by angle with the horizontal as shown in Fig. 14.5. The angle β is called the angle of surcharge. In determining the active earth pressure for this case by Rankine's theory, an additional assumption is made. It is assumed that the vertical lateral stresses are conjugate. It may be shown that, if the stress a given point is parallel to another plane, then, the stress on the later plane at the same point must be parallel to the first plane. Such planes are termed as conjugate planes and the stresses are termed as the conjugate stresses.

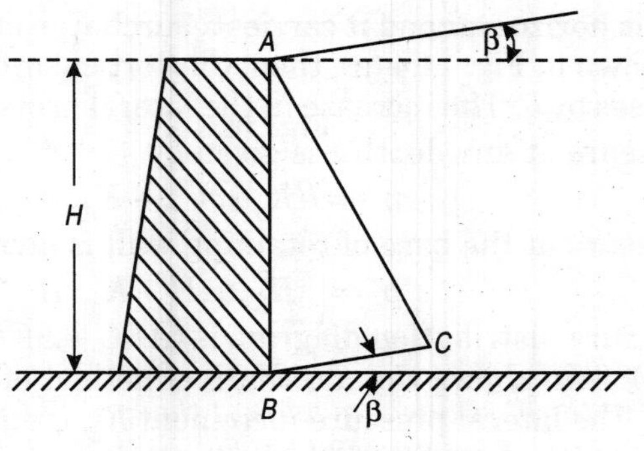

Fig. 14.5

The lateral earth pressure at any depth z in this case is given by

$$p' = K_a' . w . z \qquad ...(14.18)$$

where,
$$K_a' = \cos\beta . \left(\frac{\cos\beta - [\cos^2\beta - \cosh^2\phi]^{\frac{1}{2}}}{\cos\beta - [\cos^2\beta - \cosh^2\phi]^{\frac{1}{2}}} \right)$$

The lateral earth pressure at height H, i.e., at the base is given by

$$p' = K_a' w . H \qquad ...(14.19)$$

when
$$\beta = 0, K_a' = K_a = \left(\frac{1 - \sin\phi}{1 + \sin\phi} \right)$$

The total earth pressure is given by

$$P' = \frac{1}{2} Ka' . wH^2 \qquad ...(14.20)$$

14.8 LATERAL EARTH PRESSURE ON RETAINING WALL WITH INCLINED BACK AND SURCHARGE

Figure 14.6 shows a retaining wall with an inclined back supporting a backfill with horizontal ground surface. The total lateral pressure calculated on a vertical plane BC passing through the heel, (point B). The total pressure is the resultant of the horizontal pressure P and the weight of the wedge ABC.

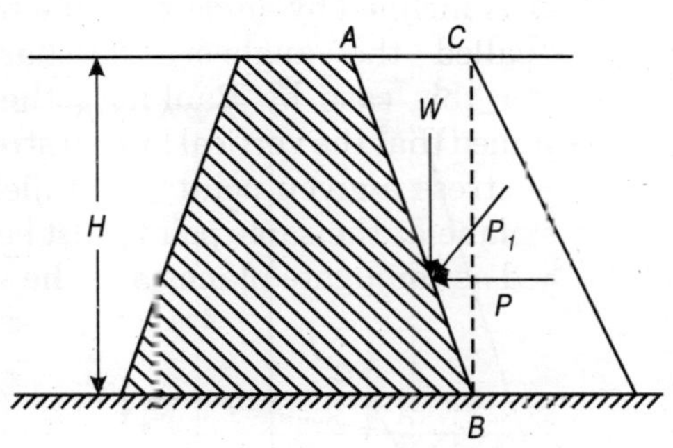

Fig. 14.6

$$\therefore \qquad P = (P^2 \cdot W^2) \qquad ... (14.21)$$

where,
$$P = \left(\frac{1}{2} K_a \cdot wH^2 \right)$$

Example 14.1 *A retaining wall is 6 m high and the level of earth is horizontal. The wall is 2 m wide at the base and 1 m wide at the top. The unit weight of earth is 16 kN/m² and that of the masonry is 24 kN/m³ and the back of wall is vertical. If the angle of repose for the earth is 30 °, determine the stability of wall.*

Solution

Step 1: Lateral earth pressure

The lateral earth pressure is given by

$$P = \frac{wH^2}{2}\left(\frac{1-\sin\phi}{1+\sin\phi}\right) \qquad [\because \phi = 30°]$$

or

$$P = \left(\frac{16\times9\times6}{2}\right)\times\left(\frac{1-0\cdot5}{1+0\cdot5}\right) = 96.00 \text{ kN}$$

Step 2: Gravity forces

The trapezoidal section of retaining wall as shown in Fig. 14.7 is divided into a rectangle *ABCE* and a triangle *AED*. The area of masonry section of retaining wall

$$= 1\times6+\frac{1}{2}\times1\times6 = 9 \text{ m}^2$$

The weight of masonry of 1 metre length of the retaining wall

$$W = 1 \times 9 \times 24 = 216 \text{ kN}$$

Let the vertical line of section of *W* pass through point *F*. Let this point be at a distan $\bar{x}$ from *C*, horizontally. Then by taking the moment of areas about *C*.

$$6 \times 0.5 + 3 \times \left(1+\frac{1}{3}\right) = 9 \times \bar{x}$$

$$\bar{x} = \frac{7}{9} = 0.777 \text{ m}$$

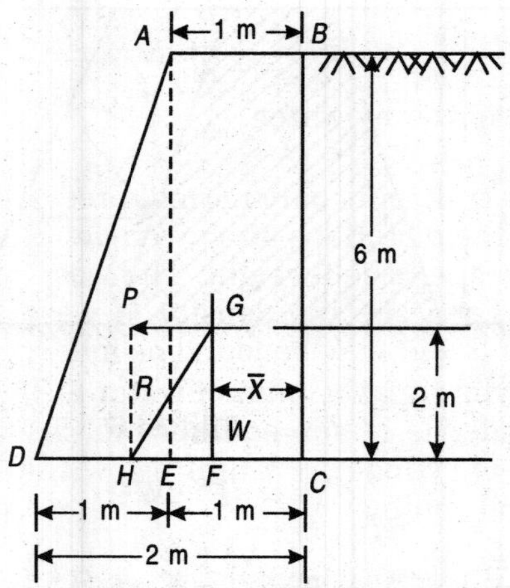

Fig. 14.7

Step 3: Resultant-force

The resultant force *R* is found as follows :

$$R = (P^2+W^2)^{1/2}$$

or

$$R = [(96.00)^2 + (216.00)^2]^{1/2} = 236.20 \text{ kN}$$

The resultant force R passes through the point H. The force P acts at a height $\left(\frac{H}{3}\right)$ = 2 m above the base.

The distance FH is found as follows :

$$\frac{FH}{FG} = \left(\frac{P}{W} \cdot FH\right) = \left(\frac{96 \cdot 00}{216 \cdot 00} \times 2\right) = 0.888 \text{ m}$$

$$\therefore \quad HC = (EC + FH) = 0.777 + 0.888 = 1.665 \text{ m}$$

The distance of point H from the centre of section

$$= \left(1 \cdot 655 - \frac{1}{2} \times 2\right) = \mathbf{0.665 \text{ m}}$$

The width of base of cross-section B is 2 m

$$\left(\frac{B}{2}\right) = 0.6667 \text{ m}$$

The resultant R passes within the middle third of the cross-section. Hence, the retaining wall is stable.

14.9 HYDROSTATIC PRESSURE OF WATER

The force transmitted between water and a solid surface is considered as a *normal pressure*. The force exerted on any given area is known as *water pressure*. When the water is at rest, then the water pressure is known as *hydrostatic pressure of water*. The intensity of pressure exerted within a fluid or on the walls of a structure is measured by the force per unit area. The intensity of pressure is given by

$$p = \left(\frac{P}{A}\right) \qquad \text{...(14.22)}$$

where, P = Force exerted by water

A = Area

The stationary water transmits only normal pressure. Therefore the normal pressure is equal in all the directions at a given depth in the water. The water remains under pressure due to self-weight. The water pressure increases with the increase of depth. Let w be the density of water. The relation between pressure and depth in a liquid may be found as follows :

Consider a vertical cylindrical column of water at rest. The surface of water is horizontal. The cylinder is in the equilibrium. The cross-sectional area of cylinder is A. The base of cylinder is h unit below the free water surface. The volume of this cylindrical column

$$V = (A \times h) \qquad \text{...(i)}$$

The weight of this cylindrical column

$$W = (w.V) = (w.A.h) \qquad \text{...(ii)}$$

The upward force exerted by water at the base

$$R = p \, . \, A \qquad \text{...(iii)}$$

The forces exerted on the sides of cylindrical column are self-balanced. Therefore

$$R = W$$

or $$p \cdot A = w \cdot A \cdot h$$

or $$p = w \cdot h \qquad \text{...(14.23)}$$

Equation 14.23 show that the intensity of pressure at a given depth h below the free surface is given by the weight of a cylindrical column of height h on a unit area. It is to further note that the intensity of pressure in a liquid at rest is equal in all the directions.

The total pressure on any horizontal surface such as the bottom of a tank of area A (above atmospheric pressure) is given by

$$P = w \cdot h \qquad \text{...(iv)}$$

When any plane surface is inclined and it is not horizontal, then, the intensity of pressure varies continuous across the surface according to varying depth.

Consider a vertical submerged surface of total area A as shown in Fig. 14.8 This surface is divided into a large member of parallel strips. The areas of the strips are a_1, a_2, a_3 etc. The depths of these strips below the free water surface are h_1, h_2, h_3 etc respectively. Total pressure on the vertical surface is given by

$$P = w.(h_1 a_1 + h_2 a_2 + h_3 a_3 + \ldots + \text{etc.}) \qquad \text{...(14.24)}$$

The terms $h_1 a_1$, $h_2 a_2$, $h_3 a_3$ constitute the moment of area A about an axis which is the line of intersection of the vertical plane of the submerged surface with the surface of water. Therefore,

$$h_1 a_1 + h_2 a_2 + h_3 a_3 + \ldots + \text{etc.} = \bar{h} \cdot A_1$$

$$P = w \cdot A \cdot \bar{h} \qquad \text{...(14.25)}$$

where, $\bar{h}$ = Depth of the centroid of area

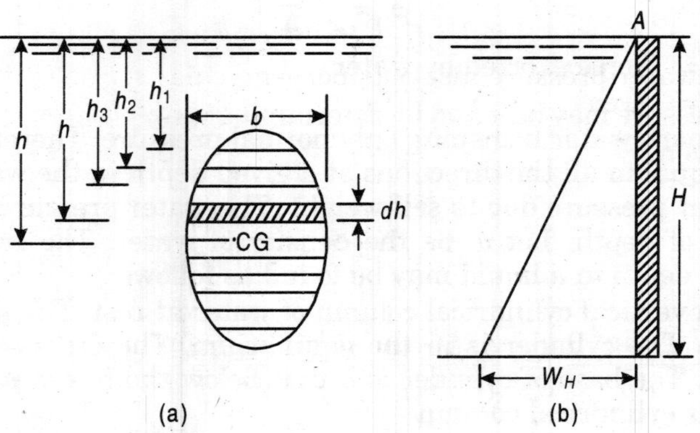

Fig.14.8

The total pressure in a vertically immersed surface may also be found by considering elementary strip of thickness dh and width b at depth h below the free surface of water. The pressure on this strip

$$\delta P = (wh \times b \times dh)$$

The pressure on the surface

$$\int dP = \int (wh \times b \times dh)$$

$$\therefore \quad P = w\int h \cdot bdh \quad \text{or } P = w\,.\,Ah \qquad \text{...(v)}$$

Figure 14.8 (b) shows a vertical surfaces submerged in the water. The top of surface is at the free surface of water. The intensity of pressure at the top

$$p_1 = 0 \qquad \text{...(vi)}$$

The intensity of pressure at the bottom

$$p_2 = w.H \qquad \text{...(vii)}$$

The pressure distribution diagram is a triangle ABC as shown in Fig. 14.8 (b) Total pressure on the plane is equal to the area of this triangular diagram

$$P = \frac{1}{2}wH \cdot H = \left(\frac{wH^2}{2}\right) \qquad \text{...(14.26)}$$

From Eq. 14.25

$$p = w\,.\,Ah = w\,(H \times 1) \times \left(\frac{H}{2}\right) = \frac{wH^2}{2} \qquad \text{...(viii)}$$

This resultant pressure acts through the centre of gravity of triangle ABC in a direction perpendicular to the surface AB. The centre of pressure is situated at *two-thirds* of the depth AB below the free surface or at *one-third*, of the depth AB above the bottom.

The total pressure on any inclined surface submerged in the water is also given by Eq 14.25, i.e.,

$$P = w\,.\,Ah \qquad \text{...(ix)}$$

But this water pressure also acts perpendicular to the inclined surface. The centre of pressure may be found by determining its depth from free water surface, which is given by

$$\bar{y} = \left(\frac{\text{Second moment of area}}{\text{First moment of area}}\right)$$

$$\therefore \bar{y} = \left(\frac{I_{00} \cdot \sin^2\theta}{A\bar{h}}\right) = \left(\frac{Ak_{00}^2 \cdot \sin^2\theta}{A\bar{h}}\right) = \left(\frac{K_{00} \cdot \sin^2\theta}{\bar{h}}\right) \qquad \text{...(x)}$$

$$\text{or} \quad \bar{y} = \left(\frac{K_{C \cdot G^2} + h^2}{\bar{h}}\right) \cdot \sin^2\theta = \left(h + \frac{K_{C \cdot G^2}}{\bar{h}}\right) \cdot \sin^2\theta \qquad \text{...(14.27)}$$

where, k_{00} = Radius of gyration about free water surface

k_{CG} = Radius of gyration about the axis passing through CG

14.10 MASONRY DAMS

The masonry dams are further classified as *gravity dams* and *arched dams.* The gravity dams resist the water pressure by action of the dead or self-weight only. The arched dams are constructed on the sites where a reliable rock abutment is provided by the sides of valley across which the dam is built. The arched dams are curved in plan and resist the pressure behind them by acting as the horizontal arches and are therefore of much lighter section than the gravity dams. The arched dams are not often constructed and are subjected to complicated stress distributions. The stability of gravity dams has only been discussed here. A dam is virtually a retaining wall for the water. But the magnitude of water pressure may be accurately determined as compared to that of earth pressure on the retaining wall. Therefore, the design of dam is capable of more rigid treatment than that of a retaining wall. Further, the dams are work of much greater magnitude than the retaining walls and it is more necessary to provide the minimum quantity of material consistent with the safety. The dams possessing the minimum section consistent with the safety are said to be designed of economical section.

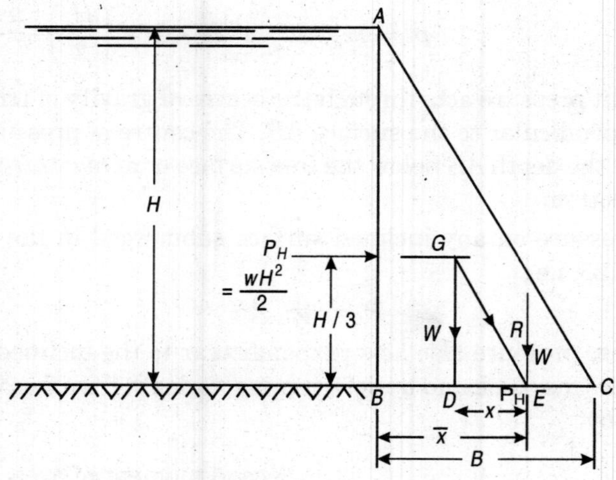

Fig. 14.9

The stability of any proposed dam is to be examined under two conditions :

1. When the reservoir is empty, and material of dam is subjected to the action of self-weight only.

2. When the reservoir is filled with water and it is subjected to combine effect of water pressure and weight of masonry.

The line of action of resultant under both the conditions should pass within the *middle-third* at any horizontal section through the dam, including at the base.

A right angled-triangle *ABC* as shown in Fig. 14.9 is an ideal theoretical section. It gives the basis to deduce a practical section. Let the water level be upto the top as shown in Fig 14.9. Consider 1 metre length of dam. The total water pressure acting on the face *AB* of dam

$$P_H = \frac{wH^2}{2} \quad \text{...(i)}$$

The line of action of water pressure is at height $\frac{H}{3}$ from the bottom of dam.

The weight of masonry of dam

$$W = \frac{1}{2} \cdot p \cdot B \cdot H \quad \text{...(ii)}$$

where, p = Unit weight of masonry

B = Width of the dam at the bottom

The weight of masonry acts at $\frac{1}{3}$rd width of the base from *B*. *When the reservoir is empty*, then, the water pressure is zero. The eccentricity at the base of dam

$$e = \left(\frac{B}{2} - \frac{B}{3}\right) = \frac{B}{6} \quad \text{...(iii)}$$

The weight of dam acts at the middle third point from the face (i.e., on the heel side). The pressure at the toe, from Eq. 11.2

$$f = \frac{W}{A}\left(\frac{x \cdot d_t}{k^2} - 1\right) \qquad \left(\because x = \frac{k^2}{d_t}\right)$$

$$\therefore \quad f = \text{zero} \quad \text{... (iv)}$$

The pressure at the heel is given by Eq. 14.1

$$f = \frac{W}{A}\left(1 + \frac{x \cdot d_c}{k^2}\right) \qquad \left(\because x = \frac{k^2}{d_c}\right)$$

$$\therefore \quad f = \left(\frac{2W}{A}\right) = \left(\frac{2W}{B \times 1}\right) = \left(\frac{2W}{B}\right) \quad \text{...(v)}$$

or

$$f = \left(\frac{2}{B} \times \frac{1}{2} \times p \times B \times H\right) = \rho.H \quad \text{...(vi)}$$

When the reservoir is full, then, the resultant of water pressure P_H and the weight of the masonry of dam *W* passes through *E*. The point *E* may be located by considering the triangle of forces *GDE*.

$$\frac{DE}{GD} = \left(\frac{P_H}{W}\right), \; x = DE = \left(\frac{P_H}{W} \times GD\right)$$

$$x = \left(\frac{P_H}{W} \times \frac{H}{3}\right)$$

or $$x = \left(\frac{WH^2}{2} \times \frac{H}{3}\right) \times \left(\frac{1}{\frac{1}{2} \times P \times B \times H}\right)$$

or $$x = \left(\frac{1}{3} \cdot \frac{w}{p} \cdot \frac{H^2}{B}\right) \qquad \text{...(14.28)}$$

Alternatively. The distance x may be found by taking the moments of both the forces about E. Therefore,

$$P_H \times \frac{H}{3} = W.\, x, \text{ or } x = \left(\frac{1}{3} \cdot \frac{w}{p} \cdot \frac{H^2}{B}\right) \qquad \text{...(viii)}$$

The point E may be at a distance $\frac{B}{3}$ on the toe side at the most, so that the resultant is at the middle third point.

$$\therefore \quad \frac{B}{3} = \left(\frac{1}{3} \cdot \frac{w}{p} \cdot \frac{H^2}{B}\right)$$

$$\therefore \quad B = \left[\frac{w}{\rho}\right]^{1/2} \qquad \text{...(14.29)}$$

Equation 14.29 gives the minimum width necessary for an idea section.

Taking $w = 10 \text{ kN/m}^2$ and $\rho = 20 \text{ kN/m}^3$

$$\therefore \quad B = 0.71\, H \qquad \text{...(14.30)}$$

The resultant pressure at the bottom, R is proportional to side GE of the triangle of forces GDE. The vertical and horizontal components of this resultant force R are W and P_H, respectively. These components also act at E. The vertical component gives normal pressure at the bottom. The horizontal component P_H tends to slide the dam horizontally. The maximum pressure at the toe

$$f_{max} = \left(\frac{2W}{B}\right) = \rho \,.\, H \qquad \text{...(ix)}$$

The co-efficient of friction, from Eq. 14.7

$$\mu > \frac{P_H}{W}, \text{ or } \mu > \frac{1}{2} wH^2 \times \left(\frac{1}{\frac{1}{2} \rho \cdot B \cdot H}\right)$$

or $$\mu > \left(\frac{w}{p} \cdot \frac{H}{B}\right) \qquad \text{...(14.31)}$$

or $$\mu > \left(\frac{10}{20} \times \frac{0.71B}{B}\right), \mu > 0.355 \qquad \text{...(x)}$$

Alternatively. The position of point E may be found by equating the moments of P_H and W and to the moment of resultant, R about the heel i.e., about toe point B in Fig. 14.9. In this method, the distance of point E, $EB = \bar{x}$ is found directly Therefore,

$$P_H + \frac{H}{3} + W \times \frac{B}{3} = W \times \left(x + \frac{B}{3}\right) + P_H \times O$$

or
$$\left(\frac{wH^2}{2} \times \frac{H}{3}\right) + \left(\frac{\rho \cdot B \cdot H}{2} \times \frac{B}{3}\right) = W \cdot \bar{x} \quad ...(14.32)$$

In a practical section, the following modifications of the theoretical triangular section become necessary. A free board is always necessary above the supply level. The upper edge is given a reasonable practical width to the triangular section at the top. Many dams carry a roadway along the top. The base width is increased by providing a sloping outer surface or a series of batters or a continuous curve.

14.10.1 Width of Base of Trapezoidal Dam

Let $ABCD$ be a trapezoidal dam section with a vertical face as shown in Fig 14.10 (a). Let width of top of dam AB be a, and width of bose of dam CD be b, and the height of dam be h. The width of the base of the dam section is determined

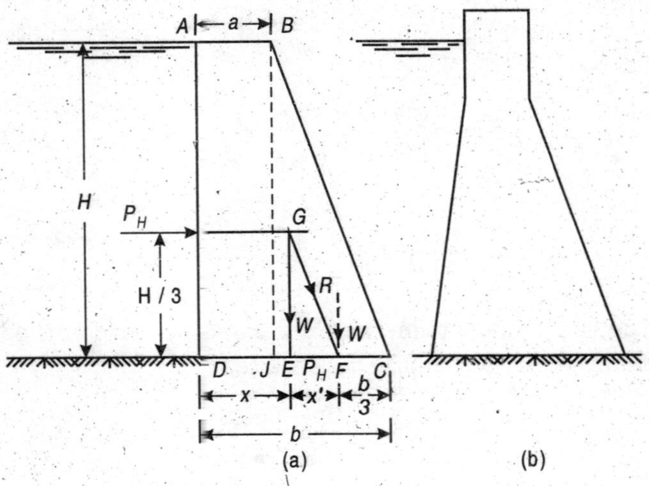

Fig. 14.10

such that the line of action of resultant R strikes the base within the middle third of the section. The point G represents the centroid of the trapezoidal dam section. The vertical through G intersects the base at the point E. Let the distance DE be a from D. The trapezoidal dam section is divided into a rectangle $ABJD$ and a triangle BJC. Taking the moments of these areas about D.

$$x\frac{(a+b)}{2}\cdot h = k\cdot a\cdot\frac{a}{2}+\frac{1}{2}h(b-a)\left[a+\frac{1}{3}(b-a)\right]$$

or
$$x\frac{(a+b)}{2} = \frac{h}{2}\left[a^2+\frac{(b-a)(b+2a)}{3}\right]$$

or
$$x\frac{(a+b)}{2} = \frac{h}{6}(b^2+ab+a^2)$$

$$x = \left[\frac{a^2+ab+b^2}{3(a+b)}\right] \quad \text{...(xi)}$$

Total water pressure per metre length

$$P_H = \frac{wH^2}{2} \quad \text{...(xii)}$$

Total weight of masonry per metre length

$$W = \rho(a+b)\frac{H}{2} \quad \text{...(xiii)}$$

From the triangle of forces *GEF*

$$\frac{EF}{GE} = \frac{P_H}{W} = \frac{wH^2}{2}\times\left(\frac{2}{\rho(a+b)H}\right)$$

or
$$\frac{EF}{GE} = \left(\frac{w\cdot H}{\rho(a+b)}\right) \quad (\because GE = \frac{H}{3})$$

or
$$EF = x' = \left(\frac{w\cdot H^2}{3\rho(a+b)}\right) \quad \text{...(xiv)}$$

The line of pressure *GF* is to be within the middle third of the base, so that there is no tensile stress in the base. Therefore,

$$x + x' = \frac{2b}{3}$$

or
$$\frac{a^2+ab+b^2}{3(a+b)}+\frac{wH^2}{3p(a+b)} = \frac{2b}{3}$$

$$a^2+ab+b^2+\frac{wH^2}{p} = 2b\,(a+b)$$

or $$a^2 + ab + b^2 + \frac{wH^2}{\rho} = 2ab + 2b^2$$

or $$b^2 + ab - \left(a^2 + \frac{wH^2}{2}\right) = 0$$

$$\therefore \quad b = \frac{-a + \left[a^2 + 4\left(a^2 + \frac{wH^2}{\rho}\right)\right]^{\frac{1}{2}}}{2}$$

$$\therefore \quad b = -\frac{a}{2} + \left[\frac{5a^2}{4} + \frac{wH^2}{\rho}\right]^{\frac{1}{2}} \qquad \text{...(14.33)}$$

It is to note that Eq. 14.33 gives the minimum width of the base required for the trapezoidal dam section. The maximum compressive stress is given by

$$f_{max} = \left(\frac{2W}{b \times 1}\right) = \left(\frac{\rho H (b-a)}{p}\right) \qquad \text{...(14.34)}$$

In actual practice, a steep batter is also given on the upstream side of the dam section as shown in Fig 14.10 (b).

Example 14.2 *A masonry dam of trapezoidal section is 75 m high and 12 m thick at the top. The water face of dam is vertical. The free water level is upto the top of the dam. The unit weight of masonry is 24 kN/m³. Determine the minimum width of the base of the dam, so that there is no tensile stress anywhere at the base. Also determine maximum compressive stress at the base.*

Check the stability of dam against sliding. The coefficient of friction between the base of dam and earth is 0.7.

Solution

Step 1. Minimum width of base

The minimum width of base is given by Eq. 14.33 as

$$b = -\frac{a}{2} + \left[\frac{5a^2}{4} + \frac{wH^2}{\rho}\right]^{\frac{1}{2}}$$

or $$b = -\frac{12}{2} + \left[\frac{5}{4} \times (12)^2 + \frac{1000}{2400} \times 75 \times 75\right]^{\frac{1}{2}} \text{ m}$$

or $$b = \mathbf{44.23\ m}$$

Step 2. Max. Compressive stress

From Eq 14.34, the maximum compressive stress at the base

$$f_{max} = \frac{\rho \cdot H(b+a)}{b} = \left(\frac{24 \times 75(44 \cdot 23 + 12)}{44 \cdot 23}\right)$$

$$f_{max} = 2288.36 \text{ kN/m}^2 \text{ or } f_{max} = \mathbf{2.288\ N/mm^2}$$

Step 3. Check for stability of dam against sliding

$$\frac{P_H}{W} = \left(\frac{wH^2}{2} \times \frac{2}{\rho \cdot H(b+a)}\right) = \left(\frac{w}{\rho}\frac{H}{(b+a)}\right) = \left(\frac{10 \times 75}{24 \times 56 \cdot 23}\right) = 0.556 < 0.7.$$

Hence, safe

PROBLEMS

14.1. The height of a masonry retaining wall is 9 m. The width at top of retaining wall is 1 m and that at bottom is 7 m. The wall is used to retain earth. The face of retaining wall is vertical. The unit weight of earth is 16 kNm^2 and that of masonry is 24 kN/m^3. The angle of friction of earth particles is 30°. The coefficient of friction between the masonry and the masonry is 0.5. The bearing capacity of earth is 120 kN/m^2. Determine the total lateral pressure, the total vertical load and the eccentricity of resultant force. Check the stability of the retaining wall against (a) tension, (b) compression, (c) sliding, and (d) overturning. Also find the maximum and minimum stresses at the base of the retaining wall.

14.2. A masonry dam of trapezoidal section is 50 m high and 8 m thick at the top. The water face of the dam is vertical. The free water level is at the top of dam. The unit weight of masonry is 24 kN/m^3. Determine the minimum width of the base of the dam, so that there is no tension anywhere at the base. Also determine the maximum compressive stress at the base. Check the stability of dam against sliding. The coefficient of friction between the base of dam and earth is 0.65.

14.3. A masonry retaining wall of trapezoidal section has its top width equal to 0.75 m and height 5 m. Its face which is in contact with the retained earth is vertical. The earth retained is level at top. The soil weighs 16 kN/m^3 and its angle of internal friction is 30°. The masonry weighs 24 kN/m^3. Determine the minimum width of the base to avoid tensile stresses and determine the maximum compressive stress for this base width.

If the coefficient of friction between base and soil is 0.60 check the stability of retaining wall.

14.4. Check the stability of the masonry retaining wall with concrete base loaded as shown and determine maximum pressure under the base.

Take unit weight of masonry as 20 kN/m^3 of concrete as 24 kN/m^3 and that of soil as 16 kN/m. The angle of internal friction of soils is 30° and coefficient of friction between base and soil is 0.5.

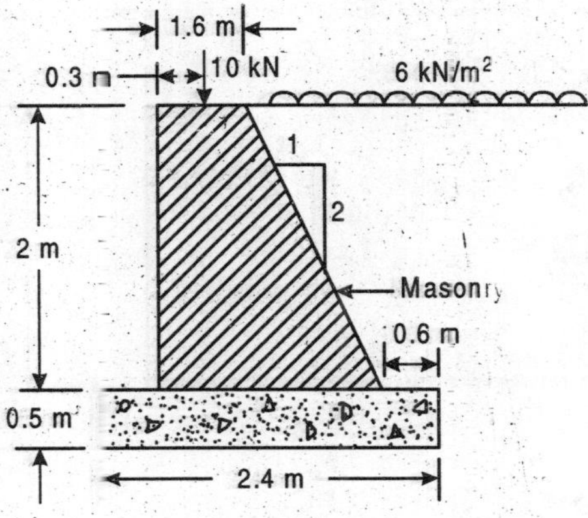

Fig. P 14.4

14.5. A masonry dam has a triangular cross-section with 15 m base width as shown in Fig. P 14.5. The reservior is filled with water. The water level is 1 m below the top. Determine the maximum height of the dam such that the dam remains structurally safe.
Take unit weight of masonry = 20 kN/m^3
Permissible stress in compression in masonry = 1.5 N/mm^2
No tensile stresses are permitted.

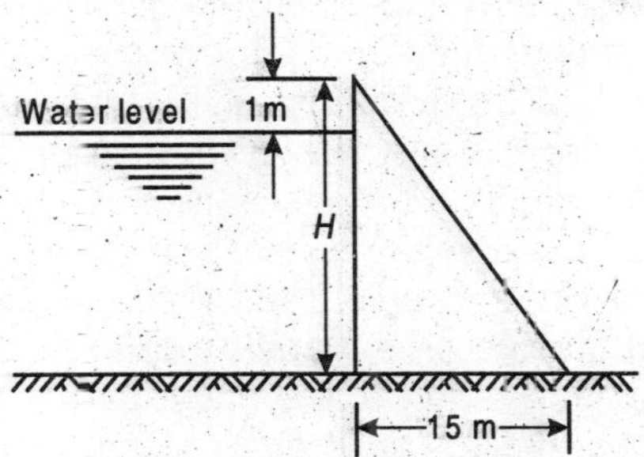

Fig. P 14.5

Chapter 15

Design of Masonry Chimneys

15.1 INTRODUCTION

The chimneys are used to escape the gases to such a height that the gases do not contaminate surrounding atmosphere. The cross-sectional area of chimney is kept large enough to allow the passage of burnt gases. The cross-sectional area of a chimney for the passage of burnt gases depends on the type of fuel and quantity of fuel to be used in a plant, available draft for carrying the burnt gases set up in the chimney and losses due to friction within the chimney. The height of chimney is kept to provide required draft. The draft is defined as the pressure available for producing a flow of burnt gases. The draft depends on the height of chimney above sea level, the type of fuel to be burnt, the type of furnace and the temperature of burnt gases. When the gases in the chimney are heated, then the hot gases expand. The hot gases occupy larger volume than before. The weight of gases per cubic metre becomes less. As a result of this, the pressure at the bottom of chimney due to weight of hot gases also becomes less than the pressure due to weight for cold air outside the chimney. The difference between the two pressure result in the flow of the burnt gasses up the chimney. For the purpose of the structural design, the height and diameter of a chimney at the top are given.

The masonry chimneys are generally built in brick masonry. The brick masonry chimneys are built in common bricks or of special radial bricks, which are usually perforated and moulded to different sizes suitable for use in various diameters. The masonry chimneys may have square, hexagonal, octagonal or circular shapes in plan. The masonry chimneys of circular shape in plan are usually preferred over the other shapes.

15.2 LINING FOR MASONRY CHIMNEYS

The material used for lining should be capable of withstanding high temperature upto 2000° F. The lining is required from below the breech opening to the height where the heat of gases does not damage the chimney. The fire bricks are used for lining in the brick masonry chimneys. The fire brick lining must be free to expand and contract independently of the main chimney. The height of lining depends on the particular purpose of chimney. The height of chimney may be as low as 6 m to 10 m but it is commonly 16 m to 26 m high in lofty chimneys. The height of lining is kept $\frac{1}{4}$th to $\frac{1}{5}$th as a common rule.

An efficient air space should be provided between the fire brick line and the outer shaft, suitably covered or protected from corebelled course of brickwork projecting from the inside of the outer shaft and leaving sufficient clearance for maximum expansion of the lining.

15.3 VARIOUS FORCES ACTING ON MASONRY CHIMNEYS

The various forces acting on the masonry chimney are as follows :

1. Self weight of the masonry chimney.
2. Weight of lining.
3. Wind pressure.
4. Seismic force.

15.3.1 Self-weight of the Masonry Chimney

The self-weight of masonry chimney W_s acts vertically. The masonry chimneys of circular cross-section in plan are built tapering from top towards bottom. The portion of a circular masonry chimneys between two horizontal sections may be considered to be frustum of a hollow cone. The volume of a hollow frustum of a cone is given by

$$V_P = \frac{\pi \cdot H}{3} [R_1{}^2 + r_1{}^2 + 2R_1r_1) - (R_2{}^2 + r_2{}^2 + 2R_2r_2)]] \quad ...(15.1)$$

where, R_1 = Outer radius of the cone at the bottom,

r_1 = Outer radius of the cone at the top

R_2 = Inner radius of the cone at the bottom

r_2 = Inner radius of the cone at the top

H = Height of the frustum of hollow cone.

The weight of masonry of the chimney

$$W_s = \rho . V$$

or

$$W_s = \rho \frac{\pi \cdot H}{3} [(R_1^2 + r_1^2 + 2R_1r_1) - (R_2^2 + r_2^2 + 2R_2r_2)]$$

where, ρ = unit weight of the material used in masonry

or

$$W_s = \rho \frac{H}{3} [\pi (R_1^2 - r_1{}^2) + \pi (r_1^2 - r_2^2) + (2 \pi R_1 r_1 - 2 \pi R_2 r_1)$$

or $$W_s = \rho \frac{H}{3}\left(A_1 + A_2 + \sqrt{A_1 A_2}\right) \qquad ...(15.2)$$

The compressive stress due to weight of masonry of the chimney at any horizontal cross-section

$$f_1 = \frac{W_s}{A}$$

or $$f_1 = \left(\frac{W}{\pi\left(R_1^2 - R_2^2\right)}\right) \qquad ...(15.3)$$

where, R_1 = Outer radius of the chimney

R_2 = Inner radius of the chimney.

15.3.2 Weight of Lining

The weight of lining in the masonry chimney, W_L also acts vertically. The weight of lining is given by

$$W_L = (\rho_2 \,.\, \pi \,.\, d \,.\, t_L \,.\, H_L) \qquad ...(15.4)$$

where, ρ_2 = Unit weight of brick lining = 20 kN/m^3

d = Mean diameter of the lining

t_L = Thickness of lining

H_L = Height of the lining.

The lining is supported directly at the base.

15.3.3 Wind Pressure

The wind pressure acts horizontally. The wind pressure acting on a structure depends on the shape of the structure, the height of structure, the location of structure, and the climatic conditions. The wind pressure per unit area increase with the height of structure above the ground level. In order to simplify the design, the chimney is divided into number of segments. The intensity of wind pressure over each segment may be assumed as uniform. The intensity of wind pressure corresponding to mid-height of each segment may be noted from IS : 875 –1984. The wind pressure acting on the chimney may be found as below :

$$P = k \,.\, p. \text{ (Projected area of chimney)} \qquad ...(15.5)$$

where, k = Shape factor

p = Intensity of wind pressure.

The shape factors for various shapes in plan of the chimney shaft are given in Table 15.1 as per IS : 875 –1984.

Therefore, the wind pressure acting on the chimney circular in plan

$$P = 0.7 \times p \times \text{(Projected area)} \qquad ...(15.6)$$

Table 15.1 *Shape (in plan) factors*

Plan shape	*Factor, k*		
	Ratios of Ht to base with (0 – 4)	*Ratio of Ht to base with (4 – 8)*	*Ratio of Ht to base with 8 or over*
Circular	0.7	0.7	0.7
Octagonal	0.8	0.9	1.0
Square (wind perp. to diagonal)	0.8	0.9	1.0
Square (wind perp. to face)	1.8	1.15	1.3

The projected area of a masonry chimney circular in plan and tapering from top to the bottom is a trapezium.

The wind pressure has an overturning effect on the chimney. The overturning moment M_w due to wind pressure causes compressive stress on the leeward side of the chimney and the tensile stress on the windward side of the chimney. The maximum compressive stress and tensile stress on the extreme fibre of chimney are equal and these are given by

$$f_w = \left(\frac{M_w}{I} \cdot R_1\right) = \left(\frac{M}{Z}\right) \qquad ...(15.7)$$

where, R_1 = Outer radius of the chimney at the section under consideration

I = Moment of inertia of the section

$$I = \frac{\pi}{64}\left(R_1^4 - R_2^4\right) = \frac{\pi}{64}(D_1^4 - D_2^4) \qquad ...(15.8)$$

It is to note that the circular shape in plan for the masonry chimney realizes the greatest stability with the minimum quantity of brickwork. It is the most effective section of flue. It is equally resistant to the wind pressure applied in any direction.

In addition to the overturning effect due to wind pressure, the wind has also aerodynamic effect. This effect of wind has not been taken into consideration for the design of chimney.

15.3.4 Seismic Forces

The seismic forces also act horizontally. The seismic forces act on a structure, when the structures are located in the seismic areas.

The worst effect out of effect due to seismic (earthquake) forces and wind effect is only taken into consideration. Only one effect is considered for the design of structure out of these two effects.

15.4 STABILITY OF MASONRY CHIMNEYS

A masonry chimney of circular shape in plan tapering from top towards bottom is shown by a diagrammatic sketch in Fig. 15.1.

The stability of tall masonry chimney depends on the weight of the masonry work of the chimney shaft W_s, and the wind pressure P_H acting on the outer face. In the calm atmosphere, the centre of pressure at any horizontal section

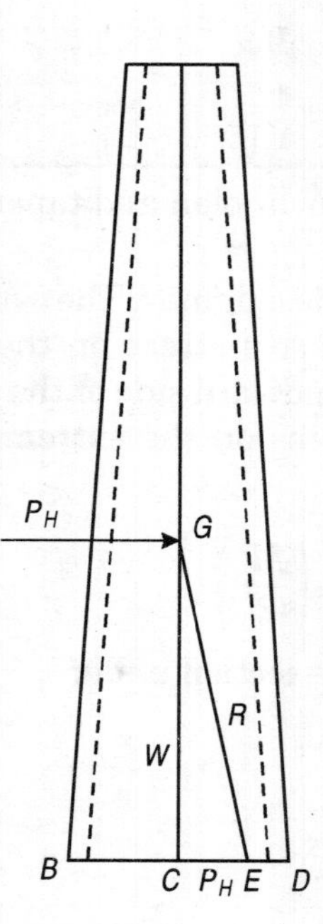

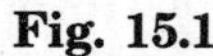

Fig. 15.1

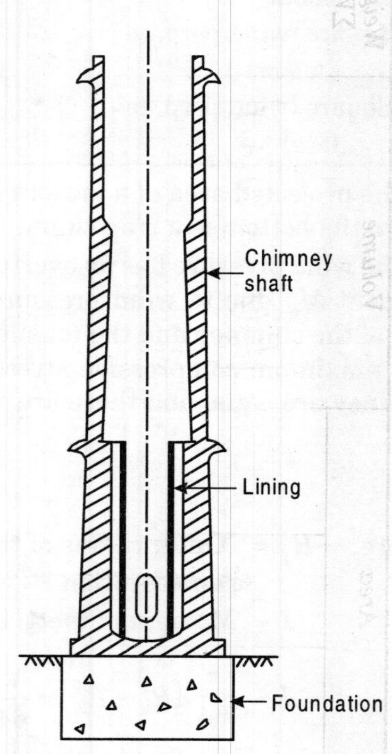

Fig. 15.2

coincides with centre of gravity of the section and the uniform intensity of compression exists over the entire section. The application of lateral wind pressure P_H causes the displacement of the centre of pressure at every horizontal section. The displacement should be restricted within such units as to prevent the development of tensile stress in the mortar joints.

Let W_s, be the weight of the chimney and P_H be the lateral wind pressure action on the chimney. In Fig. 15.1 the line of action of force P_H, is produced to meet the line of action of W_s. In the triangle of forces, *GC* represents the weight W_s, and *CE* represents the lateral wind pressure. Then, *GE* represents the line of action of the resultant of these two forces, *R*. When the point *E* falls outside the core (middle-quarter in case of circular shape in plan) then, there will be tension in the mortar at the point *R* on the windward side. When the point *F* falls inside the core, the chimney is quite stable.

Table 15.2

Section	Diameter		Area			Volume	Weight ΣW	Direct Compression W/A_2
	Outside	Inside	A_1	A_2	$\sqrt{A_1 A_2}$	$\left[\frac{H}{3}\left(A_1 + A_2\sqrt{A_1 A_2}\right)\right]$		
1	2	3	4	5	6	7	8	9
	m	m	m^2	m^2	m^2	m^2	kN	N/mm^2
$X X$	290	2.50	1.68	1.83	1.76	10.54	210.80	
$X_1 X_1$	3.15	2.75 2.55						0.1153
			2.68	2.92	2.80	16.80	336.00	
$X_2 X_2$	3.40	2.80 2.60						0.1873
			3.77	4.08	3.92	29.54	470.80	
$X_3 X_3$	3.65	2.85 2.65						0.2495
			4.94	5.34	5.14	30.84	616.80	0.3063
$X_4 X_4$	3.90	2.90 2.70						
			6.22	6.69	6.45	38.72	774.40	
$X_5 X_5$	4.15	2.95 2.75						0.3602
			7.59	8.12	7.85	47.12	942.40	
$X_6 X_6$	4.40	3.00 2.80						0.4128
			9.05	9.68	9.80	57.06	1141.20	
$X_7 X_7$	6.65	3.05 2.85						0.4641
			10.60	11.30	10.94	65.68	1313.60	
$X_8 X_8$	4.90	3.10						0.5159

Table 15.3

Section	Moments of Inertia $I = \frac{\pi}{4}(D_1^4 + D_2^4)$	Modulus of Section $Z(I/R_1)$	Projected Area A	Wind Force $P_H = 0.7 \times p \times \Sigma A$	Height from top h	Bending Stress $f_b = \frac{M}{Z} = \frac{P \cdot h}{Z}$	$\frac{P}{A} + \frac{M}{Z}$	$\frac{P}{A} - \frac{M}{Z}$	Remarks
1	2 m^4	3 m^3	4 m^2	5 kN	6 m	7 N/mm^2	8 N/mm	9 N/mm^2	10
XX	—	—	—	—	—	—	—	—	—
			18.15	19.06	6				
X_1X_1	2.16	1.336				0.0413	0.1566	0.074	No tension
			19.65	39.69	12				
X_2X_2	3.59	2.112				0.1127	0.3000	0.0746	–do–
			21.15	61.90	18				
X_3X_3	5.56	3.047				0.1827	0.4322	0.0668	–do–
			22.65	85.68	24				
X_4X_4	8.06	4.113				0.2492	0.5555	0.571	–do–
			24.15	120.97	30				
X_5X_5	10.82	5.216				0.3192	0.6794	0.4410	–do–
			25.65	137.97	36				
X_6X_6	14.32	6.509				0.3815	0.7933	0.0313	–do–
			27.15	166.48	42				
X_7X_7	18.90	8.128				0.4405	0.8946	0.0313	–do–
			28.15	166.48	42				
X_8X_8	23.50	9.583				0.4669	0.9828	0.0490	–do–

Notes : 1. The maximum compressive stress at the base is 0.9828 N/mm^2. It is less than allowable value of 1 N/mm^2.
2. There is no tension on the windward face. Hence, the design is safe.

The combined compressive stress at any horizontal section is given by

$$f = \left(\frac{W_s}{A} + \frac{M_w}{I} \cdot R_1\right) \quad \text{...(15.9)}$$

The combined compressive stress is maximum at the extreme fibre at the leeward side. The maximum combined compressive stress should be less than maximum allowable compressive stress in the masonry.

15.5 ARCHITECTURAL TREATMENT OF MASONRY CHIMNEYS

The chimneys are the structures which are emitting clouds of black smoke. Therefore, the chimneys are not handsome structures from their nature. However, the appearance of the chimneys may be improved by giving consideration and attention to the details. It is generally accepted that the shaft should be tapering. Panelling and projecting the arches are the devices which are frequently used in the brick chimneys. The enlarged top of a chimney may take one of various forms, but it should bear some suitable proportions to the size of the entire structure. The excessive ornamentation in a chimney is objectionable.

The strongly projecting coping caps string courses are undesirable, since, they offer increased resistance to the wind, the former being at the maximum leverage above the base.

15.6 SPECIFICATIONS FOR DESIGN OF MASONRY CHIMNEYS

The heights of a square chimney shall not exceed *ten times* the width the base section measured immediately above the footings. The height of circular chimneys must not exceed *twelve-times* the diameter clear of the footings. The height of octagonal chimneys should not exceed *eleven-times* the diameter.

The shaft is to batter (as shown in Fig. 15.2) 1 in 48 and not less than 1 in 50.

The projections should not extend beyond the face of the shaft to a greater distance than the thickness of wall at that level.

The minimum thickness of the uppermost section is to be one brick thick and the thickness is to increase by half the brick length at least every 6 m from the top downward.

The footings are to project beyond the basal section in all sides to a minimum distance equal to the thickness of the wall at the base.

The masonry chimneys are designed by following the above specifications and the stability of the chimney dimensions is then checked.

Example 15.1 *Design a brick masonry chimney of 48 metres height with 2.5 m diameter at the top. The horizontal wind pressure acting on the chimney is 1.50 kN per square metre. The unit weight of bricks is 20 kN/m*^{3}*. The safe compressive stress for brick masonry is 1.0 N/mm*^{2}*.*

Solution

Design. The diameter of brick masonry chimney at top is 2.5 m. Let the thickness of masonry at top be 200 mm. The chimney shaft is given a batter of 1 in 48.

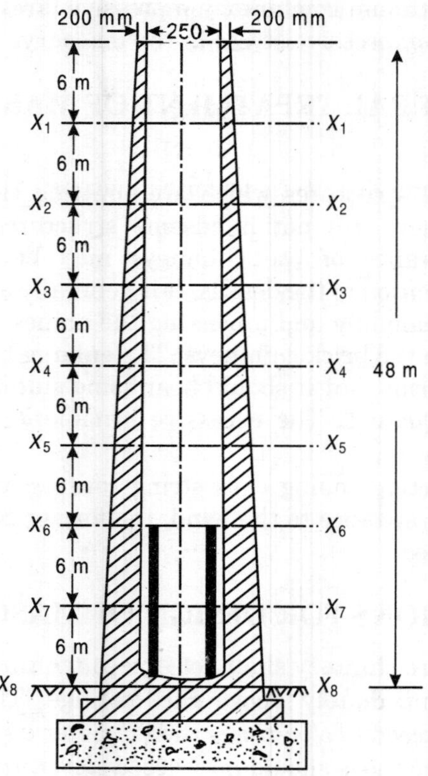

Fig. 15.3

The projections at every 6 m height from top are given. The projections of 100 mm are given. The thickness of lining is kept 200 mm. The height of lining is kept $\frac{1}{4}$ th the height of chimney (i.e., $\left(\frac{1}{4}\right) \times 48 = 12$ m]. The dimensions and data for the calculations are given in Tables 15.2 and 15.3. The dimensions of brick masonry chimney are shown in Fig. 15.3.

The combined stress due to vertical loading and moment due to wind forces have been found on leeward side $\left(\frac{P}{A}+\frac{M}{Z}\right)$ and on windward side $\left(\frac{P}{A}-\frac{M}{Z}\right)$ in Table 15.3, columns 8 and 9, respectively. The maximum compressive stress at the base does not exceed the allowable value and there is no tension at any section. Hence, the design is safe. The lining is supported on foundation directly.

The load is transferred to the foundation by pedestral footing. The cylindrical concrete foundation is provided under the chimney. The foundation is designed such that the compressive stress does not exceed the bearing capacity of the soil and there is no tension.

PROBLEM

15.1 Design a brick masonry chimney with the following data :

Height of chimney	=	42 m
Diameter at top	=	2.5 m
Unit weight of bricks	=	20 kN/m^3
Wind pressure	=	1.50 kN/m^2

The safe compressive stress for brick masonry is 1.0 N/mm^2.

Index

F

G

H

I

J

K

L

M

N

P

R

S

T

U

V

W

Y

Z